Plant Cytogenetics

Third Edition

Plant Cytogenetics

Third Edition

Ram J. Singh

USDA-Agricultural Research Service
Soybean/Maize Germplasm, Pathology, and Genetics Research Unit
Department of Crop Sciences
University of Illinois, Urbana-Champaign
Urbana, Illinois, USA

CRC Press
Taylor & Francis Group
Boca Raton London New York

CRC Press is an imprint of the
Taylor & Francis Group, an **informa** business

CRC Press
Taylor & Francis Group
6000 Broken Sound Parkway NW, Suite 300
Boca Raton, FL 33487-2742

First issued in paperback 2021

Version Date: 20160509

ISBN 13: 978-1-03-209750-3 (pbk)
ISBN 13: 978-1-4398-8418-8 (hbk)

Library of Congress Cataloging-in-Publication Data

Names: Singh, Ram J., author.
Title: Plant cytogenetics / author: Ram J. Singh.
Description: Third edition. | Boca Raton : Taylor & Francis, 2017. | Includes bibliographical references and index.
Identifiers: LCCN 2016017670 | ISBN 9781439884188 (alk. paper)
Subjects: LCSH: Plant cytogenetics.
Classification: LCC QK981.35 .S56 2017 | DDC 572.8/2--dc23
LC record available at https://lccn.loc.gov/2016017670

Visit the Taylor & Francis Web site at
http://www.taylorandfrancis.com

and the CRC Press Web site at
http://www.crcpress.com

To my wife, boys, and grandson (Kingston)

Contents

Preface

Johann Gregor Mendel (1822–1884), a monk in the Augustinian monastery at Brün, Austria (now Brno, the Czech Republic), conceived and delivered to the scientific world in 1865 the science of genetics using a garden pea. His results were presented in a clear-cut and straightforward fashion, and they were fairly easy to understand. Unfortunately, the invaluable treasure was buried in libraries until 1900, when three scientists, Carl Correns (Germany), Hugo de Vries (the Netherlands), and Erich von Tschermak (Austria), independently unearthed Mendel's Laws of Heredity. Three years later, Walter Sutton and Theodor Boveri (1903) proposed that the hereditary particles are borne by chromosomes and developed the chromosomal theory of inheritance (Sutton–Boveri Hypothesis), creating the science of cytogenetics. The pioneering cytogenetic investigations in maize, *Datura*, and *Drosophila* during the early twentieth century showed that an individual has a fixed number of chromosomes. The genes are located on the chromosomes and are inherited precisely by cell divisions (mitosis and meiosis). Alteration in the genetic material induces mutations. Since its inception, cytogenetics has exploded in many disciplines, involving organisms ranging from virus to mammals. The dawn of the twenty-first century saw the sequencing of the entire genomes of bacteria, many crops, animals, and humans.

Plant cytogenetics has been progressing at an extremely rapid pace since the author composed the first edition of *Plant Cytogenetics* that was published by CRC Press in 1993. The second edition that was published in 2003 contained several changes such as the mode of reproduction in plants and transgenic plants and this was updated in all the chapters. It has been more than a decade since the second edition saw the light of day and was widely accepted by plant cytogeneticists. The time is right to update this book (third edition) in an encyclopedic fashion that includes new chapters on Mendelian genetics and the role of cytogenetics in plant breeding. Mendel's original paper in English was added in Appendix A. All chapters have been updated with new, important information.

The book's introduction flows directly into the stream of classical and modern cytogenetic techniques after extensive exposure to Mendelism and major chronological discoveries in the science of cytology, genetics, cytogenetics, and molecular genetics. The handling of plant chromosomes assembled in Chapter 2 has undergone monumental progress. Precise identification and nomenclature of plant chromosomes from classical aceto-carmine and Feulgen staining techniques have progressed to Giemsa banding, fluorescence *in situ* hybridization, genomic *in situ* hybridization, fiber fish, and sorting and karyotyping of chromosomes by flow cytometry. Several protocols for determining nuclear DNA content by flow cytometry have been formulated and perfected. Chapter 3, Methods in Plant Cytogenetics, a new chapter, describes Mendelian genetics including gene interaction and expression, test of independence, and classical mapping of qualitative traits on the chromosomes. Chapter 4 (Cell Divisions—Mitosis and Meiosis) and Chapter 5 (Genetic Control of Meiosis) describe mitosis and meiosis in the higher plants. The mode of reproduction (sexual, asexual, and sex chromosomes, incompatibility-self, and cross pollination) in plants is described in Chapter 6. Chapter 7 includes chromosome nomenclature based on the kinetochore position on chromosomes, karyotyping by aceto-carmine, and Giemsa C- and N-banding methods. Chromosome numbers of some economically important plants are presented in Appendix B.5. Chapter 8 discusses extensively chromosomal aberrations (structural and numerical changes). Utilization of primary, secondary, and tertiary gene pools for crop improvement depends on the comprehension of genomic relationships between the cultigen and allied species and genera. Established genomic relationships based on classical taxonomy, cytogenetics, and verified by biochemical and molecular methods are presented in Chapter 9. The cause of morphological aberrations generated through cell and tissue cultures and genetic transformation are described in Chapter 10. The dazzling progress achieved since 1986 in producing genetically modified crops by private and public institutions is briefly discussed in Chapter 11. Finally, Chapter 12 summarizes the relationship of cytogenetics in plant breeding.

I have provided an adequate literature review; it covers publications from ancient time to the present.

I am optimistic that this book will be accepted by plant cytogeneticists and will encourage students to receive training in the exciting field of plant cytogenetics, particularly in wide hybridization.

Acknowledgments

During my MSc (Agricultural Botany) final year at Government Agricultural College, Kanpur, India, I was introduced to plant cytogenetics by Professor Rishi Muni Singh. Fortunately, I joined the barley cytogenetics laboratory of the late Professor Takumi Tsuchiya, a world-renowned barley cytogeneticist, at Colorado State University, Fort Collins, Colorado. I became the beneficiary of his expertise of aneuploidy in barley, which broadened my knowledge and expertise and, by coincidence, I became associated with Professor G. Röbbelen, University of Göttingen, Germany and Dr. G. S. Khush at the International Rice Research Institute, Las Baños, the Philippines. I am thankful to Dr. Theodore Hymowitz who introduced me to the genus *Glycine* in February 1983. I never looked back to barley, wheat, and rice cytogenetics. I have been devoted to soybean and its distantly-related wild perennial cousin, the *Glycine* species.

My sincere appreciation is to E. B. Patterson, W. F. Grant, G. S. Khush, M. P. Maguire, G. Röbbelen, R. T. Ramage, and D. F. Weber for their critical reviews of the first edition of *Plant Cytogenetics*. I am profoundly grateful to F. Ahmad, D. E. Alexander, K. Arumuganathan, J. P. Braselton, R. Bothmer, C. L. Brubaker, B. Burson, C. Carvalho, M. J. Cho, G. Chung, J. Dolžel, C. doVale, J. J. Doyle, T. R. Endo, M. Eubanks, B. Friebe, K. Fukui, Govindjee, W. F. Grant, J. Greilhuber, W. W. Hanna, S. A. Jackson, P. P. Jauhar, J. Jiang, G. S. Khush, T. M. Klein, R. J. Kohel, F. L. Kolb, K. P. Kollipara, A. Labeda, U. C. Lavania, M. P. Maguire, R. Obermayer, J. Orellana, R. G. Palmer, E. B. Patterson, R. T. Ramage, A. L. Rayburn, G. Röbbelen, I. Schubert, R. M. Skirvin, R. M. Stupar, C. Taliaferro, T. Tsuchiya, and D. F. Weber for their contributions, constructive suggestions, and comments, which helped improve the clarity of this third edition.

I am very grateful to various scientists and publishers who permitted me to use several figures and tables. However, I am solely responsible for any errors, misrepresentations, or omissions that may appear in this book. I sincerely acknowledge the patience of my wife, Kalindi, who let me spend numerous weekends, nights, and holidays completing this third edition of *Plant Cytogenetics*.

Ram J. Singh

Author

Ram J. Singh, PhD, is a plant research geneticist in the USDA-Agricultural Research Service, Soybean/Maize Germplasm, Pathology, and Genetics Research Unit, Department of Crop Sciences, at the University of Illinois at Urbana-Champaign. Dr. Singh earned his PhD in plant cytogenetics under the guidance of a world-renowned barley cytogeneticist, the late Professor Takumi Tsuchiya from Colorado State University, Fort Collins, Colorado. Dr. Singh isolated monotelotrisomics and acrotrisomics in barley, identified them by Giemsa C- and N-banding techniques, and determined chromosome arm-linkage group relationships.

Dr. Singh conceived, planned, and conducted excellent quality pioneering research related to many cytogenetic problems in barley, rice, wheat, and soybean and is published in highly reputable national and international prestigious journals, including *American Journal of Botany*, *Caryologia*, *Chromosoma*, *Critical Review in Plant Sciences*, *Crop Science*, *Euphytica*, *Genetic Plant Resources and Crop Evolution*, *Genetics*, *Genome*, *International Journal of Plant Sciences*, *Journal of Heredity*, *Plant Molecular Biology*, *Plant Breeding*, and *Theoretical and Applied Genetics*. In addition, he has summarized his research results by writing 18 book chapters. Dr. Singh has presented research findings as an invited speaker at national and international meetings. He has edited a book series titled *Genetic Resources, Chromosome Engineering, and Crop Improvement*. This series includes *Grain Legumes* (Volume 1), *Cereals* (Volume 2), *Vegetable Crops* (Volume 3), *Oil Seed Crops* (Volume 4), *Forage Crops* (Volume 5), and *Medicinal Plants* (Volume 6).

Dr. Singh assigned genome symbols to the species of the genus *Glycine* and established genomic relationships among species by cytogenetic approach. He constructed and introduced a soybean chromosome map based on pachytene chromosome analysis and established all possible 20 primary trisomics associating 11 of the 20 molecular linkage maps to the specific chromosomes. Dr. Singh produced fertile plants with $2n = 40$, 41, and 42 chromosomes, for the first time, from an intersubgeneric cross of soybean cv. "Dwight" ($2n = 40$) and *Glycine tomentella*, PI 441001 ($2n = 78$). The screening of derived lines showed that useful genes of economic importance from *G. tomentella* have been introgressed into soybean. He has U.S. patents on methods for producing fertile crosses between wild and domestic soybean species.

His books on plant cytogenetics (first edition, 1993; second edition, 2003) are ample testimony to his degree of insight in research problems and creative thinking. His books are widely used worldwide by students, scientists, universities, industries, and international institutes. He is chief editor of *International Journal of Applied Agricultural Research* and editor of *Plant Breeding and Biological Forum—An International Journal*. He has won many honors and awards such as the Academic Professional Award for Excellence: Innovative & Creativity, 2000; Academic Professional Award for Excellence, 2007; College of ACES Professional Staff Award for Excellence—Research, 2009; The Illinois Soybean Association's Excellence in Soybean Research Award—2010; and Foreign Fellow, National Academy of Agricultural Sciences, India—2011. He continues to hold collaborative projects with many countries.

1 Introduction

The foundation of the science of genetics was laid in 1865 (published in 1866) by Johann Gregor Mendel when he reported principles of segregation and an independent assortment based on his careful selection of seven contrasting traits and controlled artificial hybridization experiments with the garden pea (*Pisum sativum* L.). Unfortunately, the importance of Mendel's classic paper was not widely known or clearly appreciated until 1900 when three botanists, Carl Correns, Hugo deVries, and Erich von Tschermak, rediscovered Mendel's laws of inheritance as a result of their own investigations (Peters, 1959).

Mendel formulated two basic principles of heredity from his experiments with garden peas: (1) the law of segregation and (2) the law of independent assortment. He selected pure-line parent varieties differing from each other by clear-cut contrasting traits which helped him to classify the F_1 (first filial) generation and to score segregating progenies accurately in F_2 (second filial) and later generations. Although Mendel was unaware of the gene-chromosome relationship, his method of experimentations, keen observations, and statistical approach to analysis of data led him to discover and develop basic principles of heredity.

Mendel's first law relates to members of a single gene pair. In diploid species, factors (genes) occur in pairs, one member contributed from each of the parents. In hybrid individuals, in which the two members of a gene pair differ, there is no blending but rather each of the two differing gene forms is transmitted unchanged to progeny. In the formation of sex cells (gametes), later shown to occur as a result of meiosis (followed by gametophyte generations in higher plants), members of factor pairs typically behave as if they were discrete particles which separate (segregate) clearly and are delivered in equal numbers to gametes. He postulated that when contrasting members of factor pairs are delivered in equal numbers to gametes and then are combined randomly at fertilization in zygotes in an F_2 generation, the progeny segregate (or may be separated into) the two grand parental types in a ratio of three-fourths dominant trait expression: one-fourth recessive trait expression. Mendel's knowledge of probability, his use of contrasting traits, and a clear testing of the genetic constitutions of parents, F_1 and F_2 generations, enabled him to formulate the *law of segregation*.

Mendel also made crosses in reciprocal directions in which parents differ by two unrelated traits (two pairs of genes → dihybrid). The four phenotypic categories occur in nearly equal frequencies in test crosses. That is, from the dihybrid parent, round (*R*) and wrinkled (*r*) seed traits are transmitted in equal frequencies, as are yellow (*Y*) and green (*y*) seed traits (demonstrating the law of segregation for both the *Rr* and *Yy* gene pairs). In addition, *RrYy* individuals produce four gametic genotypes in equal proportions and in F_2 segregate in a phenotypic ratio of 9 (round, yellow; *R-Y-*) :3 (wrinkled, yellow; *rr Y-*) :3 (round, green; *R- yy*) :1 (wrinkled, green; *rryy*). This proves that during meiosis copies of *R* and *r* are assorted (distributed) to the meiotic products (and eventually to gametes) independently of copies of *Y* and *y*. This is the law of *independent assortment*.

Mendel's laws are not universal when nonallelic genes are linked [usually by virtue of being located on the same chromosomes but less than 50 centiMorgans (cM is a map unit) apart], segregation still occurs, but assortment is not independent. The frequencies of gametic genotypes can be estimated by observing phenotypes of zygotes either from F_2 or testcross populations. The closer the distance between the genes, the stronger the linkage. The absence of crossover products (also known as recombinants) suggests complete linkage. The strength of linkage expresses itself primarily in the production of gametes and that depends on the distance between two genes on the same chromosome. Discovery and exploitation of linkage and crossing over led to detailed chromosome mapping in numerous organisms, an activity that continues to the present day. In many species, linkage groups have now been assigned to each of the individual chromosomes of the genomes, especially through the use of chromosomal aberrations.

Shortly after the rediscovery of Mendel's laws of inheritance, Sutton (1903) reported the chromosomes in heredity and in the paper he wrote, "It appears from a personal letter that Boveri had noted the correspondence between chromosomic behavior as deducible from his experiments and the results on plant hybrids." (Boveri, Th. 1902. Ueber Mehrpolige Mitosen als Mittel zur Analyse des Zellkerns, *Verb. d. Phys.–Med. Ges. zu Würzburg*, N. F., Bd. XXXV.) Wilson (1928, p. 923) wrote, "Cytological Basis of the Mendelian Phenomena. The Sutton-Boveri Theory" gave credit to Sutton and Boveri with "Sutton Boveri Hypothesis." Since then, it is known as Sutton–Boveri hypothesis (also known as chromosome theory of inheritance or Sutton–Boveri theory). However, currently, it is written in the literature that Sutton and Boveri, independently, discovered chromosomal theory of inheritance. However, Peters (1959) gives equal credit to Sutton and Boveri in the summary with the statement that Sutton and Boveri published their findings "in the same year." But, Martins (1999) examined the facts of papers published in 1902–1903, and concluded that credit should be given only to Sutton and not to Boveri because Boveri did not publish any hypothesis of that kind during the relevant period, 1902–1903. Cytogenetics flourished once it was established that chromosomes are the vehicles of genes. Classical cytogenetic studies in plants and animals quickly led to numerous extensions of genetic information and interpretation. Cytogenetic studies have made landmark contributions to knowledge in biology. The combined discipline has been especially powerful in explaining hereditary phenomena that would have been extremely difficult to decipher by alternative approaches. In spite of a proliferation of powerful new disciplines such as molecular biology, and an ever-increasing level of complexity and performance of new apparatus, there remains a need for well-trained cytogeneticists to address questions that are most readily and economically resolved by cytogenetic approaches. Cytogenetics is a hybrid science that combines cytology (the study of chromosomes and other cell components) and genetics (the study of inheritance). The science includes handling chromosomes (chromosome staining techniques), function and movement of chromosomes (cell division; mitosis, meiosis), numbers and structure of chromosomes (karyotype analysis), and numerous modifications of structure and behavior as they relate to recombination, transmission, and expression of genes. Succeeding chapters will summarize and document much of the cytogenetic information in higher plants that has been reported in recent decades. It is hoped that this information and literature citations will prove useful to practicing cytologists and cytogeneticists, as well as to members of other disciplines.

A Brief Historical Chronology of the Progress in Genetics, Cytogenetics, and Genetic Engineering

Year	Discoverer	Discovery
		1. Pre-Mendelian Period
300 BC	Euclid	Discovery of optical properties of curved surfaces
−1285	Degli	Invention of spectacles
1590	Janssen	Discovery of operational compound microscope
1610	Galileo	Construction of first microscope by combining lenses in a tube of lead
1625	Faber	Application of microscope for anatomical studies
1651	Harvey	Development of theory of Epigenesis
1661	Malpighi	Discovery of capillaries in the lungs of animals
1665	Hooke	Discovery of cell
1672	de Graafe	Discovery of follicles
1677	Leeuwenhoek	Discovery of sperms "wild animalcules"
1679	Swammerdam	Development of preformation theory

Continued

A Brief Historical Chronology of the Progress in Genetics, Cytogenetics, and Genetic Engineering

Year	Discoverer	Discovery
1682	Grew	Stamens as the male organs of the flower
1694	Camerarius	Description of sexual reproduction in plants
1707	Linnaeus	Father of plant taxonomy
1761	Koelreuter	Equal hereditary contribution of both parents
1809	Lamarck	Acquired inheritance "Lamarckism"
1824	Amici	Discovery of pollen tube
1828	Brown	Discovery of cell nucleus
1835	Mohl	Creator of modern plant cytology
1838	Schleiden	Founder of cell theory
1839	Schwann	Father of modern cytology
1941	Remak	The phenomenon of amitosis or direct division
1842	Nägeli	Publication of first drawings of chromosomes
1845	von Siebold	Protozoa are animals consisting of a single cell
1848	Hofmeister	Illustration of tetrad formation, identification of two groups of cell at the opposite poles of the embryo sac but failure to distinguish between the synergids and the egg
1858	Darwin and Wallace	Theory of evolution through natural selection
1859	Darwin	The origin of species

2. Post-Mendelian Period

Year	Discoverer	Discovery
1865	Mendel	Discovery of the laws of heredity
1866	Darwin	Theory of pangenesis (mechanism of heredity)
1971	Miescher	Isolation of nucleic acid and nucleoprotein
1872	Hertwig	Fertilization is the union of sperm and egg nuclei (general statement)
1879	Flemming	Discovery of mitosis
1879	Strasburger	Description of megasporogenesis in plants
1879	Schleicher	Indirect division is called karyokinesis
1880	Hanstein	Coined the term protoplast; followed the sequence of early cell divisions in the development of the embryos
1881	Focke	Xenia; the pollen produced a visible influence on the hereditary characters of those parts of the ovule that surround the embryos
1882	Flemming	Discovery of tiny threads inside salamander, later known as chromosomes; mitosis and chromatin
1884	Strasburger	Discovery of actual process of syngamy
1890	Waldeyer	Fundamental fact in mitosis is the formation of the nuclear filaments named chromosomes
1892	Hertwig	Protoplasm (proto, first and plasma formation) theory
1896	Wilson	Publication of *the Cell in Development and Inheritance* (1st ed.)
1898	Nawaschin	Discovery of double fertilization in angiosperm
1900	deVries, Tschermak, and Correns	Rediscovery of Mendel's laws (independently)
1900	Juel	Parthenogenesis
1901	McClung	Sex determination is related to the chromosomes
1902–1903	Sutton, Boveri	Independently, chromosome theory of inheritance
1902	Bateson	Coined terms allelomorph (=allele), F_1, and F_2, homozygotes, heterozygote and epistasis
1905	Johannsen	Coined the terms gene, genotype, and phenotype
1905	Wilson	Coined the term X-chromosomes
1905–1908	Bateson and Punnett	Discovery of first linkage

Continued

A Brief Historical Chronology of the Progress in Genetics, Cytogenetics, and Genetic Engineering

Year	Discoverer	Discovery
1906	Montgomery	Distinction between heterochromosomes (X and Y) and autosomes
1906	Bateson	Coined term genetics (1906)
1909	Jenssens	Chiasma theory of crossing over
1910	Morgan and Colleagues	Sex linkage and genes are located on chromosomes
1912	Morgan	Linkage map of *Drosophila*
1912	Baur	Sex-linked gene in *Silene alba* (=*Melandrium album*)
1913	Sturtevant	Linear arrangement of genes on *Drosophila* chromosomes
1913–1916	Bridges	Nondisjunction as a proof of the chromosome theory of inheritance
1913, 1917	Carothers	Correlation of Mendelian segregation and the distribution of heteromorphic homologues meiosis
1917, 1919	Bridges	Recognition of deficiencies, duplications, and translocations by genetic tests
1920	Lysenko	Hereditary factors are not in the nucleus
1921	Blakeslee	Discovery of trisomics in *Datura stramonium*
1921	Sturtevant	Recognition of an inversion by genetic tests
1922	L.V. Morgan	Attached X-chromosomes in *Drosophila*
1924	Feulgen and Rossenbeck	Specific stain for DNA (Feulgen)
1925	Morgan	Publication of *The Theory of the Gene*
1926	Kihara and Ono	Introduction of terms auto- and allopolyploidy
1927	Longley	Discovery of B-chromosomes in maize
1927	Muller	X-ray causes gene mutation
1928	Heitz	Distinction between euchromatin and heterochromatin
1928	Griffith	Transformation (a harmless strain of bacteria into a virulent one)
1928	Gowen	Discovery of a gene that affects meiosis
1929	Darlington	Coined the term chiasma terminalization
1930	Kihara	Establishment of genome analysis
1931	Creighton and McClintock	Cytological proof of crossing over in maize
1931	Stern	Cytological proof of crossing over in *Drosophila*
1931	Belling	Classical model of crossing over
1932	Darlington	Publication of Recent Advances in Cytology (1st ed.)
1933	Heitz and Bauer	Discovery of the giant chromosomes in dipteran insects
1933	Painter	Discovery of salivary gland (polytene) chromosomes in *Drosophila melanogaster*
1933	Brachet	DNA is found in the chromosomes and the RNA is present in the cytoplasm of all cells
1934	McClintock	Recognition of nucleolus organizer in maize
1936	Caspersson	Development of ultraviolet photomicrography by the study of nucleic acid within nucleus
1936	Stern	Discovery of mitotic crossing over
1936	Sturtevant	Preferential segregation in *Drosophila*
1936	Sturtevant and Beadle	Relation of inversions to crossing over and disjunction
1937	Blakeslee and Avery	Chromosome doubling by colchicine
1937	Sax	Beginning of x-ray cytology and cytogenetics
1939	Thom and Steinberg	Induction of mutations by nitrous acid
1941	Beadle and Tatum	Development of one gene–one enzyme concept
1944	Avery, MacLeod, and McCarty	DNA is the hereditary material, not protein
1946	Lederberg and Tatum	Genetic exchange in bacteria

Continued

A Brief Historical Chronology of the Progress in Genetics, Cytogenetics, and Genetic Engineering

Year	Discoverer	Discovery
1947	Auerbach and Robson	Induction of mutations by nitrogen and sulfur mustard
1949	Delbruck and Bailey	Genetic exchange in bacteriophage
1949	Barr and Bertram	Discovery of small stainable body in nondividing nuclei of female and absent in male-sex chromatin or Barr body
1950	Levan and Tjio	Use of 8-hydroxyquinoline for pretreatment of chromosomes
1950	McClintock	Discovery of activator-dissociation system in maize
1951	Horowitz and Leupoid	One gene–one enzyme hypothesis
1952	Chargaff	Discovery of A-T and G-C ratio in DNA
1952	Franklin	X-ray diffraction data of DNA
1953	Watson and Crick	Announcement of the double-helix structure of DNA
1955	Benzer	Fine structure of genetic region in bacteriophage
1956	Tjio and Levan	Determination of accurate chromosome number 2n = 46 of humans
1960	Barski	Somatic cell fusion
1960	Tsuchiya	Establishment of primary trisomics in barley
1962	Ris and Plaut	Discovery of chloroplast chromosomes
1963	Nass and Nass	Discovery of mitochondrial chromosomes
1963	Marmur	DNA/DNA hybridization *in vitro*
1964	Britten and Waring	Discovery of repetitious DNA in eukaryotes
1965	Du Praw	Discovery of whole-mount electron microscopy
1968	Arber, Smith, and Nathan	Restriction of modification enzymes
1968–1970	Purdue and Gall	DNA/DNA and RNA/RNA hybridization *in situ*
1969	Jonathan and Colleagues	Synthesis of artificial gene
1972	Khorana and colleagues	Synthesis of artificial gene
1972	Boyer's research group (Berg and Colleagues)	Production of artificial gene
1974	Thomas Jr. and Wilson	Discovery of the widespread occurrence of the hair-pin-like structures (palindromes) resulting from inverted repetitious DNA, located at intervals along the chromatids of eukaryotic chromosomes.

3. Genetic Engineering Period

Year	Discoverer	Discovery
1973	Cohen, Boyer and colleagues	Insertion of a gene; DNA cloning
1974		Monsanto developed Roundup® brand agricultural herbicides
1976	Genentech	Establishment of the first genetic engineering company
	Sanger and Coulson; Maxam and Gilbert	Rapid sequencing of DNA
1983	Gusella and colleagues	Location of Huntington's disease on human chromosome 4
1984	Jeffreys	Development of genetic finger printing
1985		Plant Genetic Systems (now part of Bayer Crop Science) developed BT tobacco.
1986		Initiation of the Human Genome Projects
1987		Production of the first comprehensive human genetic map
1988		Creation of the national Center for Human Genome Research
1990		The launch of the Human Genome Project; the first gene therapy on a four-year-old girl with an immune disorder known as ADA deficiency, but procedure did not work; Mary-Clair King demonstrated that a gene on chromosome 17 causes an inherited form of breast cancer and increases the risk of ovarian cancer

Continued

A Brief Historical Chronology of the Progress in Genetics, Cytogenetics, and Genetic Engineering

Year	Discoverer	Discovery
1992		Production of rough human genetic map by international collaboration
1995		The establishment of The Institute of Genomic Research
1995		Publication of the first complete DNA sequence of the genome of a free-living organism—the bacterium *Haemophilus influenza*
1995		BT potato produced by Monsanto was approved
1996		Monsanto commercialized Roundup Ready® soybeans followed by alfalfa, corn, cotton, spring and winter canola, and sugar beets
1996		Transgenic maize, potatoes, and cotton expressing BT toxins grown in the United States on a large scale
1998		The first genome of a multicellular organism—the 97 megabase DNA sequence of the roundworm
2000		Celera Genomics published the first draft of the sequence of 90% of the human genome
2000		The Institute of Genomic Research completed the *Arabidopsis* genome map
2001		Syngenta and Myriad Genetics completed rice genome map
2002		Syngenta announced the complete Japonica rice genome sequence
2002		Beijing Genomic Institute announced the complete Indica rice genome sequence

2 Handling of Plant Chromosome

2.1 INTRODUCTION

Cytologists have devised cytological techniques from time to time to obtain precise information on chromosome numbers, chromosome structures, size and shape, and to examine the mechanism of cell division in plant species. These properties are studied by cytological techniques. The basic principles for handling the mitotic and meiotic chromosomes of all plant species are similar and consist of collection of specimens, pretreatment, fixation, and staining. However, cytological procedures are modified depending on crop species, objective of the experiments, available facilities, and, above all, personal preference of the cytologists (Sharma and Sharma, 1965; Darlington and La Cour, 1969; Jahier, 1996).

2.2 SQUASH TECHNIQUE: MITOTIC AND MEIOTIC CHROMOSOMES

2.2.1 COLLECTION OF ROOTS

Place the seeds on a moistened filter paper in a Petri dish (Figure 2.1a). For the analysis of cereal chromosomes (e.g., wheat, oat, barley, rye, and others), keep the Petri dish with seeds in a dark cold room or refrigerator (0–4°C) for 3–5 days. The cold treatment facilitates uniform and rapid seed germination. Remove the Petri dish with seeds from the refrigerator and leave at room temperature (20–25°C) for germination. Collect roots that are 1–2 cm long (Figure 2.1a, b). It should be kept in mind that roots should touch the filter paper; otherwise the mitotic index will be very low. When it is desired to collect roots from plants growing in pots, care should be taken not to break the actively growing roots. Transfer roots to vials containing cold water after washing soil from the roots. Keep the vials in ice cold water in an ice chest.

For soybean chromosome analysis, germinating seeds are kept at 30°C for 4 h before collecting the roots. It is not required when roots are collected from a sand bench or vermiculite in the greenhouse. If seeds are geminated in a sand bench in the greenhouse, seedlings are ready to collect roots after about one week depending on the temperature of the room. Figure 2.1c shows one-week-old soybean seedlings in a send bench and this is the stage to collect the roots. Seedlings are very carefully uprooted without damaging the actively growing roots (Figure 2.1d). Roots are healthy and root tip is unbroken and is of cream color (Figure 2.1d; arrow). Transfer seedlings in a container with cold tap water (Figure 2.1e). Eppendorf tubes 1.5 mL are numbered in the laboratory and 1 mL ddH$_2$O is added. Root tips (3–5) are collected and transferred immediately in vials and the same number is assigned to the seedling label (Figure 2.1f). Remove water from each tube and add 1 mL of 8-hydoxyquinoline and transfer tubes on a heating block kept in a refrigerator at 15°C for 2–3 h. Actively growing roots can be collected either from germinating seeds in a Petri dish or from seedlings growing in pots or in the field. Generally roots are collected in the laboratory after germinating the seeds in a Petri dish.

2.2.2 PRETREATMENT OF ROOTS

Pretreatment of roots is an essential step for study of somatic chromosomes. It performs several purposes: stops the formation of spindles, increases the number of metaphase cells by arresting the

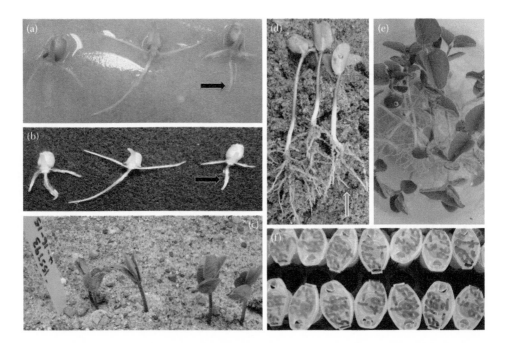

FIGURE 2.1 (a) Germinated wheat seedling in a Petri plate on a moist filter paper-3 days old (arrow shows actively growing root). (b) Germinated wheat seedling mostly three roots after 3 days ready to harvest place on black background (arrow shows creamy color root tip). (c) One week old soybean seedlings being grown in a san bench in the greenhouse. (d) Uprooted seedling showing actively active growing healthy roots (arrow). (e) Soybean seedlings in a container with water to remove the sand. (f) Labeled 1.5 mL Eppendorf tube with 1 mL distilled water. ID number assigned to each tube corresponds to the seedling identification number.

chromosomes at the metaphase plate, contracts the chromosome length with distinct constrictions, and increases the viscosity of the cytoplasm. Numerous pretreatment agents, described below, have been developed.

2.2.2.1   Ice-Cold Water

Pretreatment with ice-cold water (0–2°C) is very effective and is widely used for cereal chromosomes. This particular pretreatment is preferred over other pretreatments when chromosomes of a large number of plants must be studied.

Keep roots in numbered vials two-thirds filled with cold water. Transfer vials with roots into a container which will allow cold water to cover the top of the vials, and further cover the vials with a thick layer of ice (Tsuchiya, 1971a). Keep the container in a refrigerator for 12–24 h depending on the materials and experimental objectives. The recommended pretreatment lengths for barley and wheat chromosomes are 16–18 and 24 h, respectively. Longer pretreatment will shorten the chromosome length considerably.

2.2.2.2   8-Hydroxyquinoline

Tjio and Levan (1950) were the first to recognize the usefulness of 8-hydroxyquinoline (C_9H_7NO) for chromosome analysis. Aqueous solution is prepared by dissolving 0.5 g of 8-hydroxyquinoline in 1L double distilled water (ddH_2O) at room temperature (RT) on cold stir plate; requires 24 h or more. Make sure chemical is dissolved. Filter in a colored bottle and store in a refrigerator. Pretreat roots in 8-hydroxyquinoline for 2–3 h at 15–16°C. Warmer temperatures often cause sticky chromosomes. Pretreatment with 8-hydroxyquinoline has been very effective for plants with small-size chromosomes. It makes primary (kinetochore) and secondary constrictions (nucleolus organizer region) very clear.

2.2.2.3 Colchicine

Colchicine is a toxic natural product and secondary metabolite, originally extracted from plant *Colchicum autumnale* (meadow saffron). It is commonly used to treat gout. Since colchicine in higher concentrations induces polyploidy, a low concentration (0.1–0.5% for 1–2 h, at RT) is recommended for pretreatment. A treatment time of 1 1/2 h gives the best results for soybean chromosomes. Colchicine inhibits microtubules polymerization by binding to tubulin known as "mitotic poison"; arrests chromosomes at the metaphase stage, giving a high proportion of cells with metaphase chromosomes. This helps chromosome count very efficiently. Roots should be washed thoroughly after colchicine pretreatment. This pretreatment facilitates better penetration of fixative at the subsequent stages of chromosome preparation. Since colchicine is a toxic chemical (carcinogenic and lethal poison), necessary precautions should be taken when it is used.

2.2.2.4 α-Bromonaphthalene

The effect of α-bromonaphthalene is almost the same as that of colchicine. It is sparingly water soluble. A saturated aqueous solution is used in pretreatment for 2–4 h at RT. This pretreatment is very effective for wheat and barley chromosomes.

2.2.2.5 Para-Dichlorobenzene

Pretreatment with para-dichlorobenzene (PDB; $C_6H_4Cl_2$) is an organic compound and is colorless with strong odor. Like α-bromonaphthalene, it has a low solubility in water and is carcinogenic. It is very effective for plants with small-size chromosomes. Weigh 3 g of para-dichlorobenzene, and add to it 200 mL distilled water. Incubate overnight at 60°C, and then cool. Some crystals may remain undissolved. Shake thoroughly before using. Pretreat roots for 2–2 1/2 h at 15–20°C or at RT (Palmer and Heer, 1973).

2.2.3 FIXATION

The science of chromosome study depends on good fixative. The function of a fixative is to fix, or stop, the cells at a desired stage of cell division without causing distortion, swelling, or shrinkage of the chromosomes. The most widely used fixatives described by Sharma and Sharma (1965) are as follows:

2.2.3.1 Carnoy's Solution 1

- 1 part glacial acetic acid
- 3 part ethanol (95%–100%)

Note: This fixative is prepared fresh each time and is used for the fixation of roots and anthers. For anthers, 1 g of ferric chloride ($FeCl_2 \cdot 6H_2O$) is added to the fixative as a mordant, if aceto-carmine and propionic-carmine are used as a stain. The material should be kept in the fixative at least 24 h at RT.

2.2.3.2 Carnoy's Solution 2

- 1 part glacial acetic acid
- 3 part chloroform
- 6 part ethanol (95%–100%)

A modification of 1:3:4 has also been used for wheat chromosomes.

2.2.3.3 Propionic Acid Alcohol Solution

- 1 part propionic acid
- 3 part ethanol (95%–100%)
- 1 g/100 mL fixative (for meiotic materials only) ferric chloride

This fixative is very good for plants with small chromosomes. It provides clear cytoplasm and optimal staining for chromosomes (Swaminathan et al., 1954). After 24 h pot-fixation, roots are transferred to staining solution. Inflorescence and flower buds are first washed with two changes of 70% ethanol. Materials to be stained for observation of meiosis should be stored in 70% ethanol under refrigeration (4–5°C). Materials can be stored for a long term at –20°C.

2.2.4 STAINING OF CHROMOSOMES

Chromosome number and morphology and structure can be visualized through a microscope after optimal staining of the chromosomes. A good quality staining agent stains specifically the chromosomes, differentiates euchromatin and heterochromatin, and provides clear cytoplasm with stained nucleoli (but not with Feulgen stain). The following staining methods have been used extensively for staining plant chromosomes.

2.2.4.1 Aceto-Carmine Staining

2.2.4.1.1 Ingredient

For 1% stain:

- 1 g carmine powder (use a lot of carmine certified by the Biological Stain Commission)
- 45 mL glacial acetic acid
- 55 mL ddH$_2$O water

2.2.4.1.2 Preparation of Stain

1. Heat to boil 100 mL 45% acetic acid under a fume hood.
2. Add carmine powder to the boiling 45% acetic acid.
3. Boil for 5–10 min, with occasional stirring until color becomes dark red.
4. Cool and filter into a colored bottle and store in a refrigerator.

Note: For preparing aceto-orcein, carmine is replaced by orcein in the same procedure.

2.2.4.1.3 Staining of Root

1. Aceto-carmine stain has been very effective for somatic chromosomes of barley. Roots are transferred directly into stain after pretreatment.
2. The stain acts like a fixative and is routinely used for the analysis of F$_1$ and F$_2$ aneuploid populations of barley. Cells are fragile when slides are prepared immediately (Tsuchiya, 1971a).

2.2.4.2 Feulgen Staining

2.2.4.2.1 Ingredient

- 1 g basic fuchsin (use a stain commission type especially certified for use in Feulgen technique).
- 200 mL ddH$_2$O
- 30 mL 1 N HCl
- 3 g potassium metabisulphite (K$_2$S$_2$O$_5$)
- 0.5 g activated charcoal

2.2.4.2.2 Preparation of Stain

Method I:

1. Dissolve 1 g basic fuchsin gradually in 200 mL of boiling ddH$_2$O and shake thoroughly.
2. Cool to 50°C, and filter.

3. Add to filtrate 30 mL 1 N HCl and then 3 g potassium metabisulfite (should not be from old stock).
4. Close mouth of the container with a stopper, seal with Para film, wrap container with aluminum foil and store in a dark chamber at RT for 24 h.
5. If solution shows faint straw color, add 0.5 g activated charcoal powder, shake thoroughly and keep overnight in a refrigerator (4°C).
6. Filter and store the stain in a colored bottle in a refrigerator.

Method II: (J. Greilhuber, personal communication, 2002)

1. Dissolve g basic fuchsin in 30 mL 1 N HCl at RT (3 h, stirring).
2. Add 170 mL dH$_2$O + 4.45 g K$_2$S$_2$O$_5$ (3 h or overnight, stirring), tighten vessel.
3. Add 0.5 g activated charcoal (15 min, stirring), filter by suction. The reagent is colorless. Avoid evaporation of SO$_2$ during all steps.

Note: After discoloration, the solution has lost the red tint, but may be brown such as Cognac. These compounds are removed by the charcoal treatment. The use of aged K$_2$S$_2$O$_5$ can result in precipitation of white crystal needles (leuco-fuchsin), which are insoluble in SO$_2$ water but soluble in dH$_2$O. This can cause serious problems with washing out residual reagent from the sample. The SO$_2$ water washes (0.5 g K$_2$S$_2$O$_5$ in 100 mL dH$_2$O + 5 mL 1 N HCl) should always be done for 30–45 min for removing nonbound reagent from the stained tissue. Plasmatic background can so be avoided.

Note: Ready-to-use Feulgen stain is also available: Sigma, Schiff reagent, Fuchsin-sulfite reagent (lot # 51H5014).

2.2.4.2.3 Staining of Root

1. Wash fixed root tips once with ddH$_2$O to remove the fixative. If formaldehyde has been used in the fixative, wash roots in running tap water thoroughly for 1 h or wash in three times for 15 min in ddH$_2$O.
2. Hydrolyze roots in 1 N HCl (60°C) for 6–10 min, depending upon the specimen. Instead of 1 N HCl at 60°C, 5 N HCl at 20°C for 60 min may be used. The latter is recommended for quantitative Feulgen staining (J. Greilhuber, personal communication, 2002).
3. After hydrolysis, rinse root tips in ddH$_2$O, remove excess water and transfer root tips to Feulgen stain.
4. Adequate staining can be achieved after 1 h at RT.
5. Cut stained region of root, place on a clean glass slide, add a drop of 1% aceto-carmine or 1% propionic-carmine, place a cover glass and prepare slide by squash method.

Tuleen (1971) modified the above procedures for barley chromosomes. After pretreatment, transfer roots to 1 N HCl for approximately 7 min at 55°C. Transfer hydrolyzed roots immediately to Feulgen stain. Keep roots in stain at RT for 2 h. If it is desired to keep materials for a longer period, store in a refrigerator. Squash in 1% aceto-carmine.

Palmer and Heer (1973) developed a cytological procedure by using Feulgen stain for soybean chromosomes after comparing several techniques. Their modified version is as follows:

1. Germinate seeds until roots are 7–10 cm long. Collect root tips after first 3 h of the 30°C period. Seed germinated in a sand box or bench in a greenhouse yields a large number of secondary roots after 5–7 days of germination. It is advised to germinate seeds of soybean or large seeded legumes in a sand bench or in vermiculite in the greenhouse.
2. Excise 1 cm of root tips, slit last 1/3 with a razor blade. Pretreat tips in covered vials in saturated solution PDB 15°C for 2 h. Pretreatment of roots with 0.1% colchicine for 1 1/2 h at RT proved to be better than PDB treatment. Pretreatment with 8-hydroxyquinoline at

16–18°C for 2–3 h has been found to be very efficient to arrest a large number of metaphase cells for soybean.

3. Wash root tips with ddH_2O, fix in freshly prepared 3:1 (95% ethanol: glacial acetic acid) for at least 24 h in covered vials.
4. Wash root tips in ddH_2O, drain the water and hydrolyze in 1 N HCl 10–12 min at 60°C.
5. Wash root tips once in ddH_2O, place in Feulgen stain in covered vials for 1 1/2–3 h at RT.
6. Wash root tips in ice-cold ddH_2O.
7. Place root tips in pectinase in spot plates for 1–2 h, at 40°C. After treatment in pectinase, root tips may be stored in 70% ethanol in covered vials at 4°C in a refrigerator, or slides may be made immediately.
   Note: This step may be omitted. After appropriate staining, wash roots with chilled distilled water once and store in cold water in a refrigerator. Chromosome staining is not distorted as long as roots are in chilled ddH_2O.
8. Place a root tip on a clean slide; remove root cap with a razor blade. Place less than 1 mm of root tip (dark purple region only) in a drop of propionic-carmine or aceto-carmine stain. Place a cover glass and prepare slide by squash method.

2.2.4.2.4   Softening of Root

Pectinase has been found to be very effective to soften root tips and has been used in various combinations and concentrations depending on the materials:

1. 500 mg pectinase, 500 mg cellulase, 10 mL ddH_2O, 6 drops 1 N HCl; before use wait 24 h after solution is made. The solution can be stored in a refrigerator (2°C) for 1–2 months. Soften roots for 1–1 1/2 h at RT (Kimber et al., 1975).
2. 2% cellulase (Onozuka-Rl0), 2% pectinase (Fluka): Soften roots for 5–8 h at RT (Hadlaczky and Kalmán, 1975).
3. 2% cellulase, 2% pectinase, adjusts pH 4.5 with 0.2 N HCl. Soften roots for 1–4 h at RT (Stack and Comings, 1979).

2.2.4.3   Alcoholic-Hydrochloric Acid-Carmine

Snow (1963) developed this stain from the study of mitotic and meiotic chromosomes of several plant species.

2.2.4.3.1   Ingredient

- 4 g (certified) carmine powder
- 15 mL dH_2O
- 1 mL concentrated HCl
- 95 mL (85%) ethanol

2.2.4.3.2   Preparation of Stain

1. Add 4 g certified carmine to 15 mL dH_2O in a small beaker.
2. Add 1 mL concentrated HCl. Mix well and boil gently for about 10 min with frequent stirring.
3. Cool, add 95 mL 85% ethanol and filter.

2.2.4.3.3   Staining of Root

1. For staining, wash the fixed materials with 2–3 changes of 70% ethanol, allowing at least 1 h for each change.
2. The specimen can be stored in the last change of ethanol in a refrigerator.
3. Place the drained specimen (roots or anthers) in stain for at least 24 h. Sometimes, several days or weeks are needed to stain compact tissues (e.g., heads of Compositae).
4. Pour used stain back into stock bottle and it can be reused.

5. Rinse the specimen with dH_2O or with 70% ethanol.
6. Maceration of tissues should be done in 45% acetic acid at 60°C and prepare slides by squash method.

Note: Aceto-carmine stain is not required and materials should not be hydrolyzed in 1 N HCl because HCl bleaches the stain.

2.2.4.4 Lacto-Propionic-Orcein

Dyer (1963) used this particular stain for a large number of crops. This stain is useful for plants with small or numerous chromosomes. Chromosomes are intensely stained and cytoplasm remains clear.

2.2.4.4.1 *Ingredient*

- 2 g orcein
- 50 mL lactic acid
- 50 mL propionic acid

2.2.4.4.2 *Preparation of Stain*

1. Add 2 g natural orcein to 100 mL of a mixture of equal parts of lactic acid and propionic acid at RT.
2. Filter and dilute the stock solution to 45% with dH_2O.

2.2.4.4.3 *Staining of Root*

Stain specimens in lacto-propionic-orcein stain for 2 min and prepare slide by squash method.

2.2.4.5 Carbol Fuchsin Staining

This technique was modified from Darlington and LaCour (1969).

2.2.4.5.1 *Ingredient-1*

1. Solution A
   a. 3 g basic fuchsin in 100 mL 70% ethanol
   b. 10 mL 3% basic fuchsin
   c. Add 90 mL of 5% phenol in distilled water

Note: Solution A can be stored in a refrigerator for a long time

2. Solution B
   a. 55 mL solution A
   b. 6 mL glacial acetic acid
   c. 3 mL formalin

Note: Solution B can be stored only for 2 weeks in a refrigerator

2.2.4.5.2 *Staining Solution (100 mL)*

- 20 mL solution B
- 80 mL 45% acetic acid
- 1.8 g sorbitol

2.2.4.5.3 *Ingredients-2*

This recipe has been formulated to reduce the concentration of formalin. The higher concentration of formalin is toxic.

1. Solution A
   a. 20 mL 3% basic fuchsin in 70% ethanol
   b. 80 mL of 5% phenol in ddH_2O

Note: Solution A can be stored in a refrigerator for a long time

2. Solution B
   a. 55 mL solution A
   b. 6 mL glacial acetic acid
   c. 3 mL formalin

Note: Solution B can be stored only for 2 weeks in a refrigerator

2.2.4.5.4   Staining Solution (100 mL)
- 10 mL solution B
- 90 mL 45% acetic acid
- 1.8 g sorbitol

Caution! Handle carbol fuchsin stain under a fume hood, store stain for at least 2 weeks at RT before use (aging of stain), and store working stain in a refrigerator.

2.2.4.5.5   Staining of Root
1. Stain roots for 14–16 h in a refrigerator.
2. Wash roots in cold ddH$_2$O (store water bottle in a refrigerator) at least 3 × in order to remove carbol fuchsin stain.
3. Store roots in cold ddH$_2$O in a refrigerator.
4. Squash a root tip under cover slip in a drop of 45% acetic acid.

Note: Roots can be stored for a long time in a refrigerator (Singh et al., 1998b; Xu et al., 2000a). Soybean metaphase chromosomes stain well with clear cytoplasm in carbol-fuchsin stain (Figure 2.2). In this cell, only a pair of satellite chromosomes (arrows) is clearly distinguished from the remaining soybean chromosome complement.

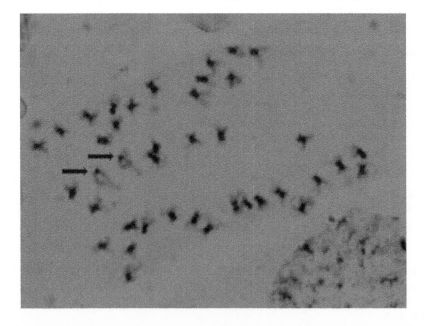

FIGURE 2.2   Somatic metaphase chromosomes of "Dwight" soybean with $2n = 40$ after carbol fuchsin staining. All chromosomes look similar except a pair of satellite chromosomes (arrows).

2.2.4.6   Giemsa Staining

2.2.4.6.1   Giemsa C-Banding Technique

The Giemsa C-banding technique has facilitated the identification of individual chromosomes in many plant species by its characteristic C-banding (constitutive heterochromatin) patterns. Several minor modifications in C-banding procedure have been made to obtain the maximum number of C-bands in the chromosomes. Procedures of seed germination and pretreatments described earlier can be followed. Fix roots in 3:1 (95% ethanol:glacial acetic acid) for at least 24 h. Soften cereal roots in 45% acetic acid or in 0.5% aceto-carmine. Pectinase and cellulase can also be used to soften the roots. Prepare slide by the squash method. Remove cover glass by dry-ice method (Conger and Fairchild, 1953).

2.2.4.6.1.1   Dehydration   The majority of researchers placed slides in 95%–100% ethanol for 1 h to obtain C-bands in *Secale* chromosomes (Thomas and Kaltsikes, 1974; Kimber et al., 1975; Singh and Lelley, 1975; Singh and Röbbelen, 1975; Bennett et al., 1977; Seal and Bennett, 1981) and *Hordeum vulgare* chromosomes (Linde-Laursen, 1975). However, Stack and Comings (1979) did not dehydrate rye chromosomes and obtained satisfactory C-banding patterns. In contrast, Limin and Dvořák (1976) and Fujigaki and Tsuchiya (1985) placed rye chromosomes slides in ethanol for 2.25 and 2 h, respectively, and observed C-banding patterns similar to those observed after 1 h dehydration. Ward (1980) treated maize chromosomes slides in 75% (10 min), 95% (1.5 h) and 100% ethanol (20 min). On the other hand, Mastenbroek and de Wet (1983) rinsed maize and *Tripsacum* chromosome slides once in 90% ethanol and twice in 100% ethanol.

After dehydration, keep air dried slides at RT overnight (Singh and Röbbelen, 1975). Linde-Laursen (1975) kept barley chromosome slides in a desiccator over silica gel for 1–2 weeks at 18°C. Air dried slides can be processed through BSG (Barium hydroxide/Saline/Giemsa) method.

2.2.4.6.1.2   Denaturation   Prepare fresh saturated solution of barium hydroxide [5 g Ba $(OH)_2 \cdot 8H_2O$ + 100 mL dH_2O)]. Solution is filtered in a Coplin jar. Keep Coplin jar in a water bath (50–55°C) or at RT. An alternative is to use hot dH_2O and cool down to 50–55°C or 20°C. Filtering is not necessary. Replace Ba $(OH)_2$ solution by cold water and rinse slides. (J. Greilhuber, personal communication).

Darvey and Gustafson (1975) prepared barium hydroxide by adding deionized distilled water at 80°C to remove CO_2 from the crystal of barium hydroxide. They shook the solution a few seconds and poured off the supernatant. Deionized water was then added to the crystals of barium hydroxide remaining in the bottom of the flask. Their protocol was the following:

Treat slides in barium hydroxide for 5–15 min at 50–55°C depending on the materials. Wash slides thoroughly in three or four changes of dH_2O for a total of 10 min and air dry the slides or transfer slides directly into 2 × SSC. Merker (1973) skipped denaturation step and still produced satisfactory C-bands in rye chromosomes.

2.2.4.6.1.3   Renaturation   Incubate slides in 2 × SSC (0.3 M NaCl + 0.03 M $Na_3C_6H5O_7 \cdot 2H_2O$), pH 7 to 7.6 at 60–65°C in a water bath or oven for 1 h (Merker, 1973; Thomas and Kaltsikes, 1974; Verma and Rees, 1974; Hadlaczky and Kalmán, 1975; Kimber et al., 1975; Singh and Röbbelen, 1975; Weimarck, 1975; Fujigaki and Tsuchiya, 1985). Incubation periods and temperatures are variable and can be determined by experimentation. After saline sodium citrate (2 × SSC) treatment, wash slides in 3 changes of dH_2O for a total of 10 min and air dry the slides.

2.2.4.6.1.4   Staining   Stain slides for 1–2 min with Giemsa stain, 3 mL stock solution + 60 mL Sörensen phosphate buffer (0.2 M), pH 6.9 (30 mL KH_2PO_4 + 30 mL $Na_2 HPO_4 \cdot 2H_2O$), freshly mixed. Monitor staining regularly. Leishman stain also produces results similar to Giemsa stain (Thomas and Kaltsikes, 1974; Darvey and Gustafson, 1975; Ward, 1980; Chow and Larter, 1981). After optimal staining, place slides quickly in dH_2O, air dry, store in xylene overnight, air dry again and mount cover glass in Euparal or Canada Balsam or Permount.

Preparation of Giemsa stock solution from powder (Kimber et al., 1975):

- 1 g Giemsa powder
- 66 mL glycerine
- 66 mL methanol

Dissolve Giemsa powder in the glycerin at 60°C for 1 h with constant stirring. Add methanol and continue stirring at 60°C for 1 day (24 h). Filter and keep in a refrigerator. It can be kept for one or two months.

Noda and Kasha (1978) did not use BSG method for banding barley chromosomes and developed the following protocol:

1. Collect barley roots, pretreat and fix as described earlier.
2. Wash roots in tap water.
3. Hydrolyze in 1 N HCl at 60°C for 7 min.
4. Treat roots in 1% pectinase at 37°C for 13 min.
5. Transfer roots in tap water and prepare slides by squash technique in a drop of 30% acetic acid.
6. After dehydration and maturation of slides, place slides in 1 N HCl at 60°C for 5 min, rinse slides in tap water, air dry at RT for 1/2 day.
7. Immerse slides in 0.07 N NaOH for 35 sec.
8. Stain in 60× diluted Giemsa solution in 0.067 M Sörensen phosphate buffer (pH 6.9) for about 30 min. After appropriate staining, rinse slides in tap water, air dry slides at RT, and mount cover glass in Canada balsam.

Limin and Dvořák (1976), working with rye, demonstrated that the combination of barium hydroxide and saline sodium citrate (SSC) treatment is necessary for obtaining clearly-differentiated dark-staining bands.

The C-banding procedure of Giraldez et al. (1979) for rye chromosomes is as follows:

1. Immerse slides in 0.2 N HCl at 60°C for 3 min.
2. Wash slides in tap water, treat in saturated solution of barium hydroxide [Ba $(OH)_2 \cdot 8H_2$] at RT for 10 min.
3. Wash in tap water, treat slides in 2 × SSC at 60°C for 1 h, stain in 3% Giemsa in phosphate buffer pH 7.0.
4. Wash in tap water, air dry, immerse in xylene for 5 min and mount in DPX.

Giemsa C-banding technique exhibit dark terminal heterochromatin and minor heterochromatic bands regardless of which technique we use. Figure 2.3a shows Giemsa C-banding pattern for the mitotic metaphase chromosomes of rye cultivar "Prolific" from the protocol described by Endo (2011).

Carvalho and Saraiva (1993) modified C-banding technique for maize chromosomes:

1. Treat 2-week-old slides with freshly prepared 5% Ba $(OH)_2 \cdot 8H_2O$ solution at 56°C with continuous agitation for 10–15 sec.
2. Wash slides in 70% ethanol (two changes), and then in 100% ethanol, and transfer slides to methanol: acetic acid (8:1).
3. Dry slides on a hot-plate for few min and stain with 3% Giemsa (MERCK) in phosphate buffer, pH 6.8, for 8–10 min, wash slides twice in dH_2O and air dry.

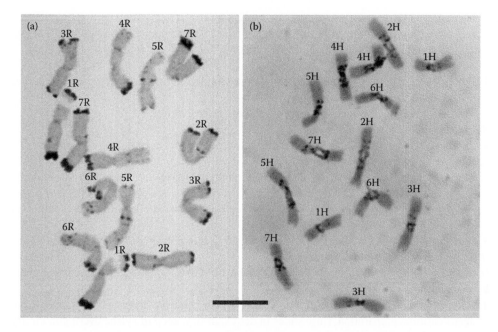

FIGURE 2.3 (a) Mitotic metaphase chromosomes of "Prolific" rye after Giemsa-Banding staining showing terminal dark heterochromatin with faint intercalary bands. (Adapted from T. R. Endo, personal communication, March 11, 2016.) (b) A mitotic metaphase chromosomes of "Shinebisu" barley after Giemsa N-banding showing centromeric heterochromatin with faint intercalary bands. (Adapted from T. R. Endo, personal communication, March 11, 2016.) Compare banding patterns of rye and barley chromosomes with C- and N-banding techniques.

2.2.4.6.2 Giemsa N-Banding Technique

This technique was originally developed to stain nucleolus organizing regions of mammalian and plant chromosomes (Matsui and Sasaki, 1973; Funaki et al., 1975).

1. Incubate slides at $96 \pm 1°C$ for 15 min in 1 N NaH_2PO_4 (pH 4.2 ± 0.2), adjust pH with 1 N NaOH.
2. Rinse thoroughly in distilled water and stain in Giemsa (dilute 1:25 in 1/15 M phosphate buffers, pH 7.0) for 20 min.
3. Rinse slide in tap water and air dry.

Gerlach (1977) modified the Giemsa N-banding technique of Funaki et al. (1975) for the staining of wheat chromosomes.

1. Incubate air dried slides in 1 M NaH_2PO_4 (pH 4.15) for 3 min at $94 \pm 1°C$.
2. Rinse slides in distilled water.
3. Stain for 30 min with a solution of 10% Gurrs Giemsa R66 in 1/15 M Sörensen's phosphate buffer (pH 6.8).
4. Rinse slides in tap water, mount in immersion oil.

Giemsa N-banding technique has been used with slight modifications following Gerlach (1977) to identify chromosomes of barley (Islam, 1980; Singh and Tsuchiya, 1982a, b), wheat (Jewell, 1981; Endo and Gill, 1984), and lentil (Mehra et al., 1986).

Singh and Tsuchiya (1982b) suggested that the Giemsa banding technique should be considered as a qualitative tool to identify individual chromosomes, while conventional staining methods should be used as a quantitative approach to establish the standard karyotype. The combination of aceto-carmine and Giemsa staining suggested by Nakata et al. (1977), used by Singh and Tsuchiya (1982a, b) for barley and Endo and Gill (1984) for wheat chromosomes, demonstrated that karyotype analysis could be conducted with greater precision than was previously possible.

1. Barley roots can be collected and fixed according to the procedures described earlier (Singh and Tsuchiya, 1982b).
2. Transfer fixed roots to aceto-carmine (0.3%) for about 2–3 h.
3. Prepare slides by squash method.
4. Photograph the cells with well-spread chromosomes by phase contrast.
5. Remove cover glass by dry-ice method.
6. Place slides in 96% ethanol for 2–4 h. Endo and Gill (1984) treated wheat chromosome slides with hot (55–60°C) 45% acetic acid for 10–15 min.
7. Air dry slides overnight at RT. Linde-Laursen (1975) placed slides in a desiccator over silica gel for 2–4 weeks.
8. Incubate slides in 1 M NaH_2PO_4, pH 4.15 for 1.5–2 min (5 min for barley) at 94 ± 1°C.
9. Rinse slides briefly in dH_2O, air dry.
10. Stain for 20 min to 24 h in 1% Giemsa (Sigma No. G04507) in 1/15 M Sörensen's phosphate buffer.
11. After optimal staining of the chromosomes, which were previously photographed from aceto-carmine preparation, rinse slides in dH_2O, air dry, keep in xylene overnight, air dry again, and mount in Permount.

Endo (2011) modified Giemsa N-banding technique described in this chapter. However, observed similar banding patterns for barley chromosomes (Figure 2.3b) than those reported by Singh and Tsuchiya (1982a).

2.2.4.6.3 HCl-KOH-Giemsa (HKG) Technique

Shang et al. (1988a, b) claimed that HKG technique produces well-separated and sharply-banded chromosomes, including centromeric bands of wheat chromosomes and that the results are more highly reproducible than C- or N-banding techniques.

1. Pretreat roots for 2.5 h with an aqueous solution of α-bromonaphthalene (0.01 mL stock solution in 10 mL water; stock solution: 1 mL α-bromonaphathalene in 100 mL absolute ethanol).
2. Wash roots twice in distilled water and hydrolyze with 5 N HCl for 20 min at RT.
3. Wash roots twice in distilled water, store material in 45% acetic acid.
4. Clean slides in 95% ethanol and wipe dry.
5. Prepare slide by taking 1 mm meristematic part of the root in 45% acetic acid.
6. Remove cover glass by liquid nitrogen. Cover glass can be removed by placing prepared slides in liquid nitrogen, by dry ice or by placing in an −80°C freezer. The purpose is not to lose the specimen or to distort the chromosome spread.
7. Air dry slide and store for 3–7 days.
8. Treat air dried slide with 1 N HCl at 60°C for 6 min and wash slides 4 times in distilled water for a total of 10 min at RT.
9. Air dry slide for half a day, dip slide into fresh 0.07 N KOH for 20–25 sec followed by dipping into 1/15 M Sörensen's phosphate buffer (pH 6.8) for 5–10 sec with shaking.
10. Stain in 3% Gurr's improved Giemsa stain solution (3 mL stain in 100 mL 1/15 M Sörensen's phosphate buffer, pH 6.8) for 1–2 h or until proper staining. The slides can be kept in the stain for 1–2 days without over staining.
11. Rinse slide in distilled water, air dry and mount in a synthetic resin (Preserveaslide, Mathesin, Coleman and Bell Co.).

2.2.4.6.4   Modified HKG-Banding Technique
The following explains the modified HKG-banding technique (Carvalho and Saraiva, 1993):

1. Hydrolyze 1–5 day old slides in 1 N HCl at 60°C for 4 to 6 min.
2. Wash slides four times in dH$_2$O for a total of 10 min.
3. Immerse slides briefly in 0.9% NaCl, plunge 10 times in 70% ethanol, dry slides on a hot plate (surface temperature 50°C) for few seconds.
4. Immerse slides in 0.06 N KOH solution for 8–12 s with continuous agitation at RT.
5. Wash slides in 70% ethanol (two changes), 100% ethanol and transfer to methanol:acetic (8:1).
6. Air dry slides on a hot plate for few minutes and stain according to procedure described for the modified C-banding procedure.

2.2.4.6.5   Giemsa G-Banding Technique
The G-banding technique was originally developed to identify human chromosomes. Kakeda et al. (1990) perfected G-banding method for maize chromosomes.

1. Germinate maize seeds in small pots filled with moist vermiculite for 2 days at 32°C under continuous light.
2. Excise 3 root tips about 1 cm long from each seed and pretreat either with a 0.05% colchicine solution or 0.05% colchicine solution containing either 10 ppm actinomycine D or ethidium bromide for 2 h at 25°C.
3. Dip root tips in the Ohnuki's hypotonic solution (55 mM KCl, 55 mM NaNO$_3$, 55 mM CH$_3$COONa, 10:5:2) for 30 min to 1 h at 25°C.
4. Fix root tips with methanol-acetic (3:1) for 1–4 days in a freezer (–20°C) or for at least 2 h at 25°C to actinomycine D-pretreated ones.
5. Remove meristem cells with a tweezer in a drop of fresh fixative or macerate enzymatically. **Note:** For enzyme maceration; wash fixed root tips for about 10 min and macerate in an enzyme mixture [2% Cellulase RS (Yakult Honsa Co., Ltd, Tokyo) and 2% MacerozymeR-200 (Yakult Honsa), pH 4.2] for 20–60 min at 37°C in a 1.5-mL Eppendorf tube. Rinse root tips with dH$_2$O two or three times. Pick up macerated roots tips with the help of a Pasteur pipette and place it on a glass slide.
6. Cut actively growing tip into small pieces with a sharp-pointed tweezer with the addition of fresh fixative.
7. Observe slide under phase contrast microscope, select slide with well-spread chromosomes and air dry for about 2 days in an incubator at 37°C.
8. Stain samples, prepared by the actinomycine D, directly in 10% Wright solution diluted with 1/15 M phosphate buffer (pH 6.8) for 10 min at 25°C, wash, and air dry slides.
9. Slides prepared by enzyme maceration, fix sample again in a 2% glutaraldehyde solution diluted with phosphate buffer for 10 min at 25°C and wash.
10. Immerse postfixed slides either in 2% trypsin (MERCK, Art. 8367) dissolved in PBS (pH 7.2) for 10 min at 25°C or in 0.02% SDS-dissolved TRIS-HCl buffer (20 mM, pH 8.0) for 2–25 min at 25°C.
11. Wash slides briefly and air dry.
12. Stain slides in 5% Wright solution in 1/30 M phosphate buffer (pH 6.8) for 5 min.

Endo (2011) compiled methodologies (C- and N-banding and FISH) for wheat chromosomes, described below:

1. Chromosome preparation: Fixative; one part glacial acetic acid, three parts (95%–99%) ethanol. Store at RT, need not to be freshly prepared. *This is a departure from the routinely fresh mixed fixative.*

2. Aceto-carmine stain solution: Dissolve 1 g carmine powder (Merck) in 100 mL 45% acetic acid and boil for 24 h, using a reflux condenser to prevent the solution from being boiled dry. Transfer to a bottle without filtration and store at RT temperature. Use the clear layer on the top. *This method of preparing aceto-carmine stain is different than those described by Tsuchiya (1971).*

3. For chromosome banding: He recommends Wright stain solution (Muto Pure Chemicals Co. LTD., Japan). Store the stain at RT.

4. Phosphate buffer: Prepare a stock solution with 0.1 M Na_2HPO_4 (14.2 g/1000 mL) and 0.1 M KH_2PO_4 (13.6 g/1000 mL).

5. 5% Ba $(OH)_2$ solution: Place 50 g Ba $(OH)_2$ in a glass container and pour hot tap water up to 1000 mL. The container should be stopped tightly for storage at RT. *This protocol is departure from the author used for banding staining.*

6. 2x SSC: Prepare a 20x SSC stock solution with 3 M NaCl and 0.3 M $C_6H_5Na_3O_7 \cdot 2H_2O$. Dilute 10-fold with distilled water for use.

Chromosome preparation:

1. Place root tips in a small vial filled with distilled water and immerse them in ice water for 16–20 h to collect metaphase cells.

2. Fix the root tips in the fixative for 1 day, stain them in aceto-carmine solution for 12 h, return to the original fixative, and store for 2–3 days for chromosome banding for 3–6 days for FISH. Perform all procedures at RT. The fixed and stained root tips can be stored at −20°C until use.

3. Stain the fixed and stain root tips again in the aceto-carmine solution for 10–20 min.

4. Make chromosome preparations by the squash method from the stained root tips and immediately store them at −70°C or below until use.

5. Remove the cover slip quickly from the frozen slide, using a razor blade, immerse the slide in 45% acetic acid at 40–45°C for 2–3 min, and air-dry the slide at RT. The air-dried slide can be used immediately or stored in an airtight container at −20°C up to several months.

Giemsa N-banding:

1. Incubate the air-dried slide in 1 M NaH_2PO_4 solution for 1.5 min at 92–95°C.

2. Wash the slide briefly with hard tap water.

3. Place the wet slide into the staining solution at RT until appropriate staining is achieved—about 2 h.

Giemsa C-banding:

1. Place the air-dried slide in a container with a lid.

2. Pour 5% Ba $(OH)_2$ solution into the container, put the lid on, and keep it for about 5 min at RT.

3. Take out slide, quickly wash with hard tap water, and incubate in 2x SSC for 10 min at 42–45°C.

4. Wash the slide briefly with hard tap water.

5. Place the wet slide into staining solution at RT until appropriate staining is achieved—usually 50 min.

6. Wash the slide briefly with hard tap water.

7. Air-dry the slide using a puffer (a camera tool).

8. Mount slide in immersion oil with or without cover slip.

2.3 SMEAR TECHNIQUE FOR PLANT CHROMOSOMES

The removal of cover glass is a prerequisite for Giemsa C- and N-banding techniques for plant chromosomes. To avoid this step, attempts are being made to prepare plant chromosome slides by smear technique, as used for the mammalian chromosomes. By smear technique, chromosome slides have been prepared from the roots (Kurata and Omura, 1978; Rayburn and Gold, 1982; Pijnacker and Ferwerda, 1984; Geber and Schweizer, 1988) and protoplasts (Mouras et al., 1978; Malmberg and Griesbach, 1980; Hadlaczky et al., 1983; Murata, 1983) of several plant species.

2.3.1 CHROMOSOME PREPARATION FROM ROOTS

1. Germination of seeds, collection and pretreatment of roots should be conducted as described earlier.
2. Cut only meristematic region (1 mm) of the roots and treat roots with hypotonic 0.075 M KCl for 15 min (potato—Pijnacker and Ferwerda, 1984), 20 min (rice—Kurata and Omura, 1978) at RT. Geber and Schweizer (1988) did not treat roots with hypotonic solution.
3. Fix roots in absolute ethanol/glacial acetic acid (3:1) or methanol/glacial acetic acid (3:1) for a minimum of 1 h (Geber and Schweizer, 1988) or 2 h or up to several months in a deep freeze (Pijnacker and Ferwerda, 1984).
4. Rinse fixed roots in 0.1 M citric acid–sodium citrate buffer, pH 4.4–4.8.
5. Enzyme treatment is quite variable: 10% pectinase (Sigma P-5146) + 1.5% cellulase (Onozuka R-10) in citrate buffer, at 37°C for 30 min (Pijnacker and Ferwerda, 1984); 6% cellulase (Onozuka R-10) + 6% pectinase (Sigma), adjust pH 4.0 with HCl, treat roots for about 60 min at 35°C (Kurata and Omura, 1978).

A mixture of pectinase, 20%–40% (v/v), (Sigma, from *Aspergillus niger*, P-5146, obtained as glycerol-containing stock solution) + 2 to 4% (w/v) cellulase (Calbiochem 21947 or Onozuka R-10) in 0.01 M citric acid-sodium citrate buffer was used for softening root tissues of *Sinapsis alba* (Geber and Schweizer, 1988). According to Kurata and Omura (1978), wash the roots in distilled water for 5–10 min at about 20°C to remove the enzyme. Place root meristem on a clean slide with a drop of fresh fixative (3 parts methanol: 1 part acetic acid) and break the root meristem into fine pieces with a needle. Add a few drops of the fixative and flame dry the slide.

Pijnacker and Ferwerda (1984) prepared slides of potato (*Solanum tuberosum*) by the following smear method:

1. Transfer one root meristem to a clean slide.
2. Remove the excess buffer; add a drop of 60% acetic acid, heat slide (without boiling) over an alcohol flame and leave for 2–5 min.
3. Suspend cells in a drop of acetic acid with the help of a fine needle, leave for one min, and heat slightly again with tilting of the slide.
4. Add Carnoy's solution to suspension when slide is cooled down (some seconds), add three more drops of Carnoy's solution on the top of the suspension, air dry the slides, and store overnight or longer and stain as needed.

Geber and Schweizer (1988) used the following procedure:

1. Spin the cells at about 4000 rpm in small conical centrifuge tube (10 mL volume) at each step to change the solution.
2. Remove the supernatant carefully with a Pasteur pipette and resuspend the pellet in approximately 5 mL liquid.
3. Transfer fixed roots to buffer, collect only meristem tissues, rinse twice in buffer.

4. Soften the tissue in enzyme solution at 37°C for 1–2 h depending upon plant material.
5. Wash twice in buffer and finally suspend the pellet in a drop of buffer to prevent sticking together of protoplasts.
6. Add an excess of fixative (freshly prepared) and change it twice, suspend the pellet in a small amount of fixative and drop this suspension onto ice-cold tilted slides, air dry.
7. Age air dried slides overnight or longer before further processing.

Note: In addition to *Sinapsis alba*, the above procedure produced an excellent chromosome spread of *Vicia faba, Pisum sativum, Crepis capillaris, Calla palustris,* and *Spirodela polyrrhiza.*

2.3.2 Chromosome Preparation from Cell Suspension and Callus

Murata (1983) developed an air dry method, described below, to study chromosomes from suspension and callus cultures. The procedure has been divided into three steps: (1) pretreatment to accumulate metaphase cells, (2) cell wall digestion and protoplast isolation, and (3) application of air drying technique.

2.3.2.1 Suspension Culture of Celery (*Apium graveolens*)
1. Add 1 mL of 0.5% colchicine solution to 9 mL of cell suspension 2–3 days after subculture in a 100 × 15 mm Petri dish and place on a gyratory shaker (50 rpm) for 2 h.
2. Transfer 2 mL of cell suspension to 2 mL of enzyme solution in 100 × 15 mL Petri dish [enzyme solution: 2% cellulysin (Calbiochem, San Diego, CA), 1% macerase (Calbiochem) and 0.6 M sorbitol, (pH 5.5–5.6)]. Substitution of 1% pectolyase Y-23 (Kikkoman) for maceration produced comparatively faster protoplast isolation.
3. Seal the Petri dish with Parafilm and place on the gyratory shaker (50 rpm) for 3–4 h at RT (25°C).
4. Filter the cells and enzyme mixture through 60 μm nylon mesh into a 15 mL centrifuge tube and centrifuge (65 × g) for 3 min.
5. Rinse twice with 0.6 M sorbitol and suspend in 5 mL hypotonic solution (0.075 M KCl), and allow to stand for 7 min at RT (25°C).
6. Remove the supernatant following centrifugation (65 × g) for 5 min, gradually add fresh fixative 3 (95% ethanol):1 (glacial acetic acid) up to 5 mL, and allow to stand for 1 h.
7. Resuspend in fresh fixative following centrifugation; repeat twice and make the final volume of the fixed cells 0.5–1.0 mL.
8. Put 5–6 drops of the fixed cells by using a Pasteur pipette onto a wet cold slide and air or flame dry.
9. Stain 3–4 min with 4% Giemsa (Gurr's R66, Bio/medical Special) diluted with 1/15 M phosphate buffer (pH 6.8), rinse in phosphate buffer and distilled water, and air dry.
10. Mount slide in DePeX mounting medium (Bio/Medical Special).

2.3.2.2 Callus Culture of *Brassica carinata*
1. Put 10–20 mg calluses 5–7 days after subculture into a 15 mL centrifuge tube with 5 mL of Murashige and Skoog (1962) (MS) liquid medium containing 0.5% colchicine and allow to stand for 5 h.
2. Discard the liquid medium following centrifugation (100 × g) for 5 min and suspend in the fresh Carnoy's fixative (3:1) for 1 h.
3. Rinse twice with dH₂O and add 5 mL of enzyme solution as in step 2 of method I.
4. Place the centrifuge tube, sealed with cap, horizontally on the gyratory shaker (100 rpm) for 2 h.
5. Filter the cell suspension through 60 μm nylon mesh into another centrifuge tube.
6. The subsequent procedures are as described in steps 6–10 of celery.

Pijnacker et al. (1986) studied the chromosomes of leaf explants of potato by following smear technique:

1. Fix leaf explants (with developing callus) either directly in 3:1 (ethanol 100%: glacial acetic acid) at 4°C for about 24 h or pretreat in a saturated α-bromonaphthalene for 3 h at RT before fixation.
2. Rinse leaf pieces in distilled water, incubate in 15% (v/v) pectinase (Sigma 5146) + 1.5% (w/v) cellulase R 10 (Yakult) in citrate buffer pH 4.8 for 45 min at 37°C, rinse, and then keep in distilled water for a minimum of 2 h.
3. Transfer one leaf piece to a clean slide and add a drop of 60% acetic acid.
4. Make leaf pieces into a fine suspension.
5. Surround this suspension with cold fixative (3:1), and then add about 3 drops of fixative on the top of the suspension, and air dry the slide.
6. Stain chromosomes by the fluorescent and Giemsa techniques.

2.3.3 CHROMOSOME PREPARATION FROM FLOWERS

Murata and Motoyoshi (1995) developed a cytological procedure for *Arabidopsis thaliana* floral tissues of young buds.

1. Dissect out sepals from young buds (1.5–2 mm in length) under a dissecting microscope.
2. Transfer buds without sepals into a 1.5 mL microtube with 1 mL dH_2O, keep at 0°C in iced water for 24 h to allow accumulation of metaphase cells.
3. Fix buds in 3:1 (99% methanol:glacial acetic acid), and store at –20°C.
4. For slide preparation, rinse buds well with dH_2O.
5. Digest buds with an enzyme solution containing 2% (w/v) cellulase Onozuka R10 (Kinki Yakult) and 20% (v/v) pectinase (Sigma); incubate for 1.5 h at 30°C.
6. Suspend tissues with micropipette, rinse suspended cells with dH_2O and fix again.
7. After three changes of fresh fixative, drop fixed cells onto wet, cold slides and flame dry.
8. Stain slides with 4% (v/v) Giemsa solution, which is diluted with 1/15 M phosphate buffer, pH 6.8 for 20–30 min, and examine with a microscope.

2.3.4 CHROMOSOME PREPARATION FROM SHOOTS

Ma et al. (1996) developed a cytological procedure to study mitotic metaphase chromosomes from shoots of roses.

1. Collect actively growing terminal shoot (2–4 mm) in the morning during the spring time burst of growth. Place 5–10 shoot tips in one 5-mL tube and keep on ice water (0°C) for transport to the laboratory.
2. Remove the young outside leaves, cut terminal 2–4 mm portion of the shoot apex and place immediately in a pretreatment solution (0.1% colchicine + 0.001 M 8-hydroxyquinoline) for 4 h in the dark at ≈25°C.
3. Fix shoot tips in a freshly prepared mixture of 2 acetone:1 acetic acid (v/v) + 2% (w/v) polyvinylpyrrolidone (MW 40,000). Fix tissues for 24 h at ambient pressure.
4. Soak shoot tips in dH_2O for 1–24 h to elute the fixative and soften the tissues.
5. Hydrolyze shoot tips in 1 N HCl for 20 min at ≈25°C. Remove 1 N HCl, wash 2× in dH_2O and soak for 10 min in dH_2O.
6. Incubate shoot tips in 0.075 M KCl for 30–60 min. (Alternative: 0.01 M sodium citrate, pH 4.6, for 15–30 min).

7. Digest the shoot tips in 5% cellulase R10 (Yakult Honsha Pharmaceutical, Tokyo) + 1% pectolyase Y23 (Seishin, Tokyo) + 0.01 M sodium citrate at pH 4.6. Use 2–4 μL of enzyme mix per shoot tip. Digest shoot tip in enzyme for 3–4 h at 37°C.
8. Prepare slides either by one-slide maceration or by suspension and spreading of protoplasts.
9. One-slide maceration:
    a. Place one shoot tip on a microscope slide, blot excess water.
    b. Add 2–3 drops of 3:1 or 1 ethanol:1 acetic acid fixative.
    c. Disperse tissues gently with forcep tips, examine briefly at ×160, until most of the cells are settled onto the surface of the slide.
    d. Wash cytoplasmic debris with additional drops of fixative. Remove excess fixative by filter paper, air dry, stain, and cover with a cover slip.
10. Suspension and spreading of protoplast:
    a. Add about 480 μL dH_2O to microcentrifuge tube containing the shoot tips and enzyme mix, vortex tube vigorously for 30–60 s to break up the shoot tips. Discard un-dissociated pieces.
    b. Centrifuge at ≈700g_n fixed suspension protoplast for 5 min at 2–4°C and discard the supernatant.
    c. Re-suspend the pellet in ≈500 μL dH_2O. Transfer supernatant cell suspension to a new microcentrifuge tube.
    d. Centrifuge the suspension at ≈700g_n for 5 min at 2–4°C, discard the supernatant, and re-suspend pellet in ≈500 μL of 3 ethanol:1 acetic acid.
    e. Centrifuge the suspension at ≈700g_n for 5 min at 2–4°C, discard the supernatant and re-suspend the pellet in 15–20 μL freshly made 3 ethanol:1 acetic acid per shoot tip.
    f. Apply 5–10 μL of the protoplast suspension to a scrupulously clean microscope slide and allow it to air-dry (≈ 30%–50% relative humidity).
    g. For FISH, allow slides to air dry for 2–3 days at low humidity to improve cell retention.

2.4 POLLEN STAINING

2.4.1 POLLEN FERTILITY

Pollen viability or fertility can be ascertained after staining the mature pollen with aceto-carmine (1%), I_2-KI (potassium iodide; 1%), methylene blue and fuchsin, fluorescein diacetate, and DAPI.

Pollen fertility after aceto-carmine staining: Determination of pollen fertility by staining in aceto-carmine is routinely used for cereal crops. Collect flowers during anthesis. Dust pollen grains on a clean slide and add a drop of aceto-carmine, place a cover glass and heat the slide over a low flame but be sure not to boil. Heating dissolves the starch grains and facilitates the staining of sperm nuclei and the vegetative nucleus.

In barley, rye, maize, oats, and wheat, pollen grains with two well-developed sperm nuclei and one vegetative nucleus are considered functional pollen grains (Figure 2.4). Degenerated pollen grains are those containing two deformed sperm nuclei, two or one underdeveloped nuclei or empty pollen grains that are shrunken and without cytoplasm and nucleus.

Staining pollen grains with I_2-KI does not require heating, and fertile pollen grains turn black while sterile pollen grains are colorless.

Sometimes pollen fertility can be estimated by the pollen germination test. Dust fresh pollen grains onto the surface of 1 mL of germination medium in depression wells, cover to reduce evaporation, and observe after 1–3 h. Germination medium: 12.5%, 20%, or 25% sucrose in distilled water + 0.01% borate (Coleman and Goff, 1985).

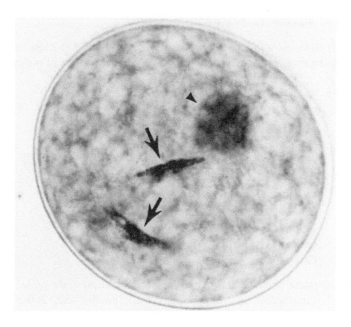

FIGURE 2.4   A fertile pollen grain with two sperm nuclei (arrows) and a vegetative nucleus (arrow head) in *Secale cereale* ($2n = 14$).

2.4.2   CHROMOSOME COUNT IN POLLEN

Stain:

Solution A: Dissolve 2 g hematoxylin in 100 mL 50% propionic acid. Allow the solution to age for about 1 week. The solution can be kept indefinitely in a stoppered brown bottle at RT (Kindiger and Beckett, 1985).

Solution B: Dissolve 0.5 g ferric ammonium sulfate [$FeNH_4 (SO_4)$] in 100 mL 50% propionic acid. The solution can be kept indefinitely in a stoppered brown bottle at RT.

Note: Mix equal volumes of solution A and B. The mixture, which turns dark brown, is ready for use immediately. The stain remains good for about 2 weeks.

The following protocol was developed to count chromosomes from maize pollen grains:

1. Collect fresh maize tassels (when about 3 cm of the tassel becomes visible above the leaf whorl) in a waxed short bag or other moisture-retaining container.
2. Fix tassel in a 14:1 mixture of 70% ethanol: formaldehyde and store in a refrigerator. This fixative produces better results than 3:1 (95% ethanol:glacial acetic acid).
3. Pretreatment is unnecessary for the fresh material. However, anthers can be pretreated with α-bromonaphthalene with 1 drop of DMSO for 30–40 min, fix anthers in glacial acetic acid.
4. Dissect out anthers and place in a drop of 45% acetic acid on a clean slide, add chloral hydrate crystal, remove debris, add a drop of stain, macerate and stain for a min, apply coverslip, heat (but not to boil), remove excess stain by filter paper by applying gentle pressure. Heat helps to darken the chromosomes and nuclei and clears the cytoplasm.
5. To make slide permanent, seal the edges of the coverslip with permount and allow to air dry. Stain begins to fade after 3 months.

2.4.3   DIFFERENTIAL STAINING OF POLLEN

A stain was developed by Alexander (1969) for examining aborted to nonaborted pollen grains. The pollen staining depends upon the concentration of stain, thickness of pollen walls, and pH of stain.

2.4.3.1   Ingredient

- 10 mL   95% ethanol
- 10 mg   Malachite green (1 mL of 1% solution in 95% ethanol)
- 50 mL dH_2O
- 25 mL   Glycerol
- 5 g   Phenol
- 5 g   Chloral hydrate
- 50 mg   Acid fuchsin (5 mL of 1% solution in dH_2O)
- 5 mg   Orange G (0.5 mL of 1% solution in dH_2O)
- 1–4 mL   Glacial acetic acid

Note: The amount of glacial acetic acid depends upon the thickness of the pollen walls: 1 mL (thin-walled pollen); 3 mL (thick-walled and spiny pollen); 4 mL (nondehiscent anthers).

2.4.3.2   Staining of Pollen with Thin Walls

1. Thin walled pollen occurs in *Phaseolus, Sorghum, Oryza, Triticum, Hordeum, Zea,* and *Lycopersicon.*
2. Stain pollen in one drop of stain, cover with a cover slip, heat over a small flame and examine microscopically.
3. If differentiation is not satisfactory, keep slide in an oven at 50°C for 24 h.
4. Mounted pollen can be stored for a week without fading of stain.

2.4.3.3   Staining of Pollen with Thick and Spiny Walls, or Both

2.4.3.3.1   Nonsticky Pollen

1. Acidify about 100 mL of the stain with 3 mL glacial acetic acid.
2. Add a small quantity of pollen into a small vial; pour enough stain to cover pollen.
3. Keep in a 50°C oven for 24–48 h.
4. Examine after 24 h for differentiation.
5. Add one drop of 45% acetic acid into the specimen vial and mix if aborted and nonaborted pollen are not well differentiated.

2.4.3.3.2   Sticky and Oily Pollen

1. Acidify about 100 mL stain with 3 mL glacial acetic acid.
2. Fix mature but nondehisced anther for 24 h in a fixative (3 ethanol:2 chloroform:1 glacial acetic acid). This fixative removes sticky materials.
3. Transfer through an ethanol-water series to water.
4. Remove the excess water by placing anthers between filter papers.
5. Cover 1 or 2 anthers with 1–2 drops of stain on a slide, split the anther wall with a needle to release pollen, remove debris.
6. Place a cover slip over the stain and store at 50°C oven for 24–48 h.
7. After 24 h, add 1–2 drops of stain along the sides of the cover slip to replace the amount of stain lost during evaporation.
8. Examine slide after 24 h. If the green color dominates, increase the acidity of stain mixture. This stain is suitable for *Hibiscus* and *Cucurbita.*

2.4.3.3.3   Staining of Pollen inside Nondehiscent Anther

1. Acidify 100 mL stain with 4 mL glacial acetic acid.
2. Collect anthers immediately after anthesis.
3. Fix for 24 h in a fixative:
   a. 60 mL   Methanol
   b. 30 mL   Chloroform
   c. 20 mL dH_2O
   d. 1 g   Picric acid
   e. 1 g $HgCl_2$
4. Transfer through 70%, 50%, and 30% ethanol, 30 min each, change gradually to hydrate the anthers, and finally rinse with dH_2O.
5. Remove excess water between two filter papers.
6. Stain anthers and keep at 50°C oven for 24–48 h.
7. Remove excess stain with blotting paper and examine. If over stained, remount in 25% glycerol containing 4% chloral hydrate and seal.

2.5   POLLEN-STIGMA INCOMPATIBILITY

2.5.1   STAINING PISTILS BY ANILINE BLUE

To ascertain compatibility in wide crosses, pollen germination, pollen tube growth, and fertilization are determined by fluorescence microscopy. An intersubgeneric hybrid between soybean and a wild perennial species, *Glycine clandestina*, of the subgenus *Glycine* is cited as an example.

1. Fix fertilized gynoecia in 1:1:18 [1 formaldehyde (HCHO approx. 37%):1 propionic acid: 18 ethanol (70%)] 24 h postpollination.
2. After 24 h fixation, wash gynoecia once with ddH_2O and transfer to 1 N NaOH at RT (25°C) for 24 h to soften the tissues.
3. Wash softened gynoecia twice, 2 min each, with ddH_2O and stain for 24 h in 0.1% water soluble aniline blue solution prepared in 0.1 N K_3PO_4 (Kho and Baër, 1968).
4. Squash gynoecia gently in a drop of 80% glycerin under a cover glass on a clean slide and observe by a fluorescence microscope. In a compatible cross of *G. clandestina* x *G. max*, pollen tube had already reached the ovules 24 h postpollination (Figure 2.5).

2.5.2   CLEARED-PISTIL TECHNIQUE

Young et al. (1979) developed a protocol known as cleared-pistil and thick-sectioning technique for detecting apomixis in grasses.

1. Collect inflorescences either from field or greenhouse grown plants at the time of stigma exertion.
2. Fix in formalin + glacial acetic acid + alcohol [known as FAA; 95% ethanol:water:40% formalin:glacial acetic acid; 40:14:3:3 (v/v)].
3. Excise fixed pistils and place in 50% ethanol in a 15×60 mm screw-cap vial.
4. Dehydration and clearing step:
   a. 70% ethanol → 85% ethanol → 100 ethanol (three changes)
   b. 1 (ethanol):1 (methyl salicylate) → 1 (ethanol):3 (methyl salicylate) → 100% methyl salicylate (two changes).
   Note: Place one excise pistil in a separate vial with 1 mL liquid for 30 min. Later, wrap 10–20 pistils from an inflorescence in a moist envelope from 12×12 mm square of single-thickness Kim wipes tissue, place in a vial, and clear using 2 mL solution for 2 h each step.

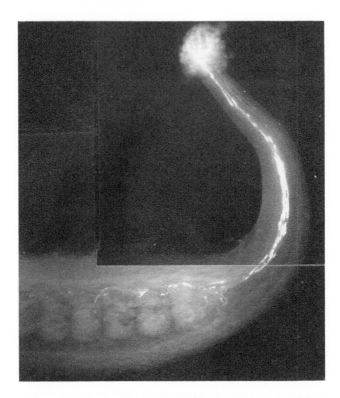

FIGURE 2.5 Pollen tube growth in a gynoecium following intersubgeneric hybridization in *Glycine clandestina* (2n = 40) × soybean (*G. max*) (2n = 40), 24 h postpollination. Pollen tubes are already in contact with ovules. (From Singh, R. J. and T. Hymowitz. 1987. *Plant Breed.* 98: 171–173. With permission.)

5. Change solution with a Pasteur pipette, store cleared pistils in methyl salicylate in vials and examine with an interference contrast microscope.
6. Mount cleared pistils in methyl salicylate under an unsealed coverslip on a microscope slide.
7. Record components of embryo sac and photograph with 35-mm Kodak Technical Pan Film, ASA 100.
8. After examination by interference contrast microscopy, transfer each pistil to a vial with 50% ethanol for embedding in plastic.
9. The infiltration steps are as follows:
   a. 50% ethanol → 75% ethanol → 100% ethanol (three changes) → 100% propylene oxide (two changes) → 50% Spurr's low viscosity embedding medium → 100% Spurr's (four changes). Steps through 50% Spurr's are 30 min each and after 50% Spurr's are 1 h each. At each Spurr's medium step, agitate the vials briefly on a vortex mixture, and evacuate.
   b. Spurr's low viscosity embedding medium (Spurr, 1969):
      i. Epoxy resin (ERL 4206: vinyl, cyclohexene dioxide; MW 140.18): 10 g
      ii. Flexibilizer (DER-736: curing agent (low viscosity), an epoxide MW 175–205, av. 380: 6 g
      iii. NSA (Nonenyl succinic anhydride): 26 g
      iv. S-1 or DMAE (Dimethylaminoethanol, curing agent): 0.2 g
   c. Cure schedule 16 h at 70°C.
10. Remove styles after the final changes, align ovaries in a size 00, square-tipped BEEM capsules, and cure the resin for 16 h at 70°C.

11. For sectioning, coat untrimmed blocks on the top and bottom edges with a mixture of Elmer's contact cement and toluene (1:2 v/v), dry overnight at RT and section at 1.8 μm with a glass knife on a Sorvall MT 2-B ultramicrotome.

12. Collect ribbons, 6–10 sections long, on a teflon-coated slide then transfer to a large drop of freshly boiled dH₂O on a slide coated with 0.5% gelatin. Expand ribbons, dry on a hot plate at 65°C.

13. Heat slide carefully for 20 sec over an alcohol flame to secure the sections.

14. Stain sections by floating the slide with 1% toluidine blue O in 0.05 M borate buffer at pH 9 for 4–8 min at 65°C.

15. Rinse slide briefly with each of dH₂O, 95% ethanol, 100% ethanol, and xylene.

16. Affix a cover slip with Permount.

2.5.3 MODIFIED CLEARED-PISTIL TECHNIQUE

Note: Pistils should remain in each following solutions at least for 30 min.

1. Fix inflorescence in 6:3:1 (v/v/v/) 95% ethanol:chloroform:glacial acetic acid for 24 h (Crane and Carman, 1987; C. B. do Valle, personal communication, 2002).

2. Store in 75% ethanol at −20°C.

3. Transfer pistils in 95% ethanol → 2:1 (v/v) 95% ethanol:benzyl benzoate → 1:2 (v/v) 95% ethanol:benzyl benzoate → 2:1 (v/v) benzyl benzoate:dibutyle phthalate (tetratogen)—clearing solution.
   Caution! Avoid contact with skin!

4. Remove solution carefully from pistil-containing vials by a fine bent Pasteur pipette. Some solution may remain with the pistil, but remove as much as possible without sucking the pistil into pipette.

5. Place cleared pistils on a microscope slide in a sagital optical section between two cover slips.

6. Fill spaces with the clearing solution.

7. A third cover slip is then placed on top, resting on the two other cover slips. Avoid formation of air bubbles.

Note: Time in clearing solution may be species specific. *Brachiaria* requires longer time than *Panicum*.

2.6 FLUORESCENCE *IN SITU* HYBRIDIZATION (FISH)

Fluorescence *in situ* hybridization is a very powerful technique for the detection of specific nucleic acid sequences and to localize highly repetitive DNA sequences in the specific regions of the chromosomes. Since the publication of Gall and Pardue (1969), numerous modifications and refinements of techniques have facilitated the diagnosis and identification of chromosomal aberrations particularly for human and animal chromosomes (Viegas-Pequignot et al., 1989; Lichter et al., 1990).

The application of *in situ* hybridization technique in plants has lagged behind compared to its use in mammalian cytogenetics. The main handicap of utilizing this technique for plant chromosomes is to obtain a high frequency mitotic metaphase cells without cell wall and cytoplasmic debris. These obstacles hinder hybridization of low-copy-number sequences to the chromosomes (Gustafson et al., 1990). *In situ* hybridization technique has been used to identify chromosomes in several plant species, particularly in wheat and its allied genera (Bedbrook et al., 1980; Hutchinson and Miller, 1982; Teoh et al., 1983; Rayburn and Gill, 1985a, b; Schubert and Wobus, 1985; Shen et al., 1987; Huang et al., 1988; Visser et al., 1988; Bergey et al., 1989; Le et al., 1989; Maluszynska and Schweizer, 1989; Mouras et al., 1989; Schwarzacher et al., 1989; Skorupska et al., 1989; Gustafson et al., 1990;

Mukai et al., 1990; Dhaliwal et al., 1991; Friebe et al., 1991a; Griffor et al., 1991; Lapitan et al., 1991; Leitch et al., 1991, 1996; Mukai and Gill, 1991; Schwarzacher and Heslop-Harrison, 1991; Wang et al., 1991; Hayasaki et al., 2001; Singh et al., 2001).

Singh et al. (2001) developed a procedure to detect repetitive DNA sequences on somatic metaphase chromosomes for soybean ($2n = 40$) root tips using fluorescence *in situ* hybridization after consulting above published results for the mammalian as well as plant chromosomes.

2.6.1 CHROMOSOME PREPARATION

2.6.1.1 Collection, Pretreatment, and Fixation of Roots

1. Germinate soybean seeds in a sand box or in vermiculite in the greenhouse (for cereals, seeds can be germinated on a moist filter paper in a Petri dish). Sand box should be kept moist (Murata, 1983; Ambros et al., 1986; Geber and Schweizer, 1988; Griffor et al., 1991; Singh et al., 2001).
2. After about 7 days, collect actively growing root tips (3–5 mm) from secondary roots.
3. Pretreat root tips in 0.1% colchicine for 2 h at RT. Often metaphase chromosomes are found to be clumped. Sometimes, 0.05% 8-hydroxyquinoline at 16°C for 4–5 h yields a higher mitotic index. For cereal chromosomes, pretreatment with ice cold water 18 h (barley) to 24 h (wheat) is recommended.
4. Fix root tips in a freshly prepared fixative methanol, acetic acid, and chloroform (3:1:1) for 2–3 h at –20°C or in 3 (95% ethanol):1 (propionic acid) or 3 (95% ethanol):1 (glacial acetic acid). However, roots can be stored for 1–2 weeks at –20°C.

2.6.1.2 Preparation of Buffer

2.6.1.2.1 Preparation of CA-SC Buffer

- 0.01 M citric-sodium citrate buffer pH 4.5–4.8 (CA-SC)
- Prepare from 10× stock solution consisting of three parts 0.1 M tri-sodium citrate and two parts 0.1 M citric acid, check pH. It should be in range of 4.5–4.8.

2.6.1.2.2 Preparation of SSC Buffer

1. 20 × SSC
   a. 70.12 g NaCl
   b. 52.8 g Na citrate
   c. 4 L ddH$_2$O, mix, filter sterilize, and adjust pH 7.4. (Stock)
2. 4 × SSC + 0.2% Tween-20 (designated as 4 × SSC)
   a. 400 mL 20 × SSC
   b. 1596 mL ddH$_2$O
   c. 4 mL Tween-20
3. 2 × SSC
   a. 100 mL 20 × SSC
   b. 900 mL ddH$_2$O

2.6.1.3 Enzyme Solution

The enzyme solution should be prepared fresh before use.

- 0.02 g cellulase (2%) cellulase Onozuka R-10 (from Yakult, Tokyo, Japan).
- 0.01 g pectinase (1%) pectinase, Sigma (cat. # P-2401).
- Dissolve in 1.0 mL 0.01 M CA-SC buffer in an Eppendorf tube.

2.6.1.4 Preparation of Slides

In between washes, cells are spun down at 2500 rpm for 5 min in an Eppendorf tube. Remove supernatant carefully with a Pasteur pipette.

1. Wash root tips 3 ×, 5 min each in 0.01 M CA-SC buffer in a spot plate at RT.
2. Place a Petri cover on some crushed ice, and dissect only 1–2 mm in length meristematic region of the root tips (cream colored at the very tip).
3. Transfer the tips into an Eppendorf tube with 0.01 M CA-SC buffer on ice.
4. Spin down in a microfuge for a few seconds. Withdraw the supernatant.
5. Add 1.0 mL of enzyme solution, seal with parafilm, and incubate in a 37°C water bath for 1 h and 30 min.
6. Aspirate the tips in the enzyme solution by drawing them into a Pasteur pipette several times. The tips should break apart and the solution should appear cloudy.
7. Again, incubate in a 37°C water bath for 10–15 min.
8. Spin down as stated above.
9. Add about a milliliter of 0.01 M CA-SC buffer and let it sit on ice for 35–45 min.
10. Spin (3000 rpm, 3 min), resuspend the pellet in 75 mM KCl (150 mM KCl has also produced good results) and let it sit for 10 min at RT.
11. Spin (5000 rpm, 3 min) and suspend in fresh fixative (three parts 100% ethanol: one part glacial acetic acid) and let it stand 17–20 min at RT.
12. Spin (5000 rpm, 3 min), suspend pellet in approximately 20–25 drops of fixative from a Pasteur pipette.
13. Place one drop of this suspension per slide using a Pasteur pipette.
    Note: Use slides that are ice-cold from −70°C freezer. Drop from about 3 cm above on the flat slide and blow very gently on slide to aid in spreading.
14. Air dry slides, once dried, can be stored in slide box sealed in a bag at −70°C for several weeks.
    Note: Slides can be prepared by squash method. Cut root tip and squash on a clean slide under a clean cover slip with a drop of 45% acetic acid.

2.6.2 Fluorescence *In Situ* Hybridization Procedure (FISH)

Check slides for good metaphase cells by mounting in 0.2 µg/mL Sigma DAPI in 1 × PBS, pH 7.0 or in 0.4 µg/mL Sigma propidium iodide in 1 × PBS, pH 7.0. Mark the position of cells from the scale on the microscope stage. Place the slides in a −80°C freezer for 5 min and remove the cover glass with a razor blade. Destain the slides by immersing in 45% acetic acid for 1–2 sec, rinse in water and air dry. *Scan slides for prometaphase and metaphase spreads by a phase-contrast lens. Remove cover slip after dipping slides in liquid nitrogen for few seconds. Cover slips can be also removed by dry ice and prepare* eight slides.

2.6.2.1 Prehybridization Method

1. Treat slides in a staining jar in 45% acetic acid for 10 min at RT and air dry for 1–2 h or overnight at RT.
2. Turn on water bath at 70°C, prepare humid chamber.
3. Take 24 µL (from 1 mg/mL stock) RNase stock solution + 776 µL 2 × SSC (800 mL). Add 100 µL RNase to each slide and cover with colored plastic cover slip (24 × 30 mm). Incubate at 37°C for 1 h in a humid chamber, drain slides (**Caution!** Slide should not get dry).
4. Treat slides in 70% formamide for 2 min at 70°C, drain slides.
5. Dehydrate slides at −20°C in 70% (5 min), 80% (5 min), 95% (5 min), and 100% (30 sec) ethanol. Air dry at RT for 1–2 h.

2.6.2.2 Hybridization

2.6.2.2.1 Probe (50 µL)

200 ng rDNA; [Internal transcribed spacer (ITS) region of nuclear ribosomal DNA probe (ITS1, 5.8S, ITS2; approximately 700 nucleotides)] (5 µL) + 19 µL dH$_2$O sterilized H$_2$O.

- Denature on PCR at 95°C for 5 min or in boiling water
- Keep on ice 5 min (immediately)
- Add 5 μL 10× dNTP + 20 μL 2.5 × random primer solution
- Mix (tapping by bottom)
- Add 10 μL Klenow from the freezer
- Mix (tapping by bottom)
- Incubate at 37°C for 3 h (Bio Prime™ DNA Labeling System, Life Technologies, cat. No.: 18094-011; lot no. EHDOO1)

2.6.2.2.2 Preparation of Hybridization Solution

Each slide needs 50 μL of solution.

- 200 μL 50% formamide
- 80 μL 50% dextron sulfate
- 53 μL TE buffer
- 40 μL 20 × SSC
- 17 μL salmon sperm (ss) DNA
- 10 μL probe

1. Mix hybridization solution in 1.5 mL tube, give quick spin, and transfer into three PCR tubes (two contains 150 μL and third contains 100 μL).
2. Denature at 80°C for 10 min (during this break prewarm humidity chamber), keep on ice for 10 min, give quick spin, mix, and keep on ice.
3. Apply 50 μL denatured hybridization solution on each slide, cover, and keep at 80°C for exactly 8 min, move humid chamber with slides quickly to 37°C for overnight for hybridization [at this time, prepare 4 × SSC + Tween-20 (will be known as 4 × SSC) and 2 × SSC].

2.6.2.2.3 Posthybridization (Next Day)

1. Turn on water bath (40°C), place two Coplin jars (designated as jar) with 2 × SSC, one with 35 mL 2 × SSC, one tube containing 35 mL formamide, and one bottle with 400 mL 2 × SSC.
2. Remove cover slip and drain slides.
3. Wash slides in 2 × SSC at 40°C for 5 min twice (discard used 2 × SSC and fill with fresh 2 × SSC). (During second wash, prepare 5% BSA: dissolve 0.24 g albumin bovine in 4.8 mL 4 × SSC), drain slides.
4. Treat slides in 50% formamide for 10 min at 40°C, drain slides.
5. Wash in 2 × SSC for 5 min twice at 40°C, wash third time in 2 × SSC at RT, drain slides.
6. Wash slides in 4 × SSC at RT for 5 min (during this time prepare humidity chamber), drain slides.
7. Add 200 μL 5% BSA, cover, treat for 5 min at RT (during at this time prepare 4 μL fluorescein-Avidin, DCS (FITC) + 796 μL 5% BSA = 800 μL; 10 μg/mL), drain slides.
8. Add 100 μL FITC to each slide, cover, incubate for 1 h at 37°C; (during this time, place three jars with 4 × SSC and one bottle with 4 × SSC in water bath at 40°C), drain slides.
9. Wash slide three times at 40°C in 4 × SSC for 5 min each [during this period, prepare 1600 μL in 2 mL tube (80 μL goat serum + 1520 μL 4 × SSC)], drain slides.
10. Add 200 μL goat serum to each slide, incubate for 5 min at RT in humid chamber [during this period, prepare 800 μL: 8 μL biotinylated antiavidin (10 μg/mL) + 792 μL 4 × SSC], drain slides.
11. Add 100 μL biotinylated antiavidin, cover, incubate at 37°C for 1 h, drain slides.
12. Wash slide four times in 4 × SSC at 40°C for 5 min, drain slides.
13. Repeat step 7.

14. Repeat step 8.
15. Repeat step 9 [during second wash, prepare propidium iodide (PI): 16 µL PI (1 mg/ mL) + 784 µL 2 × SSC].
16. Add 100 µL PI, cover, incubate in dark for 20 min at RT in humid chamber, drain slides.
17. Rinse slides briefly in 2 × SSC, drain, wipe the back of slides.
18. Add 50 µL (two drops) Vectashield mounting medium for fluorescence (Vector H-1000), cover with 22 × 32 mm cover glass, and store in dark for overnight.

2.6.2.3 Observation and Photography

Use either Zeiss Axioskop or Olympus fluorescence microscope. For Zeiss Axioskop:

1. Observe slides by using D (dark)-field.
2. Turn on UV power supply
3. Use 63 × oil lens.
4. Two levels of UV: 50% for general observation, 100% for photography by using 100 × oil lens.
5. Use neutral filter 0.1 m (−90% of light) to 0.3 m (60% light).
6. Use 400 ASA slide Fuji film.

2.6.3 GENOMIC *IN SITU* HYBRIDIZATION (GISH)

The genomic *in situ* hybridization technique is a powerful tool to distinguish genome of one parent from the other by preferential labeling of chromosome of one parent. It has proved boon to cytoge-neticists to identify precisely the inserted region in the recipient parent from the alien species and to examine evolutionary relationship of crops.

The protocol for GISH essentially is the same as for the FISH except for the blocking genomic DNA. The GISH protocol is defined to eliminate most of the cross-hybridization between total genomic DNA from the two species. Optimum results are obtained when blocking DNA exceeds the concentration of probe DNA by one hundred fold. Blocking genomic DNA fragments of 100–200 bp length are obtained by autoclaving the total genomic DNA from each parental species.

For GISH mixture, the yield of labeled probe can be increased by use of carrier DNA. The role of block and carrier DNA are combined in addition of unlabeled total genomic blocking DNA at 35 × the concentration of probe DNA immediately after stopping the NT reaction with EDTA. The BIONICK Labeling System is used for biotin-labeled DNA probes by nick translation. [BIONICK™ Labeling System (cat. no. 18247-015; lot no. KKF712)].

FISH\GISH method of Endo (2011):

1. Immerse the air-dried slide in the 0.15 N NaOH/ethanol solution for 5 min at RT.
2. Transfer the wet slide into a series of two 70% and one 99% ethanol solution for 5 min at RT
3. Dry the slide quickly with a puffer.
4. Apply the denatured hybridization mixture (10 µL per slide) onto the slide, and place a cover slip on it and incubate the slide in a moistened chamber for 6–24 h at 30°C.
5. Remove the cover slip with a pair of forceps (when the cover slip is firmly stuck on the slide, do not apply force to remove it; instead dip the slide in 2 × SSC and let the cover slip fall off) and immerse the slide for 3 min in 2 × SSC at RT.
6. Wash the slide briefly with distilled water and blow off water using a puffer.
7. Apply the detection mixture (10 µL/slide), place a cover slip on the slide and incubate in a wet chamber for about 1 h at 30°C.
8. Apply the counter staining solution (5 µL/slide) and cover with a cover slip for fluorescence microscopic observation.

FISH/GISH after chromosome banding (Endo, 2011):

1. After recording the images of chromosome banding, wash off immersion oil from the slide by dropping a mixture of xylene/99% ethanol for at least 10 min and 90% ethanol for at least 5 min at RT.
2. Dip the slide in 70% ethanol for at least 10 min and 90% ethanol for at least 5 min at RT.
3. Air-dry the slide using puffer.
4. Treat the slide as described for FISH/GISH.

2.6.4   MULTICOLOR GENOMIC *IN SITU* HYBRIDIZATION (McGISH)

This protocol was provided by K. Fukui (Personal communication, 2002). Multicolor GISH is an excellent cytogenetic tool for simultaneous discrimination of each genome and identification of diploid progenitors of allopolyploids, simultaneous mapping of different DNA sequences, physical ordering of multiple probes in a single chromosome, genome allocation of the gene of interest, detection of chromosomal aberrations, and examining chromosome organization in interphase nuclei (Mukai, 1996). Shishido et al. (1998) developed multicolor genomic *in situ* hybridization to identify somatic hybrids between *Oryza sativa* cv. "Kitaake" (AA, $2n = 24$) and *O. punctata* (BBCC, $2n = 48$) and the progeny rescued from embryo culture. The procedure is as follows:

1. Label total DNA of diploid rice species, *O. sativa* (AA), *O. punctata* (BB), and *O. officinalis* (CC) with biotin-16-dUTP (Boehringer Mannheim) or digoxigenin-11-dUTP (Boehringer Mannheim) by the standard random primed labeling protocol.
2. Denature hybridization mixture (100 ng labeled probe/slide + equal parts of 50% formamide and $2 \times$ SSC) for 10 min at 90°C and cool immediately on ice (0°C).
3. Denature chromosome spreads in 50% formamide/$2 \times$ SSC for 6 min at 70°C with the hybridization mixture (Fukui et al., 1994) and then hybridize for 3–4 days at 37°C.
4. Wash twice in $2 \times$ SSC, once in 50% formamide/$2 \times$ SSC, and once in $4 \times$ SSC each for 10 min at 40°C.
5. Apply a drop of Fluorescin-avidin (1% Vector laboratories, CA) + 1% bovine serum albumin (BSA) in BT buffer (0.1 M sodium hydrogen carbonate, 0.05% Tween-20, pH 8.3) on chromosome spread.
6. Incubate at 37°C for 1 h.
7. Wash slides three times (10 min each) at 37°C.
8. Amplify again (repeat step 5).
9. Apply a drop on chromosome spread antidigoxigenin-rhodamine in 5% goat biotinylated anti-avidin and antidigoxigenin-rhodamine in 5% goat serum in BT buffer, incubate at 37°C for 1 h.
10. Wash with BT buffer three times at 37°C (5 min each).
11. Apply a drop of 1% fluorescin-avidin and antisheep-Texas Red (1% Vector) in 1% BSA in BT buffer to chromosome spread.
12. Incubate at 37°C for 1 h.
13. Wash twice with BT buffer, and once with $2 \times$ SSC at 40°C (10 min each).
14. Counterstain chromosomes with 1 µg/mL 4′,6-diamidino-2 phenylindole (DAPI) in an antifadant solution (Vector Shield, Vector).

Note: Block chromosome spread with 5% bovine serum or goat serum albumin in BT buffer at 37°C for 5 min before each immunocytochemical step.

2.6.5   PRIMED *IN SITU* (PRINS) DNA LABELING

This protocol was provided by I. Schubert (Personal communication, 2002). PRINS DNA labeling technique, first described by Koch et al. (1989), is an alternative to FISH for the detection of nucleic acid. This technique involves labeling chromosomes by annealing an oligonucleotide DNA primer to the denatured DNA of chromosomes spread on slide glass and extending it enzymatically *in situ* with incorporation of labeled nucleotides. PRINS technique has used, extensively, in human cytogenetics for mapping of repetitive and low copy sequence, for chromosome identification, for detection of aneuploidy in sperm cells, the analysis of the human chromosome complement of somatic hybrids, for specifying chromosome rearrangements by combination of PRINS labeling of chromosome-specific alphoid sequences and chromosome painting (Pellestor et al., 1996; Gosden, 1997).

The reliable and reproducible detection of single-copy sequence below 10 kb in plants with large genome has been difficult by FISH. Menke et al. (1998) developed a procedure of PRINS for detection of repetitive and low-copy sequences on plant chromosomes. For chromosome preparation, follow the protocol used for FISH.

1. Prior to PRINS, wash slides three times in $2 \times$ SSC for 5 min at RT.
2. Treat slides with RNase (50 µg/mL in $2 \times$ SSC) for 40 min at 37°C. Subsequently a fill-in-reaction is carried out to reduce background signals caused by nicks within the chromosomal DNA, which may induce polymerase activity at sites of free 3' OH-ends.
3. To reduce background signals, wash slides first in $2 \times$ SSC and equilibrate in *Taq*-polimerase buffer (10 mM Tris-HCl, pH 8.4, 50 mM KCl, 1.5 mM $MgCl_2$) for 5 min.
4. Reaction mixture (20 µL)
   a. $1 \times$ PCR buffer (Boehringer)
   b. 1.5 mM $MgCl_2$
   c. 100 µM of each dATP, dCTP, dGTP
   d. 100 µM of 2',3'-dideoxy (dd)TTP
   e. 2U of *Taq*-DNA polymerase (Boehringer)
5. Drop reaction mixture on slide, cover with a cover slip and seal with Fixogum rubber cement.
6. Heat slides at 93°C for 90 s followed by 72°C for 20 min in a wet chamber (Zytotherm, Schutron). Before labeling, wash chromosomes and equilibrate again.
7. Labeling mixture (25 µL)
   a. $1 \times$ PCR buffer
   b. 1.5 mM $MgCl_2$
   c. 100 µM of each dATP, dCTP, dGTP, 75 µM dTTP
   d. 25 µM of digoxygenin-11-dUTP or fluorescein-12-dUTP (Boehringer)
   e. 4 µM of each of the corresponding oligonucleotide primers and 2.5 U of *Taq*-DNA polymerase
8. Seal the cover slip, denature chromosomes for 2–3 min at 93°C. Anneal primers at 55–60°C for 10 min and extend at 72°C for 40 min.
9. Wash slides twice in $4 \times$ SSC, 0.1% Tween-20 for 5 min at 42°C in order to stop the reaction.
10. Counterstain preparation labeled with FITC-dUTP with propidium iodide/4', 6-diamidino-2-phenylindole dihydrochloride (PI/DAPI, 1 µg/mL each in antifade) and examine immediately.

Detect Dig-dUTP with anti-Dig-FITC-Fab fragments from sheep (2 µg/mL, Boehringer):

11. Incubate samples in blocking solution ($4 \times$ SSC, 0.1% Tween-20, 3% BSA) at 37°C for 30 min.
12. Subsequently, apply Fab-fragments in detection buffer ($4 \times$ SSC, 0.1% Tween-20, 1% BSA) for 50 min at 37°C.
13. Remove un-specifically bound conjugates by washing slides in wash buffer three times for 5 min at 42°C.
14. After counterstaining, described above, examine slides using Zeiss epifluorescence microscope.

2.6.6 Fluorescence *In Situ* Hybridization on Extended DNA Fibers: Fiber-Fish

This protocol was contributed by S. A. Jackson, R. M. Stupar and J. Jiang, 2002. Fiber-FISH technique is a very powerful cytological tool to analyze large repetitive regions and increases the resolution of FISH analyses down to a few kb in the higher eukaryotic genomes (Stupar et al., 2001). It can be used to gauge the distances between adjacent clones up to ~500 kb and to measure repetitive loci up to ~1.7 mb. Combined with metaphase and interphase nuclei analysis, this tool allows molecular cytogeneticists to map loci to specific chromosomes and determine the distance between loci from a few kb up to several mb.

2.6.6.1 Protocol-I: Fiber-FISH on Extended Nuclear DNA Fibers

2.6.6.1.1 *Isolation of Plant Nuclei*

Isolation of plant nuclei is the same as for the preparation of high-molecular weight DNA embedded in agarose plugs. However, we suspend in 50% glycerol to store at −20°C. Almost all of the published protocols call for a 20 or 30 μm filtrations, but this can be omitted it with most of plant materials in order to obtain the maximum number of nuclei. If there is too much debris in the suspension or on slides, the last (20 μm) filtration can be added.

1. Freeze 2–5 g fresh leaf material in liquid nitrogen and grind to a fine powder with a pre-cooled (−20°C) mortar and pestle.
2. Transfer powder to a 50 mL centrifuge tube, add 20 mL chilled nuclei isolation buffer (NIB), and mix *gently* (make sure to break-up clumps) on ice for 5 min (in an ice bucket on a shaker). [NIB: 10 mM Tris-HCl pH 9.5, 10 mM EDTA, 100 mM KCl, 0.5 M sucrose, 4.0 mM spermidine, 1.0 mM spermine, 0.1% mercapto-ethanol. Prepare a large stock and store it in a refrigerator (4°C). Mercapto-ethanol should not be included in the stock. It can be added just before use.].
3. Filter through nylon mesh: 148 and 48 μm sequentially, into cold (on ice) 50 mL centrifuge tubes using cooled funnel (Nylon filters obtained from Tetko Inc, P.O. Box 346, Lancaster, NY, 14086, tel. 914-941-7767).
4. Add 1 mL NIB containing 10% (v/v) Triton x-100 (premixed) and *gently* mix the filtrate. The final concentration of Triton x-100 should be 0.5%. It removes any chloroplast contamination. Centrifuge at 2000 × g for 10 min at 4°C, decant the supernatant. If pellet is very small, skip the further cleaning steps and move directly to step 7. Otherwise, re-suspend the large pellet in 20 mL NIB (with mercaptoethanol added).
5. Filter through nylon mesh: 48 and 30 μm sequentially (optional), into cold (on ice) 50 mL centrifuge tubes.
   Note: A 22 μm filter step will lose a lot of nuclei. However, this makes the nuclei cleaner.
6. Add 1 mL NIB containing 10% (v/v) Triton x-100 (premixed) and *gently* mix the filtrate, as in step 4. Centrifuge at 2000 × g for 10 min at 4°C.
7. Decant the supernatant and resuspend the pellet in 200 μL to 5 mL of 1:1 NIB: 100% glycerol (neither mercaptoethanol nor Triton x-100 added) depending on the amount of nuclei harvested (Concentration ~5 × 10^6 nuclei/mL can be checked by staining with DAPI and examining under a microscope). Store at −20°C.

2.6.6.1.2 *Extension of DNA Fibers*

Extending the fibers is a critical step in Fiber-FISH. There are several methods of extending the fibers. The dragging method with a cover slip seems to give the most uniform results. When dragging, it is imperative that it be done slowly and smoothly. Poly-L-lysine slides obtained from Sigma can be used. These slides are treated so as to promote the adhesion of one or both ends of the DNA molecule. Silinated slides can also be used but seem to generate too much background signal. The calibration of the method should be checked occasionally by using BACs or cosmids of a known length as probes.

1. Identify the nuclei portion in the nuclei stock. The nuclei tend to settle near the bottom of the tube and the settling process can take a day or longer. The pellet may appear white and the nuclei often times sit right above this bottom film. The color of the nuclei is variable across species and samples, but normally clean nuclei have a gray/white coloration. Any layers above the nuclei tend to contain debris. Some people like to mix the nuclei stock prior to slide preparation by gently inverting the Eppendorf tube several times. It is not desirable as it mixes the debris with the nuclei.
2. With a cut P20 pipette tip, pipette 1–10 µL nuclei suspension (1–5 µL/slide depending on the suspension concentration) into ~100 µL NIB (minus mercaptoethanol and Triton) in an Eppendorf tube to dilute the glycerol. Gently mix the nuclei with the buffer and centrifuge at 3000–3600 rpm for 5 min. Remove carefully the supernatant with a pipette leaving only the nuclei pellet.
3. Re-suspend the nuclei in PBS (the final volume is 2 µL per slide). [PBS: 10 mM sodium phosphate, pH 7.0, 140 mM NaCl].
4. Pipette 2 µL suspension across one end of a clean poly-L-lysine slide (Sigma, Poly-Prep, Cat # P0425) and air dry for 5–10 min. The nuclei should dry to the point where it appears sticky; neither wet nor dry.
5. Pipette 8 µL STE lysis buffer on top of the nuclei and incubate at RT for 4 min. [STE: 0.5% SDS, 5 mM EDTA, 100 mM Tris, pH 7.0].
6. Slowly drag the solution down the slide with the edge of a clean cover slip held just above the surface of the slide, do not touch the cover slip to the slide surface as this will drag the nuclei completely off the slide. Air dry for 10 min.
7. Fix in fresh 3:1 100% ethanol: glacial acetic acid for 2 min.
8. Bake at 60°C for 30 min.
9. Slides can be used immediately but it can be stored in a box for several weeks. But it is suggested to use them immediately.

2.6.6.1.3   Probe Application

1. Apply 10 µL probe to the slide, then cover with a 22 × 22 mm cover slip and seal with rubber cement.
2. After the cement is dried, place the slide in an 80°C oven for 3 min in direct contact with a heated surface, then for 2 min in a wet chamber prewarmed in the 80°C oven.
3. Transfer wet chamber, with the slides, immediately to 37°C overnight. It is recommended to incubate at 37°C for longer periods, up to 3 or 4 days, especially for difficult probes.

2.6.6.1.4   Probe Detection

Three layer detection gives much stronger signal than does the single layer of antibodies. All antibody layers are composed of the antibodies diluted in the appropriate buffers at the concentration specified below:

Apply 100 µL antibody to each slide and a 22 × 40 cover slip is gently placed upon the antibody solution to promote an even spreading. All antibody layers are incubated in a 37°C wet-chamber for a minimum of 30 min. The first layer is often incubated for up to 45–60 min.

Note: The blocking step using 4 M buffer seems to help reduce some of the background noise. Dry bovine milk from Sigma works the best in the 4 M buffer, other substitutes (i.e., Carnation dry milk) tend to reduce the amount of signal. The 4 M and TNB buffers can be prepared at 5× and stored at −20°C. The wash solutions, 4T and TNT, can be prepared at 20× and 10×, respectively, and stored at RT.

1. One-Color Detection Protocol   Time, min (total)
   a. Wash in 2 × SSC — 5
   b. Wash in 2 × SSC 42°C — 10
   c. Wash in 2 × SSC — 5
   d. Wash in 1 × 4T — 5
   e. Incubate at 37°C in 4 M — 30
   f. Wash in 1 × 4T — 2
   g. Incubate FITC-Avidin (1 μL antibody stock/100 μL TNB buffer) — 30
   h. Wash three times in 1 × TNT — 5 (15)
   i. Incubate Biotin antiavidin (0.5 μL/100 μL TNB buffer) — 30
   j. Wash three times in 1 × TNT — 5 (15)
   k. Incubate FITC-Avidin (1 μL/100 μL TNB buffer) — 30
   l. Wash three times in 1 × TNT — 5 (15)
   m. Wash two times in 1 × PBS — 5 (10)
   n. Add 10 μL Prolong (Molecular Probes) or Vectashield (Vector Labs), cover with a 22 × 30 mm cover slip and squash

Notes:
- All 30 min incubation periods are at 37°C
- Antibodies:
  - FITC-avidin, 1 μL per 100 μL buffer
  - Biotin antiavidin, 0.5 μL per 100 μL buffer
  - Mouse anitdig, 1 μL per 100 μL buffer
  - Dig antimouse, 1 μL per 100 μL buffer
  - Rodamine antidig, 1–2 μL per 100 μL buffer
- Solutions:
  - 4 M: 3%–5% nonfat dry milk [Sigma, Cat # M7409] in 4 x SSC
  - 4T: 4 × SSC, 0.05% Tween 20
  - TNB: 0.1 M Tris-HCl pH 7.5, 0.15 M NaCl, 0.5% blocking reagent (Boehringer Mannheim)
  - TNT: 0.1 M Tris-HCl, 0.15 M NaCl, 0.05% Tween 20, pH 7.5
  - PBS: 0.13 M NaCl, 0.007 M Na_2HPO_4, 0.003 M NaH_2PO_4

2. Two-Color Detection Protocol   Time, min (total)
   a. Wash in 2 × SSC — 5
   b. Wash in 2 × SSC 42°C — 10
   c. Wash in 2 × SSC — 5
   d. Wash in 1 × 4T — 5
   e. Incubate at 37°C in 4 M — 30
   f. Wash in 1 × 4T — 2
   g. Incubate FITC antidig (1 μL antibody stock/100 μL TNB buffer) + Texas Red streptavidin (1 μL/100 μL), 37°C for 30 min
   h. Wash three times in 1 × TNT — 5 (15)
   i. Incubate FITC antisheep (1 μL/100 μL TNB buffer) + Biotinylated antiavidin (1 μL/100 μL), 37°C for 30 min
   j. Wash three times in 1 × TNT — 5 (15)
   k. Incubate Texas Red streptavidin (1 μL/100 μL TNB buffer), 37°C for 30 min.
   l. Wash three times in 1 × TNT — 5 (15)
   m. Wash two times in 1 × PBS — 5 (10)
   n. Add 10 μL Vectashield (Vector Labs), cover with a 22 × 30 mm cover slip and squash

Notes:
- Preparation of antibody stocks:
  - FITC antisheep comes ready to use and is stable at 4°C
  - FITC antidig needs 1 mL water resuspension.

3. Poly-L-Lysine Slide Preparation
   a. Boil slides in 5 M HCl for 2–3 h
   b. Rinse thoroughly with dH_2O then air dry
   c. Incubate overnight in filtered 10^{-6} g/mL poly-D-lysine (MW = 350,000, Sigma)
   d. Rinse thoroughly

4. Slide Silanation
   a. 30 min. 1:1 HCl:methanol
   b. Overnight 18 M sulfuric acid
   c. 8–10 washes in ddH_2O
   d. 10 min boiling ddH_2O
   e. 1 h 10% 3-aminopropyltriethoxy silane in 95% ethanol
   f. Rinse several times in ddH_2O
   g. Wash in 100% ethanol
   h. 80–100°C overnight before use

Note: The Poly-Prep slide from Sigma (Cat # P 0425) is ready to use, and is available and can be used instead of the above procedure.

2.6.6.2 Protocol-II: Fiber-FISH Using BAC and Circular Molecules as Targets

1. Prior to Preparing Slides
   a. Label appropriate probes (biotin and dig).
   b. Mini-prep BAC DNA (Solution I, II, III method followed by IPA precipitation; use 20 µL water for resuspension).
   c. Silanize 22 × 22 cover slips by dipping in Sigmacote for 10 min, then air dry.

2. Slide Preparation
   a. Prepare wet-chamber at 37°C, turn slide warmer up to 60°C.
   b. Dilute BAC DNA (w/cut P20 pipette tips) to appropriate level (dilute 1 µL BAC into 9 µL water). Add all 10 µL of diluted BAC to Poly-Prep slide (Sigma #P0425).
   c. Add 15 µL of *FISH lysis buffer** to BAC drop. Allow drop to spread. Let this sit at RT for ~5 min. Add water to the slide if it dries.
   d. Gently place ("drop") a silanized cover slip directly over the liquid (use tweezers to avoid air bubbles).
   e. Transfer slides to slide warmer. Allow slides to "bake" for 15 min. At this point, one should see the liquid begin to recede.
   f. Place slides in 3:1 (ethanol:glacial acetic acid), wait 1 min; gently shake slide to promote removal of the cover slip. Once cover slip falls off, transfer slides to new container of 3:1 and incubate for 1 min 30 sec. Transfer slides back to slide warmer for an additional 15 min.
   g. Add probe, denature, and detect as in nuclear fiber-FISH.

---

* FISH lysis buffer
  - 2% Sarkosyl
  - 0.25% SDS
  - 50 mM Tris (pH 7.4)
  - 50 mM EDTA (pH 8.0)

2.6.6.3   Protocol-III: Staining Fibers (Yo-Yo Staining)

1. Prepare slides for fiber analysis (as in BAC or nuclear fiber-FISH protocols).
2. Dilute Yo-Yo (molecular probes) in PBS following manufacturer's direction.
3. Add 100 μL of the Yo-Yo dilution to the slide and add a cover slip. Store in a dark place at RT for 10–20 min.
4. Wash three times in PBS (5 min each).
5. Short dry. Add antifade (Vectashield).
6. View slides.

2.6.6.4   Source of Chemicals

See Appendix B.1.

2.7   TOTAL DNA EXTRACTION: PLANT GENOMIC DNA

2.7.1   Protocol-I

2.7.1.1   Solutions

Miniprep salts buffer, pH 8.0 per 1 L. For more information on plant DNA preparation, see Dellaporta et al. (1983)

- 29.02 g   500 mM NaCl
- 12.1 g   100 mM Tris base
- 18.6 g   50 mM EDTA disodium
to 1000 mL final vol, pH 8.0 with HCl, and autoclave

Complete buffer (per sample) make just before use

- 50 mL Miniprep salts buffer
- 10 mM o-phenanthroline (Mol wt 198; is prepared by dissolving 90 mg o-phenanthroline in 500 μL ethenol)
- 0.5 g SDS (heat gently at 37°C to dissolve)

Immediately before use add

- 35 μL ß-mercaptoethanol
- 100 μL of 10 mg/mL ethidium bromide stock solution (remember this amount is needed for each sample, so multiply each amount by the number of samples extracted).
- 5 M potassium acetate (Mol wt 98) 98 g to 200 mL final volume in distilled water and no pH adjustment to 500 mL final volume, pH 8.0 with HCl, autoclave

Wash Buffer	Per 500 mL
• 1.51 g	• 25 mM Tris base
• 1.86 g	• 10 mM EDTA disodium

2.7.1.2   Procedure

1. Take ice, plastic bags, and collect fresh 2 g leaves either from the growth chamber or green house grown seedlings.
2. Grind young leaves or seedlings into fine powder using a mortar and pestle in liquid nitrogen.
3. Add 25 mL of complete buffer to mortar and continue grinding until homogenized. Pour into sterile GSA bottle. Wash pestle with additional 25 mL aliquot of complete buffer. Pour into bottle and cap.

4. Heat bottle at 65°C for 10 min, swirl occasionally.
5. Add 17 mL 5 M potassium acetate to bottle. Swirl to mix. Place in ice for 15 min.
6. Centrifuge at 9000 rpm for 10 min in JA14 rotor.
7. Filter supernatant through mira cloth into a clean (sterile) GSA bottle containing 33 mL cold isopropanol. Place in −20°C freezer for 20 min (should see DNA spool out).
8. Spool out DNA, if possible, using a glass sterring rod. Dissolve DNA in 9 mL of wash buffer (in an autoclaved plastic tube or sterile 30 mL COREX tube). If DNA will not spool, spin for 10 min at 9000 rpm, decant supernatant, dry pellets under air flow in hood and re-dissolve in 9 mL wash buffer on ice. Be sure pellets are re-dissolved well before proceeding.
9. Add 500 µL ethidium bromide (10 mg/mL stock).
10. Add 10 g CsCl, cover tightly with parafilm, invert to mix. CsCl must go into solution. Incubate 1 h at 4°C.
11. If there is any debris, centrifuge at 8000 rpm 15 min in the JA17 rotor. Rinse the rotor after using with CsCl.
12. If there is floating debris, filter supernatant through miracloth, collect in 15 mL graduated conical tubes. Protect from light. Adjust refractive index to 1.3870–1.3885 using the refractometer. Add wash buffer if the index ix too high, add solid CsCl if it is too low. (It is usually high.)
13. Fill and balance 13.5 mL polyallomer ultracentrifuge tube (fit the type 65 or 75 Ti rotors) as follows: Pipette the solution into each tube using an appropriate syringe with an 18 gauge needle. (Don't use Pasteur pipette because the tip breaks off too easily.) Balance the tube using the syringe needle to add or remove small amounts of the solution. If you don't have enough sample, fill and balance the tubes with a blank CsCl-TNE solution of the proper density. Leave a small air bubble just below the neck of the tube and have the tubes as usual, check the seal and recheck the balance.
14. Load the tubes properly in the Type 75 Ti rotor with the red tube spacer on top.
15. Spin at 20°C and 50,000 rpm for 40 h or you can use 60,000 rpm for 22 h but don't exceed 60,000 rpm. Use deceleration rate 9.
16. Remove tubes from rotor. Clean the rotor and parts after the run by rinsing out well with tap and distilled water, and drain. Cesium chloride is corrosive.
17. Visualize band with hand UV light if necessary. Collect band with 5 mL syringe and 18 gauge needle. Transfer to sterile 15 mL COREX tube.
18. Remove the ethium bromide from the DNA by extracting with butanol saturated with CsCl. (Butanol saturated with CsCl is prepared by adding a few grams of solid CsCl to 100% butanol. As the butanol is used up more can be added to the CsCl.) Add two volumes of the butanol to the DNA and mix well by vortexing. Spin at 1000 rpm for 1 min in either the clinical centrifuge or the JA 17 rotor. Remove the upper butanol layer and discard into the organic waste. Repeat the extraction until all of the ethidium bromide purple color has been removed. This will be about 3–4 extractions.
19. Can dialyze the DNA against 1 × TNE to remove the CsCl, but it is faster to ethanol precipitate:
    a. Add 2 volumes of distilled water. (This will dilute the CsCl so that it does not precipitate out.)
    b. Add 6 volumes of ethanol (i.e., 6 volumes of the original DNA volume).
       Note: Na acetate is not needed as CsCl salt concentration is already high.
    c. Incubate 1 h at −70°C unless the DNA spools out immediately.
    d. Centrifuge 15 min at 8000 rpm or in microfuge.
    e. Add 80% ethanol to remove residual salts from the tube, spin decant, dry briefly in air, and re-dissolve DNA in 500 µL of 1 × TE (10 mM Tris, 1 mM EDTA) in a microtube

(to be sure that DNA is in the Na salt form and there is not a lot of Cs that may inhibit enzymes):

    f. Reprecipitate the DNA by adding 1/10th volume of Na acetate and two volumes of ethanol; mix, freeze, spin, wash, and dry.

20. Dry pellet and re-suspend in 500 μL 1 × TNE or more if needed. Be sure all of the DNA is re-suspended evenly, this may take 10–15 min on ice to resolve, with gently inverting. Scan a 1/100 dilution. Adjust to desired concentration with 1 × TNE, rescan, and store at −20°C.

2.7.2 PROTOCOL-II

2.7.2.1 Solutions

- Sol-I: 15% sucrose, 50 mM Tris-HCl (pH 8.0), 50 mM Na$_2$EDTA, 250 mM NaCl
- Sol-II: 20 mM Tris-HCl (pH 8.0), 10 mM Na$_2$EDTA
- TE: 20% SDS, 7.5 M NH$_4$OAc, isopropanol, ethanol

2.7.2.2 Procedure

1. Grind the leaf tissue (50–100 mg) in a 1.5 mL tube with "pellet pestle" or glass rod in liquid nitrogen.
2. Add 1 mL of Sol-I to the ground sample, mix, and keep on ice for a few min.
3. Spin for 5 min at 5 K and discard the supernatant (nuclei in the pellet!).
4. Re-suspend the pellet in 0.3 mL of Sol-II by gently tapping and add 20 μL of 20% SDS, invert to mix.
5. Incubate the sample at 70°C for 15 min (remove parafilm because it will melt at 70°C).
6. Add 170 μL of 7.5 M NH$_4$OAc and mix thoroughly cover with parafilm inside out and keep on ice for 30 min.
7. Sediment the NH$_4$OAc-SDS pellet at 15 K for 5 min.
8. Transfer supernatant to new tube and add 0.6 vol (300 μL) of isopropanol and incubate on ice for 15 min, mix well and whitish cotton material will precipitate, keep on ice.
9. Pellet the DNA at 15 K for 5 min.
10. Wash the pellet with cold 80% ethanol, dry, and re-suspend in 50 μL of TE.

Note: Clean up; re-suspend in 100 μL TE + 50 μL 7.5 M ammonium acetate ~30 min on ice, transfer supernatant to a new tube + 300 μL ethanol, wash with 80% ethanol.

2.7.3 PROTOCOL-III

This protocol was taken from Rogers and Bendich (1985).

2.7.3.1 Buffer/Tissue

Weight of Tissue	<10 mg	10–50 mg	50–500 mg	>500 mg
2 × CTAB buffer	1 μL/mg	1 μL/mg	1 μL/mg	1 μL/mg
1 × CTAB buffer	Bring to 20 μL total volume (including time)	(a) None	(a) None	(a) None
		(b) 1 μL/mg	(b) 1 μL/mg	(b) 1 μL/mg
		(c) 2 μL/mg	(c) 2 μL/mg	(c) 2 μL/mg
Grinding	0.5 mL MT	0.5 mL or 1.5 mL MT	1.5 mL MT or M & P	M & P
Vessel for further manipulations	0.5 mL MT	0.5 mL MT	0.5 mL or 1.5 mL MT	CT

Note: To use this table, find appropriate column for the weight of the tissue.

Note that the amount of 1x CTAB buffer added depends on the tissue type:

1. Leaves, roots, shoots, seedlings, and tissue culture cells.
2. Diploid cotyledons.
3. Polyploid cotyledons (legumes), embryos, endosperm, whole seeds, whole grains, and pollen grains.
4. For polyploid cotyledons, weighing 30 mg, the tissue would be ground with dry ice in a 0.5 or 1.5 mL microcentrifuge tube, followed by addition of 30 µL of 2x CTAB buffer and 60 µL of 1x CTAB buffer.
5. For sample weighing 3 mg, the tissue would ground in a 0.5 mL centrifuge tube, followed by addition of 3 µL of 2x CTAB buffer and 14 µL of 1x CTAB buffer.
6. For dried specimens, add 5 µL of $1 \times$ CTAB buffer per mg of tissue and no $2 \times$ CTAB buffer.

Vessel abbreviation: MT, microfuge tube; M & P, mortar and pestle; CT, centrifuge tube.

2.7.3.2 Procedure

1. Weigh and grind tissue into a fine powder in dry ice in the appropriate vessel (see table above).
2. When grinding in microfuge tube, 3 mm glass or stainless steel rods, fil at one end so that it can fit into the bottom of the tube and use as pestles.
3. Proceed grinding, warm vessel slowly in order to evaporate the dry ice.
4. Immediately after all of the dry ice evaporated, 65°C $2 \times$ CTAB extraction buffer (2% [CTAB (w/v), 100 mM Tris (pH 8.0), 20 mM EDTA (pH 8.0), 1.4 M NaCl, 1% PVP (polyvnylpyrrolidone), MW 40, 000] and add $1 \times$ CTAB extraction buffer according to above table.
5. Heat the mixture in 65°C water bath for 1–3 min.
6. If mortar and pestle is used, transfer the mixture to a microfuge tube using a rubber policeman.
7. If the weight of tissue is less than about 50 mg, or genome size is small, add 25–50 µg of yeast tRNA to precipitate nucleic acid.
8. Add an equal volume of chloroform/isoamyl alcohol (24:1).
9. After gentle but thorough mixing, the tube is centrifuged at $11,000 \times g$ for 30 s.
10. Transfer the top phase to a new tube.
11. Add one volume of CTAB precipitation buffer [1% CTAB, 50 mM Tris (pH 8.0), 10 mM EDTA (pH 8.0)]. Mix gently.
12. Centrifuge the tube for 10 s (for fibrous precipitates) to 60 s (for fine precipitates).
13. Rehydrate the pellet in high salt TE buffer [10 mM Tris (pH 8.0), 1 mM EDTA (pH 8.0), 1 M NaCl].
14. Heat to 65°C for 5–10 min that facilitates rehydration of some of the pellets.
15. After complete rehydration, reprecipate nucleic acids with two volumes of ethanol.
16. Pellet the nucleic acids for 1 min by centrifugation, wash with 80% ethanol and dry in a desiccator.
17. Rehydrate nucleic acids pellet in a small volume of 0.1 xTE buffer [1 mM Tris (pH 8.0), 0.1 mM EDTA (pH 8.0)].

2.7.3.3 Determination of DNA Yields and Susceptibility to Restriction Enzymes

1. Treat nucleic acids with ribonuclease A (100 µg/mL) and T1 (10 units/mL) for 30 min at 37°C.
2. Treat parts of some preparation with DNase I (100 units/mL) for 1 h at 37°C.

3. Run a small sample of each nucleic acid solution to electrophoresis on 0.5% or 0.6% (for fresh tissue) or 1% or 1.5% (for herbarium and mummified specimens) agarose gels and compare to a standard amount of bacteriophage lambda DNA by densitometry of the fluorogram negative.

4. Estimate DNA length by comparison to intact lambda DNA and set of size marker (1 kb ladder from BRL).

5. Include standard amounts of unlabeled plasmid on the blot. Hybridize blot to a ^{32}P-labeled plasmid (pBD[4]) containing the yeast (*Saccharomyces cerevisiae*) ribosomal RNA genes at 30°C below the thermal denaturation point.

6. Analyze the autoradiogram of this blot by inspection.

7. For EcoRI and BamHI assays, incubate 1 μg of the DNA overnight with 5 units of the enzyme in the appropriate buffer.

8. Examine sample by gel electrophoresis.

2.7.4 PROTOCOL-IV

This protocol for DNA extraction was developed by modifying protocol III by Jeff and Jane Doyle, published first in *Phytochemical Bulletin* in 1987 and again in *Focus* 12:13–15, 1990. The DNA extraction procedure, described below, was kindly provided by Jeff Doyle (March 4, 2016).

2.7.4.1 Background

This procedure was developed in 1984 as an alternative to the lengthy, expensive, and low-yield cesium chloride-ethidium bromide ultracentrifugation procedure that required the isolation of nuclei, and which produced only degraded DNA from soybean leaves in our hands. In looking for a method that used small amounts of tissue, we found a paper by Saghai-Maroof et al. (PNAS 81:8014–8019, 1984) that was in turn modified from a procedure developed in the laboratory of Arnold J. Bendich (U. of Washington), which used the strong detergent CTAB. However, the Saghai-Maroof et al. procedure used lyophilized tissue, and there was no lyophilizer in the lab of Roger Beachy at Washington University in St. Louis, where we were working (JJD was a postdoc, JLD was a technician). We tried the Saghai-Maroof procedure exactly as described, but this failed with soybean leaves. We therefore modified the procedure by doubling the concentration of components of Saghai-Maroof et al.'s extraction buffer to compensate for the greater water content of fresh tissue. This very simple modification worked beautifully.

We did not publish the procedure because it was such a simple modification of existing procedures, until asked to do so by the editor of the *Phytochemical Bulletin*.

This was at the time that DNA variation was just beginning to be used as a source of characters for plant systematics studies, and interest was high; we were sending the procedure to many colleagues, so this seemed reasonable. *Phytochemical Bulletin* was the publication of the Phytochemical Section of the Botanical Society of America, and in 1987 it was a set of typewritten sheets stapled together. It does not appear to have been accessioned in any libraries, so it is impossible to find. In 1987, JJD and EE Dickson tested the procedure on herbarium specimens and preserved tissues, and published the results in the plant systematics journal, Taxon, but this paper is rarely cited even though it gives the same procedure.

Because of the inaccessibility of the original protocol, and given the ever-growing need for plant DNA minipreps, we were asked by the editor of *Focus* in 1990 to publish the protocol there, and did so. Unfortunately, *Focus* was also not a "real" scientific journal, but instead was the trade publication of GIBCO-BRL, Inc., so it cannot be found in libraries, either. Ironically, we discovered in 1990, while on a seminar trip to Texas A&M University, that Dr. Brian Taylor of that university had published an identical procedure, in *Focus*, in 1982 (Taylor & Powell, *Focus* 4:4–6), of which we were completely unaware! He had developed it directly from a Bendich procedure he had learned while at the University of Washington.

In 1990 we taught the technique to a large group of people at a NATO workshop on "molecular taxonomy" held in Norwich, England. The proceedings were published, so that the procedure has now appeared four different times:

Doyle, J. J. and E. E. Dickson. 1987. Preservation of plant samples for DNA restriction endonuclease analysis. *Taxon* 36:715–722.

Doyle, J. J. and J. L. Doyle. 1987. A rapid DNA isolation procedure for small quantities of fresh leaf tissue. *Phytochemical Bulletin* 19:11–15.

Doyle, J. J. and J. L. Doyle. 1990. A rapid total DNA preparation procedure for fresh plant tissue. *Focus* 12:13–15.

Doyle, J. J. 1991. DNA protocols for plants. pp. 283–293 in: G. Hewitt, A. W. B. Johnson, and J. P. W. Young (eds.), *Molecular Techniques in Taxonomy.* NATO ASI Series H, Cell Biology, Vol. 57.

2.7.4.2 Buffer/Tissue

1. CTAB isolation buffer (2% hexadecyltrimethylammonium bromide [CTAB; Sigma H-5882])
2. 1.4 M NaCl
3. 0.2% 2-mercaptoethanol
4. EDTA 100 mM Tris-HCl, pH 8.0)
5. Glass centrifuge tube

2.7.4.3 Procedure

1. Preheat 5–7.5 mL of CTAB isolation buffer (2% hexadecyltrimethylammonium bromide [CTAB: Sigma H-5882], 1.4 M NaCl, 0.2% 2-mercaptoethanol, 20 mM EDTA, 100 mM Tris-HCl, pH 8.0) in a 30 mL glass centrifuge tube to 60°C in a water bath.
2. Grind 0.5–1.0 g fresh leaf tissue in 60°C CTAB isolation buffer in a preheated mortar.
3. Incubate sample at 60°C for 30 (15–60) min with optional occasional gentle swirling.
4. Extract once with chloroform-isoamyl alcohol (24:1), mixing gently but thoroughly.

    This produces two phases, an upper aqueous phase which contains the DNA, and a lower chloroform phase that contains some degraded proteins, lipids, and many secondary compounds. The interface between these two phases contains most of the "junk"—cell debris, many degraded proteins, etc.
5. Spin in clinical centrifuge (swinging bucket rotor) at room temperature to concentrate phases. We use setting 7 on our IEC clinical (around $6000 \times g$) for 10 min.

    This is mainly to get rid of the junk that is suspended in the aqueous phase. Generally the aqueous phase will be clear, though often colored, following centrifugation, but this is not always the case.
6. Remove aqueous phase with wide bore pipet, transfer to clean glass centrifuge tube, add 2/3 volumes cold isopropanol, and mix gently to precipitate nucleic acids.

    A wide bore pipet is used because DNA in solution is a long, skinny molecule that is easily broken (sheared) when it passes through a narrow opening. Gentleness also improves the quality (length) of DNA. In some cases, this stage yields large strands of nucleic acids that can be spooled out with a glass hook for subsequent preparation. In most cases, this is not the case, however, and the sample is either flocculent, merely cloudy-looking, or, in some instances, clear. If no evidence of precipitation is observed at this stage, the sample may be left at room temperature for several hours to overnight. This is one convenient stopping place, in fact, when many samples are to be prepped. In nearly all cases, there is evidence of precipitation after the sample has been allowed to settle out in this manner.

7. If possible, spool out nucleic acids with a glass hook and transfer to 10–20 mL of wash buffer (76% EtOH, 10 mM ammonium acetate).
   a. Preferred alternative: Spin in clinical centrifuge (e.g., setting 3 on IEC) for 1–2 min. Gently pour off as much of the supernatant as possible without losing the precipitate, which will be a diffuse and very loose pellet. Add wash buffer directly to pellet and swirl gently to resuspend nucleic acids.
   b. Last resort: Longer spins at higher speeds may be unavoidable if no precipitate is seen at all. This will result, generally, in a hard pellet (or, with small amounts, a film on the bottom of the tube) that does not wash well and may contain more impurities. Such pellets are difficult to wash, and in some cases we tear them with a glass rod to promote washing at which point they often appear flaky.
       Nucleic acids generally become much whiter when washed, though some color may still remain.
8. Spin down (or spool out) nucleic acids (setting 7 IEC, 10 min) after a minimum of 20 min of washing. The wash step is another convenient stopping point, as samples can be left at room temperature in wash buffer for at least two days without noticeable problems.
9. Pour off supernatant carefully (some pellets are still loose even after this longer spin) and allow to air dry briefly at room temperature.
10. Resuspend nucleic acid pellet in 1 mL TE (10 mM Tris-HCl, 1 mM EDTA, pH 7.4).
    a. Although we commonly continue through additional purification steps, DNA obtained at this point is generally suitable for restriction digestion and amplification, so we'll stop here.
    b. If DNA is to be used at this stage, pellets should be more thoroughly dried than indicated above.
    c. Gel electrophoresis of nucleic acids at this step often reveals the presence of visible bands of ribosomal RNase as well as high molecular weight DNA.
11. Add RNAase A to a final concentration of 10 µg/mL and incubate 30 min at 37°C.
12. Dilute sample with 2 volumes of distilled water or TE, add ammonium acetate (7.5 M stock, pH 7.7) to a final concentration of 2.5 M, mix, add 2.5 volumes of cold EtOH, and gently mix to precipitate DNA.
    DNA at this stage usually appears cleaner than in the previous precipitation. Dilution with water or TE is helpful, as we have found that precipitation from 1 mL total volume often produces a gelatinous precipitate that is difficult to spin down and dry adequately.
13. Spin down DNA at high speed (10,000 × g for 10 min in refrigerated centrifuge, or setting 7 in clinical for 10 min).
14. Air dry sample and resuspend in appropriate amount of TE.

2.8 DETERMINATION OF NUCLEAR DNA CONTENT OF PLANTS BY FLOW CYTOMETRY

2.8.1 INTRODUCTION

Determination of accurate amount of nuclear DNA content is extremely important to understand the hereditary constituent of an organism. Flow cytometry, originally developed for medical studies, is an easy, rapid, accurate, and convenient tool for estimating plant genome size, ploidy level, assessing DNA content and analyzing the cell cycle (Doležel et al., 1989, 1991; Awoleye et al., 1994; Bharathan et al., 1994; Baranyi and Greilhuber, 1995, 1996; Greilhuber, 1998; Winkelmann et al., 1998). A voluminous information on the nuclear DNA contents in plants is being published by Bennett and his colleagues (Bennett and Leitch, 1995, 1997, 2001; Bennett et al., 1998, 2000; Hanson et al., 2001).

In addition to determining the nuclear DNA content, flow cytometry in higher plants is used for studying plant protoplast (protoplast size, cell wall synthesis, chlorophyll content, alkaloid content, RNA content, protein content, protoplast-microbe interaction, sorting of protoplast fusion products), and chromosomes (chromosome size, centromeric index, sorting of large quantities of chromosomes of single type for gene isolation and mapping) (Doležel et al., 1999; www.ueb.cas.cz/olomoucl). Several protocols have been developed, described below, to estimate nuclear DNA contents in the higher plants.

2.8.2 Protocol-I

This protocol was provided by Arumuganathan and taken from Arumuganathan and Earle (1991).

2.8.2.1 Stock Solutions

1. $MgSO_4$ buffer:
   a. Dissolve in ddH_2O
   b. 0.246 g (10 mM) $MgSO_4 \cdot 7H_2O$
   c. 0.370 g (50 mM) KCl
   d. 0.120 g (5 mM) Hepes
   Adjust volume to 100 mL with ddH_2O, adjust pH 8.0 (Arumuganathan and Earle, 1991).
2. Triton x-100 stock (10% w/v): 1 g Triton x-100 in 10 mL ddH_2O
3. Propidium iodide (PI) stock (5 mg/mL): PI (Calbiochem, 537059): 5.0 mg/mL ddH_2O. Cover with aluminum foil and store in refrigerator
4. RNase (DNase free): (Roche Molecular Biochemicals, Cat # 1 119 915)
5. Alsever's Solution: Dissolve in ddH_2O
   a. 0.055 g citric acid
   b. 2.05 g glucose
   c. 0.42 g sodium chloride
   d. 0.8 g sodium citrate
   Adjust volume to 100 mL with ddH_2O, adjust pH to 6.1, and autoclave.
6. Chicken red blood cells (CRBC):
   Fresh chicken blood diluted with Alsever's solution to a concentration of ~10^7 CRBC/mL.[*] Store in refrigerator (can be used up to 4 weeks).

2.8.2.2 Final Solutions

The following solutions should be freshly prepared using the stock solutions, protected from light with aluminum foil, and keep on ice.

1. Solution A: (25 mL, enough for 15 samples):
   a. 24 mL $MgSO_4$ buffer (ice cold)
   b. 25 mg dithiothreitol (Sigma, D-0632)
   c. 500 µL PI stock
   d. 625 µL Triton x-100 stock
2. Solution B: (7.5 mL, enough for 15 samples):
   a. 7.5 mL solution A
   b. 17.5 µL RNase (DNase free)

2.8.2.3 Procedure

Note: The entire procedure should be done on ice or in a cold room

---

[*] CRBCs are used as an internal standard. Fresh chicken blood is collected into heparnized tubes from chicken that are 2 to 3 weeks old. Usually 1:100 dilution in Alsever's solution gives ~10^7 CRBC/mL. Human leucocytes (HLNs) [(6.5 ± 0.21) pg/2C] or red blood cells from rainbow trout (TRBC) [(5.05 ± 0.18) pg/2C] can also be used as internal standards; the values of these standards were determined by comparison with CRBCs in more than 20 experiments. Suspensions of nuclei from plants whose nuclear DNA content has previously been determined can also be used as internal standards.

2.8.2.3.1   Preparation of Suspension of Nuclei from Plant Tissues

1. Excise young leaves (without mid rib) from healthy plants.
2. Place pieces of leaves (~20 mg from test plant and ~20 mg from standard plant)[*] in plastic Petri dishes (35 × 10 mm) on ice.
3. Add 1 mL solution A and slice the tissues into thin pieces (≤0.5 mm pieces) by a sharp scalpel.
4. Filter the homogenate through 30-μm nylon mesh into microcentrifuge tube.
5. Centrifuge at high speed (15,000 rpm) for 15–20 sec and discard the supernatant.
6. Re-suspend pellet in 200–400 μL solution B.
7. Incubate for 15 min at 37°C.
8. Run the sample on a flow cytometer.

2.8.2.3.2   Preparation of Suspension of Nuclei from Plant Protoplast

1. Isolate protoplasts from callus,[†] cell suspensions or tissues by standard methods using wall-degrading enzymes.
2. Add 1 mL solution A to the pellet of protoplasts (~10^5) and incubate on ice for 10 min.
3. Filter the lysate through 33-μm nylon mesh into microcentrifuge tube.
4. Follow steps 5–8 described above for plant tissues.

2.8.2.3.3   Flow Cytometric Analysis of Suspensions of Nuclei

1. For estimation of DNA content of nuclei, measure the relative fluorescence of nuclei by using an EPICS PROFILE flow cytometer[‡] (Coulter Electronics, Hialeah, FL) with an argon ion laser operating at a wavelength of 488 nm. Filters included a 457–502 nm laser blocking filter and a 610 nm long-pass absorbance filter in front of the photomultiplier tube. The instrument is aligned and analog digital converter linearity is checked according to the procedure recommended by Coulter Electronics.
2. Objects are analyzed for logarithmic (LFL) and linear integral fluorescence (FL) and forward-angle light scatter (FS), a parameter related to particle size. Instrument protocol is set to generate three histograms:
   a. LFL versus FS (64 × 64 channel)
   b. Frequency versus LFL (256 channel, 3-decade log scale)
   c. Frequency versus FL (256 channel)

2.8.2.3.4   Analysis Procedure

1. Aspirate about 50 μL of stained nuclear suspension into EPICS PROFILE flow cytometer.
2. Adjust the high voltage of the photomultiplier tube (PMT) so that the signals corresponding to the populations of intact nuclei ($G_0 + G_1$ nuclei, $G_2 + M$ nuclei, CRBC nuclei or internal standard and nuclei of the highest ploidy) fall within the scale of the logarithmic fluorescence intensity.[§] *CRBC may be problematic as standard for plant samples* (R. Obermayer, personal communication).
3. Set a gating region around the zone containing signals from the intact nuclei on the bivariate histogram to eliminate extraneous signals from the conditioned histogram of fluorescence.

---

[*] Avoid the use of more than 50 mg tissue/mL solution A for chopping. Larger amounts of tissue not only increase the time spent for chopping and dilute the stain, but also increase the amount of debris and decrease the frequency of intact nuclei. Wash tissues excised from greenhouse or field-grown plants to remove debris and chemical residues that may fluoresce.

[†] It is sometimes easier to obtain clean preparations of nuclei from callus.

[‡] Preference depends upon the availability of flow cytometer. (Commercially available flow cytometers, two basic optical parameters of particles can be analyzed: light scatter and fluorescence emission).

[§] The logarithmic histogram can accommodate signals varying in amplitude by at least a factor of 1000. This is useful for detecting nuclear peaks corresponding to very large or very small genomes.

4. Adjust the gain of the fluorescence amplifier (and/or the high voltage of the PMT if necessary) so that the modes of the peaks corresponding to the CRBC nuclei or internal standard and the $G_0 + G_1$ nuclei of the test sample fall at desired positions within the scale of the histogram of the linear fluorescence intensity.[*]

5. Set analysis regions about peaks to calculate frequency of nuclei, mean relative fluorescence, standard deviation, and coefficient of variation (CV) for each peak.

6. Repeat the above steps for replicate measurements from samples.

2.8.2.3.5   Estimation of Nuclear DNA Content

Compare the mean position of the peaks due to the plant nuclei with the mean peak position of the internal standard nuclei.

$$\text{Nuclear DNA amount} = \left[\frac{\text{mean position of unknown plant nuclear peak}}{\text{mean position of known nuclear peak}} \right]$$
$$\times \text{DNA content of known standard}$$

Note: Suspensions of plant nuclei whose nuclear DNA content has previously been determined can also be used as internal standards.

2.8.3   PROTOCOL-II

This protocol was provided by Obermayer (2000).

2.8.3.1   Flow Cytometry after DAPI Staining

2.8.3.1.1   Chemicals
- Citric acid 1-hydrate ($C_6H_8O_7 \cdot H_2O$)—MERCK
- Triton x-100 ($C_{34}H_{62}O_{11}$)—MERCK
- Anhydrous di-sodium hydrogen phosphate (Na_2HPO_4)—MERCK
- Diamidino-2-phenylindole (DAPI)
- Sulforhodamine 101 (SR)

2.8.3.1.2   Stock Solutions
- DAPI: 0.1 mg/mL dH_2O, store at $-20°C$
- SR: 0.3 mg/mL dH_2O, store at $4°C$
- Triton x-100: 5% Triton (dissolve in dH_2O by incubation at $37°C$ and store at $4°C$)

2.8.3.1.3   Isolation Buffer
- 0.1 M citric acid 1-hydrate
- 0.5% Triton x-100 from 5% stock solution in dH_2O (e.g., 60 mL isolation buffer contains 1.26 g citric acid 1-hydrate + 6 mL (5%) Triton stock + 54 mL dH_2O)
- Store isolation buffer at $4°C$ for 3 weeks (maximum)

2.8.3.1.4   Staining Solution
1. Dissolve 0.4 M Na_2HPO_4 at $37°C$ (water bath) in dH_2O.
2. Cool at RT and mix with DAPI-stock in proportion 1:20 (5 μg/mL DAPI) and with SR-stock in proportion 1:100 (3 μg/mL SR). (Thus, 100 mL staining solution contains 5.679 g Na_2HPO_4, 5 mL DAPI- stock, 1 mL SR-stock and 94 mL dH_2O).
3. The staining solution can be stored at RT for about 3 months

---

[*] Peaks for the internal standard are kept at highest possible position between channel 32 and 224 of 256-channel histogram. This is usually done by keeping amplifier gain at 5 and adjusting the PMT voltage. Gain setting of less than 5 on the PROFILE causes nonlinear gain amplification. A simple statistical procedure using confidence intervals is used to compare mean DNA content of the strains.

2.8.3.1.5   Sample Preparation

1. Pour 0.3 mL isolation buffer into a small Petri dish (5 cm diameter), add 10 mg young leaves and cut into thin pieces with a sharp razor blade (use each side only twice). Slope the Petri dish and rinse down with 0.3 mL isolation buffer.
2. Filter 0.4 mL of the suspension through a 48 μm nylon tissue into a test tube (55 × 12 mm).
3. Add 2 mL staining solution (proportion 1 + 5) and mix carefully.
4. Measure the sample at the flow cytometer immediately or within few hours.

2.8.3.2   Flow Cytometry after Propidium Iodide (PI) Staining

2.8.3.2.1   Chemicals

1. Citric acid 1-hydrate ($C_6H_8O_7 \cdot H_2O$)—MERCK
2. Triton x-100 ($C_{34}H_{62}O_{11}$)—MERCK
3. Anhydrous di-sodium hydrogen phosphate (Na_2HPO_4)—MERCK
4. Ribonuclease A (RNase): This is important for good digestion—SIGMA Ribonuclease (Ribonuclease I; EC 3.1.27.5) Type I-AS; From Bovine Pancreas (9001-99-4) EC No. 232-646-6.
5. Tri-sodium citrate dihydrate ($C_6H_5Na_3O_7 \cdot 2H_2O$)—MERCK
6. Anhydrous sodium sulfate (Na_2SO_4)—MERCK
7. Propidium iodide (PI): SIGMA (95%–98% purity); P-4170, 25535-16-4

2.8.3.2.2   Stock Solutions

1. PI: Dissolve PI 1 mg/mL in dH_2O and store at 4°C; normally, prepare 120 mL and store in 10 fractions.
2. Triton x-100: Dissolve 5% Triton in dH_2O by incubation at 37°C and store at 4°C (e.g., 1.5 mL Triton + 28.5 mL dH_2O).
3. RNase stock solution: Dissolve 3 mg RNase/mL dH_2O, heat to 80°C (to destroy DNAse), store at −20°C in small portions in 2 mL Eppendorf tubes. The RNase stock solution is added after the isolation of the nuclei in a proportion of 1:20, which results in a final concentration of 0.15 mg/mL RNase. (0.060 g RNAse for 20 mL stock solution).
4. 10 × stock solution: Sodium citrate (100 mM) and sodium sulfate (250 mM): 5.882 g sodium citrate + 7.102 g sodium sulfate for 200 mL 10× stock) and store at 4°C.

2.8.3.2.3   Isolation Buffer

1. 0.1 M citric acid 1-hydrate, 0.5% Triton x-100 from 5% stock solution in dH_2O.
   (The 60 mL isolation buffer contains 1.26 g citric acid 1-hydrate and 6 mL 5% Triton stock solution in 54 mL dH_2O).
2. The isolation buffer can be stored at 4°C for 3 weeks.

2.8.3.2.4   Staining Solution

1. Dissolve 0.4 M Na_2HPO_4 in dH_2O.
2. Mix with 10 × stock of sodium citrate (100 mM) and sodium sulfate (250 mM) in a proportion of 1:10. Finally, add PI stock in a concentration of 0.06 mL stock/mL staining solution. (100 mL staining solution contains 5.679 g Na_2HPO_4 + 6 mL PI-stock + 10 mL 10× stock of sodium citrate and sodium sulfate + 84 mL dH_2O.)
3. The staining solution can be stored at RT for about 3 months in a dark bottle.
4. When a staining solution has a concentration of 60 μg/mL, a proportion of 1 part (0.4 mL) nuclei suspension + 5 parts (2 mL) staining solution results in a final PI concentration of 50 μg/mL.

2.8.3.2.5 Sample Preparation

1. Transfer 0.55 mL isolation buffer into a small Petri dish, add about 15 mg young leaves and cut with a sharp razor blade. Slope the Petri dish and rinse down with 0.55 mL isolation buffer.
2. Filter 0.95 mL of the suspension through a nylon tissue into a test tube (55×12 mm).
3. Add 0.05 mL RNase stock solution.
4. Digest at 37°C for 30 min in the water bath. (If necessary, sample can be stored overnight at 4°C.)
5. Divide the suspension of each sample into two test tubes, each containing 0.4 mL.
6. Add 2 mL staining solution to each test tube (proportion $1 + 5$) and mix carefully.
7. Incubate at least for 20 min at RT.
8. Measure the sample by the flow cytometer after 20 min or within a few hours.

2.8.4 PROTOCOL-III

This protocol has been developed by Professor J. Doležel and colleagues (www.ueb.cas.cz/olomoucl). The main advantages of flow cytometric analysis are as follows:

- Rapidity, precision, and convenience
- No need for dividing cells
- Nondestructive (requires small amount of tissue)
- Analysis of large population of cells (detection of mixoploidy)

2.8.4.1 Procedure

1. Chop 20 mg (higher quantities are usually needed for callus and cultured cells) leaves with a sharp scalpel in 1 mL of ice cold LB01 buffer in a Petri dish (Doležel et al., 1989).
2. It is preferable to include a DNA fluorochrome [DAPI ($2 \mu g/mL$) or PI ($50 \mu g/mL + 50 \mu g/mL$ RNase)] in the buffer prior to chopping.
   Note: A modified version; isolate protoplast, re-suspend in ice-cold LB01 to a concentration of 10^5–10^6/mL. The concentration of Triton x-100 in LB01 buffer should be increased to 0.5% (v/v). This improves the release of the nuclei from the protoplast. The protoplast should be 90%–100% viable.
3. Filter the suspension through 42 μm nylon mesh.
4. Store on ice prior to analysis (a few min to 1 h).
5. Analyze the relative DNA content by flow cytometry.

2.8.4.2 Fluorescent Staining of Nuclear DNA

Flow cytometry can quantify precisely the nuclear DNA content of cells by a suitable fluorescent dye. Nuclear DNA content can be estimated using several stains that bind specifically and stoichiometrically to DNA. Some of them quantitatively intercalate between pairs of double stranded nucleic acids, others bind selectively to DNA either to AT- or GC-rich regions.

The selection of fluorescent stains is important. The fluorescent molecules absorb and emit light at a characteristic wave length. Thus, fluorescent probes can be selectively excited and detected even in a complex mixture of fluorescent molecular species. The emitted light has always less energy because some of the absorbed energy is lost as heat. This phenomenon is called "stokes shift" that facilitates discrimination between the exciting light and the emitted fluorescence.

2.8.4.3 Cell Cycle in Higher Plants

The cell cycle in higher plants (eukaryotes) is divided into cellular (growth phase and division phase-cytokinesis) and nuclear events. The nuclear event includes mitosis (M) and interphase

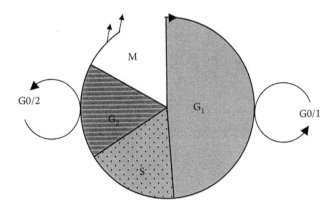

FIGURE 2.6 Diagrammatic figure of the cell cycle in plants. The interphase includes G_1, S, and G_2 phase. (M, mitosis). (Redrawn from www.ueb.cas.cz/olomoucl)

periods. Interphase is divided into a synthetic (S) phase, and two gaps, G_1 and G_2. Nonproliferating cells may leave the cell cycle either in G_1 or G_2 (Figure 2.6).

2.8.4.4 Nuclear DNA Content and the Cell Cycle

A diploid cell has a 2C nuclear DNA content during G_1 phase (i.e., two copies of nuclear genome). Nuclear DNA content doubles from 2C to the 4C level during the S phase of the cell cycle. The S phase is followed by G_2 phase (second period of cell growth), during which the DNA content is maintained at the 4C level. The DNA content returns suddenly to 2C level as G_2 phase moves into mitosis which generates two identical daughter nuclei (Figure 2.7).

2.8.4.5 Determination of Nuclear Genome Size

$$2C\ DNA\ (pg) = \frac{Sample\ G_1\ Peak\ Mean \times Standard\ 2C\ DNA\ Content\ (pg)}{Standard\ G_1\ Peak\ Mean}$$

A comparison of relative positions of G_1 peaks corresponding to the sample nuclei and the nuclei isolated from a plant with known DNA content, respectively, provide precise estimation of the "unknown" DNA content.

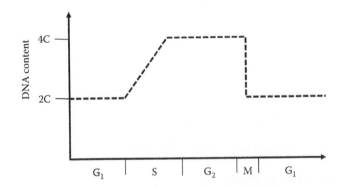

FIGURE 2.7 A graph showing changes of nuclear DNA content during the cell cycle. The G_1 phase constitutes 2C DNA content and G_2 4 C DNA content. (Redrawn from www.ueb.cas.cz/olomoucl)

2.8.4.6 Resolution of DNA Content Histograms

The resolution of histogram of nuclear DNA content is usually estimated by coefficient of variation of the peaks corresponding to cells in G_1 or G_2 phase of the cell cycles.

$$CV = \frac{\text{Standard Deviation}}{\text{Mean} \times 100}$$

The CV reflects the variation in nuclei isolation, staining, and measurement. The smaller the CV, the more accurate is the analysis of DNA content.

2.8.4.7 Instrument Calibration

The precision of flow cytometric measurement can be controlled using instrument standards (e.g., fluorescently labeled microspheres, nuclei isolated from chicken or fish red blood cells). Adjustment of the flow cytometer to give a maximum signal amplitude and minimal CV using instrument standard defines the optimal operating conditions.

2.8.4.8 Preparation of Fixed CRBC Nuclei for Instrument Alignment

1. Mix 1 mL fresh chicken blood with 3 mL CRBC buffer I. Centrifuge at 50 g for 5 min.
2. Discard the supernatant, add 3 mL CRBC buffer I and gently mix. Centrifuge at 50 g for 5 min.
3. Discard the supernatant, re-suspend the pellet in 2 mL CRBC buffer II, and vortex briefly.
4. Immediately add 2 mL CRBC buffer III, and mix briefly.
5. Centrifuge at 250 g for 5 min. Discard the supernatant. Add 2 mL CRBC buffer III, and mix gently.
6. Centrifuge at 120 g for 5 min, and discard the supernatant. Transfer the pellet to a clean tube, add 2 mL CRBC buffer III, and mix gently.
7. Centrifuge at 90 g for 5 min, and discard the supernatant.
8. Re-suspend the pellet, add 2 mL ice-cold fresh fixative (3 ethanol: 1 glacial acetic acid), and vortex briefly.
9. Leave overnight at 4°C, do not shake.
10. Gently remove the fixative, and resuspend the pelleted nuclei.
11. Add 6 mL of ice cold 70% ethanol, vortex briefly, and syringe through a 30G needle, three times.
12. Filter the nuclear suspension through a 42 μm nylon filter to remove large clumps.
13. Store in aliquots of 2 mL at −20°C. If the concentration of the nuclei is too high, dilute it using ice cold 70% ethanol.

Note: Isolated nuclei can be stored for up to several years without any sign of deterioration. Fixed nuclei are not suitable as a standard for estimation of nuclear DNA content in absolute units (genome size).

2.8.4.9 CRBC Buffers

2.8.4.9.1 CRBC Buffer I

- 1.637 g   140 mM NaCl
- 588.2 mg   10 mM sodium citrate
- 24.23 mg   1 mM Tris

Adjust volume to 200 mL, and adjust pH to 7.1.

2.8.4.9.2 CRBC Buffer II

- 0.41 g   140 mM NaCl
- 2.5 mL   5% (v/v) Triton x-100

Adjust volume to 50 mL.

2.8.4.9.3 CRBC Buffer III

- 54.77 g   320 mM sucrose
- 1.85 g   15 mM $MgSO_4 \cdot 7H_2O$
- 530 μL   15 mM ß-mercaptoethanol
- 60.57 mg   1 mM Tris

Adjust volume to 500 mL, and pH to 7.1.

2.8.4.10 Determination of Ploidy in Plants by Flow Cytometry

Flow cytometry helps to determine the ploidy level in plants without counting chromosomes. Nuclear DNA content of G_1 reflects the ploidy level in cells and that can be used to distinguish polyploid plants from the diploids.

Ploidy	DNA Content (G_1 phase)
$2n = x$	1C
$2n = 2x$	2C
$2n = 4x$	4C

2.8.4.11 Identification of Interspecific Hybrids

Flow cytometery may help to identify interspecific F_1 hybrids where parents differ by nuclear DNA content. If applicable, flow cytometry permits screening of large number of hybrids without counting chromosomes.

2.8.4.12 Identification of Aneuploid Lines

Flow cytometry can be used to identify aneuploid lines based on quantity of nuclear DNA content. It depends on the precision of the measure (characterized by coefficient of variation of G_1 peaks) and on the difference in DNA content between diploid and aneuploid plants. Basically, there are two approaches for detection of aneuploid plants using nuclear DNA flow cytometry. In both cases, internal standards should be used.

1. The use of euploid plant of the same species as an internal standard.
2. The use of a different species as in internal standard.

$$D = \frac{(G_1 \text{ Peak Ratio 2}) - (G_1 \text{ Peak Ratio 1})}{(G_1 \text{ Peak Ratio 1})} \times 100 \, (\%)$$

3. D = the relative difference in DNA content between euploid and aneuploid plant.

2.8.5 Protocol-IV

This protocol was provided by A. L. Rayburn; personal communication, 2002).

2.8.5.1 Chemicals

- Hexylene glycol (2-methyl-2, 4-pentanediol; Aldrich cat # 11,210-0)
- $MgCl_2$
- PEG 6000
- Propidium iodide (1 mg/mL stock)
- RNase

- Sodium chloride
- Sodium citrate (trisodium salt)
- Triton x-100

2.8.5.2   Extraction Buffers

- dH$_2$O   425 mL
- Hexylene glycol   65 mL
- 1 M Tris, pH 8.0   5 mL
- 1 M MgCl$_2$ 5 mL

2.8.5.3   Stain Solution A (10 mL)

- PEG 6000 (3% w/v)   0.3 g
- ddH$_2$O   4.0 mL
- 8 mM citrate buffer   4.4 mL
- PI (100 µg/mL)   1.0 mL
- RNase-180 unit/mL   0.5 mL
- Triton x-100 (0.1%)   0.1 mL

2.8.5.4   Stain Solution B (10 mL)

- PEG 6000 (3% w/v)   0.3 g
- ddH$_2$O   4.2 mL
- 0.8 mL NaCl   4.7 mL
- PI (100 µg/mL)   1.0 mL
- Triton x-100 (0.1%)   0.1 mL

2.8.5.5   Isolation and Staining of Nuclei

1. Germinate maize seeds in a 1:1 mixture of perlite and vermiculite. Expose plants to a day length of 14 h. The irradiance is approximately 100 µM photons m^{-2} S^{-1}. These conditions result in tissues that are most conducive to nuclear isolation.
2. Harvest seedlings after 2 weeks, remove 2.5 cm section of the seedling starting at the first mesocotyl, remove coleoptile.
3. Slice seedling section into disks by a clean razor blade of approximately 3 mm in length and place into an extraction buffer.
   Note: Keep sample at 4°C at this and all subsequent steps. Remember to turn on the water bath and the centrifuge. Also prepare the stain and salt solutions and place under the box on the bench so they are not exposed to light.
4. Place sample beakers in a beaker that is a little larger and full of ice. Turn the homogenizer (Biospec products, Bartlesville, OK) on and grind sample for 25 sec at 4500 rpm. Wash homogenizer with ddH$_2$O on high power for several seconds, then dry it by Kimwipes, and repeat this after every sample.
5. Filter homogenate through 250 µm mesh into the second beaker and then filter the sample through 53 µm into the third beaker. Place the sample in the labeled test tube.
6. Centrifuge the filtrate in 15-mL Corex tube at 500 g, 4°C for 15 min.
7. Remove the supernatant, resuspend the pellet in 300 µL of PI stain solution A and vortex for 10 sec.
8. Place in the water bath at 37°C for 20 min.
9. Add 300 µL stain solution B to each microfuge tube.
10. Place on ice, in the dark, in the refrigerator for at least 1 h.

Note: This method yields a large number of nuclei (maize; 250,000–30,000 nuclei per isolation).

2.8.6 VALIDITY OF GENOME SIZE MEASUREMENTS BY FLOW CYTOMETRY

Despite numerous and basically correct assertions that the flow cytometry is a fast and precise method for nuclear DNA content determination, published results obtained with this method have not always been reproducible (Greilhuber and Ebert, 1994; Baranyi and Greilhuber, 1996; Greilhuber and Obermayer, 1997; Doležel et al., 1998; Greilhuber, 1998; Temsch and Greilhuber, 2000; and Vilhar et al., 2001; J. Greilhuber, personal communication). Greilhuber and Obermayer (1997) reinvestigated, applying flow cytometry with DAPI and ethidium bromide as DNA stains and Feulgen densitometry, the genome size in nineteen of the twenty soybean cultivars examined by Graham et al. (1994), for which intercultivar variation up to 11% had been claimed. No reproducible genome size differences were found among these cultivars with either technique, or a correlation of genome size with maturity group as reported by Graham et al. (1994) was not confirmed. These discrepancies were attributed to the omission of internal standardization by Graham et al. (1994). Temsch and Greilhuber (2000) re-evaluated genome size variation in *Arachis hypogaea* reported by Singh et al. (1996) and failed to reproduce their results. Bennett et al. (2000) examined nuclear DNA contents of *Allium cepa* cultivars, an out breeder with telomeric heterochromatic, from diverse geographical and environmental origin and found no large difference in C-value (nuclear DNA content) between different cultivars.

Discrepancies in the genome size estimation from various laboratories may be attributed to technical variation, and lack of reliable methods for genome size estimation. There are two most widely used methods, Feulgen densitometry and flow cytometry, for determining DNA content. Flow cytometry is more convenient and rapid and has become an extremely more popular method than Feulgen densitometry. The simplicity of flow cytometry may be deceiving and may also, like densitometry, lead to the generation of flawed results as have been clearly demonstrated for the soybean, ground nut, pea, and *Allium*. It has been shown that fluorochromes which bind preferentially at AT or GC rich regions of DNA are not suitable for genome size estimation in plants. Thus, preference of DAPI over PI is not a wise choice. The proper concentration of fluorochrome is critical (Doležel, 1991).

Second, plants such as *Helianthus annuus* may contain so-called "fluorescence inhibitors," that is, still unidentified secondary metabolites which probably bind to chromatin and reduce the fluorescence yield (Price et al., 2000). This stoichiometric error in flow cytometry is comparable to the "self-tanning error" in Feulgen densitometry (Greilhuber, 1986, 1988), that is, the diminishing of the Feulgen reaction by tannins and related compounds that are contained in many plant tissues.

In order to test the reproducibility of flow-cytometric data and promote better standardization procedures, Doležel et al. (1998) conducted collaborative experiments among four laboratories and compared inter-laboratory results on plant genome size. Each laboratory used a different buffer/or procedure for nuclei isolation. Two laboratories used arc-lamp-based instruments and two used laser-based instruments. The results obtained after nuclei staining with PI (a DNA intercalator) agreed well with those obtained using Feulgen densitometry. The data recorded after nuclei staining with DAPI did not agree with those obtained using Feulgen densitometry. Small, but statistically significant, differences were observed between data obtained with individual instruments; differences between the same types of instruments were negligible while larger differences were observed between lamp- and laser-based instruments.

Also of relevance for the evaluation of the reproducibility of flow-cytometric results is the study by Vilhar et al. (2001) who examined the reliability of the nuclear DNA content measurements in two laboratories. Nuclear DNA content obtained by image cytometry was comparable to photometric cytometry and flow cytometry. Image cytometry exhibited little variation among repeated experiments within each laboratory or among different operators using the same instruments. Image cytometry produced accurate and reproducible results and may be used as an alternative to photometric cytometry in plant nuclear DNA measurements. They proposed two standards for quality control of nuclear DNA content measurements by image cytometry: (1) the coefficient of variation of the peak should be lower than 6% and, (2) the 4C/2C ratio should be from 1.9 to 2.1.

2.9 KARYOTYPING AND SORTING OF PLANT CHROMOSOMES BY FLOW CYTOMETRY

Sorting of individual chromosomes by flow cytometry is a valuable cytological tool in physical gene mapping, identification of chromosomes, isolation of molecular markers, and construction of chromosome-specific DNA libraries (Doležel et al., 1999). However, this technique has lagged behind in plants because of symmetric chromosomes of some plant species (Doležel et al., 1994). However, a range of tissues and culture types should permit selection of proper system for synchronization of cell division and collection of high yield of metaphase chromosomes from root tips of each plant species (Doležel et al., 1994; Kaeppler et al., 1997). Karyotyping of symmetrical chromosomes is feasible by flow cytometer (Pich et al., 1995). Mitotic index (MI) is estimated as: (number of cells in mitosis/total number of cells) × 100 (%).

2.9.1 PROTOCOL-I

2.9.1.1 Accumulation of Mitotic Metaphase

This protocol is highly efficient for the large seeded legumes (*Vicia faba, Pisum sativum*) and cereals (*Hordeum vulgare, Secale cereale, Triticum aestivum*). Four prerequisites are: (1) accumulation of cells at metaphase, (2) Preparation of chromosome suspensions, (3) Flow analysis and sorting of chromosomes, (4) Processing of sorted chromosomes. Cell-cycle synchronization and metaphase chromosome accumulation (Doležel et al., 1999).

Note: Perform all incubations at 25 ± 0.5°C in darkness and all solutions are aerated. Keep aeration stones and tubing clean to avoid bacteria and fungi contamination.

2.9.1.1.1 Large Seeded Legumes (Faba Bean, Garden Pea)
2.9.1.1.1.1 Materials for Flow Sorting
- Seeds
- Perlite or vermiculite
- 1× Hoagland's nutrient
- Hydroxyurea (HU)
- Amiprophos-methyl (AMP)
- Aquarium bubbler with tubing and aeration stones
- 4-liter plastic tray (25 cm long × 15 cm wide × 11 cm high)
- 750 mL plastic tray (14 cm long × 8 cm wide × 10 cm high) including an open-mesh basket to hold germinated seeds
- Biological incubator (heating/cooling) with internal temperature adjusted to 25 ± 0.5°C

2.9.1.1.1.2 Germination of Seeds
1. Imbibe 30–35 seeds (needed to prepare one sample) for 24 h in dH₂O with aeration.
2. Wet an inert substrate with 1× Hoagland's nutrient solution and put it into a 4-liter plastic tray.
3. Wash seeds in dH₂O, spread them over the surface of the wet substrate, and cover them with a 1-cm layer of wet substrate.
4. Cover the tray with aluminum foil and germinate seeds at 25 ± 0.5°C in a biological incubator in the dark. Optimum root length is ~4 cm and it takes about 2–3 days.
5. Remove seedlings from the substrate and wash in dH₂O.

2.9.1.1.1.3   Accumulation of Metaphase Cells in Root Tips
1. Select ~30 seedlings with primary roots of similar length.
2. Thread seedling roots through the holes of an open-mesh basket placed in a 750 mL plastic tray filled with dH_2O.
3. Transfer the basket with seedlings to a second plastic tray containing HU treatment solution.
4. Incubate seedlings of faba bean with main roots about 2 cm long for 18.5 h or 18 h (garden pea) at $25 \pm 0.5°C$.
5. Wash roots in several changes with dH_2O.
6. Immerse in HU- free 1× Hoagland's nutrient solution for 4.5 h at $25 \pm 0.5°C$.
7. Transfer the basket with seedlings to a tray filled with APM solution and incubate for 2 h at $25 \pm 0.5°C$.

2.9.1.1.2   Cereals

2.9.1.1.2.1   Additional Materials
- Cereal seeds
- Glass Petri dishes 18 cm diameter
- Paper towels cut to 18 cm diameter
- Filter papers cut to 18 cm diameter

2.9.1.1.2.2   Germination of Seeds
1. Place several layers of paper towel into an 18 cm Petri dish and top them with a single sheet of filter paper.
2. Moisten the paper layers with dH_2O.
3. Spread the seeds on the filter paper. Approximately 50 seedlings are required to prepare one sample (1 mL chromosome suspension).
4. Cover the Petri dish and germinate the seeds at $25 \pm 0.5°C$ in a biological incubator in the dark. (Optimum root length is 2–3 cm and is achieved in 2–3 days.)
5. Select ~50 seedlings with roots of similar length and process through HU and AMP treatment as described in accumulation of metaphase cells in root tips for large seeded legumes (1–7). For HU treatment, incubate seedlings for 18 h. For HU-free Hoagland's nutrient solution, incubate for 6.5 h.
6. Transfer the basket with seedlings to a plastic tray filled with an ice water bath (1–2°C).
7. Place the container in a refrigerator and leave for overnight.

2.9.1.2   Analysis of the Degree of the Metaphase Synchrony

2.9.1.2.1   Materials
- Root tips synchronized at metaphase
- 3:1 (v/v) ethanol/glacial acetic acid, freshly prepared
- 70% and 95% (v/v) ethanol
- 5 N HCl
- Schiff's reagent SIGMA (Lot # 79H5076)
- 45% (v/v) acetic acid
- Fructose syrup
- Xylene
- DePeX (Serva)
- Microscope slides
- 18 × 18 mm coverslips
- Coplin jars
- Microscope

2.9.1.2.2 Fixation and Staining

1. Collect 1 cm root tips in dH_2O; five roots may be enough.
2. Fix root tips in 3:1 ethanol/glacial acetic acid overnight at 4°C.
3. Remove fixative with washes (3×) in 70% ethanol. Fix root tips may be stored in 70% ethanol at 4°C for up to 1 year.
4. Wash root tips in several changes of dH_2O.
5. Hydrolyze root tips in 5 N HCl at RT for 25 min.
6. Wash in dH_2O and incubate in Schiff's reagent for 1 h at RT.
7. Wash root tips in dH_2O and soften for ~1 min in 45% acetic acid at RT.
8. To make slide for immediate count, cut purple dark stained meristem tip and squash it in a drop of fructose syrup between microscope slide and 18 × 18 mm cover slip. Prepare at least five different slides.
9. Analyze about 1000 cells from each slide and determine the proportion of cells in metaphase.
10. To make permanent slide, squash the purple dark stained meristem tip in a drop of 45% acetic acid and immediately place the slide on a block of dry ice or dip in the liquid nitrogen, remove cover slip, dehydrate in two changes of 96% ethanol in a Coplin jar and air dry overnight, dip slide in xylene and mount on a drop of DePeX or Permount, analyze about 1000 cells from each slide, and determine the proportion of cells in metaphase.

2.9.2 PROTOCOL-II

2.9.2.1 Suspensions of Plant Chromosomes

2.9.2.1.1 Materials

- Root tips synchronized in metaphase
- Formaldehyde fixative
- Tris buffer
- LB01 lysis buffer
- 0.1 mg/mL DAPI stock solution
- 5°C water bath
- 5 mL polystyrene tubes (Falcon 352008; Becton Dickinson)
- Mechanical homogenizer (Polytron PT 1300D with a PT-DA 1305/2E probe; Kinematica)
- 50 µm (pore size) nylon mesh in 4 × 4 cm squares
- 0.5 mL tubes for polymerase chain reaction (PCR)
- Microscope slide
- Fluorescence microscope with 10× to 20× objectives and DAPI filter set

2.9.2.1.2 Preparation of Chromosome Suspension

1. Cut root tips (1 cm) immediately after the APM treatment, rinse in dH_2O.
2. Fix in formaldehyde fixative at 5°C for 30 min (faba bean, garden pea, rye) or 20 min (barley).
3. Wash root tips three times (5 min each) in 25 mL Tris buffer.
4. Excise root meristem (1.2–2.0 mm) of 30 roots with a sharp scalpel in a glass Petri dish and transfer them to 5 mL polystyrene tube containing 1 mL LB01 lysis buffer.
5. Isolate chromosomes by homogenizing at 9500 rpm using a Polytron PT 1300D for 15 sec (faba bean and garden pea) or 10 sec (barley and rye).
6. Pass the suspension of released chromosomes and nuclei into a 5 mL polystyrene tube through 50-µm-pore-size nylon filter to remove large tissues and fragments.
7. Store the suspension on ice. It is recommended to analyze chromosomes on the same day, although can be stored overnight.

8. Transfer 50 µL chromosome suspension in a 0.5 mL PCR tube.
9. Add 1 µL of 0.1 mg/mL DAPI stock solution, place a small drop (~10 µL) of DAPI-stained suspension on a microscope slide.
10. Observe the suspension under low magnification (10× or 20×) using a fluorescence microscope. Do not cover with a coverslip.

Note: The suspension should contain intact nuclei and chromosomes. The concentration of chromosomes in the sample should be $\geq 5 \times 10^5$/mL. If the chromosomes are damaged (broken and/or appear as long extended fibers), the formaldehyde fixation is too weak and should be prolonged. If the chromosomes are aggregated and/or the cells remain intact, the fixation is too strong and should be shortened.

2.9.2.2 Alignment of Flow Cytometer for Chromosome Analysis and Sorting

The alignment of the flow cytometer is crucial to achieve the highest purity in the sorted chromosomes fraction. This protocol describes suitable setups and fine tuning of the instrument for chromosome analysis and sorting, for example, FACS Vantage. (For operation and alignment of the flow cytometer, read manufacturer's instructions.)

2.9.2.2.1 *Additional Materials*

Calibration beads (Polysciences): BB beads (univariate analysis) or YG beads (bivariate analysis) 530 ± 30 nm and 585 ± 42 nm band-pass filters (bivariate analysis).

2.9.2.2.2 *Alignment of Flow Cytometer*

1. Switch on the laser (s):
   a. Univariate analysis; operate the argon ion laser in multi-UV mode (351.1–363.8 nm) with 300 mW output power.
   b. Bivariate analysis; operate the first argon laser in multi-UV mode (351.1–363.8 nm) with 300 mW output power, and the second argon ion laser at 457.9 nm with 300 mW output power.
2. Allow the laser(s) to stabilize for 30 min. Peak the laser optics for maximum light output.
3. Empty the waste container and fill the sheath container with sterile sheath fluid SF50.
4. Adjust sheath fluid pressure (10 psi for FACS Vantage instrument) and leave the fluid running to fill all plastic lines and filters in the instrument.
5. Install a nozzle (70 µm orifice) and check for the air bubbles.
6. Install appropriate optical filters for alignment:
   a. Univariate analysis: Use a 424 ± 44 nm band-pass filter in front of the DAPI detector.
   b. Bivariate analysis: Use a 530 ± 30 nm band-pass filter in front of the DAPI detector and a 585 ± 42 nm band-pass filter in front of the mithramycin detector. Use a half mirror to split the fluorescence from the first and second lasers.
7. Trigger on forward scatter (FS) and select a threshold level.
8. Run fluorescent beads at a flow rate of 200 particles/sec, using BB beads for univariate analysis and YG beads for bivariate analysis.
9. Display the data on a dot plot of FS versus DAPI, and on one parameter histograms of FS and DAPI fluorescence.
10. Align the instrument to achieve maximum signal intensity and minimum coefficient of variation of FS and DAPI signals.
11. For bivariate analysis, use one parameter histogram of mithramycin fluorescence and align the second laser to achieve maximum intensity and the lowest coefficient of variation of the mithramycin signal. Change only settings specific for the second laser; do not adjust the controls. Adjust dual-laser delay and dead-time parameters as needed.

2.9.2.2.3 *Adjust Sorting Device*

1. Switch on the sorting device and warm up the deflection plates for 30 min.
2. Run calibration beads at a flow rate of 200 particles/sec. It is important that adjustment of sorting device be done with the sample running.
3. Switch on the test sort mode.
4. Adjust the drop drive frequency and drop drive amplitude to break the stream at a suitable distance from the laser intercept point (check for satellite drops).
5. Adjust the drop drive phase to obtain single side streams.
6. Adjust the position of side streams so that they enter the collection tubes.
7. Switch off the test sort mode.
8. Calculate drop delay and perform its optimization:
    a. Define sorting region for single beads (avoiding doublets and clumps).
    b. Select one-droplet sort envelope (number of deflected droplets) and sort mode giving the highest purity and count precision (counter mode in FACS Vantage instrument).
    c. Sort 20 beads onto a microscope slide.
    d. Check the number of beads using a fluorescence microscope (do not cover the drop with a cover slip).
    e. If the number of sorted beads is not correct, change the drop delay by a factor of 0.25 and repeat step c to e until the drop delay setting results in the highest number of sorted beads. To determine optimal drop delay, it may be convenient to sort 20 beads (at 0.25 step settings) on the same slide.

2.9.3 PROTOCOL-III

2.9.3.1 Univariate Flow Karyotyping and Sorting of Plant Chromosomes

This protocol describes the analysis and sorting of plant chromosomes stained with DAPI. The flow cytometer must be equipped with a UV light source to excite this dye.

2.9.3.1.1 *Materials*

- Chromosome suspensions
- 0.1 mg/mL DAPI stock solution
- LB01 lysis buffer
- Collection liquid
- Sheath fluid SF50 for flow cytometric analysis; 40 mM KCl/10 mM NaCl (sterilize by autoclaving)
- Computer with spreadsheet or other software for theoretical flow karyotypes (available from Professor Dole_el)
- 20 µm pore size nylon mesh in 4 × 4 cm squares
- Flow cytometer and sorter (e.g., Becton Dickinson FACS Vantage) with a UV argon laser (Coherent Innova 305) and a 424 ± 44-nm band-pass filter
- Microscope slides
- Fluorescence microscope with DAPI filter set

2.9.3.1.2 *Preparation of Theoretical Flow Karyotypes*

1. Prepare theoretical flow karyotypes using either a spreadsheet or dedicated computer software.
2. Predict the assignment of chromosomes to chromosome peaks on a flow karyotype.
3. Determine the resolution (coefficient of variation) of chromosome peaks needed to discriminate individual chromosome types.

Note: Theoretical flow karyotypes can be modeled based on relative length or DNA content of individual chromosomes, and are very useful in planning experiments with chromosome analysis. The model predicts the complexity of the analysis and limitations of univariate flow karyotyping. It may be used to predict positions of peaks representing specific chromosomes on a flow karyotype, and to study the effect of resolution (coefficient of variation) of chromosomes peaks on discrimination of individual chromosome types.

2.9.3.1.3 Performing Flow Cytometry

1. Stain a chromosome suspension (~1 mL/sample) by adding 0.1 mg/mL DAPI stock solution to a final concentration of 2 μg/mL. Analysis can be conducted immediately after addition of DAPI without incubation. If necessary, the stained suspension can be kept on ice.
2. Filter the suspension through a 20 μm nylon mesh.
3. Make sure that the flow cytometer is properly aligned for univariate analysis and that a 424 ± 44 nm band-pass filter is placed in front of the DAPI fluorescence detector.
4. Run a dummy sample (LB01) lysis buffer containing 2 μg/mL DAPI to equilibrate the sample line. This ensures stable peak positions during analysis and sorting.
5. Introduce the sample and let it stabilize at appropriate flow rate (200 particles/sec). If possible, do not change the flow rate during the analysis. Significant changes in the flow rate during the analysis may result in peak shifts.
6. Set a gating region on a dot plot of forward scatter (FS) and DAPI peak/pulse height to exclude debris, nuclei, and large clumps.
7. Adjust photomultiplier voltage and amplification gains so that chromosome peaks are evenly distributed on a histogram of DAPI signal pulse area/integral.
8. Collect 20,000 to 50,000 chromosomes and save the results on a computer disk.

2.9.3.1.4 Sorting of Chromosomes

1. Make sure that the sorting device is properly adjusted.
2. Run the sample and display the signals on a dot plot of DAPI signals pulse width versus area/integral.
3. Adjust the DAPI pulse width amplifier gain and width offset as needed to achieve optimal resolution of the width signal.
4. Check for the stability of the break-off point and the side streams.
5. Define sorting region for the largest chromosome on the dot plot of DAPI pulse width versus DAPI pulse area/integral.
6. Select one-droplet sort envelope (number of deflected droplets) and sort mode giving the highest purity and count precision (counter mode in FACS Vantage instrument).
7. Sort an exact number of chromosomes (e.g., 50) onto a microscope slide.
8. Check the number of chromosomes using a fluorescence microscope (do not cover the drop with a coverslip).
9. If the number is not correct, repeat adjustment of the sorting device using fluorescent beads.
10. Define a sorting region for the chromosome to be sorted on the dot plot of DAPI pulse width versus DAPI pulse area/integral.
11. Select sort mode and sort envelope according to required purity, number of chromosomes to be sorted, and desired volume for the sorted fraction. (Consult manufacturer's instruction for explanation of sort modes and sort envelopes.)
12. Sort the required number of chromosomes into polystyrene tube containing the appropriate amount of collection liquid.
    Note: The amount and composition of collection liquid depends on the number of sorted chromosomes and on their subsequent use. For PCR, use a small volume (20–60 μL) of sterile, dH_2O in a 0.5 mL PCR tube.

13. Microcentrifuge the tube for 5–10 sec at RT.
14. Sort chromosomes onto a microscope slide for determination of purity.

2.9.3.2   Bivariate Flow Karyotyping and Chromosome Sorting

This protocol provides information for chromosome isolation and bivariate analysis after dual staining with DAPI and mithramycin with preferentially bind AT-rich and GC-rich regions of DNA, respectively.

2.9.3.2.1   Additional Materials

- 100 mM $MgSO_4$ solution (filter through a 0.22 µm filter; store at 4°C)
- 1 mg/mL mithramycin stock solution
- A 424 ± 44-nm band-pass filter, and a 490-nm long-pass filter

2.9.3.2.2   Performing Flow Cytometry

1. To a chromosome suspension (~1 mL/sample), add 100 mM $MgSO_4$ solution to a final concentration of 10 mM.
2. Stain chromosomes by adding 0.1 mg/mL DAPI stock solution to a final concentration of 1.5 µg/mL and 1 mg/mL mithramycin stock solution to a final concentration of 20 µg/mL.
3. Allow to equilibrate for 30 min on ice.
4. Make sure that the flow cytometer is properly aligned for bivariate analysis. Use a half mirror to split the DAPI fluorescence through 424 ± 44-nm band-pass filter and the mithramycin fluorescence through 490-nm long-pass filter.
   Note: Because of the optical design of the dual-laser FACS Vantage instrument (which employs spatially separated beam geometry), a half mirror is used to reflect all light from the second laser at 90° toward the mithramycin detector. The DAPI fluorescence (excited by the first laser) is not reflected and enters the DAPI detector directly.
5. Run dummy sample (LB01 lysis buffer containing 1.5 µg/mL DAPI and 20 µg/mL mithramycin) to equilibrate the sample line. This ensures stable peak positions during analysis and sorting.
6. Filter the sample through a 20 µm nylon mesh.
7. Run the sample and let it stabilize at the appropriate flow rate (200 particles/sec).
8. Set a gating region on a dot plot of FS versus DAPI peak/pulse height. Gating is used to exclude small debris and large clumps and nuclei. Use this gate to display other parameters (DAPI pulse area/integral and mithramycin pulse area/integral).
9. Adjust photomultiplier voltage and amplification gains so that chromosome peaks are evenly distributed on histograms of DAPI pulse area/integral and mithramycin pulse area/integral.
10. Display the data on a dot plot of DAPI pulse area/integral versus mithramycin pulse area/integral.
11. Collect 20,000 to 50,000 chromosomes and save the results on a computer disk.

2.9.3.2.3   Chromosome Sorting

1. Make sure that the sorting device is properly adjusted.
2. Run the sample and display the signals on a dot plot of DAPI pulse area/integral versus mithramycin pulse area/integral.
3. Check for stability of break-off point and the side streams.
4. Define sorting region for the largest chromosome on the dot plot of DAPI pulse area/integral versus mithramycin pulse area/integral.
5. Proceed with cell sorting as described (step 6–19 in flow sorting univariate flow karyotyping), but define a sorting region on the dot plot of DAPI pulse area/integral versus mithramycin pulse area/integral.

2.9.4   PROTOCOL-IV

2.9.4.1   Physical Mapping of DNA Sequences Using PCR

This protocol describes the use of PCR on flow-sorted chromosomes for physical mapping of DNA sequences to individual chromosomes or their regions. Experiments involving PCR require extremely careful technique to prevent contamination.

2.9.4.1.1   Materials
- PCR premix
- Loading buffer
- 1.5% (w/v) agarose gel
- 1 × TAE electrophoresis buffer
- DNA molecular weight markers
- 0.5 µg/mL ethidium bromide solution
- 0.5 mL PCR tubes
- Thermal cycler
- Horizontal gel electrophoresis apparatus and power supply
- UV transilluminator and gel documentation system

2.9.4.1.2   Sorting Chromosomes
1. Prepare 0.5 mL PCR tubes containing 19 µL sterile dH$_2$O. The final volume after sorting will be ~20 µL.
2. Sort 500 chromosomes into each tube.
3. Freeze the tube and store at −20°C for up to 6 months. It is important to freeze the tubes even if the reaction is to be performed on the same day of sorting.

2.9.4.1.3   Perform PCR
1. Thaw a chromosome fraction and add 30 µL PCR premix. Vortex and microcentrigauge briefly.
2. Place tube in a thermal cycler and perform PCR amplification using the following cycles:

Initial steps:	2 min	94°C (denaturation)
35 cycles:	1 min	94°C (denaturation)
	1 min	58°C (annealing)
	2 min	72°C (extension)
1 cycle:	10 min	72°C (extension)
Final step:	Indefinitely	4°C (hold)

Annealing temperature must be optimized for a given primer pair and template.

2.9.4.1.4   Analyze PCR Products
1. Take equal amounts of PCR products (5–10 µL) from each tube and add 1–2 µL loading buffer.
2. Load samples onto a 1.5% agarose gel bathed in 1 × TAE electrophoresis buffer. Also load DNA molecular weight markers.
3. Run electrophoresis at a constant voltage of 4–5 V/cm until the bromophenol blue reaches a point 3 cm from the edge of the gel.
4. Stain the gel with 0.5 µg/mL ethidium bromide solution.
5. Photograph the gel and analyze the presence of products in individual lanes.

2.9.4.2 Two-Step Sorting

This procedure is developed to sort chromosomes when their frequency in the original suspension is too low. This is frequently the case for large chromosomes. During the first sort, the sample is enriched for the required chromosome. During the second sort, the chromosomes are sorted with a high purity. In some cases, it may be practical to enrich the sample for more than one chromosome. Individual chromosomes are sorted during the second sort.

2.9.4.2.1 Additional Materials

1.5 mL polystyrene cup (Delta Lab, Cat # 900022)
1.5 mL polystyrene PCR tube

2.9.4.2.2 Enrich Sample for Desired Chromosome

1. Make sure that the sorting device is properly adjusted.
2. Run the sample and display the signals on a suitable distribution.
3. Perform trial sorting onto a microscope slide and check the number of chromosomes using a dot plot of DAPI pulse width versus DAPI area/integral to define the sorting region for univariate analysis, or using DAPI area/integral versus mithramycin area/integral for bivariate analysis.
4. Select the sort mode and sort envelope that allow for the highest recovery (enrich mode and three deflected droplets in FACS Vantage instrument).
5. On a suitable distribution, define a sorting region for the chromosomes to be sorted.
6. Sort ≥100,000 chromosomes into 400 μL LB01 lysis buffer in a 1.5 mL polystyrene cup.

Note: The actual number of chromosomes that should be sorted depends on the number of chromosomes that will be sorted during the second sort. It is recommended to sort at least five times more chromosomes than the final number required.

2.9.4.2.3 Perform Second Sort

1. Add fluorescent dye to reach recommended final concentrations.
2. Run the sample and define a sorting region for the chromosome to be sorted.
3. Select sort mode and sort envelope according to required purity, number of chromosomes to be sorted, and desired volume for the sorted fraction. (Consult manufacturer's instructions for explanation of sort modes and sort envelopes.)
4. Sort the required number of chromosomes into a 1.5 mL polystyrene PCR tube containing the appropriate amount of collection liquid. [For PCR, use a small volume (20–60 μL) of sterile, dH_2O in 0.5 mL PCR tube].
5. Microcentrifuge the tube for 5–10 sec at RT.
6. Sort chromosomes onto a microscope slide for determination of purity.

2.9.4.3 Reagents and Solutions

See Appendix B.3

Remarks

1. It may take a week from the preparation of chromosome suspension to their analysis and chromosome sorting by flow cytometer.
2. Although protocols described above are not complicated but require careful planning, and expertise in several disciplines.
3. A large number of chromosomes can be sorted by the flow cytometer.
4. Chromosomes isolated from the above procedures have been suitable for scanning electron microscopy, fluorescence *in situ* hybridization, PRINS, and immune-localization of chromosome proteins (see Doležel et al., 1999).

2.10   PRODUCTION OF WIDE HYBRIDS THROUGH *IN VITRO* TECHNIQUE

The media used for immature embryo or seed rescue differ from crop to crop. However, plants have been obtained from explants of cotyledon, hypocotyl, stem, leaf, shoot apex, root, young inflorescences, flower petals, patioles, ovular tissues or embryos derived from calluses established either through embryogenesis or organogensis. Modification in media composition depends entirely on source of explants and crops. The rescue of plants from immature seed harvested from 19- to 21-day postpollination is described here as it is extremely difficult to obtain plants from wide crosses in the genus *Glycine*. The protocol described has been derived from the modification of several media formulations (Appendix 2-IV).

Disinfection and excision of mature or embryo and immature embryo and seed:

1. Surface sterile seeds or pods with either with 5% sodium hypochlorite or 50% commercial bleach for 15–30 min followed by washing three to four times with sterile distilled water in hood.
2. Dissect the pods (soybean) and culture the immature seeds on initial medium.
   All media are prepared fresh by adding the compounds in order.
   a. Adjust pH 5.8 using 0.1 or 1.0 N KOH prior autoclaving.
   b. Autoclave at 1.46 kg/cm^2 for 20 min at 121°C.
   c. Add filter sterilized (Millipore HA 0.45 μm) Glutamine (160 mL H$_2$O + 7.305 g Glutamine at 18°C and IAA.
   d. Dispense under aseptic condition.
3. All cultures are incubated at 25 ± 1°C, with a 16 h photoperiod under cool white fluorescent tubes (ca45μEM^{-2}S^{-1}) in a Percival model LVL incubator.
4. Subculture every 2 w for the first 6 w followed by 4 w interval (total 10 w). Observe regularly for the seed germination and also for callus formation.
   a. Transfer germinating seedling to step 5.
   b. Remove embryo with cotyledonary tissue from nongerminating seeds and place on the same medium for 2 weeks.
5. Place germinating seed on germination medium and subculture at 2-w intervals until shoot is completely elongated. Transfer shot to rooting medium if roots have not developed.
6. Transfer germinated seedlings to sterile vermiculite and peat moss (3:1) in a closed chamber (GA 7 vessels, Magenta Company) with enough moisture to maintain humidity.
7. Acclimatize seedlings to greenhouse condition by slowly displacing the lid of Magenta box.
8. Transfer plantlets to 15 cm clay pots in the greenhouse. The potting soil consists of 1:1:1 mixture of clay, peat, and sand.
9. Fertilize plants twice a week with a half strength Hoagland's solution.

3 Methods in Plant Cytogenetics

3.1 INTRODUCTION

The science of genetics relates to heredity and variation-continuity of the life. Cell division, mitosis and meiosis, and precise DNA replication predict the inheritance of a particular trait from parents to the progenies. A clear understanding of the mode of inheritance is revealed by the science of genetics and its foundation was laid by J. G. Mendel in 1865 based on experiments of the garden pea (*Pisum sativum*) (Appendix A). The selection of garden pea was a wise choice because: (1) it is a self-pollinated plant and pollination can be controlled and selfing of F_1 posed no problem; (2) it is easy to cultivate this plant and requires only a single growing season; (3) this plant has many distinguishing sharply defined inherited traits. However, Mendel's laws were unrecognized until 1900 and this may be due to a number of special reasons (Strickberger, 1968): The variability among the F_2 and further hybrid generations could be traced to the original variability in the first parental cross. The factors that could be traced were followed but did not change during the period of observation but only expressed themselves in new and different combinations among the offspring.

Mendel used discontinuous traits while at that time scientists were looking for continuous variation (Galton, Darwin, and others). Mendel's approach with probability events and mathematical ratios was an unfamiliar idea to biology. Mendel was in constant correspondence with Nägeli but he did not appreciate Mendel's work because he was working with *Hieracium* (hawkweed), an apomictic plant and F_1 hybrid did not segregate and all F_2 plants were identical to the maternal plant.

Mendel was fortunate not to run into the complication of linkage during his experiments. He selected seven genes of the pea which has $2n = 14$ chromosomes (seven linkage groups) and these genes are on different chromosomes. Mendel worked with two genes in chromosome 1, three genes in chromosome 4 and 1 gene each in chromosomes 5 and 7 (Table 3.1; Blixt, 1975). Blixt (1975) assumed that out of 21 dihybrid combinations, Mendel theoretically could have studied, no more than four (*a-i, v-fa, v-le, fa-le*) ought to have linkage seeing the current genetic map of pea. It has been found that *a* and *i* in chromosome 1 are so distantly located and no linkage is normally detected and the same is true for *v, le,* and *fa* for chromosome 4. However, *v* and *le* should have shown linkage. Mendel either did not publish results or did not make appropriate cross. Thus, Mendel did not run into the linkage complication.

Unfortunately, Mendel's work was ignored and virtually forgotten for 34 years after its publication. However, Mendel's work was discovered when three scientists, DeVries, Correns, and Tschermak independently found the same results reported by Mendel in 1866. Now, it is known as Mendelian inheritance, Mendel's laws, Mendelian factors, and others.

3.2 MENDELIAN GENETICS

3.2.1 MONOHYBRID INHERITANCE

1. *Complete dominance:* Mendel selected seven contrasting characters individually. He hybridized with a variety containing one different character. For example, he hybridized round seed coat plant with wrinkled seed shape. The F_1 was only one type (round-seed). He allowed F_1 plants to self-pollinate. In F_2, plants produced 5474 round seeds and 1850 wrinkled. This suggests that these F_2 ratios are close to 3:1 and can be also expressed as

TABLE 3.1

Relationship between Modern Genetic Terminology and Character Pairs Used by Mendel

Character Pair Used by Mendel	Alleles in Modern Terminology	Located in Chromosomes
Seed color: yellow-green	*I-i*	1
Seed coat and flowers: colored-white	*A-a*	1
Mature pods: smooth expanded-wrinkled	*V-v*	4
Inflorescences: from leaf axils— umbellate in top of plant	*Fa-fa*	4
Plant height: >1 m-around 0.5 m	*Le-le*	4
Unripe pods: green-yellow	*Gp-gp*	5
Mature seeds: smooth-wrinkled	*R-r*	7

Source: Adapted from Blixt, S. 1975. *Nature* 256: 206.

3/4:1/4, or 75:25 or 75%:25% (Figure 3.1). Based on the results of all seven contrasting characters, Mendel concluded:

a. F_1 plants expressed one type of character.
b. The result was always the same regardless of which parent was hybridized.
c. The trait hidden in F_1 reappeared in F_2.

This is known as the principle or law of segregation. As the science of genetics progressed, it was established that each trait is controlled by a gene and each trait is controlled by a pair of genes.

2. *Incomplete dominance:* Since Mendel's law of segregation, geneticists discovered that heterozygotes F_1 and F_2 express partial or incomplete dominance. For example, when

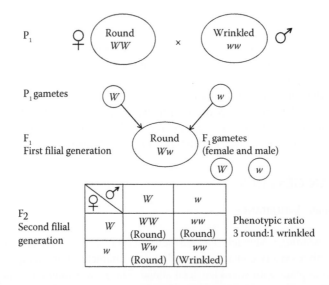

FIGURE 3.1   Monohybrid segregation showing round seed coat is complete dominant on wrinkled seed coat. F_1 plants produced seeds with round seed coat and F_2 plants segregated for 3 round and 1 wrinkled seed coat. This is Mendel's law of segregation. (Redrawn from Sinnott, E. W., L. C. Dunn, and Th. Dobzhansky. 1950. *Principles of Genetics.* McGraw-Hill Book Company, Inc., New York.)

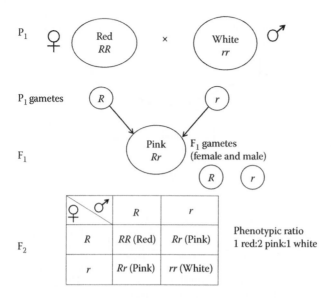

FIGURE 3.2 Sometimes, when red-flowered plants crossed with a white-flowered plant, F_1 plants produce pink flowers and F_2 plants segregate for 1 red, 2 pink, and 1 white flowers. This is a departure from the 3:1 ratio. However, the segregation pattern shows that red is incomplete dominant over white and heterozygous plants (Rr) produce pink flowers. (Redrawn from Sinnott, E. W., L. C. Dunn, and Th. Dobzhansky. 1950. *Principles of Genetics*. McGraw-Hill Book Company, Inc., New York.)

red-flowered (*RR*) plant was hybridized with white-flowered (*rr*) plants, the F_1 plants (*Rr*) produced pink-flowered plants. After selfing F_1 plants, the F_2 ratio segregated for 1 red (*RR*):2 pink (*Rr*):1 white (*rr*) plants (Figure 3.2).

3.2.2 DIHYBRID INHERITANCE

The monohybrid hybridization (parents differing for a single gene pair) led Mendel to discover the principle of segregation. Mendel proposed the law of independent assortment by mating two parents differing in two pairs of genes (dihybrid crosses). Mendel hybridized plants with yellow-round and green-wrinkled seeded plants. The F_1 plant produced yellow-round seeds. In F_2, of the 556 seeds the segregation was: 315 yellow and round, 108 yellow and wrinkled, 101 green and round, and 32 green and wrinkled. This suggests that these numbers are very close to a 9:3:3:1 ratio (Figure 3.3). Thus, smooth and wrinkled and yellow and green genes do not interfere with and two gene pairs inherit independently—*independent assortment*. We can obtain the ratio of each phenotypic combination by multiplying the probabilities of the individual phenotypes giving 9:3:3:1 segregation pattern (phenotypic ratio):

3/4 round × 3/4 yellow	= 9/16 round yellow
3/4 round × 1/4 green	= 3/16 round green
1/4 wrinkled × 3/4 yellow	= 3/16 wrinkled yellow
1/4 wrinkled × 1/4 green	= 1/16 wrinkled green

A question is raised about the genotypes of the segregating plants in F_2 population. The genotype of smooth and yellow plants is *SSYY* and wrinkled green is *ssyy*. The plant with *SSYY* genotype will produce gametes with *SY* and *ssyy* plant will produce gametes with *sy*. The resulting f1 plants will be smooth and yellow with genotype *SsYy*. The F_1 plant will produce eggs and

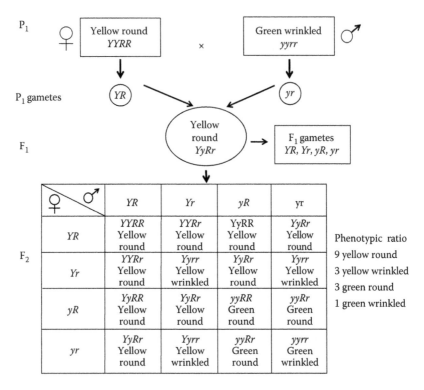

FIGURE 3.3 When two parents differ by two contrasting characters (yellow round vs. green wrinkled), F₁ plants produce yellow round seeds and in F₂ a 9:3:3:1 ratio is observed. This is the law of independent segregation of Mendel. (Redrawn from Sinnott, E. W., L. C. Dunn, and Th. Dobzhansky. 1950. *Principles of Genetics*. McGraw-Hill Book Company, Inc., New York.)

sperms with genotype *SY*, *Sy*, *sY*, and *sy* in equal proportion (1:1:1:1). In F₂, genotypes of plants are expected as:

SSYY = 1	Smooth yellow
SSYy = 2	Smooth yellow
SsYY = 2	Smooth yellow
SsYy = 4	Smooth yellow
Ssyy = 1	Smooth green
SSyy = 1	Smooth green
Ssyy = 1	Smooth green
ssYY = 1	Wrinkled yellow
ssYy = 2	Wrinkled yellow
ssyy = 1	Wrinkled green

Mendel hybridized plants differing three genes (trihybrid cross) such as:

1. Smooth and wrinkled seed shape (*S* and *s*)
2. Yellow and green seed color (*Y* and *y*)
3. Violet and white flower (*V* and *v*)

He hybridized plants smooth, yellow, and violet (*SSYYVV*) to wrinkled, green, and white plants (*ssyyvv*). The F₁ plants expressed the phenotype of the dominant parents. The F₁ was allowed to self-pollinate and produced F₂ seeds. The F₁ hybrids produced gametes with genotype *SYV*, *SYv*, *SyV*,

Syv, sYV, sYv, syV, and *syv.* Based on gamete combinations, $8 \times 8 = 64$ combinations are expected. The F_2 plants produced 8 phenotypes and 27 assumed genotypes:

1. 27 smooth yellow violet = 8 genotypes
2. 9 smooth yellow white = 4 genotypes
3. 9 smooth green violet = 4 genotypes
4. 9 wrinkled yellow violet = 4 genotypes
5. 3 smooth green white = 2 genotypes
6. 3 wrinkled yellow white = 2 genotypes
7. 3 wrinkled green violet = 2 genotypes
8. 1 wrinkled green white =1 genotype

It is interesting to note that the number of gene-pair differences is more than three, the number of possible combinations between them is considerably increased as summarized below:

Characteristics of segregation and independent assortment of crosses involving *n* pairs of alleles:

Number of Heterozygous Allelic Pairs	Number of Kinds of Gametes	Number of Phenotypes Testcross	Number of Genotypes in F_2	Number of Phenotypes in F_2[a]
1	2	2	3	2
2	4	4	9	4
3	8	8	27	8
4	16	16	81	16
n	2^n	2^n	3^n	2^n

[a] Assuming complete dominance in each allelic pair.

3.2.2.1 Gene Interaction and Expression

Mendel's laws of segregation and independent assortment are precisely correct in case gene action expressed complete dominance and *lack of interference* between different genes did not occur. However, the exceptions were when traits inherited through cytoplasm, and interaction and expression of genes modified Mendelian ratios. Sometimes, dominance is not observed in some crosses, such as red (*RR*) and white (*rr*) flower plants. The F_1 plants with genotype *Rr* produce pink flowers and F_2 plants segregate for red (*RR*) 1:2 pink (*Rr*):1 (*rr*) white. Thus, genotypic ratios are phenotypic ratios.

1. *Lethal genes*: The lethal effect of a gene occurs in a particular genotype under a particular set of environmental conditions. It may be dominant or recessive gene lethal as long as lethality depends on its presence in homozygous condition. In house mouse, Cuénot observed that the yellow mice never breed true and concluded that yellow mice are heterozygous and hybrid. The crossing of two yellow mice produced only yellow and black and can be shown as:

   Yy (yellow) $\times$ *Yy* (yellow) gametes *YY, Yy, Yy,* and *yy.* The progenies with *YY* mice die and only mice with genotype *Yy* (yellow) and *yy* (black) survive. Such mechanism has been demonstrated in several organisms including human, plant, and animals. It is also known as balanced lethal.

2. *Complementary genes*

   a. *Duplicate recessive gene (9:7):* In soybeans, parents with white flowers were hybridized. The F_1 plants produced purple flowers and F_2 plants segregated for 9/16 purple and 7/16 white flowers. Figure 3.4 shows a diagrammatic presentation of a checkerboard. The genotypes of white parents are assumed as *CCpp* and *ccPP* and the genotype of F_1 plant is *CcPp* (purple). In F_2, genotypes of white flower plants are *CCpp,*

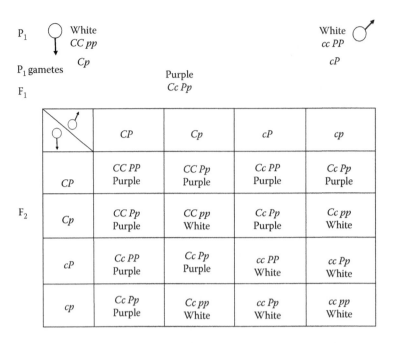

FIGURE 3.4   When parent white flowers are hybridized, F_1 plants produce purple flowers and F_2 plants segregate for nine purple to seven white flowers; modification of Mendel's Law of Segregation. This also suggests that both white flower parents have different white flower genes. (Redrawn from Sinnott, E. W., L. C. Dunn, and Th. Dobzhansky. 1950. *Principles of Genetics*. McGraw-Hill Book Company, Inc., New York.)

Ccpp, ccPP, ccPp, Ccpp, ccPp, and *ccpp.* Occasionally the genes participating in such interactions are called complementary genes.

b. *Epistasis*

   i. *Dominant epistasis (12:3:1):* When dominant allele *A* masks the expression of *B*, *A* is an epistatic gene of *B. A* can express itself only in the presence of a *B* or *b* allele. Therefore, it is known as dominant epistasis; *B* expresses only when aa is present. This phenomenon of interaction between nonallelic genes is known as epistasis. In summer squashes, three common fruit colors—white, yellow, and green—are always found. In crosses between white and yellow and between white and green, white is always found to be dominant; and in crosses between yellow and green, yellow is always found to be dominant. Yellow thus acts as a recessive in relation to white but as a dominant in relation to green. Thus, there is evidently a gene, *W*, which is epistatic to those for yellow and green; and so long as it is present, no color is produced in the fruit, regardless of whether genes for color are present. In plants where gene for white is lacking (*ww*), the fruit color will be yellow if gene *Y* is present and green if it is absent. Green-fruited plants by *wwYY* and white-fruited ones are either by *WWYY* or by *WWyy*. This assumption is based on that there are two independent gene pairs, one epistatic over the other, and can be tested by crossing a homozygous white that also carries yellow, *WWYY*, with a green, *wwyy*. The F_1 plants, *WwYy*, are white fruited. This plant will produce four kinds of gametes, *WY*, *Wy*, *wY*, and *wy*. Figure 3.5 shows expected F_2 plants; 12/16 plants will carry *W* and will be white regardless of other *Y* genes. Thus, *W* masks everything which is hypostatic to it, so that *Y*, which segregates quite independently of white, produces a visible effect only, white, in 3/16 of the plants lacking *W*. Plants (1/16) with *wwyy* gene will be green. Thus, the expected ratio is 12 white:3 yellow:1 green (Figure 3.5).

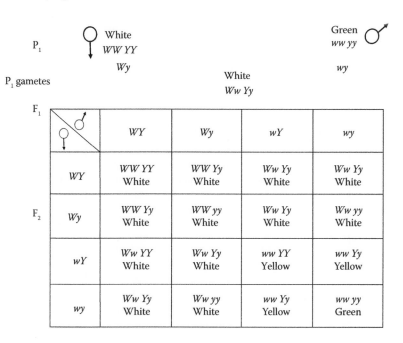

FIGURE 3.5 This checkerboard shows when white fruited is dominant over green fruited, F_2 plants segregate 12 white:3 yellow:1 green fruited. In this case, W gene masks everything which is hypostatic to it, Y, which segregates independently of white and expresses only when fruit lacks Y. (Redrawn from Sinnott, E. W., L. C. Dunn, and Th. Dobzhansky. 1950. *Principles of Genetics*. McGraw-Hill Book Company, Inc., New York.)

    ii. *Dominant suppression epistasis (13:3):* When the dominant gene locus (*I*) in homozygous (*II*) and heterozygous (*Ii*) condition and homozygous recessive alleles *cc* of another gene locus (*C*) produce the same phenotype, the F_2 phenotypic ratio becomes 13:3; this case is the result of the influence of dominant genes that do not permit certain dominant genes to function normally. The genotype *IICC, IiCC, IIcc, Iicc,* and *iicc* produce one type of phenotype (13) and genotype *iiCC, iiCc* will produce another type of phenotype (3) (Figure 3.6).

    iii. *Color reversion (9:3:4):* Complete dominance at both gene pairs, but one gene, when homozygous recessive, is epistatic to the other. The classical example is mice coat color. When black (*CCaa*) mice are crossed with albino (*ccAA*) mice, the progenies are all agouti color. When these F_1 agoutis are inbred, their progenies consist of 9/16 agouti, 3/16 black, and 4/16 albino (Figure 3.7).

    iv. *Complete dominance at both gene pairs, but either gene, when dominant, epistatic to the other—duplicate genes (15:1):* Genes with the same expression are known as duplicate genes. The classical example is the inheritance of capsule or pod form in the shepherd's-purse Bursa. One race has triangular capsules where as in another race, capsules are ovoid or top-shaped. When two capsule forms were crossed, F_1 showed triangular capsules suggesting that triangular trait is dominant over ovoid capsule form. In F_2, a 15:1 ratio is obtained; a deviation from 9:3:3:1 Mendelian ratio.

3. *Test of Independence:* A clear understanding of the probability is of fundamental importance to understand the laws of heredity: (1) to appreciate the operation of genetic mechanism, (2) predicting the likelihood of certain results from a particular hybridization, and (3) determining how well an F_1 phenotypic ratio fits a particular postulated genetic

P₁   ◯ White leghorn   White Wyandotte ♂
              II CC   *ii cc*
            │ *IC*   *ic*

P₁ gametes

                                        White
                                       Ii Cc

F₁

♂ / ♀	*IC*	*Ic*	*iC*	*ic*
IC	*II CC* White	*II Cc* White	*Ii CC* White	*Ii Cc* White
Ic	*II Cc* White	*II cc* White	*Ii Cc* White	*Ii cc* White
iC	*Ii CC* White	*Ii Cc* White	*ii CC* Colored	*ii Cc* Colored
ic	*Ii Cc* White	*Ii cc* White	*ii Cc* Colored	*ii cc* White

F₂ *(label at left)*

FIGURE 3.6 This checkerboard shows 13:3 ratio which tells that presence if I gene is in recessive condition, dominant gene C expresses color; thus one (I) is suppressing the expression of C gene. (Redrawn from Sinnott, E. W., L. C. Dunn, and Th. Dobzhansky. 1950. *Principles of Genetics*. McGraw-Hill Book Company, Inc., New York.)

P₁   ◯ Black   Albino ♂
              CC aa   *cc AA*
P₁ gametes │ *Ca*   Agoutii   *cA*
                               Cc Pp
F₁

♂ / ♀	*CA*	*Ca*	*cA*	*ca*
CA	*CC AA* Agouti	*CC Aa* Agouti	*Cc AA* Agouti	*Cc Aa* Agouti
Ca	*CC Aa* Agouti	*CC aa* Black	*Cc Aa* Agouti	*Cc aa* Black
cA	*Cc AA* Agouti	*Cc Aa* Agouti	*cc AA* Albino	*cc Aa* Albino
ca	*Cc Aa* Agouti	*Cc aa* Black	*cc Aa* Albino	*cc aa* Albino

F₂ *(label at left)*

FIGURE 3.7 This checkerboard shows 9:3:4 ratio showing the presence of C is necessary to express any other color. The gene C produces black color mice when the albino gene is in recessive condition. (Redrawn from Sinnott, E. W., L. C. Dunn, and Th. Dobzhansky. 1950. *Principles of Genetics*. McGraw-Hill Book Company, Inc., New York.)

mechanism. Based on transmission of two pairs of genes which are assumed to be on separate chromosome pairs, the chance of the simultaneous occurrence of two or more independent events is equal to the product of the probability that each will occur separately. The application of the probability to genetics permits demonstration of the two independent, nongenetic events, the binomial expression, and to determine the goodness of fit.

a. *Two independent nongenetic events—two coin tosses:* This can be elucidated by tossing 50 times two coins simultaneously. The outcome of any single toss will be HH (head), HT (head and tail), and TT (tail). Of the 50 tosses, the results were:

HH	12
HT	27
TT	11

We can assume three possibilities: (1) the deviation from the predicted result is within limits set by chance alone and fall of coins is *unbiased*, (2) if two coins are tossed simultaneously, does each one of them have an equal chance of coming to rest heads or tails? In this case, there is a remote chance that the ultimate fall of one coin affects the other. Thus, the second hypothesis will be that the coins themselves are *independent* of each other, (3) the third assumption may be, at least at the outset, that *successive tosses are also independent* of each other. This would tell that having obtained two heads the first toss does not in any way affect the outcome of the second toss or any other tosses.

We expect to see 1/4HH + 1/2HT + 1/4 TT; thus, for two independent events of known probability can be stated as $a^2 + 2ab + b^2$. The a represents head and b represents tail and can be simplified as $(a + b)^2$. We can summarize the observed and expected values of tossing two coins of heads and tails:

Class	Observed	Expected
HH	12	12.5
HT	27	25.5
TT	11	12.5
	50	50.0

b. *Four independent nongenetic events—four coin tosses*:
    In four coin tosses, the possible combinations of heads and tails actually obtained in an experiment are as follows:

Class	Observed
HHHH	9
HHHT	32
HHTT	29
HTTT	25
TTTT	5
	100

In a two-coin toss, the expansion of $(a + b)^2$ gave an expected ratio of $a^2 + 2ab + b^2$. Therefore, by expending $(a + b)^4$, we expect: $a^4 + 4a^3b + 6\ a^2b^2 + 4ab^3 + b^4$. By substituting the numerical values, we will obtain:

$$\left(\frac{1}{2}\right)^4 + 4\left[\left(\frac{1}{2}\right)^3 \cdot \frac{1}{2}\right] + 6\left[\left(\frac{1}{2}\right)^2 \cdot \left(\frac{1}{2}\right)^2\right] + 4\left[\left(\frac{1}{2}\right) \cdot \left(\frac{1}{2}\right)^3\right] + \left(\frac{1}{2}\right)^4 = 1:$$

or $1/16 + 4/16 + 6/16 + 4/14 + 1/16 = 1$

We can compare observed results with the calculated results for expectancy in a four-coin toss:

Class	Observed	Calculated
HHHH	9	6.25 (= 1/16 of 100)
HHHT	32	25.00 (= 4/16 of 100)
HHTT	29	37.50 (= 6/16 of 100)
HTTT	25	25.00 (= 4/16 of 100)
TTTT	5	6.25 (= 1/16 of 100)
	100	100.00

c. *Determination of goodness of fit-nongenetic events:* Chi-square test is used to accept the observed and expected assumptions. The formula of chi-square is:

$$\chi^2 = \sum \frac{[(o-e)^2]}{e}$$

o = observed frequencies

e = calculated frequencies

Σ = summed for all classes

The calculated chi-square of coin tosses (1:2:1 expectation) can be expressed as:

Class	Observed O	Expected E	Deviation o–e	Squared Deviation (o–e)^2	$\frac{(o-e)^2}{e}$
HH	12	12.5	−0.5	0.25	0.02
HT	27	25.0	+2.0	4.00	0.16
TT	11	12.5	−1.5	2.25	0.18
	50	50.0	0		$\chi^2 = 0.36$

How often, by chance, can we expect a value of $\chi^2 = 0.36$ in a 1:2:1 ratio? Thus, we can accept that the observed deviation was produced only by chance. To answer this question, we should consult a table of chi-square (Table 3.2). In order to use this table, it is necessary to know the degrees of freedom; it is one less than the number of classes involved and represents the number of independent classes for which chi-square value is calculated. Based on probability values and degree of freedom presented in Table 3.2, we can conclude that observed chi-square 0.36 with two degrees of freedom fits probability values between 0.95 and 0.80. This suggests that for an expected ratio of 1:2:1, we can expect a deviation as large as or larger than we experienced in between 80% and 95%. Such deviation could be due to chance and assumption is good and a good fit between observed results and our expected values.

The chi-square test is very useful for obtaining an objective approximation of goodness of fit. However, this is reliable when the observed and expected frequency in any class is five or more. The method provides useful information for an earlier judgmental answer.

3.2.2.2 Genetic Mapping of Chromosomes[*]

In case two genes do not follow Mendelian inheritance, these genes are lined and on the same chromosomes. Bateson and Punnett (1905–1908) could not explain the pea in the following F_2 results:

---

[*] From G. W. Burns 1969.

P	Homozygous purple long (*R Ro/R Ro*) × red round (*r ro/r ro*)
F$_1$	All purple long (*R Ro/r ro*)
F$_2$	Purple long (*R Ro/r ro*) = 296
	Purple round (*R Ro/r ro*) = 19
	Red long *r ro/Ro Ro*) = 27
	Red round (*r ro/r ro*) = 85

These results did not fit the 9:3:3:1 ratio and Bateson and Punnett could not figure out the cause of this segregation. This now suggests that these two genes are linked. These genes may be arranged in a chromosome in two ways: (1) the two dominants, *R* (purple) and *Ro* (round), may be located on one member of the chromosome pair and two recessive, *r* and *ro*, on the other chromosome pair. This arrangement is known as *cis;* (2) the dominant of one pair and the recessive of the other may be located on one chromosome of the pair, with the recessive of the first gene pair and dominant of the second gene pair on the other chromosome. This arrangement is known as *trans* (Figure 3.8).

The classical genetic maps using qualitative genes are based on the extent of recombination frequencies observed either from the test cross or by selfing the F$_1$ plants. In test cross, F$_1$ plants are backcrossed to the recessive homozygous parent and progenies should segregate in a ratio of 1:1:1:1 if genes are more than 50 cM apart on the same chromosome or on different chromosomes. A departure from this ratio suggests that two genes are linked and frequency depends on how close

TABLE 3.2
Chi-Square Table

Degrees of Freedom	Probability										
	Nonsignificant							Significant			
	0.95	0.90	0.80	0.70	0.50	0.30	0.20	0.10	0.05	0.01	0.001
1	0.004	0.02	0.06	0.15	0.46	1.07	1.64	2.71	3.84	6.64	10.83
2	0.10	0.21	0.45	0.71	1.39	2.41	3.22	4.60	5.99	9.21	13.82
3	0.35	0.58	1.01	1.42	2.37	3.66	4.64	6.25	7.82	11.34	16.27
4	0.71	1.06	1.65	2.20	3.36	4.88	5.99	7.78	9.49	13.28	18.47
5	1.14	1.61	2.34	3.00	4.35	6.06	7.29	9.24	11.07	15.09	20.52
6	1.63	2.20	3.07	3.83	5.35	7.23	8.56	10.64	12.59	16.81	22.46
7	2.17	2.83	3.82	4.67	6.35	8.38	9.80	12.02	14.07	18.48	24.32
8	2.73	3.49	4.59	5.53	7.34	9.52	11.03	13.36	15.51	20.09	26.12
9	3.32	4.17	5.38	6.39	8.34	10.66	12.24	14.68	16.92	21.67	27.88
10	3.94	4.86	6.18	7.27	9.34	11.78	13.44	15.99	18.31	23.21	29.59

Source: From R. A. Fisher and F. Yates, *Statistical Tables for Biological, Agricultural and Medical Research,* 6th ed., Table IV, Oliver & Boyd. Ltd., Edinburgh, 1963, by permission of the authors and publishers.

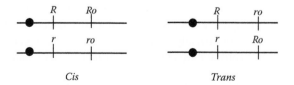

FIGURE 3.8 Genes arranged in *cis* and *trans* position. (Redrawn from Burns, G. W. 1969. *The Science of Genetics.* N. H. Giles, and J. G. Torrey, Eds., The Macmillan Biology Series., The Macmillan Company, Collier Macmillan Limited, London.)

the genes are. If Bateson and Punnett would have conducted testcross, the following segregation is expected:

P	Homozygous purple long ($R\ Ro/R\ Ro$) × red round ($r\ ro/r\ ro$)
F_1	All purple long ($R\ Ro/r\ ro$)
Testcross progeny	Purple long ($R\ Ro/r\ ro$) = 192
	Red round ($r\ ro/r\ ro$) = 182
	Purple round ($R\ Ro/r\ ro$) = 23
	Red long ($r\ ro/Ro\ Ro$) = 30
	427

These results widely depart from 1:1:1:1 ratio. Two groups resemble parents while two groups, with far less frequencies, are nonparental types and this deviation is too large than the expected 1:1:1:1 ratio. To check this hypothesis, the chi-square test for a 1:1:1:1 expectation shows:

Phenotype	o	e	o–e	$(o-e)^2$	$\dfrac{(o-e)^2}{e}$
Purple long	192	106.75	85.25	7267:563	68.08
Red round	182	106.75	75.25	5662.563	53.05
Purple round	23	106.75	–83.75	7014.063	65.70
Red long	30	107.75	–76.75	5890.563	55.91
	427	427.00	0.00		$\chi^2 = 242.01$

Testcross progeny: The following table of testcross progenies shows that 87.6% plants are parental types and 12.4% are nonparental types.

Phenotype	Genotype	Number	Frequency	Type
Purple long	$R\ Ro/r\ ro$	192	0.4496	Parental
Red round	$r\ ro/r\ ro$	192	0.4262	Parental
Purple round	$R\ ro/r/ro$	23	0.0538	Nonparental
Red long	$r\ Ro/r/ro$	30	0.0702	Nonparental
		427	1.0000	

If R and Ro genes are not linked, it is expected to have 50% each of parental and nonparental types because parents should have produced gametes in equal frequency. Bateson and Punnett recognized that their population was segregating as 7 purple long, 1 purple round, 1 red long, and 7 red round. However, it should have segregated in a ratio of 9:3:3:1 if genes are unlinked.

3.2.2.2.1   Chromosome Mapping

Crossing over and crossover frequency: If genes are arranged in *cis* linkage, it can be easily determined based on frequency of the four possible types of F_1 gametes in unequal frequency. Depending on the closeness of genes on the same chromosome, the occurrence of chiasma during meiotic prophase-I and separation of chromosomes at anaphase-I suggest that 87.6% gametes contained parental type chromosomes and 12.4% crossover (nonparental chromosome type). The more closely the genes, the frequency of crossover product is reduced and is described as a map unit. A map unit is equal to 1% of crossing-over; this represents the linear distance within which 1% crossing-over takes place. Thus, for sweet peas, the distance from R and Ro would be 12.4 map units or cM.

Classical chromosome mapping is easily conducted based on the crossover frequencies observed from the testcross than the data observed from the F_2 population. The association of genes using dihybrid testcrosses in *Drosophila melanogaster*, the fruit fly, has been well known; normal is designated as +:

Gene Symbol	Phenotypes
+	Normal wing (dominant)
cu	Curled wing (recessive)
+	Normal thorax (dominant)
sr	Striped thorax (recessive)
+	Normal bristles (dominant)
ss	Spineless bristles (recessive)

Testcross 1 in cis arrangement:

P	Female		Male
	Normal wing normal thorax	×	Curled wing striped thorax
	+ +/*cu sr*		*cu sr/cu sr*

Progenies:

F$_1$ Phenotypes	Maternal Chromosomes	Numbers	Percentage
Normal normal	+ +	436	43.6 Parental
Curled wing and striped	*cu sr*	468	46.8 Parental
Normal striped thorax	+ *sr*	42	4.2 Crossover
Curled wing normal	*cu* +	54	5.4 Crossover

% parental types: 90.4; % crossover types: 9.6.

This suggests that cu and sr are linked and the distance between the two genes is 9.6 cM.

Testcross 2 in Tans arrangement:

P	Female		Male
	Normal wing normal thorax	×	Curled wing striped thorax
	+ *sr/cu*/+		*cu sr/cu sr*

Progenies:

F$_1$ Phenotypes	Maternal Chromosomes	Numbers	Percentage
Normal normal	+ +	49	4.9 Parental
Curled wing and striped	*cu sr*	47	4.7 Parental
Normal striped thorax	+ *sr*	442	44.2 Nonparental
Curled wing and normal	*cu* +	462	46.2 Nonparental

Parental types: 9.6%; nonparental type; 90.4%.

It should be noted that the percentage of recombinant is 9.6%, but with the *trans* linkage in the female parent, the parental types in the F$_1$ now constitute the smaller class.

Testcross 3 in trans arrangement:

P	Female		Male
	Normal wing and normal bristles	×	Curled wing and spineless bristles
	++/*cu ss*		*cu ss/cu ss*

Progenies:

F$_1$ Phenotypes	Maternal Chromosomes	Numbers	Percentage
Normal wing and normal bristles	+ +	450	45.0 Parental
Curled wing and spineless bristles	*cu ss*	465	46.5 Parental
Normal wing and spineless bristles	+ *ss*	39	3.9 Nonparental
Curled wing and normal bristles	*cu* +	46	4.6 Nonparental

Parental types: 91.5%; nonparental types: 8.5%.

The *trans* arrangement in the female gives compatible results:

Testcross 4 in cis arrangement:

P	Female		Male
	Normal normal	×	Curled wing and spineless bristles
	+ ss/cu +		cu ss/cu ss

F₁ Phenotypes	Maternal Chromosomes	Numbers	Percentage
Normal wing and normal bristles	+ +	45	4.5 Parental
Curled wing and spineless bristles	cu ss	40	4.0 Parental
Normal wing and spineless bristles	+ ss	461	46.1 Nonparental
Curled wing and normal bristles	cu +	454	45.4 Nonparental

Parental types: 8.5%; nonparental types: 91.5%.

At this stage, we have the following distance cM among the genes based on crossing over and can be represented as follows:

cu-sr	9.6
cu-ss	8.5

Based on the above information, we can assume that *cu, ss*, and *sr* genes are on the same chromosomes and their arrangement may be either a or b. To determine the correct alternatives, we should know the cross-over frequencies between ss and sr, which is 9.6 + 8.5 = 18.1 or 9.6−8.5 = 1.1.

3.2.2.2.2 Three-Point Test cross

Three-point test cross provides more precise distance among three genes than those obtained from two-point test cross (Figure 3.9a). In *Drosophila*, three pairs of genes: a recessive homozygote cu, ss, and sr fly were testcrossed with a fly with dominant (normal) phenotype. Based on segregation in testcross progenies, we can determine the correct gene sequence.

P	Female		Male
	Normal normal normal	×	Curled spineless striped
	+ ++/cu ss sr		cu ss sr/cu ss sr

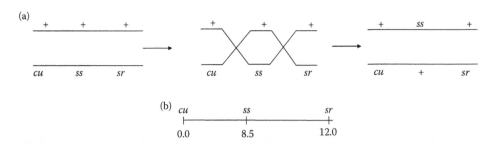

FIGURE 3.9 (a) Double crossing-over; (b) Arrangement of genes on the chromosome. (Redrawn from Burns, G. W. 1969. *The Science of Genetics*. N. H. Giles, and J. G. Torrey, Eds., The Macmillan Biology Series., The Macmillan Company, Collier Macmillan Limited, London.)

F₁ Phenotypes	Maternal Chromosomes	Numbers	%	Percentage
Normal normal normal	+ + +	440	44.0	Parental
Curled spineless striped	cu ss sr	452	45.2	Parental
Normal spineless striped	+ ss sr	40	4.0	cu-ss single c.o.
Curled normal normal	cu + +	33	3.3	cu-ss single c.o.
Normal normal striped	+ + sr	11	1.1	ss-sr single c.o.
Curled spineless normal	cu ss +	12	1.2	ss-sr single c.o.
Normal spineless normal	+ ss +	7	0.7	Double c.o.
Curled normal striped	cu + sr	5	0.5	Double c.o.

Parental: 89.2%
Single c.o. (*cu-ss*): 7.3%
Single c.o. (*ss-sr*): 2.3%
Double c.o.: 1.2%
Based on data observed in the above table, we can establish the following conclusions:

1. Double crossover: The double crossover frequency is the smallest (0.7%; 0.5%) and may determine the gene sequence. Since genes in the female parent are in *cis* arrangement, then + *ss* + and *cu* + *sr* individuals must be the product of double crossover.
2. Thus, the gene sequence on the map will be *cu-ss-sr* (Figure 3.9a,b).
3. The precise distance between *cu* and *ss* is 7.0 + 1.2 = 8.5 (single crossover + double crossover) (Figure 3.9b).
4. The true distance between ss and sr is 2.3 + 1.2 = 3.5 (single crossover + double crossover).
5. The correct distance between *cu* and *sr* is 7.3 + 1.2 + 2.3 + 1.2 = 12.0 (*cu-ss* single crossover; *ss-sr* single crossover + twice the double crossovers). The double crossover implies the crossing over between *cu* and *ss* and between *ss* and *sr*. In two-point crossover, the distance between *cu* and *sr* is 9.6, which is lower than those observed in the three-point testcross (12.0) (Figure 3.10a). This is due to the inability to detect double crossovers without a third marker between *cu* and *sr*.
6. It has been established based on classical cytogenetics that the number of known linkage groups never exceeds the number of pairs of homologous chromosomes in diploid organisms.
7. Assuming this is the first linkage group, three genes can be placed as shown in Figure 3.10b.

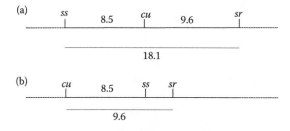

FIGURE 3.10 Arrangement and order of genes (a) on the chromosomes. (b) Change in the order after detecting double crossover. (Redrawn from Burns, G. W. 1969. *The Science of Genetics*. N. H. Giles, and J. G. Torrey, Eds., The Macmillan Biology Series., The Macmillan Company, Collier Macmillan Limited, London.)

8. In *Drosophila*, when a female homozygote fly (*cu ss sr/cu ss sr*) is crossed using F_1 heterozygote (*++ +/cu ss sr*) as a male, only two parental phenotypes will be recovered because
   crossing over does not occur in male flies; dipterans are unusual in this aspect.

3.2.2.2.3 *Interference and Coincidence*

Once a crossing over occurs in one chromosome segment, the probability of another crossing over
in the adjacent region is reduced. This phenomenon is known as interference. Interference appears
to be unequal in different parts of a chromosome and among chromosomes of a given complement. In general, interference appears to be greatest near the centromere and at the distal ends of
a chromosome. The degrees of interference are commonly expressed as coefficient of coincidence.
Coincidence values ordinarily vary from 0 and 1; absence of interference results in a coincidence
value of 1, whereas complete interference results in a coincidence of 0. Coincidence is usually quite
small for a short map distance.

In conclusion, knowledge of Mendelian genetics and its modified ratios is the foundation of modern genetics, cytogenetics, plant breeding, and molecular biology.

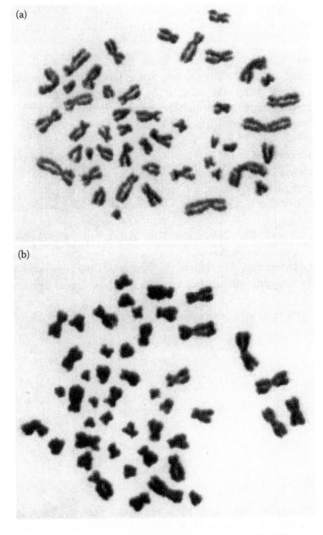

FIGURE 3.11 Metaphase chromosome of human from embryonic lung fibroblasts grown *in vitro*. Adapted
from Tjio, J. H. and A. Lavan. 1956. *Hereditas* 42 (1–2): 1–6).

3.3 RISE AND DECLINE OF PLANT CYTOGENETICS

Shortly after the rediscovery of Mendel's laws of inheritance, Walter Sutton (1903) proposed the chromosome in heredity based on studies in *Brachystola* (grasshopper):

1. The chromosome group of the presynaptic germ cells is made up of two equivalent chromosome series, and that a strong ground exists for the conclusion that one of these is paternal and the other one is maternal.
2. The process of synapsis (pseudo-reduction) consists of the union in pairs of the homologous numbers (i.e., those that correspond in size) of the two series.
3. The first post-zygotic or maturation mitosis is equational and hence results in no chromosome differentiation.
4. The second post-zygotic division is a reducing division, resulting in the separation of the chromosomes which have conjugated in synapsis, and their relegation to different germ cells.
5. The chromosomes retain a morphological individuality throughout the various cell divisions.

On the basis of the above observations, Sutton demonstrated that chromosomes are the vehicles for the genes used by Mendel in his study on the pea. Thus, cytogenetics is a hybrid of cytology and genetics. Several organisms have been used in cytogenetic studies, including drosophila (Morgan, Sturtevant, Bridges; 1910–1922), datura (Blakeslee, Avery, Satina, Rietsema, and associates; 1912–1956), maize (McClintock, Randolph, Rhoades, and others), wheat (Kihara and colleagues, Sears and colleagues), barley (Tsuchiya and colleagues), tomato (Rick and Khush and colleagues), and others. Cytogenetic chromosome maps for barley, tomato, rice, and maize have been developed by using aneuploid stocks such as primary trisomics. Unfortunately, when these pioneering cytogeneticists either retired or expired, these positions were filled by molecular geneticists designating a field of molecular cytogenetics. Thus, classical cytogenetics is slowly fading away. Cytogenetics is being taught by plant breeders without using a cytology laboratory. My personal experience is that students are graduating in plant breeding without seeing chromosomes.

Contrastingly, revolution in human cytogenetics occurred when Joe Hin Tjio and Albert Lavan (1956) determined the accurate ($2n = 46$) chromosome number in humans (Figure 3.11). Prior to this year, human chromosome number was considered to be $2n = 48$.

4 Cell Division

4.1 INTRODUCTION

Cell division is a continuous process that occurs in all living organisms. It has been divided into two categories: mitosis and meiosis. Both forms of nuclear division occur in eukaryotes and these processes comprise the cell cycle: G_1 (growth) $\rightarrow$ S (synthesis of DNA) $\rightarrow$ G_2 (growth) $\rightarrow$ M (mitosis or meiosis) $\rightarrow$ C (cytokinesis) (Smith and Kindfield, 1999). Mitosis occurs in somatic tissues where each chromosome is divided identically into halves, both qualitatively and quantitatively, producing genetically identical to the parent nucleus. In contrast, meiosis takes place in germ cells with the consequence that nuclei with haploid chromosome numbers are produced. Both types of cell division play an important role in the development and hereditary continuity of a eukaryotic organism.

4.2 MITOSIS

4.2.1 PROCESS OF MITOSIS

The term mitosis is derived from the Greek word *mitos* for thread; coined by Flemming in 1879 (see Chapter 1). The synonym of mitosis is karyokinesis, that is, the actual division of a nucleus into two identical parental daughter nuclei. It is also known as equational division because the exact longitudinal division of each chromosome into identical chromatids and their precise distribution into daughter nuclei leads to the formation of two cells; identical to the original cell from which they were derived.

The process of mitotic cell division has been divided into six stages: (1) interphase, (2) prophase, (3) metaphase, (4) anaphase, (5) telophase, and (6) cytokinesis.

4.2.1.1 Interphase

Two more terms, resting stage and metabolic stage, have been used to identify interphase cells. However, interphase cells should not be described as being in a "resting stage" because their nuclei are very active as they prepare for cell division. The DNA replication and transcription occur during interphase (Manuelidis, 1990). Interphase consists of three phases: G_1 (gap 1; pre-DNA synthesis) phase, S phase (DNA synthesis), and G_2 (gap 2; post-DNA synthesis). The duration of mitotic division is short compared to time required for the cells going through interphase (Figure 4.1). Thus, "metabolic stage" is a more appropriate term for the interphase cells. The interphase nucleus contains one or more prominent nucleoli and numerous chromocenters depending on the heterochromatic nature of the chromosomes. Chromosomes cannot be traced individually and they are very lightly stained (Figure 4.2a).

4.2.1.2 Prophase

All the chromosomes are beginning to become distinct and they are uniformly distributed in the nucleus. In early prophase, chromonemata become less uniform. The chromosomes are more or less spirally coiled and seem to be longitudinally double (Figure 4.2b). The two longitudinal halves of a chromosome are known as chromatids. As the prophase stage advances to mid- and late-prophase, chromosomes become thicker, straighter, and smoother (Figure 4.2c). The two chromatids of a chromosome become clearly visible. The nucleoli begin to disappear in late-prophase. It was formerly believed that chromosomes in a prophase nucleus are arranged in a haphazard fashion throughout the nucleus, but this is not true. Several studies have shown that, in interphase

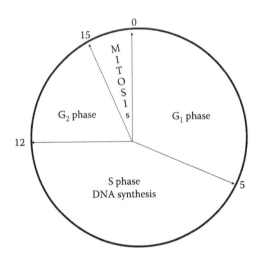

FIGURE 4.1 Mitotic cycle.

and prophase nuclei, kinetochores (centromeres) or primary constrictions are oriented toward one pole while telomeres face opposite to the kinetochores and are attached to the nuclear membrane. This orientation suggests that chromosomes maintain substantially the previous telophasic position (Figures 4.2b and c).

4.2.1.3 Metaphase

During metaphase, kinetochores move to the equatorial plate. Nucleoli and nuclear membrane disappear. Chromosomes are shrunk to the minimum length. Kinetochores are attached to spindle fibers, while chromosome arms may float on either side in the nucleus (Figure 4.2d). Karyotype analysis of a species is generally studied at metaphase after pretreatment of specimen cells.

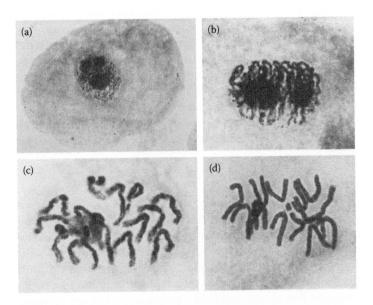

FIGURE 4.2 Mitosis in barley $(2n = 14)$ without pretreatment and with aceto-carmine staining. (a) Interphase. (b) Early prophase. (c) Late prophase. (d) Metaphase.

4.2.1.4 Anaphase

As the anaphase stage ensues, kinetochores become functionally double and the two sister chromatids of each chromosome separate (Figure 4.3a). It appears that the kinetochores (spindle fiber attachment) regions of the two sister chromatids are being pulled to opposite poles by the spindle fibers. Chromatids are more slender and densely stained. Each daughter chromosome moves to a polar region (Figure 4.3b). By the end of anaphase, spindle fibers disappear and the compact groups of chromosomes at the two poles are of identical genetic constitution.

Based on studies in frog eggs and egg extracts, Karsenti and Vernos (2001) suggested that mitotic spindle is a molecular machine capable of distributing the chromosomes to the daughter cells with stunning precision. The spindle is composed of microtubules that facilitate movement of chromosomes precisely during cell division. Microtubules in the spindle are arranged in two antiparallel arrays with their plus end at the equator and their minus end to the poles. Three kinds of microtubules are present in the spindle:

1. Astral microtubules originate from the poles and radiate out into cytoplasm.
2. Some microtubules connect the poles to a specific site (kinetochore) on the chromosomes.
3. A variable number of microtubules originate from the spindle poles and overlap in an anti-parallel way at spindle equator or interact with the chromosome arms.

Most plants do not have morphologically distinct microtubules (Marc, 1997). However, plants contain functionally similar microtubule-organizing centers. The major key element of microtubules in plants is γ-tubulin, which is omnipresent in eukaryotes including plants. The initiation

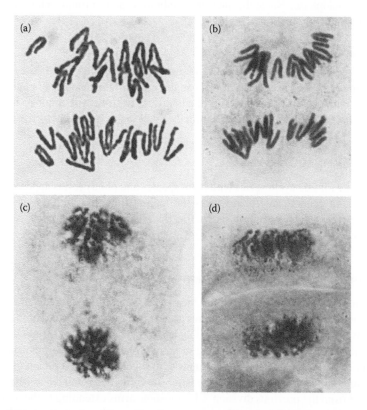

FIGURE 4.3 Mitosis in barley ($2n = 14$) without pretreatment and with aceto-carmine staining. (a) Anaphase. (b) Late anaphase, chromosomes have already reached their respective poles. (c) Telophase. (d) Cytokinesis.

of microtubules during cell division is initiated by γ-tubulin. Microtubule-organizing centers are dispersed in the cell. However, microtubules can assemble and self-organize into bipolar spindle without centriolar centrosome but their regulation is not yet established.

Forer and Wilson (1994) presented four models for chromosome movement during postmetaphase of mitosis:

1. The microtubules and filaments are attached to the kinetochore.
2. At the onset of anaphase, motor molecules fixed in a spindle filament matrix push poleward on the kinetochore microtubules.
3. Chromosome arms are pushed poleward in anaphase by forces independent of the forces that push the kinetochore fiber poleward; the forces on the chromosome arms also may arise from motor molecules associated with the spindle matrix.
4. The addition of a tubulin subunit to kinetochore microtubules occurs at the kinetochore. The removal of tubulin subunits from kinetochore microtubules occurs at the pole and at the kinetochore. Both polymerization and depolymerization are regulated at least in part by "compression" and "stretching" forces on the kinetochore microtubules. The compression of kinetochore microtubules arises when motor molecules push the kinetochore fiber into the kinetochore and chromosomes into the kinetochore fiber. The stretching force arises by motor molecules pulling kinetochore fiber out of the kinetochore.

4.2.1.5 Telophase
Chromosomes contract and form a dense chromatid ball. Chromosomes of the two daughter nuclei reorganize at the telophase. Nucleoli, nuclear membranes, and chromocenters reappear, and the chromosomes lose their stainability (Figure 4.3c).

4.2.1.6 Cytokinesis
The division of cytoplasm and its organelles between daughter cells is called cytokinesis, which begins during the late telophase stage at or near the equatorial (metaphase) plate (McIntosh and Koonce, 1989; Figure 4.3d). Cytokinesis in plants differs from that in animals. In plants, it takes place by the formation of a cell plate but in animals, cytokinesis begins by furrowing.

4.2.2 DURATION OF MITOSIS

The duration of the mitotic cycle varies with plant species and ploidy levels appear to have no effect at all (Table 4.1). In general, interphase takes the longest time and prophase is shorter than the remainder of the cycle. The results have shown that the greater the DNA content per cell, the longer the mitotic cycle (Van't Hof and Sparrow, 1963; Evans and Rees, 1971).

4.2.3 CHROMOSOME ORIENTATION AT INTERPHASE AND PROPHASE

After mitotic anaphase, chromosomes remain localized during interphase and eventually reappear in the same position (kinetochores and telomeres are located at opposite sides in the nucleus) during the next prophase (Rable model). Singh and Röbbelen (1975) demonstrated after Giemsa staining of chromosomes of several species of *Secale* that in an interphase cell the chromocenters (telomeres) mostly lay at one side of the nucleus, while opposite to it generally lay a region of very dense filamentous structures called the kinetochore. The latter site could be visualized as being a spindle pole position retained from the previous division. Chromosome arms extending to the region of chromocenters represent their telomeric ends (Figure 4.4a). Therefore, chromosomes may have a relatively fixed position in the nucleus (Comings, 1980). Chromosomes are more closely associated during interphase and become more loosely oriented as mitosis progresses. The chromocenters gradually

TABLE 4.1
Mitotic Cycle Time (h) in Several Plant Species

Species	2n	Ploidy Level	Mitotic Cycle	Authority
1. *Haplopappus gracilis*	4	2x	10.50	Sparvoli et al. (1966)
2. *Crepis capillaris*	6	2x	10.75	Van't Hof (1965)
3. *Trillium erectum*	10	2x	29.00	Van't Hof and Sparrow (1963)
4. *Tradescantia paludosa*	12	2x	20.00	Wimber (1960)
5. *Vicia faba*	12	2x	13.00	Van't Hof and Sparrow (1963)
6. *Impatiens balsamina*	14	2x	8.80	Van't Hof (1965)
7. *Lathyrus angulatus*	14	2x	12.25	Evans and Rees (1971)
8. *Lathyrus articularis*	14	2x	14.25	Evans and Rees (1971)
9. *Lathyrus hirsutus*	14	2x	18.00	Evans and Rees (1971)
10. *Avena strigosa*	14	2x	9.80	Yang and Dodson (1970)
11. *Secale cereale*	14	2x	12.75	Ayonoadu and Rees (1968)
12. *Allium cepa*	16	2x	17.40	Van't Hof (1965)
13. *Allium fistulosum*	16	2x	18.80	Van't Hof (1965)
14. *Hyacinthus orientalis*	16	2x	24.00	Evans and Rees (1971)
15. *Zea mays*	20	2x	10.50	Evans and Rees (1971)
16. *Melandrium album*	22	2x	15.50	Choudhuri (1969)
17. *Lycopersicon esculentum*	24	2x	10.60	Van't Hof (1965)
18. *Tulipa kaufmanniana*	24	2x	23.0	Van't Hof and Sparrow (1963)
19. *Avena strigosa*	28	4x	9.90	Yang and Dodson (1970)
20. *Pisum sativum*	28	4x	12.00	Van't Hof et al. (1960)
21. *Triticum durum*	28	4x	14.00	Avanzi and Deri (1969)
22. *Allium tuberosum*	32	4x	20.60	Van't Hof (1965)
23. *Helianthus annuus*	34	2x	9.00	Van't Hof and Sparrow (1963)
24. *Triticum aestivum*	42	6x	10.50	Bennett (1971)

are distributed into the net-like structure of the condensing chromosomes. At early prophase, fused chromocenters can be clearly seen and such association is still visualized during mid-prophase (Figure 4.4b). Thus, the orientation of chromosomes in interphase nuclei is loosened during later stages of the mitotic cycle and may be due to normal chromosome movement. Differences in squashing of material may also contribute to this effect.

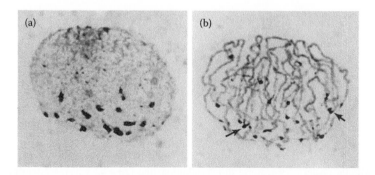

FIGURE 4.4 Mitotic nuclei of rye (2n = 14) after Giemsa C-banding staining. (a) Interphase, with chromocenters (telomeric bands) on the lower portion of nucleus and spindle pole (kinetochores) on the upper side of the nucleus. (b) An early prophase nucleus with fused telomeres (arrows). (From Singh, R. J. and G. Röbbelen. 1975. *Z. Pflanzenzüchtg.* 75: 270–285. With permission.)

4.2.4 SOMATIC ASSOCIATION

Nonrandom arrangement of chromosomes during mitosis has been reported even by earlier cytologists (Vanderlyn, 1948). Somatic association of homologous chromosomes has been observed in a large number of plants (Avivi and Feldman, 1980). It has been speculated that in plants homologous chromosomes are very closely associated during interphase (Kitani, 1963; Feldman et al., 1966; Wagenaar, 1969). However, due to lack of suitable materials and techniques, all the somatic association studies in plants, reviewed by Avivi and Feldman (1980), are from observations or measurements of distances between homologous chromosomes at the metaphase stage, except for the observations of Singh et al. (1976). Singh et al. (1976) demonstrated that homologous chromosomes are more closely associated during interphase of somatic cells than at metaphase. Recently, nonrandom chromatin arrangements in intact mitotic interphase nuclei of barley by confocal fluorescent microscope were recorded by Noguchi and Fukui (1995). Centromere were clustered at one site, and subtelomeric regions dispensed or close to the nuclear membrane on the opposite site (polarity). FISH and EM studies also supported polarity results. In *Chrysanthemum multicore*, they recorded homologous satellite chromosome (SAT) pair fuse with each other during telophase. The SAT fusion in a pair of homologous chromosomes occurs after the nuclear fusion. A similar result was reported in *Plantago ovata* (Dhar and Kaul, 2004).

Somatic association of homologous chromosomes of rye (*Secale cereale*) was studied by the Giemsa banding technique at interphase in wheat-rye addition lines. Telomeres of the rye chromosomes, appearing as chromocenters, showed close somatic association in disomic addition lines (Figures 4.5a and b), but they were distributed at random in double monosomic additions. This demonstrates directly that somatic association of homologs at interphase is even closer in nondividing nuclei than in metaphase cells. Feldman and Avivi (1973) suggested that the positions occupied by chromosomes, when examined at metaphase, were in fact interphase positions. However, it is unlikely that such measurements in metaphase determine interphase positions. Studies in other crop plants at metaphase also showed higher mean distances between homologous chromosomes than those reported for interphase nuclei (Table 4.2). The differences surely reflect dislocation of chromosomes from interphase to metaphase by chromosome contraction and spindle-mediated movement, reducing the association of homologous chromosomes.

In contrast, evidence of somatic association of homologous chromosomes has been rejected by Darvey and Driscoll (1972), Dvořák and Knott (1973), Therman and Sarto (1977), and Heslop-Harrison and Bennett (1983).

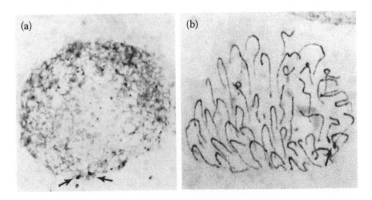

FIGURE 4.5 Mitotic nuclei of wheat-rye di-telo-addition line ($2n = 42$ wheat + 2 telo rye) after Giemsa C-banding staining. (a) Interphase with two chromocenters (arrows) representing telomeres of the rye telocentric chromosomes. (b) An early prophase with dark telomeric bands (arrow) showing somatic association. (From Singh, R. J., G. Röbbelen, and M. Okamoto. 1976. *Chromosoma* (Berlin) 56: 265–273. With permission.)

TABLE 4.2

Comparison of Mean Distances between Homologous Chromosomes Observed in Wheat, Oat, and Barley

Crops	Mean Distance[a]		Authors
	Homologous	Nonhomologous	
	Somatic Metaphase		
Hexaploid wheat	0.285	0.445	Feldman et al. (1966)
Hexaploid wheat	0.341	0.368	Darvey and Driscoll (1972)
Diploid oat	0.374	0.388	Sadasivaiah et al. (1969)
Hexaploid oat	0.332	0.440	Thomas (1973)
Barley	0.321	0.485	Fedak and Helgason (1970b)
	Somatic Interphase		
Wheat-rye addition lines	0.250	0.387	Singh et al. (1976)

[a] Theoretical random distribution, 0.452.

Precise spatial position of chromosomes during pro-metaphase of human fibroplasts and He la cells was recorded by FISH (Nagele et al., 1995). Chromosomes were arranged in wheel-shaped (rosette) and homologues were consistently positioned on opposite sides of the rosette, but the mechanism is unknown. Chromosome 16 is usually positioned adjacent to those of the X chromosome. This observation does not support somatic association.

Studies of various plant species have elucidated that somatic association is a prerequisite for meiotic pairing (Brown and Stack, 1968; Chauhan and Abel, 1968; Stack and Brown, 1969; Loidl, 1990). Palmer (1971) observed no evidence for a nonrandom association of homologous chromosome in the premeiotic or ameiotic mitoses of homozygous ameiotic plants or in the premeiotic mitosis of normal sibs and supported the classical view that homologous chromosomes do not synapse until zygonema. Furthermore, the formation (during zygonema stage) and function of the synaptonemal complex do not necessarily favor somatic association (Moens, 1973).

Despite numerous examples that favor somatic association of homologous chromosomes (Avivi and Feldman, 1980), it is still a highly controversial issue. Therefore, a definite conclusion that the phenomenon of somatic chromosome association is common in all organisms cannot be drawn (Loidl, 1990).

4.3 MEIOSIS

Reduction of the somatic chromosome number to the haploid number occurs in the germ cells of both plants and animals. Meiosis, known as reduction division, consists of two nuclear divisions. The first division is disjunctional. This division includes the initiation and maintenance of homologous chromosome association in prophase and metaphase and subsequent movement of homologous kinetochores of each bivalent to opposite poles during anaphase-I, thus reducing the chromosome number by half in each daughter nucleus. Mitosis occurs in cycle II producing four haploid cells from each of the diploid cells (meiocytes) that undergo meiosis. Meiotic stages have been divided into two cycles. The first cycle consists of stages responsible for producing daughter nuclei with the haploid chromosome number, and after the second cycle four haploid nuclei will have been produced.

4.3.1   Process of Meiosis

4.3.1.1   Cycle 1

4.3.1.1.1   Prophase-I

4.3.1.1.1.1   Leptonema   The chromonemata become more distinct from one another at leptonema and appear as very long and slender threads. The chromosomes may be oriented with kinetochores toward the same side of the nucleus, forming a so-called "bouquet." Leptotene cells have
large nucleoli and distinct nuclear membranes (Figure 4.6a).

4.3.1.1.1.2   Zygonema   The mechanism of synapsis of homologous chromosomes is a complex
phenomenon and is described by light microscope, electron microscope (synaptonemal complex-
SC), and molecular experiments. The light microscope view is presented here.

Synapsis starts initially from one or more regions of homologous chromosomes and gradually extends at several secondary sites zipper-like until it is complete (Burnham et al., 1972).
Chromosomes begin to shorten and thicken (Figure 4.6b). A physical multiple interstitial inter-
homologue connection is a prerequisite for homologous chromosome pairing and recombination
(Kleckner, 1996). The site of chromosome pairing initiation is still debated. The intimate association of homologous chromosomes in the synaptonemal complex, alignment of homologous chromosomes, and recombination has been critically reviewed by Santos (1999). Recombination should be
required for synapsis was concluded.

Centric heterochromatin contains multiple pairing elements. These elements act initiating the
proper alignment of achiasmate chromosomes early in meiosis (Karpen et al., 1996). This suggests that heterochromatin in eukaryote is one of the extremely mysterious components. It is distinguished from euchromatin by sparsely populated with genes, inhibits the function of euchromatic

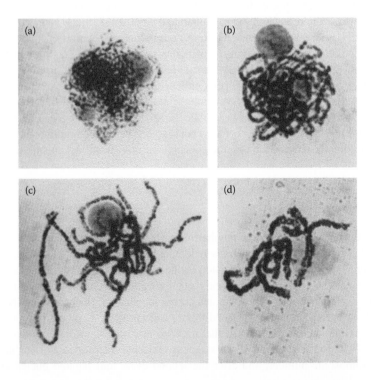

FIGURE 4.6   Meiosis in barley ($2n = 14$) after propiono-carmine staining. (a) Leptonema. (b) Zygonema.
(c) Pachynema. (d) An early diplonema.

gene (position effect variegation), replicates late in the S phase, and is rich in tandemly repeated satellite sequences (Karpen et al., 1996).

However, in trisomics and in triploids, three homologous chromosomes associate in a trivalent synapsed arrangement.

4.3.1.1.1.3 Pachynema The synapsis of homologous chromosomes is completed. Chromosome regions unpaired in pachynema will remain usually unpaired. Chromosomes are noticeably thicker and shorter than those in the leptonema and paired entities visibly represent the haploid number of that species. Each pair of chromosomes is termed a bivalent. The term bivalent is used for the paired homologous chromosomes. Some scientists still use an outdated term tetrad. Tetrad should be used only for the four meiotic products such as the tetrad after meiosis in anthers. The unpaired chromosome is a univalent and three paired homologous chromosomes are trivalent, etc.

Nucleoli are clearly visible and certain chromosomes may be attached to them. These chromosomes are known as nucleolus organizer chromosomes or satellite (SAT) chromosome (Figure 4.6c).

4.3.1.1.1.4 Diplonema Chromatids composing bivalents begin to separate, or repel, at one or more points along their length (Figure 4.6d). Each point of contact is known as a chiasma (pl, chiasmata). At each point of contact, two chromatids have exchanged portions by crossing over. The number of chiasmata varies from organism to organism. In general, longer chromosomes have more chiasmata than shorter chromosomes.

Chiasmata may be interstitial or terminal. Interstitial chiasmata may be found anywhere along the length of a chromosome arm but terminal chiasmata are located at a chromosome tip (telomere). In most cases, a terminal chiasma occupies an interstitial position earlier but as cell division progresses, its position moves to the tip of the chromosome arm in which it occurred (Figure 4.7a). This phenomenon has been termed terminalization.

Torrezan and Pagliarini (1995) examined the influence of heterochromatin on chiasma localization and terminalization in maize. Chiasma frequency was higher during the diplonema with a predominance of interstitial chiasmata located in euchromatin. During diakinesis, reduction in chiasma frequency was recorded in euchromatin with a consequent increase in the heterochromatin. The chiasmata localized in euchromatin appear to move to the region of the heterochromatin blocks. It appears that heterochromatin serves as a barrier against terminalization since terminalization beyond the bands was not recorded in any heteromorphic bivalent.

4.3.1.1.1.5 Diakinesis Chromosomes continue to shorten and thicken. In a squash preparation, compact and thick chromosomes lie well-spaced in the nucleus, often in a row near the nuclear membrane. This is a favorable stage to count the chromosomes. The number of chiasmata is reduced due to terminalization. The nucleolus begins to decrease in size (Figure 4.7b).

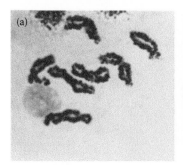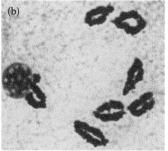

FIGURE 4.7 Meiosis in barley (2n = 14) after propiono-carmine staining. (a) Late diplonema, showing 7 bivalents. (b) Diakinesis, a bivalent associated with nucleolus.

4.3.1.1.2 Metaphase-I

As soon as cells reach the metaphase stage, the nuclear membrane and nucleoli disappear, and spindle fibers appear. Bivalents are arranged at the equatorial plate with their kinetochores facing the two poles of the cell. Chromosomes reach their maximum contractions (Figure 4.8a).

4.3.1.1.3 Anaphase-I

Homologous chromosomes begin to separate toward the opposite poles of the spindle (Figure 4.8b). The dissociation of chromosomes initiates on schedule leaving unpaired chromosomes behind. Chromosome movement to their respective poles is either coupled with the shortening of spindle microtubules or kinetochore motors chew microtubules as they drag chromosomes to the poles (Zhang and Nicklas, 1996). This generates two groups of dyads with the haploid chromosome number (Figure 4.8c). The meiotic anaphase-I chromosomes are much shorter and thicker than the chromosomes of mitotic anaphase.

4.3.1.1.4 Telophase-I

At the end of anaphase-I, chromosomes reach their respective poles and polar groups of chromosomes become compact (Figure 4.8d). Nuclear membranes and nucleoli start to develop and eventually two daughter nuclei with haploid chromosomes are generated. The chromatids are widely separated from each other and show no relational coiling. As a result of crossing over, each chromatid derived from a particular bivalent may be different genetically from either of the parental homologs that entered the bivalent.

4.3.1.1.5 Interkinesis

Interkinesis is the time gap between division I and division II. Generally it is short or may not occur at all. At the end of telophase-I, cytokinesis does not invariably follow division I. In many vascular plants, two daughter nuclei lie in a common cytoplasm and undergo second division.

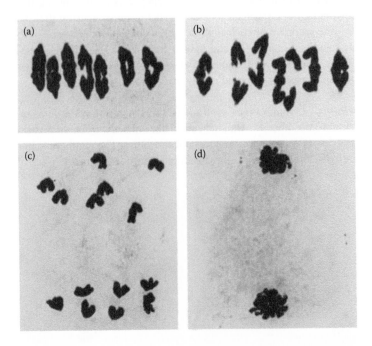

FIGURE 4.8 Meiosis in barley ($2n = 14$) after propiono-carmine staining. (a) Metaphase-I showing 7 bivalents. (b) An early anaphase-I, chromosomes are beginning to disjoin. (c) Anaphase-I, showing 7-7 chromosome migration. (d) Telophase-I.

4.3.1.2   Cycle 2

In the second division of meiosis, prophase-II (Figure 4.9a), metaphase-II (Figure 4.9b), anaphase-II (Figure 4.9c), and telophase-II resemble similar stages of mitotic divisions. This second division yields a quartet (tetrad) of uninucleate cells (Figure 4.9d) called microspores (in male) produced from pollen mother cells (PMC) and megaspores (in female) produced from megaspore mother cells (MMC). Tetrad analysis is a powerful scientist's tool to determine gamete development, cell division, chromosome dynamics, and recombination (Copenhaver et al., 2000).

The difference between mitosis and meiosis has been shown diagrammatically in Figure 4.9. Mitosis occurs in all growing vegetative tissues; $2n$ cells produce only $2n$ cells or $1n$ cells produce only $1n$ cells during mitosis while meiosis is confined only specialized in reproductive tissue; $2n$ tissues produce $1n$ cells and after fertilization with $1n$ spore produces $2n$ cells (Figure 4.10a). Chromosomes usually do not pair in mitotic prophase; 1-chromatid chromosomes replicate and produce 2-chromatid chromosomes and that separates at anaphase while in meiosis synapsis and exchange of 2-chromatid homologue in early prophase leads to the reduction of chromosome number (Figure 4.10b). During mitosis, kinetochores lie on the metaphase plate but in meiosis kinetochores lie on either side of the metaphase plate. Kinetochores divide in mitotic anaphase while kinetochores do not divide during meiotic anaphase I. Mitosis concludes with the production of two identical diploid ($2n$) daughter cells, in absence of mutation, while the end of meiosis results in production of four haploid (n) daughter nuclei: two identical and two altered (resultant of synapses, crossing over, recombination, and independent assortment of nonhomologous chromosomes). When egg and sperm unite, a zygote is produced and has the same amount of genetic materials as their parents maintaining the continuity of chromosome number but variation in genotypic constitution and phenotypic appearance is evolved by meiosis.

Although most people use mitosis in reference to division II of meiosis, division II differs from mitosis in three ways. (1) Mitosis always is preceded by an interphase in which DNA synthesis occurs; meiotic division II is not. (2) The chromatids within meiotic division II chromosomes are

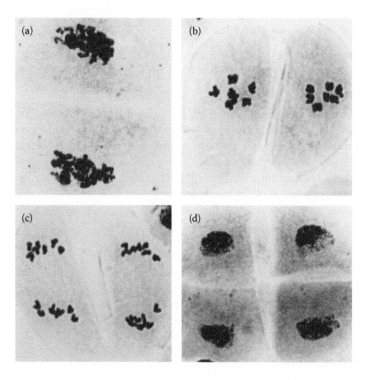

FIGURE 4.9   Meiosis in barley ($2n = 14$) after propiono-carmine staining. (a) Prophase-II. (b) Metaphase-II. (c) Anaphase-II. (d) A quartet or tetrad cell.

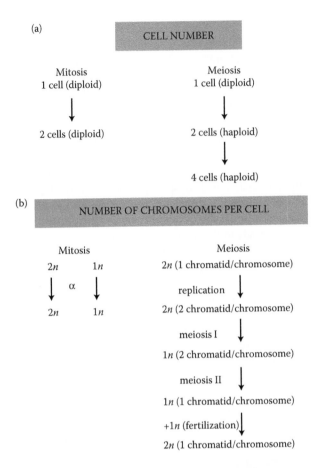

FIGURE 4.10 Diagrammatic explanation of the product of mitosis and meiosis. (a) Cell number. (b) Number of chromosomes per cell. (From Smith, M. U. and A. C. H. Kindfield. 1999. *Am. Biol.* Teach. 61: 366–371. With permission.)

not necessarily identical (due to crossing-over) but in mitosis they are. (3) Chromatids within a chromosome of mitotic prophase and metaphase are tightly wound around each other; in meiotic division II, the chromatids appear as if they are repelling each other in late prophase and metaphase. [This could be a carry-over from the anaphase I (Figure 4.8c) when the chromatids appear to be repelling each other (e.g., four chromatid arms for a metacentric chromosome in the anaphase I)]. Division II could be described as superficially appearing as a mitotic division.

The following characteristic features of meiosis suggest that cell division is under genetic control (Baker et al., 1976):

1. The replication of bulk DNA (premeiotic S-phase)
2. Synapsis of homologous chromosomes, chiasma formation, and crossing over
3. Segregation of homologous kinetochores at anaphase-I followed by segregation of sister kinetochores to opposite poles at anaphase-II
4. Formation of nuclei with the haploid chromosome number

4.3.2 DURATION OF MEIOSIS

Bennett (1971) reviewed published results on meiotic duration. The results showed that in diploid species the duration is positively correlated with the DNA content per nucleus and with the mitotic

TABLE 4.3
Duration of Meiosis (h) in Diploid Species[a]

Species	2n	Meiotic Cycle	DNA per Cell (in Picograms)
1. *Antirrhinum majus*	16	24.0	5.5
2. *Haplopapus gracilis*	4	36.0	5.5
3. *Secale cereale*	14	51.2	28.7
4. *Allium cepa*	16	96.0	54.0
5. *Tradescantia paludosa*	12	126.0	59.0
6. *Tulbaghia violacea*	12	130.0	58.5
7. *Lilium henryi*	24	170.0	100.0
8. *Lilium longiflorum*	24	192.0	106.0
9. *Trillium erectum*	10	274.0	120.0

Source: From Bennett, M. D. 1971. *Proc. Roy. Soc. Lond., Ser.* B. 178: 277–299.
With permission.

[a] Results from several authors, data taken at different temperatures, however, provide convincing evidence.

cycle time (Table 4.3). Meiotic duration is also influenced by chromosomal organization, DNA structure, and the developmental pattern of the plant species.

4.3.3 GAMETOGENESIS

The end products of meiosis in higher plants are four microspores in the male and four megaspores in the female containing the haploid (*n*) chromosome number. Two mitotic divisions occur in microspores. The first division produces a tube nucleus and a generative nucleus. The second division occurs only in the generative nucleus, producing 2 sperm cells (Figure 4.11a).

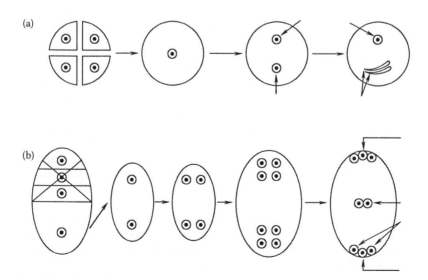

FIGURE 4.11 Diagrammatic explanation of gametogenesis in higher plants. (a) Formation of pollen grains. (b) Development of embryo sac.

Megaspores are produced in the ovules from a megaspore mother cell. Of the four megaspores produced by meiosis, only one normally survives. Three mitotic divisions occur during mega-gametogenesis. In the first division, a megaspore nucleus divides to give rise to the primary micropylar and primary chalazal nuclei. The second mitosis produces two nuclei each at the micropylar and at the chalazal region of the embryo sac. The third mitosis results in four nuclei at each of the opposite poles of an embryo sac. One nucleus from each pole, known as polar nuclei, moves to the middle of the embryo sac and the two nuclei fuse to give rise to a secondary nucleus, or polar fusion nucleus, with $2n$ chromosome constitution. At the micropylar end of the embryo sac, of the three nuclei remaining one nucleus differentiates into an egg cell and the remaining two become synergids (member of the egg apparatus). The three nuclei at the chalazal end of the embryo sac are called antipodal nuclei (Figure 4.10b). Cell wall formation around the antipodals, synergids, and egg occurs. At maturity, the embryo sac consists of the following cells: antipodals (3), synergids (2), egg (1), and the large central cell that contains the polar nuclei. There are many variations to embryo sac development detailed in Maheshwari (1950). The present text covers a typical gametogenesis of many plants, particularly the grasses so important in plant breeding.

During pollination and fertilization in higher plants, pollen grains fall onto the stigma. The tube nucleus of a pollen grain directs growth of the pollen tube down the style, its travel through the micropyle, and its entrance into the nucellus. The two sperms are released into the embryo sac. One sperm nucleus unites with the egg nucleus to give rise to a $2n$ zygote (embryo). The other sperm fuses with the secondary nucleus to form an endosperm nucleus with triploid chromosome number. From this double fertilization, a seed develops. Higashiyama et al. (2001) demonstrated by laser cell ablation in flowering plants that two synergid cells adjacent to the egg cell attract pollen tubes. Once fertilization is completed, the embryo sac no longer attracts the pollen tube and this cessation of attraction might be involved in blocking polyspermy.

4.4   DIFFERENCES BETWEEN MITOSIS AND MEIOSIS

Mitosis	Meiosis
1. Mitosis occurs in somatic cells of an organism	Meiosis takes place in germ cells of an organism
2. Mitosis occurs in both sexually and asexually reproducing organism	Meiosis takes place in sexually reproducing organism
3. Cells divide only one time	Cells divide two times; the first and second meiotic divisions
4. Interphase ensues prior to each mitotic division	Interphase occurs only meiosis I; does not occur prior to meiosis II
5. Synthesis of DNA occurs during interphase I	Synthesis of DNA occurs during interphase I but not in interphase II
6. Duration of prophase is short; probably few hours	Duration of prophase is longer; may be few days
7. Prophase is simple	Prophase is completed and is divided into leptonema, zygonema, pachynema, diplonema, and diakinesis
8. Usually, homologous chromosome pairing does not occur	Synapsis of homologous chromosome pairing occurs during prophase I
9. Two chromatids of a chromosome do not exchange chromosome during prophase	Chromatids of two homologous chromosomes exchange chromosome segments during crossing over in prophase I
10. The arms of the chromatids in prophase are close to one another	The arms of chromatids are widely separated in prophase II of meiosis
11. Chromosomes are duplicated at the beginning of prophase	When prophase I starts, chromosomes appear single; although DNA replication has taken place in interphase I
12. Bouquet arrangement of interphase and prophase chromosomes is not recorded	It is recorded in prophase I (convergence in animal and some plants)
13. Synaptonemal complex is absent	Synaptonemal complex is observed during homologous chromosome pairing

Continued

Mitosis	Meiosis
14. Crossing over is absent	Crossing over is present
15. Chiasmata are absent	Chiasmata are present
16. All the centromeres are aligned in same plate (metaphase plate)	Kinetochores are lined up in two planes; parallel to one other
17. Chromosome arms are away from metaphase plate	Chromosome arms are attached with their homologues at the metaphase I plate
18. Two chromatids of a chromosome are genetically identical	Two chromatids of a chromosome are genetically different due to crossing over
19. The genetic constitution of each daughter nuclei is identical	The genetic constitution of daughter nuclei is different from the parent; mixture of maternal and paternal genes
20. Division of kinetochores occurs at anaphase	The kinetochores move to the opposite poles and division of kinetochores occurs during anaphase II
21. Separation of chromosomes occurs simultaneously at anaphase	Usually short chromosomes (without interstitial chiasmata) separate first and separation of long chromosomes is delayed
22. Anaphase chromosomes are single stranded	Anaphase I chromosomes are double stranded and anaphase II chromosomes are single stranded
23. Similar chromosomes move toward the opposite poles during anaphase I	Dissimilar chromosomes move toward the opposite poles during anaphase I and anaphase II
24. Spindle fibers disappear at telophase	Spindle fibers do not disappear completely at telophase I
25. Nucleoli reappear at telophase	Nucleoli do not reappear at telophase
26. Cytokinesis occurs at the end of telophase resulting in two precise daughter diploid nuclei	Cytokinesis usually does not occur after telophase I but occurs after telophase II resulting in four new haploid nuclei
27. The chromosome number remains constant at the end of mitosis	The chromosome number reduced precisely half (haploid) from diploid
28. Mitosis helps in multiplication of cells-growth	Meiosis does not help in multiplication of cells
29. Mitosis helps in repair and healing	Meiosis helps in the formation of gametes and maintains chromosome number of an individual

5 Genetic Control of Meiosis

5.1 INTRODUCTION

Meiosis is a complex process that occurs in all sexually reproducing eukaryotic organisms. It is a unique cell division that produces haploid gametes from diploid parental cells and during sexual reproduction union of two haploid spores restores the diploid chromosome complement of an organism. Thus, meiosis helps maintain chromosome numbers constant from generation to generation, and ensures the operation of Mendel's laws of heredity. The steps of gametogenesis are premeiotic, meiotic (pairing of homologous chromosomes, assembly of the synaptonemal complex [SC], formation of chiasma [formation of stable connections between homologs formed at the site of cross-overs], recombination, and creation of haploid meiotic products), and postmeiotic mitosis (gametophytogenesis—the equational division). The meiotic events are divided into a series of substages based on changes in chromosome morphology (Table 5.1).

The entire process of meiosis is under control of a large number of genes (Darlington, 1929, 1932; Prakken, 1943; Rees, 1961; Katayama, 1964; Riley and Law, 1965; Baker et al., 1976; Maguire, 1978; Golubovskaya, 1979, 1989; Gottschalk and Kaul, 1980a, b; Koduru and Rao, 1981; Kaul and Murthy, 1985; Kaul, 1988; Roeder, 1997; Zickler and Kleckner, 1999; Pagliarini, 2000; Villeneuve and Hillers, 2001). Grishaeva and Bogdanova (2000) estimated more than 80 genes specifically controlling meiosis and meiotic recombination in *Drosophila melanogaster*.

The precise sequence of meiosis is sometimes disturbed by mutations. Meiotic mutants drastically change the normal behavior of chromosomes beginning with the initiation of premeiotic DNA synthesis (premeiotic mutants), during prophase-I (synaptic mutants), anaphase-I to telophase-II (disjunction mutants) (meiotic mutants) and after the completion of the second division of meiosis (postmeiotic mutants) (Figure 5.1). Meiotic mutants are rather common in the plant kingdom, and are distributed in a large number of crop species covering a wide range of crop families. In the literature, meiotic and postmeiotic mutants predominate while premeiotic and disjunction mutants are relatively rare.

Meiotic mutants have been identified chiefly on the basis of cytological observations, genetic evidence, and pollen or ovule abortion. Sometimes, they exhibit changes in general growth habit of plants. Meiotic mutants occur by spontaneous origin in natural populations, may be induced by mutagenesis, or may result from interspecific hybridization (Tables 5.2 through 5.4).

5.2 SYNAPTIC MUTANTS

It has been established by classical cytology that pairing of homologous chromosomes is initiated during zygonema by a longitudinal zipping up. It has been demonstrated in many plants that pairing sites are numerous (multiple sites) and fairly uniformly distributed along chromosomes. This phenomenon is not universal. In nematode (*Caenorhabitis elegans*), each chromosome contains a single site-homologue recognition region (HRR) and in every case it is located at one end of the chromosome. This region promotes homologous chromosome pairing (Roeder, 1997). By mid-pachynema, pairing is completed. At the end of metaphase-I, chromosomes are highly contracted, the repulsion of homologues is maximum and chromosome association is maintained only by chiasmata (Maguire, 1978). Meiotic events are precisely checked at each checkpoint (station) to ensure that one event does not occur until the preceding event has been completed. Two checkpoints operate in meiosis. The recombination checkpoint guards the cells that do not exit pachynema until recombination intermediates have been resolved. The metaphase checkpoint impedes cells from exiting metaphase-I until all chromosome pairs have been properly oriented on the metaphase plate (Roeder, 1997). Asynaptic and desynaptic (asyndetic) mutations ignore the checkpoints (Baker et al., 1976).

TABLE 5.1

Major Events in Meiotic Prophase

Meiotic Prophase	Chromosome Morphology	Bouquet Formation	DSB Repair	Cytological Signs
1. Leptonema	Axial elements begin to develop	Telomeres begin to cluster	DSBs appear	Early nodules
2. Zygonema	Chromosome synapsis initiates	Telomeres tightly clustered	DSBs disappear	Early nodules
3. Pachynema	Chromosomes fully synapsed	Telomeres disperse	DHJ	Late nodules
4. Diplonema	SC disassembled	–	Mature recombinants	Chiasma
5. Diakinesis	Further chromosome contraction	–	–	Chiasma

Source: From Roeder, G. S. 1997. *Genes & Develop.* 11: 2600–2621. With permission.
Note: DSB, Double-strand break; DHJ, Double holliday junction; SC, synaptonemal complex.

Rhoades (1956) identified an ameiotic mutant (*am1*) in maize in which meiosis did not occur and pollen mother cells degenerated. The trait is a monogenic recessive and homozygous plants carrying the *am1* gene are completely male sterile and partially female sterile. Palmer (1971) studied *am1* mutants cytologically and recorded normal premeiotic mitoses.

5.2.1 DISTRIBUTION OF SYNAPTIC MUTANTS

Since the discovery of the first synaptic mutants in maize by Beadle and McClintock (1928), the occurrence of such mutations has been recorded in a large number of plant species. In 1964, Katayama reported synaptic mutants in 20 families, consisting of 50 genera and about 70 species of higher plants. Koduru and Rao (1981) estimated synaptic mutants to have been observed in 126 species belonging to 93 genera. The estimates of Koduru and Rao (1981) are subject to some reduction. For example, they considered *Triticum vulgare* and *T. aestivum* to represent different species. Similarly they considered red clover and *Trifolium pratense* to be distinct species. Therefore, they counted four species instead of two.

The family Gramineae comprises the largest number of species with identified synaptic mutants, followed by Leguminosae, Liliaceae, Solanaceae, and Malvaceae. The majority of synaptic mutants reported in higher plants are in diploid species.

Although desynaptic mutants have been reported in a large number of plant species, their origin, inheritance, gene symbol designation, and meiotic chromosome behavior have been examined in only a few plant species (Table 5.2).

5.2.2 ORIGIN OF SYNAPTIC MUTANTS

Synaptic mutants of spontaneous origin have been isolated from natural populations. Similar mutations may also be induced by mutagenesis or derived from species hybrids. Generally, these mutants are identified in segregating generations, in which they are distinguished from normal plants because of pollen or ovule abortion. Meiotic mutants mostly exhibit monogenic recessive inheritance (Table 5.2; Koduru and Rao, 1981). However, the reported exception is a digenic recessive inheritance in cotton (Beasley and Brown, 1942; Weaver, 1971). Katayama (1964) and Koduru and Rao (1981) cited Hollingshead (1930a, b) in which she reported a dominant gene for asynapsis in *Crepis*. Katayama (1964) cited Hollingshead (1930b). I could not conclude from the paper that a dominant gene was controlling pairing. Koduru and Rao (1981) cited Hollingshead (1930a). However, I was unable to locate a claim of mono-factorial dominant inheritance. These two citations appear to deliver misinformation.

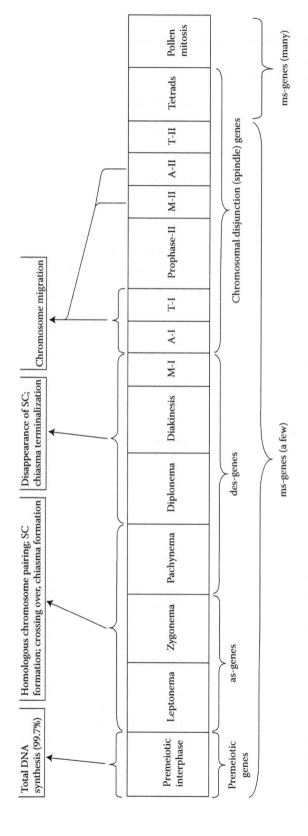

FIGURE 5.1 A diagrammatic sketch showing meiotic mutants observed at subsequent stages of meiosis. (Revised and redrawn from Golubovskaya, I. N. 1979. *Int. Rev. Cytol.* 58: 247–290. With permission.)

TABLE 5.2
Salient Characteristic Features of a Few Synaptic Mutants

Species	2n	Gene Symbol	Origin	Genetics	Description	Authority
					Asynaptic Mutants	
Brassica campestris	20	as, as3	I[a]	3:1	Chromosome pairing was not apparent during pachynema, diakinesis, and MI.	Stringam (1970)
Datura stramonium	24	bd	S	3:1	A complete lack of homologous chromosome pairing (24I) during first meiosis.	Bergner et al. (1934)
Hordeum vulgare	14	as	I	3:1	Absence of pachytene chromosome pairing. At MI, 14 univalent were obtained.	Sethi et al. (1970)
Oryza sativa	24	–	I	3:1	Absence of chromosome pairing and 24I at diakinesis and MI.	Kitada and Omura (1984)
Pisum sativum	14	as	I	3:1	Chromosome pairing lacked completely at pachynema and MI.	Gottschalk and Klein (1976)
Secale cereale	14	sy1	S	3:1	Absence of meiotic chromosome pairing	Sosnikhina et al. (1992)
Solanum commersonii	24	sy2	I	3:1	Absence of chromosome pairing at pachynema, diplonema, diakinesis, and MI.	Johnston et al. (1986)
Sorghum vulgare	20	as4	S	3:1	Characterized by lack of normal chromosome association in prophase.	Stephens and Schertz (1965)
Triticum durum	28	–	I	3:1	Majority of the sporocytes showed 28I at MI; lacked SC (La Cour and Wells,1970).	Martini and Bozzini (1966)
Zea mays	20	as	S	3:1	Twenty univalent were most frequent and 18I + 1II were less frequent.	Beadle (1930)
Zea mays	20	as	S	3:1	Bivalent ranged 0–9.95 at MI.	Miller (1963)
					Desynaptic Mutants	
Allium ascalonicum	8	–	S	?	At pachynema, chromosome pairing is complete; at MI, bivalent ranged from 0 to 4.	Darlington and Hague (1955)
Avena strigosa	14	ds	S	3:1	Chromosomes were synapsed in earlier stages of meiosis; bivalent ranged from 0 to 7.	Dyck and Rajhathy (1965)
Collinsia tinctoria	14	c[st]	I	3:1	Seven bivalent were observed in 26.87% cells while 14I in 12.8% sporocytes.	Mehra and Rai (1972)
Glycine max	40	st5	S	3:1	Almost complete pachytene chromosome pairing; very little pairing at MI.	Palmer and Kaul (1983)
Hordeum vulgare	14	des1 – des 15	S, I	3:1	Pachytene chromosome pairing is normal; chromosomes undergo desynapsis during diplonema.	Ramage (1985)
Lathyrus odoratus	14	–	I	3:1	Frequencies of bivalent varied from plant to plant and bivalents ranged from 0 to 7.	Khawaja and Ellis (1987)
Lycopersicon esculentum	24	as1 – as5	S	3:1	Variable chromosome pairing at pachynema, diakinesis and MI.	Soost (1951)
Pennisetum glaucum	14	–	I	3:1	Chromosome pairing at pachynema is normal; bivalents at MI ranged from 0 to 7.	Subba Rao et al. (1982)
Pisum sativum	14	ds	I	3:1	Mean bivalents ranged 0–6; the most frequent (31.5–57.6%) was 14I.	Gottschalk and Baquar (1971)
Secale cereale	14	sy7, sy10	S	3:1	Variable chromosome pairing at pachynema.	Fedotova et al. (1994)

(Continued)

TABLE 5.2 (*Continued*)
Salient Characteristic Features of a Few Synaptic Mutants

Species	2n	Gene Symbol	Origin	Genetics	Description	Authority
Triticum aestivum	42	*ds*	S	3:1	Chromosome synapsis occurred in normal fashion but fell apart after pachynema.	Li et al. (1945)
Avena sativa	42	*syn*	I	3:1	Homologous chromosomes were paired in early meiosis but disassociated prematurely at late prophase-I.	Rines and Johnson (1988)

[a] I = induced by mutagenesis; S = spontaneous origin.

TABLE 5.3
Synaptic Mutants Induced by Mutagens

Types of Mutagen	Crops	2n	Authors
1. X-rays	*Capsicum annuum*	24	Morgan (1963)
	Collinsia tinctoria	14	Mehra and Rai (1970)
	Nicotiana sylvestris	24	Goodspeed and Avery (1939)
	Oryza sativa	24	Katayama (1961)
	Pisum sativum	14	Gottschalk and Klein (1976)
	Zea mays	20	Morgan (1956)
2. Gamma rays	*Allium cepa*	16	Konvi_ka and Gottschalk (1974)
	Brassica oleracea	18	Gottschalk and Kaul (1980b)
	Hordeum vulgare	14	Sethi et al. (1970)
	Vicia faba	12	Sjödin (1970)
3. Fast neutrons	*Triticum durum*	28	Martini and Bozzini (1966)
	Vicia faba	12	Sjödin (1970)
4. Colchicine	*Capsicum annuum*	24	Panda et al. (1987)
	Lathyrus odoratus	14	Khawaja and Ellis (1987)
	Lathyrus pratensis	14	Khawaja and Ellis (1987)
5. Diethyl sulphate	*Hordeum vulgare*	14	Prasad and Tripathi (1986)
6. Ethyleneimine	*Brassica campestris*	20	Stringam (1970)
	Oryza sativa	24	Singh and Ikehashi (1981)
7. Ethylenemethane sulphonate	*Nigella sativa*	12	Datta and Biswas (1985)
	Vicia faba	12	Sjödin (1970)
8. *N*-nitroso-*N*-methyl urea	*Oryza sativa*	24	Kitada et al. (1983), Kitada and Omura (1984)
	Zea mays	20	Golubovskaya (1979)

TABLE 5.4
Synaptic Mutants of Hybrid Origin

Hybrids	2n	Gene Symbol	Genetics	Authors
Avena abyssinica × *A. barbata*	28	*ds2*	3:1	Thomas and Rajhathy (1966)
Gossypium hirsutum × *G. barbadense*	52	*as*	15:1	Beasley and Brown (1942)
Nicotiana rustica × *N. tabacum*	42	*as*	3:1	Swaminathan and Murty (1959)
Sorghum durra × *S. candatum*	20	*as1*	3:1	Krishnaswamy and Meenakshi (1957)
Triticum monococcum × *T. aegilopoides*	14	–	15:1	Smith (1936)

TABLE 5.5
Desynaptic Genes in Barley

Desynaptic Genes	Origin	Chromosome Behavior (Range Bivalents)	Female Fertility (%)	Chromosome	Inheritance
des1	X-rays	7 O II to 5 rod II + 4 I	45	1	3:1
des2	X-rays	7 O II to 2 rod II + 10 I	1	3	3:1
des3	S	7 O II to 4 O II + 2 rod II + 2 I	1–33	–	3:1
des4	S	7 O II to 3 rod + 8 I	18	1	3:1
des5	S	7 O II to 14 I	7	1	3:1
des6	S	7 O II to 14 I	16	1	3:1
des7	S	7 O II to 14 I	33	2	3:1
des8	S	7 O II to 14 I	22	–	3:1
des9	S	7 O II to 5 rod II + 4 I	90	–	3:1
des10	S	7 O II to 3 O II + 2 rod II + 4 I	60–80	–	3:1
des11	S	6 O II + 1 rod II to 2 rod II + 10 I	40	–	3:1
des12	S	7 O II to 1 rod II + 12 I	35	–	3:1
des13	S	7 O II to 4 O II + 2 rod II + 2 I	20–30	–	3:1
des14	S	7 O II to 1 rod II + 12 I	35	–	3:1
des15	S	5 O II + 2 rod II to 1 rod II + 12 I	25	–	

Source: Adapted from *des* 1–8, Hernandez-Soriano, J. M., R. T. Ramage, and R. F. Eslick. 1973. *Barley Genet. Newsl.* 3: 124–131; *des* 9–14, Hernandez-Soriano, J. M. and R. T. Ramage. 1974. *Barley Genet. Newsl.* 4: 137–142; and *des* 15, Hernandez-Soriano, J. M. and R. T. Ramage. 1975. *Barley Genet. Newsl.* 5: 113.

Synaptic mutants of spontaneous origin predominate in the literature. In barley, of the 15 desynaptic genes identified, 13 were of spontaneous origin and 2 were induced by irradiation with X-rays (Table 5.5). Furthermore, synaptic mutants have been induced by various mutagens, with or without intention.

Singh and Ikehashi (1981) isolated two desynaptic mutants of rice by ethyleneimine treatment. They identified 93 sterile and partial sterile segregating M_2 families among 1924 M_2 lines. Cytological observations revealed chromosomal interchanges, triploids, desynapsis, and genic male sterility (Table 5.6).

TABLE 5.6
Cytological Observation and Pollen Fertility of Sterile and Partially Sterile Lines from Two Ethyleneimine Treatments

Cytological Observation	Ethyleneimine Treatment			
	1 h 0.4%	3 h 0.2%	Total	Pollen Fertility (%)
Reciprocal translocations	39	22	61	1.6–80.0
Triploids	1	1	2	0.9–2.1
Desynaptic plants	1	1	2	2.1–22.1
Meiosis not studied	4	7	11	33.0–89.0
Fertile plant	4	9	13	80–100
Normal meiosis	2	2	4	0.0–4.0
Total	51	42	93	

Source: From Singh, R. J. and H. Ikehashi. 1981. *Crop Sci.* 21: 286–289. With permission.

5.2.3 CYTOLOGICAL BEHAVIOR OF SYNAPTIC MUTANTS

The term asynapsis originally was proposed by Randolph (1928) to describe the absence of normal chromosome pairing during first meiotic division. The meiotic mutant was first described by Beadle and McClintock (1928) in maize who reported that the majority of the sporocytes at metaphase-I showed 20 univalents (I) and rarely 10 bivalents (II). Mutant plants showed variable pollen and ovule abortion. Beadle (1930) assigned the gene symbol *as* to this particular asynaptic mutant.

Asynaptic mutants do not exhibit normal pairing of homologous chromosomes at pachynema (failure to synapse in the first place) while failure to maintain association after first synapsis is known as desynapsis. In asynapsis, most or all of the chromosomes remain univalents at diakinesis and metaphase-I (Table 5.2). This behavior indicates that the asynaptic mutant of maize described by Beadle (1930) is probably asynaptic rather than desynaptic. Miller (1963) examined *as* mutants cytologically and observed zero to nearly complete synapsis of homologous chromosomes at early pachynema and at metaphase-I. The range of bivalents was 0–9.95. In the desynaptic (*dy*) mutant of maize, bivalents were mostly loosely associated (Figure 5.2). The maize asynaptic (*as*) mutant also induces polyploid meiocytes, elongated and curved spindles, misdivision of univalents and monod centromere, and partial or complete failure of cytokinesis after the meiotic divisions (Miller, 1963).

Asynaptic mutants have been reported in several plants (Table 5.2). In *Brassica campestris* (2n = 20), Stringam (1970) isolated two asynaptic mutants. Mutants *as* and *as3* displayed no chromosome pairing at pachynema and no bivalents were recorded at metaphase-I. The *as2* mutants showed a variable number of bivalents (up to 4 loosely paired) at diakinesis. Kitada and Omura (1984) identified one asynaptic (MM-19) and two desynaptic (MM-4, MM-16) mutants from *N*-methyl-*N*-nitrosourea-treated rice seeds. In the MM-19 mutant, the homologous chromosomes lacked complete synapsis at pachynema, resulting in 24I at metaphase-I. In contrast, mutants MM-4

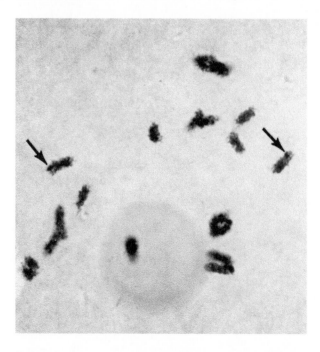

FIGURE 5.2 Photomicrograph of a microsporocyte at diakinesis from a desynaptic maize plant showing eight bivalents [one of which is nearly separated to univalents and two pairs of widely separated univalents. In one pair of univalents, there is an equationally separated distal knob (arrows), which indicates that a crossover has occurred between the knob and the kinetochore, so that in each univalent there is a knob-carrying and knobless chromatid] and 4 univalents. (From Maguire, M. P., A. M. Paredes, and R. W. Riess. 1991. *Genome* 34: 879–887. With permission.)

and MM-16 showed variable chromosome pairing from zygonema to pachynema and displayed both univalents and bivalents at metaphase-I.

Chromosome disjunction at anaphase-I to telophase-I is highly irregular in asynaptic mutants. The second division is essentially normal but cells inherit chromosomal abnormalities resulting from the first meiotic division. Asynapsis produces chromosomally unbalanced male and female spores, resulting in a high level of pollen and ovule abortion. Gottschalk (1987) found in X-ray-induced pea desynaptic mutants that *ds* genes influence microsporogenesis more strongly than megasporogenesis.

It has been demonstrated by electron microscopy (EM) that in asynaptic mutants, formation of the synaptonemal complex (SC) is blocked (La Cour and Wells, 1970; Golubovskaya and Mashnenkov, 1976). La Cour and Wells (1970) examined two synaptic mutants of *Triticum durum* ($2n = 4x = 28$) by light and electron microscopes. The light microscope showed suppression of chromosome pairing at zygonema and pachynema and EM revealed the absence of SC. When SC is eliminated in the bivalents, the two lateral elements (cores) are set free (Figure 5.3). Maguire et al. (1991)

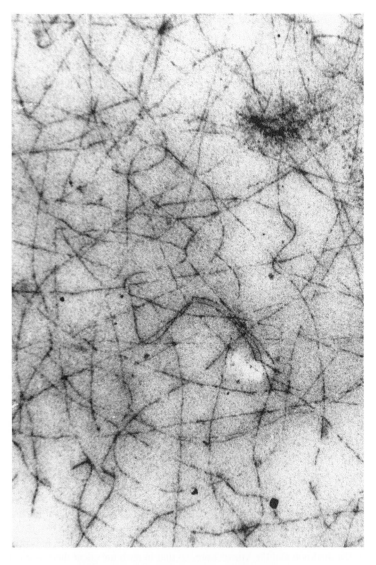

FIGURE 5.3   Electron micrograph of asynaptic (*as*) nucleus at mid- to late pachynema. Note widespread lack of homologue pairing, short triple association. (Courtesy of Dr. M. P. Maguire.)

observed normal crossing over followed by failure of chiasma maintenance in a desynaptic mutant of maize. They examined normal and desynaptic stocks by EM and found statistically significant wider dimension of the SC central region and less twisting of synapsed configuration at pachynema in desynaptic mutants compared to normal. Chromosomes undergo desynapsis after pachynema to diakinesis and by metaphase-I, desynapsis is completed. The SC is, apparently, rapidly disintegrated following pachynema (Maguire, personal communication). A reduction of chiasma frequency or complete failure of chiasma formation occurs at diplonema to metaphase-I, resulting in various frequencies of univalents and bivalents. The action of desynaptic genes differs among sporocytes as well as among plants.

Since pachynema chromosome analysis is not feasible for a large number of plant species, the action of desynaptic genes is often ascertained on the basis of studies of diakinesis and metaphase-I. Chromosome associations at these later stages of meiosis appear to correlate well with the amount of pachynema chromosome synapsis. The degree of desynapsis is reflected by the number of bivalents at metaphase-I and the frequency of chiasmata per cell. Chiasmata are normally not randomly distributed among cells, chromosomes, and bivalents. They also vary between genotypes and between and within cells (Rees, 1961; Jones, 1967, 1974). Chiasmata in desynaptic plants are mostly terminal at metaphase-I and are rarely interstitial (Li et al., 1945).

Prakken (1943) classified desynaptic mutants depending on their expressivity: weakly desynaptic (several univalents), intermediate desynaptic (many univalents), and completely desynaptic (exclusively univalents and rarely any bivalents).

Bivalents move to the equatorial plate at metaphase-I while univalents tend to be distributed at random in the cytoplasm. The number of univalents varies within different microsporocytes in the same plant. This suggests that within a chromosome complement of a species, there may be differences among the different chromosomes concerning their requirements for the initiation of pairing (Rees, 1958; Swaminathan and Murty, 1959; Koduru and Rao, 1981).

Disjunction of bivalents at anaphase-I is usually normal. Univalents sometimes move to the poles at random without dividing while in other cases they divide equationally (Soost, 1951; Miller, 1963). Univalents that fail to move to either pole remain as laggards at the equatorial plate. At telophase-I, those chromosomes that reach the poles organize dyad nuclei, while laggards often form micronuclei. The second meiotic division is essentially normal, and irregularities are restricted to the first meiotic division. As a consequence of meiotic irregularities in asynaptic and desynaptic mutants, chromosomally unbalanced male and female microspores and megaspores are generated, resulting in reduced pollen and ovule viability.

5.2.4 SYNAPTIC MUTANTS AND RECOMBINATIONS

Homologous chromosome pairing, synaptonemal complex formation, crossing over, chiasma formation, and gene recombination in normal plants occur during zygonema and pachynema (Henderson, 1970; Palmer, 1971; Gillies, 1975). One expects to observe a reduction in recombination frequencies in synaptic mutants because of the variable level of chromosome pairing. However, conflicting views prevail in the literature. Enns and Larter (1962) recorded reduction in recombination percentage (40%–43% to 14%–16%) between marked loci on chromosome 2 in homozygous desynaptic plants of barley. The reduction in recombination was positively associated with reduced chiasma formation. Nel (1979) demonstrated in maize that recombination was reduced between the marked loci on chromosome 3 in the asynaptic heterozygotes and recombination between the same loci was reduced to a greater extent in asynaptic homozygotes. This, however, contradicts his earlier report (Nel, 1973).

Instances have also been reported in which recombination values between two genes are not affected by the synaptic mutants. In maize, the crossing over value was similar in the $sh1$-wx region in asynaptic and nonasynaptic plants (Beadle, 1933; Dempsey, 1959). In two asynaptic mutants (as_1, as_4) of tomato, Soost (1951) recorded apparently no significant difference in percent crossing overs between the d_1 (dwarf growth habit) and Wo (wooly) marker genes in as and As plants.

A higher recombination between certain marked loci in asynaptic (*as*) plants was reported in maize (Miller, 1963) and tomato (Moens, 1969). Miller (1963) proposed that higher than normal rates of recombination occurring in *as* plants are due to the fact that the genetic markers he tested are in distal segments of the short arm or in segments near the centromere where chromosomes pair more often than in intercalary regions. The increased recombination in synaptic mutants may be due in some cases to compensation for the loss of recombination in some other parts of the genome (Miller, 1963; Sinha and Mohapatra, 1969; Omara and Hayward, 1978).

5.2.5 FACTORS INFLUENCING PAIRING IN SYNAPTIC MUTANTS

The degree of chromosome pairing in synaptic mutants may be influenced by temperature, humidity, and chemicals (Prakken, 1943; Ahloowalia, 1969; Koduru and Rao, 1981). Furthermore, it also varies from plant to plant, day to day, year to year and between specimens collected at different times during the same day (Prakken, 1943; Soost, 1951). Moreover, the degree of expression of each synaptic gene is variable (Table 5.2). Goodspeed and Avery (1939) reported with regard to an asynaptic mutant of *Nicotiana sylvestris* that high temperature and low humidity greatly increased asynapsis and high temperature and high humidity decreased asynapsis. Ahloowalia (1969) recorded in a desynaptic mutant of rye grass ($2n = 14$) that at lower temperature (11°C) mean bivalent/cell was 7.71 ± 0.040 but at 28°C desynapsis (5.39 ± 0.068/cell) was observed. In contrast, Li et al. (1945) observed a greater extent of pairing at higher temperature and decreased pairing at lower temperature in desynaptic mutants of wheat.

Magnard et al. (2001) examined a temperature sensitive, *tardy asynchronous meiosis* (*tam*), mutant in *Arabidopsis* isolated from EMS-mutagenized M_2 population. Although *tam* allele is temperature sensitive, *tam* plants are fertile even at the restrictive temperatures. The *tam* mutant is a recessive trait, nonallelic to *mei1*, and regulates delayed and asynchronous meiosis I and II. The range of aberrant tetrads at 22°C in *tam* plants was 5%–90%, and it reached to 100% at 27°C and 12–24 h exposure at 27°C induced total aberrant tetrads. It is unique that the percentage of this aberration decreased when the mutant was returned at 22°C for 24 h. Chromosome abnormalities during meiosis in *tam* mutant include condensed chromosome, chromosome lagging or scattering that resulted in the formation of aberrant tetrads with more than four spores and sometimes more than one nucleus in a cell. The delayed and asynchronous meiosis produced mostly dyads in *tam* mutant. By contrast, wild type (WT) plants were not temperature sensitive as meiosis was completely synchronous.

Experiments have shown that an increase in ion content of potassium (Law, 1963) and phosphate (Bennett and Rees, 1970; Fedak, 1973) increases the number of bivalents in desynaptic plants. Ahloowalia (1969) recorded an increased bivalent frequency in a desynaptic mutant of rye grass treated with 5-ethyl, 5-phenylbarbituric acid, and 5,5-diethylbarbituric acid. It was hypothesized that hydrogen bonds keep chromosomes paired after their initial pairing. The desynaptic mutants may be defective in a thermosensitive compound-controlling hydrogen bonds in the condensation of chromosomes. These studies suggest that desynaptic mutants may be deficient in certain ions required for normal synapsis, and that when these chemicals are added chromosome pairing is enhanced. This aspect needs to be investigated further.

5.3 GENETIC CONTROL OF RECOMBINATION

The universal rule of meiosis is to segregate parental and recombinant chromosomes into gametes precisely in equal (50%) frequencies. The foundation of Mendelism is laid on this principle. The non-Mendelian segregation is recorded during meiosis in the yeast (*Saccharomyces carevisiae*), which is regulated by an *ntd 80* gene (Allers and Lichten, 2001a). Two types of meiotic recombination in yeast were discovered: (1) reciprocal crossing over and (2) nonreciprocal crossing over (gene conversion). Both mechanisms are highly correlated and their relationships are altered in mutants such as *zip1, msh 4, msh 5,* and *mer 3* where frequencies of gene conversion are normal, but reciprocal crossing over is reduced about twofold (Storlazzi et al., 1996; Allers and Lichten, 2001a).

5.4 GENES RESPONSIBLE FOR CHROMOSOME DISJUNCTION

Clark (1940) reported a meiotic divergent spindle mutation (*dv*) induced in maize pollen treated with UV light. The *dv* gene disrupts the structure and function of spindles. The mutation is inherited as a simple Mendelian recessive, showing partial male and female sterility. The *dv* mutant exhibits normal chromosome behavior until metaphase-I (bivalents assume a normal equatorial orientation). The spindles do not converge at each pole as is observed for normal maize. The migration of chromosomes at anaphase-I is impaired, since chromosomes disjoin irregularly in several groups. As a result, single telophase-I nuclei are not formed. The second meiotic division is synchronous in all the nuclei of both daughter cells; instead of normal tetrads of microspores, one to several small nuclei (polyads) are formed. Golubovskaya and Sitnikova (1980) induced three mutants (*ms28, ms43, mei025*) in maize by *N*-nitroso-*N*-methylurea. These mutations are similar to *dv*, being inherited as single recessive genes causing complete pollen sterility.

A meiotic mutant that shows precocious centromere division (*pc*) in *Lycopersicon esculentum* was described by Clayberg (1959). Chromosome pairing is normal until metaphase-I. The precocity first appears at anaphase-I in some bivalents which often lag and undergo premature centromere division. The centromeres of those chromosomes not lagging in the first division divide in most cases by prophase-II. All the chromosomes were regularly oriented at metaphase-II plate. The precociously divided chromosomes move to the poles at random without further division. Many chromosomes lag in the second division and frequently form restitution nuclei. Irregular chromosome segregation results in gametes of unbalanced chromosome numbers as revealed by the appearance of trisomics progeny. The mutation segregates as a single recessive gene (*pc*) and shows 0%–10% pollen fertility.

Beadle (1932) identified a recessive partially fertile mutant of maize (*va*). The homozygous plant (*va/va*) exhibits normal prophase-I but cytokinesis is often absent at telophase-I resulting in gametes with diploid ($2n = 20$) and tetraploid ($2n = 40$) chromosome constitutions. A failure in cytokinesis may occur either at the first or at the second meiotic division. In *Datura stramonium*, Satina and Blakeslee (1935) observed a mutant that displayed a completely normal first meiotic division but the second division did not occur, and instead of tetrads, dyads were formed with doubled ($2n = 24$) chromosome numbers. Ploidy increased with each subsequent generation. The mutant was designated by the gene symbol *dy*.

Golubovskaya and Mashnenkov (1975) isolated a recessive mutant, absence of the first meiotic division (*afd*), by treating dry maize seeds with 0.012% solution of *N*-nitroso-*N*-methylurea. Meiotic prophase-I was absent in the mutant. Homologous chromosomes failed to pair and remained as 20I at metaphase-I. Chromosomes resembled C-mitotic chromosomes. At anaphase-I, 20 chromatids migrated to each pole. Since the kinetochores divided in the first division, chromatid movement to the poles at anaphase-II was random, leading to complete male and female sterility.

A failure in chromosome migration may result in the formation of multiploid microsporocytes, that is, mixoploidy (incomplete formation of cell walls following nuclear division). An example was reported by Smith (1942) in barley. The mutant segregated as a recessive gene, designated *mu*. The number of bivalents in the *mu* mutant varied in different metaphase plates from fewer than 7 to more than 100 multiples of seven. Cytokinesis was suppressed in some premeiotic divisions and in other groups, migration of chromosomes took place in the formation of the prometaphase plate. Pollen grains were of various sizes and devoid of starch. Sharma and Reinbergs (1972) also isolated a mutant, *ms-au*, in barley with comparable effects. Takahashi et al. (1964) reported a dwarf barley plant, controlled by a recessive gene and an intensifier gene. The dwarf plant carried cells with chromosome numbers ranging from $2n = 28$ to 210, together with cells carrying a normal chromosome complement of $2n = 14$.

A triploid inducer gene (*tri*) in barley was reported by Ahokas (1977) and studied by Finch and Bennett (1979). The mutant apparently suppresses at random the second meiotic division in about half of the megaspore mother cells. This results in the formation of embryo sacs with diploid nuclei. The triploid inducer gene does not appear to affect microsporogenesis.

Rhoades and Dempsey (1966) found a mutant for elongate chromosomes (*el*) in the open pollinated maize variety Hays Golden. The mutant segregated as a monogenic recessive. The despiralization of the chromosomes occurred at both meiotic anaphases. The most significant effect of the *el* gene was the production of unreduced eggs with variable chromosome composition and in varying frequencies. Ears borne on *el/el* plants had plump and shriveled seeds, as well as aborted ovules. Chromosome counts revealed that plump seeds were diploid ($2n = 20$). A considerable frequency of shriveled seeds did not germinate, but chromosome analysis of 825 shriveled seeds showed 82% $2n = 30$ chromosomes and 18% aneuploid chromosome numbers ranging from $2n = 25$ to 33. It is interesting to note that all the offspring (961) were normal diploids ($2n = 20$) in crosses between normal female plants and *el/el* male plants.

The *el* gene generates a series of polyploid plants. Rhoades and Dempsey (1966) obtained vigorous plants up to the $5x$ ($2n = 50$) level, but beyond this level the plants were dwarfed and female sterile. It was determined that the unreduced eggs originated by omission of the second meiotic division.

Beadle (1929) identified a meiotic mutant in maize that showed apparently normal first and second meiotic divisions of microsporogenesis, but microspore quartet cells underwent a series (at least 4) of meiosis-like divisions in which the chromosomes were distributed to the two poles at random and without splitting. Cells with only one chromosome in the spindle were seen, and fragmentation of chromosomes was rather common. Plants were completely pollen sterile and partially female sterile. Beadle (1931) named the mutant polymitotic and assigned the gene symbol *po*. The polymitotic trait inherited as a simple Mendelian recessive and was not allelic to the *as* (asynaptic) gene.

5.5   OTHER MEIOTIC MUTANTS

Genes for "long" and "short" chromosomes have been induced by X-ray in barley. The gene for long chromosome causes a high frequency of elongated rod bivalents at metaphase-I (Burnham, 1946). On the other hand, Moh and Nilan (1954) found a recessive mutant for short chromosomes among the progenies derived from barley seed that had been subjected to atom bomb irradiation. In this mutant, chromosomes paired at early meiotic prophase and there was a considerable shortening of the chromosomes during prophase, especially at diakinesis; the bivalents usually underwent precocious terminalization of all chiasmata, resulting in 14 univalents at metaphase-I.

A recessive gene for chromosome stickiness (*st*) has been recorded in *Zea mays* (Beadle 1937) and *Collinsia tinctoria* (Mehra and Rai, 1970). Chromosomes were clumped at prophase-I and frequently were associated. Thus, chiasma formation was impaired and chromosomes did not orient at the equatorial plate. Chromosomal fragmentation was common at anaphase-I and plants homozygous for the sticky gene were male and female sterile. Stout and Phillips (1973) did not record any difference in histone composition between normal plants and *st/st* mutant maize plants.

Meiotic mutants in maize have been extensively studied. They are summarized by Carlson (1988) as follows:

1. Absence of meiosis: Ameiotic (*am1*, *am2*).
2. Absence or disruption of synapsis: Asynaptic (*as*), desynaptic (*dy*, *dys1*, *dys2*) and absence of first division (*afd*).
3. Changes in structural organization of chromosomes: Elongate (*el*) and sticky (*st*).
4. Improper meiotic segregation and/or defective meiotic spindle: Divergent spindle (*dv*, *ms28*, *ms43*, *mei025*).
5. Failure of cytokinesis and/or irregularities of cell shape: Variable sterile (*ms8*, *ms9*).

6. Extra divisions following meiosis: Polymitotic (*po*).

7. Mutants with several effects on meiosis: Plural abnormalities of meiosis: *pam2, ms17*.

5.6 ROLE OF HETEROCHROMATIN IN CHROMOSOME PAIRING

Heterochromatin plays an important role in chromosome pairing and chromosome alignment. The heterochromatin of centromeric regions, supernumerary chromosomes (B-chromosomes or accessory chromosomes), nucleolus organizer regions, and knobs (e.g., knob 10 in maize) consists of highly repetitive DNA. Distal euchromatin of chromosomes is constituted of mainly unique sequences interspersed with moderately repeated sequences that have regulatory and functional roles in transcription, replication, synapsis, and recombination (Gillies, 1975; Loidl, 1987).

Several reports show that heterochromatin (the highest degree of DNA synthesis-S phase) increases chiasma frequency (Rees and Evans, 1966), increases the frequency of nonhomologous association at diplonema (Church, 1974), increases crossing over (Nel, 1973), and lengthens meiotic prophase. Thus, heterochromatin prolongs the time during which crossing over takes place (Rhoades and Dempsey, 1972). Supernumerary chromosomes suppress homoeologous chromosome pairing in *Triticum aestivum* (Sears, 1976) and in *Lolium* (Evans and Macefield, 1973). On the other hand, Romero and Lacadena (1982) observed a promoter effect of rye B-chromosomes on homoeologous chromosome pairing in hexaploid wheat. This suggests that the mechanism of heterochromatin-mediated homologous chromosome pairing is undetermined (Roeder, 1997).

5.7 DIPLOID-LIKE MEIOSIS IN ALLOPOLYPLOIDS

Diploid-like meiosis is under genetic control and is a common occurrence in allohexaploids of several crop plants of the family Gramineae, for example, *Triticum aestivum* (Okamoto, 1957; Sears and Okamato, 1958; Riley and Chapman, 1958a), *Avena sativa* (Rajhathy and Thomas, 1972; Jauhar, 1977), *Festuca arundinacea* (Jauhar, 1975), and *Hordeum parodii* (Subrahmanyam, 1978).

5.7.1 THE 5-B SYSTEM IN WHEAT ($2n = 6x = 42$)

Meiotic chromosome pairing under genetic control in the hexaploid wheat cultivar Chinese Spring was first demonstrated by Okamoto (1957). He studied meiotic chromosome pairing in the F_1 between plants which were monosomic for a telocentric chromosome 5 BL ($2n = 40 + 1$ telo 5BL) and AADD ($2n = 28$; derived from amphidiploid *T. aegilopoides* x *Ae. squarrosa*) plants. Two kinds of F_1 plants were expected: (1) $2n = 34$ chromosomes without the 5BL chromosome and (2) $2n = 35$, with the 5BL chromosome. Chromosome pairing data in 34-chromosome plants revealed a low frequency of univalents and an increased number of bivalents and multivalents compared to 35-chromosome plants (Table 5.7). Based on these results, Okamoto suggested that chromosome

---

TABLE 5.7

Asynaptic Effect of 5-B Chromosome in Chinese Spring Wheat

	No. of PMCs	Uni-Valents	Bivalents		Tri-Valents	Quadri-Valents	5-Valents	6-Valents	7-Valents
			Closed	Open					
34-chromosome plant	200	8.24	5.07	3.445	0.735	0.685	0.19	0.22	0.02
35-chromosome plant	200	23.82	1.025	4.115	0.320	0.005	0	0	0

Source: From Okamoto, M. 1957. *Wheat Inf. Sev.* 5: 6. With permission.

5BL carries a gene or genes for asynapsis. Subsequently, Riley and Chapman (1958a) reported similar results. Wall et al. (1971) assigned the gene symbol *Ph* (homoeologous pairing suppressor). A single *Ph* gene was postulated to control meiotic pairing of homoeologous chromosomes. Mello-Sampayo (1972) suggested two loci, acting additively, and Dover (1973) reported two linked loci for *Ph*. It has been demonstrated that the three genomes (A, B, and D) are genetically very closely related, and *Ph* controls only homologous chromosome pairing. When *Ph* is removed or its activity is suppressed, not only do homoeologous chromosomes pair but they also pair with the chromosomes of related species and genera, making alien gene transfer possible (Sears, 1975, 1976).

Since the discovery of *Ph*, several chromosome pairing suppressor and promoter genes in Chinese Spring wheat have been identified (Sears, 1976). In addition to *Ph*, there are three more minor pairing suppressors located on chromosomes 3AS, 3DS (Mello-Sampayo and Canas, 1973) and 4D (Driscoll, 1973); the chromosome 4D gene is as effective as the gene on chromosome 3AS, while the gene located on chromosome 3DS is only half as effective as the *Ph* gene.

Besides suppressors, there are several promoters of chromosome pairing located on chromosome arms 5DL, 5AL, 5BS (Feldman, 1966; Riley et al., 1966b; Feldman and Mello-Sampayo, 1967; Riley and Chapman, 1967), and 5DS (Feldman, 1968) and 5AS (Dvořák, 1976). Furthermore, genes on 2AS, 3BL, and 3DL may also be considered promoters for pairing (Sears, 1977). It has been observed in Chinese Spring wheat that asynapsis occurs at low temperature (15°C or below), when chromosome 5D is absent (nulli-5D tetra-5B) and a lesser reduction at 20°C. But pairing is more or less normal at 25°C (Riley, 1966).

Diploid-like meiosis under genetic control has been suggested in allopolyploid species of several genera: *Chrysanthemum* (Watanabe, 1981), *Glycine* (Singh and Hymowitz, 1985a), *Gossypium* (Kimber, 1961), *Pennisetum* (Jauhar, 1981), and *Solanum* (Dvořák, 1983a).

5.7.2   ORIGIN OF THE *Ph* GENE

Three theories have been postulated regarding the origin of the *Ph* gene:

1. It was already present in the diploid B-genome species (Okamoto and Inomata, 1974; Waines, 1976). Ekingen et al. (1977) suggested that the homoeologous pairing suppressor of chromosome 3D of hexaploid wheat already existed in *Ae. squarrosa*. However, this possibility needs to be investigated further.
2. The *Ph* gene was transferred to chromosome 5B from an accessory chromosome. It has been shown that B-chromosomes appear to have a suppressive effect on chromosome pairing very similar to that of the 5B chromosome. Furthermore, the accessory chromosomes of rye have a slight suppressive effect on pairing in hybrids with *T. aestivum*, whether 5B is present or not (Dover and Riley, 1972).
3. The *Ph* gene arose as a single mutation following formation of the AABB amphidiploid (Riley and Chapman, 1958a). Riley et al. (1973) suggested "an activity like that of the *Ph* allele must have occurred in the first hybrids and allotetraploid from which polyploid wheat was evolved." According to Sears (1976) "an origin of *Ph* by mutation in the newly formed AABB tetraploid seemed a more likely possibility."

5.8   HAPLOIDY

Haploids have been isolated in diploid as well as polyploid species of higher plants as spontaneous occurrence in natural populations (Kimber and Riley, 1963), by induction in anthers, pollen, and ovule culture (Sunderland, 1974; Maheshwari et al., 1980; Morrison and Evans, 1988) and after interspecific crosses (Gupta and Gupta, 1973; Kasha, 1974; Rowe, 1974; Choo et al., 1985). The genetics and mechanisms of barley haploid ($2n = x = 7$) and potato dihaploid ($2n = 2x = 24$)

production have been extensively studied and well documented (barley—Symko, 1969; Kasha and Kao, 1970; potato—Hougas and Peloquin, 1957; Hougas et al., 1958).

5.8.1 Mechanism of Chromosome Elimination

The gradual and selective elimination of *Hordeum bulbosum* (wild species) chromosomes in diploid ($2n = 14$) interspecific hybrids of *H. vulgare* (cultigen) and *H. bulbosum* (alien) is attributed to chromosome fragmentation, micronuclei formation, and degradation of chromatin (Subrahmanyam and Kasha, 1973; Thomas, 1988), lagging chromosomes and bridges (Lange, 1971; Bennett et al., 1976), noncongressed chromosomes at metaphase or the failure of chromosome migration to anaphase poles (Bennett et al., 1976). These chromosomal abnormalities have been postulated to relate to *vulgare* (V) and *bulbosum* (B) genome ratios (Subrahmanyam and Kasha, 1973), asynchronous cell cycle phases and mitotic rhythms (Lange, 1971; Kasha, 1974), differential amphiplasty (Bennett et al., 1976; Lange and Jochemsen, 1976), inactivation of alien (B) DNA by nuclease (Davies, 1974) and formation of multipolar spindles and asynchrony in nucleoprotein synthesis (Bennett et al., 1976).

It has been reported by Subrahmanyam and Kasha (1973) and Bennett et al. (1976) that normal double fertilization occurs in V × B crosses. Both reports recorded selective elimination of *bulbosum* chromosomes in embryos on the third day. Subrahmanyam and Kasha (1973) found 93.69% haploid nuclei ($2n = 7$) 11 days after pollination (DAP). In contrast, Bennett et al. (1976) recorded 93.6%–100% haploid nuclei 5 DAP; the discrepancy in results was attributed to different genetic backgrounds (Bennett et al., 1976).

The mechanism of dihaploid ($2n = 24$) isolation from *Solanum tuberosum* ($2n = 4x = 48$; female) and *S. phureja* ($2n = 2x = 24$; pollinator) is different from that observed in *vulgare–bulbosum* crosses. Wagenheim et al. (1960) observed that developing dihaploid embryos were associated with hexaploid endosperms, suggesting that both sperm nuclei fused to form a restitution nucleus and then combined with the secondary nucleus. Montelongo-Escobedo and Rowe (1969) recorded restitution sperm nuclei (single male gametes) in over 30% of the pollen tubes of the superior pollinators while an inferior pollinator had two sperm nuclei in 97% of the pollen tubes. They suggested that the production of pollen grains containing a single male gamete was under genetic control.

5.8.2 Genetics of Chromosome Elimination

The entire process of the elimination of *bulbosum* chromosomes in interspecific hybrids of *H. vulgare* and *H. bulbosum* is under genetic control and the stability of chromosomes in hybrids is influenced by the balance of parental genomes. Furthermore, haploids ($n = 7$) are produced in high frequency (98.2%) from diploid ($2n = 2x = 14$) VV × BB crosses (Table 5.8). Chromosome elimination does not occur in true hybrids with the genome ratio of 1V:2B (Davies, 1974). Subrahmanyam and Kasha (1973) concluded based on development and chromosomal abnormalities that endosperm stability could be placed in the following descending order: 1V:4B > 1V:2B > 1V:1B > 2V:1B. The more stable embryos (VBB) were much larger and better developed at the time of embryo culture than the other embryos regardless of the endosperm constitution (Kasha, 1974).

Ho and Kasha (1975) demonstrated from primary and telotrisomic analysis that chromosomes 2 (both arms) and 3 (long arm) most likely carry a gene or genes, located on *H. vulgare* chromosomes, for the elimination of *H. bulbosum* chromosomes (Table 5.9). They crossed seven primary trisomic stocks ($2n = 2x + 1 = 15$) of barley with pollen from autotetraploid ($2n = 4x = 28$) *H. bulbosum*. Since primary trisomics generate 8- and 7-chromosome gametes, two kinds of hybrid progenies are expected, namely: (1) 21-chromosome triploid hybrids and (2) 22-chromosome hyper-triploid hybrids. The 21-chromosome triploids should be similar to the relatively stable hybrids produced from the cross *H. vulgare* ($2x$) × *H. bulbosum* ($4x$).

TABLE 5.8

Types and Frequencies of Progeny Obtained from Interspecific Crosses between *Hordeum vulgare* (V) and *H. bulbosum* (B)

Cross-Combination (_ × _)	No. of Plants	Genotype and Chromosome Number					Expected Genomic Constitution	
		V 7	VV 14	VB 14	VBB 21	VVBB	Embryo	Endosperm
VV × BB	1544	1517		26	1		1V:1B	2V:1B
BB × VV	35	35					1V:1B	1V:2B
VV × BBBB	87				87		1V:2B	2V:2B
BBBB × VV	6				6		1V:2B	1V:4B
VVVV × BB	4		4				2V:1B	4V:1B
VVVV × BBBB	79		76			3	2V:2B	4V:2B
BBBB × VVVV	34		34				2V:2B	2V:4B

Source: Modified from Kasha, K. J. 1974. In *Haploids in Higher Plants. Advances and Potential. Proc. Ist Int. Symp.* K. J. Kasha, Ed., Ainsworth Press, Canada, pp. 67–87.

The frequency of 22-chromosome hybrids should be similar to the usual transmission frequency of the extra chromosome through female gametes in trisomic plants. However, if a specific chromosome carries a factor or factors controlling chromosome elimination, the 22-chromosome plant is not expected to survive or its frequency will be much lower in crosses with a plant trisomic for that chromosome (Table 5.9). This is because the controlling genes on this chromosome will be in the ratio of 1V:1B, and the *bulbosum* chromosomes will be eliminated. Therefore, it is expected to find 8-chromosome ($2n = x + 1$) haploid plants of *H. vulgare* if they are able to survive following chromosome elimination.

TABLE 5.9

Chromosome Number Segregation in Progenies of Crosses between Seven Primary Trisomics ($2n = 2x + 1 = 15$) of Barley and autotetraploid ($2n = 4x = 28$) *Hordeum bulbosum*

Trisomic Type	No. of Florets Pollinated	Total Progeny Obtained	Frequency and (%) of Progeny with $2n =$			Expected Trisomic Transmission	χ^2 Value
			21 Chromosomes	22 Chromosomes	Others[a]		
1	2,537	81	69 (85.2)	11 (13.6)	1 (1.2)	15.0	0.06
2	2,843	57	55 (96.5)	1 (1.8)	1 (1.8)	22.5	10.73[b]
3	1,015	75	72 (96.0)	2 (2.7)	1 (1.3)	20.3	11.44[b]
4	975	39	28 (71.8)	10 (25.6)	1 (2.6)	29.6	0.19
5	1,651	60	52 (86.7)	8 (13.3)	0 (0.0)	20.0	1.31
6	2,283	99	80 (80.0)	18 (18.2)	1 (1.0)	16.1	0.17
7	1,924	95	79 (83.2)	14 (14.7)	2 (2.1)	20.0	0.60
Total	13,228	506	435 (86.0)	64 (12.6)	7 (1.4)		

Source: From Ho, K. M. and K. J. Kasha. 1975. *Genetics* 81: 263–275. With permission.

[a] Aneuploids.
[b] Significant at 1% level.

5.8.3 Haploid-Initiator Gene in Barley

A partially dominant haploid-initiator gene (*hap*) was identified by Hagberg and Hagberg (1980) in barley. The haploid-initiator gene controls the abortion or the survival of abnormal embryos and endosperms. Plants homozygous for the *hap* gene produce progeny that include from 10% to 14% haploids. It is clearly shown in barley that male sperm nuclei reach the synergid cells about 1 h after pollination. One of the two male sperm nuclei reaches the two polar nuclei and forms a triploid endosperm. In plants of *hap/hap* and *hap/+* genotypes, the egg cell is not always reached by the other sperm nucleus. Thus, in a *hap/hap* plant evidently about half of the eggs stay unfertilized and some of these develop into haploid embryos. The frequency of haploid occurrence is highly influenced by the genotype and also by the environment. Using the *hap* system with marker genes, the breeders do not need to use embryo culture technique. However, they have to make a large number of crosses—a greater number than is needed using the *H. bulbosum* technique (Hagberg and Hagberg, 1987).

5.9 MALE STERILITY

Male sterility is common in higher plants. It occurs spontaneously in natural populations and may be induced in mutation experiments. Kaul (1988) estimates that in about 175 species male sterility (*ms* genes) has arisen spontaneously and in about 35 species male sterile genes have been induced by mutagenesis.

The *ms* genes cause complete breakdown of microsporogenesis but macrosporogenesis typically remains completely uninfluenced. By contrast, prophase-I and prophase-II meiotic mutants affect both micro- and macrosporogenesis, resulting in chromosomally unbalanced spores causing male and female sterility. On the other hand, both male and female spores are functionally normal in plants with self-incompatibility systems, but self-fertility is genotypically controlled. Compared to normal fertile plants, male sterile plants possess smaller anthers, shrunken and nondehiscent.

5.9.1 Classification of Male Sterility

In general, male sterility is divided into three different types: (1) genetic, (2) cytoplasmic, and (3) cytoplasmic-genetic. Genetic male sterility produces nonfunctional androecium or absence of pollen in plants in several ways. One of these is termed structural and includes structural malformation of stamens into pistils, nondehiscence of anthers because of the absence of anther pores or the absence of stamens. In a second type, termed functional, viable pollen is formed but normal anther dehiscence is prevented by barriers such as faulty, or lack of, exine formation. A third basis, termed sporogenous, is the result of an extreme scarcity of pollen, due to the abortion of microsporogenous cells in premeiotic, meiotic, and postmeiotic mutants (Figure 5.1).

Genetic male sterility exhibits Mendelian inheritance. The majority of *ms* genes represent monogenic recessive genes. In some cases, male sterility is controlled by a single dominant gene and occasionally is due to several recessive genes (Jain, 1959; Gottschalk and Kaul, 1974; Driscoll, 1986; Kaul, 1988). The soybean contains 9 (*ms1* to *ms9*) genetic male sterile (female sterile and male fertile) lines. All are recessive in nature (Palmer, 2000).

Cytoplasmic male sterility is inherited maternally and regardless of the nuclear constitution, plants carrying sterile cytoplasm are male sterile (Edwardson, 1970). Cytoplasmic-genetic male sterility involves the interaction of cytoplasm and nuclear factors. The progenies of cytoplasmic-genetic male sterile plants can be fertile when certain genetic stocks carrying nuclear fertility-restorer genes are used as pollinators (Allard, 1960).

Male sterility that is not genetically inherited is also produced by chemicals (male gametocides) or by changing the atmospheric temperature, light intensity, soil conditions, and growing conditions (field vs. greenhouse).

5.9.2 MECHANISM OF MALE STERILITY

Most male sterility in plants has a sporogenous basis. On the other hand, functional and structural *ms* genes are relatively less prevalent (Kaul, 1988). The *ms* genes act with remarkable precision at definite meiotic stages (premeiotic, meiotic, and postmeiotic as shown in Figure 5.1) and impair the normal development of sporogenous tissues, tapetal cells, pollen mother cells, and microspores.

A relatively smaller number of premeiotic *ms* genes act on sporogenous cells by causing degeneration of anthers (anthers remain rudimentary in the flowers), archesporial tissues and pollen mother cells. Some genes are effective during meiotic divisions. They disturb the normal sequence of meiosis or affect a particular stage of meiosis, causing abnormal chromosome behavior and microspore formation (Kaul, 1988).

Many *ms* genes act during postmeiosis, particularly immediately after the tetrad stage. Their action typically blocks normal pollen formation, resulting in empty, nonfunctional pollen grains. At the same time, the tapetal cell layer begins to disorganize and much of the tapetal cytoplasm is replaced by vacuoles. The tapetum plays an important role in microsporogenesis, especially by transporting nutrients to developing microspores. A deformed tapetum provides an insufficient supply of nutrients to microspores, thus starving them (Cooper, 1952; Rick and Butler, 1956; Filion and Christie, 1966; Mian et al., 1974; Chauhan and Kinoshita, 1979; Albertsen and Phillips, 1981; Dundas et al., 1982; Graybosch and Palmer, 1988; Kaul, 1988).

6 Mode of Reproduction in Plants

6.1 INTRODUCTION

The mode of reproduction in plants may be sexual (amphimixis), asexual (apomixis or agamospermy), or by specialized vegetative structures (vegetative reproduction). Sexual reproduction involves pollination, germination of pollen grains on the stigma, growth of pollen tubes down the style, the entrance of pollen tube to the ovule through the micropylar opening, the discharge of two sperm nuclei into the embryo sac and fusion of a sperm nucleus with the egg nucleus producing an embryo and conjugation of a second sperm nucleus with the two polar nuclei (secondary nucleus) forming a triploid endosperm. Thus, seeds develop as the result of double fertilization and the majority of plants reproduce sexually (Figure 6.1). The disruption of the sexual process may lead to apomixis.

Apomictic plants of some species produce seed directly from chromosomally unreduced megaspore mother cells or from somatic cells of the nucellus or ovule without fertilization. In addition, several plants propagate by specialized vegetative structures like bulbs, corms, cuttings, runners, rhizomes, tubers, and by grafting.

This chapter reviews the salient cytogenetic information on the mode of reproduction in plants (Stebbins, 1950; Maheshwari, 1950; Nygren, 1954; Fryxell, 1957; Bashaw, 1980; Grant, 1981; Nogler, 1984; Hanna, 1991, 1995; Asker and Jerling, 1992; Ramachandran and Raghavan, 1992; Chapman and Peat, 1992; den Nijs and van Dijk, 1993; Naumova, 1993; Koltunow et al., 1995; Jefferson and Bicknell, 1996; Ramulu et al., 1999). The genetic basis of apomixis, its prevalence in polyploidy, and exploitation to enhance yield of major crops such as wheat, rice, and maize (apomixis-revolution) is discussed.

6.2 SEXUAL REPRODUCTION

Sexual reproduction is a complex biological activity that facilitates genetic diversity and speciation. In flowering plants, the reproductive organs are in the flower. Meiosis and fertilization are two essential processes in the sexual cycle of higher plants. Sexual reproduction consists of two generations—sporophytic and gametophytic. The sporophytic generation begins when an egg nucleus unites with a sperm nucleus producing an embryo and the second sperm nucleus fuses with the polar nuclei (secondary nucleus) producing triploid endosperm, and continues with the development of seed, seedling, mature plant, and flowers. Sporophytic or somatic tissues contain diploid ($2n$) chromosome number. The flower contains spore-forming organs called anthers and ovary through meiosis. Anthers and the ovaries produce, respectively, haploid (n) microspores and megaspores. Thus, alternation of sporophytic and gametophytic generations is a rule in sexual reproduction (Figure 6.1).

6.3 ASEXUAL REPRODUCTION

Asexual reproduction (uniparental) in plants may occur either without sexual fusion of the spores (apomixis) or by vegetative organs (specialized vegetative structures). Apomixis is common in forage grasses, citrus, mango, blackberries, guayule, and ornamental shrubs. Vegetative propagation occurs by stem modification (bulbs, corms, runners, rhizomes, and tubers), root modification (tuber roots), and flowers and by inflorescence modification (vivipary). Furthermore, generation of plants by grafting and cell and tissue cultures is also included in vegetative propagation.

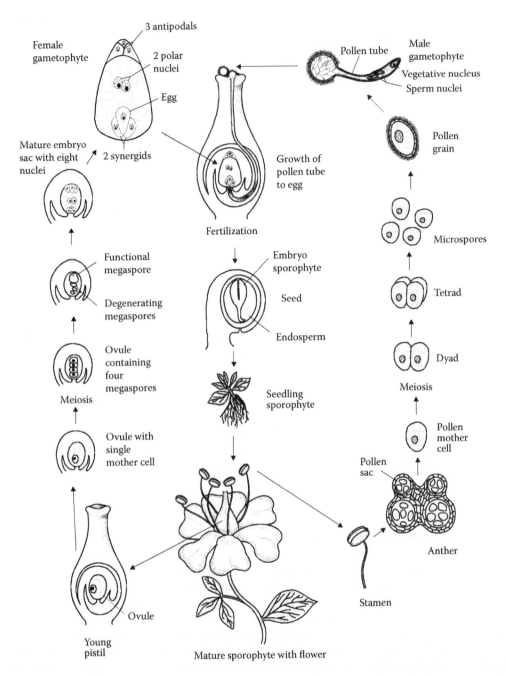

FIGURE 6.1 A diagrammatic sketch of the sexual life cycle (amphimixis) in plants. (Nelson, G. E., G. G. Robinson, and R. A. Boolootian: *Fundamental Concepts of Biology*, 2nd edition. p. 189. 1970. John Wiley & Sons, Inc., New York. Copyright Wiley-VCH Verlag GmbH & Co. KGaA. Reproduced with permission.)

6.3.1 APOMIXIS

The term "apomixis" was coined by Winkler (1908) and replaced the misused term apogamy. Apomixis is a type of asexual reproduction in which seed (sporophyte) is produced from the female gametophyte without fertilization. The offspring of apomictic plants are exact genetic replicas of the maternal plant. Apomixis is synonymous with asexual seed formation or agamospermy (Asker and Jerling, 1992). Solntseva (1976) defined apomixis in angiosperms as a method of seed production in which the embryo develops from the cells of the gametophyte with various disruptions of the sporogenesis and sexual processes.

Apomixis is widely distributed in flowering plants (Grant, 1981; http://www.apomixis.uni-goettingen.de) and the literature is inundated with reviews, research articles, and proceedings of conferences and symposia. Darlington (1939a, 1958) considered apomixis a reproductive mechanism for escaping sterility but an evolutionary blind alley; an escape guided in one or several steps by natural selection; but it is an escape that leads to extinction. Stebbins (1950) proposed in *Variation and Evolution in Plants* that there is no evidence that apomicts have ever been able to evolve a new genus or even a subgenus. In this sense, all agamic complexes are closed systems resulting in an evolutionary blind alley. Their proposal can be accepted under a single dominant gene hypothesis and at the same ploidy level. The predominance of apomixis in polyploid apomictic populations strongly supports this model (Grimanelli et al., 1998b).

However, the current knowledge of apomixis neither supports Darlington's blind alley concept nor Stebbins' speciation statement. On the contrary, de Wet and Harlan (1970), based on experimental evidence from *Capillipedium*, *Dichanthium* and *Bothriochloa* and a compilospecies *B. intermedia*, proposed that sexual and asexual polyploid species are in a very active stage of evolution. Evolutionary mechanisms like hybridization, recombination, polyploidy, clonal selection, and somatic mutation are constantly generating new sexual and apomictic forms or species.

6.3.1.1 The Compilospecies Concept

Harlan and de Wet (1963) stated: "The term compilospecies is taken from the Latin compilo: to snatch together and carry off, to plunder, or to rob. A compilospecies is genetically aggressive, plundering related species of their heredities, and in some cases it may completely assimilate a species, causing it to become extinct." The compilospecies concept is based on the behavior of the *Bothriochloa intermedia* and cytogenetic structure of *Bothriochloa–Dichanthium–Capillipedium* apomictic complex (Figure 6.2). In this complex, sexual and asexual reproductions are independent and genetically active. Their habitats are always contiguous producing a number of intertaxonomic gene flow interactions among species and even genera.

The major agamospecies are tetraploid, and all are connected either directly or indirectly with the compilospecies *B. intermedia* (Figure 6.2). The tetraploid cytotype predominates, and cytogenetically behaves like segmental alloploids. Preferential bivalent formation ensures production of genetically and physiologically balanced and hence functional male and female sexual spores even among the most apomictic biotypes. It is likely that chromosome pairing in alloploids may be under genetic control. The extensive introgression generates a high degree of heterozygosity and polymorphism. The entire complex (18 species) has an extremely dynamic cytogenetic system, ploidy level $2x–6x$, and is in a progressive state of evolution (de Wet and Harlan, 1970).

6.3.1.2 Sources of Apomixis

Apomixis is widespread throughout the plant kingdom and is most common in members of the families Poaceae, Compositae, and Rosaceae (Nygren, 1954; Asker and Jeling, 1992; Ramulu et al., 1999). The majority of wild polyploid perennial relatives of cultivated species harbor the apomixis and only a small proportion of cultivated crops in natural populations are apomictic. Apomixis can be induced by chemicals and irradiation and is isolated by wide hybridization. The occurrence of apomixis is very rare in the diploid species and more than 90% of the apomictics are usually polyploids (Grant, 1981; Asker and Jerling, 1992).

Alien species are the most reliable source for isolating apomixis. Wild relatives of wheat [e.g., *Elymus scabrus* (Hair, 1956), *Elymus rectisetus* (Liu et al., 1994)]; maize [e.g., *Tripsacum dactyloides* (Farquharson, 1955; de Wet, 1979; Dewald and Kindiger, 1994; Leblanc et al., 1995a,b)], and pearl millet [e.g. polyploid *Pennisetum* species (Dujardin and Hanna, 1984; Chapman and Busri, 1994)] carry genes for apomixis. Furthermore, many tropical grass genera such as *Panicum*,

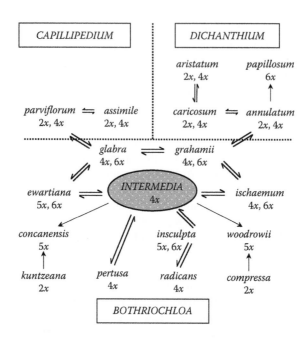

FIGURE 6.2 A diagrammatic sketch showing three major agamospecies (*Capillipedium*, *Dichanthium*, *Bothriocloa*) are directly or indirectly connected with the compilospecies *intermedia*. The gene flow is continuous introgression (double line with half arrow) and hybridization without further introgression (solid line). (Redrawn from de Wet, J. M. J. and J. R. Harlan. 1970. *Evolution* 24: 270–277. With permission.)

Paspalum, Cenchrus, Eragrostis, Poa, Elymus, Hordeum, Brachiaria, Setaria, Saccharum, and *Sorghum* are also a rich source of apomixis (Chapman, 1992). A number of apomictic species have been identified in the *Bothriochloa–Dichanthium–Capillipedium* complex (de Wet and Harlan, 1970). Some species of *Malus, Rubus*, and wild species of strawberry (*Fragaria*) are apomictic (Nogler, 1984). Apomixis predominates in cross-fertilizing plants but has not developed in self-fertilizing plants.

Certain types of apomixis can be induced by mutagenesis in some crops. Ramulu et al. (1998) developed various strategies to induce and isolate mutants for apomixis in *Arabidopsis* and *Petunia* by treatment with ethylemethanesulfonate (EMS) and by transposon mutagensis. Ohad et al. (1996) screened ≈50,000 EMS-treated M_1 plants of *Arabdopsis thaliana* and identified a total of 12 lines from M_2 and M_3 families that displayed elongated siliques in the absence of fertilization-independent endosperm (*fie*). The allele *fie* (female gametophyte) is the source of signals that activate sporophytic fruit and seed coat development, not transmitted by the female gametophyte. The inheritance of *fie* allele by the female gametophyte results in embryo abortion, even when the pollen carries the wild type *FIE* allele (essential for female reproductive development). Ohad et al. (1996) concluded that *fie* gene has the genetic potential of inducing autonomous endosperm formation in *Arabidopsis thaliana* as observed in certain apomictic plants.

A similar approach by Chaudhury et al. (1997) in *Arabidopsis thaliana* was pursued. They isolated six putative fertilization-independent seed mutants (*fis*) by treating the stamenless pistillata (*pi*) mutant with EMS. The *pi* mutant is characterized by short siliques without seed. The *fis* mutant in the *pi* background bears long siliques with developing seeds, though anthers are pollenless. Three mutant alleles (fis_1, fis_2, fis_3) are independent and are associated with three different chromosomes. They suggested that fis_3 and *fie* (described by Ohad et al., 1996) may be allelic. The gametophytic nature of *fis* mutants indicates that their time of action is downstream from the point at which apomixis operates in plants. In normal sexual reproduction, *fis* genes are likely to play a key regulatory role in the development of seeds after normal pollination and fertilization.

Facultative apomixis has been induced in *Pennisetum glaucum* (Hanna and Powell, 1973; Arthur et al., 1993) and *Sorghum bicolor* (Hanna et al., 1970). The expression of apomixis in these experiments was variable and low. These examples indicate that apomixis in diploid plants can be induced by mutagenesis.

Apomixis can be produced through hybridization of sexual and apomictic plants and may be transferred to the cultigen through cytogenetic or molecular methods (Hanna and Bashaw, 1987). The introgression of genes for apomixis from hexaploid ($2n = 6x = 54$) perennial East African grass species *P. squamulatum* to synthetic autotetraploid pearl millet has been extensively attempted (Dujardin and Hanna, 1983, 1984, 1987; Hanna et al., 1993; Chapman and Busri, 1994). Hanna et al. (1993) isolated seven apomictic BC_4 plants. The expression of the apomictic trait showed that most highly apomictic plants had $2n = 27$ or 29 chromosomes. No BC_4 plants were obligately apomictic, although one plant produced 89% maternal type and 6 of the 7 BC_4 plants produced a few offspring that formed only aposporous embryo sacs. This study failed to transfer the apomictic trait from a noncultivated wild perennial species to the cultigen. Lubbers et al. (1994) isolated molecular markers that may facilitate the identification of genes for apomixis. It may be possible to transfer these genes to pearl millet through nonconventional methods, such as somatic hybridization and genetic transformation.

No genes for apomixis have been found in the major economically important crops (wheat, barley, rice, maize, and soybean). Polyembryony is common in several cultivars of citrus (Koltunow et al., 1996). Apomixis is being utilized to produce disease-free scion and uniform root stocks in citrus.

6.3.1.3 Distribution of Apomixis

Apomixis is distributed throughout the plant kingdom from algae to angiosperms (Asker and Jerling, 1992; Koltunow et al., 1995). It has been estimated that apomixis occurs in about 300 genera belonging to 80 families (Khokhlov, 1976). Apomixis has uneven taxonomic distribution occurring mostly in polyploid species of *Poaceae*, *Rosaceae* and *Asteraceae*, and is often associated with fertile hybrids that otherwise would have been sterile (den Nijs and van Dijk, 1993). In the citrus group, however, apomixis is not related to polyploidy, but it is related to hybridization.

Hojsgaard et al. (2014) summarized the occurrence of apomixis in angiosperms. This mechanism was found to be taxonomically widespread with no clear tendency to specific groups and to occur with sexuality at all taxonomical levels. Adventitious embryony was the most frequent form (148 genera) followed by apospory (110) and diplospory (68). Numbers of genera containing sporophytic or gametophytic apomicts decreased from the tropics to the arctic, a trend that parallels general biodiversity. While angiosperms appear to be predisposed to shift from sex to apomixis, there is also evidence of reversions to sexuality. Such reversions may result from genetic or epigenetic destabilization events accompanying hybridization, polyploidy, or other cytogenetic alterations.

6.3.1.4 Types of Apomixis

Two main types of apomixis are known: gametophytic apomixis (gametophytic agamospermy) and adventitious embryony (nucellar) (Figure 6.3). The two forms of apomixis are unrelated (Asker and Jerling, 1992). In gametophytic apomixis, an egg cell develops parthenogenetically from an unreduced embryo sac producing only maternal-type offspring. In contrast, adventitious embryony does not have a gametophytic stage, but only sporophytic stage, and is connected with parthenogenesis. Embryos develop from somatic cells of the nucellus or integument. Thus, alternation of generation is eliminated (Grant, 1981).

6.3.1.4.1 Gametophytic Apomixis

Gametophytic apomixis has been divided into diplospory and apospory (Figure 6.3). Both forms maintain the alternation of generations, but the gametophytic and sporophytic stages contain the same level of ploidy (Nogler, 1984). Meiosis and fertilization are circumvented in both forms; an unreduced egg cell develops asexually by parthenogenesis. Although the predominant mode of

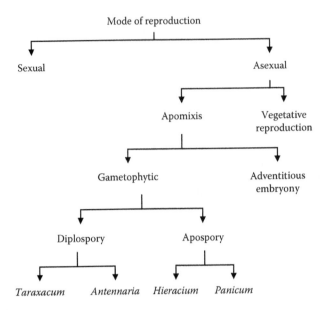

FIGURE 6.3 Classification of the mode of reproduction in plants.

reproduction in gametophytic apomixis is obligate, it may sometime be combined with sexuality. Such type is known as facultative apomixis and frequency is influenced by several factors.

6.3.1.4.1.1 Diplospory Diplospory (generative apospory) occurs in plants where meiosis is absent or development of a linear tetrad of megaspores does not occur. It is subdivided into two forms named after the genera in which they were first discovered (Figure 6.3).

Taraxacum Type: The megaspore mother cell enters into the meiotic prophase-I but chromosomes remain as univalents resulting in restitution nuclei with full somatic chromosome complement. The second division is normal producing dyads with unreduced chromosome number. Usually the chalazal dyad cell gives rise to an eight-nucleate embryo sac after three mitoses (Figure 6.4).

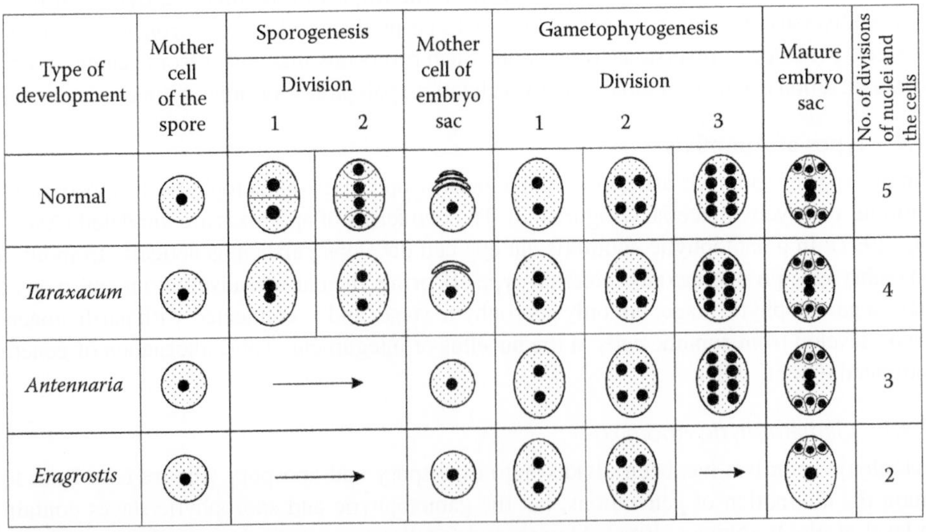

Type of development	Mother cell of the spore	Sporogenesis Division 1	Sporogenesis Division 2	Mother cell of embryo sac	Gametophytogenesis Division 1	Gametophytogenesis Division 2	Gametophytogenesis Division 3	Mature embryo sac	No. of divisions of nuclei and the cells
Normal	●	●	●	●	●	●	●	●	5
Taraxacum	●	●	●	●	●	●	●	●	4
Antennaria	●	→		●	●	●	●	●	3
Eragrostis	●	→		●	●	●	→	●	2

FIGURE 6.4 Megagametogenesis in amphimixis and various apomictics.

There are four divisions of nuclei and cells instead of the five observed in normal embryogenesis (Figure 6.4). The *Taraxacum* type of apomixis, in which embryo and endosperm develop without fertilization, is known as mitotic diplospory or autonomous apomixis and is found among genera of the Compositae (*Taraxacum, Erigeron, Chondrilla*), rarely in *Arabis holboellii, Agropyron scabrum*, and in certain *Paspalum* species (Nogler, 1984; Asker and Jerling, 1992).

Antennaria Type: The megaspore mother cell does not undergo meiosis but directly undergoes three mitotic divisions producing an eight-nucleate unreduced (diploid) embryo sac (Figure 6.4). The *Antennaria* type is also called mitotic diplospory and has a wide taxonomic distribution (Nogler, 1984).

Other Types of Diplospory: Deviation from the *Taraxacum* and *Antennaria* types of diplospory has been observed.

Eragrostis: It is similar to the *Antennaria* type, but with the difference that the embryo sac formed has four nuclei because the third mitosis is omitted. Thus, cells and nuclei have two divisions (Figure 6.4).

Ixeris: An asyndetic meiotic chromosome pairing produces a restitution nucleus, as observed in the *Taraxacum* type, which then undergoes a second meiotic division without cytokinesis. This results in two unreduced instead of four reduced nuclei. The two mitoses produce an eight-nucleate embryo sac. The *Ixeris* type of diplospory is found in *Ixeris dentata*.

Allium: Premeiotic chromosome doubling by endomitosis or endoreduplication in the female produces unreduced nuclei. Meiosis is normal and yields a tetrad with unreduced nuclei of parental genotypes. Two subsequent mitoses in the chalazal dyad result in an eight-nucleate embryo sac. It is observed in *Allium nutans* and *A. odorum*.

6.3.1.4.1.2 Apospory   Apospory, earlier termed "somatic apospory" (Nogler, 1984), is characterized by the development of unreduced embryo sacs directly from the somatic (vegetative) cells located in the center of the nucellus. There may be multiple embryo sacs in an ovule but only one of them matures into an aposporous embryo sac (Nogler, 1984; Asker and Jerling, 1992). Aposporous apomixis is of two main types: the bipolar type *Hieracium* and the monopolar *Panicum* type (Figure 6.3).

Hieracium type: The initial cell produces an eight-nucleate bipolar embryo sac by mitotic divisions along with the development of the Polygonum type observed in the related sexual taxa. Thus, an ovule may contain two or more eight-nucleate embryo sacs, one is reduced which may degenerate and others are unreduced (Khokhlov, 1976). Apospory is widely distributed in several families, for example, in the genera of *Hypericum, Poa, Ranunculus, Crepis, Hieracium, Hierochloe*, and *Beta* (Nogler, 1984).

Panicum type: The unreduced being four-nucleate monopolar embryo sac is produced after the second mitosis; the third mitosis is absent. This type of apomixis was discovered in *Panicum maximum* (Warmke, 1954) and is found in grasses belonging to Panicoideae, and Andropogoneae (Nogler, 1984; Asker and Jerling, 1992). It is widely distributed in the genera of the *Bothriochloa–Dichanthium–Capillipedium* complex, *Cenchrus, Chloris, Digitaria, Eriochloa, Heteropogon, Hyparrhenia, Panicum, Paspalum, Pennisetum, Sorghum, Themeda*, and *Urochloa* (Nogler, 1984). An exceptional case is recorded within the Panicoideae. The unreduced aposporous sacs are usually four-nucleate and the reduced ones are eight-nucleate and bipolar.

6.3.1.4.2 Adventitious Embryony

The terms nucellar or adventive embryony are synonymous with adventitious embryony (Asker and Jerling, 1992). Adventitious embryony is an asexual reproduction in which adventitious embryos develop from one or more of the nucellar and integumentary cells of the ovule.

Adventitious embryony differs from diplospory and apospory in that it lacks embryo sacs and produces viable seeds. Thus, these embryos do not contain polar nuclei or endosperm. Double fertilization is apparently absent and, even if present, it is influenced by environmental conditions (Naumova, 1993). Meiosis is usually normal with reduced chromosome numbers in embryo sacs.

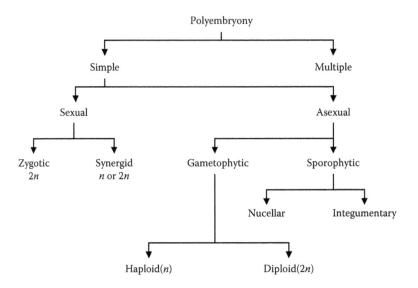

FIGURE 6.5 Classification of polyembryony in plants.

Adventitious embryos can be induced in plants experimentally through tissue culture. The production of somatic embryos (embryoids, i.e., embryo-like structures) *in vitro* is a prerequisite for regeneration of plants; however, adventitious embryos and embryoids are not always produced in a similar way (Naumova, 1993).

Adventitious embryony is widespread in economically important crops belonging to the genera *Citrus, Euphorbia, Mangifera, Malus, Ribes, Beta*, and several genera of grasses. It has been identified in more than 250 species of 121 genera belonging to 57 families of flowering plants (Naumova, 1993).

Polyembryony: Polyembryony, the occurrence of more than one embryo in a seed, may be simple or multiple (Figure 6.5). Simple polyembryony is divided into sexual and asexual. Polyembryony from sexual reproduction may be zygotic-zygotic polyembryony, develop from suspensor-suspensor polyembryony or synergid cells-synergid polyembryony (Lakshmanan and Ambegaokar, 1984). Polyembryony is controlled by several genes. It may be autonomous where no induction is necessary, or induced when pollination is required. Asexual polyembryony may arise from gametophyte or sporophyte. The gametophyte polyembryony can be haploid or diploid, and the sporophytic form is from nucellar and integumentary cells (Figure 6.5).

6.3.1.5 Endosperm Development in Apomixis

6.3.1.5.1 Autonomous Endosperm Development

Plants with autonomous apomixis diploid parthenogenesis do not require pollination, and neither the egg nucleus nor the polar nuclei are fertilized. The central cell (polar nuclei) develops parthenogenetically producing endosperm without fertilization resulting in autonomous endosperm development (Asker and Jerling, 1992). Autonomous endosperm usually contains variable chromosomes due to fusion of polar nuclei and endomitosis. Autonomous endosperm development is rather common in most apomictic Compositae (Nogler, 1984) and Asteraceae (Asker and Jerling, 1992).

6.3.1.5.2 Pseudogamous Endosperm Development

Fertilization of the polar nuclei is required for endosperm development. According to Asker and Jerling (1992), pseudogamy includes cases where embryo formation takes place independently of pollination, that is, embryos develop before pollination, which is necessary for the production of endosperm and germinable seed. Pseudogamous apomicts often exhibit poor seed set. Seed development depends on a precise ratio of maternal (m) and paternal (p) genomes in the endosperm (Haig and Westoby, 1991). This ratio is called endosperm balance ratio (EBN) and is usually 2m:1p. The failure in crosses between autotetraploid and diploid related species is attributed to the departure from EBN.

Parthenogenic embryo may develop in association with a viable endosperm. Thus, the development of an embryo into a seed is a preadaptation for apomixis. Morgan et al. (1998) isolated a tetraploid, partially male-fertile, aposporous apomictic line 169-46 from *Pennisetum glaucum* ($2n = 4$ $x = 28$) x *P. squamulatum* ($2n = 6x = 54$). This interspecific F_1 hybrid ($2n = 42$; expected 41) was crossed with a colchicine-induced amphidiploid ($2n = 6x = 42$) between pearl millet and Napier grass (*P. purpureum*, $2n = 4x = 28$). This double-cross apomictic hybrid was crossed as a male to tetraploid pearl millet three times. In each generation, Morgan et al. (1998) selected apomictic and partially fertile plants and eventually identified a tetraploid apomictic pearl millet line 169–46. The central-cell nuclei (secondary nucleus) of 169–46 are unreduced ($2n = 4x = 28$) and on fusion with 14 chromosome gametes from pearl millet, produce an endosperm with 4m:1p genomic ratio. The maternal and paternal genomes are similar in $4x$ pearl millet, and thus m:p ploidy ratio in the endosperm should equal the m:p endosperm balance ratio (2:1). Therefore, 4m:1p endosperm imbalance causes poor seed set in 169–46. By BC_3 autotetraploid pearl millet exhibits meiotic irregularities but has a fairly good seed set. Apomictic diploid pearl millet has not yet been identified (W. W. Hanna, personal communication).

6.3.1.6 Identification of Apomixis

Apomixis can be identified by morphological, cytological, genetic, biochemical, and molecular methods:

1. Uniform progeny from a single plant, that is, matromorphic progeny.
2. Maternal type progeny appears in the F_1; thus, no genetic variation is expected in the F_2.
3. In crosses between a recessive genotype as a female parent and dominant genotype as a male, exclusively recessive genotype is obtained.
4. A high degree of polyploidy is usually associated with apomixis and is constant from parent to the progeny.
5. An unusually high degree of fertility in aneuploids, triploids, wide crosses, or other plants expected to be sterile is an indication of apomixis.
6. Multiple embryos, seedlings, stigmas, and ovules suggest an indication of apomixis (Hanna, 1991).
7. Cytohistological examinations are useful tools in identifying and verifying various types of apomixis.
8. Both biochemical and molecular techniques are currently being used to screen apomixis at the seedling stage in *Pennisetum* species (Lubbers et al., 1994) and *Poa pratensis* (Mazzucato et al., 1995). Mazzucato et al. (1996) developed an auxin test to determine frequency of parthenogenesis.

6.3.1.7 Classification of Plants Based on Natural Crossing

1. Dioecious plants: Staminate and pistillate flowers are produced on separate plants of a species; entirely cross-pollinated.
2. Apomictic plants: Union of spores is not involved; cross-pollination is required.
3. Self-incompatible plants: Both male and female spores are produced at the same time and are functional but fail to fertilize and produce seeds; entirely cross-pollinated.
4. Dichogamous plants: Plants include protandry and protogyny ensuring cross-pollination.
5. Cleistogamous plants: Flowers remain closed and are completely self-fertilized.

6.3.1.8   Regulation of Apomixis

Despite numerous publications, the inheritance of apomixis is not clearly demonstrated nor is genetics of apomixis clearly understood. Although apomixis is genetically controlled, it is also influenced by environmental factors like temperatures and light regimes changes (Gerstel et al., 1955; Bashaw, 1980; Hanna and Bashaw, 1987; Hanna, 1991, 1995; Asker and Jerling, 1992; Ramachandran and Raghavan, 1992; den Nijs and van Dijk, 1993; Lutts et al., 1994). Hanna et al. (1973) observed that in guineagrass sexuality was controlled by a dosage effect of two or more dominant alleles present at different loci or at a single locus. It has been demonstrated that apomixis is controlled by qualitative traits mainly by recessive and dominant genes and in some species where apomixis is facultative, polygenes may be involved. An intermediate to recessive inheritance with quantitative dosage effects for apomixis has been suggested in polyploid species of *Rubus* (Chirsten, 1952; Berger, 1953), but a dominant gene for apomixis in apple was reported (Sax, 1959). Based on inheritance of apospory in buffelgrass, Sherwood et al. (1994) postulated a two-locus model for tetrasomic transmission in which the dominant allele A of one locus is required for apospory, but is hypostatic to the dominant allele B of the second locus which confers sexuality. Bicknell et al. (2000) recorded monogenic dominant inheritance in *Hieracium piloselloides* ($2n = 3x = 27$), and *H. aurantiacum* ($2n = 3x + 4 = 31$). They could not recover diploid apomictic plants from these crosses due to selection against the survival of diploid zygotes. The dominant genes in both species are either closely linked or possibly allelic. This shows that apomixis has a genetic basis, but how it is regulated is still unresolved. Genetic analysis of *Paspalum simplex* showed that apomixis to be under the control of a single dominant gene (Pupilli et al., 2001). den Nijs and van Dijk (1993) concluded that different genetic mechanisms for apomixis may exist even within the same species. More critical studies are needed to clearly understand these mechanisms.

Grimanelli et al. (1998a) supported a one dominant-gene hypothesis for the failure of meiosis in *Tripsacum*. However, the apomictic process may be controlled by a cluster of linked genes inherited in a tetrasomic fashion. van Dijk et al. (1999) did not find a single locus control for apomixis in *Taraxacum* and suggested that several genes were involved in the control of apomixis. In a subsequent study, Grimanelli et al. (1998b) proposed non-Mendelian transmission of apomixis in maize-*Tripsacum* hybrids. A gene (or genes) controlling apomixis in *Tripsacum* was linked with a segregation distortion-type system, promoting the elimination of the apomixis allele when transmitted through haploid gametes. This protects the diploid level from being invaded by apomixis.

Occasional occurrence of unreduced egg cells or parthenogensis in sexual plants may not be related to apomixis. Three sets of genes, acting at different reproductive stages of plants, may be responsible for apomixis: (1) failure of reduction in number of chromosomes, (2) failure of fertilization, and (3) development of nonreduced unfertilized egg-cells.

The mode of reproduction in facultative apomictic plants is often influenced by environmental factors such as day length, light intensity, temperature shocks, plant age, nutrition supply, exposure to certain growth hormones, or growing environments (greenhouse vs. field). Knox and Heslop-Harrison (1963) reported that in *Dicanthium aristatum* an 8-h light treatment produced inflorescence with 79% aposporous embryo sacs, but a 16-h photoperiod produced only 47%. Chapman (1992) recorded 60% apomictic embryos in the same species with a photoperiod excess of 14 h while more than 90% were found with less than 14 h. Contradictory results were observed in *Poa ampla* where a 20-h daylight regime produced more sexuality (Williamson, 1981). Sexuality was promoted in *Poa pratensis* when plants were grown and flowered in early spring in the glasshouse, but it was not recorded in plants grown in the field (Grazi et al., 1961). Temperature and seasonal variation have been found to have significant influence on the expression of apomixis in *Malus* (Schmidt, 1977). Similar results were reported in *Cenchrus ciliaris* (syn. *Pennisetum ciliare*) where frequency of sexual pistils differed between sampling dates in one field, indicating that environment may influence development of embryo sac type (Sherwood et al., 1980).

In *Brachiaria decumbens* reproductive behavior appeared similar between greenhouse- and field-grown plants, but in *B. brizantha* a much lower level of sexuality was observed in field-grown plants. Sexual reproduction was more frequent in adverse climatic conditions in *Paspalum cromoyorrhizon* (Quarin, 1986).

Occasional isolation of apomixis in plants is due to pollinations: condition of pollination, choice of pollinators, condition of pollen applied (fresh vs. old), or time of pollination (early vs. delayed). Martínez et al. (1994) observed that an early pollination by $2x$ and $4x$ pollen in an apomictic tetraploid ($2n = 4x = 40$) *Paspalum notatum* sometimes prevented parthenogenetic development of the unreduced egg cell in an aposporous embryo sac allowing fertilization. Wide hybridization often induces apomictic seed development. It may be due to slow pollen tube growth or sperm nuclei may fail to fertilize the egg. In sexual plants, wide hybridization, pollination with irradiated pollen, or chemically treated pollen promote haploid formation (Asker and Jerling, 1992).

To conclude, apomixis is under complex genetic control; however, eight genetic models for the inheritance of apomixis can be proposed from the published results:

Model 1: One disomic locus, apospory recessive (*aa*).
Model 2: One disomic locus, apospory dominant (*AA, Aa*).
Model 3: Two disomic loci, additive gene action, requires at least two dominant alleles for sexuality (*Aabb, Aabb, aaBb, aabb* genotypes apomictic).
Model 4: Two disomic loci, dominant allele *A* is required for apospory, dominant allele *B* confers sexuality and is epistatic to dominant gene *A* (*Aabb, Aabb* → apomictic).
Model 5: One tetrasomic locus, apospory recessive (*aaaa*) to sexuality.
Model 6: One tetrasomic locus, dosage effect (*Aaaa, AAAa,* and *AAAA*); sexual genotype *aaaa*.
Model 7: One tetrasomic locus, diplospory dominant (*Aaaa, Aaaa,* and *AAAa, AAAA*).
Model 8: One disomic dominant but controlled by a cluster of linked loci.

6.3.1.9 Apomixis in Crop Improvement

Apomixis helps fix heterosis in a desired heterozygous gene combination. Extreme heterozygosity is a characteristic of plants reproducing by apomixis. It fixes new heterozygous genotypes with valuable agronomic traits and resistance to pests and pathogens (Asker and Jerling, 1992; Nassar, 1994). Although apomixis is a potentially powerful tool in plant breeding, its application in crop improvement is limited to turf and forage grasses (Hanna, 1991; Voigt and Tischler, 1994).

Obligate apomixis may suppress sexuality completely. Thus, a significant reduction in cost of hybrid seed production could be avoided. Since outcrossing within a population in the commercial field with obligate apomicts is not a problem, it can be used for other vegetative propagated crops. Purity can be maintained by controlling mechanical mixture during harvest. Farmers will not be required to buy hybrid seed every year because they could use seeds indefinitely without the risk of contamination from recombination. Apomixis would be disadvantageous to seed industries because it would result in a loss of control of commercial hybrids and reduction in seed sales. In obligate apomixis, progeny testing for stability and field isolation are not needed in commercial seed production. It can be effective in a breeding program provided that cross-compatible sexual and partial sexual plants are available to allow generation of new gene combinations.

Introgression of the apomixis from *Tripsacum dactyloides* ($2n = 2x = 36$; $2n = 4x = 72$) to maize (*Zea mays*) was extensively attempted by a number of research groups (Harlan and de Wet, 1977; Grimanelli et al., 1998a, 1998b; Hoisington et al., 1999; Blakey et al., 2001). Two pathways were used:

1. Maize $2n = 20$ x *Tripsacum* $2n = 2x = 36 \rightarrow 2n = 28 \rightarrow 2n = 20$ pathway (the most common way).

2. Maize $2n = 20$ x *Tripsacum* $2n = 4x = 72 \rightarrow 2n = 46 \rightarrow 2n = 56 \rightarrow 2n = 38 \rightarrow 2n = 20$ pathway (Harlan and de Wet, 1977).

Harlan and de Wet (1977) isolated highly tripsacoid maize lines with $2n = 20$ chromosomes. Dominant resistance to six maize diseases was found in BC_8 populations but no apomictic line was recovered.

As of March 1999, 10 patents related to apomixis, including for maize, pearl millet, and rice, have been issued (Bicknell, 1999). Despite enormous research and financial resources from public, private, and international institutes like CIMMYT, apomictic maize has not been commercialized.

Facultative apomixis has been identified in some plant groups for utilization in breeding programs. The best source of apomixis is the distantly related species of the cultigens. The main emphasis is on isolation, identification, and incorporation of either obligate or facultative apomixis in several major crops such as *Citrus*, berries, apple, fodder grasses, rubber plants (*Parthenium*), pearl millet, maize, wheat, barley, rice, *Sorghum*, sugar beet, *Brassica*, *Pea*, soybean, alfalfa, cotton, potato, and tobacco (Asker and Jerling, 1992). In rice, wild relatives failed to induce apomixis (Rutger, 1992); however, Zhou et al. (1993) discovered a rice line SAR-1 that showed a high degree of pollen sterility but seed set under isolation was 55.33%. Cyto-embryological examination revealed division of eggs without fertilization, or adventitious embryo developed from cells of ovary wall and produced seed. Endosperm developed normally to provide nutrients for embryo development. The apomixis trait is heritable and is being incorporated for rice varietal improvement.

Several factors should be examined carefully before apomixis can be successfully used to produce new cultivars. These are: facultative behavior, number of genes involved, modifiers, environmental factors, ploidy levels, and seed sterility. Although apomixis provides a unique opportunity to develop and maintain superior genotypes, its use is restricted because of lack of basic understanding of the mechanism of apomixis (Asker and Jerling, 1992; Hanna, 1995). Studies on synteny among grass species have revealed homoeologous regions in chromosomes. For example, a chromosome segment conditioning apomixis in *Paspalum simplex* is homoeologous to the telomeric region of the long arm of rice chromosome 12 (Pupilli et al., 2001). Ortiz et al. (2001) discovered that *Paspalum notatum* ($2n = 20$) linkage groups 1, 3, 4, 5, 6, 8, 10 have synteny regions with maps of maize and rice. Apomictic genes could be isolated and transferred to desired sexual diploid crops through transformation provided such a system is available.

6.3.2 VEGETATIVE REPRODUCTION

Vegetative reproduction is an asexual form of reproduction in higher plants by which new identical individuals are generated from a single parent without sexual reproduction. The offspring of asexual propagation is known as clone. The principal method of asexual reproduction in higher plants is by vegetative propagation. It is characterized by mitosis which occurs in the shoot and root apex, cambium, intercalary zones, callus tissues, and adventitious buds.

Vegetative propagation facilitates easy, rapid multiplication of economically valuable heterozygous plants without alteration and does not induce genetic diversity. The perpetuation of seedless oranges, bananas, and grapes are excellent examples.

1. Vegetative Reproduction by Specialized Vegetative Organs
   Many plants multiply through specialized vegetative structures by modifying stems. Common examples are as follows:
   a. Bulbs
      Bulbs are a short basal, underground stem surrounded with thick, fleshy leaves, common in the onion, daffodil, and hyacinth (family Liliaceae). Bulbils or bulblets are miniature bulbs used in propagation.

b. Corms
    Corms are short, upright, hard, or fleshy bulb-like stems usually covered with papery, thin, dry leaves and do not contain fleshy leaves. It is common in gladiolus, crocus, and water chestnut. Cormels are miniature corms. Fleshy buds develop between the old and new corms.

c. Runners (Stolons)
    Runners are a horizontal aboveground stem that usually produces plants by rooting at nodes. Strawberry reproduces vegetatively via stolons.

d. Rhizomes
    A rhizome is a horizontal, prostrate, or underground stem that contains nodes and internodes of various lengths and readily produces adventitious roots. Species with rhizomes are easily propagated by cutting the rhizomes into small pieces that contain a vegetative bud. Examples of species with underground rhizomes are Johnson grass, brown grass, and hops.

e. Tubers
    A tuber is fleshy portion of a rhizome, underground storage stem. Potato is the best example of a species with tubers. A small piece of potato with a bud (known as an eye) is planted to produce more potatoes. This ensures uniformity in the next generation as each eye generates a replica of its parent.
    Vegetative propagation is also common by tuberous roots like sweet potato. Tuberous roots of some species may contain shoot buds at the "stem end" as part of their structure. Other examples include dahlia and begonia. The primary tap root develops into an enlarged tuberous root that can be propagated by dividing into several portions, each with a bud (Janick et al., 1981).

2. Vegetative Reproduction by Adventitious Roots and Shoots
   Reproduction of an entire plant from a buried branch or stem is called layering or layerage. Once new roots and shoots emerge, plantlets are separated from the mother plants. Cutting is one of the most important methods of vegetative reproduction. Small pieces of stems (cuttings) are used by horticulturists and nurserymen for multiplying and reproducing ornamental crops.

3. Vegetative Reproduction by Grafting
   Reproduction of an entire plant by union of a small actively growing shoot (scion) grafted onto root-stock that is resistant to pathogens and pests is an invaluable tool in plant propagation. Grafting is quite common for a large number of domestic fruit crops to produce disease-free crops.

4. Vegetative Reproduction by Tissue Culture
   Plant propagation through cell and tissue cultures is termed micropropagation. It involves regeneration of plants aseptically from cells (including protoplasts) and tissues (immature embryos, leaves, roots, and stems) in artificial cultures. A single protoplast can regenerate an entire whole plant. Thus, a large number of plantlets can be generated via embryogenesis and organogenesis from a small piece of the stock plant. Micropropagation can result in the isolation of disease-free plants and has proven efficient for orchid propagation where natural propagation rate is very slow.

5. Advantages of Vegetative Reproduction
   Vegetative reproduction (propagation) has numerous advantages such as efficient commercial crop production (potato and sugarcane), exploitation of heterosis (hybrid vigor is not lost), avoidance of dormancy and juvenile period (grafting on older root stocks allows new wood of seedling to produce fruit sooner than if it remains on its own root stock), maintenance of sterile or lethal genotypes, facilitation of physiological and genetic studies, and increases plants of unique genotypes in breeding programs.

Despite many advantages, vegetative reproduction has a serious problem. All vegetative propagated plants from the same source are genetically uniform. This means that genetic vulnerability exists. If the genotype is susceptible to a pest or pathogen, or if a new pest or a pathogen develops that can infect the genotype, then all plants of the clone will be susceptible. If a disease strikes the members of a clone, production of disease-free seed or shoot is extremely difficult.

6.4 THE CHROMOSOMAL BASIS OF SEX DETERMINATION IN PLANTS

The majority of flowering plant species are hermaphroditic having flowers with stamens and carpels. In a relatively small number of plant species, sex is determined by the presence or absence of a pair of sex chromosomes (chromosomal mechanism of sex determination). The species may be heterogametic (produce two kinds of gametes and offspring with two sexes) or homogametic (one kind of gamete and offspring). The sex chromosomes (heteromorphic sex pair = heterochromosomes; X and Y) are distinct from all other chromosomes (autosomes). In certain plants, sex determination is under genetic control where discrete sex-chromosome systems are not discovered and usually influenced by autosomal genes (Burnham, 1962). Heteromorphic sex chromosomes are clearly defined in only a few species (Table 6.1).

6.4.1 SYSTEM OF SEX DETERMINATION

6.4.1.1 Male Heterogametic (Female XX; Male XY = X-Y System)

The female is homogametic and all eggs carry an X chromosome. The male is heterogametic and 50% spores carry an X chromosome and 50% will contain a Y chromosome. Random fertilization ensures a sex ratio of 1 female:1 male in every generation. The X-Y system is common in *Cannabis sativa* (hemp), *Humulus lupulus* (hop), *Rumex angiocarpus*, *Silene latifolia* (syn. *Melandrium album*), and *M. rubrum* (Table 6.1). The X-Y system is also prevalent in mammals including humans, *Drosophila*, and many other species.

TABLE 6.1
Established Heteromorphic Sex Chromosomes in Plants

Species	2n Female	2n Male
Cannabis sativa	$18 + XX$	$18 + XY$
Humulus lupulus	$18 + XX$	$18 + XY$
Humulus lupulus var. *cordifolius*	$16 + X_1X_1X_2X_2$	$16 + X_1Y_1X_2Y_2$
Humulus japonicus	$14 + XX$	$14 + XY_1Y_2$
Rumex angiocarpus	$12 + XX$	$12 + XY$
Rumex tenuifolius	$24 + (XX) XX$	$24 + (XX) XY$
Rumex acetosella s. str.	$36 + (XXXX) XX$	$36 (XXXX) XY$
Rumex graminifolius	$48 (XXXXXX) XX$	$48 (XXXXXX) XY$
Rumex hastatulus	$6 + XX$	$6 + XY_1Y_2$
Rumex acetosa	$12 + XX$	$12 + XY_1Y_2$
Rumex paucifolius	$24 + (XX) XX$	$24 + (XX) XY$
Melandrium album[a]	$22 + XX$	$22 + XY$
Melandrium rubrum	$22 + XX$	$22 + XY$

Source: Adapted from Westergaard, M. 1958. *Adv. Genet.* 9: 217–281.

[a] *Silene latifolia*.

6.4.1.2 Male Heterogametic (Female XX; Male XO = X-O System)

The male produces two types of gametes: one with the X chromosome and the other without a sex chromosome (O). The female produces gametes only with the X chromosome. The X-O system is found only in *Vallisneria spiralis* ($2n = 20$), and *Dioscorea sinuata* (yam). The X-O system is common in grasshoppers, crickets, and roaches.

6.4.1.3 Male Heterogametic (But with One Extra Chromosome)

This system is proposed in *Phoradendron flavescens* (American mistletoe) and *Phoradendron villosum* [more evidence is needed (Burnham (1962)].

6.4.1.4 Female Heterogametic (Female XY (ZW); Male XX (ZZ) (Z-W System))

The female is heterogametic and the egg determines the sex. The male is homogametic and produces only one kind of gamete. The Z-W system is found in $6x$ ($2n = 42$) strawberry (*Fragaria elatior*). This system is absent in all animals, but found in birds and some insects including butterflies and moths.

6.4.1.5 Compound Chromosomes (Interchanges among X, Y, and Autosomes)

These chromosomes are found in certain races of common hop and in garden sorrel (*Rumex acetosa*) where a chain of three chromosomes is observed in the heterogametic sex, arranged at meiotic metaphase-I as $Y_1 X Y_2$ with the X oriented to pass to one pole and $Y_1 Y_2$ migrate to the other pole. In certain cases, there is a chain of five chromosomes in the male. This may be due to interchanges between autosomes and sex chromosomes (Burnham, 1962).

Several dioecious species of plants are without sex chromosomes but sex expression is under genetic control. These are: *Spinacia oleracia, Ribes alpium, Vitis vinifera, Carica papaya, Asparagus officinalis,* and *Bryonia dioica*. The sex determining gene in the spinach is on chromosome 1.

6.4.2 DETERMINATION OF HETEROGAMETIC SEX IN PLANTS

6.4.2.1 Cytological Identification

A classic example of a heteromorphic pair of sex chromosomes (XX female; XY male) that has been widely studied in plants is white campion, *Silene latifolia* (syn. *Melandrium album*). Cytological investigation establishes that the Y chromosome is larger than the X, and both are larger than the autosomes. Westergaard (1958) divided X and Y chromosomes into four hypothetical segments. Segment IV is homologous in X and Y chromosomes, and these ends pair during meiosis. The other segments of X and Y are nonhomologous (Figure 6.6). In the Y chromosome, segment I carries the gene that suppresses the development of female sex organs, segment II initiates anther development, and segment III controls the last stages in anther development. The sex expression gene in the Y chromosome is present on the segment that is nonhomologous with the X chromosome and is never separated by crossing over (Westergaard 1958).

Doležel and Göhde (1995) analyzed nuclear DNA content of *S. latifolia* by high-resolution flow cytometry. They found lower DNA content in female than male plants due to sex chromosome heteromorphism. Kejnoský et al. (2001) identified, for the first time, male-specific genes, *MROS*, expressed in *S. latifolia*. The *MROS* genes were located on chromosomes using the flow-sorted X chromosomes and autosomes as a template for PCR with internal primers. Their results indicate that at least two copies of the *MROS3* gene are located in tandem on the X chromosome with additional copies on autosomes, while *MROS1, MROS2,* and *MROS4* are exclusively located on autosomes. They conclude that *MROS3* is a low-copy gene family that is connected with the normal pollen development, present in dioecious and other dicots. For example, *MROS3* homologoues are also discovered in a nonrelated *Arabidopsis thaliana* genome. This suggests an ancient origin of the *MROS3* gene. It is possible that the X chromosome of *S. latifolia* has some regions homologous to chromosomes III and IV of *A. thaliana*.

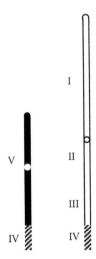

FIGURE 6.6 A diagrammatic sketch of sex chromosomes (XY) of *Silene latifolia* (syn. *Melandrium album*). Segments I, II, and III are different in Y, segment IV is the homologous region and segment V is the differential region of X. (Redrawn from Westergaard, M. 1958. *Adv. Genet.* 9: 217–281. With permission.)

Another dioecious crop, hemp (*Cannabis sativum*), was extensively investigated and has a sexual system similar to *S. latifolia*. Female plants contain two X chromosomes whereas male plants have one X and one Y chromosome. The Y chromosome is much larger than the X chromosome and autosomes. By using flow cytometry, Sakamoto et al. (1998) reported a genome size of 1638 Mbp for diploid female (2*n* = 18 + XX), and 1683 Mbp for diploid male (2*n* = 18 + XY) plants. Karyotype analysis revealed that the X chromosome was submetacentric and the Y chromosome subtelocentric. The Y chromosome has the longest long arm with a satellite in the terminal of its short arm (Figure 6.7).

6.4.2.2 Sex Linkage Inheritance

In *Lychnis alba*, broad-leaved female (normal) × narrow-leaved male produce F_1 plants in which all female plants are broad-leaved but part of the males are broad-leaved and part are narrow-leaved

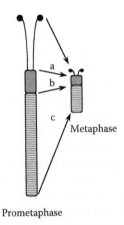

FIGURE 6.7 A diagrammatic sketch of specific condensation of the long arm and satellite of the Y chromosome of *Cannabis sativa*. a, NOR region; b, short arm; c, long arm. (Redrawn from Sakamoto, K. Y. et al. 1998. *Cytologia* 63: 459–464. With permission.)

(Burnham, 1962). The sex determining gene is on the X chromosome and no allele is present on the Y chromosome.

6.4.2.3 Crosses between Dioecious and Monoecious *Bryonia* Species

The segregation in crosses between dioecious and monoecious *Bryonia* species is shown below:

1. *Bryonia dioica* (dioecious) female x *Bryonia alba* (monoecious) male → All female
2. *Bryonia alba* (monoecious) female x *Bryonia dioica* (dioecious) male → 1 female:1 male
3. *Bryonia dioica* (dioecious) female x *Bryonia macrostylis* (monoecious) male → all female
4. *Bryonia macrostylis* (monoecious) female x *Bryonia dioica* (dioecious) male → 1 female: 1 male
5. *Bryonia dioica* (dioecious) female x *Bryonia multiflora* (dioecious) male → all monoecious
6. *Bryonia multiflora* (dioecious) female x *Bryonia dioica* (dioecious) male → 1 female:1 male
7. *Bryonia macrostylis* (monoecious) female x *Bryonia alba* (monoecious) male → All hermaphrodite
8. *Bryonia alba* (monoecious) male x *Bryonia macrostylis* (monoecious) female → 1 female: 1 male

6.4.2.3.1 Conclusion

1. Dioecious × monoecious species produce 1 female:1 male. This suggests the male is the heterogamete of the dioecious species.
2. Crosses between dioecious × dioecious species may give monoecious progeny.
3. Crosses between monoecious × monoecious species may give dioecious species.
4. Competition tests in pollination
   In *Silene*, pollination by excess pollen produces excess female progeny, but sparse pollination gives 1 female:1 male (male heterogametic). By contrast, in *Cannabis sativa* (hemp), sparse pollination increases the proportion of male.
5. Self-pollination of plants that are normally unisexual
   In *Asparagus* ($2n = 20$), only a few seeds are obtained from a male plant, and they segregate into 3 male:1 female. No sex chromosome has been identified cytologically, but sex determining genes must be on the chromosomes. The XY male [⊗] → 1XX female:2XY male:1YY male; XX female × YY male → All XY and male. This suggests that a dominant gene for maleness is on the Y chromosome.
6. Crosses between diploid and autotetraploids
   When the male is heterogametic, chromosome doubling in dioecious species (XX and XY) will produce autotetraploids with genotypes XXXX and XXYY. When the female is homogametic, crosses between XXXX female × YY male → 1 XXX female:1 XXY male; XX female × XXYY male → 1 female:5 male. When the female is heterogametic, crosses between XXYY × XX → 4 XXY:1 XYY:1 XXX (5 female:1 male). In *Spinacia oleracea* (spinach) and also in *Silene latifolia*, male plants are heterogametic.
7. Sex ratio among progeny of primary trisomics
   In spinach, a diploid female × primary trisomic (Triplo 1) male cross produces an F_2 population that segregates in a trisomic ratio of 2 female:1 male. This suggests that a gene for sex determination is dominant and is located on chromosome 1.

6.4.3 SEX EXPRESSION

Table 6.2 compares sex expression in *Silene latifolia* and *Drosophila melanogaster*. Investigations on sex expression in diploid, triploid, and tetraploid with different dosages of X and Y chromosomes show that a plant is male when one or more Y chromosomes are present and a plant is female when Y is absent. The expression of female traits is possible when the ratio of X:Y reaches 1:4. Plants with XX chromosomes are females.

TABLE 6.2

Sex Expression (the Ratio of Sex Chromosomes and Autosomes) in *Silene latifolia* (syn. *Melandrium*) and *Drosophila*

Chromosome Constitution	*Melandrium*	*Drosophila*
1. 2A + XX	Female	Female
2. 2A + XXX	Female	Female
3. 3A + X	–	Male
4. 3A + XX	Female	Hermaphrodite
5. 3A + XXX	Female	Female
6. 4A + XX	Female	Male
7. 4A + XXX	Female	Hermaphrodite
8. 4A + XXXX	Female	Female
9. 4A + XXXXX	Female	–
10. 2A + XY	Male	Male
11. 2A + XYY	Male	Male
12. 2A + XXY	Male	Female
13. 2A + XXYY	–	Female
14. 3A + XY	Male	–
15. 3A + XXY	Male	Hermaphrodite
16. 3A + XXXY	Male	Female
17. 4A + XY	Male	–
18. 4A + XXY	Male	–
19. 4A + XXYY	Male	–
20. 4A + XXXY	Male	–
21. 4A + XXXYY	Male	–
22. 4A + XXXXY	Male →Hermaphrodite	–
23. 4A + XXXXYY	Male	–

Source: Adapted from Westergaard, M. 1958. *Adv. Genet.* 9: 217–281.

Sex expression on the same plant or on different plants is controlled either by qualitative or quantitative genes confined to sex chromosomes (X or Y or both) that, through interaction with sex genes on the autosomes, determine which sex will be expressed (Westergaard, 1958). Terminologies have been assigned for sex expression in plants based on presence or absence of sex organs (Table 6.3).

Based on the system of sex determination in diploid and polyploid plants, two kinds of mechanisms are evident:

1. The active Y chromosome plays a decisive role as an enhancer for maleness and a suppressor for gynoecism in determining the sex.
2. The sex deciding genes are in the X chromosome, and the sex depends on the X chromosome/autosome ratios, the Y chromosome being inactive (Westergaard 1958).

6.4.4 THE EVOLUTION OF DIOECISM

It has been demonstrated that in higher plants hermaphroditism is the original form of sexuality. Dioecism originated from bisexual or monoecious species through mutation and natural selection. The unisexual condition developed as a result of a trigger mechanism that suppresses the potentialities of the opposite sex in males and females. The forward evolution is from bisexual to unisexual, and the reverse evolution is from dioecism to bisexuality.

TABLE 6.3

Terminologies for Sex Expression in Plants

Type of Plants	Expression of Sex
1. Hermaphrodite	Perfect flower (bisexual; a flower with androecium and gynoecium)
2. Monoecious	Separate male and female flowers on the same plant (not synonymous with imperfect and should never be applied to individual flower)
3. Dioecious	Separate male and female flowers on different plants
4. Andromonoecious	Perfect and male flowers on the same plant
5. Gynomonoecious	Perfect and female flowers on the same plant
6. Trimonoecious	Perfect, female, male flowers on the same plant (polygamous)
7. Androdioecious	Perfect and male flowers on different plants
8. Gynodioecious	Perfect and female flowers on different plants

Source: Adapted from Westergaard, M. 1958. *Adv. Genet.* 9: 217–281.

6.5   INCOMPATIBILITY

6.5.1   SELF- AND CROSS-INCOMPATIBILITY

Morphological modification in flowers ensures cross-pollination and is widespread in families of the crop plants. However, in certain plants physiological mechanisms play an important role in cross-pollination. Self- and cross-incompatibilities are selective physiological reactions in fertilization that prevent (a) self-fertilization in certain hermaphrodites and (b) cross-fertilizations between certain of these self-incompatible individuals. The simplest incompatibility was found in *Nicotiana sanderae* and has been referred to as *Nicotiana type incompatibility*. In this system, incompatibility is determined by a multiple allelic series of *s* genes at one locus. Diploid pollen contains one *s* gene and there are two genes in the somatic tissues of the style. Pollen tube growth is inhibited in styles containing an allele in common with that of the pollen. Since the discovery of Nicotiana type incompatibility, very large numbers of different S alleles have been postulated. Personate incompatibility has been reported in a number of *Trifolium* species. In this system, plants are self-incompatible, and the various reactions are identified.

1. When plants of the same genotype are hybridized, there is no fertility because both sperm and style tissues and eggs have the same allele (Figure 6.8a).
2. When plants differ in one S allele, pollen grain with S_3 allele will fertilize eggs with S_1 and S_2 but pollen grain with S_2 allele will not germinate through the style and will fail to fertilize eggs with S_1 and S_2 (Figure 6.8b).
3. When the parents differ in both alleles (Figure 6.8c), four groups of progenies are produced: S_1S_3, S_1S_4, S_2S_3, and S_2S_4. Members of each group are cross-fertile in six combinations among the four progeny groups.
4. The induced autotetraploid from diploid self-incompatible species usually breaks down the genetic balance of the newly synthesized lines (Figure 6.9a–d).
5. Various modifications of *Nicotiana* and personate types of incompatibilities have been identified where pollen tube reaction is determined by the genotype of the sporophyte rather than genotype of the pollen tube. A great complexity of incompatibility relationships is possible because of the variability of allelic interactions and these are
   a. Frequent occurrence of reciprocal cross differences.
   b. The occurrence of incompatibility with the female parent.
   c. A family can consist of three incompatibility groups.
   d. Homozygotes are a normal part of the system.
   e. An incompatible group may contain two genotypes.

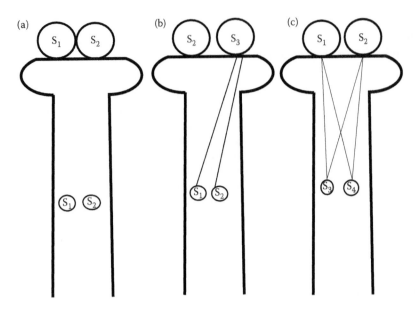

FIGURE 6.8 Diagrammatic sketch of pollen tube interaction in self-incompatibility behavior in plants. (a) Selfing or outcrossing pollen tube will not germinate; (b) Pollen with S_3 gene will travel through stigma and fertilize S_1 and S_2 female eggs; (c) Pollen with S_1 and S_2 genes will be able to fertilize eggs with S_3 and S_4 genes. (Redrawn from Elliott, F. C. 1958. *Plant Breeding and Cytogenetics*. McGraw-Hill Book Company, New York.)

6.5.2 MICROSPOROGENESIS

The end product of meiosis, in anthers, is the primary sporogenous cells contained in the microsporangia that gives rise to microspore mother cells (pollen mother cells; PMCs). Microspores with half chromosome number are produced through meiosis. The pollen grain is a mature microspore.

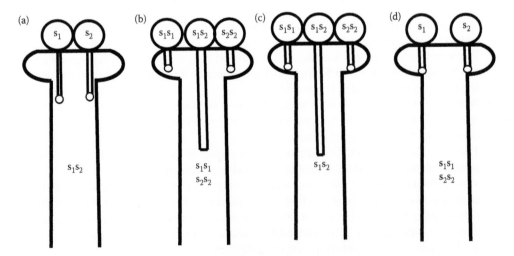

FIGURE 6.9 The incompatibility behavior of diploid and tetraploid plants. (a) Diploid, self-incompatible; (b) Tetraploid self, compatible; (c) Diploid × tetraploid, compatible; (d) Tetraploid × diploid, incompatible. (Redrawn from Elliott, F. C. 1958. *Plant Breeding and Cytogenetics*. McGraw-Hill Book Company, New York.)

6.5.3 GAMETOGENESIS

Nucleus of microspores divides to form the generative nucleus and tube nucleus. In many crop plants, generative nucleus divides again prior to dehiscence of the pollen grains from the anther. The pollen grain germinates on the stigma, the protoplast extruding through a pore of the wall in the form of a tube which grows down the style tissue and directs the pollen tube growth. The generative nucleus divides to produce sperm nuclei and follow the tube nucleus.

6.5.4 NATURE OF POLLINATION

When pollen grains fall onto the stigma of the same flower of the same plant, self-pollination occurs and the union of male and female nuclei generates zygote which is known as self-fertilization or autogamy. Cleistogamous flowers ensure self-pollination. By contrast, when pollens of one plant fertilize eggs of another plant, it is called cross-fertilization or allogamy. In allogamous plants, wind and insects are the major pollination dissemination agents. Wind-pollinated flowers usually lack petals, nectar, and odor but constitute enlarged or feathery stigmas adapted to the interception of airborne pollen. Wind-pollinated flowers produce a large number of anthers.

6.5.5 FLOWER MODIFICATION FOR CROSS-POLLINATION

Various floral parts modification facilitates cross-pollination: (1) The separation of plants by sexes; (2) the separation of staminate and pistillate inflorescences on the same plant; (3) the flower structure prevents self-pollination, such as unequal maturation of stamens and pistils, flower shape and the arrangement of stigma and anthers, and protective film over stigma surface; (4) anthers and pistils are of different lengths in separate plants known as heterostyly; and (5) self-incompatibility mechanisms. Heterostyly flowers are termed pin flowers (styles are long and stamens are short) and thrum flowers (styles are short and stamens are long).

6.6 BARRIERS TO CROSSABILITY IN PLANTS

1. Geographical separation.
2. Timing of flowering due to day length.
3. Species flowering at different length.
4. Varietal differences preventing hybridization.
5. Self- and cross-incompatibility.
6. Separation of sexes.
7. Failure of pollen tube to germinate.
8. Bursting of pollen tubes.
9. Pollen tube growth too slow to reach ovary.
10. Pollen tube reaching to ovary but fertilization does not take place.
11. Fertilization occurs but embryo development is arrested.
12. Embryo develops for few days but viable seeds are not formed.
13. Disharmony in parents produces unviable and sterile plants.

7 Karyotype Analysis

7.1 INTRODUCTION

An individual displays its characteristic $2n$ chromosome number—half maternal and half paternal. A deviation in either direction (+ or −) results in chromosomal imbalance. Cytological techniques, described in Chapter 2, determine the chromosome constitution of an organism and facilitate recognition of the individual chromosomes. Three terms, namely, karyotype, karyogram, and idiogram, are often referred to in the identification of chromosomes. Karyotype is the number, size, and morphology of a chromosome set of a cell in individual or species (Battaglia, 1994). Karyogram is the physical measurement of the chromosomes from a photomicrograph where chromosomes are arranged in descending order (longest to shortest). An idiogram represents a diagrammatic sketch (interpretive drawing) of the karyogram (Figure 7.1). The classification of chromosomes is based on physical characteristics, such as size of chromosomes, features of telomere, position of kinetochore, secondary constriction, size and position of heterochromatic knobs, and relative length of chromosomes (Figure 7.1).

Karyotype analysis is usually based on somatic mitotic metaphase chromosome measurement. This can be estimated with three assumptions:

1. The exact length (image parameter).
2. Chromosomes within a complement exhibit an even condensation–condensation pattern (Fukui and Kakeda, 1994).
3. Metaphase chromosomes (maximum condensation) are obtained after ice cold water pretreatment and aceto-carmine staining (Figure 7.2). However, differential condensation among chromosome arms may be the cause of variation in relative length values.

Occasionally, conventional staining techniques (aceto-carmine or Feulgen) do not distinguish chromosomes of similar morphological features. In this case, Giemsa C- and N-banding techniques have helped identification of these chromosomes (Singh and Tsuchiya, 1981b; Schlegel et al., 1987; Gill et al., 1991). Furthermore, the recent adaptation of the *in situ* hybridization technique for plant chromosomes, particularly for cereals, has also facilitated to distinguish morphologically similar chromosomes. Pachynema chromosomes identify small and similar mitotic metaphase chromosomes such as soybean (Singh and Hymowitz, 1988).

7.2 NOMENCLATURE OF CHROMOSOMES

The kinetochore (centromere) position is a very useful landmark for the morphological identification and nomenclature of chromosomes (Battaglia, 1955; Levan et al., 1964; Matérn and Simak, 1968; Naranjo et al., 1983). Battaglia (1955) disagreed to use the term kinetochore for the centromeric region of the chromosome body. Monocentric chromosomes contained one centromere and nomenclatured chromosomes based on position of centromere in the chromosomes:

1. *Median centromere* (isobrachial chromosomes): Centromere is situated in the middle of the chromosome resulting in an arm ratio of 1:1.
2. *Submedian centromere* (heterobrachial chromosome): Centromere is located near the middle of the chromosome resulting in an arm ratio of more than 1:1 but less than 1:3 (from 1:1 to 1:2.9).

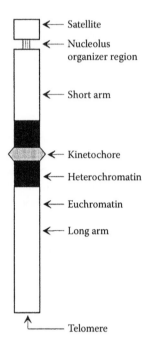

FIGURE 7.1   An idiogram of a metaphase chromosome.

3. *Subterminal centromere* (hyperbrachial chromosome): Centromere is near one extremity of the chromosome resulting in ratio of 1:3 or more.
4. *Terminal centromere* (monobrachial chromosome): Centromere is situated at one extremity of the chromosome resulting in an arm ratio of 0:1.

Battaglia (1955) considered satellite as a part of the chromosome distal to a nucleolar constriction and is universally accepted that for each satellite there is one nucleolus. It is an established fact that a satellite is a spheroidal body and diameter that is either the same or smaller than the diameter of the chromosomes, situated at one extremity connected to the chromosome body by a thin thread

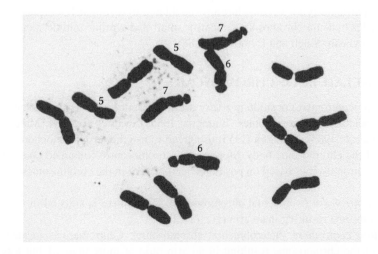

FIGURE 7.2   Somatic mitotic metaphase chromosomes of barley after aceto-carmine staining showing 2n = 14 chromosomes. (Adapted from Singh, R.J. 1974. Ph. D. thesis, Colorado State University, Ft. Collins, CO.)

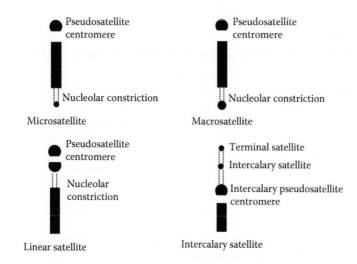

FIGURE 7.3 Diagrammatic sketch of the variability of chromosome morphology based on location of secondary nucleolar constriction. (From Battaglia, E. 1955. *Caryologia* 8: 179–187. With permission.)

(Figure 7.3). The position of the satellite may be terminal (a satellite between its nucleolar constriction and one extremity) or intercalary (a satellite between two nuclear constrictions). Satellite chromosome is designated based on size and location (Figure 7.3):

1. Microsatellite: A spheroidal satellite of small size, that is, having a diameter equal or less than one half the chromosomal diameter.
2. Macrosatellite: A spheroidal satellite of large size, that is, having diameter greater than one half the chromosomal diameter.
3. Linear satellite: A satellite having the shape of a long chromosomal segment.

Levan et al. (1964) discussed thoroughly the nomenclature of chromosomes, and only salient points of their paper will be taken into consideration in this chapter. The relative lengths of the long arm (l) and short arm (s) are shown by the arm ratio ($r = l/s$). Based on arm ratio, Levan et al. (1964) grouped chromosomes in six categories (Table 7.1). Naranjo et al. (1983) proposed a modified version of Levan et al. (1964) nomenclature. They divided chromosomes into 8 equal units. Like Levan et al. (1964), they felt difficulties in assigning chromosomes to a particular type when the arm ratio was exactly 5:3, 6:2, or 7:2, which were three boundary ratios separating the four intermediate groups m (median region), sm (submedian), st (subterminal), and t (terminal). They suggested that the chromosomes with these arm ratios must be classified as m-sm, sm-t, or st-t. In a random sample of chromosomes, only a few will fit into these categories and most of the chromosomes will fall in one of the six types identified in Table 7.1.

7.3 KARYOTYPE ANALYSIS BY MITOTIC METAPHASE CHROMOSOMES

Karyotype analysis has played an important role in the identification and designation of chromosomes in many plant species. Barley (*Hordeum vulgare* L.) is cited here as an example. Barley is a basic diploid, contains $2n = 2x = 14$ chromosomes and the chromosome-linkage group relationship has been established based on cytogenetics and molecular studies.

In barley, chromosome 5 is the smallest chromosome and chromosomes 6 and 7 are the nucleolus organizer chromosomes and are morphologically distinct. Based on conventional staining techniques, chromosomes 1–4 are difficult to distinguish (Figure 7.2). Chromosome designations of Tjio

TABLE 7.1

Nomenclature of Chromosomes Based on Levan et al. (1964)

Centromere Position	Arm Ratio l/s	Chromosome Designations	
Median sensu stricto	1.0	M	
Median region		m	(Metacentric)[a]
Submedian	1.7	sm	(Submetacentric)
Subterminal	3.0	st	(Subtelocentric)
Terminal	7.0	t	(Acrocentric)
Terminal sensu stricto	∞	T	(Telocentric)

Source: Adapted from Levan, A., K. Fredga, and A. A. Sandberg. 1964. *Hereditas* 52: 201–220.

[a] Recommended terms to discard. Since these terms are often used in chromosome nomenclature, I suggest not abandoning them.

and Hagberg (1951) were accepted by barley cytogeneticists until Tuleen (1973) and Künzel (1976) questioned the identity of chromosomes 1, 2, and 3. Both authors observed, based on the results of multiple translocation analysis, that chromosome 2 is the longest chromosome in the barley complement. The application of the Giemsa C-banding (Figure 7.4) and N-banding technique helped to identify all the barley chromosomes while it was not possible by conventional staining techniques (Linde-Laursen, 1975; Singh and Tsuchiya, 1981b, 1982a, b; Kakeda et al., 1991). Furthermore, the combination of aceto-carmine and Giemsa staining technique applied to the same cell (Nakata et al., 1977) helped to construct karyotype analysis better than either of techniques alone.

The salient features of the seven barley chromosomes based on conventional and Giemsa C- and N-banding techniques (Singh and Tsuchiya, 1982a, b), and homoeologous (in parenthesis) groups (Costa et al., 2001) with wheat are described next:

Chromosome 1 (7H)

This is the third longest chromosome and is a metacentric (Figure 7.5; Table 7.2). In four cells (cells 1, 3, 7, 9) out of ten cells measured, the long and short arms of chromosome 1 were equal in

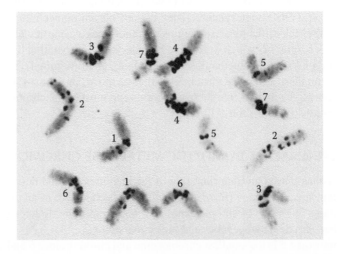

FIGURE 7.4   Somatic mitotic metaphase chromosomes of barley after Giemsa C-banding. (From Singh, R. J. and T. Tsuchiya. 1981b. *Z. Pflanzenzüchtg.* 86: 336–340. With permission.)

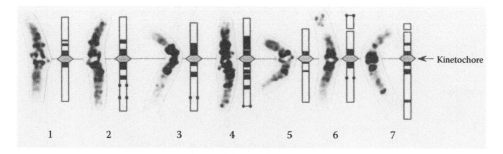

← Kinetochore

1   2   3   4   5   6   7

FIGURE 7.5 Karyogram and idiogram of Giemsa N-banded chromosomes of barley. (Redrawn from Singh, R. J. and T. Tsuchiya. 1982b. *J. Hered.* 73: 227–229. With permission.)

length. In three cases (cells 4, 5, 10), the long arm was longer than the short arm and in three other instances (cells 2, 6, 8) the short arm was longer (Table 7.2). Since both arms were almost equal in measurement, their designation in karyogram and idiogram depended on the morphological effects of telotrisomic plants, gene-chromosome arm relationships, and also on the Giemsa N-banding pattern (Singh and Tsuchiya, 1982a, b).

Chromosome 2 (2H)
This is the longest chromosome among the five nonsatellite chromosomes of the barley complement. It carries its kinetochore at the median (l/s = 1.26) region (Tables 7.2, 7.3; Figure 7.5). Similar results were reported by Tuleen (1973) and Künzel (1976).

Chromosome 3 (3H)
Tjio and Hagberg (1951) identified chromosome 3 as a median (arm ratio = 1.09) chromosome. If chromosomes 1 and 3 of Tjio and Hagberg (1951) are switched, their results will agree with the results presented in Table 7.2. Chromosome 3 showed a dark centromeric band. The band on the short arm appeared as a large block at metaphase. The long arm had a dark interstitial band (close to the kinetochore) and a faint dot on each chromatid in the middle of the long arm (Figure 7.5).

Chromosome 4 (4H)
This chromosome contains its kinetochore at the median region (l/s = 1.21), and was correctly identified in all the studies. Conventional staining techniques do not distinguish chromosome 4 from chromosomes 1, 2, and 3. However, based on Giemsa C- and N-banding techniques, chromosome 4 was easily distinguished from the rest of the chromosomes because it is the most heavily banded in the barley complement; about 48% of the chromosome is heterochromatic. Sometimes it is difficult to locate the centromere position in condensed Giemsa banded metaphase chromosomes (Figure 7.4). However, the appearance of a diamond-shaped centromere position and the use of aceto carmine stained Giesma N-banding technique facilitated the precise localization of the kinetochore (Figure 7.5).

Chromosome 5 (1H)
This chromosome is the shortest among the five nanosatellite chromosomes of barley and has an arm ratio (1.42) similar to chromosome 3 (Tables 7.2, 7.3). It has a centromeric band and an intercalary band on the long arm and a band on the short arm that is darker than those of the long arm (Figure 7.5).

Chromosome 6 (6H)
This chromosome has a larger satellite than chromosome 7 and has an arm ratio of 1.66 (without the satellite). A similar observation was also recorded by other workers (Table 7.3). Chromosome 6 showed a dark centromeric band in both arms, a faint intercalary band on the long arm and a faint dot on each chromatid on the telomere of the satellite (Figure 7.5).

TABLE 7.2
Relative Chromosome Arm Length (%), Mean Arm Ratios of *Hordeum vulgare* cv. Shin Ebisu (SE 16)

Chromosome No. and Relative Arm Length (%)

Cell No.	1 l	1 s	2 l	2 s	3 l	3 s	4 l	4 s	5 l	5 s	6 l	6 s	7 l	7 s	6 Sat	7 Sat
1	7.69	7.69	9.23	6.92	9.23	6.15	7.69	6.92	7.69	5.38	7.69	4.61	8.46	4.61	2.30	1.53
2	7.69	8.09	8.90	7.29	9.72	6.47	8.90	5.67	7.29	4.85	7.29	4.04	9.72	4.04	2.42	1.62
3	6.98	6.98	9.56	8.08	8.46	6.61	8.08	6.99	7.35	5.15	6.61	4.42	9.56	5.14	2.20	1.47
4	8.43	7.23	8.43	7.23	9.04	6.62	7.83	6.63	7.23	5.42	7.23	4.21	9.64	4.82	2.41	1.80
5	8.41	7.96	8.85	7.52	8.85	6.19	7.08	6.19	7.08	5.31	7.08	4.42	10.62	4.42	1.77	1.77
6	7.14	7.93	9.52	7.14	9.52	6.74	7.93	5.95	6.74	4.76	7.14	4.37	9.92	5.16	2.38	1.58
7	7.41	7.41	9.05	6.58	9.47	6.58	7.41	6.58	7.41	4.94	8.23	4.94	9.05	4.94	2.47	1.65
8	6.36	7.07	9.19	7.06	9.89	7.06	7.77	7.36	7.77	5.65	7.42	4.24	9.54	4.59	2.82	2.12
9	7.64	7.64	8.92	7.32	9.55	7.01	7.64	6.37	7.64	5.09	7.01	4.45	8.92	4.77	2.22	1.27
10	8.62	7.66	8.62	6.71	8.62	5.75	7.66	6.70	7.66	5.75	6.70	3.83	10.92	4.78	1.91	1.91
Mean	7.64	7.57	9.03	7.19	9.24	6.52	7.80	6.44	7.39	5.23	7.24	4.36	9.64	4.73	2.29	1.67
95% confidence limit	±0.51	±0.27	±0.26	±0.29	±0.34	±0.28	±0.34	±0.29	±0.23	±0.23	±0.34	±0.21	±0.53	±0.24	±0.21	±0.17
Relative length[b]	65.89	65.28	77.88	62.06	79.68	56.24	67.29	55.53	63.72	45.14	62.46	37.57	83.12	40.78	19.75	14.42
95% confidence limit	±4.39	±2.34	±2.23	±2.56	±2.95	±2.48	±2.93	±2.56	±1.97	±2.04	±2.90	±1.85	±4.58	±2.07	±1.81	±1.48
Arm ratio (l/s)	1.01		1.26		1.42		1.21		1.41		1.66[a]		2.04[a]			

Source: From Singh, R. J. and T. Tsuchiya. 1982a. *Theor. Appl. Genet.* 64: 13–24. With permission.

Note: l = long arm, s = short arm.

[a] Arm ratios do not include satellite.

[b] Based on 100 units for both arms of chromosome 6.

TABLE 7.3
Comparison of Relative Chromosome Arm Length and Arm Ratios in Barley Observed by Several Authors

Authors	1		Arm Ratio	2		Arm Ratio	3		Arm Ratio	4		Arm Ratio	5		Arm Ratio	6		Arm Ratio	7		Arm Ratio
	l	s	l/s	l	s	l/s	l	s	l/s	l	s	l/s	l	s	l/s	l	s	l/s	l	s	l/s
Tjio and Hagberg (1951)	78.4	58.5	1.34	71.7	61.1	1.17	63.7	58.6	1.09	67.2	51.8	1.30	60.7	44.3	1.37	62.1	37.9	1.64	78.5	32.2	2.44
	63.7	58.6	1.09				78.4	58.5	1.34												
Tuleen (1973)	64.5	61.3	1.05	75.0	60.5	1.24	74.5	56.8	1.31												
Künzel (1976)	66.5	64.2	1.04	73.9	63.0	1.17	72.4	58.5	1.24	68.2	58.0	1.18	63.6	44.4	1.43	62.1	36.7	1.69	79.8	37.7	2.12
Singh and Tsuchiya (1982a)	65.9	65.3	1.01	77.9	62.1	1.25	79.7	56.2	1.42	67.3	55.5	1.21	63.7	45.1	1.41	62.5	37.6	1.66	83.1	40.8	2.04

Source: From Singh, R. J. and T. Tsuchiya. 1982a. *Theor. Appl. Genet.* 64: 13–24. With permission.

Chromosome 7 (5H)

This chromosome has the longest long arm in the barley karyotype and carries a submedian kineto-chore (Tables 7.2, 7.3). It showed an equally dense centromeric band at the distal portion of the long arm and a faint intercalary band was also observed in the short arm (Figure 7.5).

The literatures on karyotype studies are voluminous. It has been shown here that combination of several techniques facilitates construction of a karyogram of a crop species better than one technique alone.

7.4 KARYOTYPE ANALYSIS BY PACHYTENE CHROMOSOMES

Sometimes pachytene chromosomes are used when somatic chromosomes do not show distinguishing landmarks. The classical examples for conducting karyotype analysis based on pachytene chromosomes are in maize (McClintock, 1929b), tomato (Barton, 1950), *Brassica* (Röbbelen, 1960), and rice (Khush et al., 1984). Singh and Hymowitz (1988) constructed a karyogram (idiogram) for the soybean (an economically important oil seed crop) pachytene chromosomes for the first time. The soybean contains a high chromosome number ($2n = 40$), small and similar (symmetrical) chromosome size (1.42–2.84 μm), and lack of morphological landmarks. Individual somatic chromosomes of the soybean have not been clearly distinguishable; however, only a pair of satellite chromosomes (arrows) is occasionally visible (Singh et al., 2001).

No reliable techniques are known that consistently produce a high degree of success with squash preparations of soybean meiocytes (Palmer and Kilen, 1987). Singh and Hymowitz (1988) were not possible to trace all 20 pachytene chromosome bivalents in a single cell but the isolation of 1–3 bivalents, observations on euchromatin and heterochromatin differentiation, and chromosome measurements (Figure 7.6) facilitated the identification and construction of pachytene chromosome map (an idiogram) of the soybean for the first time (Figure 7.7). The heterochromatin is distributed proximal to and on either side of the centromeres on the long and short arms, and 6 of the 20 short arms are totally heterochromatic. This latter feature makes soybean pachytene chromosomes rather unique.

7.5 KARYOTYPE ANALYSIS BY FLOW CYTOMETRY

Flow cytogenetics may be defined as the use of flow cytometry to short and analyze individual chromosomes and physical mapping of genes of economic importance using FISH technology. However, this technique has not been universally successful in plants because of lack of quality chromosomes and an inability to resolve a single chromosome on flow karyotype.

Lucretti et al. (1993) invented a procedure to sort only metacentric chromosomes of *Vicia faba* by flow cytometry. They located rDNA locus by FISH. Preparation of high quality chromosome suspensions is a prerequisite for successful chromosome sorting and karyotyping. They listed several factors for poor quality chromosomes in suspensions, such as the splitting of metaphase chromosomes into chromatids, chromosome breakage, chromosome clumping, presence of interphase nuclei, and presence of cellular and chromosomal debris.

Flow cytometry may be an extremely valuable tool if we can distinguish aneuploid from diploid plants based on relative surplus or deficit of DNA content. With this view in consideration, Samoylova et al. (1996) examined relative surplus of DNA content in *Arabidopsis* primary and telotrisomics in interphase nuclei measured by flow cytometer to distinguish diploid (wild-type) from trisomic plants. They measured differences in nuclear fluorescence intensity between diploid and trisomics. The relative surplus of genomic DNA recorded by primary and telotisomics was attributed to the extra chromosome (Figure 7.8). However, flow karyotype contradicts cytological observation. Cytologically chromosomes 5 and 3 are larger than chromosomes 4 and 2 (the smallest) but

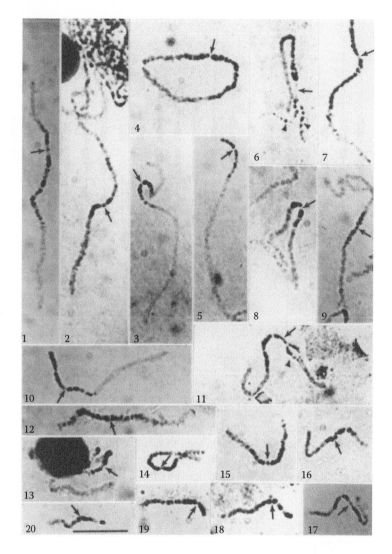

FIGURE 7.6 Photomicrographs of the pachytene chromosome complement of *Glycine max* x *G. soja* F₁ hybrid. Each figure shows a different chromosome. For example; 1, chromosomes 1 to 20, chromosome 20. Arrows indicate centromere location. (From Singh, R. J. and T. Hymowitz. 1988. *Theor. Appl. Genet.* 76: 705–711. With permission.)

relative DNA content (%) was found lesser for chromosomes 5 and 3 than 4 and 2 (Figure 7.8). At this stage, we may conclude that flow karyotyping for plants needs perfection.

7.6 KARYOTYPE ANALYSIS BY IMAGE ANALYSIS

The beginning of chromosome image analysis goes back to early 1980s when computer systems were at the cradle stage to handle huge digital data of images. There were only a few expensive image analyzing systems available and imaging techniques suitable for plant chromosomes analysis were under development. Some trial studies were carried out using human chromosomes especially in the field of semi-automatic identification of human chromosomes (Casperson et al., 1971; Castleman and Melnyk, 1976; Lundsteem et al., 1980).

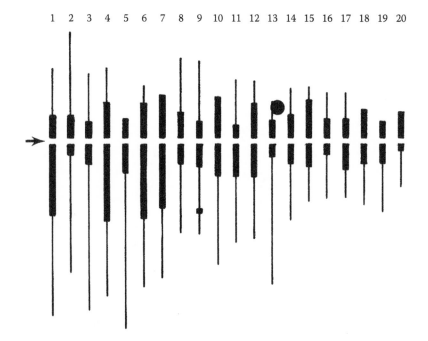

FIGURE 7.7   Proposed idiogram, based on Figure 7.6, of the pachytene chromosomes of the soybean. Arrow indicates centromere location. (From Singh, R. J. and T. Hymowitz. 1988. *Theor. Appl. Genet.* 76: 705–711. With permission.)

In 1985, the first comprehensive chromosome image analyzing system (CHIAS) with software fulfilling the basic requirements of cytologists and cytogeneticists was developed (Fukui, 1986). Then further development of imaging methods such as quantifying chromosome morphology and its band patterns in barley (Fukui and Kakeda, 1990), quantifying uneven condensation patterns appearing at the prometaphase chromosomes in rice (Fukui and Iijima, 1991), and simulating human

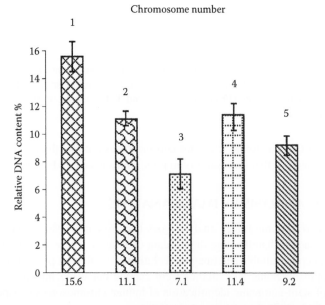

FIGURE 7.8   Diagrammatic representation of the flow karyotype of *Arabidopsis thaliana* interphase chromosomes. (From Samoylova, T. I., A. Meister, and S. Miséra. 1996. *Plant J.* 10: 949–954. With permission.)

vision for identifying and quantifying chromosome band patterns in *Crepis* (Fukui and Yamisugi, 1995) followed. Imaging method for quantification of pachytene chromomeres was soon released.

No personal computers with enough imaging capability allow image analysis in every cytology and cytogenetics laboratory. The basic points that should be in mind when the application of imaging methods are as follows:

1. Importance of the quality of chromosome images. No imaging method can create new information that is not originally included in the chromosome images.
2. The information of the images is basically reduced by each application of image manipulation. Imaging methods present the essence of the chromosome image of the original image as visible and thus in a perceptible way.
3. Imaging methods can present the image information by numerical data.

The standard and basic procedures for image analyses and the manuals (Kato et al., 1997) can be obtained either by written form or via the Internet (http://mail.bio.eng.osaka-u.ac.jp/cell/).

7.7   CHROMOSOME NUMBER OF ECONOMICALLY IMPORTANT PLANTS

The comprehension of chromosome number of plants is important for taxonomists, cytogeneticists, evolutionists, plant breeders, and molecular geneticists. Chromosome number of plants is compiled in two volumes in *Chromosome Atlas of Flowering Plants* (Darlington and Wylie, 1955), in several journals, such as *Taxon, Rhodora, American Journal of Botany, International Organization of Plant Biosystematists Newsletter (ISPOB), Annals of the Missouri Botanical Garden, New Zealand Journal of Botany,* and *Systematic Botany.* Chromosome number of a few plants is listed in Appendix B.5.

8 Chromosomal Aberrations

8.1 STRUCTURAL CHROMOSOME CHANGES

8.1.1 DEFICIENCIES

8.1.1.1 Introduction

The loss of a segment from a normal chromosome is known as deficiency (Df). The term deletion is often used as a synonym of deficiency. Deficiencies indicate any chromosomal loss and the term deletions should be confined to a deficiency involving an internal region of a chromosome (McClintock, 1931). Thus, deficiency may be either intercalary (Figure 8.1a) or terminal (Figure 8.1b). The deficiency method, which is also known as the pseudo-dominant method, has been effectively utilized for locating genes in the chromosomes of maize (McClintock, 1931, 1941a; Chao et al., 1996) and tomato (Khush and Rick, 1967b, 1968a).

8.1.1.2 Origin and Identification of Deficiencies

Induced deficiencies may be generated by X-raying pollen carrying normal (wild type) alleles and applying the pollen to the stigma of female flowers. The female parent carries recessive alleles at loci in the genome that will hopefully be represented in segments of induced deficiency. A majority of the plants show dominant phenotype but occasionally a few plants with the homozygous recessive phenotype appear in the progenies. Cytological examination of these plants with recessive phenotype at pachynema often reveals normal association along the entire length of each chromosome except in one region of one chromosome. This observation is a strong indication that the recessive allele may be carried in hemizygous condition. A loop is generally observed if the deficiency is long enough and is located in an interstitial region. A terminal deficiency results in an unpaired end region.

X-ray-induced deficiencies occur at nonrandom positions in chromosomes. Two reports provided conflicting results in tomato. Gottschalk (1951) found 73.2% centromeric breaks, 18.4% breaks in heterochromatin and 8.4% in the euchromatin. Khush and Rick (1968a) recorded the highest frequency (60%) of breaks in the heterochromatin. The frequency of breaks in kinetochores was intermediate (20%) and only 15% of breaks occurred in euchromatin.

Fast neutron also induces breaks in chromosomes in a nonrandom fashion (Khush and Rick, 1968a). However, fast neutron was found to be much more efficient than X-rays in inducing breaks in euchromatin in tomato. Both types of radiation produced breaks preferentially in heterochromatin. A terminal deficiency with one break in heterochromatin was the most frequent; a terminal deficiency in euchromatin was not observed. Terminal deficiencies were observed for only those arms whose losses were tolerated by gametophytes or sporophytes. Broken ends in a kinetochore or euchromatin did not heal without reuniting with other broken ends. The most frequent type of interstitial deficiency obtained in tomato was the type that resulted from both breaks in the euchromatin. All such breaks were obtained from fast neutron treatment of pollen, and none was induced by X-ray treatment.

In maize the *r-X1* deficiency, a small intercalary deficiency located in the long arm of chromosome 10 that includes the *R1* locus, induces terminal deficiencies, monosomics and trisomics, and nondisjunction of chromosomes in the early embryos (Weber, 1983; Lin et al., 1990; Weber and Chao, 1994).

Several *Aegilops* species such as *Ae. caudata* ($2n = 14$; CC), *Ae. cylindrica* ($2n = 28$; CCDD), *Ae. geniculata* ($2n = 28$; UUMM), *Ae. longissima* ($2n = 14$; $S^l S^l$), *Ae. sharonensis* ($2n = 14$, $S^{sh} S^{sh}$), *Ae. speltoides* ($2n = 14$; SS), and *Ae. triuncialis* ($2n = 28$, UUCC) contain gametocidal gene (*Gc*) that induces chromosome aberrations in common Chinese Spring wheat ($2n = 42$, AABBDD) (Endo, 1988, 1990; Endo and Mukai, 1988; Kota and Dvořák, 1988; Ogihara et al., 1994;

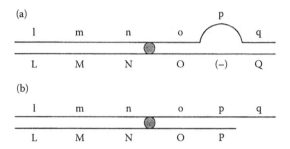

FIGURE 8.1 Diagrammatic sketch showing types of deficiencies. (a) Intercalary; (b) terminal.

Endo and Gill, 1996; Nasuda et al., 1998). The *Gc* gene in *Aegilops* species was identified during the production of alloplasmic chromosome alien addition lines between crosses of common wheat and *Aegilops* species (Endo, 1990). Taxonomic nomenclature of *Aegilops* species by van Slageren is used as suggested by Dr. B.S. Gill (personal communication; Table 8.2).

The following procedures (Figure 8.2) generated 436 deficiencies in Chinese Spring wheat by using gametocidal gene (Endo, 1990; Endo and Gill, 1996):

1. Back cross the monosomic alien addition and translocation lines as female to euploid Chinese Spring wheat as a male.
2. Examine the chromosome constitutions of the progeny by C-banding technique.
3. Select plants with deficiency or deficiencies, and without alien chromosome.
4. Self-pollinate deficiency heterozygote plants. Screen the self-progeny cytologically for deficiency homozygous plants with the least degree of aberrations in the other chromosomes. For example, translocations and aneuploidy.
5. In case deficiency homozygote is not found among 10 or more offsprings from the structural heterozygotes, cross the deficiency heterozygote as female with an appropriate nullisomic-tetrasomic, or an appropriate ditelosomic of Chinese Spring to create the deficiency hemizygotes in the F_1 progeny.
6. Screen F_2 progeny for deficiency homozygote.

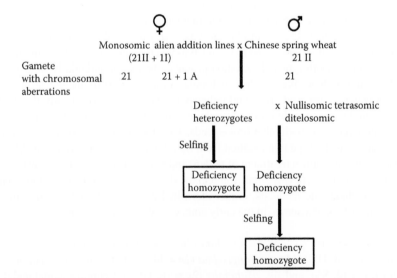

FIGURE 8.2 Diagrammatic sketch showing the production of the deficiency stocks in common wheat. The "A" represents the *Aegilops* chromosome causing chromosomal aberrations.

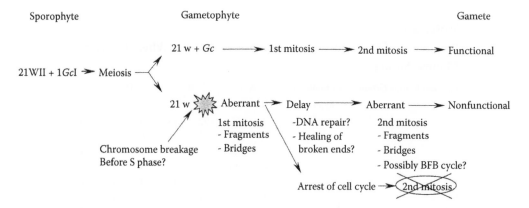

FIGURE 8.3 Diagrammatic scheme showing the mode of action of *Gc* gene (gametocidal chromosome) in Chinese Spring wheat background causing semi-sterility. (Redrawn from Nasuda, S., B. Friebe, and B. S. Gill. 1998. *Genetics* 149: 1115–1124. With permission.)

Endo (1988) observed deficiencies and translocations in almost half the progeny of a monosomic addition line of common wheat which carried a chromosome of *Ae. cylindrica*. Chromosome breaks induced by *Gc* gene (gametocidal) occur in various regions of all the wheat chromosomes, and also in the *Ae. cylindrica* chromosomes (Endo and Gill, 1996; Tsujimoto et al., 2001). The frequency of chromosome structural changes was far less both in the self-progeny of disomic addition plants and in the F_1 monosomic alien addition line progeny derived from reciprocal crosses with common wheat. Nasuda et al. (1998) observed that *Gc* gene-induced chromosomal breakage probably occurs prior to first pollen mitosis and during first and second pollen mitosis (Figure 8.3). The schematic diagram of Gc action in common wheat carrying different types of GC chromosomes 3C, 2S, and 4S is shown in Figure 8.4.

Kota and Dvořák (1988) found deficiencies, translocations, ring chromosomes, dicentric chromosomes and paracentric inversions during the production of a substitution of chromosome $6B^S$ from *Ae. speltoides* for chromosome 6B of Chinese Spring wheat; 49 of the 138 plants contained chromosome aberrations. Chromosome rearrangements were recorded in both wheat and *Ae. speltoides* chromosomes. The B genome chromosomes showed the highest frequency of structural changes, followed by A genome and D genome chromosomes. The chromosome aberrations were nonrandom in the B genome. Chromosomes 1B and 5B were the most frequently involved. Chromosome rearrangements were also frequent for the $6B^S$ chromosome of *Ae. speltoides*. Chromosome aberrations are induced prior to syngamy and are seldom transmitted through the gametophyte if the parent is

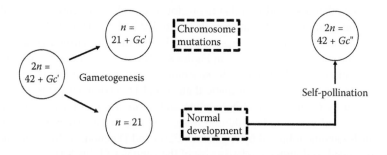

FIGURE 8.4 Diagrammatic scheme showing the mode of action of *Gc* stands for "gametocidal chromosome." (Redrawn from Endo, T. R. 2015. *Advances in wheat Genetics: From Genome to Field*, Y. Ogihara et al. (eds.); DOI: 10.1007/978-4-431-55675-6_8.)

TABLE 8.1

Frequency of Deficiencies in Chromosome Arms in Wheat Cultivar Chinese Spring

Homoeologous Group	Genome →	A		B		D		
	Arms →	S	L	S	L	S	L	Total
1		5	6	22	18	5	8	64
2		9	6	13	11	6	12	57
3		4	8	10	12	9	3	46
4		4	13	9	14	5	15	60
5		11	23	9	18	4	12	77
6		5	8	11	15	7	11	57
7		13	25	6	16	6	9	75
Total		**140 (32%)**		**184 (42%)**		**112 (26%)**		**436**

Source: From Endo, T. R. and B. S. Gill. 1996. *J. Hered.* 87: 295–307. With permission.

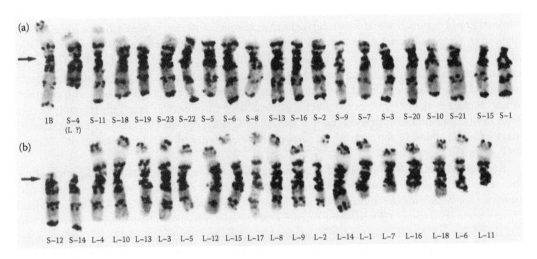

FIGURE 8.5 A series of deficiencies in chromosome 1B (normal chromosome 1B extreme left) detected by Giemsa C-banding technique. The horizontal line represents the kinetochore (Reproduced from Endo, T. R. and B. S. Gill. 1996. *J. Hered.* 87: 295–307. With permission.)

used as a male. The other assumption is that the factor responsible for genomic instability is active only in the maternal germ line, different from that recorded by Endo (1988).

The number of deficiencies predominates (42%) for B-genome chromosomes (Table 8.1). A series of deficiencies in chromosome 1B is shown in Figure 8.5. Breakage occurs generally adjacent to the heterochromatic regions. The B genome chromosomes are more vulnerable to breakage because these chromosomes are more heterochromatic than A and D genome chromosomes. Deficiencies are being identified by Giemsa C- and N-banding and by *in situ* hybridization techniques or by GISH in wheat–rye and wheat-barley addition lines. The rate of deficiency was highest (42%) in chromosomes of B genome followed by A genome (32%) and D genome (26%). The distribution of break points correlated well with the relative size of the genomes (Table 8.1).

In another study, Tsujimoto et al. (2001) induced a total of 128 chromosomal aberrations (terminal deficiencies, 110; translocations, 7; dicentrics, 8; insertions, 1; highly chimeric, 2) in 1B chromosome of common wheat by *Gc* gene. These lines were produced by crossing a monosomic alien

TABLE 8.2
Frequency of Plants with Aberrant Barley Chromosomes

Barley Chromosome	Selfed Progeny		Back-Cross Progeny	
Number of Plants	Number of Plants Examined (%)	Number of Plants with Aberrations	Number of Plants Examined (%)	Number of Plants with Aberrations
2H	15	1 (6.7)	22	3 (13.6)
3H	35	2 (5.7)	60	9 (15.0)
4H	48	16 (33.3)	20	3 (15.0)
5H	7	1 (14.3)	18	3 (16.7)
6H	43	8 (18.6)	51	7 (16.7)
7H	28	6 (21.4)	156	26 (16.7)
Total	176	34 (19.3)	327	51 (15.6)

Source: From Shi, F. and T. R. Endo. 1999. *Genes Genet. Syst.* 74: 49–54. With permission.

chromosome addition line of Chinese Spring carrying chromosome 2C of *Ae. cylindrica* (abbreviated as CS +2C) with nullisomic 1B-tetrasomic 1D.

Barley and rye are diploid ($2n = 14$) and unable to tolerate deficiencies. Production of deficiencies in wheat-barley and wheat–rye addition lines by *Gc* factor will enhance the gene mapping in barley and rye. By using the *Gc* gene from *Ae. cylindrica*, Shi and Endo (1999) produced chromosome aberrations for barley chromosomes 2H, 3H, 4H, 5H, and 7H (Table 8.2). They identified a total of 31 deficiencies, 26 translocations, and 2 isochromosomes. Serizawa et al. (2001) produced 7 deficiencies and 15 translocations in barley chromosome 7H. The breakpoints of the deficiencies and translocations by N-banding, FISH, and GISH suggest that they are localized in general to the distal and proximal regions of barley chromosomes.

Friebe et al. (2000) produced 56 deficiencies for rye chromosomes by *Gc* gene. They crossed seven disomic alien addition lines (DAALs) of CS wheat-"Imperial" rye ($2n = 44$; 21W II + 1R II) as a female with DAALs of CS wheat–*Ae. cylindrica* ($2n = 44$; 21W II + 2C II) as male. The double monosomic alien addition line ($2n = 44$; 21W II + 1R I + 1C I) is expected to produce four types of gametes: 21W, 21W + R, 21W + 2C, and 21W + R + C. Since univalents in wheat are expected to be eliminated in about 3/4 of the gametes, the expected frequencies are in the proportion of 9/16, 3/16, 3/16, and 1/16, respectively. It has been established that chromosome containing *Gc* gene induces chromosome structural changes in gametes without *Gc* chromosome. The target gametes for producing the deficiencies in rye chromosome in this case are 3/16 which is considerably low. To enhance the screening of higher frequencies of deficiencies, they backcrossed the double MAALs ($2n = 21$ W II + R I + 2C I) with the corresponding wheat–rye DAALs and screened the BC_1 progenies cytologically to identify disomic plants for a given rye chromosome and monosomic for the gametocidal chromosome ($2n = 45$, 21 II + R II + 2C I). All gametes of these plants have the target rye chromosome and 3/4 of the gametes are expected to be without 2C (21 W + R) and, thus are not subjected to chromosome aberrations. These plants were either backcrossed with the corresponding wheat–rye addition line or selfed. The derived lines were screened by C-banding for chromosomal structural changes in the rye chromosomes. By using this procedure, 33 deficiencies, 22 wheat–rye dicentrics, and 7 wheat–rye and rye–rye translocations were recovered.

8.1.1.3 Meiotic Chromosome Pairing in Wheat Deficiency Line

The degree of meiotic chromosome association at metaphase-I between homologous chromosome arms of wheat is drastically influenced by a deficiency in one arm. Curtis and Lukaszewaski (1991) examined meiotic metaphase-I pairing in 4AL arm containing relative length of deficiency from 6% to 50%. The pairing frequency continued to decline as the deficiency length increased (Table 8.3).

TABLE 8.3

Pairing Frequencies at Metaphase-I of Deficient Chromosomes 4AL with Complete Homologous and Telosome Chromosomes

| | | Pairing Frequency (%) of Deficient Arm with | | | |
| | | Complete Homologue | | Complete Telosome | |
Chromosome	Relative Length of Deficiency	S	L	S	L
Df4A06L	6	90.0	32.0	–	–
Df4A08L	8	84.0	64.0	85.2	29.8
Df4A11L	11	89.5	50.0	89.0	16.4
Df4A17L	17	70.6	11.8	76.1	5.4
Df4A23L	23	82.7	6.9	70.2	3.8
Df4A34L	34	83.0	3.0	68.6	0.0
Df4A36L	36	87.0	0.0	77.8	0.0
Df4A39L	39	–	–	41.7	0.0
Df4A50L	50	84.0	0.0	–	–

Source: From Curtis, C. A. and A. J. Lukaszewaski. 1991. *Genome* 34: 553–560. With permission.
Note: S, short arm; L, long arm.

Pairing frequency in 4AL deficient for 34% reduced to 3% and was absent (0%) when deficiency reached to 36%. The lack of pairing between deficient arm and its normal homologue is probably related to an inability of homologues of unequal length to initiate pairing and this may result in no recovery of recombinant chromosomes (Curtis and Lukaszewaski, 1991). Hohmann et al. (1995) recorded pairing reduction of 60% in the deficiency of the most distal 1% of chromosome arm 7AL.

Deficiencies in wheat, rye, and barley chromosomes induced by *Gc* gene mechanism are mostly terminal. Progeny of deficiency stocks of wheat breed true suggesting that broken chromosomal ends either heal after breakage by the synthesis of telomeric sequences or undergo fusions to produce dicentric or translocated chromosomes. Dicentric chromosomes undergoing breakage-fusion-bridge (BFB) cycles in first few divisions of sporophyte are particularly healed before germ line differentiation and ends are totally healed in the ensuing gametophytic stage (Friebe et al., 2001).

8.1.1.4   Transmission of Deficiencies

Deficiencies in maize were transmitted through eggs, but not through pollen (McClintock 1938a; Rhoades and Dempsey, 1973). The morphologically and cytologically detectable deficiencies were rarely transmitted to the next generation in tomato. Small cytologically undetectable deficiencies were transmitted through male and female normally. In the case of a deficiency of the entirely heterochromatic 2S arm, only one individual of 1416 plants examined showed transmission of the deficiency. No tomato euchromatic deficiencies were known to be transmitted unless they involved heterochromatin (Khush and Rick, 1967b). Even small deficiencies are lethal during gametophyte development in *Vicia faba* (Schubert and Rieger, 1990).

A total of 338 deficiencies homozygous lines of common wheat (289 single, 39 double. 7 triple, and 3 quadruple—deficiency lines) are maintained as they were more or less fertile (B. S. Gill, personal communication). About 67% of the wheat deficiencies are homozygous and they transmit normally (Endo and Gill, 1996). Some deficiency homozygotes either are sterile or could not be obtained at all. These lines have to be maintained as heterozygotes as they behave just like a monosomic chromosome and the progeny must be screened cytologically for homozygotes. For example, all homozygous plants for 2AS deficiencies were highly male and female sterile because all deficiencies were larger than that of 2AS-5. All 4BS deficiency homozygotes were totally male sterile. These stocks are being maintained producing new disomic stocks for the 2AS or 2BS deficiency chromosomes and monosomic for the 2A or 4B short-arm telocentric chromosomes. They recovered deficiency homozygotes at a high rate in the progeny of these stocks. Most of the barley deficiencies in wheat are heterozygous and transmitted to offspring like monosomic addition in wheat. Barley deficiencies and translocations in wheat-barley addition lines are differentiated by GISH (Shi and Endo, 1999).

8.1.1.5 Genetic Studies

Deficiencies have been utilized very effectively to locate marker genes by the pseudo-dominant technique in maize (McClintock, 1941a), tomato (Rick and Khush, 1961; Khush and Rick, 1967b, 1968a), and wheat (Gill et al., 1996; Sutka et al., 1999; Tsujimoto et al., 2001). Khush and Rick (1968a) located 35 marker genes belonging to 18 arms of the 12 tomato chromosomes by the induced-deficiency technique. In addition to gene location, induced deficiencies helped better the understanding of tomato genomes in ways briefly listed below:

1. *Association of a linkage map with its respective chromosome*: Rick and Khush (1961) associated the unlocated *a-hl* linkage group with chromosome 11 by induced deficiencies.
2. *Identification of unmarked chromosome*: By induced deficiencies, Khush and Rick (1966) associated the marker *alb* with chromosome 12.
3. *Location of a marker with a particular arm*: Numerous induced deficiencies were used to assign markers to particular arms of the tomato chromosomes.
4. *Location of markers to a specific region of a chromosome*: Several markers, such as *sy*, *ru*, *bls*, *sf*, *ra*, *var*, *ag*, and *t^v*, were located to a specific segment of a tomato chromosome by induced deficiencies.
5. *Assignment of unlocated markers to respective chromosomes*: Induced deficiencies helped to assign unlocated markers *clau*, *alb*, *lut*, and *fd* to their respective chromosomes.
6. *Orientation of the linkage maps*: Proper arm assignments of linkage maps of chromosomes 2, 4, 6, 8, and 11 were made.
7. *Location of centromere positions*: Induced deficiencies facilitated the precise locations of the centromeres of nine chromosomes and the approximation of the other three chromosomes of tomato.
8. *Production of aneuploids*: Induced deficiencies were utilized to generate tertiary trisomics, secondary trisomics, and telotrisomics.

Deficiency stocks are an excellent tool for physical mapping of molecular and qualitative trait loci in wheat (Gill et al., 1996; Sutka et al., 1999; Tsujimoto et al., 2001). Gill et al. (1996) developed physical maps by locating 80 DNA and 2 morphological markers using 65 deficiency lines for homoeologous group 5 chromosomes. The maps were constructed for chromosome 5B in wheat and 5D in *Aegilops tauschii*. They recorded suppression of recombination in the centromeric regions and was prominent in the gene-rich regions.

8.1.1.6 Use of Deficiencies in Hybrid Maize Breeding

Patterson (1973) proposed a system for producing hybrid maize by utilizing nuclear male sterility genes (*ms*) in conjunction with chromosome deficiencies in the background of a normal (nonsterile) cytoplasm. Male sterile stocks are normally maintained by *ms/ms* (female) × *ms/+* (male) crosses. The progeny are expected to segregate 1/2 male sterile (*ms/ms*):1/2 male fertile (*ms/+*) (Figure 8.6a). On the other hand, all plants from this testcross are expected to be sterile if pollen grains carrying the + allele are nonfunctional in fertilization. Detasseling could be greatly reduced or eliminated in female rows in hybrid production if this result could be approximated. As shown in Figure 8.6b, this differential transmission of *ms* locus alleles might be achieved by a male fertile allele carried immediately proximal to a terminal deficiency that is female transmissible, but not male-transmitted. Differential transmission could equally well be achieved by a male fertile allele carried immediately adjacent to an

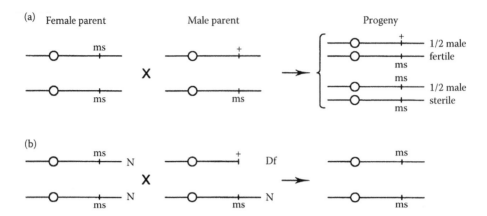

FIGURE 8.6 Use of chromosome deficiencies to produce male sterile progeny in maize. (a) Expected segregation in male sterile (female) × male fertile (male); (b) A proposed method to produce male sterile progeny in maize. (Redrawn from Patterson, E. B. 1973. *Proc. 7th Meeting Maize and Sorghum Sect.* Secretariat, Institute for Breeding and Production of Field Crops, Zagreb, Yugoslavia.)

internal deficiency showing the same transmission characteristics. In maize, however, very few simple deficiencies showing the required transmission characteristics have been recognized and saved.

Patterson (1973) demonstrated that the required level of differential transmission of *ms* alleles could be accomplished by utilizing female-transmissible duplicate-deficient (Dp-Df) chromosome complements derived by adjacent-1 disjunction from some heterozygous reciprocal translocations. In suitable Dp-Df complements, the deficient segment is closely linked in coupling phase to the + allele of an *ms* gene in *ms/ms* (female) by Dp-Df +/*ms* or Dp-Df + *ms/ms* (male) crosses. As shown in Figure 8.7, there are two options with the same chromosome structures. In Figure 8.7 (a) is shown the wild type allele (+) of a chromosome 6 male sterile gene located just proximal to the interchange point in the 6^9 chromosome; it is thus located just proximal to a terminal deficiency for the tip of the long arm of chromosome 6. In Figure 8.7 (b) is shown the alternative option of using the wild type allele of a chromosome 9 male sterile gene located just distal to the interchange point on the 6^9 chromosome; in this position the locus is part of a chromosome 9 segment that becomes triplicated in the derived Dp-Df plant. Suitably-marked Dp-Df plants will have received the recessive *ms* allele from the male parent. The interchange point on the 6^9 chromosome represents the proximal terminus of a deficiency for the tip of the long arm of chromosome 6. A locus on either side of that interchange point is thus linked genetically and physically to the same terminal deficiency. Since the 6^9 chromosome cannot be male-transmitted in the absence of the 9^6 chromosome, a wild type allele of an *ms* gene carried on it cannot be transmitted to progeny except if it is transferred to a normally arranged homologue as a result of crossing over between the locus and the adjacent interchange point.

In commercial use, Dp-Df stocks would be derived in inbred lines destined to be used as female parents, or components of female parents, in production of hybrid maize. In initial crosses, suitably marked Dp-Df plants like those shown in Figure 8.7 are self-pollinated to propagate plants of the parental Dp-Df constitution and are crossed as male parents to male sterile plants of the same line. The latter cross confirms the constitution of the Dp-Df male parent and at the same time produces seeds that will yield virtually all male sterile plants for use in female rows of foundation or production fields. In routine foundation field practice, male fertile Dp-Df plant produced by self or sib open pollination would serve as pollinators in male rows. Female rows would consist of male sterile plants; seed produced on these plants is suitable, in turn, for planting additional male sterile rows in foundation or production fields. Male rows in production fields would consist of standard inbred line, since such lines normally carry wild type alleles of recessive nuclear male sterile genes and thus function as natural restorers of male fertility in hybrid seed sold to farmers.

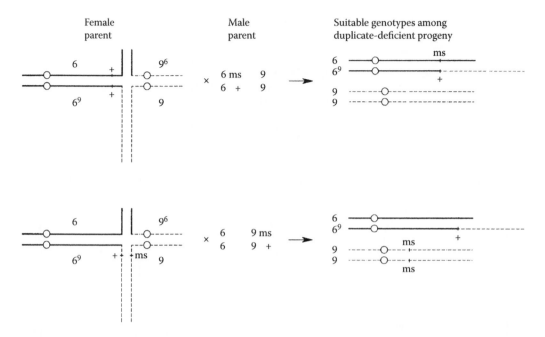

FIGURE 8.7 The derivation of suitably marked duplicate-deficient progeny from a reciprocal translocation in maize. (Redrawn from Patterson, E. B. 1973. *Proc. 7th Meeting Maize and Sorghum Sect.* Secretariat, Institute for Breeding and Production of Field Crops, Zagreb, Yugoslavia.)

Four different Dp-Df complements were identified (Patterson, 1973) that when used in suitable combinations with three different male sterile genes in testcrosses each gave progenies in which only 1 or 2 plants per 1000 were male fertile. Commercial procedures require that Dp-Df plants be used as pollinators in foundation fields. Thus far, Dp-Df complements showing strong linkage with suitable *ms* loci have not met maize industry requirements for agronomic performance with respect to pollen and seed production.

8.1.1.7 Ring Chromosomes

Ring chromosomes have been found in several plant species including *Zea mays*, *Nicotiana tabacum*, *Antirrhinum majus*, *Petunia hybrida* and *Hordeum vulgare*, as well as in *Drosophila* and man. Cytological behavior of ring chromosomes is unique. McClintock (1931, 1938b, 1941a,b) studied thoroughly the mitotic and meiotic behaviors of ring chromosomes in maize and developed the breakage-fusion-bridge (BFB) cycle hypothesis to explain changes in chromosome sizes (duplications and deficiencies). She also followed these chromosome aberrations in genetic studies (Figure 8.8). The breakage-fusion-bridge cycle phenomenon has been confirmed by Morgan (1933) and Braver and Blount (1950) in *Drosophila*, by Stino (1940) in tobacco, Schwartz (1953a,b) and Fabergé (1958) in maize, Michaelis (1959) in *Antirrhinum majus*, Tsunewaki (1959) in wheat-*Agropyron* hybrids, Frost et al. (1959) in *Matthiola incana*, Maizonnier and Cornu (1979) in *Petunia hybrida*, and Singh and Tsuchiya (1981e) in barley. Most of the ring chromosomes studied were in the disomic condition where a ring chromosome compensated for a deficiency in the standard homologues.

Ring chromosomes may occur spontaneously due to meiotic irregularities or they may be induced by X-ray or gamma ray treatment. Only ring chromosomes are produced from the ring chromosomes. According to McClintock (1941a) ring chromosomes originate from a univalent chromosome by nonhomologous chromosome synapsis of two arms followed by crossing over. In barley, it is believed that the ring chromosomes obtained in the progenies of Triplo 7 (semi-erect) may have originated in a similar fashion (Singh and Tsuchiya, 1981e).

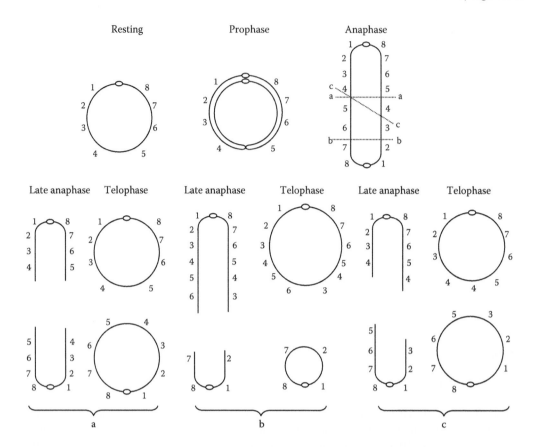

FIGURE 8.8 Origin of ring chromosomes of various sizes by breakage-fusion-bridge cycle. Breakage points (a–a; b–b; c–c), shown at anaphase-I, in a dicentric bridge produce ring chromosomes of various sizes. (Redrawn from McClintock, B. 1941a. *Genetics* 26: 542–571. With permission.)

Various sizes of ring chromosomes have been observed in disomic wheat–rye addition lines by BFB cycle (Lukaszewaski, 1995). However, this study could not discover mechanism like transposable element in wheat as was discovered in maize.

Ring chromosomes are not stable during cell division and often are eliminated during cell divisions. Their numbers and sizes are changed in successive cell cycles depending on the position of the breakage in dicentric double-sized ring chromosomes (Figure 8.8). During mitosis, small ring chromosomes exhibit the following features:

1. Reduced frequency with which double-sized or interlocking rings originate.
2. Frequent loss of ring chromosomes from the nuclei.
3. Changes in size of the ring chromosomes less frequent.
4. Increase in the number of rings.

McClintock (1938b) utilized a ring chromosome to locate a marker gene in a particular region of chromosome 5. The ring chromosome carried the locus *Bml* (allele bml shows brown midrib when homozygous or hemizygous). The rod chromosomes were of three types: both copies of chromosomes lacked the *Bml* locus, one copy was deficient for the locus while the other carried bml, or both carried *bml*. The plants expressed variegation for *Bml* and *bml* expression through the frequent loss of the ring chromosomes from somatic tissues.

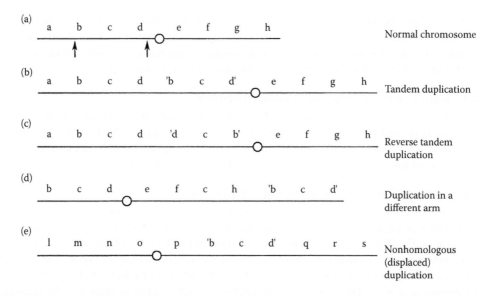

FIGURE 8.9 (a to e) Diagrammatic sketch of the types of duplications. Arrows show a segment of chromosome duplicated in several fashions.

8.1.2 DUPLICATIONS

8.1.2.1 Introduction

An extra piece of chromosome segment attached to either the same homologous chromosome or transposed to one of the nonhomologous members of the genome is termed duplication. Based on transposition of the segment, Burnham (1962) classified duplications shown in Figure 8.9 as follows:

a. Normal chromosome (Figure 8.9a)
b. Tandem duplication (Figure 8.9b)
c. Reverse tandem duplication (Figure 8.9c)
d. Duplication in a different arm (Figure 8.9d)
e. Non-homologous (displaced) duplication (Figure 8.9e)

Duplications are more frequent and less lethal to organisms than are deficiencies. Combined duplications and deficiencies, abbreviated as (Dp-Df) for duplicate and deficient, produce changes in the amount of genetic material in spores, pollen, gametes, seeds, and plants.

8.1.2.2 Origin of Duplications and Deficiencies

1. Duplications occur in nature and are also produced experimentally. In several diploid crops, duplicate genes, determined by the F_2 segregation (15:1 or 9:7), have been recorded. Furthermore, the formation of occasional rod- or ring-shaped bivalents at meiosis in haploid suggests the presence of duplications (Burnham, 1962).
2. Generally, ionizing types of radiation break chromosomes, producing duplications and deficiencies.
3. Viable Dp-Dfs are isolated in the progenies of chromosomal interchanges in maize (Gopinath and Burnham, 1956; Phillips et al., 1971; Patterson, 1978; Carlson, 1983), barley (Hagberg, 1962), and cotton (Menzel and Brown, 1952, 1978; Contolini and Menzel, 1987; Menzel and Dougherty, 1987).
4. Dp-Dfs are produced from breakage of dicentric bridges at meiotic anaphases-I and -II, and spore mitotic anaphase (Figure 8.10) in the progenies of inversion heterozygotes

(McClintock, 1938a; Rhoades and Dempsey, 1953). The observed pseudo-alleles near the end of short arm of the chromosome 9 in maize suggested to McClintock (1941c) the presence of a series of similar genes which arose through duplications. A similar situation can originate with regard to duplications from breakage-fusion-bridge cycles.

8.1.2.3   Identification of Duplications and Deficiencies

In favorable species Dp-Df plants may be identified cytologically at pachynema by an unpaired segment that is present in single copy and by pairing patterns displayed by a segment present in triplication. In maize Dp-Df plants may be recognized by pollen phenotypes because they produce equal numbers of usually distinguishable normal and Dp-Df pollen grains.

8.1.2.4   Transmission of Duplications and Deficiencies

The frequency of Dp-Df individuals in progenies depends on: (1) frequencies of Dp-Df megaspores and (2) viability of Dp-Df megaspores, megagametophytes, gametes, and zygotes. In maize, Dp-Df kernels borne on Dp-Df plants often comprise less than one-third of the total kernels (Patterson, 1978). Dp-Df complements are transmitted almost exclusively through the female in maize. The frequencies are variable, depending especially on the lengths of the deficient and duplicated segments. There was no transmission of Dp-Df chromosomes through the pollen in material studied by Rhoades and Dempsey (1953). However, Patterson (personal communication) has evidence that at least three Dp-Df complements derived from reciprocal translocations in maize may be male-transmitted even under conditions of pollen competition. The usual lack of pollen transmission may be attributed either to pollen inviability or the inability of Dp-Df pollen to compete with chromosomally balanced pollen in effecting fertilization. By utilizing B-A translocations, Carlson (1986) and Carlson and Roseman (1991) produced heritable segmental (proximal and distal) duplications in maize. In cotton, Dp-Dfs obtained from adjacent-1 disjuction of heterozygous translocation tetravalents were usually ovule-viable and also occasionally pollen-viable. Adjacent-2 Dp-Dfs were rarely recovered (Menzel and Dougherty, 1987).

8.1.2.5   Use of Duplications and Deficiencies in Genetic Studies

Duplications can modify genetic ratios because of inviable gametes or alter rates of transmission, and they may change phenotypes due to dosage effects of particular alleles. The Dp-Df stocks in maize are

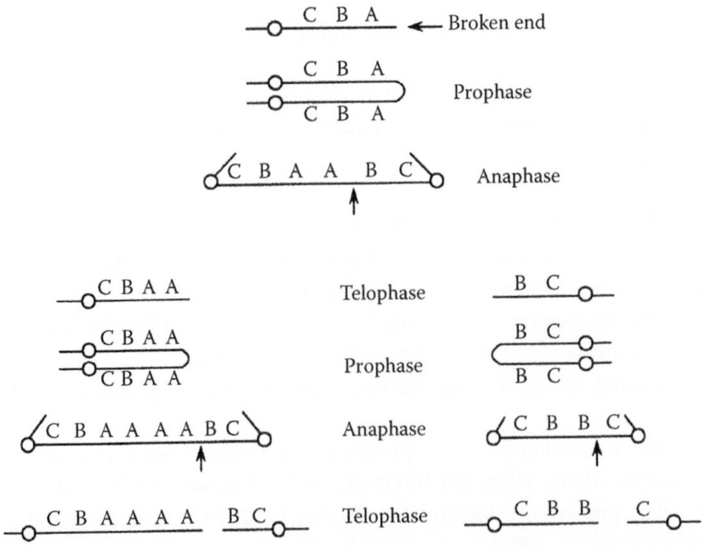

FIGURE 8.10 Production of Dp-Dfs from breakage-fusion-bridge cycle. Breakage points are shown by arrows. (Redrawn from McClintock, B. 1941c. *Genetics* 26: 234–282. With permission.)

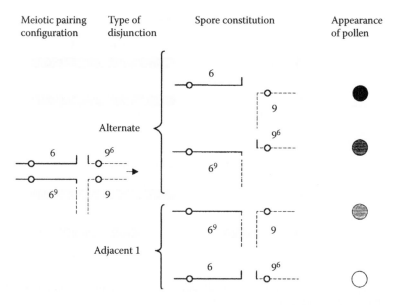

FIGURE 8.11 Chromosome configurations and spore constitutions after meiosis in a maize plant hetero-zygous for a reciprocal translocation. Appearance of pollen grains with Dp-Df chromosome constitutions reflects unbalance of spores. (Redrawn from Patterson, E. B. *Maize Breeding and Genetics.* 693–670. 1978. Copyright Wiley-VCH Verlag GmbH & Co. KGaA. Reproduced with permission.)

useful in localizing and sequencing genes in chromosomes. Distortion in genetic segregation ratios of linked genes may be used to determine the positions of translocation points with respect to gene loci and the sequence of gene loci with respect to each other (Patterson, 1978). In the simplest application translocation, heterozygotes homozygous for a counterpart dominant allele are used as females and are pollinated by plants homozygous for (or carrying) a recessive mutant allele. In this initial extraction of a female-transmissible Dp-Df complement arising from adj.-1 disjunction, Dp-Df plants in maize may often be identified by pollen phenotype (Figure 8.11). If the gene is located in the Df segment, a recessive allele will be in a hemizygous condition and will be expressed. If the gene is not located in the Df segment, Dp-Df plants carrying the recessive allele are crossed as males to a tester stock homo-zygous for the mutant allele. Recombination frequency of the mutant locus with the interchange point is measured and this recombination frequency can be compared with the independently or simultane-ously measured recombination frequencies of various mapped genes in the same chromosome with the same interchange point in Dp-Df plants. If the mutant locus is in fact linked to the deficiency, this procedure will furnish an approximate positioning of the new mutant locus in the chromosome map. Further elaborations of Dp-Df mapping techniques were also discussed (Patterson, 1978).

Modern techniques which sequence RFLP's with respect to each other and to marker genes have found various unbalanced chromosome complements useful in these studies. Particularly useful have been Dp-Df complements derived from heterozygous reciprocal translocations and inver-sions, simple monosomics, tertiary trisomics, and heterozygous deficiencies generated from B-A translocations (Carlson and Curtis, 1986).

8.1.3 INTERCHANGES

8.1.3.1 Introduction

Interchanges are known as segmental chromosomal interchanges, reciprocal translocations, or simply translocations (Burnham, 1956; Ramage, 1971). Translocations are the result of the recipro-cal exchange of terminal segments of nonhomologous chromosomes (Figure 8.12).

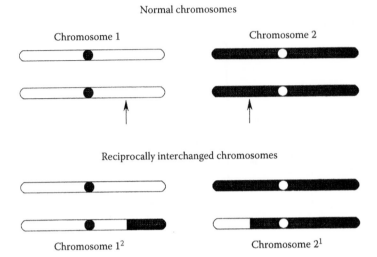

FIGURE 8.12 Diagrammatic sketch showing origin of an interchanged (reciprocal translocation) chromosomes (arrows show the breakage points).

Translocations were first observed by Gates (1908) in *Oenothera rubrinervis* where more than two chromosomes were attached to form a ring. Subsequently, Cleland (1922) suggested in studies based on *Oenothera* species that chromosomes in a ring are associated in a specific and constant order. Belling and Blakeslee (1926) put forth an hypothesis of segmental interchange from chromosome pairing studies in *Oenothera* and suggested that the formation of a circle is due to reciprocal translocation. However, McClintock (1930) was the first to provide cytological evidence of interchanges between two nonhomologous chromosomes in maize.

8.1.3.2 Identification of Interchanges

An interchange heterozygote may be identified cytologically, by its effects on partial pollen and seed sterility or by genetic tests. A reciprocal translocation is obtained when each member of two nonhomologous chromosomes is broken, as shown in Figure 8.12, and the two terminal segments exchange reciprocally, producing the interchanged chromosomes, designated as chromosomes 1^2 and 2^1. Thus, an organism carrying chromosomes 1, 1^2, 2^1, 2 is known as an interchanged heterozygote or a translocation heterozygote. Interchange heterozygote chromosomes form a cross-shaped configuration at pachynema of meiosis (Figures 8.13a, 8.14, and 8.15). A group of four chromosomes, known as a quadrivalent or tetravalent, coorient at diakinesis and metaphase-I in such a fashion that several types of configurations are observed depending on the positions of kinetochores (Figure 8.13b).

In coorientation configuration, the chromosomes are distributed in equal numbers to the opposite poles. In both alternate (zig zag) and adjacent-1 (open) segregations, homologous kinetochores move to opposite poles (Figure 8.13b). In contrast, in adjacent-2 (open) segregation, homologous kinetochores move the same pole (Figure 8.13b).

The region between an interchange point and a kinetochore is termed an interstitial region. If there is no crossing over in an interstitial region, alternate chromosome migration results in viable male and female gametes: $1 + 2$ standard normal or $1^2 + 2^1$ translocated, but balanced, chromosomes. Meiotic products from adjacent disjunctions are typically inviable due to duplications and deficiencies. However, crossing over in the interstitial segments will transfer blocks of genes between the interchanged and normal chromosomes, altering the types of products from alternate and adjacent-1 disjunctions, but all products will be abortive from adjacent-2 disjunction (Figure 8.13a; Table 8.4).

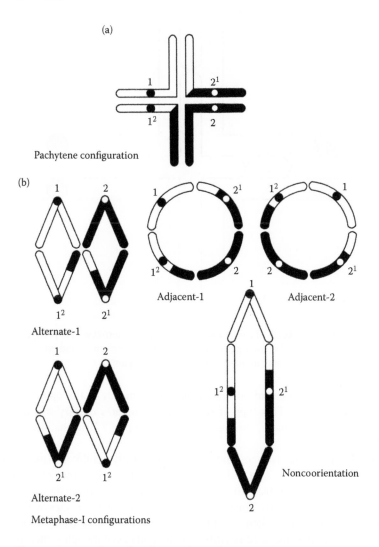

FIGURE 8.13   Chromosome configurations in a reciprocal translocation heterozygote. (a) A cross-shaped configuration at pachynema. (b) Possible chromosome configurations at metaphase-I.

When an interchange heterozygote orients on the meiotic metaphase plate, an approximate ratio of 1 alternate:1 adjacent arrangement of chromosomes is frequently observed because of the random behavior of two pairs of cooriented kinetochores (Endrizzi, 1974; Chochran, 1983; Rickards, 1983). In alternate orientations either the two standard chromosomes or the two interchanged chromosomes may orient to a given pole. The two configurations may be distinguished from each other by chromosome morphological traits (Endrizzi, 1974; Lacadena and Candela, 1977).

Chromosome configurations in interchange heterozygotes can be identified in cotton because of differences in chromosome lengths of A and D genomes (Endrizzi, 1974).

The deficiency in the adjacent-2 class in a T4-5 interchange is most likely due to the knob or may be attributed to nonterminalized chiasmata that may have caused a higher frequency of coorientation of homologous kinetochores to opposite poles (Table 8.5).

In noncooriented arrangement, normal chromosomes carrying nonhomologous kinetochores are on opposite sides of a quadrivalent located equidistant from the equatorial plate. The two noncooriented interchanged chromosomes are stretched in the middle and are not attached to the poles (Figure 8.13b). At anaphase-I, 1 cooriented and 2 noncooriented chromosomes move to the

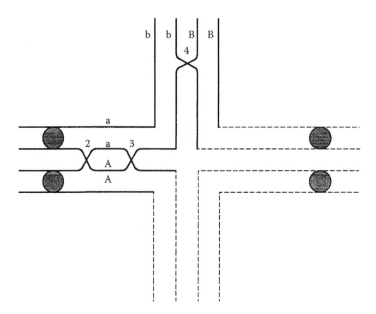

FIGURE 8.14 Diagrammatic sketch of a reciprocal translocation heterozygote at pachynema showing crossing overs between kinetochore and marker gene *a* (2), between marker gene *a* and breakage point (interstitial region-3) and between breakage point and marker gene *b* (distal region-4)

one pole while the other cooriented normal chromosome moves to the opposite pole, producing 3-1 chromosome segregation. This type of segregation generates tertiary trisomics and monosomics in the progenies. Furthermore, chromosome segregation of 2:2 results in unbalanced gametes (deficiency-duplication).

In diploid plant species, heterozygous interchanges generally exhibit approximately 50% pollen sterility. In the *Glycine max × G. soja* cross, studied by Singh and Hymowitz (1988), pollen fertility ranged from 49.2% to 53.3% (Table 8.6). This result suggests that alternate (zig-zag) and adjacent (open) configurations are in nearly equal proportions. Other plant species with 50% sterility in translocation heterozygotes have been listed by Burnham (1962). Interchanged heterozygotes usually do not show 50% pollen and seed sterility in polyploids even though sometimes parents differ

TABLE 8.4
Nature of Crossing Over, and Types of Gamete Formations in Three Types of Chromosome Coorientation

		Types of Orientation	
Nature of Crossing Over	Alternates 1 and 2	Adjacent-1	Adjacent-2
1. Noncrossing over	Original combinations are viable nonviable	All gametes are nonviable	All gametes are nonviable
2. Crossing over between centromere and marker gene *a*	Original combinations are viable	Only cytological crossover products are viable	All gametes are nonviable
3. Crossing over within interstitial segment; between marker gene *a* and breakage point	Original combinations are viable	Only cytological and genetical crossover products are viable	All gametes are nonviable
4. Crossing over in the distal region; between breakage point and marker gene *b*	All combinations are viable	All combinations are nonviable	All gametes are nonviable

TABLE 8.5

Frequency of Chromosome Configurations in the Interchange Heterozygotes of *Gossypium hirsutum*

Translocation Stocks	Types of Chromosome Orientation			
	Adj. 1[a]	Adj. 2[a]	Alt. 1[b]	Alt. 2[b]
T4-5	48	16	45	37
T10-19	68	31	60	30
AG184	20	6	20	13

Source: From Endrizzi, J. E. 1974. *Genetics* 77: 55–60. With permission.

[a] Adj., adjacent.

[b] Alt., alternate.

TABLE 8.6

Meiotic Chromosome Configuration at Diakinesis and Pollen Fertility (%) in *Glycine max* x *G. soja* and in Its Reciprocal Crosses

Hybrids	2n	II	IV	Total PMC	Pollen Fertility (%)
Bonus[a] × PI 81762[b]	40	18.3	0.9	38	49.2
PI 81762 × Bonus	40	18.2	0.9	25	52.0
Essex[a] × PI 81762	40	18.2	0.9	48	53.3
PI 81762 Essex	40	18.4	0.8	25	51.2

Source: From Singh, R. J. and T. Hymowitz. 1988. *Theor. Appl. Genet.* 76: 705–711. With permission.

[a] *Glycine max.*

[b] *G. soja.*

by two interchanges. Tetraploid cotton (Brown, 1980), hexaploid wheat (Baker and McIntosh, 1966), and hexaploid oat (Singh and Kolb, 1991) have been extensively studied. In hexaploid oat, pollen fertility in intercultivar F_1 hybrids ranged from 92.0% to 99.7% and in parents it ranged from 94.1% to 95.5% (Table 8.7). These results suggest that male and female spores containing Dp-Df in polyploids are as competitive as normal spores.

8.1.3.3 Robertsonian Translocations

The kinetochore is an important part in the chromosome and is responsible for a remarkable and accurate movement of meiotic prometaphase homologous chromosomes to metaphase plate. It participates in spindle checkpoint control and moves chromosomes poleward at anaphase (Yu and Dawe, 2000). Meiotically unaligned (unpaired-univalents; nonhomologous) univalents remain laggard at ana-/telophase-I and often misdivide at the kinetochore. Centric misdivision followed by the fusion of broken arms from different chromosomes produces Robertsonian translocation (Robertson, 1916). Robertsonian translocation is found in plants, animals, and humans and plays a key role in the karyotype evolution (see Zhang et al., 2001). Two different fused kinetochores can be distinguished by FISH.

8.1.3.4 Interchanges in Genetic Studies

Interchange stocks have been used to identify linkage groups, to associate new mutants to specific chromosomes and to construct new karyotypes. This technique may be superior to the gene-marker

TABLE 8.7

Meiotic Chromosome Configurations at Diakinesis in Intercultivar Hybrids of Hexaploid Oat

| F₁ Hybrids | Frequency of Chromosome Associations[a] at Diakinesis (No.) | | | | | | | (%) | |
	2III	1IV + 19II	1III + 19II + 2I	1IV + 18II + 2I	2IV + 17II	1VI + 18II	Total Pollen Mother Cells	1IV + 19II	Pollen Fertility
Andrew/Gopher	10	22	–	–	–	–	32	68.8	99.7
Andrew/Hazel	3	40	1	1	7	–	52	76.9	92.0
Andrew/PA12422	4	16	1	–	2	–	23	69.6	–
Andrew/Otee	15	5	–	–	–	–	20	25.0	99.7
Hazel/Gopher	72	69	1	–	–	–	142	48.6	–
PA12422/Gopher	88	25	–	–	–	–	113	22.1	–
Hazel/Otee	87	40	1	–	–	–	128	31.3	–
Otee/PA12422	74	14	–	–	–	–	88	15.9	–
Hazel/Ogle	25	–	–	–	–	–	25	0.0	94.7
Hazel/PA12422	56	46	–	–	–	11	113	40.7	91.6

Source: From Singh, R. J. and F. L. Kolb. 1991. *Crop Sci.* 31: 726–729. With permission.

[a] I = univalent, II = bivalent, III = trivalent, IV = quadrivalent, VI = hexavalent.

stock method because an interchange, in contrast to gene markers, usually does not affect the expression of other traits. Semi-sterility is often more clearly expressed than genetic markers and two chromosomes may be tested for linkage instead of only one (Lamm and Miravalle, 1959).

The interchange-gene-linkage analysis is based on the association of contrasting traits with partial sterility. An interchange behaves as a dominant marker for partial sterility located simultaneously in the two interchanged chromosomes at the points where the original breakage and exchange have occurred. For example, translocated chromosomes are designated T and their normal chromosome homologues are symbolized as N. An individual heterozygous for a translocation (T/N) and for a gene pair (Aa) produces four kinds of gametes AT, AN, aT, and aN. In an F_2 generation, plants that are T/T (translocation homozygote) or N/N (normal homozygote) genotypes are fertile (F), while plants with T/N (translocation heterozygote) are partial sterile (PS). Thus, four classes in progenies are expected in F_2:

- A PS (T/N) = Normal, partial sterile, translocation heterozygote (six)
- A F (T/T N/N) = Normal, fertile, translocation homozygote or normal homozygote (six)
- a PS (T/N) = Mutant, partial sterile, translocation heterozygote (two)
- a F (T/T, N/N) = Mutant, fertile, translocation homozygote or normal homozygote (two)

Based on the above information, Tuleen (1971) used translocation tester stocks of barley carrying unequal interchanged chromosome pieces which could be readily identified at somatic metaphase. By using qualitative seedling mutants, four phenotypic classes can be recognized in F_2 by identifying N/N and T/T individuals. A linkage can be detected by studying the recessive fraction of F_2 population; plants with T/Naa genotype are generated from the union of a nonrecombinant gamete (aN) and a recombinant gamete (aT) while aaT/T and aa N/N plants are generated from the union of two recombinant or two nonrecombinant gametes, respectively. In F_2, a deviation from 1(N/N):2 (T/N):1(T/T) in recessive homozygotes can be used to determine the linkage. The mutants' glossy seedling and virido-xantha are located on chromosome 4 of barley because both mutants showed linkage with T4-5e and independent segregation with T1-5a. Furthermore, both mutants are close to a breakage point, because no recombinant (T/N) genotype was recovered (Table 8.8).

This technique is efficient and quite useful because only a small sample of recessive F_2 seedlings, classified for the translocation by root tip squashes, is needed to detect linkage. However, a limitation of this technique is that it is time-consuming because classification for translocation by the root tip method is tedious, applicable mainly to plants with large identifiable chromosomes and requires expertise in cytological technique.

Interchanges have been very useful to test the independence of linkage groups established genetically. This was demonstrated by translocation analysis in barley. Kramer et al. (1954) observed that chromosome 1 of barley carried two linkage groups (III and VII). They used 9 translocation

TABLE 8.8
Interchange-Gene Linkages for T4-5e and T1-5a, F$_2$ Recessive Seedlings in Barley Classified Cytologically for the Interchanges

Mutant	T4-5e				T1-5a			
	NN	NT	TT	x^2	NN	NT	TT	x^2
Glossy seedling	7	0	0	14.0[a]	1	5	1	0.0
Virido-xantha	10	0	0	20.0[a]	2	4	1	0.3

Source: Adapted from Tuleen, N. A. 1971. In *Barley Genetics II. Proc. 2nd. Int. Barley Genet.* Symp. R. A. Nilan, Ed., Washington State University Press, Pullman, pp. 208–212.

[a] Significant at the 1% level.

lines involving 6 chromosomes (a-b, b-d, c-d, a-e, c-e, c-b, b-f, e-f) and crossed with marker genes representing seven linkage groups of barley. Three interchange testers showed that linkage groups III and VII are not independent. With this information, Haus (1958) studied gene a_{c2} (white seedling) of linkage group III and y_c (virescent seedling) of linkage group VII of barley. The measured recombination value between these two genes was $28.14 \pm 1.59\%$ from combined F_2 and F_3 data, confirming that these two genes, a_{c2} and y_c, belong to the same linkage group rather than to two different linkage groups.

8.1.3.5   Principles of Producing Interchange Testers

Complete interchange tester sets have been established and are being utilized in cytogenetic studies in several plant species, such as barley (Burnham et al., 1954), garden pea (Lamm and Miravalle, 1959), maize (Burnham, 1954), tomato (Gill et al., 1980), rye (Sybenga and Wolters, 1972; Sybenga et al., 1985; Sybenga, 1996), cotton (Ray and Endrizzi, 1982), and pearl millet (Minocha et al., 1982).

Chromosomal interchanges occur spontaneously and may also be induced by irradiation (X-ray, gamma ray, fast neutron, thermal neutron, ultra violet light) and chemical mutagens (ethyleneimine, ethyl methane sulfonate). Translocations have also been reported in the progenies of maize carrying the sticky gene (*st*) (Beadle, 1937) as well as activator and dissociation-controlling elements (McClintock, 1950). They have also been found in aged seeds of barley and wheat (Gunthardt et al., 1953).

The chromosomes involved in different translocations may be identified by studying the F_1's generated by intercrossing them with translocations involving known chromosomes, then observing the F_1's at diakinesis or metaphase-I. The three possibilities are as follows:

- The occurrence of two quadrivalents (IV) suggests that both translocation testers involve different chromosomes.
- The presence of a hexavalent (VI) in F_1 sporocytes indicates that one of the chromosomes involved in the two translocation stocks is common.
- When both translocation stocks involve the same chromosomes, F_1 hybrids may show bivalents only.

8.1.3.6   Identification of Interchanged Chromosomes

Unidentified interchanged stocks can be identified cytologically by studying chromosome configurations at pachynema (Figure 8.15), diakinesis and metaphase-I in an F_1 hybrid involving known translocation testers and aneuploid stocks. For example, reciprocal translocations in various wheat cultivars have been recorded in intervarietal F_1 hybrids and the chromosomes involved in interchanges are identified using aneuploids of wheat cv. Chinese Spring. Since Chinese Spring wheat is considered the most primitive cultivar with standard chromosomes (Sears, 1953b), it is hybridized with other wheat cultivars. Chromosome pairing for chromosomal structural changes at meiotic metaphase-I is analyzed in F_1 plants. The observation of 1IV + 19II suggests that the parental cultivars differ from Chinese Spring by one reciprocal translocation. Similarly, the chromosome association of 2IV + 17II suggests that two independent interchanges are involved. In cases where two interchanges are observed, the frequency of cells with 1IV is higher than cells with 2IVs. This suggests that interchanged segments are different (Vega and Lacadena, 1982). Furthermore, the higher the frequency of quadrivalents, the larger the chromosome segments involved in translocation. Similar results were observed in intervarietal F_1 hybrids of oat (Singh and Kolb, 1991).

Singh and Kolb (1991) observed three types of multivalent associations in all intercultivar F_1 hybrid combinations among six parental lines (Andrew, Gopher, Hazel, Ogle, Otee, and a breeding line PA 12422) of oat ($2n = 6x = 42$): 1IV + 19II (one interchange), 2 IV + 17II (two independent interchanges), and 1VI + 18II (progressive or successive interchanges involving three chromosomes).

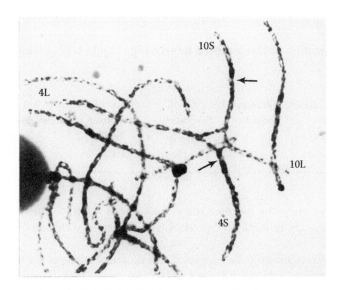

FIGURE 8.15 A photomicrograph of a cell at pachynema showing a cross-shaped configuration in an heterozygous reciprocal translocation in maize. Chromosomes involved in the interchange are numbered and kinetochores are shown by an arrow. (Courtesy of Dr. D. F. Weber).

The telomeres of chromosomes were designated arbitrarily as follows: 1.2, 3.4, 5.6, 7.8, 9.10 … 41.42.

Andrew and Gopher are differentiated by a single interchange because 68.8% of the sporocytes showed a 1IV + 19II chromosome configuration (Table 8.7). Suppose a reciprocal translocation occurred between the chromosomes of Andrew designated 5.6 and 7.8, the resultant chromosomes designated 5.7 and 6.8 would correspond to the chromosome structures of Gopher; however, Andrew and Gopher are not related.

Based on results presented in Table 8.7, it is evident that Gopher and Otee have a similar chromosome arrangement, that is, 1.2, 3.4, 5.7, 6.8, 9.10 … 41.42. Also Hazel and Ogle cultivars carry a similar chromosome constitution. Since Hazel differs from Gopher and Otee by a single interchange and from Andrew by two independent interchanges, a reciprocal translocation occurred between chromosomes 1.2 and 3.4 that resulted in 1.3, 2.4. 5.7, 6.8, 9.10 … 41.42 chromosome ends for Hazel and Ogle. It should be mentioned that the frequency of sporocytes with the 1IV + 19II configuration in the Andrew/Hazel hybrid was higher (76.9%) compared with the sporocytes (13.4%) with 2IV + 17II (Table 8.7). This suggests that the chromosome segments involved in the second interchange may be small. The recorded chain quadrivalent configurations in the Andrew/Hazel and Hazel/Gpoher F_1 hybrids also suggest that the second interchange (1.3, 2.4) is a small segment.

The F_1 hybrids between Hazel/PA 12422 showed 1VI + 18II, together with 1IV + 19II and 21II configurations. If it is assumed a reciprocal translocation occurred between chromosomes 3.4 and 9.10 of Otee, then PA 12422 has 3.9 and 4.10 interchanged chromosomes, giving 1.2, 3.9, 4.10, 5.7, 6.8 … 41.42 chromosome ends in Pa 12422. Thus, the chromosome arrangement in a hexavalent configuration observed in Hazel/PA 12422 cross can be 1.2-2.4-4.10-10-9-9.3-3.1. Furthermore, when any four of the six chromosomes (e.g., 1.2-2.4-4.10-10.9) associated in a quadrivalent configuration, the configuration should be an open chain because the two ends of the quadrivalent will be nonhomologous with each other.

The low frequency of 1IV + 19II configuration in PA 12422/Gopher (22.1%) and Otee/PA12422 (15.9%) hybrids suggests that both the interchanged segments (3.9 and 4.10) of PA 12422 are short. The small homologous segments are not large enough to form chiasmata with

TABLE 8.9

Cytological Identification of Unidentified Interchange Stocks with Various Known Aneuploid Stocks

Unidentified Translocation Stocks	Translocation Lines	Primary Trisomics	Telotrisomics	Monosomics	Monotelodisomics	
Tra. stock A	VI + II	V + II	V^t + II	III + II	IV^t + II	Associated
Tra. stock B	2IV + II	IV + III + II	IV + III^t + II	IV + I + II	1IV + 1^tII	Independent

homologous counterparts or by the time the chromosomes reach diakinesis and metaphase-I, the chiasmata are already terminalized producing an open ring, a chain quadrivalent, or two bivalents.

Once a cultivar is recognized for a reciprocal translocation, attempts are made to identify the interchanged chromosome with the help of monosomics or monotelodisomics. Generally, aneuploids are used as a female parent and cultivars are crossed as a pollen parent. The F_1 population will segregate for disomic and monosomic or monotelosomic plants. In a critical combination, if a cultivar differs by one reciprocal translocation, the majority of sporocytes will show 1III + II (in monosomic) or $1IV^{het}$ + II (in monotelosomic) chromosome association. This suggests that the monosomic chromosome is involved in an interchange. However, in a noncritical combination, a chromosome configuration of 1IV + II + 1I (in monosomic) and 1IV + 1^tII (in monotelodisomic) will be recorded (Table 8.9). It should be indicated that the frequency of trivalents or quadrivalents depends on the size of chromosome segments involved in a translocation; the higher the frequency of quadrivalents, the larger the segments interchanged. Furthermore, the occurrence of chains or open quadrivalent configurations suggests that one of the chromosome segments involved in an interchange is shorter than the others.

8.1.3.7 B-A Interchanges

B-A interchanges or translocations are very useful for studying the breakage-fusion-bridge cycle (Zheng et al., 1999) and genetic analysis (Beckett, 1978; Carlson, 1986) in maize. These interchanges have been obtained by translocation between a member of the standard, "A" chromosome, set and a type of supernumerary chromosome, "B" chromosome, found in certain strains of maize. After reciprocal exchanges between A and B chromosomes, A^B and B^A chromosome types are produced (Figure 8.16a).

Dominant and recessive traits can be located on A chromosomes by a pseudo-dominance technique, because B chromosomes are believed to be largely genetically inert. Roman and Ullstrup (1951) were the first to locate a gene (*hm*) for reaction to *Helminthosporium carbonum* using B-A interchanges.

For locating a recessive gene (*r*), homozygous recessive (*rr*) plants are used as female and are pollinated with a B-A interchange carrying a dominant gene (*R*) for aleurone and plant color on the B^A chromosome. As expected, each egg and polar nucleus will have *r* genotype and the female contribution to the endosperm will be of *rr* genotype. In contrast, hyper- and hypoloid sperm nuclei will be generated in B-A interchanges because of nondisjunction of a B^A chromosome at the second mitotic division in the microspore (Figure 8.16b). When an egg (*r*) is fertilized by a hyperploid (*RR*) sperm, the constitution of embryo tissue will be colored while aleurone will be colorless because it lacks a B^A chromosome and thus an *R* allele. By contrast, the embryo will be colorless when an egg is fertilized by a hypoploid sperm and aleurone cells of the endosperm will be colored after the polar nuclei are fertilized by a hyperploid sperm carrying an *R* allele on each of two B^A chromosomes. In maize, endosperm and embryo color traits may be scored on unshelled

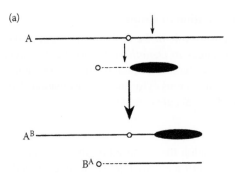

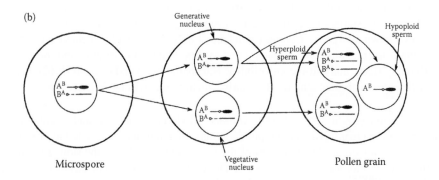

FIGURE 8.16 Diagrammatic sketch of B-A translocations. (a) Production of a B-A (A^B, B^A) translocations in maize (arrows show breakage points). (b) Development of hypoploid and hyperploid sperms due to nondisjunction of the B^A chromosome during second mitosis of a microspore. (Redrawn from Beckett, J. B. 1978. *J. Hered.* 69: 27–36. With permission.)

ears and if a gene is located on a B^A chromosome, the kernels will segregate into colored and colorless endosperm types, reflecting the frequencies of transmission and functioning of hyper- and hypoploid sperms. In the example given here, noncorrespondence between aleurone and embryo color is itself evidence of nondisjunction. On the other hand, if the dominant allele of a gene is not located on a particular B^A chromosome, hypoploidy for that chromosome will not uncover the relevant recessive phenotype.

A dominant gene can also be located on a chromosome by a further test. A dominant gene in hemizygous condition will be expressed in F_1 plants. If selfed or testcrossed to recessive plants, all progeny will be dominant. To test whether the dominant gene is located on an A^B chromosome proximal to the interchange point, the hypoploid is selfed or backcrossed to recessive plants. A locus just proximal to the interchange point in A^B may yield all, or nearly all, dominant progeny. Positions on an A^B farther from the interchange point will indicate by recombination values whether there is linkage and its strength. If a gene is not located on either interchanged chromosome of a B-A translocation, a 3:1 or a 1:1 ratio will be observed in F_2 or test cross progenies of hypoploid plants.

B-A interchanges may be maintained in the heterozygous condition by crossing B-A translocation heterozygotes with pollen from plants carrying normal karyotype. About 1/4 of the plants in progenies are expected to be heterozygous for the translocation. These plants can be identified by pollen sterility, or by chromosome analysis. A modern maintenance procedure consists of developing stocks of homozygous B-A translocations. These are maintained by selfing or sibbing. For use in research, these homozygous stocks may be used directly. Alternatively, they may be crossed as male parents to vigorous standard strains. All nonhypoploid progeny will be either heterozygous or hyperploid, suitable in either case for locating genes (Beckett, 1978).

8.1.3.8   Role of Interchanges in Evolution of Crops

Interchanges in nature have played a major role in the evolution and speciation of several crop species. Earlier investigations on interchanges in *Oenothera* led to the discovery of several genetic systems. Among cultivated crop species, the role of reciprocal translocations in the speciation of *Secale* species is a classic example and numerous cytogenetic investigations have been conducted to determine the progenitor of cultivated rye, *Secale cereale*.

8.1.3.8.1   Oenothera *Species*

Hugo de Vries (1901) formulated a mutation theory based on the peculiar behavior of *Oenothera lamarkiana* and developed the *theory of Intracellular Pangenesis* to explain the mutants. However, it was proved later that the majority of de Vries mutants were not the result of gene mutation but were due to chromosomal aberrations: trisomics, triploids, tetraploids, and interchanges. Furthermore, all the de Vries lines studied bred true and behaved like pure lines because they were permanent heterozygotes known as *Renner Complex* (Cleland, 1962). This peculiar behavior is attributed to several mechanisms including gametophytic and zygotic lethals, microspore and megaspore competition (Renner effect), and self-incompatibility. Burnham (1962) and Cleland (1962) discuss cytogenetics of *Oenothera* at great length. Thus, only salient points will be discussed here.

8.1.3.8.1.1   *Gametophytic and Zygotic Lethals*   Both gametophytic and zygotic lethals are found in most of the *Oenothera*. Gametophytic lethals are either megaspore lethal or pollen lethal. In such cases, sperms do not fertilize eggs of the same genotype due to the slow growth of pollen tubes or the slow development of embryo sacs. Therefore, only gametes of differing genotypes are compatible, producing only heterozygous plants. In *O. lamarkiana*, one gamete complex is called *velans* and the other is called *gaudens*. Homozygous *velans* or *gaudens* are lethal in the zygotes and do not survive; thus only heterozygous *velans/gaudens* survives. This particular mechanism is known as a balanced lethal (Cleland, 1962).

8.1.3.8.1.2   *Microspore and Megaspore Competition*   Megaspore competition probably occurs in all the *Oenothera*. The pattern of female transmission is due to genetic compatibility or incompatibility regarding embryo sac development. The distortion in genetic ratio depends on problems related to interacting effects of megaspore and pollen tube growth. In *O. hookeri*, the embryo sac develops regularly from the micropylar spore. In contrast, the chalazal cell often develops in *O. muricata* to form an embryo sac. Virtually all the eggs carry "rigens" spores and thus "curvans" spores rarely reach the egg. Therefore, "rigens" finds itself in the mycropylar spore and this particular cell develops into an embryo sac (Cleland, 1962).

8.1.3.8.1.3   *Chromosome Designation in* Oenothera   Chromosomes in *Oenothera* were designated based on their meiotic associations. Attempts have not been made to conduct karyotype analysis because somatic chromosomes are small and morphologically similar. Considering the "hookeri" complex primitive and having standard and original chromosome end arrangement, the chromosomes of other *Oenothera* species were designated. It has been observed that reciprocal translocation was not limited to the 14 ends of the chromosome complement, but every one of the 14 ends has been found associated with every other end among the chromosome arrangements. Thus, there are 91 possible associations of 14 ends by twos and they have been observed among the complexes where segmental arrangements have been fully determined.

O. hookeri:	1.2	3.4	5.6	7.8	9.10	11.12	13.14	
[Standard]	1.2	3.4	5.6	7.8	9.10	11.12	13.14	
O. hookeri:	1.2	3.4	5.6	7.10	9.8	11.12	13.14	$= \bigcirc 4 + 5\mathrm{II}$
[Johansen arrangement]	1.2	3.4	5.6	7.10	9.8	11.12	13.14	

This arrangement is quite common among the races of *O. hookeri* found in California and adjacent areas.

O. lamarkiana:	1.2	3.4	5.6	7.8	9.10	11.12	13.14	Hookeri
	1.2	3.4	7.6	5.8	9.10	11.12	13.14	Velans
	1.2	3.4	5.6	7.8	9.10	11.12	13.14	Hookeri
	1.2	3.9	5.6	7.11	14.8	4.12	13.10	Gaudens

Thus, there have been 15 possible types of arrangement of chromosomes into circle pairs. Chromosomes in all the *Oenothera* are equal in size, with median centromeres leading to the regular separation of adjacent chromosomes to opposite poles. Although repeated interchanges occurred in the genomes of *Oenothera*, most of the chromosomes still carry their unaltered chromosome morphology. Interchanges were of equal length and breaks occurred near centromeres. In contrast, chromosome morphologies in *Rhoeo* differ significantly from each other, resulting in uneven spacing of centromeres in a circle at metaphase-I. This arrangement leads to a high degree of irregularity at anaphase-I when chromosomes move to their respective poles.

8.1.3.8.1.4 Secale *Species* The genus *Secale* L. consists of a cultigen, rye (*S. cereale*) and four major wild species namely, *S. vavilovii, S. africanum, S. montanum,* and *S. silvestre*. All contain $2n = 14$ chromosomes. Wild and cultivated species are separated from each other by an effective reproductive isolation and reciprocal translocations have played an important role in the evolution of cultivated rye.

Of the four wild species of *S. cereale, S. silvestre* is a unique species. It differs from others morphologically and cytogenetically (Khush, 1962, 1963) and shows the least Giemsa C-banding pattern (Singh and Röbbelen, 1975; Bennett et al., 1977). Furthermore, crossability rate with other species is extremely low. Khush (1962) obtained 0%–0.8% seed set in *S. cereale* × *S. silvestre* but Singh (1977) did not record mature seed set from any combinations involving *S. silvestre* (Figure 8.17). In these crosses seed developed for about 15 days but collapsed later. Plants could have been obtained through embryo culture. It is interesting to note that *S. silvestre* and *S. vavilovii* did not produce hybrid seed, although both are annual and self-pollinating. These results support the suggestion of Khush (1963) that *S. silvestre* should be placed in a separate section, *Silvestria*.

The nature of the barriers among *Secale* species is partly geographical, and the isolated habitats (Stutz, 1972) rarely allow natural hybridization (Kranz, 1963, 1976). On the other hand, translocations have introduced limitations of genetic exchange by chromosomal sterility.

Secale montanum is assumed to be the oldest within the four major species, *S. silvestre, S. africanum, S. vavilovii,* and *S. cereale* (Riley, 1955; Khush and Stebbins, 1961; Khush, 1963). Among the different attempts to describe the translocation system within *Secale* (Riley, 1955; Khush, 1962; Kranz, 1963), the first consistent evidence on the relationships between these species based on Giemsa C-banding patterns (chromosomal differentiation) was presented by Singh and Röbbelen (1977).

Interspecific hybrids of *S. montanum* with *S. vavilovii* and with *S. africanum*, respectively, showed a ring of four chromosomes. But in the two cases different chromosomes are involved. Evolution apparently occurred separately from these two species through one different translocation each. The same translocation which distinguishes *S. montanum* from *S. vavilovii* was found in *S. cereale* but differed for the second interchange resulting in 1IV + 3II configuration in F_1 of *S. cereale* and *S. vavilovii*. Additional translocations occurred to the extent that *S. cereale* is now separated from *S. vavilovii* by two and from *S. africanum* and *S. montanum* and *S. silvestre* by three interchanges (Figure 8.18). As expected, the F_1 involving *S. africanum* and *S. vavilovii* showed two quadrivalents (Figure 8.18a). In *S. cereale* × *S. africanum*, but more frequently in crosses with *S. montanum*, the configuration of 1VIII + 3II was observed (Figure 8.18b), but this association was reduced in certain nuclei to 1VI + 4II (Singh and Röbbelen, 1977). Obviously one of the three

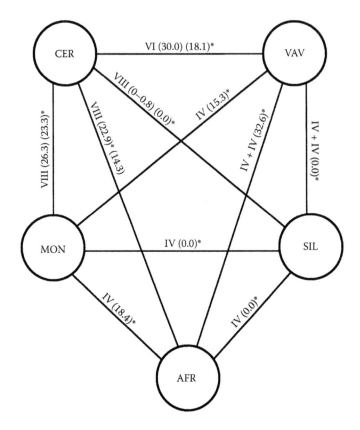

FIGURE 8.17 Cytogenetic relationships among five major *Secale* species. Abbreviations: AFR, *africanum*; CER, *cereale*; MON, *montanum*; SIL, *silvestre*; VAV, *vavilovii*. Crossability rate (%) in parentheses is from Khush, G. S. 1962. *Evolution* 16: 484–496. With permission. In parentheses with an * is from Singh, R. J. 1977. *Cereal Res. Commun.* 5: 67–75. With permission.

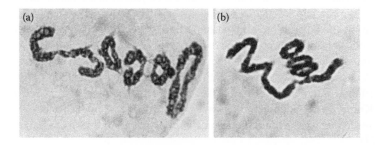

FIGURE 8.18 Meiotic metaphase-I configurations in *Secale* interspecific F$_1$ hybrids. (a) *S. africanum* × *S. vavilovii* (2 IV + 3 II); (b) *S. cereale* × *S. montanum* (1VIII + 3 II).

translocations carried only a rather small segment. Riley (1955) also found a chain of eight chromosomes in *S. cereale* × *S. delmaticum* hybrids, though with even smaller frequency (1/70). *Secale delmaticum* and *S. montanum* are similar in gross chromosome structure (Riley, 1955). Stutz (1972), on the other hand, described 7II in hybrids of *S. cereale* × *S. vavilovii*, supposing complete structural similarity between these species. For these differences, the only plausible explanation is that the material of Stutz was of different origin and that the easy pitfall of misclassification of species was not effectively avoided (Khush, 1963).

Secale vavilovii and *S. africanum* showed 2IV + 3II, but the 1IV + 5II configuration was in combinations found of *S. montanum* and *S. vavilovii* (Figure 7.16; Khush, 1962). *Secale montanum* and *S. africanum* also gave 1IV + 5II. Thus, three species differ from each other with respect to the translocated chromosomes. It is evident that the same chromosomes cannot be involved to distinguish *S. cereale* from *S. vavilovii* and *S. montanum* as Heemert and Sybenga (1972) postulated.

In none of the above interspecific hybrids studied was the nucleolus organizer chromosome (SAT-chromosome) included in translocations. This confirms previous findings (Jain, 1960; Kranz, 1963; Heemert and Sybenga, 1972) and suggests that the SAT-chromosome is less subject to structural changes, possibly due to functional restrictions. However, the SAT-chromosome is very vulnerable to structural changes when incorporated into a wheat nucleus, producing translocations and substitution lines (Zeller, 1973).

8.1.4 INVERSIONS

8.1.4.1 Introduction

A change in the linear sequence of the genes in a chromosome which results in the reverse order of genes in a chromosome segment is called an inversion. Inversions are widespread in plants, insects and mammals, and are found in nature or induced by radiations and chemical mutagens. Chromosomal breakage followed by healing produces new chromosomes. Inversions are responsible for speciation (Stebbins, 1950). In plants, McClintock (1931) was the first to show cytologically in maize that when a long inversion is heterozygous, a large loop is seen at pachynema, and bridge(s) and acentric fragment(s) may occur at meiotic anaphases-I and -II.

8.1.4.2 Types of Inversions

Based on the positions of the two breaks of an inversion in relation to the kinetochore of a chromosome, inversions are of two types: (1) paracentric inversion and (2) pericentric inversion.

8.1.4.2.1 Paracentric Inversion

In a paracentric inversion both breaks occur in the same arm so the inverted region does not include a kinetochore. In a heterozygous inversion a loop is observed at pachynema, and chromatin bridges and acentric fragments may be found at anaphases-I and -II. A dicentric bridge and acentric fragment at anaphase-I may arise due to an error in the normal process of crossing over (C.O.). However, if the bridge and fragment is moderately constant, then this is most likely due to a paracentric inversion (Sjödin, 1971).

8.1.4.2.1.1 Cytological Behavior Different configurations of dicentric bridges and fragments at anaphases-I and -II are expected depending on the types of crossing over within the loop (Figure 8.19; Table 8.10). A single crossing over within the loop will produce a BF (bridge fragment); 2-strand double crossing over in the loop will yield 2 normal (*in*) and 2 inversion (*IN*) chromatids, and B and F will not be observed; a 3-strand double C.O. will produce B and F together with one *in* and one *IN* chromatids. A 4-strand double will give rise to 2 B and 2 F. This will result in all the unbalanced chromatids (Table 8.10). It gets more complicated when simultaneous crossing over occurs at position 6 and there is a single C.O. in the loop. This will produce LF (loop and fragment) and BF together with *in* and *IN* chromatids. Thus, 50% of the chromatids are expected to be genetically balanced (Table 8.10). Furthermore, the situation gets even more complicated following a triple C.O. (one at 6 and a double C.O. in the loop). Crossing over at 6 and 1 and 2 (3-strand) will produce only *in* and *IN* chromatids. Crossing over 3 (2-strand) and 4 (4-strand) and simultaneously at 6 will generate only LLFF chromatids and likewise triple C.O. at 6, 1, and 5 will produce BBFF chromatids. In both cases all the produced gametes are unbalanced (McClintock, 1938a; Burnham, 1962; Ekberg, 1974; Table 8.10).

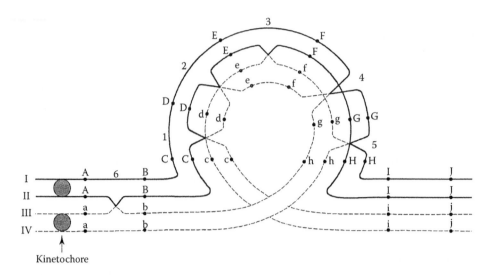

FIGURE 8.19 Schematic diagram of a paracentric inversion heterozygote at pachynema. The numbers represent the positions of crossing overs. Chromatids are designated by numbers I, II, III, and IV. Chromatids I and II contain dominant gene loci and their corresponding recessive alleles are on chromatids III and IV. Crossover products are in Table 8.10.

TABLE 8.10
Chromosome Constitution in a Paracentric Inversion

Crossing over (C.O.) in Inversion	Chromatid Constitution		Anaphase I	Anaphase II	Balanced Chromatids (%)
	Normal in	Inversion IN			
I C.O. in Inversion Loop at:					
1,2 = 2-Strand double	2 in	2 IN	in	No bridge	100
1,3 = 3-Strand double	1 in	1 IN	BF	No bridge	50
3,5 = 3-Strand double	1 in	1 IN	BF	No bridge	50
3,4 = 4-Strand double			2B, 2F	No bridge	0
II C.O. at 6 and Single C.O. in Inversion Loop at:					
3 = 2 Strand	1 in	1 IN	LF	Bridge in 1 cell	50
1 = 3 Strand	1 in	1 IN	BF	No bridge	50
5 = 3 Strand	1 in	1 IN	BF	No bridge	50
4 = 4 Strand	1 in	1 IN	LF	Bridge in 1 cell	50
III C.O. at 6 and Double C.O. in Version Loop at:					
1,2	2 in	2 IN	in	No bridge	100
1,3	1 in	1 IN	BF	No bridge	50
1,4	1 in	1 IN	LF	Bridge in 1 cell	50
1,5			2B2F	No bridge	0
3,4			2L2F	Bridge in both cells	0

Note: in, normal; IN, inversion; B, dicentric bridge; F, acentric fragment; L, dicentric loop; LF, loop and fragment.

Anaphase configurations, where chromosome 4 in maize contained an inverted segment (paracentric), were examined thoroughly by McClintock (1938a). The frequencies (%) of various configurations were as follows: 1B + 1F = 41.4; 2B + 2F = 3.1; F = 2.6. Anaphase-II configurations depended very much on the preceding anaphase-I configurations: no B + 1F = 42.7%; no B + 2F = 1.3%; 1B + 1F = 3.0%. Anaphase-II configurations with 1B + 1F in each daughter nucleus can be readily recognized and this configuration is due to a triple crossing over. The observed frequency suggests that this rarely occurs. This conclusion can be confirmed from the observation of Russell and Burnham (1950). They recorded no bridges at anaphase-II in a paracentric inversion heterozygote for chromosome 4 of maize. They observed the higher frequencies of cells with B and F (average 11.4%) at the second meiotic division.

Ekberg (1974) examined different configurations at anaphases-I and -II in three paracentric inversion heterozygotes of barley (Table 8.11). Two inversion stocks (NI 501, CI 501) showed similar configurations while XI 501 differed significantly showing 76% normal sporocytes. The frequencies of BBFF and LLFF were low (0%–3%). A crossover within the inversion accompanied by a proximal crossover was the most common event.

In a *Vicia faba* paracentric inversion, the frequencies (%) of anaphase-I (anaphase-II), studied by Sjödin (1971), were as follows: in = 27 (21); BF = 40(24); LF = 30(52); BBFF = 2(0.2); LLFF = 0.4(2). Again, the frequencies of BBFF and LLFF were the lowest.

Das (1955) examined four X-ray-induced paracentric inversions in barley. The frequencies of 1B + 1F at anaphase-I ranged from 49.99% (D59-14) to 60.61% (D58-15). In contrast, only a few cells (approximately 2% of the sporocytes) of hybrids involving barley cultivar OAC 21 with B and F were detected (Powell and Nilan, 1968). This suggests that the long sections of chromosomes around kinetochore are not involved in crossing over. Thus, this particular paracentric inversion cannot be detected by B and F. Therefore, the occurrence of localized chiasmata in barley may be the reason for the rarity of reported inversions.

Detection of inversions is influenced by temperature. Swanson (1940) observed in *Tradescantia* F_1 hybrids that warmer temperatures increased chiasma formation, particularly in the interstitial regions, with a correlated increase in bridge frequency (Table 8.12). Detection of inversions is also influenced by temperature in barley. Powell and Nilan (1963) found optimal C.O. in an inversion heterozygote of barley between 15°C and 21°C. Thus, under field conditions when temperatures are adverse, crossing over in short inversions in varietal hybrids may be too low for detection.

The breakage of dicentric bridges occurs at various positions in different sporocytes at telophases-I and -II. Fragment chromatids remain in the cytoplasm and can be distributed at random to either pole at telophase-I (dyad) or telophase-II (quartet). Fragments not included in nuclei may form micronuclei.

TABLE 8.11

Normal and Aberrant Sporocytes (%) at Anaphases-I and -II in Three Heterozygous Paracentric Inversions in Barley

	N	BF	LF	BBFF	LLFF	No. PMCs Studied
NI501 AI	41	34	22	3	1	959
AII	38	35	22	3	2	333
CI501 AI	42	36	20	2	1	1327
AII	35	34	29	1	1	315
XI501 AI	76	13	10	0.3	0.3	689
AII	78	14	8	0	0	100

Source: From Ekberg, I. 1974. *Hereditas* 76: 1–30. With permission.
Note: N, normal sporocytes; BF, bridge fragment; LF, loop fragment.

TABLE 8.12
Effect of Temperature on Production of Bridges in Harvard Hybrid *Tradescantia*

Temperature °C	Interstitial Chiasmata per Cell	Terminal Chiasmata per Cell	Univalent (%)	Bridges (%)
12–15	1.6	6.07	78.5	17.5
19–22	2.88	7.7	54.0	26.1
27	3.8	7.62	53.0	47.8

Source: From Swanson, C. P. 1940. *Genetics* 25: 438–465. With permission.

Paracentric inversions have been studied in several plant species, including maize (McClintock, 1931; Morgan, 1950; Russell and Burnham, 1950; Rhoades and Dempsey, 1953), barley (Smith, 1941; Das, 1955; Holm, 1960; Powell and Nilan, 1963, 1968; Kreft, 1969; Ekberg, 1974; Yu and Hockett, 1979), *Tradescantia* (Swanson, 1940), *Vicia faba* (Sjödin, 1971), and *Lilium formosanum* (Brown and Zohary, 1955). The widespread occurrence of inversions in *Tradescantia* suggests that inversions have played a major role in speciation. Paracentric inversion is rather common in *Drosophila* because crossover products, dicentric bridges, and fragments are selectively eliminated into polar bodies during oogenesis by preferential segregation (Sturtevant and Beadle, 1936).

8.1.4.2.1.2 Fertility The characteristic feature of chromosomal interchanges and inversions is that these chromosomal aberrations produce varying degrees of pollen and ovule abortion. Paracentric inversion causes pollen and ovule abortion because single crossovers, certain double crossovers in the inversion loop, and triple crossovers (rare) generate bridges and fragments at anaphases-I and -II, producing spores containing chromatids with duplications and deficiencies. These spores generally abort and the degree of pollen abortion can be predicted from the frequencies of anaphases-I and -II cells containing bridges and fragments. If crossing over does not occur in the inversion loop, all the spores would be functional because two spores would contain normal chromatids and two spores would have inverted chromatids.

Morgan (1950) observed 28.2% pollen abortion and 4% ovule abortion in In4a paracentric inversion of maize. The normal sibs showed 2.9% aborted pollen. Russell and Burnham (1950) studied a paracentric inversion involving chromosome 2 (In2a). They did not find ovule abortion but pollen abortion ranged from 12.4% to 21.6% with an average of 16.5%. Pollen abortion in In3a of maize ranged from 11.5% to 27.6% (Rhoades and Dempsey, 1953). They stated that the observed percentage of abortion was always less than that indicated by the cytological observations.

In a paracentric inversion of *Vicia faba*, Sjödin (1971) observed an average of 37% (range 28%–59%) pollen abortion, which fully agreed with the predicted (38%) value. In barley, Ekberg (1974) studied pollen and seed fertility of three paracentric inversion stocks. The degree of pollen fertility and seed set did not differ significantly (Table 8.13).

According to McClintock (1938a) if a fragment chromosome is included in a tube nucleus and if the fragment compensates a deficiency portion, the tube nucleus would contain a complete genomic complement. Thus, it could be expected to function in the growth of a pollen tube with a deficient chromosome 4 into an embryo sac. The resulting zygote would be heterozygous for a broken chromosome 4 with a terminal deficiency of the long arm. Such plants were recovered.

8.1.4.2.1.3 Linkage Studies Paracentric inversions have been used in locating genes in specific segments of chromosomes in maize (Morgan, 1950; Russell and Burnham, 1950; Rhoades and Dempsey, 1953), barley (Ekberg, 1974), and *Drosophila* (Burnham, 1962).

Parental genotype gametes containing *in* and *IN* chromosomes are mostly viable in paracentric inversions. Therefore, inversions are often known as "crossover suppressors." Generally, an

TABLE 8.13

Pollen Fertility, Seed Set, and Segregation Ratios for Three Heterozygous Paracentric Inversion Stocks in Barley

Inversion Stock	Pollen Fertility (%)	Seed Set (%)	Segregation Ratio F:PS	χ^2 for 1:1 Segregation
NI 501	78.2 ± 2.3	78.3 ± 1.1	101:91	0.52
XI 501	83.6 ± 2.8	83.3 ± 0.9	233:209	1.30
CI 501	74.6 ± 4.6	78.2 ± 1.1	47:33	2.4

Source: From Ekberg, I. 1974. *Hereditas* 76: 1–30. With permission.
Note: F, Fertile; PS, Partial Sterile.

inversion homozygote (*IN IN*) carrying a dominant allele, for example *X*, is pollinated by a genetic stock which contains a recessive allele *x* and a normal karyotype (*in*). In inversions and translocations, a breakage point is considered a dominant locus that behaves as a single factor. F_1 (inversion heterozygote) plants are either selfed to produce an F_2 or are backcrossed to the recessive parent. Reciprocal crosses can also be made.

There are two possibilities. A marker gene may be located outside an inversion loop or inside the inversion. In independent assortment, four types of gametes are expected from an F_1 (*X IN*, *X in*, *x IN*, *x in*) in an equal proportion of 1:1:1:1. Partial sterility is used as a marker in the linkage test to identify members of the plant population with the inverted chromosome. Recombinants in BC_1 are scored to determine the association between a marker gene and a break point. If no, or a few, recombinants (*X in*, *x IN*) are found, the break point is located close to the *x* locus. On the other hand, if a higher frequency of recombinants is found, the locus *x* is located farther away.

The other possibility is that the marker locus *x* is located within the inversion loop. A single crossover within the inversion is expected to result in transmission of only parental type gametes. Recombinants are produced following 2-strand and 3-strand double crossovers, and in equal proportions. Furthermore, simultaneous crossing over at the proximal region 6 (Figure 8.19) and double crossover or triple crossover within the inversion will not alter the proportions of recombinants.

Rhoades and Dempsey (1953) investigated breakage points of the inversion In3a in relation to genes *Rg*, *lg2*, *A1*, and *et*. All genes lie in the long arm of chromosome 3 in the order given, with *Rg* nearest to the kinetochore. The standard recombination values between these loci are: $Rg\text{-}lg_2 = 16\%$, $lg_2\text{-}A1 = 34\%$, $A1\text{-}et = 12\%$, and *Rg* is 15 map units from the kinetochore. F_1 plants heterozygous for the inversion and for these genetic markers were test crossed to structurally normal plants homozygous for the respective recessive alleles. Table 8.14 shows 16.3% recombination between *Rg* and Lg_2 in the inversion heterozygote and 15.4% recombination in normal plants. The standard value is 16%. Reduction in recombination (0.6%) occurred between Lg_2 and *A1* in the *IN/in* fraction, while the recombination value in the *in/in* fraction was 24.9%. Based on the reduction of genetic recombination in the $Lg_2\text{-}A1$ region, Rhoades and Dempsey (1953) suggested that these loci are located in the inverted segment, while *Rg* is in the proximal region.

Dobzhansky and Rhoades (1938) wrote whether an inversion causes the genes in a chromosome to be inherited as a unit depends directly on certain types of crossover.

8.1.4.2.2 Pericentric Inversion

In a pericentric inversion, the two breaks occur in opposite arms of a chromosome (Figure 8.20). Dicentric bridges and acentric fragments at anaphases-I and -II are not observed in pericentric inversions.

Pericentric inversions have been found in maize (Morgan, 1950; Zohary, 1955), *Vicia faba* (Sjödin, 1971), *Allium thunbergii* (Watanabe and Noda, 1974), *Scilla scilloides* (Noda, 1974), and in many grasshopper species (White and Morley, 1955). Pericentric inversions are rare in *Drosophila*

TABLE 8.14

Test Cross Data from a Plant of $\dfrac{rg^{(1)}Lg_2^{(2)}A1}{Rg\ lg_2a1}$ **Constitution**

	IN/in	in/in	Recombinations
$rg\ Lg_2\ A1\ (0)$	612	182	
$Rg\ lg_2\ al\ (0)$	470	162	Rg-Lg_2 = IN/in = 16.3%
$rg\ lg_2\ al\ (1)$	108	32	= in/in = 15.4%
$Rg\ Lg_2\ A1\ (1)$	111	35	Lg_2-$A1$ = IN/in = 0.6
$rg\ Lg_2\ al\ (2)$	0	59	= in/in = 24.9%
$Rg\ lg_2\ A1\ (2)$	3	60	
$rg\ lg_2\ A1\ (1.2)$	2	4	
$Rg\ Lg_2\ al\ (1.2)$	3	13	

Source: From Rhoades, M. M. and E. Dempsey. 1953. *Am. J. Bot.* 40: 405–424. With permission.

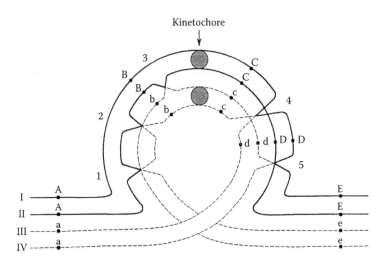

FIGURE 8.20 Schematic diagram of a pericentric inversion heterozygote at pachynema. (See Figure 8.19 for explanations). Crossover products are in Table 8.15.

because single crossovers between normal and inverted segments in heterozygotes produce Dp and Df chromatids which are lethal in zygotes or the embryos.

8.1.4.2.2.1 Cytological Behavior   Any single crossover at any point and 3-strand double cross-overs in pericentric inversion heterozygotes produce chromatids consisting of one normal (*in*), one inversion (*IN*) and two Dp and two Df constitution (Figure 8.20). A 2-strand double crossover generates two normal and two inverted chromatids. However, a 4-strand double crossover (at 1, 5) results in all four chromatids with duplications and deficiencies (Table 8.15). Pericentric inversions in Orthoptera are usually characterized by nonhomologous pairing between normal and inverted chromosomes, which prevents crossing over in the inverted region. Thus, pericentric inversions are also known as "crossover *suppressors*," since recombinant strands are not recovered (White and Morley, 1955; Cabrero and Camacho, 1982).

TABLE 8.15
Chromosome Constitutions in a Pericentric Inversion after Crossing Over

Crossing Over in Inversion	Position of C.O.	Chromatid Constitution after C.O.		
		Normal in	Inversion IN	Dp + Df
1. Any single C.O.	Any point	1	1	Dp + Df (2)
2. Double C.O.				
2 strand	1, 2	2	2	–
3 strand	1, 3 or 1, 4	1	1	Dp + Df (2)
4 strand	1, 5	0	0	Dp + Df (4)

Pericentric inversions may shift the kinetochore position in a chromosome and result in a change of arm ratio. This feature facilitates the identification of inverted chromosomes by karyotype analysis at mitotic and meiotic stages. At metaphase-I of meiosis, inverted chromosomes may contribute to asymmetrical bivalents.

Morgan (1950) identified two pericentric inversions in maize at pachynema. In *Vicia faba*, Sjödin (1971) identified two pericentric inversions at mitotic metaphase, and in both cases chromosome 1 was involved in the inversion.

8.1.4.2.2.2 Fertility Heterozygous pericentric inversions cause pollen and ovule abortion because spores are produced that contain chromosomes with deficiencies and duplications. In *Vicia faba*, Sjödin (1971) observed about 50% pollen abortion in two heterozygous pericentric inversions. Pollen abortion in heterozygous pericentric Inversion 2b in maize ranged from 18.3% to 25.6% (19.1% average), and ovule abortion was 20.1%. Normal plants showed 2.7% to 4.1% pollen abortion. In pericentric Inversion 5a in maize, the average ovule abortion was 12.5% and pollen abortion was 28.3%. The differences between ovule and pollen abortion are related to the higher frequencies of crossing over in male than in female flowers (Morgan, 1950). In the natural population, four pericentric inversion heterozygotes of *Scilla scilloides* (2n = 18) showed an average of 46.4% pollen abortion (Noda, 1974).

8.2 NUMERICAL CHROMOSOME CHANGES (HETEROPLOIDY)

8.2.1 EUPLOIDY

8.2.1.1 Introduction

An individual carrying chromosome numbers other than true monoploid or diploid numbers is called heteroploid (Sharp, 1934). Heteroploidy is divided into euploidy and aneuploidy. In the euploid, an individual carries an exact multiple of the basic chromosome number while in the aneuploid the chromosome number is some number other than an exact multiple of the basic set.

In polyploids, x is the basic (monoploid) chromosome number, n is the gametic chromosome number of chromosomes and 2n is the zygotic or somatic chromosome number. For example, the genomic formula of *Triticum aestivum* is 2n = 6x = 42 and *Hordeum vulgare* is 2n = 2x = 14. In both cases, the basic chromosome number x is seven. The basic set of chromosomes in a diploid is called a genome.

8.2.1.2 Classification of Euploidy

Euploidy is divided into auto- and allopolyploidy. In the autopolyploid, the genomes are alike because one basic genome is multiplied (x, monoploid; 2x, diploid; 3x, triploid; 4x, tetraploid; 5x, pentaploid; 6x, hexaploid; 7x, heptaploid; 8x, octoploid). In allopolyploids two or more genomes derived from different genomically unlike, distinct species are present (Figure 8.21).

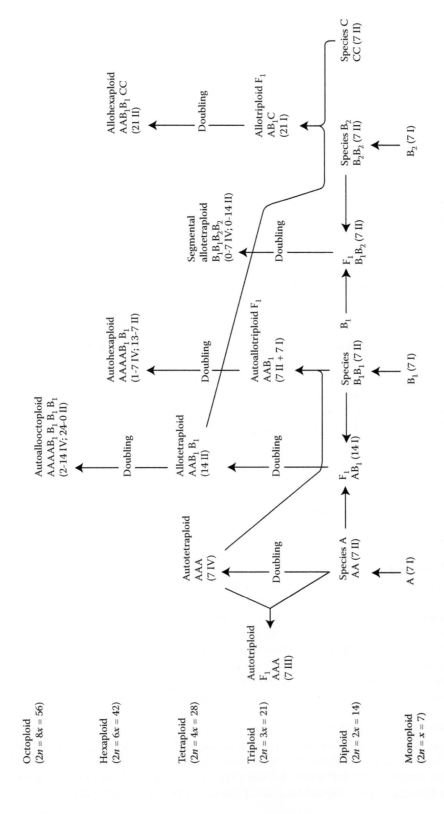

FIGURE 8.21 Diagram showing ploidy levels, species, genomes, and meiotic pairing at metaphase-I. The flow chart shows interrelationships between auto- and allo-polyploids and path of their origin. Diploid chromosome numbers of species A, B1, B2, and C are assumed as $2n = 2x = 14$. (Modified from Stebbins, G. L. Jr. 1950. *Variation and Evolution in Plants*. Columbia University Press, New York.)

Stebbins (1950) recognized two additional euploid categories, namely segmental allopolyploid and auto-allopolyploid combinations. Segmental allopolyploids carry genomes intermediate in degree of similarity and generally exhibit preferential pairing. For example, a segmental allotetraploid with genomes $B_1 B_1 B_2 B_2$ usually forms bivalents and occasionally quadrivalents. Auto-allopolyploidy is confined to hexaploidy and higher levels of polyploidy (Stebbins, 1950). The term amphiploidy or amphidiploidy denotes polyploidy derived after hybridization between two or more genomically dissimilar species separated by chromosomal sterility (Figure 8.21).

Polyploidy has played a major role in the speciation of higher plants. In angiosperms, 30%–35% (Stebbins, 1950, 1971) to 47% (Grant, 1971) of the species is of polyploid origin. The species in which the basic chromosome number is $x = 10$, or higher, evolved by polyploidization.

8.2.1.2.1 Autopolyploidy

8.2.1.2.1.1 Autotriploids (3x, AAA) An individual possessing three basic sets of homologous chromosomes is known as autotriploid.

1. Sources of Autotriploids
   Autotriploids arise from the progenies of diploid ($2x$) species by spontaneous origin at low frequencies. They originate from the sexual fusion of an unreduced ($2n$) egg and haploid (n) male sperm. Harlan and de Wet (1975) recorded polyploid individuals in 68 genera that originated from unreduced gametes. Triploids are identified and distinguished from diploids in the natural population because of their more vigorous vegetative growth, profuse tillering, *gigas* phenotypes, and high ovule abortion.

   Spontaneously-induced triploids have been isolated and identified in numerous sexually and vegetatively propagated crops such as *canna* (Belling, 1921), *Datura* (Belling and Blakeslee, 1922), maize (McClintock, 1929c), tomato (Lesley, 1928; Rick, 1945), rice (Ramanujam, 1937; Rao and Reddi, 1971; Khush et al., 1984), barley (Sandfaer, 1975), pearl millet (Dujardin and Hanna, 1988; Rao et al., 1988), Sorghum (Schertz and Stephens, 1965), and others (Burnham, 1962; Balog, 1979, 1984).

   Rick (1945) found in tomato that triploids appeared spontaneously at the rate of 0.08% (45/55,000 plants). Sandfaer (1975) observed 0.05% (179/381,563 seeds) triploids of spontaneous origin in barley. Seeds of 39 barley cultivars were examined to determine the frequency of autotriploids. Barley seeds were divided into three groups: group A included fully developed seeds that constituted 95.8%, group B contained shriveled seed (2.0%), and group C had very light "empty" seeds (2.2%). Seeds of group A were assumed to be diploid ($2n = 2x = 14$) and chromosomes were not determined. Germination of group B seeds was 73.9% and the frequency of triploid plants ($2n = 3x = 21$) was 2.86%. The germination rate of group C seeds was only 2.4% but its members consisted of 10.29% triploid plants (Table 8.16). The frequency

---

TABLE 8.16
Summary of the Total Material Studied from 39 Barley Cultivars

Number of Spikes	Number of Seeds	Seed Size Groups	1000 Kernel Weight (g)	Number of Seeds	Germination (%)	Diploid $2n = 14$	Triploid $2n = 21$	Aneuploid	Triploid (%)
		A	41.2	365,645	–	–	–	–	–
18,719	381,563	B	13.7	7,480	73.9	5,364	158	7	2.86
		C	3.5	8,438	2.4	182	21	1	10.29

Source: From Sandfaer, J. 1975. *Hereditas* 80: 149–153. With permission.

of triploid plants may have been influenced by the genotypes of the cultivars because of the 39 cultivars examined, triploids were recorded (range 1–25 plants) in 22 cultivars.

Autotriploids are produced experimentally from tetraploid ($4x$) by diploid ($2x$) crosses. Häkansson and Ellerström (1950) suggested that seed development was better in rye when a plant with the higher chromosome number was used as female parent rather than the reverse. In the $2x \times 4x$ cross of rye, 1275 spikelets were pollinated and 36 well-developed kernels were obtained. Seventeen seeds germinated; four seedlings were triploid ($2n = 3x = 21$) and two plants were diploid ($2n = 14$). In $4x \times 2x$ hybridization, 783 pollinated spikelets yielded 20 well-developed seeds; six seeds germinated giving four triploids and two tetraploids. Diploid and tetraploid plants may have originated after selfing. Similar results were reported by Tsuchiya (1960b) in barley. In a $4x \times 2x$ cross 980 seeds were obtained from 2721 pollinations, but only 31 seeds germinated (1.13%). By contrast, triploid plants were not found in a $2x \times 4x$ cross (Table 8.17). In rice, the success rate of finding triploids depended on the crossing methods and the frequency of triploids ranged from 0.0% to 1.09% (Morinaga and Kuriyama, 1959).

In *Triticum monococcum* ($2n = 2x = 14$), Kuspira et al. (1986) obtained a total of 1750 paper-thin, shriveled seeds from 7700 pollinated florets in a $4x \times 2x$ cross (23.0% seed set). In a $2x \times 4x$ cross, 9200 pollinations resulted in 260 seeds (3% seed set). Not all seeds (2010) germinated. However, 15-day-old embryos from a $4x \times 2x$ cross that were excised aseptically produced triploid plants. This result indicates that endosperm collapse 15 days postfertilization caused the death of embryos.

Twin seedlings in some cases produce triploid. Müntzing (1938) cytologically examined a total of 2201 twin plants from 16 species. Triploid plants were the most frequent (77) followed by haploid (11) and tetraploid plants were quite rare (2).

It has been demonstrated that normal seed development depends on a correct ploidy ratio (2:3) between embryo and endosperm that occurs normally in a $2x \times 2x$ cross (Watkins, 1932; Esen and Soost, 1973). Departure from this quantitative ratio such as 3:4 ($2x \times 4x$) or 3:5 ($4x \times 2x$) between embryo and endosperm genome sets disturbs physiological and genetic balance, leading to the collapse of endosperm and then death of the embryos. In $2x \times 4x$ crosses of citrus, about 92%–99% of the seeds containing triploid embryos and tetraploid endosperm aborted at different stages of embryogenesis (Esen and Soost, 1973). Lin (1984) dismissed the hypothesis of embryo and endosperm relations based on his studies of diploid and tetraploid maize that involved a gametophyte mutant (*ig*), and suggested "development of maize endosperm evidently is affected by the parental source of its set of chromosomes." Since he used a gametophyte mutant stock (*ig*) that generates chromosome abnormalities, however, the embryo and endosperm ratio (2:3) theory should perhaps be investigated further.

A large number of autotriploids have been isolated from the progeny of different lines of homozygous recessive male-sterile (*Ms1 ms1 ms1*) soybean (Xu et al., 2000a). Autotriploids

---

TABLE 8.17
Crossing Success from Reciprocal Crosses between Tetraploid and Diploid Barley

Cross	Number of			Number of Seeds		
	Florets	Seeds	Fertility (%)	Sown	Germinated (%)	Success (%)
$4x \times 2x$	2721	980	36.01	980	31 (3.16)	1.13
$2x \times 4x$	77	58	75.32	58	0 (0.0)	0.0

Source: From Tsuchiya, T. 1960b. *Seiken Zihô* 11: 29–37. With permission.

derived from *ms1 ms1* soybean always carry the recessive *ms1* gene either in homozygous (*ms1 ms1 ms1*) conditions and have poor seed set. Chen and palmer (1985) isolated 138 plants with greater than 40 chromosomes ($2n = 44$ to $2n = 71$), but not 41 chromosomes, from 32 male fertile (*Ms1 ms1 ms1*) triploid plants.

Autotriploids can be induced by mutagens. Pantulu (1968) obtained triploids ($2n = 3x = 21$) from gamma rays irradiated pearl millet seed. Singh and Ikehashi (1981) isolated 93 totally sterile to semi-sterile plants in an M_2 population of rice treated with a chemical mutagen, ethyleneimine. Two plants were found to be triploid ($2n = 3x = 36$). Morphologically triploid rice plants were dark-green, tillered profusely and showed extremely low pollen (0.9%–2.1%) and ovule fertility. The triploids probably originated by the fertilization of unreduced female gametes ($2n$) (caused by mutagen) by n male gametes. Colchicine induces formation of tetraploids as well as triploids (Myers, 1944; Jauhar, 1970).

2. Cytological Behavior in Autotriploids

In autotriploids only two of three homologous chromosomes associate at any point during pachynema. Pairing patterns determine the frequencies and types of trivalent association at metaphase-I (Darlington and Mather, 1932; Benavente and Orellana, 1984). Various combinations of univalents, bivalents, and trivalents are possible at diakinesis and metaphase-I (Table 8.18). Although all five species shown are $2n = 3x = 21$ and have intermediate chromosome size, they exhibit highly variable trivalent frequencies. The mean numbers of trivalents per cell ranged from 3.75 (*Secale cereale*) to 6.18 (*Pennisetum glaucum*). The highest frequencies of sporocytes with 5III + 2II + 1I were observed in barley, rye, and *T. monococcum*. In autotriploid maize ($2n = 3x = 30$), 9III + 1II + 1I chromosome association was the most frequent (McClintock, 1929c).

Trivalent association is expected to be more frequent with long chromosomes than with shorter chromosomes (Darlington, 1929; Darlington and Mather, 1932; Burnham, 1962). However, it has been suggested that aside from environmental influences, trivalent frequencies are genetically determined (Kuspira et al., 1986).

Based on random chromosome pairing, crossing-over, and chiasma formation during pachynema, there are four possible trivalent configurations at diakinesis: (a) chain (V-shaped), (b) ring-rod (frying-pan), (c) triple arc (bird-cage), and (d) Y-shaped (Figure 8.22a). The V-shaped trivalent can arise by distal chiasmata involving one

---

TABLE 8.18

Frequencies of Trivalents at Metaphase-I in Autotriploids in Five Species with $x = 7$ Chromosomes

Species	Frequencies of Trivalents at Metaphase-I								Total %	Mean No./Cell				Author
	0	1	2	3	4	5	6	7	7III	No. of Cells	I	II	III	
Hordeum spontaneum	1	5	4	27	45	53	44	13	6.77	192	2.33	2.34	4.66	Tsuchiya (1952)
Lolium perenne	0	0	0	3	8	32	53	68	41.46	164	0.79	0.79	5.34	Myers (1944)
Pennisetum glaucum	0	0	0	2	0	7	9	20	52.63	38	0.82	0.82	6.18	Pantulu (1968)
Secale cereale	1	0	4	4	3	5	3	0	0.00	20	3.25	3.25	3.75	Lamm (1944)
Triticum monococcum	2	8	37	89	123	126	87	29	5.80	500	2.65	2.60	4.38	Kuspira et al. (1986)

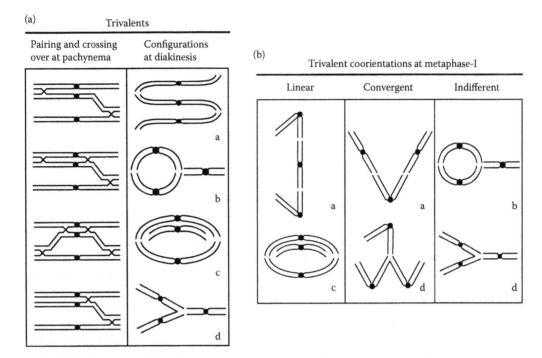

FIGURE 8.22 (a) Possible trivalent configurations at diakinesis in an autotriploid, based on crossing over and pachytene chromosome pairing. (b) Types of trivalent coorientations at metaphase-I in an autotetraploid. (Redrawn from Kuspira, J. et al. 1986. *Can. J. Genet. Cytol.* 28: 867–887. With permission.)

chromosome on both arms. A frying-pan (ring-rod) requires three crossover and three chiasmata (one interstitial and two distal). The triple arc configuration is rarely observed and arises by four reciprocal exchanges (two interstitial and two distal) in both arms. A Y-shaped trivalent needs a minimum of two reciprocal exchanges (one interstitial and one distal) involving only the same arm of all three homologues (Figure 8.22a–d).

Each group of three homologous chromosomes at diakinesis and metaphase-I can coorient in three different ways: (1) linear, (2) convergent, and (3) indifferent. The convergent pattern, particularly V-shaped trivalent coorientation at metaphase-I (Figure 8.22b-a), is the most frequent. Convergent typed (Figure 8.22b-d) is a hypothetical configuration and is generally not observed. Linear types rod-shaped (Figure 8.22b-a) and triple-arc (Figure 8.22b-c) and indifferent, that is, ring-rod (frying-pan, Figure 8.22b-b) and Y-shaped (Figure 8.22b-d) orientations are relatively rare (Table 8.19). It has been elucidated that orientation of the kinetochores within trivalents at metaphase-I towards the poles in meiocytes is random (Satina and Blakeslee, 1937; Balog, 1979; Kuspira et al., 1986).

TABLE 8.19

Frequencies of Trivalent Coorientations at Metaphase-I in Autotriploids

Species	2n	Trivalent Coorientations at Metaphase-I			Total	Author
		Convergent (%)	Linear (%)	Indifferent (%)		
Hordeum spontaneum	21	806 (90.2)	76 (8.5)	12 (1.3)	894	Tsuchiya (1952)
Triticum monococcum	21	561 (93.5)	0 (0.0)	39 (6.5)	600	Kuspira et al. (1986)

Disjunction of chromosomes at anaphase-I in trivalent is mainly determined by the orientation and coorientation of their kinetochores at metaphase-I. In convergent coorientation, two chromosomes regularly go to one pole and middle one to the opposite pole. Chromosomes of indifferent orientation often migrate in a 1:1 fashion. Occasionally a 2:1 segregation is found. Linear alignment usually chromosome segregate in a 1:1 fashion and the middle chromosome behaves like a lagging univalent. These observations indicate that chromosome migration at anaphase-I is nonrandom (Darlington, 1929; Burnham, 1962; Balog, 1979).

In autotriploid ($2n = 3x = 21$) of barley and *T. monococcum*, the highest frequency of sporocytes at anaphase-I showed 10-11 chromosome disjunction followed by 9-12, 8-13, and 7-14 (Tsuchiya, 1952; Kuspira et al., 1986). Lagging chromosomes in barley and *Lolium perenne* ranged from 0 to 6 and in *T. monococcum* ranged from 0 to 3 (Table 8.20). A contrasting difference among three autotriploids was the frequencies of sporocytes without laggards. In *T. monococcum*, 89.2% of the sporocytes did not carry laggards but the frequencies were much lower in barley and ryegrass (Table 8.20). Some of the laggards may be included in dyads. Tsuchiya (1952) found 50% telophase-I cells without laggards while anaphase-I cells showed 29.92%. Lagging chromosomes that do not reach either pole are eventually eliminated.

3. Fertility and Breeding Behavior in Autotriploids

Ovule fertility in autotriploids ranges from high sterility to complete fertility. Complete sterility has been found to be very useful for breeding seedless fruits, particularly seedless watermelon (Kihara, 1958), tuber-bearing plants, vegetatively propagated ornamentals, and fruit trees.

Although pollen fertility was high in triploid *Datura* (Satina and Blakeslee, 1937), barley (Tsuchiya, 1952), *T. monococcum* (Kuspira et al., 1986), and pearl millet (Dujardin and Hanna, 1988), seed set after selfing was very poor. In triploid *Datura*, while Satina and Blakeslee (1937) recorded pollen fertility between 50% and 60%, only 15% or less of pollen germinated and of the latter nearly 2/3 burst within or near the stigma, leaving about 5% functional pollen. These results suggest that micro gametophytes with other than haploid chromosome numbers are unviable because pollen grains with the aneuploid chromosome numbers ($n + 1$, $n + 2$, or higher) are genetically and physiologically unbalanced and cannot compete in fertilization with n chromosome spores. Pollen transmission of an $n + 1$ spore in diploid is rare.

On the female side, ovules with n, $n + 1$, $n + 2$, $n + 3$, and rarely $n + 4$ and $n + 5$ are functional. The maximum limit of tolerance of extra chromosomes is usually 3 in true diploids. Therefore, zygotes containing more than three extra chromosomes in addition to their normal chromosome complement will typically abort. However, there are a few exceptions where autotriploids set normal seed. These cases indicate that male and female spores can, sometimes, tolerate unbalanced chromosome numbers resulting in a high frequency of viable spores, and viable progeny that are produced may include diploid as well

---

TABLE 8.20

Frequencies (%) of Laggard at Anaphase-I in Three Autotriploid ($2n = 3x = 21$) Species

Species	Frequency of Cells with Laggards							Total No. of Cells	Author
	0	1	2	3	4	5	6		
Lolium perenne	37.0	26.0	17.0	9.0	9.0	1.0	1.0	1103	Myers (1944)
Hordeum spontaneum	29.92	26.26	21.26	15.35	2.36	3.15	0.79	127	Tsuchiya (1952)
Triticum monococcum	89.20	7.2	1.20	2.4	0.0	0.0	0.0	250	Kuspira et al. (1986)

as most of the possible aneuploids. In autotriploid of spinach, Janick and Stevenson (1955) observed normal seed set. Progenies of $3x \times 2x$, and the reciprocal, yielded plants ranging from diploid ($2n = 2x = 12$) to triploid ($2n = 3x = 18$) (Tabushi, 1958).

8.2.1.2.1.2 Hypertriploids

1. Sources of Hypertriploids
   Occasionally, trisomics generate a low frequency of triploids ($3x$) and near triploids, such as hypo triploids ($2n = 3x - 1$) and hypertriploids ($2n = 3x + 1$), in their progenies (Tsuchiya, 1960a, 1967; Singh and Tsuchiya, 1975a; Xu et al., 2000b). Hypertriploid ($2n = 3x + 1 = 22$) plants in barley were relatively short as compared to diploid but there was a marked increase in leaf size and stomatal length in hypertriploid compared to diploid sibs (Singh and Tsuchiya, 1975a). Hypertriploid soybean plant was more vigorous in vegetative growth with robust main branches and large, dark-green leathery leaves, matured much later than its disomic sib, a typical trait of an autotriploid plant (Xu et al., 2000b). A spontaneous hypertriploid ($2n = 3x + 1 = 61$) plant in the soybean was identified in the progeny of a cross between T31 (glabrous, $p2\ p2$) and an unidentified primary trisomic ($2n = 41$) (Xu et al., 2000b). It is assumed that hypertriploid plants probably originated by the fusion of an unreduced egg ($n = 40$) from T31 and an $n + 1$ ($n = 21$) sperm from the primary trisomic line. The frequency of hypertriploid in the soybean is extremely low. All 146 F_1 plants involving T31 (female) crosses with 14 primary trisomics contained either $2n = 40$ or $2n = 41$ chromosomes.

2. Cytological Behavior in Hypertriploids
   In hypertriploids ($2n = 3x + 1$), one chromosome is present four times while others occur in three copies. Hypertriploid plants have been found in several plant species. Singh and Tsuchiya (1975a) isolated four hypertriploid barley plants ($2n = 3x = 22$) in the progenies of Triplo 4 (Robust). Thus, it was assumed that chromosome 4 was present four times. As was expected, chromosome configurations at metaphase-I were various combinations of quadrivalents, trivalents, bivalents, and univalents. The maximum chromosome association was 1IV + 6III (Figure 8.23a). Quadrivalents were either N-shaped or with the fourth chromosome associated with tandem-V trivalent (Figure 8.23a,b). An association of 6III + 1II + 2I in some cells indicates that a quadrivalent dissociated to form 1III + 1I and 1III + 1II + 1I. The minimal chromosome association was 1IV + 1III + 5II + 5I. The frequency of univalents ranged from 0 to 5. An average frequency of chromosome association

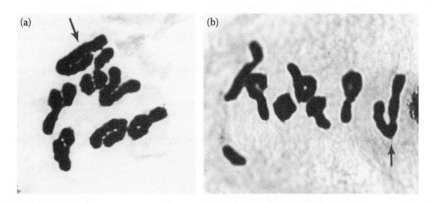

FIGURE 8.23 Meiotic metaphase-I configurations in a hypertriploid ($2n = 3x + 1 = 22$) in barley. (a) 1IV (arrow) + 6III; (b) 1IV (arrow) + 5III + 1II + 1I. (From Singh, R. J. and T. Tsuchiya. 1975a. *Caryologia* 28: 89–98. With permission.)

per cell in all hypertriploids was 0.35IV + 4.91III + 2.06II + 1.73I. Meiotic chromosome pairing in hypertriploid soybean was similar to the barley hypertriploid. Theoretically, the chromosome configurations at metaphase-I in soybean hypertriploid are mainly the following three types:

a. 1 IV + 19 III, 1 IV + 18 III + 1 II + 1 I————1 IV + 19 II + 19 I.
b. 19 III + 2 II, 18 III + 3 II + 1 I, 17 III + 4 II + 2 I————-1 III + 20 II + 18 I.
c. 20 III + 1 I, 19 III + 1 II + 2 I, 18 III + 2 II + 3 I———— 1 III + 19 II + 19 I.

Meiotic chromosome pairing in soybean hypertriploid showed above three types of chromosome association in 90% of the studied PMCs which is similar to the barley hypertriploid. In soybean hypertriploid, the second type of chromosome configuration was observed in 63% of the PMCs, suggesting that the four homologous chromosomes primarily formed two bivalents (Xu et al., 2000b). In barley hypertriploid, the frequency of the first type was marginally higher than that of the second type. These differences are probably due to the difference in chromosome size between soybean and barley. The soybean chromosomes are about 10 times smaller than those in barley.

The genetic male sterile soybean autotriploids showed chromosome association of enneavalent (association of three trivalents), hexavalent, pentavalent, and quadrivalent at diakinesis and metaphase-I (Chen and Palmer, 1985). Nonhomologous chromosome association was proposed. However, Xu et al. (2000b) recorded only quadrivalent and higher multivalents were not found in hypertriploid. Precocious chromosome separation in some bivalents and trivalents was common in some PMCs. This may be due to an early terminalization of chiasma in small soybean chromosomes.

The most frequent trivalent type in hypertriploids studied in barley was ring-rod (frying-pan) followed by tandem-V and tandem chain. The frequencies of triple arc and Y-shaped trivalents were very low and similar to the autotriploids. It is interesting to note that the tandem-V trivalent was found in the highest frequency in autotriploids of barley (Tsuchiya, 1952). It was not possible to examine shape of trivalents in the soybean hypertriploid because of small chromosomes. Chromosome disjunction at anaphase-I was fairly normal in 42.5% cells, without lagging chromosomes. The number of lagging chromosomes ranged from 0 to 4. Chromosome disjunction at anaphase-I in soybean hypertriploid was mostly 30–31 resulting meiotic products with higher chromosome numbers.

3. Fertility and Breeding Behavior in Hypertriploid

Hypertriploid plants in the soybean showed 63% fertile pollen while diploid had 98%. The first flush of flowers generally abort and that was also observed for the hypertriploid. The hypertriploid produced 98 selfed seeds, 16 seeds in hypertriploid (female) × diploid (male), and no seed in the reciprocal cross.

The self-seed of the soybean hypertriploid germinated poorly and only 43 (44%) of 98 seeds could germinate. Most of the shrunken and irregular-shaped seeds were not viable and it may be attributed to the poor endosperm development. The chromosome number ranged from $2n = 50$ to $2n = 69$ and plants with $2n = 40$ or 41 were not recovered. This suggests that soybean male and female spores tolerate higher chromosome numbers than those recorded in autotriploid and hypertriploid of barley. One more back cross is needed to isolate primary trisomics. This study clearly demonstrates that soybean is not a true diploid but is of tetraploid origin.

8.2.1.2.1.3 *Autotetraploids (4x, AAAA)*

1. Sources of Autotetraploids

Autotetraploids $(4x)$ occasionally appear spontaneously in natural diploid $(2x)$ populations by nondisjunction either in somatic tissue (meristematic chromosome doubling) or in reproductive tissues by formation of unreduced gametes (de Wet, 1980). Autotetraploids of spontaneous origin are also found among the progenies of twin seedlings and male sterile or in response to genes causing abnormal meiosis. A classic example of spontaneous origin of an autotetraploid is *Oenothera gigas* which contained $2n = 28$ chromosomes and was originally considered to be a genic mutant (Lutz, 1907). Furthermore, autotetraploids are successfully induced from temperature shocks, cell and tissue cultures, irradiations, and chemicals (colchicine, certain growth hormones, e.g., naphthalene-acetic acid) (Burnham, 1962). Chromosome doubling occurs in zygotes from fusion of an unreduced egg ($2x$) and an unreduced sperm ($2x$) and in somatic tissues from nondivision where chromosomes split but nuclear division fails, resulting in tetraploid tissues.

Colchicine: Of all the above means of obtaining tetraploids (polyploidy), colchicine has been found to be the most effective chemical to induce polyploidy in a large number of plant and animal species (Blakeslee and Avery, 1937; Blakeslee, 1939; Eigsti and Dustin, 1955; Burnham, 1962). Colchicine is an alkaloid and is a highly poisonous (carcinogenic) chemical that should not be to be absorbed by the skin (Blakeslee and Avery, 1937). Colchicine is extracted from the seeds and bulbs of the wild meadow saffron or autumn-flowering *Crocus* (*Colchicum autumnale* L.) (Eigsti and Dustin, 1955).

Colchicine acts by inhibiting spindle formation and preventing anaphase. Chromosomes stay at the equatorial plate but split longitudinally and divided chromosomes remain in a single restitution nucleus doubling the chromosome number. This process will be continued as long as the drug is present. Colchicine is also used as a pretreatment agent for counting somatic chromosomes. Colchicine appears to affect only actively dividing cells. Therefore, colchicine is applied to growing organs of plants, to rapidly germinating seeds, to growing shoots and buds and axillary nodes. Colchicine is water-soluble.

Seed treatment: Colchicine concentrations and treatment durations depend on the crop species. Blakeslee and Avery (1937) treated rapidly-germinating *Datura* seeds with colchicine concentrations ranging from 0.003125% to 1.6% for 10 days. Concentrations up to 0.1% were not effective but tetraploids were obtained in the higher concentrations (0.2%–1.6%).

Seed treatment with 0.25% aqueous colchicine was found to be the most effective in producing autotetraploids in chick pea (*Cicer arietinum* L., $2n = 16$), but seedling treatment failed (Pundir et al., 1983). Colchicine-treated seeds and seedlings should be washed thoroughly in running tap water before planting.

Seedling treatment: Actively growing shoots of young seedlings may be immersed in aqueous colchicine solution or colchicine may be applied to shoots by absorbent cotton or swab. Immersion of roots should be avoided. Sears (1941) obtained amphidiploid sectors from sterile intergeneric hybrids of Triticinae by use of colchicine. The crowns and bases of plants were wrapped with absorbent cotton and the cotton was soaked with 0.5% aqueous solution of colchicine. The plants were transferred to a high humidity chamber to maintain humidity. Treated plants can be covered with clear plastic bags. Cotton was kept wet for 2–5 days by applying colchicine twice a day. Cotton was removed and plants were transferred to the greenhouse for further growth. He obtained tetraploid sectors in 39 of the 60 surviving plants, representing 17 different hybrids. Stebbins (1949) isolated autotetraploids from 20 different species of the family Poaceae (Gramineae) by treating shoots with 0.1%–0.2% aqueous solution of colchicine for 8–24 h. Roots were washed thoroughly before planting. However, root feeding of colchicine to *Glycine* species hybrids failed to produce amphiploids but shoot treatment was successful. Jauhar (1970) obtained autotetraploid pearl millet by treating 10-day-old shoots and 2-week-old seedlings with 0.2% aqueous solution of colchicine for 11–12 h.

Schank and Knowles (1961) recorded that the most successful colchicine treatment to induce tetraploidy in safflower was a 0.1% aqueous solution applied to a cotton swab wedged between cotyledons of 3-day-old seedlings. The shoot treatment was less successful.

For potato, Kessel and Rowe (1975) attempted 10 different colchicine treatments to induce autotetraploids. The success rate was low (Table 8.21). Axillary treatment produced the best results, though only 2.8% plants were chromosomally doubled. However, the duration of colchicine treatment was 24 h. The longer treatment period (4–6 days) could have increased the frequency of tetraploids. Any of these treatments can be used to induce polyploidy in crop plants.

Thiebaut et al. (1979) recommended treating barley haploid seedlings at the three-leaf stage using a solution containing 0.1% colchicine, 2% DMSO, 0.3 mL/L (10 drops) of Twin 20, and 10 mg/L GA_3 under 25–32°C. Seedlings were treated for 5 h.

Colchicine (0.1%–0.2%) can be incorporated in liquid or solid media before autoclaving. Actively growing young shoots or axillary buds are exposed to colchicine for 7–14 days, and then are transferred to colchicine-free medium. Chromosomally doubled shoots are slow-growing with dark-green, thick leaves. Lyrene and Perry (1982) preferred colchicine treatment for blueberry in liquid medium on a rotating wheel for 24 h with 0.2% colchicine.

TABLE 8.21
Treatments, Methods, and Frequency (%) of Tetraploids Induced by Colchicine in Potato

Treatment	Methods	Chromosome Doubling (%)
1. Smearing	Lanoline paste containing 1% colchicine was smeared on the eyes of the tuber and the tubers were allowed to sprout.	1.9
2. Dropping I	Absorbent cotton with 1% colchicine was placed on the eyes of the tubers and cotton was kept moist for 5 days. Untreated eyes and sprouts should be removed.	1.2
3. Dropping II	Cotton was placed in the bottom of a Petri dish, and tubers with marked eyes were placed on the cotton. Cotton was kept moist with 1% colchicine for 5 days.	0.0
4. Soaking	The tubers were soaked in a Petri dish containing 1% colchicine for 5 days.	0.0
5. Seedling	A drop of lanoline paste containing 1% colchicine was placed on a growing point of approximately 7.5 cm tall seedling.	0.7
6. Subaxillary	Potato was grafted onto tomato root stock. After scions were 25–30 cm tall, axillary bud was removed, after 24 h cotton was wrapped in leaf axils and cotton was soaked with 1% colchicine. The treated plant was placed in a plastic bag. After 24 h, bag and cotton were removed.	2.8
7. Agar	Seeds were placed in a Petri dish on agar jelly containing 1% colchicine. Germinated seeds were washed and seedlings were transplanted to pots.	0.3
8. Colchicine solution	Seeds were germinated on filter paper in a Petri dish soaked in either 0.5% or 0.1% colchicine. Duration of treatment was not indicated. Seedlings were washed and transplanted to small pots. 0.6 (0.5 colchicine)	1.5 (0.1 colchicine)
9. Postgermination	Germinated seeds were soaked for 3–7 days in 1% colchicine.	0.9
10. Soaking	Dry seeds were soaked for 3–7 days in colchicine containing: 0.5%; 0.25%	0.4; 0.0

Source: From Kessel, R. and P. R. Rowe. 1975. *Euphytica* 24: 65–75. With permission.

Nitrous Oxide: It has been demonstrated that nitrous oxide is a chromosome doubling agent. Compared to colchicine, it induces a relatively higher frequency of chromosome-doubled plants, causes less lethality and the gas is relatively harmless. Chromosome doubling by nitrous oxide treatment is very effective for plants still in tissue culture but is unsuitable for large plants (Hansen et al., 1988)

Taylor et al. (1976) excised heads of diploid red clover containing 2 cm stems 24 h after crossing and placed them in vials containing a 2% aqueous sucrose solution. Vials with heads were placed in a gas-tight chamber and nitrous oxide was maintained to 6 bars atmospheric pressure (approximately 90 psi). After 24 h, the heads were removed to a dark incubator at 20°C for a seed maturation period of 3 weeks. A total of 226 plants were treated with nitrous oxide and 160 plants (71%) were identified as putative tetraploid ($2n = 4x = 28$) based on pollen size. Of 136 plants examined cytologically, 119 plants were tetraploids, 3 plants were diploid, 2 plants died, and 12 plants were aneutetraploid ($2n = 26$, 27, or 29). No plants with chromosomal chimaerism were found.

Hansen et al. (1988) compared the effect of nitrous oxide and colchicine on chromosome doubling of anther culture-derived young seedlings of wheat still in culture. Two colchicine treatments (0.005% and 0.01% for 24 h) were compared with two nitrous oxide treatments (24 and 48 h at 6 atm.). Both nitrous oxide treatments were as effective as 0.01% colchicine; however, colchicine treatment killed a significant proportion of the treated plants, while nitrous oxide treatment was nontoxic. The low concentration of colchicine (0.005%) showed lower chromosome doubling efficiency.

A mixture of 0.1% colchicine and 2.0% or 4.0% DMSO (dimethyl sulfoxide) proved to be more efficient in producing doubled haploids in barley than that observed with colchicine alone or by nitrous oxide treatment (Subrahmanyam and Kasha, 1975).

2. Morphological Characteristics of Autotetraploids

Generally autotetraploids are slow in growth and exhibit dark-green, large leaves. It is a common conception that autotetraploids usually produce *gigas* phenotype, plants larger than their diploid ancestors (Ramanujam and Parthasarathy, 1953; Kuspira et al., 1985). The *gigas* phenotype has been expressed in autotetraploid rye (Müntzing, 1951), pearl millet (Jauhar, 1970; Hanna et al., 1976), *Sorghum* (Schertz, 1962), safflower (Schank and Knowles, 1961), chickpea (Pundir et al., 1983) and *Triticum monococcum* (Kuspira et al., 1985), and is attributed to the increased sizes of cells, stomata, leaves, pollen, and seed. In self-pollinating species, autotetraploids are either smaller or similar in height to their diploid counterparts.

In *T. monococcum* autotetraploids ($2n = 4x = 28$), plant height was reduced by 15% compared to diploid sibs. Furthermore, the average reduction of number of tillers per plant was 37.5% (Kuspira et al., 1985). Autotetraploids are usually a few days later in flowering than the counterpart diploids.

3. Cytological Behavior in Autotetraploids

In autotetraploids, each chromosome is present four times. Thus, chromosome associations such as quadrivalents, trivalents, bivalents, and univalents are expected based on random association of four homologous chromosomes. Chromosome pairing is usually studied at diakinesis or metaphase-I of meiosis because chromosomes at these stages are in condensed form. However, the majority of chiasmata are usually terminalized.

Kuspira et al. (1985) diagrammed ten possible quadrivalent configurations for each group of the four homologous chromosomes (Figure 8.24). However, they observed only chains (convergent-a; parallel-a) and rings (convergent-b; parallel-b) of four chromosomes (Figure 8.24). Of the 272 quadrivalents examined, chain configurations comprised 70 and ring quadrivalents numbered 202. The open ring quadrivalent was the highest (197/272). The average chromosome association per cell was 0.62I + 9.86II + 0.2 III + 1.74IV.

Quadrivalent coorientations at metaphase-I

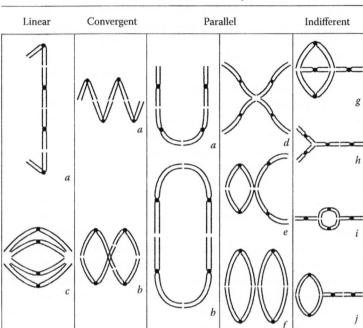

FIGURE 8.24 Expected 10 types of quadrivalent coorientations at metaphase-I in an autotetraploid. Generally, quadrivalent coorientations convergent a and b and parallel a and b are observed. (Redrawn from Kuspira, J., R. N. Bhambhani, and T. Shimada. 1985. *Can. J. Cenet. Cytol.* 27: 51–63. With permission.)

Table 8.22 compares the frequencies of chromosome configurations observed in autotetraploids of nine species containing basic chromosome number $x = 7$. The frequencies of bivalents and quadrivalents are different among species and even vary within a species. For example, Morrison and Rajhathy (1960a) recorded a higher frequency of quadrivalents (5.1 per cell) than those recorded (1.74 per cell) by Kuspira et al. (1985) in *T. monococcum*; these differences may be contributed to genetic factors.

TABLE 8.22

Mean Number of Chromosome Configurations Observed per Microsporocyte at Metaphase-I in Tetraploids of Six Species with $2n = 4x = 28$

Species	Univalents	Bivalents	Trivalents	Quadrivalents	Source
Arrhenatherum elatius	0.0	4.3	0.0	4.8	Morrison and Rajhathy (1960a)
Avena strigosa	0.1	5.0	0.1	4.4	Morrison and Rajhathy (1960a)
Hordeum bulbosum	0.1	5.6	0.2	4.00	Morrison and Rajhathy (1960a)
Hordeum vulgare	0.3	5.7	0.1	3.9	Morrison and Rajhathy (1960a)
Pennisetum americanum	2.64	8.97	0.38	1.49	Hanna et al. (1976)
Petunia hybrida	0.1	4.6	0.1	4.6	Morrison and Rajhathy (1960b)
Pisum sativum	0.3	5.21	0.16	4.20	Mercy-Kutty and Kumar (1983)
Secale cereale	0.5	6.1	0.2	3.7	Müntzing (1951)
Triticum monococcum	0.2	3.5	0.1	5.1	Morrison and Rajhathy (1960a)
Triticum monococcum	0.62	9.86	0.23	1.74	Kuspira et al. (1985)

Chromosome migration at anaphase-I in autotetraploids is determined by the coorientation of their kinetochores at metaphase-I. In *T. monococcum* autotetraploids, Kuspira et al. (1985) recorded 14-14, 16-13 + 1F (false univalent), 14-12 + 2F in 67%, 6.5% and 1.3% of sporocytes, respectively. Moreover, 15-13, 16-12, and 15-12 + 1F chromosome migrations were observed in 20.0%, 3.5%, and 1.7% of sporocytes, respectively. Based on anaphase-I chromosome movement, 70.90% balanced gametes ($n = 14$) is expected.

4. Fertility and Breeding Behavior in Autotetraploids

Autotetraploids are partially pollen and ovule abortive because of unbalanced spore constitution. The synthesized chickpea autotetraploid showed 84.4% (range 79.2%–90.2%) pollen fertility compared to 96.0% in diploids (Pundir et al., 1983).

Selection for higher fertility is accompanied by a higher frequency of bivalents in autotetraploid maize (Gilles and Randolph, 1951), pearl millet (Gill et al., 1969; Jauhar, 1970), job's tears (*Coix lacryma-jobi*) (Venkateswarlu and Rao, 1976), and safflower (Schank and Knowles, 1961). Gilles and Randolph (1951) observed fewer quadrivalents (7.46 per cell) and more bivalents in autotetraploids maize at the end of a 10-year period than were present at the beginning of the period (8.47 per cell). They suggested that the original seed stocks had a gene or genes in heterozygous condition for influencing quadrivalent formation. In safflower, Schank and Knowles (1961) examined tetraploids from C_1 to C_3 generations and found a decrease in frequency of quadrivalents and an increase in pollen fertility and seed set. Gill et al. (1969) studied autotetraploid pearl millet for five generations (C_0 to C_5) and observed a reduction in the frequency of multivalents and an increase in the frequency of bivalents. Seed fertility improved through six generations, but was still lower in tetraploids compared to diploids. Similarly, Jauhar (1970) observed a shift from multivalent to a bivalent type pairing in autotetraploid pearl millet from the C_0 to C_6 generations. Some of the quadrivalents formed in the C_5 and C_6 generations appeared to be loosely associated. He suggested that structural divergence coupled with genetic factors was the cause of sterility in raw autotetraploids.

Selection for high fertility in autotetraploids is under genetic control (Roseweir and Rees, 1962; Doggett, 1964; Bender and Gaul, 1966; Narasinga Rao and Pantulu, 1982). Selection for high fertility in autotetraploid rye was influenced by increasing the quadrivalent frequency and reducing trivalent frequency, and the bivalent association was under genetic control. However, the high frequency of quadrivalents and high fertility found in rye may not occur in all other tetraploids (Roseweir and Rees, 1962). Doggett (1964) observed variations in seed set (5%–80%) among individual heads of *Sorghum* but it was not possible to isolate lines of higher fertility after eight generations; fertility, however, was suggested to be under genetic control.

Morrison and Rajhathy (1960b) contradicted the above findings. They examined 10 autotetraploids representing different families and found no evidence of an increase in the numbers of bivalents. Approximately two-third of homologous chromosomes formed quadrivalents, and no evidence of genetic control over quadrivalent frequencies was found. Plants with small chromosomes contained as many quadrivalents as plants with large chromosomes. They concluded there is a fairly uniform number of quadrivalents for all species and that either present-day estimates of chiasmata are out of line or else their frequency has no effect on the number of quadrivalents formed.

Mastenbroek et al. (1982) studied chromosome associations in early and advanced generations of tetraploid maize ($2n = 4x = 40$). The increased yield in tetraploid maize selected for 22 years was not due to a change in chromosome associations during diakinesis. It is interesting to note that frequency of both quadrivalent and bivalent formation remained constant from generation to generation and increase in yield was associated with selection against production and functioning of cytologically unbalanced gametes. Selection for fertility was a heritable unit.

Bender and Gaul (1966) suggested that diploidization of autotetraploids could be achieved artificially by an extensive genome reconstruction by means of induced chromosome mutations and gene mutations. Based on these assumptions, Friedt (1978) selected highly fertile lines derived from the hybridization of five autotetraploid barley varieties previously irradiated up to seven times, but the results were negative.

The above results suggest that autotetraploids do not breed true, but throw a low frequency of aneuploids in their progenies because of occasional 3:1 disjunction during anaphase-I (Müntzing, 1951; Doyle, 1986).

5. Genetic Segregation in Autotetraploids

The genetics of autotetraploids is quite complex if pursued to any depth. In autotetraploids, every chromosome and gene locus is present in four copies. Quadrivalent formation and tetrasomic inheritance are the criteria to identify autotetraploids. Multivalents are also formed in reciprocal translocations but the genetic ratios distinguish autotetraploids from reciprocal translocations.

The number of possible genotypes in diploid with two alleles is AA, Aa, and aa. On the other hand, the number of genotypes in $4x$ level is five:

AAAA or A^4	$\rightarrow$	Quadruplex
AAAa or A^3a	$\rightarrow$	Triplex
AAaa or A^2a^2	$\rightarrow$	Duplex
Aaaa or Aa^3	$\rightarrow$	Simplex
aaaa or a^4	$\rightarrow$	Nulliplex

Methods of calculating theoretical genetic ratios for genotypes triplex (AAAa), duplex (AAaa), and simplex (Aaaa) in random chromosome, random chromatid, and maximum equational segregations have been described in detail by Allard (1960) and Burnham (1962) and are summarized in Table 8.23.

6. Advantage of Autopolyploidy

a. Only certain ploidy level is beneficial. For example, $4x$ apple is not desirable but $3x$ apples are good and some of the best varieties of the apple are $3x$. Red clover and turnip perform the best at $4x$ but fodder beets are generally unfavorable at this level.

b. Sterility associated with ploidy may be desirable. Certain floral plants have longer lasting blooms if sterile. Autotriploids may be beneficial if propagated vegetatively. Seedless banana and watermelon are triploid and commercially successful.

c. Among cereals, $4x$ rye is successful may be of out breeding nature. Fertility in $4x$ is low because of cytological abnormalities that results in genetic unbalance and cross pollination with $2x$ plants. This produces $3x$ seeds which abort prematurely.

d. Incompatibility system that successfully maintain cross fertilization at $2x$ may break down and allow self-fertilization at $4x$.

8.2.1.2.2 Allopolyploidy

8.2.1.2.2.1 Origin of Allopolyploidy An allopolyploid individual is derived by interspecific hybridization from two or more genomically distinct and distantly-related diploid species followed by chromosome doubling of sterile F_1 hybrids. Interspecific and intergeneric (wide hybrid) F_1 hybrids are usually sterile because genomes are highly divergent and chromosomes lack affinity or form only a small number of loosely synapsed bivalents. This type of synapsis is known as heterogenetic association. In contrast, homogenetic association occurs between chromosomes of the same genome. However, normal chromosome pairing and seed fertility are restored by doubling the chromosomes to a condition known as amphiploidy or amphidiploidy (Stebbins, 1950; Allard, 1960). The majority of allopolyploids are either tetraploids or hexaploids.

TABLE 8.23

Expected Gametic Types and F$_2$ and BC$_1$ Phenotypic Ratios in Autotetraploids for Random Chromosome, Random Chromatid, and Maximum Equational Segregations

Genotype	Gametic Types			Phenotypic Ratios (F$_2$)			Phenotypic Ratios (BC$_1$)		
	Chromosome	Chromatid	Maximum Equational Segregation	Chromosome	Chromatid	Maximum Equational Segregation	Chromosome	Chromatid	Maximum Equational Segregation
AAAa	1AA + 1Aa	15AA + 12Aa + 1aa	13AA + 10Aa + 1aa	allA	783A:1a	575A:1a	AllA-	27A-:1a	23A-:1a
AAaa	1AA + 4Aa:1aa	3AA + 8Aa + 3aa	2AA + 5Aa + 2aa	35A:1a	20.8A:1a	77A:4a	5A-:1a	3.7A-:1a	7A-:2a
Aaaa	1Aa + 1aa	1AA + 12Aa + 15aa	1AA + 10Aa + 13aa	3A:1a	2.5A:1a	407A:169a	1A-:1a	0.87A-:1a	11A-:13a

Source: Adapted from Burnham, C. R. 1962. *Discussion in Cytogenetics. Burgess*, Minneapolis, MN.

Allopolyploidization in nature has played the most important role in the evolution of the crop species wheat, oat, cotton, tobacco, brassicas, and sugarcane. The origin, cytology, and breeding behavior of a few allopolyploids have been described in a chapter on genome analysis.

The classical examples of artificially synthesized allopolyploids are *Raphanobrassica* and Triticale. Karpechenko (1927) produced an intergeneric hybrid between radish (*Raphanus sativus*, $2n = 2x = 18$; genome RR) and cabbage (*Brassica oleracea* $2n = 2x = 18$; genome BB). Chromosomes of the sterile F_1 hybrid showed absolutely no homologies. Fertility was restored by doubling the chromosomes through the formation and union of unreduced male and female gametes.

8.2.1.2.2.2 Characteristics of Allopolyploidy Plant geneticists are optimistic to create a new crop through allopolyploidization. The success of new crops depends on the following factors:

1. It is extremely difficult to predict the nature of interaction between two genomes. The interaction for any trait may be intermediate or heterotic or lethal.
2. Interaction between nuclear and cytoplasmic genes may produce cytoplasmic genetic male sterility.
3. Economically beneficial traits can be transferred to cultigen from the alien species by hybridization followed by chromosome doubling and backcrossing.

8.2.1.2.2.3 Man-Made Crop Triticales, *the first man-made cereal of great potential economic value*, are amphiploids of *Triticum* and *Secale*. Concerted cytogenetic and breeding efforts have been made for the past several decades to improve the yield and qualities of triticales particularly hexaploid ($2n = 6x = 42$) and octoploid ($2n = 8x = 56$) cytotypes. Tetraploid ($2n = 4x = 28$) and decaploid ($2n = 10x = 70$) triticales also have been studied. Triticales may be either primary (original "raw" amphiploids) or secondary (advanced lines derived from crosses between triticales and hexaploid wheats) (Gupta and Priyadarshan, 1985).

Hexaploid triticales (AABBRR) are amphiploids produced from tetraploid (AABB) wheat and diploid (RR) rye. Due to intensive breeding efforts, several secondary hexaploid triticale lines and cultivars have been released. However, cytological instability, reduced seed set, and shriveled seed are a few of the major constraints for the success of triticales. The cytological instability was attributed to tetraploid cytoplasm, allogamous rye, and the differences between duration of meiosis, DNA content between rye and wheat and interaction between the genes of wheat and rye chromosomes.

Octoploid triticales (AABBDDRR) are derived from doubling the chromosomes of F_1 hybrids of hexaploid wheat (AABBDD) and diploid rye (RR). Weimarck (1973) found a lower degree of aneuploidy, stabler meiosis, and higher fertility than with primary triticale. Fertility was not correlated with cytological disturbances. Octoploid triticales contain good winter hardiness, high protein content, good baking quality and have early flowering and seed maturity and large kernel size. However, seed sterility and early sprouting are under desirable traits that have hampered the popularity of octoploid triticale.

Tetraploid triticales are produced in three ways: (i) by hybridizing *T. monococcum* (AA) with diploid rye (RR) and doubling the chromosomes of the sterile F_1 (AR) to obtain fertile amphiploids (AARR); (ii) by isolating tetraploid triticale lines (AARR) after crossing autoallohexaploid wheat (AAAABB) with diploid rye (RR), followed by the elimination of B genome chromosomes; (iii) by hybridizing hexaploid triticale (AABBRR) with diploid rye (RR) to isolate ABRR and after selfing, to select AARR, BBRR, and (AB) (AB) RR lines. Tetraploid triticales do not have commercial value at present but can be used to improve hexaploid triticale and diploid rye (Krolow, 1973).

Decaploid triticale (AABBDDRRRR) is obtained by crossing hexaploid wheat and tetraploid rye, followed by chromosome doubling. Decaploid triticale is found to be useless because of poor vigor and seed fertility.

8.2.2 Aneuploidy: Trisomics

8.2.2.1 Primary Trisomics

8.2.2.1.1 Introduction

An organism containing a normal chromosome complement and one extra chromosome is known as a trisomic. The extra chromosome may be a primary, secondary, tertiary, telocentric, acrocentric, or compensating type (Figure 8.25). An individual with a normal chromosome complement plus an extra complete chromosome ($2n = 2x + 1$) is designated a primary trisomic, and the individual is called Triplo. Blakeslee (1921) coined the term "trisome" for a plant with addition of one member of the basic chromosome complement, and Bridges (1921) was the first to apply trisomic technique to associate a gene *ey* with the fourth chromosome of *Drosophila melanogaster*. Primary trisomics, since then, have been used extensively for the determination of gene-chromosome-linkage group relationships in several plant species (Khush, 1973; Singh, 1993, 2003).

8.2.2.1.2 Sources of Primary Trisomics

Generally triploids are considered to be one of the best and most dependable sources for establishing primary trisomic series (Table 8.24). However, primary trisomics have been isolated occasionally from the progenies of normal diploids (spontaneous origin), asynaptic and desynaptic plants, mutagen-treated progenies, interchange heterozygotes, and in the progenies of related and unrelated trisomics. The requirement to all the sources of primary trisomics is the nondisjunction of chromosomes during meiosis. This results in aneuploid ($n + 1, n + 2, n + 3, \ldots$) gametes and when these gametes are fertilized by a male sperm with n chromosome number, aneuploid plants ($2n + 1$, $2n + 2, 2n + 3, \ldots$) are originated.

8.2.2.1.2.1 Autotriploids A complete set of primary trisomics and other aneuploids have been isolated in a majority of diploid species from the progenies of autotriploids ($3x$) by diploid ($2x$) crosses (Tables 8.24 and 8.25). Triploid plants occasionally occur in natural populations by spontaneous origin. They are frequently contrasted from diploids by their taller height, more vigorous growth, and profuse tillering habits. They may show higher sterility. Autotriploid plants are also produced by crossing tetraploid and diploid plants. Generally, autotriploid plants are pollinated by the diploid of the same variety in order to isolate a complete trisomic series in a uniform genetic background (Table 8.25).

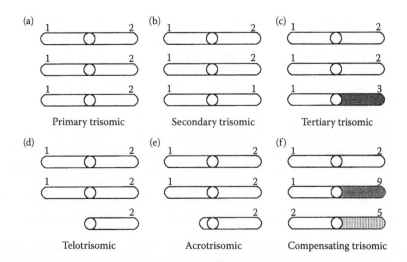

FIGURE 8.25 Diagrammatic sketch of various types of trisomics.

TABLE 8.24
Primary Trisomics Derived from Autotriploids

Species	2n	Author
1. *Agropyron cristatum*	14	Imanywohaet al. (1994)
2. *Antirrhinum majus*	16	Rudorf-Lauritzen (1958); Sampson et al. (1961)
3. *Arabidopsis thaliana*	10	Sears and Lee-Chen (1970)
4. *Avena barbata*	14	Nishiyama (1981)
5. *Beta vulgaris*	18	Levan (1942); Kaltsikes and Evans (1967); Romagosa et al. (1986)
6. *Clarkia unguiculata*	18	Vasek (1956)
7. *Collinsia heterophylla*	14	Garber (1964)
8. *Corchorus olitoris*	14	Iyer (1968)
9. *Crepis capillaris*	6	Babcock and Navashin (1930)
10. *Datura stramonium*	24	Blakeslee and Avery (1938)
11. *Fragaria bracteata*	14	Yarnell (1931)
12. *Helianthus annuus*	34	Jan et al. (1988)
13. *Hordeum chilense*	14	Cabrera et al. (1999)
14. *Hordeum spontaneum*	14	Tsuchiya (1960a)
15. *Hordeum vulgare*	14	Kerber (1954); Tsuchiya (1967)
16. *Humulus lupulus*	18	Haunold (1970)
17. *Lolium perenne*	14	Myers (1944); Meijer and Ahloowalia (1981)
18. *Lotus pedunculatus*	10	Chen and Grant (1968a, 1968b)
19. *Lycopersicon esculentum*	24	Lesley (1928, 1932); Rick and Barton (1954)
20. *Lycopersicum peruvianum*	24	Sree Ramulu et al. (1977)
21. *Medicago sativa*	16	Kasha and McLennen (1967)
22. *Nicotiana sylvestris*	24	Goodspeed and Avery (1939)
23 *Psathyrostachys juncea*	14	Wei et al. (1995)
24. *Oenothera blandina*	14	Catcheside (1954)
25. *Oryza sativa*	24	Hu (1968); Iwata et al. (1970); Iwata and Omura (1984); Khush et al. (1984); Cheng et al. (2001)
26. *Oryza glaberrima*	24	Ishiki (1991)
27. *Pennisetum glaucum*	14	Gill et al. (1970)
28. *Petunia axillaris*	14	Reddi and Padmaja (1982)
29. *Petunia hybrida*	14	Smith et al. (1975)
30. *Secale cereale*	14	Kamanoi and Jenkins 1962; Zeller et al. (1977); Melz et al. (1988); Fujigaki and Tsuchiya (1990)
31. *Solanum chacoense*	24	Lee and Rowe (1975)
32. *Solanum tuberosum*	24	Wagenvoort (1995)
33. *Sorghum bicolor*	20	Schertz (1966, 1974)
34. *Spinacia oleracea*	12	Tabushi (1958); Janick et al. (1959)
35. *Triticum monococcum*	14	Kuspira et al. (1986)
36. *Verbena tenuisecta*	10	Arora and Khoshoo (1969)
37. *Zea mays*	20	McClintock (1929c); McClintock and Hill (1931)

Various combinations of chromosome configurations, trivalents, bivalents, and univalents are observed at diakinesis and metaphase-I in triploids. Chromosomes move at random during anaphase-I and gametes with chromosome numbers ranging from n to $2n$ are expected at the end of meiosis. Thus, it is possible to obtain plants with $2x$ to $3x$ chromosome numbers in the progeny of an autotriploid plant. However, plants with $2x$ (diploid), $2x + 1$ (primary trisomic), $2x + 1 + 1$ (double

TABLE 8.25
Species, Haploid (*n*), and Frequencies of Aneuploid Plants in Triploid (3*x*) by Diploid (2*x*) Crosses

Species	*n*	2*x*	2*x*+1	2*x*+2	2*x*+3	2*x*+4	2*x*+5	2*x*+6	2*x*+7	2*x*+8	3*x*	Other	Total Nos. Plant	% 2*x*+1	Author
Datura stramonium	12	58	138	79	10									285	48.4	Satina et al. (1938)
Hordeum spontaneum	7	29	59	22	6	1						1	8	126	46.8	Tsuchiya (1960a)
Lolium perenne	7	30	52	27	6	4								119	43.7	Myers (1944)
Lotus pedunculatus	5	126	73	2									1	202	36.1	Chen and Grant (1968a)
Lycopersicon esculentum	12	303	342	131	7							2	14	799	42.8	Rick and Barton (1954)
Nicotiana sylvestris	12	1	17	20	8	3	3		1	1			3	57	29.8	Goodspeed and Avery (1939)
Oryza sativa	12	2	20	25	14	8	3							72	27.8	Khush et al. (1984)
Oryza glaberrima	12	6	11	9	5	1								32	34.4	Ishiki (1991)
Pennisetum glaucum	7	9	20									1	11	91	22.0	Gill et al. (1970)
Secale cereale	7	2	23	12	6	3							2	58	39.7	Kamanoi and Jenkins (1962)
Spinacia oleracea	6	30	39	14	8	6	15	16						128	30.5	Tabushi (1958)
Spinacia oleracea	6	90	118	23	3	9	41	43						327	36.1	Janick et al. (1959)
Triticum monococcum	7	4	59	2										135	43.7	Kuspira et al. (1986)

Frequencies of Plants with Chromosome Numbers

TABLE 8.26

Chromosome Numbers in the Progenies of Autotriploid Plants of Rice

Reference	24	25	26	27	28	29	30	2x + 1 (%)	Total
				2n					
Ramanujam (1937)	6	9	8	9	11	4	3	18.0	50
Katayama (1963)	3	6	8	1	2	1	0	28.5	21
Hu (1968)	20	42	61	14	1	0	1	30.2	139
Watanabe et al. (1969)	6	35	31	12	2	1	0	40.2	87
Khush et al. (1984)	2	20	25	14	8	3	0	27.8	72

trisomic), and $2x + 1 + 1 + 1$ (triple trisomic) chromosome numbers are frequently obtained in varying frequencies (Table 8.25). These numbers indicate that the maximum number of extra chromosomes tolerated by gametes in most diploid species is one, or two or rarely three. This limitation may be due to the fact that male and female gametes, or zygotes and embryos, with higher than three extra chromosomes abort in the progenies of triploids of the diploid species because the duplication of extra genetic material causes genetic and physiological imbalance.

The true nature of a species, whether it is a diploid or a polyploid, can be ascertained by looking at the frequency of aneuploids in progeny from autotriploids. For example, the basic chromosome number of rice is disputed by several rice scientists. Nandi (1936) proposed, based on karyotype analysis, that rice (*Oryza sativa* L.) is a balanced secondary allotetraploid with $x = 5$ as a basic chromosome number, where A and B chromosomes duplicated after hybridization between putative ancestral species. However, it was evident from the progenies of autotriploids that the maximum number of extra chromosomes tolerated by rice gametes was five and plants with $2n + 1$ and $2n + 2$ chromosomes were in the highest frequency (Table 8.26). The narrow tolerance limit of the extra chromosomes in gametes shows that rice is truly diploid with $x = 12$ basic chromosome number. Polyploid species tolerate much higher numbers of extra chromosomes (Ising, 1969; Khush, 1973).

The frequency of occurrence of each trisomic type in progenies of autotriploids varies with types of trisomics, year of planting and above all it is influenced by the genetic background of an autotriploid. Tsuchiya (1960a) reported in barley that Pale (Triplo 3) trisomic appeared most frequently in each of two years (Table 8.27) followed by Slender (Triplo 2), Purple (Triplo 6), Bush (Triplo 1), and

TABLE 8.27

Frequency of Seven Types of Primary Trisomics in the Progenies of Autotriploids and Double or Triple Trisomic Plants of Barley

Type of Trisomics	From Autotriploid Progeny			From Double and Triple Trisomics	
	1953	1954	Total		
	%	%	%	%	%
Bush	5(20.8)	3(8.6)	8(13.6)	1(5.6)	9(11.7)
Slender	3(12.5)	8(22.9)	11(18.6)	3(16.7)	14(18.2)
Pale	6(25.0)	10(28.6)	16(27.2)	3(16.7)	19(24.7)
Robust	2(8.3)	2(5.7)	4(6.8)	3(16.7)	7(9.1)
Pseudo-normal	1(4.2)	3(8.6)	4(6.8)	3(16.7)	7(9.1)
Purple	3(12.5)	6(17.2)	9(15.3)	3(16.7)	12(15.6)
Semi-erect	4(16.6)	3(8.6)	7(11.8)	2(11.1)	9(11.7)

Source: From Tsuchiya, T. 1960a. *Jpn. J. Bot.* 17: 177–213. With permission.

Semi-erect (Triplo 7). The frequencies of Robust (Triplo 4) and Pseudo-normal (Triplo 5) were lowest (6.8%) among 59 primary trisomics (Table 8.27). Similar results have been reported in *Datura stramonium* (Blakeslee and Avery, 1938), *Lycopersicon esculentum* (Lesley, 1928, 1932; Rick and Barton, 1954), *Nicotiana sylvestris* (Goodspeed and Avery, 1939), *Zea mays* (McClintock and Hill, 1931), *Oryza sativa* (Hu, 1968), and *Beta vulgaris* (Romagosa et al., 1986). Chen and Grant (1968a), working with the trisomics of *Lotus pedunculatus*, were unable to isolate Triplo 2. They suggested two possibilities: the first explanation was that Triplo 2 may have been grouped with other types. This outcome is possible because somatic chromosomes of *Lotus pedunculatus* are small and do not allow karyotype analysis. Furthermore, pachytene chromosome analysis was not conducted and genetic tests were not performed. A second possibility is that Triplo 2 may be unviable or less viable than other Triplos. A similar situation was reported in *Oenothera* (de Vries and Boedijn, 1923) and *Arabidopsis thaliana* (Steinitz-Sears, 1963).

Isolation of primary trisomics from autotriploids is preferred because the frequency of aneuploids is extremely high in the progenies of autotriploids compared to such other sources as synaptic mutants, translocation heterozygotes, induced mutations, and normal diploids. Autotriploids generally produce 30%–80% simple primary trisomic plants while other sources yield 1%–3%.

It has been observed that the initial phase of seed development is normal in autotriploids but the endosperm shrivels after a week, resulting in the death of the embryo. The failure in endosperm development is caused by an extremely unbalanced chromosome number and this is the most likely explanation for the occurrence of plants with $2x + 1$, $2x + 2$, $2x + 3$ chromosome numbers in the progenies of autotriploids such as barley, rice, maize, tomato, and others (Table 8.25). Rick and Notani (1961) reported that the primitive variety of tomato Red Cherry tolerates a significantly higher frequency of aneuploids than of a large fruited cultivated variety San Marzano. It was also reported that extra chromosomes caused far less anatomical modification in Red Cherry than in a large fruited tomato. The triploids and aneuploids of Red Cherry had also higher fertility. Similar observations were reported in barley where trisomics of wild barley (*Hordeum spontaneum*) were vigorous and fertile compared to the trisomics of cultivated barley (Tsuchiya, 1960a, 1967). In the cultivated varieties the gene balance may be upset as they represent a short time selection of individuals which are superior in one or a few special traits. It has also been suggested that primitive varieties tolerate significantly higher chromosome numbers than cultivated varieties.

8.2.2.1.2.2 Synaptic Mutants (Asynaptic and Desynaptic)   A few trisomic plants have been isolated from the progenies of mutant plants. Synaptic mutants show a high frequency of univalents at diakinesis and metaphase-I of meiosis. Their occurrence is due to the disturbance in normal bivalent pairing governed by a homozygous recessive gene. The random movement of chromosomes generates $(n + 1)$ gametes and when these gametes are fertilized by a haploid male gamete primary trisomics are produced. Koller (1938) isolated a primary trisomic in asynaptic progenies of *Pisum sativum*. Katayama (1963) obtained 9.2% (21/227) plants with $2n = 25$ chromosomes from the progeny of asynaptic rice plants. However, no attempts were made to identify trisomics morphologically, cytologically, or genetically. Palmer (1974) and Palmer and Heer (1976) obtained a few trisomics among progeny of homozygous recessive asynaptic soybean, *Glycine max*. Xu et al. (2000c) identified possible 20 primary trisomics of the soybean from the aneuploid lines obtained from the progenies of asynaptic and desynaptic, and male sterile lines and male sterile induced autotriploids.

Dyck and Rajhathy (1965) isolated six of the seven possible primary trisomics from the progeny of desynaptic *Avena strigosa*. Rajhathy (1975) further reported all the expected seven primary trisomics from the progeny of desynaptic plants. Several backcrosses were made to eliminate the desynaptic gene.

Five morphologically distinct primary trisomics of *Vicia faba* were isolated in the progenies of asynaptic mutant and were identified cytologically by translocation tester sets (Barceló and Martin, 1990).

Primary trisomics obtained from the progenies of synaptic mutants are not suitable for use in cytogenetic analysis because they include sterile plants that cause the release of unrelated trisomics in their progeny (Rick and Barton, 1954).

8.2.2.1.2.3 Mutagen Treatment Primary trisomics have been isolated sporadically from the progenies of plants treated by mutagens. Soriano (1957) isolated trisomics in *Collinsia hetero-phylla* by colchicine treatment. However, trisomics did not express contrasting morphological traits (Dhillon and Garber, 1960; Garber, 1964). Their independence was not tested cytologically and genetically. Therefore, it is likely that not all the possible (7) trisomics were induced. Parthasarathy (1938) obtained accidently a trisomic plant from the progenies of X-irradiated rice seeds. Martin (1978) isolated four primary trisomics and two double trisomics from the progeny of an X-ray induced mutant in *Vicia faba*. Simeone et al. (1985) produced fourteen possible primary trisomics in *Durum* wheat from the progenies of mutagen (X-ray, thermal or fast neutrons) treated seeds; only transmission of the extra chromosome was reported. Thakare et al. (1974) isolated morphologically distinct five primary trisomics by irradiating seeds of *Corchorus olitorius* cv. JR063 ($2n = 14$) with gamma rays and thermal neutrons. These primary trisomics were produced from chromosomal interchanges. One trisomic was secondary for chromosome 5 and another was tertiary for chromosome 2.

Chemical irradiation treatment disturbs cell division, leading to nondisjunction of chromosomes during anaphase-I. This treatment may produce female gametes with $n + 1$ chromosome constitution and may give rise to a trisomic plant after fertilization with an n chromosome sperm. However, it is possible that trisomics isolated from mutagen treatment are not always true primary trisomic types but may in some cases include tertiary trisomics.

8.2.2.1.2.4 Normal Diploids Earlier reports on the occurrence of primary trisomics were based mainly on their occasional isolation from the progenies of normal diploids: *Datura* (Blakeslee, 1924); *Matthiola* (Frost and Mann, 1924); *Crepis* (Babcock and Navashin, 1930); *Secale* (Takagi, 1935); *Hordeum* (Smith, 1941); *Lycopersicon* (Rick, 1945), and *Nicotiana* (Goodspeed and Avery, 1939). In diploid ($2n = 24$) progenies of marigold (*Tagetes erecta*), Lin and Chen (1981) isolated trisomic plants. The occurrence of primary trisomics from diploid progenies is possible if one bivalent fails to move to the metaphase plate and is included in one telophase nucleus. Trisomic plants are obtained when $n + 1$ gametes produced in this way are fertilized by normal sperm. Belling and Blakeslee (1924) demonstrated cytologically eight cases of 11–13 chromosome separation in 1137 PMC of normal disomic *Datura* (0.4% $n + 1$ pollen grains).

8.2.2.1.2.5 Other Sources Another good source of primary trisomics is among the progenies of double trisomics or multiple trisomics. Sometimes, they are an excellent source for obtaining missing trisomics. Goodspeed and Avery (1941) found the missing trisomic 12 in the progeny of a multiple trisomic while the other 11 primary trisomics of *Nicotiana sylvestris* were isolated from other sources. Tsuchiya (1960a) obtained all seven primary trisomics from double and triple trisomics in barley. The frequency varied with the trisomic types (Table 8.27). The double and triple trisomics are frequently obtained as siblings of primary trisomics in the progenies of autotriploids. They are relatively vigorous and more or less seed fertile.

Sometimes unrelated primary trisomics are obtained from the progenies of primary trisomics. They have been recorded in *Lycopersicon esculentum* (Lesley, 1928), *Datura stramonium* (Blakeslee and Avery, 1938), *Nicotiana sylvestris* (Goodspeed and Avery, 1939), and *Lotus pedunculatus* (Chen and Grant, 1968b). No unrelated primary trisomics were isolated in barley (Tsuchiya, 1960a, 1967). Avery et al. (1959) suggested from their studies of *Datura* that the presence of an extra chromosome encourages in some way nondisjunction of other chromosome sets. However, such events are rare and one should not depend on this source to isolate primary trisomics.

TABLE 8.28

Frequency of Trisomics from Interchanged Heterozygotes of Barley

Interchange	Number of Plants	Number of Trisomics	Percentage Trisomics
a + b	1008	26	2.58
b + c	1708	16	0.94
b + d	1380	53	3.84
b + g	1290	59	4.57
e + f	1629	92	5.65

Source: From Ramage, R. T. 1960. *Agron. J.* 52: 156–159. With permission.

Occasionally, progenies of interchange heterozygotes produce primary trisomics due to a 3-1 chromosome disjunction at anaphase-I from a noncooriented quadrivalent. The frequency of primary trisomics ranged from 0.94% to 5.65% (Table 8.28).

The progenies of secondary, tertiary, and telotrisomics throw primary trisomics with a low frequency and the trisomic types are those to which secondary, tertiary, or telotrisomics belong. This outcome happens when secondary, tertiary, or telocentric chromosomes disjoin from a trivalent and move to one pole while two normal chromosomes move to the opposite pole. Gametes with a complete normal extra chromosome are generated and after fertilization with normal haploid sperm primary trisomics result. Singh and Tsuchiya (1977) isolated as high as 1.5% plants with $2n = 15$ chromosomes in the progenies of seven monotelotrisomics of barley. In general, aneuploid plants of hybrid origin have a heterotic effect, which may give them a better tolerance against numerical imbalance.

8.2.2.1.3 Identification of Primary Trisomics

In general, each of the primary trisomics in a diploid species differs from its normal diploid sibs and also with each other in several distinctive traits. The differences are morphological, anatomical, cytological, physiological, or genetic. Each chromosome carries distinctive genes which affect qualitative as well as quantitative traits responsible for plant growth, vigor, and development. The modifications in expression are such that each trisomic type is distinct and easily distinguishable.

8.2.2.1.3.1 Morphological Identification The identification of a complete primary trisomic series established in *Datura*, maize, barley, diploid oat, pearl millet, rice, *Arabidopsis*, tomato, and several other crop species is based on morphological features such as growth habit, plant height, degree of branching and tillering, leaf size, leaf shape, color and texture of leaf surface, internode length, days to flower, seed fertility, and many other visible morphological traits.

The main diagnostic feature of a plant with $2n = 2x + 1$ chromosome constitution is that plant shows slower growth habit than its diploid sibs. Primary trisomic plants can be distinguished fairly easily from normal diploids at seedling, maximum tillering, flowering, or maturity stages. The nomenclature of a trisomic type is based on a most distinctive morphological feature. The 12 trisomic types of *Datura*, a classical example, were named according to the shape and size of *capsules*. The extra chromosome which modified the phenotypic appearance of the plant also contains a factor responsible for the change in capsule size and shape: Globe—shortening and widening the capsule; Cocklebur—narrow capsule; Echinus—long spines; Glossy—shiny surface; microcarpic—downy surface, and so on (Figure 8.26).

The seven primary trisomics of barley were classified into seven separate types based on distinct, easily noticeable morphological features of different trisomic plants: Bush—bushy growth habit; Slender—slender appearance; Pale—pale color leaves; Robust—vigorous growth habit;

Normal (2*n*)

2*n* + 1·2   2*n* + 3·4   2*n* + 5·6   2*n* + 7·8

2*n* + 9·10   2*n* + 11·12   2*n* + 13·14   2*n* + 15·16

2*n* + 17·18   2*n* + 19·20   2*n* + 21·22   2*n* + 23·24

FIGURE 8.26 Capsules of diploid (2*n* = 24) and 12 possible primary trisomics (2*x* + 1) in *D. stramonium*. (From Avery, A. G., S. Satina, and J. Rietsema. 1959. *Blakeslee: The Genus Datura*. Ronald Press, New York, p. 289. With permission.)

Pseudo-normal—similar to diploid; Purple—dark purple color in leaf sheaths; Semi-erect—semi-erect growth habit. Five (Bush, Slender, Pale, Pseudo-normal, Semi-erect) of the seven primary trisomics are readily distinguishable from each other as well as from diploids at an early seedling stage because they exhibit distinguishing seedling traits (Tsuchiya, 1960a, 1967). The remaining two types, Robust and Purple, are easier to identify in the time span tillering to heading.

The primary trisomics of pearl millet were significantly shorter in plant height, later to flower and narrower in leaf width than their diploid sibs. They were designated Tiny, Dark green, Lax, Slender, Spindle, Broad, and Pseudo-normal. Tiny was weak in growth and vigor while Pseudo-normal was very similar to diploid sibs (Gill et al., 1970).

Sears and Lee-Chen (1970) identified four of the five primary trisomics of *Arabidopsis thaliana* based on their easily identifiable morphological appearance of leaves: Concave (C), Round (R), Yellow (Y), and Narrow (N). The fifth trisomic type, Fragilis (F), was identified as early maturing with protruding stigmas. It is self-sterile because pollen is shed before the stigma becomes receptive.

Primary trisomics have been named in maize (Rhoades and McClintock, 1935), tomato (Rick and Barton, 1954), rice (Khush et al., 1984), and soybean (Xu et al., 2000c) based on the length of extra chromosome at pachynema. For example, in rice 12 possible primary trisomics are expected. Thus, Triplo 1 carries the longest and Triplo 12 the smallest extra chromosome. The same nomenclature is employed in maize, tomato, and soybean.

1. Effect of the Length of Extra Chromosome on Plant Morphology

   The degree of growth, vigor, and development of a trisomic plant depends roughly on the length of extra chromosome present in trisomic condition. The trisomics in which the extra chromosome is long are distinct morphologically at earlier stages of plant growth than those primary trisomics carrying a short extra chromosome. This relationship has been observed in maize (McClintock, 1929c), tomato (Rick and Barton, 1954), barley (Tsuchiya, 1960a, 1967), pearl millet (Gill et al., 1970), and rice (Khush et al., 1984) and in many other species (Khush, 1973; Singh, 1993). However, there are some exceptions. Triplo 7 and Triplo 8 of tomato are comparatively weak while Triplo 3 and 6 (longer chromosomes) are relatively vigorous (Rick and Barton, 1954).

   The soybean contains $2n = 40$ chromosomes and always expresses diploid-like meiosis. Of the 20 primary trisomics, seventeen did not express obvious differences in morphology from their disomic sib. Triplo 1 showed the largest leaves, pods, and seeds among the 20 primary trisomics, and they were about one-third larger than those of the disomic sib. Gray-colored saddle seeds in Triplo 17 distinguished it from the disomic and other primary trisomics (Xu et al., 2000c).

   Phenotypic changes in trisomic plant organs have been explained on the basis of Bridges' (1922) gene balance theory. This theory assumes that the genetic complement of a diploid permits an individual to develop and function as an integrated organism. However, when an individual chromosome is added to the normal complement of a diploid species, the gene balance is greatly disturbed. Such imbalance is reflected in physiological, morphological, and developmental deviations. Each chromosome affects the anatomy, physiology, and morphology of the plant in a distinctive way reflecting the differential gene content of the different chromosomes. The gene balance theory was formulated by Bridges to explain the *Datura* trisomics. Sinnott et al. (1934) supported the theory of Bridges and found a good fit between means of all trisomics and disomics for most of the characters studied. However, Sampson et al. (1961) and Rajhathy (1975) pointed out that the gene balance theory does not consider additive and optimum gene action as well as epistatic and compensating genes located on the other chromosomes.

   The imbalance which an extra chromosome exerts over the balanced condition may best be shown by capsules of the Globe trisomic of *Datura*. Blakeslee and Belling (1924b) observed that the Globe trisomic with two extra chromosomes ($2n = 2x + 2$) has a greater imbalance (2/24) compared with its respective primary trisomic ($2n = 2x + 1$). In tetrasomic condition, the depression of capsules is severe in expression. Similarly, the tetraploid Globes have their capsules relatively more depressed and their spines relatively stouter as we pass from $4x$ to $4x + 2$ and $4x + 3$ chromosome constitutions. A $4x + 4$ Globe was also expected but presumably was not isolated because the imbalance from the extra chromosomes would be 4/48. This is the same imbalance found in $2x + 1$ Globe. It is, therefore, evident that the tolerance limit of imbalance in *Datura* is only two extra chromosomes to its diploid chromosome complement. Tetrasomic ($2n\ 2x + 2 = 44$) soybean also exerts substantial alteration in morphology than those recorded in primary trisomics. All the available tetrasomics in the soybean are viable, slow in vegetative growth, altered morphology, and partial fertile (Singh and Xu, unpublished results).

2. Effect of a Nucleolus Organizer Chromosome on Plant Morphology

   It is interesting to note that Triplo 9 of rice, which is a principal nucleolus organizer chromosome, has stoutness in some plant characters such as thick, dark-green leaves and culms, along with a large panicle with bold grains. Triplo 10, a weak nucleolus organizer, is not distinguishable from diploids at early plant growth stages but has small grains (Khush et al., 1984). Triplo 6 and Triplo 7 of barley carry extra nucleolus organizer chromosomes and are vigorous (Tsuchiya, 1960a, 1967). In maize, Triplo 6 is not distinguishable from disomic sib on gross plant phenotype (Rhoades and McClintock, 1935). Similarly, in rye

(Kamanoi and Jenkins, 1962) and also in pearl millet (Gill et al., 1970), plants trisomic for a nucleolus organizer chromosome are pseudo-normal. Rick and Barton (1954) observed in tomato that Triplo 2, in which the extra chromosome is the second longest as well as a nucleolus organizer chromosome, has very straggly growth habit, a large terminal leaf segment, and nearly normal leaf color and flowers. The nucleolus organizer chromosome of moss verbena when in trisomic condition increases the leaf and flower size (Arora and Khoshoo, 1969). By contrast, Triplo 13 of the soybean showed shorter nodes, dark green leaves, and smaller pods and seeds than disomic (Xu et al., 2000c).

3. Effect of Genetic Background

The main morphological distinguishing features are similar for each trisomic type established in several cultivated varieties of barley (Kerber, 1954; Ramage, 1955; Tsuchiya, 1967) and almost like those of wild barley, *Hordeum spontaneum* (Tsuchiya, 1960a). Similar observations have been reported in rice where trisomic sets have been established in indica and japonica rice. The extra chromosome exerts the same modification in morphological traits, though different trisomic names were given depending on observations of the authors (Hu, 1968; Iwata et al., 1970; Iwata and Omura, 1984; Khush et al., 1984). From these results, it is suggested that trisomic plants have identical or close distinguishing characters regardless of the genetic background, provided detailed and close observations are made on many plant organs throughout the whole growing period from early seedling to maturity. However, some primary trisomics established from other than autotriploid sources do not express clear diagnostic traits in *Clarkia unguiculata* (Vasek, 1956, 1963), *Collinsia heterophylla* (Dhillon and Garber, 1960; Garber, 1964), *Solanum chacoense* (Lee and Rowe, 1975), and *Lycopersicum peruvianum* (Sree Ramulu et al., 1977). Primary trisomics in these cases were not distinguishable morphologically from each other, nor from disomic sib, but were vigorous and fertile. Furthermore, even multiple trisomics isolated from interspecific hybrids of *Solanum* did not differ morphologically from diploid sibs. This is probably due to the genetic variability present in the material (Lee et al., 1972).

The primary trisomic series established in barley were partly similar morphologically but were different in their seed fertility (Ramage, 1955, 1960; Tsuchiya, 1960a, 1967). These differences may be attributed mainly to their sources of isolation. Tsuchiya (1960a,b, 1967) obtained his trisomics from the progenies of autotriploids while Ramage's trisomic set was isolated from translocation heterozygotes. The high sterility in Ramage's trisomic series is attributed to chromosomal abnormalities, originated from X-ray-induced segmental interchanges. These may have introduced disadvantageous changes into the chromosomes and genes.

8.2.2.1.3.2 Cytological Identification The true nature of a trisomic plant, identified morphologically, is confirmed by cytological observations. Identification of the extra chromosome in trisomic plants is based first on somatic chromosome count and is verified by the analysis of meiotic stages such as pachynema, diakinesis, metaphase-I, anaphase-I, or telophase-I. At diakinesis or at metaphase-I, one trivalent plus bivalents (III + II) or one univalent plus a bivalents (I + II) chromosome configuration predominates suggesting the presence of an extra chromosome.

Somatic metaphase chromosomes have been utilized to some extent to identify primary trisomics of barley (Tsuchiya, 1960a, 1967), moss verbena (Arora and Khoshoo, 1969), diploid oat (Rajhathy, 1975), rye (Zeller et al., 1977; Fujigaki and Tsuchiya, 1990), and *Triticum monococcum* (Friebe et al., 1990b). The smallest chromosome number 5 of the five nonsatellited chromosomes of the barley complement is carried by trisome Pseudo-normal. The nucleolus organizer chromosomes 6 and 7 were easily identified from karyotype analysis as they exhibited distinctive morphology. Purple primary trisomic of barley contains three chromosome 6 with large satellite (Figure 8.27). Moreover, it has been difficult to identify chromosomes 1 through 4 in barley by standard staining methods because they are similar in length and also lack morphological landmarks. Hence, cytological identification of primary trisomics Bush, Slender, Pale, and Robust is difficult. However, in this situation identification

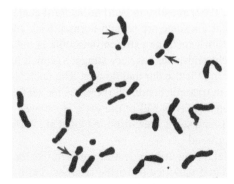

FIGURE 8.27 A photomicrograph of mitotic metaphase chromosome of a primary trisomic ($2n = 41$) for Triplo 6 (Purple) of barley showing three nucleolus organizer chromosomes. (From Tsuchiya, T. 1960a. *Jpn. J. Bot.* 17: 177–213. With permission.)

of the extra chromosome is possible by Giemsa C- and N-banding techniques (Singh and Tsuchiya, 1981b, 1982a,b). The application of the Giemsa C-banding method helped Zeller et al. (1977) to identify six of the seven possible primary trisomics of *Secale cereale*. After Giemsa staining, chromosomes of *S. cereale* exhibit telomeric heterochromatin and intercalary euchromatin differentiation. Such differentiation is not feasible by aceto-carmine or Feulgen staining techniques used routinely. Three nucleolus organizer chromosomes in of the soybean were frequently observed by Xu et al. (2000c) at mitotic metaphase by Feulgen staining while other chromosomes were indistinguishable.

Pachytene chromosome analysis has played a major role in identifying individual chromosomes in trisomic condition in maize (Rhoades and McClintock, 1935), tomato (Rick and Barton, 1954; Rick et al., 1964), and rice (Khush et al., 1984). Centromere positions in tomato have been precisely located since centromeric regions are flanked by heterochromatin. Such differentiation helped to identify individual chromosomes by their relative length (Barton, 1950). The presence of heterochromatic knobs at specific positions on certain chromosomes of maize facilitated an easy morphological identification (Rhoades and McClintock, 1935).

The three homologous chromosomes in primary trisomics compete to pair with one another, but only two of the three homologues are synapsed in a normal fashion at a given position to form a bivalent. The third one attempts to associate with the paired homologues in a random manner and may form a loose trivalent configuration or pair itself (Figure 8.28). When an unpaired chromosome

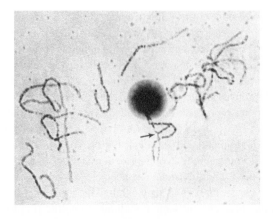

FIGURE 8.28 A trivalent configuration of Triplo 9 in rice at pachynema associated with the nucleolus. Arrow shows extra chromosome 9 partially paired nonhomologously. (From Khush, G. S. et al. 1984. *Genetics* 107: 141–163. With permission.)

does not have the opportunity for 2-by-2 association as proposed by McClintock (1932), the unpaired chromosome frequently folds back, showing nonhomologous chromosome association. Nonhomologous chromosome association (long arm paired with short arm) has been frequently observed at pachynema in primary trisomics of maize (McClintock and Hill, 1931; McClintock, 1932; Rhoades and McClintock, 1935; Rhoades, 1940), tomato (Sen, 1952; Rick and Barton, 1954; Rick et al., 1964), *Sorghum* (Poon and Wu, 1967; Venkateswarlu and Reddi, 1968), *Solanum* (Vogt and Rowe, 1968; Ramanna and Wagenvoort, 1976; Wagenvoort, 1995), rice (Khush et al., 1984), and soybean (Xu et al., 2000c). Such nonhomologous pairing could be misinterpreted as a secondary trisomic chromosome in the case of a chromosome with a median kinetochore. However, a close observation of the several pachytene trivalents for the mode of chromosome pairing leaves no doubt for correct identification.

At pachynema some chromosomes, particularly long ones, show a chance association with the nucleolus. The analysis of several well spread cells of Triplo 9 and Triplo 10 of rice indicated that chromosome 9 is the main nucleolus organizer (Figure 8.28) and chromosome 10 is a weak nucleolus organizer. In barley, chromosome 6 is the main nucleolus organizer while chromosome 7, with a smaller satellite than chromosome 6, has weak nucleolus organizer ability.

In general, trivalent + bivalents (Figure 8.29a) or univalent + bivalents (Figure 8.29b) chromosome associations are observed at diakinesis or metaphase-I in primary trisomics established in several diploid plant species. The frequency varies from crop to crop, stage of meiosis and also among primary trisomic type. In barley, the range of trivalent association (1III + 6II) at metaphase-I in seven primary trisomics was 63.1% (Pseudo-normal) to 78.6% (Slender), averaging 75.9%. The trivalent configuration was higher at diakinesis, averaging 89.1% (Table 8.29). Sears and Lee-Chen (1970), working with *Arabidopsis* primary trisomics, also observed regular trivalent formation (75%) at metaphase-I. It is expected to record about 66% trivalent configuration in primary trisomics if one chiasmata is formed in each arm.

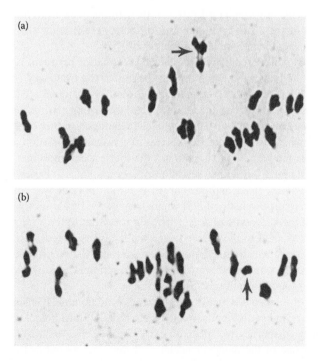

FIGURE 8.29 Meiotic chromosome configurations at metaphase-I in Triplo 20 of the soybean. (a) A cell with 1 III (arrow) + 19 II. (b) A cell with 20II + 1I (arrow). (From Xu, S. J. et al. 2000c. *Crop Sci.* 40: 1543–1551. With permission.)

TABLE 8.29

Chromosome Associations and Types of Trivalents at Diakinesis (DK) and Metaphase-I (MI) in Seven Trisomic Types of Barley (%)

Types of Trisomics	Stage	Chromosome Associations			Types of Trivalents			
		1III + 6II	7II + 1I	Others	Chain	Frying-pan	Y-type	Triple-arc
Bush	DK	82.9	17.1	–	27.6	60.3	0	12.1
	MI	70.7	28.7	0.7	67.3	31.8	0.9	0
Slender	DK	89.3	10.7	–	53.1	43.2	1.1	2.7
	MI	78.6	20.0	1.4	69.2	29.8	0.4	0.6
Pale	DK	94.3	5.7	–	40.8	56.0	0	1.0
	MI	76.0	23.8	0.2	62.4	36.1	0.9	0.6
Robust	DK	92.9	7.1	–	46.2	48.7	0	5.1
	MI	77.7	21.9	0.4	64.9	33.0	1.1	1.1
Pseudo-normal	DK	79.5	20.5	–	49.7	47.1	0.6	2.5
	MI	63.1	36.5	0.4	72.0	25.6	1.8	0.6
Purple	DK	89.0	11.0	–	41.6	53.9	0	4.5
	MI	77.7	22.3	0.1	60.2	37.8	0.5	0.1
Semi-erect	DK	93.9	6.1	–	20.3	70.7	0.8	8.1
	MI	71.7	28.2	0.2	47.1	46.7	5.7	0.4
Average	DK	89.1	10.9	–	40.6	54.1	0.4	4.9
	MI	75.9	23.7	0.5	62.5	35.5	1.3	0.02

Source: Adapted from Tsuchiya, T. 1960a. *Jpn. J. Bot.* 17: 177–213.

The types of trivalents at metaphase-I differ in their frequencies from those recorded at diakinesis. The frequency of the chain type increases at metaphase-I, while the frying-pan and other complicated types are significantly reduced (Table 8.29).

It has been generally considered that there is a good correlation between chromosome length and trivalent formation—the longer the chromosome, the higher the frequency of sporocytes with trivalent configuration. It has been observed in maize (Einset, 1943), tomato (Rick and Barton, 1954), and also in *Lotus pedunculatus* (Chen and Grant, 1968a) that a long extra chromosome has a greater opportunity of forming chiasmata with its homologue than does a shorter one. This expectation can be well documented when we compare frequencies of trivalent association among barley, maize, and tomato; barley has the longest and tomato has the shortest chromosomes. The mean of sporocytes with trivalent configuration at diakinesis for primary trisomics of barley was 89.1% (Tsuchiya, 1960a) and for tomato 48.2% (Rick and Barton, 1954). A similar trend was not observed in primary trisomics of *Avena strigosa* (Rajhathy, 1975) where only 41.3% of the sporocytes showed trivalent configuration, though the karyotype of *Avena strigosa* is similar to the karyotype of barley. These differences may be attributed to the source of trisomics. *Avena strigosa* primary trisomics were derived from desynaptic mutants while autotriploids were the source of barley primary trisomics. Although *Avena strigosa* trisomics were grown several generations, it is likely that they may not yet be completely homozygous (Rajhathy, 1975).

The disjunction of chromosomes in primary trisomics is nonrandom. Generally, two homologues ($n + 1$) disjoin from a trivalent and move to one pole while the third homologue goes to the opposite pole in a majority of sporocytes. When an extra chromosome remains as a univalent, it may either be included in one of the daughter nuclei or it may lag behind and fail to be included in either telophase nucleus. At late telophase-I, the lagging chromosome divides equationally and sister chromatids pass toward opposite poles. However, rarely do both reach these poles. When they reach the poles on time, they are included in the respective daughter nuclei. Otherwise, each may form a micronucleus. The lagging chromosome is located sometimes at the equatorial region or near one of the telophase nuclei. If present elsewhere in the cytoplasm, it is often eliminated. The second division is generally normal. If micronuclei are not lost permanently, they may be included in one of the daughter nuclei. The frequency of univalents may not provide an accurate picture for estimates on the transmission rate of an extra chromosome.

Goodspeed and Avery (1939), working with *Nicotiana sylvestris* primary trisomics, observed no apparent difference in trivalent frequencies among four different primaries. Neither did they find length differences in the extra chromosomes nor significant differences in transmission frequencies of these four primaries. In contrast, Einset (1943) reported in maize that when a short chromosome is present in triplicate fewer trivalents and more univalents are present at metaphase-I. Similar observations were recorded in triploid *Hyacinths* (Belling, 1925) and *Tulipa* (Newton and Darlington, 1929; Darlington and Mather, 1932) where shorter chromosomes form fewer chiasmata at metaphase-I than longer ones.

8.2.2.1.3.3 Identification of Primary Trisomics with Translocation Testers   In most diploid species, simple primary trisomics have been distinguished among themselves and also from diploid sib by morphological appearances and to some extent by karyomorphology. The independence of primary trisomics may also be verified by translocation tester sets. Tsuchiya (1961) used Burnham's translocation tester sets which marked all seven chromosomes of barley (Burnham, 1956; Burnham and Hagberg, 1956) to test mainly the independence of trisomics for chromosomes 1, 2, 3, and 4. These chromosomes are difficult to distinguish cytologically since they are similar in karyotype.

Generally, trisomic plants are used as female parents and are pollinated by the translocation testers because pollen with an extra chromosome does not compete successfully with normal pollen grains in fertilization. Chromosome pairing in F_1 trisomics is analyzed at diakinesis or at metaphase-I. If the extra chromosome of a certain trisomic type is partly homologous to one of the interchanged chromosomes in the testers, a chain of five chromosomes (pentavalent) plus bivalent (Figures 8.30 and 8.31), or the derivative types are obtained. On the other hand, if the extra chromosome is not partly homologous with either of the interchanged chromosomes, a quadrivalent plus a trivalent plus bivalents are observed.

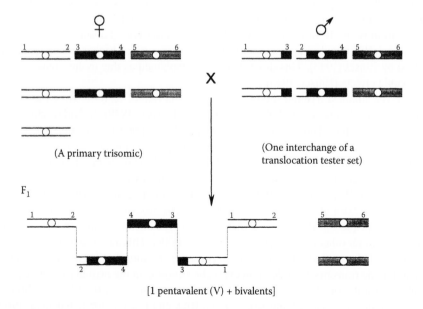

[1 pentavalent (V) + bivalents]

FIGURE 8.30 Diagrammatic sketch showing procedure to identify the extra chromosome in primary trisomics by translocation tester sets.

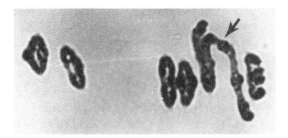

FIGURE 8.31 A meiotic metaphase-I chromosome configuration of 1V + 4II in an F₁ hybrid between primary trisomic Bush and interchange suggests that the extra chromosome in Bush is for chromosome 1. (From Singh, R. J. 2003. *Plant Cytogenetics*, 2nd edition. CRC Press, Boca Raton, FL.)

8.2.2.1.4 Transmission of the Extra Chromosome in Primary Trisomics

8.2.2.1.4.1 Female Transmission Theoretically about 50% primary trisomic plants are expected in the progenies of primary trisomics crossed as females. However, it is rarely observed. The transmission rate of the extra chromosome depends on such factors as the meiotic behavior of the extra chromosome, seed size, genetic background, growing conditions, types of primary trisomics, and number of seeds studied.

It has been established in maize that trisomics for long chromosomes (2, 3, 5) transmit at higher frequencies of primary trisomics (an average of 48%) than the trisomics for the medium (6, 7, 8; average 37%) and short (9, 10; average 25%) chromosomes (Einset, 1943). A positive relationship was observed between transmission rate of an extra chromosome and the frequency of trivalents at metaphase-I. In maize, a higher frequency of trivalents was observed in trisomics for long chromosomes than for shorter ones (Einset, 1943). This observation indicates that the extra chromosome remains as a univalent in the majority of the sporocytes in trisomics for the shorter chromosomes. The univalent lags behind in meiotic divisions and eventually gets eliminated, yielding gametes without an extra chromosome.

The above relationship, however, was not observed in primary trisomics of *Datura*, barley, tomato, rice, diploid oat, nor always even in maize. McClintock and Hill (1931) reported a transmission rate of chromosome 10 of maize to be 33.06%, but Einset (1943) observed only 28%. Rhoades (1933a) obtained 31% female transmission in trisome 5 but a much higher transmission rate (52%) was reported by Einset (1943). This discrepancy was explained as due to environmental effects or possibly to genic effects (Einset, 1943).

In *Datura*, the smallest trisome showed a transmission rate of 32.68% while the third smallest transmitted only at a frequency of 2.99% (Blakeslee and Avery, 1938). In barley, chromosome 5 is the shortest chromosome (Tjio and Hagberg, 1951) and is carried by trisomic Pseudo-normal. It showed the lowest female transmission (9.8%) rate in *H. spontaneum* genetic background but it had the highest transmission rate (31.1%) for the same trisome in *H. vulgare* (Table 8.30). No relationship was established between female transmission and length of the extra chromosome in the primary trisomics of the potato (Wagenvoort, 1995) and soybean (Xu et al., 2000c). An average (range) female transmission of the extra chromosome in 20 primary trisomics of the soybean was 41.6% [27% (triplo 20) – 59% (triplo 9)]. Primary trisomics of the soybean showed higher female transmission than those recorded for other diploid species (Xu et al., 2000c). This again suggests that soybean is of tetraploid origin. This concludes that there is no consistent relationship between the length of chromosome and female transmission rate of an extra chromosome in the primary trisomic progenies.

Generally, transmission of an extra chromosome is somewhat higher in selfed or hybrid progenies of primary trisomics (Goodspeed and Avery, 1939; Rick and Barton, 1954; Tsuchiya, 1960a, 1967; Liang, 1979; Lin and Coe, 1986; Curtis and Doyle, 1992). Tsuchiya (1960a, 1967) reported an average of 25.9% and 29.1% trisomic plants in selfed progenies of wild (*H. spontaneum*) and cultivated

TABLE 8.30

Frequency of Primary Trisomics in Progenies of Two Primary Trisomic Sets of Barley (%)

Types of Trisomics	*Hordeum Spontaneum*		*Hordeum Vulgare S.E. 16*	
	2x + 1 Selfed	(2x + 1) x 2x	2x + 1 Selfed	(2x = 1) x 2x
Bush	31.4	24.9	12.5	–
Slender	21.2	27.5	25.7	28.3
Pale	21.0	27.2	25.5	13.3
Robust	28.8	11.3	36.7	19.5
Pseudo-normal	24.6	9.8	27.9	31.1
Purple	19.3	17.4	25.9	23.4
Semi-erect	23.4	24.5	33.3	21.9
Average	25.9	22.7	29.1	20.4

Source: From Tsuchiya, T. 1960a. *Jpn. J. Bot.* 17: 177–213. With permission.

(*H. vulgare*) trisomic series, respectively. Sears and Lee-Chen (1970) reported a 25.8% transmission rate of the extra chromosome in selfed progenies of five primary trisomics of *Arabidopsis*. The range was 21%–30%. Kaltsikes and Evans (1967) studied three trisomics of *Beta vulgaris* and reported an over-all 22.07% transmission rate of the extra chromosome. Vogt and Rowe (1968) observed 20%–24% female transmission in three trisomics of potato. Khush (2010) examined female transmission (2n + 1 × 2n) in rice primary trisomics which ranged from 15.5% (Triplo 1) to 43.9% (Triplo 4).

It has been observed that the occurrence of a higher frequency of trivalents in the sporocytes of primary trisomics for the longer chromosomes is due to the fact that long chromosomes have more physical opportunity to pair with homologues than do shorter chromosomes. It is therefore expected that higher female transmission rates for trisomics of the longer chromosomes will be found because gametes with n and $n + 1$ chromosome numbers are expected to be formed almost in equal (50% each) numbers. However, such relationship has been observed only in maize primary trisomics by Einset (1943). In spite of the equal frequencies of female gametes with n and $n + 1$ chromosomes, it is assumed that gametes carrying an extra-long chromosome are more unbalanced genetically than gametes with a shorter chromosome and result in increased gametic and zygotic lethality. Gametic competition among megaspores has also been suggested to occur as has been shown for microspores. McClintock and Hill (1931) suggested that a basal megaspore with $n + 1$ chromosome constitution may not function to produce the embryo sac but be replaced by a cell with the n chromosome number. However, such megaspore substitution was not reported in later studies of maize (Singleton and Mangelsdorf, 1940; Rhoades, 1942). Rédei (1965) reported some postreduction selection between the basal megaspores of different constitutions in *Arabidopsis*. The transmission of an extra chromosome has been observed (0%–22%) through male gametes in primary trisomics. The low transmission of primary trisomics was accounted for by the preference in development of $2n$ cells over $2n + 1$ cells (Sears and Lee-Chen, 1970).

Frequently, small seeds transmit a higher frequency of primary trisomic plants than large and plump seeds (Lesley, 1928; Einset, 1943; Ramage and Day, 1960; Tsuchiya, 1960a; Liang, 1979). Trisomic seeds are narrower, thinner, and later germinating while large seeds are generally disomic. Tsuchiya (1960a) separated large and small seeds of two primary trisomics, Bush and Slender. He observed in both cases that small seeds showed poor germination but transmitted a considerably higher frequency of primary trisomic plants (Table 8.31). Ramage and Day (1960) suggested the use of a seed blower to enrich seed lots to contain a high frequency of trisomics, because trisomic seeds are lighter.

TABLE 8.31
The Number and Percentage of Trisomics in Two Seed Groups of Bush and Slender Differing in Size

Types of Trisomics	Seed Size	Number of Seeds			Trisomics Obtained	
		Sown	Germinated	%	Number	%
Bush	Large	261	247	(94.63)	44	17.81
	Small	108	88	(81.48)	71	80.68
Slender	Large	128	108	(84.37)	17	15.74
	Small	46	20	(43.47)	18	90.00

Source: From Tsuchiya, T. 1960a. *Jpn. J. Bot.* 17: 177–213. With permission.

8.2.2.1.4.2 Pollen Transmission   The transmission of extra chromosomes through pollen in general is very low because in diploid species pollen with $n + 1$ chromosome constitution is unbalanced and generally unable to compete in fertilization with pollen carrying the balanced, n, chromosome number. The extra chromosome was not transmitted through pollen in $2x \times 2x + 1$ crosses of *H. spontaneum* primary trisomics (Tsuchiya, 1960a). Tetrasomic plants were obtained in the selfed progenies of Robust, (0.5%) Pseudo-normal (0.4%), and Semi-erect (5.9%) of cultivated barley but no trisomic plants were recovered in $2x \times 2x + 1$ crosses (Tsuchiya, 1967). Male transmission of the extra chromosome has been observed in *D. stramonium* (Blakeslee and Avery, 1938), maize (McClintock and Hill, 1931), tomato (Lesley, 1928, 1932), and rice (Khush, 2010). Pollen transmission of the extra chromosome was observed in only four primary trisomics of *Datura* (Blakeslee and Avery, 1938). Pollen transmission $(2n \times 2n + 1)$ of extra chromosome in primary trisomics of rice ranged from 0% (Triplo 1, 2, 3, 6, 7) to 21.4% (Triplo 9).

Lesley (1928) reported in tomato that male transmission of the extra chromosome was as frequent as through eggs in two trisomic types. Sears and Lee-Chen (1970) observed an average male transmission of 12% in *Arabidopsis thaliana*. The value ranged from 0.0% (F trisomic) to 22.0% (N trisomic). Buchholz and Blakeslee (1922) found in Glove trisomic of *Datura* that pollen tubes with the extra chromosome are slower growing than those of normal pollen. Tetrasomic plants have been isolated from the selfed progenies of primary trisomics of the soybean. All are viable, vigorous, and partial fertile suggesting male transmission of the 21 chromosome gametes.

8.2.2.1.4.3 Progenies of Primary Trisomics   Generally, normal diploids and related primary trisomics are obtained in the progenies of primary trisomics. However, unrelated chromosomal variants are occasionally isolated, and they have been reported in tomato (Lesley, 1928; Khush, 1973), *Datura* (Blakeslee and Avery, 1938), *Nicotiana sylvestris* (Goodspeed and Avery, 1939), barley (Tsuchiya, 1960a, 1967), and *Avena strigosa* (Rajhathy, 1975). The unrelated chromosome variants from the progenies of primary trisomics are telotrisomics, acrotrisomics, hypertriploids, triploids, hypotriploids, and haploids, fragments of various sizes, ring chromosomes, secondary trisomics, tertiary trisomics, and various other chromosomal types. Einset (1943) studied the chromosomes of 1916 plants derived from $2x + 1 \times 2x$ crosses. There were 658 trisomics, 5 monosomics, 1 plant with 19 chromosomes + 1 fragment, 1 plant with 20 chromosomes + 1 fragment, 3 plants with 21 chromosomes + 1 fragment, 2 haploids, and 3 triploids.

Tetrasomic plants (an individual carrying two extra chromosomes in addition to its normal somatic chromosomes complement $[2n = 2x + 2]$ is designated as tetrasomics) are expected in the progenies of primary trisomics if an extra chromosome is transmitted through male spore. Tetrasomics are rare and die prematurely in the progenies of barley, maize, rice, and tomato primaries. However, primary trisomics of the soybean transmit tetrasomics in low frequencies. The plants are viable, and compared to their counterpart primary trisomics are slow in vegetative and reproductive growth and

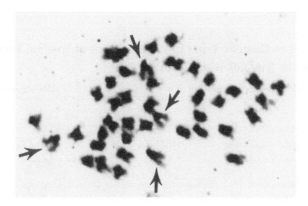

FIGURE 8.32 A mitotic metaphase cell of tetrasomic for chromosome 13 showing 4 SAT chromosomes (arrows).

partial fertile to completely fertile. Tetrasomics mostly breed true and occasionally related trisomics are identified. Similarly, the progeny of tetrasomics × disomic is primary trisomics. Tetrasomics in the soybean is a unique cytogenetic stock and that may reduce time devoted for cytological identification of tetrasomics and F_1 hybrids. Tetrasomics for chromosome 13 with four NOR are, sometimes, observed at mitotic metaphase (Figure 8.32). The isolation of viable tetrasomics in the soybean suggests tetraploid nature of the soybean genome.

8.2.2.1.5 Genetic Segregation in Primary Trisomics

Primary trisomics are very helpful for locating a gene on a particular chromosome, verifying the independence of linkage groups and also associating the genetic linkage groups with the individual chromosomes. The use of primary trisomics is often more efficient and dependable than other conventional and translocation procedures.

The principle of genetic segregation in primary trisomics has been described in detail by Burnham (1962), Hermsen (1970), and Khush (1973). When a primary trisomic is used to locate a gene on a particular chromosome, the genetic ratios are very much modified from 3:1 (F_2) or 1:1 (BC_1). The ratios encountered depend on the genotypes of the F_1 primary trisomic plants, whether duplex (AAa) or simplex (Aaa), on the type of chromosome segregation and on the female transmission rate of the extra chromosome.

8.2.2.1.5.1 Association of a Recessive Gene

A mutant with the recessive genotype aa is crossed as a pollen parent onto a primary trisomic of the genotype AAA. In the F_1, primary trisomic plants are identified morphologically and cytologically. Generally one disomic plant is saved (control) and the remaining ones are discarded. The F_1 trisomic plants are allowed to self-pollinate or are test crossed (if sterile) to the homozygous recessive parent (BC_1). Two types of gametes are produced from the F_1 primary trisomics, n and $n + 1$. If $n + 1$ gametes are not male-transmitted with 50% frequency, total phenotypic ratios of $17A:1a$ and $5A:1a$ are expected in F_2 and BC_1, respectively. However, these ratios are never recorded because although usually only n gametes function in male-transmission, $n + 1$ gametes function in less than 50% frequency in female transmission.

On the basis of 50% and 33.3% female transmission of $n + 1$ gametes through females, assuming no male transmission of the extra chromosome, the expected frequencies of gametic types, genotypic frequencies, and phenotypic ratios derived from random chromosome, random chromatid, and maximum equational segregations in primary trisomic plants with AAa genotype are shown in Tables 8.32 through 8.34, respectively. As is evident from Table 8.34, the genetic ratios are very much modified if a gene is located in the extra chromosome in primary trisomic analysis, and results are quite different from disomic F_2 ($3A:1a$) and BC_1 ($1A:1a$) ratios.

TABLE 8.32

Expected Frequencies of Gametic Types in Various Types of Possible Segregation in Primary Trisomics or Triploids with Genotype AAa

	50% Female Transmission of $n + 1$ Gametes		30.3% Male Transmission of $n + 1$ Gametes	
Type of Segregation	$n + 1$	n	$n + 1$	n
Random chromosome	$1AA + 2Aa$	$2A + 1a$	$1AA + 2Aa$	$4A + 2a$
Random chromatid	$6AA + 8Aa + 1aa$	$10A + 5a$	$6AA + 8Aa + 1aa$	$20A + 10a$
Maximum equational	$5AA + 6Aa + 1aa$	$8A + 4a$	$5AA + 6Aa + 1aa$	$16A + 8a$

Source: Adapted from Burnham, C. R. 1962. *Discussion in Cytogenetics*. Burgess, Minneapolis, MN.

1. Random Chromosome Segregation

    The expected frequencies of gametic types from a primary trisomic F_1 of duplex (AAa) genotypic constitution having 50% transmission of the extra chromosome through female are $n + 1$, $1AA + 2Aa$, and n, $2A + 1a$ (Figure 8.33; Table 8.32). The expected genotypic frequencies in F_2 are as follows: $2x + 1 = 2AAA + 5AAa + 2Aaa$ and $2x = 4AA + 4Aa + 1aa$ (Table 8.33). Phenotypic frequencies will show as: $9A\text{-}:0a:: 8A\text{-}:1a$ (Table 8.34). Similar modifications are expected in the BC_1:$3A\text{-}:0a:: 2A:1a$. These ratios indicate that no recessive homozygous plants will be obtained in the trisomic fraction.

    However, the expected 50% female transmission of $n + 1$ gametes is not observed. If we assume the female transmission of the extra chromosome is 33.3%, the frequency of an n gamete will become twice $(4A + 2a)$ and the $n + 1$ gamete's $(1AA + 2Aa)$ proportion will be unchanged (Table 8.32). A similar proportion of change will be recorded in the F_2 genotypic frequencies $(2x + 1 = 2AAA + 5AAa + 2Aaa$ and $2x = 8AA + 8Aa + 2aa)$ (Table 8.33). The F_2 phenotypic ratio in trisomic $(2x + 1)$ fraction will be unaltered but it will be changed in the disomic fraction $(16A\text{-}:2a)$. Thus, the overall phenotypic ratio modifies from $17A\text{-}:1a$ to $12.5A\text{-}:1a$ (Figure 8.33). A similar change occurs in BC_1 (Table 8.34).

2. Random Chromatid and Maximum Equational Segregation

    Sometimes homozygous recessive plants are obtained in the trisomic fraction of the F_2 and BC_1 populations. It is due to random chromatid crossing over or maximum equational segregation (Burnham, 1962). The expected gametic type frequencies in random chromatid segregation with 50% female transmission of $n + 1$ gametes are $6AA + 8Aa + 1aa$ and n gametes are $10A + 5a$ but with the 33.3% female transmission of $n + 1$ gametes the frequency of n gametes becomes $20A + 10a$ (Table 8.32). The genotypic frequencies in an F_2 population having 50% female transmission of $n + 1$ gametes are expected as follows: $2x + 1 = 12AAA + 22AAa + 10Aaa + 1aaa$, and $2x = 20AA + 20Aa + 5aa$. In F_2, $8A:1a$ phenotypic ratio is expected for the disomic fraction and $44A\text{-}:1a$ for the trisomic fraction. With 33.3% female transmission of the extra chromosome, the F_2 phenotypic ratio for the disomic fraction will be modified and the total ratio is slightly smaller $(11.27A\text{-}:1a)$ than that recorded in random chromosome assortment ($12.5\ A\text{-}:1a$). A similar proportion of changes occur in BC_1 (Tables 8.33 through 8.35).

    As in random chromatid segregation, the occurrence of recessive homozygotes in the trisomic fraction of the F_2 population can be explained based on maximum equational segregation (Burnham, 1962). Assume that only one chiasma is formed between a locus and a kinetochore in each of the three chromosomes of a trivalent (Figure 8.34). The gametic genotypes, $n + 1 = 5AA + 6Aa + 1aa$ and $n = 8A + 4a$, are expected at the end of meiosis. There is not a real difference between the results expected from random chromatid segregation and maximum equational segregation (Tables 8.33 through 8.35). In both cases, a very low frequency of recessive homozygotes is expected with similar proportions in the trisomic fraction.

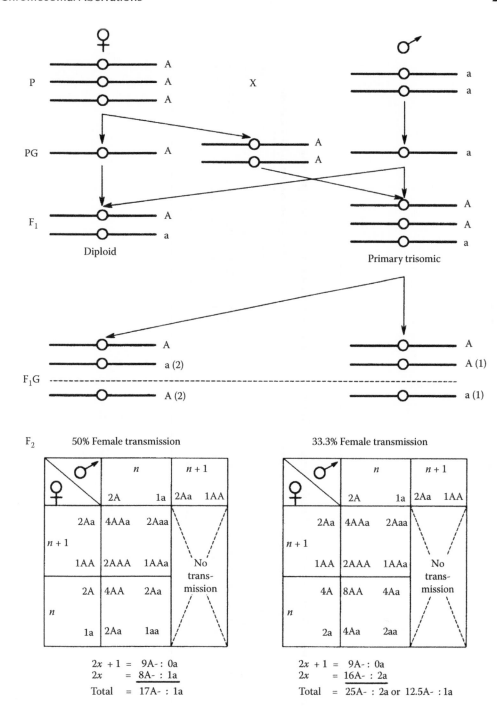

FIGURE 8.33 Diagrammatic presentation of random chromosome segregation showing association of a recessive marker (*a*) gene with a particular chromosome by primary trisomic method.

TABLE 8.33
Expected Genotypic Frequencies in Various Types of Possible Segregation in F$_2$ Primary Trisomics with Genotype AAa

Type of Segregation	50% Transmission of $n + 1$ Gametes		33.3% Transmission of $n + 1$ Gametes	
	2x + 1	2x	2x + 1	2x
Random chromosome	2AAA + 5AAa + 2Aaa	4AA + 4Aa + 1aa	2AAA + 5AAa + 2Aaa	8AA + 8Aa + 2aa
Random chromatid	12AAA + 22AAa + 10Aaa + 1aaa	20AA + 20Aa + 5aa	12AAA + 22AAa + 10Aaa + 1aaa	40AA + 40Aa + 10aa
Maximum equational	10AAA + 17AAa + 8Aaa + 1aaa	16AA + 16Aa + 4aa	10AAA + 17AAa + 8Aaa + 1aaa	32AA + 32Aa + 8aa

Source: Adapted from Burnham, C. R. 1962. *Discussion in Cytogenetics.* Burgess, Minneapolis, MN.

TABLE 8.34

Expected Phenotypic Ratio (Normal: Mutant) in Various Types of Possible Segregations in F$_2$ and Female (AAa) Back Crossed (BC$_1$) to *aa*

	50% Transmission of $n+1$ Gametes						33.3% Transmission of $n+1$ Gametes					
	F$_2$			BC$_1$			F$_2$			BC$_1$		
			Total			Total			Total			Total
Type of Segregation	2x+1	2x	Ratio A:a	2x+1	2x	Ratio A:a	2x+1	2x	Ratio A:a	2n+1	2n	Ratio A:a
Random chromosome	9:0	8:1	17:1	3:0	2:1	5:1	9:0	16:2	12.5:1	3:0	4:2	3.5:1
Random chromatid	44:1	8:1	14:0	14:1	10:5	4:1	44:1	80:10	11.27:1	14:1	20:10	3.1:1
Maximum equational	35:1	32:4	13.4:1	11:1	8:4	3.8:1	35:1	64:8	11:1	11:1	16:8	3:1

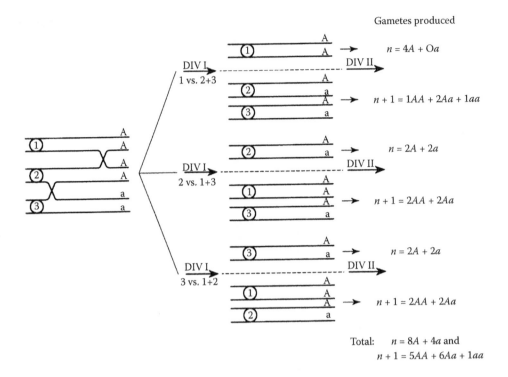

FIGURE 8.34 Diagrammatic presentation of maximum equational segregation in a primary trisomic heterozygote (*Aaa*).

8.2.2.1.5.2 Association of a Dominant Gene　In this case, the genotype of the selected primary trisomic is *aaa* and the genotype of the mutant is *AA*. The genotype of the F_1 primary trisomic is *Aaa* (simplex). The expected gametic types together with the F_2 genotypic, and phenotypic ratios based on 50% transmission of the extra chromosome in random chromosome assortment, random chromatid, and maximum equational segregations are provided in Table 8.35. It should be pointed out that separation of the trisomic and disomic fractions is advised when the genotype of the F_1 primary trisomic is in simplex (*Aaa*) condition.

8.2.2.1.5.3 Association of Codominant Gene　Primary trisomics have been used to associate isozyme markers to the specific chromosomes of tomato (Tanksley, 1983), rice (Delos Reyes et al., 1998), and faba bean (Torres et al., 1995). These markers segregate in a codominant fashion (genotypic ratios) in the F_2. In a critical combination (random chromosome), F_2 population segregates in a trisomic ratio depending on female transmission of the extra chromosome:

	Genotypic Ratio	Female Transmission
1.	5:7:1	30%
2.	5:7.5:1	33.3%
3.	6:11:1	50%

Molecular markers such as simple sequence repeats (SSR) can be associated with the chromosomes by primary trisomics. Cregan et al. (2001) associated SSR loci of molecular linkage group A_1 to triplo 5, and F to triplo 13 of the soybean. All SSR markers segregated in a ratio of 6:11:1. The SSR loci not associated with triplo 5 and triplo 13 segregated in a disomic fashion of 1:2:1 (Table 8.36).

TABLE 8.35

Expected Gametic Genotypes from Primary Trisomic Aaa Genotype, F$_2$ Genotypic Frequencies, F$_2$ and BC$_1$ Phenotypic Ratios (If Female Transmission of $n + 1$ Gamete Is 50%)

Type of Segregation	Gametic Genotype		F$_2$ Genotype		Phenotypic F$_2$ Ratio		Phenotypic BC$_1$ Ratio	
	$n + 1$	n	$2x + 1$	$2x$	$2x + 1$	$2x$	$2x + 1$	$2x$
Random chromosome	2Aa + 1aa	1A + 2a	2AAa + 5 Aaa + 2 aaa	1 AA + 4Aa + 4aa	7:2	5:4	3:2	1:2
Random chromatid	1AA + 8Aa + 6aa	10A + 20a	1AAA + 10AAa + 22Aaa + 12aaa	10AA + 40Aa + 40aa	11:4	5:4	9:6	1:2
Maximum equational	1AA + 6Aa + 5aa	4A + 8a	1AAA + 10AAa + 17Aaa + 10aaa	4AA + 16Aaa + 16aa	26:10	5:4	7:5	1:2

Source: Adapted from Burnham, C. R. 1962. *Discussion in Cytogenetics.* Burgess, Minneapolis, MN.

TABLE 8.36

Segregation Ratios in F$_2$ Generation of the SSR Markers with Triplo 5 and Triplo 13 in the Soybean

Triplos and SSR Markers	Linkage Groups	Segregation Ratio	χ² Probability for Goodness of Fit to	
			1:2:1	6:11:1
Triplo 5				
Satt276	A1	27:26:3	0.001	0.05
Satt364	A1	26:27:3	0.001	0.10
Satt471	A1	25:27:2	0.001	0.14
Satt300	A1	26:27:3	0.001	0.10
Satt155	A1	25:28:3	0.001	0.18
Satt006	L	6:11:3	0.58	
Triplo 13				
Satt569	F	24:34:5	0.0027	0.43
Satt193	F	24:34:5	0.0027	0.43
Satt030	F	24:34:5	0.0027	0.43
Satt343	F	24:34:5	0.0027	0.43
Satt657	F	20:38:5	0.0074	0.71
Satt022	N	7: 9:5	0.67	

Source: From Cregan, P. B. et al. 2001. *Crop Sci.* 41: 1262–1267. With permission.

This elucidates that primary trisomics is an excellent cytogenetic stock to associate molecular markers with the specific chromosomes by modification in the F$_2$ genetic ratio.

8.2.2.1.5.4 Association of Gene by Dosage Effect The RFLP markers are assigned to specific arms of the chromosomes by gene dosage comparison of the autoradiographs in rice. By using secondary trisomics, telotrisomics, and primary trisomics of rice cv. IR36, Singh et al. (1996) crossed with a tropical japonica variety, MaHae. Both are highly polymorphic. An F$_1$ disomic has one allele each of IR36 and MaHae and the intensities of both the autoradiographic bands in Southern blots were similar. An F$_1$ primary trisomic has two copies of the IR36 allele and one of the MaHae allele. The intensity of the IR36 band is expressed as twice that of the MaHae band. An F$_1$ secondary trisomic has three copies of the IR36 allele (located on the arm for which it is secondary) and one copy of MaHae allele. The intensity of the IR36 band for a marker present on that arm is three times that of MaHae band (Figure 8.35). If the marker in question is not located on the arm for which it is secondary trisomic, then the F$_1$ secondary trisomic shows similar intensity of IR36 and MaHae bands. The telotrisomic behaves the same way as the secondary trisomics except that the intensity of the IR36 band is twice the intensity of the MaHae band. Based on these principles, RFLP markers were assigned to specific chromosome arms and the positions of the centromeres were mapped between the nearest two markers located on opposite arms of a chromosome of rice.

8.2.2.1.5.5 Gene-Chromosome-Linkage Group Relationships Primary trisomics have been used to associate a marker gene with a particular chromosome, to determine chromosome-linkage group relationships and to test the independence of linkage groups (Burnham, 1962; Khush, 1973). Rick and Barton (1954) disproved the independence of the linkage groups X and XII and VI and VIII with the help of primary trisomic analysis in tomato. The following relationships between

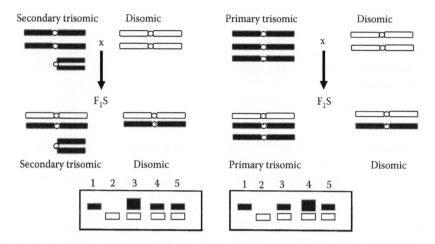

FIGURE 8.35 Diagrammatic presentation of dosage analysis in F_1 primary and secondary trisomics. Lanes 1-5 are IR 36, MaHae, F_1 secondary trisomic, F_1 primary trisomic, and F_1 disomic, respectively. (From Singh, K. et al. 1996. *Proc. Natl. Acad. Sci. USA*. With permission; Adapted from Khush, G. S. 2010. *Breeding Sci*. 60: 469–474.)

chromosomes-genes-(linkage groups) were established: 1-*y* (III); 2-*d* (I); 7-*wt* (x), *mc* (XII); 8-*l* (VI), *al* (VIII), dl; 9-*wd*; 10-H (VII). Moreover, Rick et al. (1964) analyzed 30 more marker genes with 11 of the 12 primary trisomics. Several linkage groups that were previously supposed to be independent were associated with the same chromosome and none of the tested genes showed association with Triplo 12, the shortest chromosome. In rice, three markers (gh_1, nl_1, and gl_1) belonging to three different linkage groups (VI, IX, and XII) gave trisomic segregations with Triplo 5. Thus, these three linkage groups belong to chromosome 5 (Khush et al., 1984). Similar results were obtained by Iwata and Omura (1976) who obtained trisomic ratios for d_1 (VI), nl_1, and gl_1 with their trisomic L. The gene d_2 (VII) and *I-Bf* (V) gave trisomic segregations with Triplo 9. Thus, linkage groups V and VII belong to the same chromosome.

Tsuchiya (1960a) studied 10 marker genes belonging to four linkage testers with the seven primary trisomics of barley. The observed results indicated that chromosome 6 carried no genetic linkage group previously established; two genetic linkage groups III and VII were located on chromosome 1. Thus, the linkage groups were reduced to 6. However, several new genes e_c (early), x_n (Xantha seedling), and *o* (orange lemma) were associated with chromosome 6 by reciprocal translocations. Tsuchiya (1967) verified the association of gene *o* with chromosome 6.

Lee-Chen and Steinitz-Sears (1967) studied linkage relationships in *Arabidopsis thaliana* by the primary trisomic method. Trisomic types NC and NS both gave a trisomic ratio with a marker gene *tz*. This demonstrates that NC and NS trisomics carry the same chromosome. A gene for narrow leaf (*an*) of linkage group-I did not show a trisomic ratio with any of the four tested primary trisomics, suggesting that the gene *an* may be located on the 5th trisomic type, which was not isolated.

By utilizing primary trisomics, Khush et al. (1984) identified all the possible 12 linkage groups of rice. They selected 22 marker genes, one or two genes representing each linkage group, and crossed them with the primary trisomics. As soon as a marker was located on a specific chromosome, its tests with the remaining chromosomes were discontinued. They tested 120 of the possible 264 combinations involving 22 genes before markers for all of the chromosomes were identified.

Two dominant genes, *Cl* (clustered spikelets) and *Ps1* (purple stigma-1), gave trisomic ratios with Triplo 6. The F_2 ratios in both cases were modified to a ratio of 4 normal: 5 mutant (instead of 1:3) in the diploid fraction. In the trisomic fraction of the F_2 of the Triplo 6 × *Ps1*, the ratio agreed with the expected 2 normal: 7 mutant. However, the ratio of normal to mutant in the trisomic section of the F_2 population of Triplo 6 × *Cl* was 7:2. This apparent reversal was due to the fact that *Cl cl cl* individuals have normal phenotype (as determined from trisomic F_1 phenotypes) instead of mutant. Therefore, the expected genotypic ratio of 2 *Cl Cl cl*: 5 *Cl cl cl*: 2 *cl cl cl* was modified into seven normal to two mutant.

8.2.2.2   Secondary Trisomics

8.2.2.2.1   Introduction

Secondary trisomic plants carry an extra isochromosome (both arms homologous) in addition to its normal somatic chromosome complement. Secondary trisomics have been reported in *Datura stramonium* (Blakeslee, 1924b; Blakeslee and Avery, 1938; Avery et al., 1959), *Zea mays* (Rhoades, 1933b; Schneerman et al., 1998), *Lycopersicon esculentum* (Sen, 1952; Moens, 1965; Khush and Rick, 1969; Rick and Gill, 1973), *Hordeum spontaneum* (Tsuchiya, 1960a), *Avena strigosa* (Rajhathy and Fedak, 1970), and *Oryza sativa* (Singh et al., 1996). Secondary trisomics have been studied morphologically and cytologically in *Datura*, maize, tomato, and rice but were utilized extensively in genetic and linkage studies in tomato (Khush and Rick, 1969) and rice (Singh et al., 1996).

8.2.2.2.2   Sources of Secondary Trisomics

The secondary chromosome (isochromosome) might have arisen directly or progressively by way of an unstable telocentric fragment which would itself undergo a further misdivision or segregation without division (Darlington, 1939b).

In *Datura*, secondary trisomics have appeared spontaneously from unrelated primary trisomics occasionally, but rarely from unrelated secondaries or diploid parents. They most frequently appear spontaneously in the progenies of related primary trisomics. Rhoades (1933b) isolated a secondary trisomic plant for the short arm of chromosome 5 from the offspring of an unrelated primary trisomic for chromosome 6. In tomato, Sen (1952) identified two secondaries (2n + 8L.8L; 2n + 9L.9L) from the progenies of a mutation experiment and Khush and Rick (1969) obtained 9 of the possible 24 secondaries among the offspring of haplo-triplo disomics, double iso-compensating trisomics, tertiary monosomics, and segmental deficiencies. Rick and Gill (1973) isolated three more new secondaries (2n + 5L.5L; 2n + 7L.7L; 2n + 11L + 11L) from the progenies of 2x + 1 × 2x crosses. Singh et al. (1996) isolated secondary trisomics for both arm of chromosomes 1, 2, 6, 7 and one arm of chromosomes 4, 5, 8, 9, and 12 from the progenies of a large population of each respective primary trisomics of rice (Table 8.37). The rarity of secondaries may be due to the poor viability of gametes and zygotes carrying an extra secondary chromosome.

8.2.2.2.3   Identification of Secondary Trisomics

8.2.2.2.3.1   Morphological Identification   In general, primary trisomics are morphologically, with a few exceptions, intermediate between those of their secondaries. In *Datura*, the seedlings of the 2n + 1.1 secondary showed very narrow leaves due to factors in the .1 half but the 2n + 2.2 secondary carried relatively broad leaves because of factors located on the half chromosome .2. The primary trisomic 2n + 1.2 was intermediate in leaf width (Blakeslee and Avery, 1938). In tomato, secondary trisomics for the long arms (6L, 7L, 9L, 10L) showed much slower growth rates at all stages of plant growth than those recorded for the secondaries composed of short arms (3S, 7S, 9S). Secondary trisomics for the short arms were indistinguishable morphologically from disomics. The secondary 2n + 12L.2L was quite distinct at all stages of growth but it showed only about half of the traits of its primary (Khush and Rick, 1969). In rice, most of the secondary trisomics resembled their counterpart primary trisomics for several morphological traits. In general, secondary trisomics showed slower vegetative growth rate and lower seed fertility than the corresponding primary trisomics. Some of the morphological features of primary trisomics were exaggerated in the secondary trisomics while other secondary trisomics were indistinguishable from the normal diploid

TABLE 8.37

Frequency of Secondary Trisomics in the Progenies of Rice Primary Trisomics

Trisomics	Total Plants Grown	Secondary Trisomics		Frequency (%)
		Short Arm	Long Arm	
Triplo 1[a]	–	1	1	–
Triplo 2	1632	2	1	0.18
Triplo 3[a]	–	0	0	–
Triplo 4	1812	1	0	0.05
Triplo 5	1536	1	0	0.13
Triplo 6	2112	3	1	0.19
Triplo 7	3300	2	1	0.09
Triplo 8	1608	0	3	0.25
Triplo 9	2127	0	3	0.19
Triplo 10	1776	0	0	–
Triplo 11	600	1	1	3.33
Triplo 12	1632	2	0	0.12

Source: From Singh, K., D. S. Multani, and G. S. Khush. 1996. *Genetics* 143: 517–529. With permission.

[a] Triplo 1 and Triplo 3 are highly sterile and large populations could not be grown. $2n + 1S.1S$ and $2n + 1L.1L$ and .3L were selected from the progenies of primary trisomics.

sibs. For example, secondary trisomic $2n + 7L.7L$ was very weak and sterile, $2n + 7S.7S$ was partial fertile, and $2n + 4S.4S$ was morphologically like diploid sibs (Singh et al., 1996).

8.2.2.2.3.2 Cytological Identification Secondary trisomics are identified by the observation of a ring trivalent at diakinesis or metaphase-I (Figure 8.36a). At pachynema, based on euchromatin and heterochromatin differentiations, chromosome arm length, and pairing pattern, Khush and

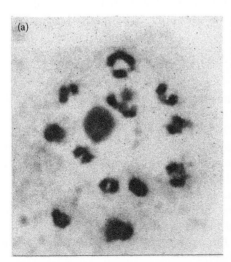 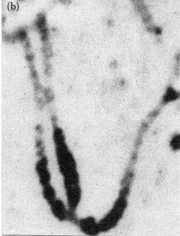

FIGURE 8.36 (a and b) Chromosome configurations in secondary trisomics of tomato. (a) A ring trivalent configuration at diakinesis in secondary $2n + 10L.10L$. (b) A trivalent configuration at pachynema in secondary $2n + 12L.12L$; one chiasma can be seen in the 12 Ls. (From Khush, G. S. and C. M. Rick. 1969. *Heredity* 24: 127–146. With permission.)

TABLE 8.38

Chromosome Association at Diakinesis in Secondary Trisomics of Tomato

Secondary Trisomics	Chromosome Association (% Sporocytes)		
	1II + 1III	12II + 1I	11II + 3I
$2n + 3S.3S$	44	56	0
$2n + 7S.7S$	32	68	0
$2n + 9S.9S$	40	56	4
$2n + 6L.6L$	68	32	0
$2n + 8L.8L$	58	42	0
$2n + 9L.9L$	62	38	0
$2n + 10L.10L$	60	40	0
$2n + 12L.12L$	72	28	0

Source: Adapted from Khush, G. S. and C. M. Rick. 1969. *Heredity* 24: 129–146.

Rick (1969) identified nine secondaries of tomato. Figure 8.36b shows a trivalent configuration at pachynema for a secondary trisomic. When a secondary chromosome remains as a univalent, it forms a "doughnut" (ringlet) or pair with the other homologous arms of the two normal chromosomes to form a trivalent. Ring trivalents in tomato secondaries for long arms are more frequent than those for the short arms (Table 8.38). In maize secondary $2n + 5S.5S$, the highest frequency of sporocytes showed a ring trivalent (43.6%), followed by a ring univalent (32.3%) (Rhoades, 1933b). Similarly in *Datura*, the highest (44.9%) frequency was for the ring trivalent, followed by a V-shape trivalent (22.0%) and a ring univalent (16.9%) (Belling and Blakeslee, 1924).

Secondary trisomics in rice were identified precisely at pachynema and that facilitated to pinpoint the position of kinetochores with certainty (Singh et al., 1996). At diakinesis, various types of chromosome association were recorded in the secondary trisomics. Like tomato's secondaries, secondary trisomics in rice for the long arms showed higher frequency of ring trivalent than those for the short arms. The frequency of ring trivalent ranged from 0.5% ($2n + 4S.4S$) to 25.6% ($2n + 8L.8L$) (Table 8.39).

8.2.2.2.4   Transmission of the Extra Chromosome in Secondary Trisomics

The transmission rate is correlated with the size of the extra secondary chromosome. The extra chromosomes with long arms cause greater imbalance than those for the short arms (Khush and Rick, 1969). Of the 14 secondaries studied in *Datura*, the female transmission rate of the secondary chromosome ranged from 2.4% ($2n + 1.1$) to 29.3% ($2n + 19.19$).

Secondary trisomics regularly throw related primaries. The frequency in *Datura* ranged from 0.06% ($2n + 19.19$) to 9.9% ($2n + 9.9$). Occasionally, unrelated primary trisomics appear in the offspring of secondaries. This suggests that the presence of one extra chromosome in some way stimulates nondisjunction in other chromosome pairs.

The female transmission rate of the extra secondary chromosome in tomato ranged from 0.0% ($2n + 6L.6L$) to 33.6% ($2n + 7S.7S$) and the frequencies of related primaries ranged from 0.0($2n + 7S.7S$) to 14.9% ($2n + 10L.10L$). Like *Datura*, unrelated primaries were isolated in tomato, but the frequency was low (Khush and Rick, 1969).

The female transmission rates of the extra isochromosome in rice ranged from 19.5% ($2n + 6S.6S$) to 40.6% ($2n + 12S.12S$) (Table 8.40). Male transmission of isochromosome in rice was recorded for $2n + 4S.4S$ as two plants in its progeny carried two extra 4S.4S isochromosomes (Singh et al., 1996).

TABLE 8.39
Chromosome Association at Diakinesis in Secondary Trisomics of Rice

Secondary Trisomics	Total Cell Observed	Chromosome Association (% Sporocytes)				
		12 II + 1 I	1 II + 1 III			1 III + 3 I
			Ring	Chain	Others	
2n + 1S.1S	101	46.5	15.9	27.7	7.9	9
2n + 1L.1L	165	29.7	24.8	23.0	18.2	4.2
2n + 2S.2S	147	74.8	`6.8	6.8	11.6	0.0
2n + 2L.2L	60	48.3	21.7	18.3	10.0	1.7
2n + 4S.4S	215	81.4	0.5	6.5	11.6	0.0
2n + 5S.5S	127	65.4	10.2	14.2	7.1	3.1
2n + 6S.6S	110	58.2	11.8	18.2	5.5	6.4
2n + 6L.6L	142	50.7	19.7	15.5	11.3	2.8
2n + 7S.7S	156	57.7	15.4	17.9	9.0	0.0
2n + 7L.7L	141	55.3	15.6	12.1	12.1	4.9
2n + 8L.8L	125	60.0	25.6	11.2	3.2	0.0
2n + 9L.9L	118	44.1	18.6	16.9	15.3	5.1
2n + 11S.11S	146	65.1	11.6	12.3	10.3	0.7
2n + 11L.11L	115	75.5	6.9	10.1	5.2	2.6
2n + 12L.12L	93	55.9	15.1	14.0	12.9	2.1

Source: From Singh, K., D. S. Multani, and G. S. Khush. 1996. *Genetics* 143: 517–529. With permission.

TABLE 8.40
Transmission Rate (%) of the Extra Isochromosomes in the Selfed Progenies of Rice

Secondary Trisomics	Total Plants	Disomics	Secondary Trisomics	Related Primary
2n + 5S.5S	181	69.1	28.2	2.7
2n + 6S.6S	41	63.4	19.5	17.1
2n + 8L.8L	116	55.2	24.1	20.7
2n + 11S.11S	103	68.0	25.2	6.8
2n + 11L.11L	124	69.3	23.4	7.3
2n + 12S.12S	123	54.5	40.6	4.9

Source: From Singh, K., D. S. Multani, and G. S. Khush. 1996. *Genetics* 143: 517–529. With permission.

8.2.2.2.5 Genetic Segregation in Secondary Trisomics

Secondary trisomics can be used to locate genes in a particular half-chromosome in much the same way as with primary trisomics (Avery et al., 1959). If a marker gene is located in the extra chromosome, a ratio of 3:1:: all (4):0 will be recorded for disomic and secondary trisomic fractions in random chromosome segregation. This suggests that no recessive homozygous plants will be obtained in the secondary trisomic fraction (Figure 8.37). In random chromatid segregation, the possible gametes with genotypes *AAA* (11), *AAa* (12), and *Aaa* (1) are expected in secondary trisomics. Thus, in F$_2$ recessive homozygous plants will not be obtained in the secondary trisomic portion

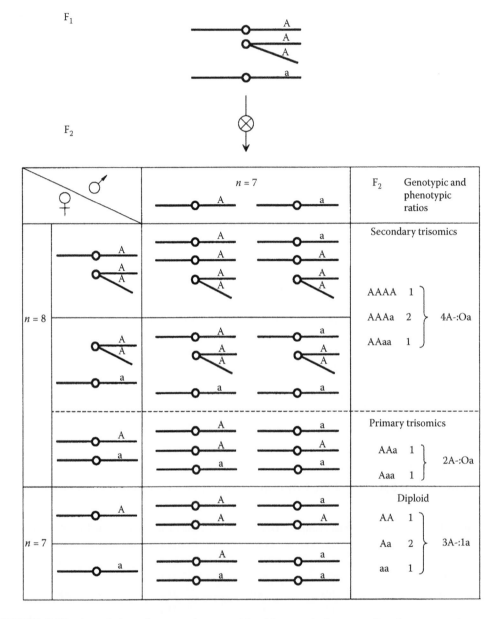

FIGURE 8.37 Association of a recessive gene (*a*) with a particular arm of a chromosome by use of secondary trisomics (random chromosome segregation).

(all *A*-:0 *a*), but a very low frequency is expected in the primary trisomic portion (27.8 *A*-:1 *a*). The disomic portion is expected to segregate in a ratio of 4.76 *A*-:1 *a* (Figure 8.38). If a gene is not located in the secondary chromosome, a disomic ratio (3:1::3:1) in F_2 and in BC_1 (1:1::1:1) are observed for both diploid and secondary trisomic fractions.

Secondary trisomics were used for the first time by Khush and Rick (1969) to associate genes with a particular arm of a chromosome. They studied genetic segregation in the back crosses of five secondaries and located precisely the centromere positions for chromosomes 8, 9 and 10, while they disproved the previously reported linkage map for chromosome 10.

Secondary trisomics in rice have been used to locate marker genes on particular arm of the rice chromosomes (Singh et al., 1996). Secondaries for the short arm for chromosome 2, 4, 5, 6, and 12

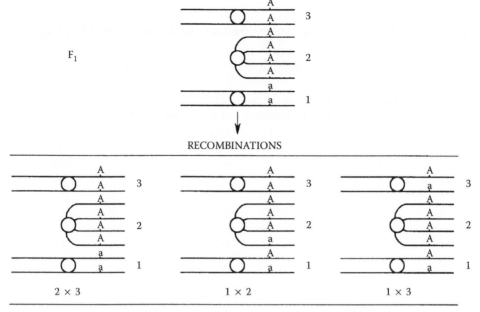

RECOMBINATIONS

Frequency of possible gametes with different genotypes:

Anaphase separation	$n = 8$						$n = 7$	
	7 + Secondary			7 + Primary				
	AAA	AAa	Aaa	AA	Aa	aa	A	a
2 × 3								
1.3 vs 2	0	0	0	0	4	0	0	0
1 vs 2.3	4	0	0	0	0	0	0	2
1.2 vs 3	0	4	0	0	0	0	2	0
1 × 2								
1.3 vs 2	0	0	0	2	2	0	0	0
1 vs 2.3	2	2	0	0	0	0	1	1
1.2 vs 3	1	2	1	0	0	0	2	0
1 × 3								
1.3 vs 2	0	0	0	1	2	1	0	0
1 vs 2.3	2	2	0	0	0	0	1	1
1.2 vs 3	2	2	0	0	0	0	1	1
Total								
$n = 7$ + Secondary	11	12	1	--	--	--	--	--
$n = 7$ + Primary	--	--	--	3	8	1	--	--
$n = 7$	--	--	--	--	--	--	7	5

FIGURE 8.38 Association of a recessive gene (*a*) with a particular arm of a chromosome by use of secondary trisomics (Random chromatid segregation).

TABLE 8.41

F$_2$ Segregation in the Crosses of Secondary Trisomics for Chromosome 11 of Rice with Marker Genes

Cross	Total Plants (Number)	Disomics			Trisomics			Arm Location of the Gene
		Normal (Number)	Recessive (Number)	χ^2 (3:1)	Normal (Number)	Recessive (Number)	Recessive (%)	
$2n$ + 11S.11S/z-1	98	60	17	0.35	21	0	0.0	11S
$2n$ + 11L.11L/z-1	91	51	16	0.04	19	52	0.8	11S
$2n$ + 11S.11S/v-4	214	100	37	0.25	62	15	19.5	11L
$2n$ + 11L.11L/v-4	207	111	28	1.75	68	0	0.0	11L
$2n$ + 11S.11S/la	214	100	37	0.29	59	18	23.4	11L
$2n$ + 11L.11L/la	207	109	30	0.87	68	0	0.0	11L
$2n$ + 11S.11S/z-2	200	119	29	2.31	42	10	19.4	11L
$2n$ + 11L.11L/z-2	245	147	39	1.61	59	0	0.0	11L

Source: From Singh, K., D. S. Multani, and G. S. Khush. 1996. *Genetics* 143: 517–529. With permission.

and long and short arm for chromosome 11 were used in locating genes on their respective arms (Table 8.41). For example, segregation of four genetic markers, z-1, v-4, la, and z-2, was analyzed in their crosses with $2n$ + 11L.11L as well as with $2n$ + 11S.11S. The segregation results show that z-1 is located on 11S and the remaining three genes are located on 11L (Table 8.41).

8.2.2.3   Tertiary Trisomics

8.2.2.3.1   Introduction

A tertiary trisomic individual consists of an interchanged nonhomologous chromosome in addition to the normal somatic chromosome complement. Tertiary trisomics have been utilized very effectively in tomato for determining centromere positions on the linkage maps and to associate a gene with a particular arm of a chromosome (Khush and Rick, 1967c). They have been used in barley for the construction of "balanced tertiary trisomics" (BTTs) for hybrid seed production (Ramage, 1965).

8.2.2.3.2   Sources of Tertiary Trisomics

In general, interchange heterozygotes that throw tertiary trisomics in their progenies have been described in maize (Burnham, 1930, 1934), pea (Sutton, 1939), *Oenothera lamarckiana* (Emerson, 1936), *Oenothera blandina* (Catcheside, 1954), *Datura stramonium* (Avery et al., 1959), barley (Ramage, 1960), tomato (Khush and Rick, 1967c; Gill, 1978), *Phaseolus vulgaris* (Ashraf and Bassett, 1987), *Pennisetum americanum* (Singh et al., 1982), *Secale cereale* (Janse, 1985, 1987), and lentil (Ladizinsky et al., 1990). Among all the published results on tertiary trisomics, the investigation of Khush and Rick (1967c) in tomato is the most thorough. They reported seven tertiaries; two were isolated from tertiary monosomics and five from the interchanged heterozygotes.

In *Datura*, the tertiaries appeared spontaneously in a very low frequency. In only six instances, tertiary trisomics occurred spontaneously among approximately two million *Datura* plants grown to adult stage. The rarity of occurrence of tertiaries and secondaries in *Datura* was attributed to the stable structure of the chromosomes (Avery et al., 1959).

Tertiary trisomics originate from an interchange heterozygote in the following way. An interchange heterozygote occasionally forms a noncooriented quadrivalent configuration at diakinesis and metaphase-I and a 3:1 random disjunction of chromosomes at anaphase-I will generate n + 1 gametes. Thus, eight possible types of $2x$ + 1 individuals are expected in the progeny of a selfed interchange heterozygote. Of the four tertiaries, two will be in homozygous background and the other two in translocation heterozygous background. For isolating tertiaries in tomatoes, Khush and Rick (1967c) hybridized interchanged heterozygotes with normal diploids. According to this procedure, the progeny should segregate in a proportion of two tertiaries (homozygous) to two primaries (interchange heterozygous) in the $2x + 1$ fraction, and one normal to one interchange heterzygote in the $2x$ fraction (Figure 8.39). A plant is designated as a primary trisomic if an extra chromosome is normal; the trisomic may be a primary trisomic, a primary trisomic interchange heterozygote or a primary trisomic interchange homozygote. On the other hand, if the extra chromosome is a translocated chromosome, a plant is designated as a tertiary trisomic (Ramage, 1960).

8.2.2.3.3   Identification of Tertiary Trisomics

8.2.2.3.3.1   Morphological Identification
Tertiary chromosomes are composed of parts of two nonhomologous chromosomes. Thus, tertiaries inherit certain morphological features of the two related primaries. Like telotrisomics for the long arm, tertiary trisomics carrying a long arm generally exert a greater influence on plant morphology than when carrying a short arm.

Furthermore, if tertiary chromosomes contain two long arms, the longer of the two arms usually exerts greater phenotypic effect, and such tertiaries frequently resemble the primary trisomics carrying the longer arm. In contrast, tertiaries for the short arms appear to contribute very little to alter the phenotypic expression of the tertiary trisomics. In tomato, tertiary trisomic $2x + 5S.7L$ resembled very closely Triplo 7, while the $2x + 7S.11L$ tertiary trisomic was similar to Triplo 11 in that the effect of the short arm was relatively minor. Tertiary trisomics for short arms such as $2x + 9S.12S$ resembled neither Triplo 9 nor Triplo 12 (Khush and Rick, 1967c). This suggests that the short arms have very little effect on phenotypes of tertiaries in tomato.

8.2.2.3.3.2   Cytological Identification
In tertiary trisomics, a maximum association of five chromosomes (pentavalent) or derived configurations, such as 1IV + I, 1III + 2I, 1III + II, 2II + 1I, 1II + 3I or 5I, are expected during diakinesis and metaphase-I. The derived configurations are attributed to the failure of chiasma formation or chiasma maintenance. Khush and Rick (1967c) analyzed chromosome associations at diakinesis in the seven tertiaries of tomato. The highest frequency of sporocytes showed 1III + 11II and the next most frequent configurations were 1V + 10II and 12II + 1I. Sporocytes with such other configurations as 1IV + 10II + 1I, 1III + 10II + 2I, 1III + 3I, and 10II + 5I were rarely recorded. The formation of a pentavalent association depends on the length of the tertiary chromosome. In tomato, tertiary chromosomes composed of short arms showed lower frequencies of the 1V + 10II configuration than those carrying long arms (Table 8.42).

The true nature of tertiary chromosomes can be established precisely by pachytene chromosome analysis. Such investigation has not been feasible in crops other than tomato, where breakage point and arm identity may be accurately determined (Khush and Rick, 1967c).

8.2.2.3.4   Transmission of the Extra Chromosome in Tertiary Trisomics

Tertiary trisomics usually throw, in addition to the parent type, small proportions of related primaries (Avery et al., 1959; Khush and Rick, 1967c). In *Datura*, a very low frequency of related secondaries was also isolated, but these were not recorded in tomato (Khush and Rick, 1967c). However, Janse (1985) observed in a tertiary trisomic of rye that 58.1% of all microspores going through pollen mitosis carried seven chromosomes and 41.9% microspores contained eight chromosomes. This suggests that up to the end of first pollen mitosis aneuploid spores are not significantly fewer than euploid spores. Therefore, the failure of $n + 1$ gametes to be transmitted through pollen presumably results from the failure of $n + 1$ pollen to compete with n chromosome pollen. In tomato, tertiary chromosomes are transmitted through females and frequencies ranged from 7.43% ($2x + 2L.10L$) to 40.78% ($2x + 7S.11L$). A high female transmission rate of tomato tertiaries was attributed to the genetically heterozygous background (Khush and Rick, 1967c). In common bean, the transmission rate of the extra tertiary chromosome after selfing ranged from 28% to 41% (Ashraf and Bassett, 1987) and in *Datura* it ranged from 14.14% ($2x + 1.18$) to 32.05% ($2x + 2.5$).

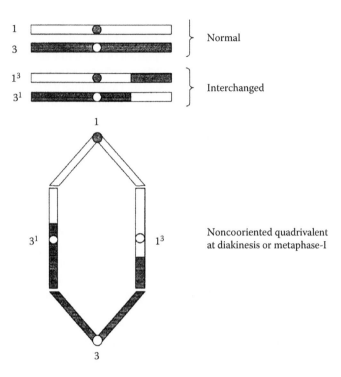

Noncooriented quadrivalent at diakinesis or metaphase-I

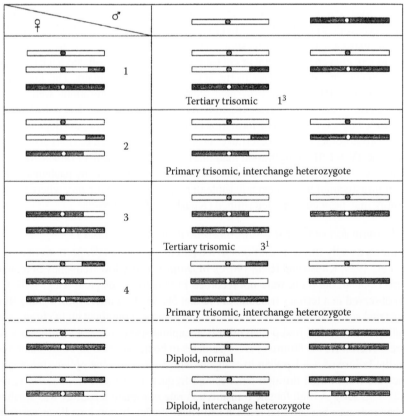

FIGURE 8.39 Expected types of plants from the progeny of an interchanged heterozygote.

TABLE 8.42

Frequencies of Various Chromosomal Associations Observed at Diakinesis in Tomato Tertiary Trisomics

Tertiary Trisomics	Number of cells	10 II + 1 V	11 II + 1 III	12 II + 1 I	10 II + 1 IV + 1 I	10 II + 1III + 2 I	11 II + 3 I	10 II + 5 I
$2n + 1L.11L$	50	15	20	13	1	1	0	0
$2n + 2L.10L$	50	12	17	12	1	6	2	0
$2n + 4L.10L$	50	22	20	6	1	1	0	0
$2n + 5S.7L$	50	13	20	15	0	2	0	0
$2n + 7S.11L$	50	5	22	21	1	0	1	0
$2n + 9L.12L$	50	14	16	19	0	0	0	1
$2n + 9S.12S$	50	11	17	17	2	2	0	1

Source: From Khush, G. S. and C. M. Rick. 1967c. *Can. J. Genet. Cytol.* 9: 610–631. With permission.

8.2.2.3.5 Genetic Segregation in Tertiary Trisomics

Genetic ratios in F_2 populations of tertiary trisomics are modified in a fashion similar to those recorded in secondary trisomics, telotrisomics, and acrotrisomics. Assuming duplex genetic constitution (*AAa*) of an F_1, no male transmission of the tertiary chromosome, 50% female transmission and a marker gene located on either arm of the tertiary chromosome, a trisomic ratio 3:1 $(2x)$::all(4):0 $(2x + 1)$ or a total of 7:1 ratio is expected. Furthermore, a 5:1 ratio should be obtained when a female transmission rate of the tertiary chromosome reaches 33.3%. On the other hand, if a marker gene is not located on the tertiary chromosome, a disomic ratio should be recorded in the F_2 for both $2x$ and $2x + 1$ portions. Tertiary trisomics have been used in associating a gene with a particular arm of a chromosome in barley (Ramage, 1965) and tomato (Khush and Rick, 1967c).

Khush and Rick (1967c) utilized five tertiary trisomics of tomato in genetic and linkage studies. Genetic segregation of tertiary trisomic 4L.10L will be cited as an example. The markers *clau, ra, di, ful,* and *w-4* are located on chromosome 4. The genes *clau* and *ful* are on 4S and *ra, di,* and *w-4* on 4L. The two markers *ag* and *tv* are on chromosome 10. As expected, markers *clau* and *ful* showed a disomic ratio and the remaining markers trisomic ratio. The occurrence of a low frequency of recessive homozygotes in the trisomic fraction can be explained by double reduction (Table 8.43).

8.2.2.3.6 Balanced Tertiary Trisomics (BTT)

Ramage (1965) proposed a scheme utilizing balanced tertiary trisomics for production of hybrid barley seed. Balanced tertiary trisomics are tertiary trisomics constituted in such a way that the dominant allele of a marker gene closely linked with the interchange break-point is carried on the tertiary chromosome. The recessive allele is carried on each of the two normal chromosomes. The dominant marker allele for a mature plant character, such as red plant color (*R*), may be carried on either the centromere portion or on the interchanged segment of the extra chromosome and should be linked with a male fertile gene (*Ms*). The two normal chromosomes should carry the corresponding recessive male sterile allele (*ms*). All balanced tertiary trisomics would be male fertile and red and all diploids would be male sterile and green. All functioning pollen produced by the balanced tertiary trisomics would carry the male sterile (*ms*) and the green plant color alleles (*r*). The self-progeny of such a trisomic would be planted in an isolation block. Diploid plants would be green and male sterile. All seed set on them would produce male sterile diploids in the next generation. The diploid plants would be harvested separately and seed produced on them would be used to plant the female rows in the hybrid seed production field. Balanced tertiary trisomics planted in isolation would be dominant and would be harvested separately. Seed produced on balanced tertiary trisomic

TABLE 8.43

Segregation Ratios in F$_2$'s of 2n + 4L.10L Tertiary Trisomic in Tomato

				Progeny					
			2n			$n+1$			
				%			%		χ^2
Gene	Chromosome	Total	Normal	Recessive	Recessive	Normal	Recessive	Recessive	(3:1)
clau	4	496	263	95	26.2	94	44	31.8	3.58
ra	4	496	272	86	24.0	138	0	0.0	46.00
di	4	496	300	58	16.2	136	2	1.4	40.81
ful	4	239	107	33	23.5	73	26	26.2	1.17
w-4	4	239	107	33	23.5	99	0	0.0	33.00
ag	10	202	91	25	21.5	83	3	3.4	21.80
tv	10	202	89	27	23.2	83	3	3.4	21.80

Source: From Khush, G. S. and C. M. Rick. 1967c. *Can. J. Genet. Cytol.* 9: 610–631. With permission.

plants would produce approximately 70% male sterile diploids and 30% balanced tertiary trisomics (Figure 8.40). This seed would be used to plant an isolation block the following year.

The prerequisites for utilizing the BTT system in hybrid seed production are that balanced tertiary trisomics produce abundant pollen and that during anthesis sufficient wind and insects are available for pollen dissemination. Several other alternatives to distinguish tertiary trisomics from diploids have been suggested, such as resistance or susceptibility to phytocides, seed size or shape and differential plant heights (Ramage, 1965).

8.2.2.4 Telotrisomics

8.2.2.4.1 *Introduction*

A telocentric chromosome consists of a kinetochore (centromere) and one complete arm of a normal chromosome. The plant with a normal chromosome complement plus an extra telocentric chromosome (2x + 1 telocentric; designated as telo) is designated as mono-telotrisomic or telotrisomics (Kimber and Sears, 1969; Tsuchiya, 1972a). The telocentric chromosome contains a terminal kinetochore (Figure 8.41).

8.2.2.4.2 *Sources of Telotrisomics*

Mono-telotrisomics originate spontaneously, often in the progenies of primary trisomics (*Zea mays*—Rhoades, 1936; *Datura stramonium*—Blakeslee and Avery, 1938; *Lotus pedunculatus*—Chen and Grant, 1968b; *Hordeum vulgare*—Tsuchiya, 1971b; Fedak et al., 1971; Singh and Tsuchiya, 1977; Siep, 1980; Furst and Tsuchiya, 1983; Shahla and Tsuchiya, 1990; Tsuchiya, 1991; *Avena strigosa*—Rajhathy, 1975; *Secale cereale*—Zeller et al., 1977; Sturm and Melz, 1982; Melz and Schlegel, 1985; Zeller et al., 1987; *Oryza sativa*—Singh et al., 1996; Cheng et al., 2001), triploids (*H. vulgare*—Tsuchiya, 1971b; Fedak and Tsuchiya, 1975), and *Lotus pedunculatus* (Chen and Grant, 1968a). Of the eleven reported telotrisomics of barley, six (Triplo 2L, 2S, 3L, 4L, 5S, 6S) appeared in the progenies of their respective primary trisomics and three (Triplo 1L, 1S, 5L) were isolated from triploids. Only, Triplo 3S and Triplo 7S originated in the progenies of plants carrying 2n = 13 + 1 acro 3L^{3S} + 1 telo 3S, and 2n = 15 + 1 telo 7S, respectively (Table 8.44).

In rice, Singh et al. (1996) isolated and identified telotrisomics for the short arm of chromosomes 1, 8, 9, and 10 and for the long arm of chromosomes 2, 3, and 5 from the progenies of their respective primary trisomics. Cheng et al. (2001) isolated all 24 telotrisomics of rice from approximately 180,000 plants derived from the trisomics and other aneuploids. *This report now*

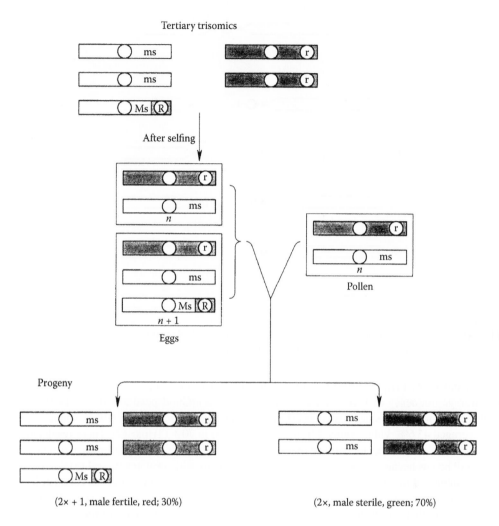

FIGURE 8.40 Utilization of balanced tertiary trisomics (BTT) to produce hybrid barley. (Redrawn from Ramage, R. T. 1965. *Crop Sci.* 5: 177–178. With permission.)

claims the establishment of a complete set of telotrisomics in an economically important diploid crop. They were able to complete the set by examining cytologically the variants morphologically distinct from the primaries and disomics. This study demonstrates that a complete set of telotrisomics can be established in any crop if a large population of primary trisomics are grown.

Khush and Rick (1968b) isolated six telotrisomics of the possible 24 of tomato from the progenies of terminal deficiency (Triplo 4L), tertiary monosomics (Triplo 3L, 8L, 7L), and compensating trisomics (Triplo 3S, 10S).

The low frequency of occurrence of telocentric chromosomes in the progenies of aneuploids of diploid crop plants such as maize, barley, tomato, and many others may be ascribed to the following reasons (Tsuchiya, 1972a).

The frequency of univalents in primary trisomics is low. The overall frequency of the sporocytes having one univalent was 23.7% (range 20.0%–36.5%) for the seven primary trisomics of *Hordeum spontaneum* and 22.5% (range 15.3%–29.3%) in *H. vulgare* (Tsuchiya, 1960a, 1967). In *Beta vulgaris*, the univalent frequency of three primary trisomics averaged 23.9%, ranging from 21.36% (Triplo 2) to 29.88% (Triplo 3) (Kaltsikes and Evans, 1967). Working with the primary trisomics of *Arabidopsis thaliana*, Sears and Lee-Chen (1970) found the 5II + 1I chromosome configuration to

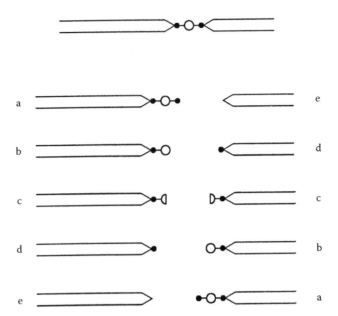

FIGURE 8.41 Diagrammatic sketch of origin of various types of telocentric chromosomes. (Redrawn from Steinitz-Sears, L. M. 1966. *Genetics* 54: 241–248. With permission.)

range from 10.3% (Triplo F) to 30.1% (Triplo R), with an average of 23.5%. In *Nicotiana sylvestris*, the average frequency of 12II + 3I was 4% (range 0%–8%) (Goodspeed and Avery, 1939). A much higher frequency of univalents was observed in tomato by Rick and Barton (1954), where 11 primaries averaged 45.6%. Values ranged from 26% (Triplo 3) to 70% (Triplo 12). These results indicate that the probability of the appearance of telocentric chromosomes in the progenies of primary trisomics will be low because only a low frequency of univalents will lag at anaphase-I to telophase-I and will be subjected to misdivision—a prerequisite for the origin of telocentrics. In contrast, in wheat monosomics the 20II + 1I configuration is observed sometimes in all the sporocytes (Person, 1956). This suggests that a univalent in wheat monosomics has greater chance of producing telocentric

TABLE 8.44
Sources of Eleven Telotrisomics of Barley

Proposed Designation of Telotrisomics	Source	Author
Triplo 1L (7HL)[a]	Autotriploid	Tsuchiya (1971b, 1972a)
Triplo 1S (7HS)	Triploid hybrid	Tsuchiya (1971b)
Triplo 2L (2HL)	Slender	Tsuchiya (1971b)
Triplo 2S (2HS)	Slender	Tsuchiya (1971b)
Triplo 3L (3HL)	Pale (F₁ hybrid)	Tsuchiya (1971b)
Triplo 3S (3HS)	Novel compensating diploid ($2n = 13 + 1$ acro 3L³ˢ + 1 telo 3S)	Singh and Tsuchiya (1981d)
Triplo 4L (4HL)	Robust	Tsuchiya (1971b)
Triplo 5L (1HL)	Triploid hybrid	Tsuchiya (1971b, 1972b)
Triplo 5S (1HS)	Pseudo-normal	Furst and Tsuchiya (1983)
Triplo 6S (6HS)	Purple (F₂)	Seip (1980)
Triplo 7S (5HS)	Semi-erect (F₁ hybrid) ($2n = 15 + 1$ telo 7S)	Shahla and Tsuchiya (1983)

[a] Homoeologous designation; From Costa, J. M. et al. 2001. *Theor. Appl. Genet.* 103: 415–424. With permission.

chromosomes by misdivision than those recorded in primary trisomics. Misdivision of a kineto-chore is a potent source of obtaining telocentric chromosomes in cotton (Brown, 1958) and wheat (Sears, 1952a,b; Steinitz-Sears, 1966).

8.2.2.4.3   Identification of Telotrisomics

8.2.2.4.3.1   Morphological Identification   In general, it has been observed in diploid species that a telotrisomic for the long arm of a chromosome has similar effect on plant morphology as the corresponding primary trisomic. Goodspeed and Avery (1939) observed that a plant with $2n + 1f$ (fragment = telocentric) chromosome constitution was similar in plant morphology to the corre-sponding primary trisomics. Rhoades (1936) discovered a plant with $2n + 1$ telo 5S in maize that resembled neither its primary trisomic nor diploid sibs but was intermediate in appearance.

The effects of the short and long arms seem to depend on the genetic material present in the extra telocentric chromosome. Khush and Rick (1968b) recorded in tomato that Triplo 4L, 7L, and 8L resembled their respective primary trisomics. Similarly, in barley Triplo 1L, 2L, 3L, 4L, and 5L were morphologically similar to their corresponding primaries (Singh and Tsuchiya, 1977). Triplo 2S resembled neither the primary trisomic, Slender, nor diploid sibs. In contrast, morphological features of Triplo 1S, 3S, 5S, 6S, and 7S of barley (Singh and Tsuchiya, 1977, 1981b; Siep, 1980; Shahla and Tsuchiya, 1983) and Triplo 3S and 10S in tomato (Khush and Rick, 1968b) were similar to their diploid sibs.

The influence of long and short arm in telotrisomic condition can be precisely compared only in rice because all 24 telotrisomics are described (Cheng et al., 2001). Triplo 4S, 5S, 7S, 9S, 10S, and 11S were morphologically indistinguishable from the disomic sibs. In this situation, chromosome count is required to establish the telotrisomic nature. Triplo 1S, 2S, 3S, 6S, 8S, and 12S expressed some characteristic morphological alterations. Telotrisomics for the long arm (Triplo 4L, 5L, 6L, 7L, 9L, 10L, and 12L) resembled their corresponding primaries. All telotrisomics were fertile even telotrisomics for Triplo 3 were highly sterile. Telotrisomics for chromosome 1, 2, and 3 neither resembled diploid sibs nor their respective primary trisomics. This study demonstrates that the short arm may contain less genetic material compared to long arm thus its influence is not that apparent on plant morphology than the long arm.

8.2.2.4.3.2   Cytological Identification   The extra telocentric chromosome is identified by somatic as well as meiotic chromosome counts. The precise arm identification is performed by karyotype analysis, either utilizing somatic metaphase or meiotic pachytene chromosomes. The somatic karyotype analysis, however, does not provide enough evidence to distinguish between long and short arms where both arms are of equal size and also lack euchromatin and heterochromatin differentiation. In the barley chromosome complement, chromosome 1, 2, 3, and 4 have almost median kinetochores and are indistinguishable by aceto-carmine or Feulgen staining techniques. Therefore, the initial identification of telotrisomics was based on morphological features and genetic studies (Fedak et al., 1971, 1972; Singh and Tsuchiya, 1977). Furthermore, translocation testers also did not elucidate the arm identity. Tsuchiya (1972a) analyzed meiotic pairing of three F_1 hybrids of Triplo 1L and translocation testers T1-4a, T1-6a, and T1-7a. He observed a 1V + 5II chromosome configuration at metaphase-I. This only demonstrated that the telocentric chromosome traced to chromosome 1.

Based on genetic analysis, Fedak et al. (1971) reported one telocentric chromosome to be the short arm of chromosome 5 because it showed trisomic ratios with genes *trd* (third outer glume) and a_t (albino seedling); these genes are located on the short arm of linkage group 5 (Robertson, 1971). From karyotype analysis, Tsuchiya (1972b) demonstrated that previously identified telo 5S was a telo 5L (Figure 8.42). On this basis, he corrected the genetic linkage map of chromosome 5.

By utilizing the Giemsa N-banding pattern, Singh and Tsuchiya (1982a,b) and Tsuchiya (1991) correctly identified and designated 11 telotrisomics (Telo 1L, 1S, 2L, 2S, 3L, 3S, 4L, 5L, 5S, 6S, and 7S) of barley. Chromosome 3 contains its kinetochore at a median region and its long arm is

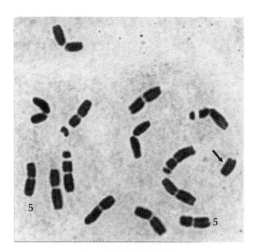

FIGURE 8.42   Mitotic metaphase chromosomes of Triplo 5L after aceto-carmine staining of barley. Compare long arm of chromosome 5 (chromosome 5 numbered) and Telo 5L (arrow). (Adapted from Singh, R. J. 1974. *Cytogenetics of Telotrisomics in Barley*. Ph.D. thesis, Colorado State University.)

the second longest in the barley complement. Telo 3L showed a centromeric (fainter than telo 3S) and an intercalary band and it also showed a faint dot on each chromatid at its distal region. Telo 3S showed a dark centromeric band only.

Similarly the N-banding technique demonstrated that the previously identified Telo 4S (Singh and Tsuchiya, 1977, 1981a; Linde-Laursen, 1978; Tsuchiya and Singh, 1982) was actually telo 4L. The morphology of correctly identified Triplo 4L is similar to Triplo 4. Giemsa N-banding technique revealed a deficiency in two telocentric chromosomes, telo 2S and 4L of barley. Telo 2S showed a 50% distal deletion while telo 4L had a 32% deficiency. The apparent loss of a distal portion of telo 2S and 4L did not affect morphology, transmission rate (Singh and Tsuchiya, 1977), or meiotic behaviors (Singh and Tsuchiya, 1981a). In this case it is likely that the broken end healed and started functioning as a normal telomere (Hang, 1981). If telotrisomic plants having deficiency in the telocentric chromosome were used in genetic-linkage analysis, a wrong conclusion could be drawn, because genes located in the deficient segment would show a disomic ratio. Telotrisomic analysis with Triplo 2S-*yst3* may be ascribed to a deficiency in telo 2S (Tsuchiya and Hang, 1979). These results indicate the importance of detailed karyotype study of telotrisomic plants.

Pachytene chromosome analysis has played a major role in maize and tomato cytogenetics. In both crops, long and short arms could be distinguished without doubt because of characteristic euchromatin and heterochromatin differentiation. Khush and Rick (1968b) were able to identify all the available six mono-telotrisomics of tomato at pachynema. Triplo 3S, 7L, and 10S possessed a truly terminal kinetochore and telo 3L, 4L, and 8L had subterminal kinetochores.

Rice telotrisomics for 9S and 10S were easily identified as they were associated with the nucleolus while pachytene chromosome failed to identify other telotrisomics (Singh et al., 1996). Cheng et al. (2001) identified telocentric nature by FISH using a rice centromeric-specific bacterial artificial chromosome (BAC) clone, 17p22, as a marker probe.

8.2.2.4.3.3   Meiotic Behavior of Telotrisomics   In telotrisomic plants, a telocentric chromosome may associate with its normal homologues, forming a trivalent (Figure 8.43a,b) or remain as a univalent. Singh and Tsuchiya (1981a) studied meiotic behavior of seven telocentric chromosomes in telotrisomic conditions. The frequencies of the different configurations are shown in Table 8.45. In the trivalent configurations, the telocentric chromosome was frequently tightly associated with an interstitial chiasma which persisted into metaphase-I (Figure. 8.43b). The average frequency of 1III + 6II chromosome configuration was 77% at diakinesis for the seven mono-telotrisomics,

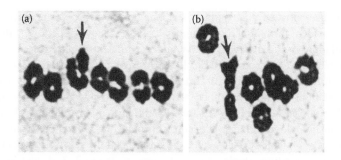

FIGURE 8.43 Chromosome configurations at metaphase-I in Triplo 3L of barley. (a) Tandom V-shape trivalent. (b) Rod-shape trivalent; interstitial chiasma (arrow) between telocentric and normal chromosome is evident. Arrows show a telocentric chromosome. (From Singh, R. J. and T. Tsuchiya. 1982a. *Bot. Gaz.* 142: 267–273. With permission.)

TABLE 8.45
Chromosome Associations at Diakinesis and Metaphase-I, Types and Frequency of the Trivalents at Metaphase-I, Separation of Chromosome at Anaphase-I, and Pollen Fertility in Seven Telotrisomics of Barley (%)

Triplo	Diakinesis		Metaphase-I		Trivalents				Anaphase-I Separation		Pollen Fertility
	1III + 6II	7II + 1I	1III + 6II	7II + 1I	V	Rod	Ring-rod	8-7	7-1-7	Others	
1L	80.0[a]	20.0	74.8	25.2	67.4	7.9	24.7	82.0	12.0	6.0	86.5
1S	80.0	20.0	59.9	40.1	59.6	31.2	9.2	72.6	12.2	15.2	93.1
2L	60.2	39.8	54.6	45.4	64.0	13.8	22.2	80.0	18.0	2.0	83.7
2S	86.0	14.0	70.0	30.0	71.4	17.2	11.4	76.0	12.0	12.0	96.9
3L	77.8	22.2	71.0	29.0	57.2	15.9	26.9	59.4	20.3	20.3	75.4
4L	66.0	34.0	63.0	37.0	60.3	27.0	12.7	75.0	19.0	6.0	95.0
5L	89.1	10.9	71.0	29.0	44.0	31.2	24.8	56.0	20.0	24.0	81.1
Average	77.0[b]	23.0	66.3	33.7	60.6	20.6	18.8	71.5	16.2	12.2	87.4

Source: From Singh, R. J. and T. Tsuchiya. 1977. *Z. Pflanzenzüchtg.* 78: 327–340. With permission.

[a] Observed χ^2-value of trivalent association among telotrisomics at diakinesis (96.88) and metaphase-I (36.58) was significant at $P < 0.001$ for 6 df.

[b] Significant (χ^2-value 48.8, $P < .001$).

ranging from 60.2% (Triplo 2L) to 89.1% (Triplo 5L). The remaining cells had 7II + 1I. As the meiotic stage proceeded from diakinesis to metaphase-I, the frequency of 1III + 6II configuration decreased with a corresponding increase of cells with 7II + 1I because of chiasma terminalization. Rhoades (1940) observed that a telocentric chromosome in maize was associated at metaphase-I with its normal homologue in 59% of the sporocytes. Smith (1947) reported the 1III + 6II configuration in 65.9% sporocytes in *Triticum monococcum* while the telocentric chromosome remained as a univalent in 34.1% cells.

The association of telocentric chromosome with its normal homologue depends on the length of the extra chromosome. Khush and Rick (1968b) observed in tomato that the telocentric chromosomes for long arms (Triplo 3L, 4L, 7L, and 8L) showed more trivalent formation than those for the short arms. Similar results were observed in the telotrisomics of rice (Singh et al., 1996). In barley, the long and short arms do not have significantly different effect on the formation of trivalents because the difference between the lengths of the long and short arms is not large enough to exert such an effect on trivalent formation. For example, Triplo 2L showed the lowest (60.2%) frequency of trivalent association at diakinesis, even though 2L is the third-longest long arm in the barley chromosome complement (Tjio and Hagberg, 1951). A great deal of difference was found in Triplo 2L studied at different times (Tsuchiya, 1971b). This result suggests that telo 2L may be sensitive to environmental conditions or that some other factors may be involved in chromosome association.

A complete chromosome in primary trisomic condition may have greater physical opportunity for synapsis with its homologues than a telocentric chromosome. This holds true for barley but not for tomato. The average metaphase-I trivalent configuration for the seven telotrisomics of barley was 66.3% while the corresponding simple primary trisomics (Triplo 2, 3, 4, and 5) showed trivalent in 77.9% sporocytes (Tsuchiya, 1967). This trend, however, was not recorded in tomato. Rick and Barton (1954) reported trivalent frequencies as: Triplo 3 = 70%; Triplo 4 = 56%; Triplo 7 = 36%; Triplo 8 = 24%; and Triplo 10 = 46%. When the respective telotrisomics are compared, it is evident that telo 7L (58.0%) and 8L (58.0%) showed a higher trivalent association than those recorded for their respective primaries (Khush and Rick, 1968b). This feature, therefore, differs from crop to crop and a definite conclusion cannot be drawn.

Different types of trivalents are observed at diakinesis and metaphase-I (Table 8.45). These are tandem V-shaped (Figure 8.43a), rod-shaped (Figure 8.43b), and ring-and-rod shaped. In barley telotrisomics, tandem V-shaped trivalents were predominant, with an average frequency of 60.6% of the sporocytes. Rod-shaped trivalents were next with an average of 20.6%. Ring-and-rod trivalents were lowest in frequency with an average of 18.8% (Table 8.45). The last type has a more complicated chiasma pattern than either of the other two types of trivalents. Rhoades (1940) also observed tandem V-shaped trivalents in the majority of the sporocytes of a maize mono-telotrisomics. The movement of these members of a trivalent is not random at anaphase-I. Among mono-telotrisomics of barley, in an average of 71.5% of the sporocytes, the telocentric chromosome moved to a pole with one of its homologues, giving 8-7 chromosome disjunction (Table 8.45; Figure 8.44a). This has been observed in Z. mays (Rhoades, 1936), T. monococcum (Smith, 1947), and L. esculentum (Khush and Rick, 1968b).

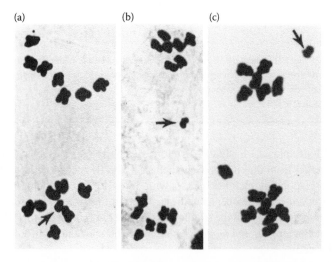

(a)   (b)   (c)

FIGURE 8.44 Separation of chromosomes at anaphase-I in Triplo 3L of barley. (a) 8-7 chromosome disjunction; arrow indicates telocentric chromosome. (b) 7-1-7 chromosome disjunction; telocentric chromosome as a laggard. (c) 8-7 chromosome disjunction; telocentric chromosome (arrow) is on the upper pole with six chromosomes and normal chromosomes on the other pole. (From Singh, R. J. and T. Tsuchiya. 1981a. *Bot. Gaz.* 142: 267–273. With permission.)

Sometimes, the telocentric chromosome does not move to either pole and remains as a laggard at the equatorial plate (Figure 8.44b). In barley, an average of 16.2% sporocytes showed 7-1-7 separation in the seven mono-telotrisomics, ranging from 12% (Triplo 1L, 2S) to 20.3% (Triplo 3L). At late anaphase-I, the lagging telocentric chromosome started to divide and as anaphase progressed, it divided giving an 8-8 separation and sometimes laggards were not included in the telophase nuclei. Occasionally, the telocentric moved to one pole and the two normal homologues passed to the other pole (Figure 8.44c). This type of separation may be responsible for the occurrence of primary trisomics with 15 complete chromosomes in the progenies of telotrisomics of barley (Singh and Tsuchiya, 1977).

At telophase-I, an average of 91.7% of the sporocytes for the seven telotrisomics showed normal separation of chromosomes and at interkinesis, normal and abnormal cells with micronuclei were found. Meiotic abnormalities of the first division were reflected in the 2nd division, which included laggards and micronuclei at the quartet stage (Singh and Tsuchiya, 1981a). Pollen fertility in seven telotrisomics of barley ranged from 75.4% (triplo 3L) to 96.9% (Triplo 2S), while the pollen fertility of normal diploid sibs was 97.8% (Table 8.45; Singh and Tsuchiya, 1977).

1. Meiotic Abnormalities
   In the progenies of telotrisomics of barley, Singh and Tsuchiya (1981a) observed multiploid cells in Triplo 3L (Figure 8.45). A similar phenomenon was observed in Triplo 3 of *H. spontaneum* (Tsuchiya, 1960a) and Triplo 4 in *H. vulgare* (Tsuchiya, 1967).

   Asynaptic cells were observed in Triplo 2L at a much higher frequency (4%) than in Triplo 2 (Tsuchiya, 1960a). Triplo 2S also showed asynaptic cells in 0.3% of the sporocytes. This indicates that both arms of chromosome 2 may have an effect on the formation of asynaptic cells.

2. Stability of Telocentric Chromosomes
   The cytological stability of a telocentric chromosome depends on the structure of its kinetochore and its behavior during cell division. Marks (1957) classified telocentric chromosomes into two categories: one group is stable and the other is unstable. The instability of a telocentric chromosome is caused by nondisjunction and misdivision of univalents.

   Rhoades (1940) indicated that telocentric chromosomes undergo structural changes in somatic cells. The loss and modification of telocentric chromosomes in somatic tissues suggests that a terminal kinetochore is unstable. Thus the stability of a telocentric chromosome depends on the kinetochore constitution. The kinetochore of most mitotic chromosomes is observed as a constriction, and in meiotic chromosomes it appears as a simple, homogeneous, translucent body (Rhoades, 1940).

   Based on her cytogenetic analysis of telocentric chromosomes in hexaploid wheat, Steinitz-Sears (1966) suggested that the relative instability of a telocentric chromosome may be attributed to the degree of completeness of its kinetochore (Figure 8.41). Generally,

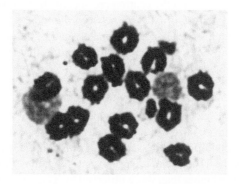

FIGURE 8.45 A multiploid sporocyte observed in Triplo 3L at diakinesis showing 14II + 2I. (Adapted from Singh, R. J., unpublished results.)

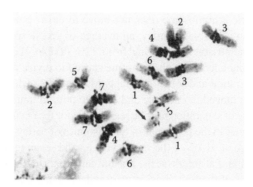

FIGURE 8.46 Giemsa N-banded mitotic mid-metaphase chromosomes of Triplo 5L of barley. Telo 5L contains half of the kinetochore (arrow). (From Singh, R. J. and T. Tsuchiya. 1982b. *J. Hered.* 73: 227–229. With permission.)

telocentric chromosomes of category d and e are unstable. Darlington (1939b) explained the absence of telocentric chromosomes in plants by their instability. It appears that his generalization of unstable telocentric chromosomes was based on the study of the d and e categories (Figure 8.41).

Barley telotrisomics are fairly stable except for Triplo 1L which shows chimaerism (Tsuchiya, 1972a). Singh and Tsuchiya (1981a) speculated that barley telocentric chromosomes contain complete kinetochores because they are highly stable. However, *Giemsa N-banding technique revealed that they contained half of the kinetochore* (Figure 8.46) and belong to category c of Steinitz-Sears (1966). The appearance of the diamond-shape kinetochore in complete chromosomes and a half-diamond in telocentric chromosomes with the N-banding technique demonstrate that breakage occurred in the middle of the kinetochore. The stability of barley telocentric (Singh and Tsuchiya, 1981b) and the lack of secondary trisomics in the progenies of mono-telotrisomics (Singh and Tsuchiya, 1977) indicate that stability of barley telocentric chromosomes does not depend on completeness of the kinetochore. As a matter of fact, there have been at least two cases of stable telocentric chromosomes in the natural population. Tsunewaki (1963) observed them in tetraploid wheat ($2n = 28$) and Strid (1968) recorded them in *Nigella doerfleri* ($2n = 12$). Stable telocentric chromosomes were also produced in grasshoppers (*Myrmeleotettix maculatus*) following centric misdivision (Southern, 1969).

Koo et al. (2015) examined 80 telosomes originated from the misdivision of the 21 chromosomes of wheat that have shown stable inheritance over many generations. They studied kinetochore size by probing with the centromere-specific histone H3 variant, CENH3. By comparing the signal, intensity of CENH3 between the intact chromosomes and derived telosomes showed that telosomes had approximately half of the signal intensity compared to that of normal chromosomes (Figure 8.47). This verifies that telosomes of wheat belong to category c of Steinitz-Sears (1966).

8.2.2.4.4   Transmission of the Extra Chromosome in Telotrisomics

The transmission of an extra telocentric chromosome depends on the length of the telocentric chromosome, its meiotic behavior, size of seeds, genetic background and also sometimes depends on environmental growing conditions. An average transmission of 34.6% (range; 30.5%—Triplo 4L to 38.4%—Triplo 1S) of the 11 telocentric chromosomes of barley through selfing was recorded (Table 8.46). These values are higher than those observed by Tsuchiya (1967) in primary trisomics. These results clearly demonstrate that transmission rates vary depending on genetic and environmental conditions.

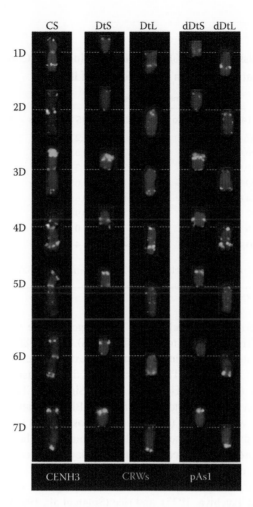

FIGURE 8.47 Immuno-FISH-based karyotype of D-genome chromosomes of wheat and their derived telosomes using CENH3 (white), CRWs (red), and pAs1 (green) as probes. CRWs (red signals) colocalized with CENH3 (white signals) in most of the chromosome except dDt1DS, 4D, Dt4DS, dDt4DS, Dt5DL, and dDt5DL. The centromeric regions of chromosome or chromosome arm were seen as pinkish red colors because the CRWs (red signals) are abundant in centromeric region and much brighter than CENH3 signals except in the above-mentioned telosomes. The dDt1DS stock contained multiple chromosome rearrangement including inversion, deletion, and centromere shift. Note that the CRWs were not detected in Dt4DS and dDt4DS, instead the pAs1 signal was overlapped with the CENH3 signal in these telosomes. A very faint pAs1 FISH site was detected in the terminal region of dDt6DS, indicating a terminal deletion. Short arm and long arm telosomes present in the ditelosomic stocks are represented as (DtS) and (DtL), respectively and short arm and long arm telosomes present in the double ditelosomic stocks are represented as (dDtS) and (dDtL), respectively. (From Koo, D-H. et al. 2015. *PLOS ONE*, DOI:10.1371/journal.pone.0137747.)

The female transmission rate of telocentric chromosomes in the progenies of telotrisomics of rice ranged from 28.6% (Triplo 2L) to 47.5% (Triplo 9S). It is evident that telocentrics for the short arm transmitted in a greater frequencies than telocentrics for the long arm (Singh et al., 1996).

The transmission of the telocentric chromosome has been observed to be a little higher, 34.6% (mean of seven telotrisomics), in a hybrid genetic background (Table 8.46). Similar results have been reported by Khush and Rick (1968b) in tomato telotrisomics.

The infrequent male transmission of the extra telocentric chromosome has been observed in maize (Rhoades, 1936), tomato (Khush and Rick, 1968b), barley (Singh and Tsuchiya, 1977;

TABLE 8.46
Female Transmission Rate (%) of the Extra Telocentric Chromosome in Selfed F_2 Population in Seven Telotrisomics of Barley

| Triplo | Chromosome Numbers (2n) | | | | | |
	14	14 + 1 telo	14 + 2 telo	15	Others	No. of Plants Studied
1L	67.4	31.4	–	1.2	–	86
1S	56.4	38.4	0.7	3.6	0.5	426
2L	60.9	37.5	0.3	1.3	0.1	1179
2S	59.8	37.6	1.1	1.0	0.6	797
3L	64.6	34.5	0.1	0.6	0.2	1935
4L	67.9	30.5	0.7	0.9	0.1	1100
5L	66.7	32.3	–	1.0	–	96
Average (total)	63.4	34.6	0.4	1.4	0.2	(5619)

Source: From Singh, R. J. and T. Tsuchiya. 1977. *Z. Pflanzenzüchtg.* 78: 327–340. With permission.

Siep, 1980), and rye (Melz and Schlegel, 1985). The range of male transmission of a telocentric chromosome was 0.0% (Triplo 2S, 5L) to 6.7% (Triplo 5S). Siep (1980) recorded 10.53% in Triplo 6S. By contrast, a high rate of male transmission of the extra telocentric chromosome 8S and 9S was recorded in the rice (Table 8.47). The occurrence of ditelotetrasomic plants in the progenies of telotrisomics of barley (Fedak and Helgason, 1970a; Singh and Tsuchiya, 1977; Tsuchiya and Wang, 1991), rice (Singh et al., 1996), and also in *Datura* (Blakeslee and Avery, 1938) indicates that sometimes male gametes with an extra telocentric chromosome can take part in fertilization. It is generally believed that unbalanced gametes would be immature at the time of anthesis and would fail to compete with matured and balanced gametes. If delayed pollination occurs, chances may be better for unbalanced gametes to take part in fertilization.

Related primary trisomics have been obtained in the progenies of tomato (Khush and Rick, 1968b), barley (Singh and Tsuchiya, 1977), and rice (Singh et al., 1996). Rhoades (1936) obtained

TABLE 8.47
Female Transmission Rate (%) of Extra Telocentric Chromosomes in Selfed or Backcross Progenies of Seven Telotrisomics of Rice

Triplo	Total Plants	Disomics	Telotrisomics	Related Trisomics
1S	80	60.0	38.8	1.2
2L[a]	56	67.8	28.6	3.6
3L[a]	150	68.7	29.3	2.0
5L	136	66.2	30.9	2.9
8S[b]	324	48.1	40.1	0.0
9S[b]	80	32.5	47.5	0.0
10S	105	65.7	31.4	2.9

Source: From Singh, K., D. S. Multani, and G. S. Khush. 1996. *Genetics* 143: 517–529. With permission.

[a] Backcross progeny.

[b] About 12% and 20% plants, respectively, contained a pair of extra telocentric chromosomes.

a secondary trisomic in the progenies of maize telotrisomics. In barley, secondary trisomic plants have not been isolated so far in the progenies of telotrisomics (Singh and Tsuchiya, 1977).

8.2.2.4.5 Genetic Segregation in Telotrisomics

Telotrisomics have been used in genetic linkage studies in several plant species such as maize (Rhoades, 1936, 1940), diploid wheat (Moseman and Smith, 1954), tomato (Khush and Rick, 1968b), barley (Tsuchiya, 1971b, 1972a,b; Fedak et al., 1972; Tsuchiya and Singh, 1982), rice (Singh et al., 1996), hexaploid wheat (Sears, 1962, 1966a), cotton (Endrizzi and Kohel, 1966), and oat (McGinnis et al., 1963). These chromosomes are very useful in associating a gene with a particular arm of a chromosome. With the use of multiple marker stocks, centromere position and gene sequence on a linkage map can be determined (Rhoades, 1936, 1940; Khush and Rick, 1968b; Reeves et al., 1968).

The principle of linkage studies with telotrisomics is different from that in simple primary trisomics. Generally, mutants are crossed as male parents and telotrisomics used as female parents. Telotrisomics are identified cytologically and morphologically in F_1 hybrids. Telotrisomics and one diploid sib (control) are saved and are allowed to self-pollinate. The chromosome number of each F_2 plant should be counted in order to separate disomics, telotrisomics, and other chromosomal types. Segregation ratios are calculated separately for the disomic and mono-telotrisomics portions of the F_2 populations. Since the extra chromosome derives from one arm of a standard chromosome, the genetic ratios are modified depending on the type of segregation and transmission rate of the extra chromosome (Reeves et al., 1968). If a gene is not located on a particular arm of a chromosome, a disomic ratio is obtained for both disomic and trisomic portions (3:1::3:1) and the result is known as a noncritical combination. If a gene is on the telocentric chromosome (critical combination), no recessive homozygotes will be obtained in the telotrisomic portion although the diploid portion will show a disomic ratio (3:1::4:0), provided that the gene is close to the centromere. When both disomic and trisomic portions are combined, a 7:1 ratio is expected in random chromosome segregation with a 50% female transmission rate of the telocentric chromosome. This ratio is further narrowed to 5:1 when the female transmission rate approaches 33.3%. Since the rate of pollen transmission of the extra telocentric chromosome is extremely low, male gametes with the extra chromosome have been considered nonfunctional.

Sometimes recessive homozygotes are obtained in the telotrisomic fraction. This is due to random chromatid crossing over and it happens when a gene is far from the centromere. It can be explained as follows. Let us assume chromosome 1 with a recessive gene b came from the male parent and chromosome 2 (complete) and 3 (telocentric) came from the female parent with dominant genes BB; chromosome 1, 2, 3 designate homologous chromosomes. There are three possibilities of recombinations: 1×3; 2×3; 1×2 (Figure 8.48). At the end of meiosis, gametes in a genotypic ratio of $11BB$: $12Bb$: $1bb$ $(n + 1$ telo) and $7BB$:$5bb$ (n) are expected to be generated. If the female transmission of the telocentric chromosome is 50%, a $119B$-:$25bb$ $(4.76B$-:$1bb)$ ratio in disomic portion, $283B$-:$5bb$ $(56.6B$-:$1bb)$ ratio in telotrisomic portion, and a total ratio of $402B$-:$30bb$ $(13.4B$-:$1bb)$ are expected. However, with 33.3% female transmission of the extra chromosome, there will be no change in the ratio of disomic and trisomic sections but segregation in the total population should be $521B$-:$55bb$ and the ratio should be $9.47B$-:$1bb$. That is 9.55% of the F_2 population should be recessive homozygotes (b) while 6.95% recessive homozygotes are expected when the female transmission rate of the extra telocentric chromosome is 50%.

Thus, the genetic ratio is modified from 3:1 to 5:1 or 7:1 (random chromosome) and 13.4:1 or 9.47:1 (random chromatid) in telotrisomic analysis when a gene is located on the telocentric chromosome. This modification, however, depends on the transmission rate of the telocentric chromosome through the female, and the distance of a gene from the centromere.

Linkage studies with telotrisomics for both arms of a chromosome are desirable whenever these stocks are available. This provided a definite association of a gene with a particular arm of a chromosome 2 of barley (Tsuchiya and Singh, 1982). The four genes f $(=lg)$, $gs6$, v, and $gs5$ were analyzed with Triplo 2L and 2S. The genes f $(=lg)$ and $gs6$ showed disomic ratios with Triplo 2L but a

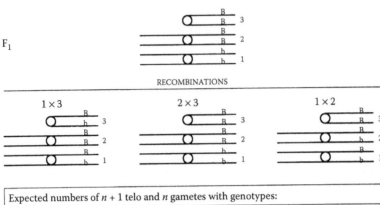

Expected numbers of $n + 1$ telo and n gametes with genotypes:					
Anaphase separation	$n = 7 + 1$ telo			$n = 7$	
	BB	Bb	bb	B	b
1×3					
1.3 vs. 2	1	2	1	2	0
1 vs. 2.3	2	2	0	1	1
2×3					
1.3 vs. 2	0	4	0	2	0
1 vs. 2.3	4	0	0	0	2
1×2					
1.3 vs. 2	2	2	0	1	1
1 vs. 2.3	2	2	0	1	1
Total $n = 7 + 1$ telo	11	12	1		
$n = 7$				7	5

F$_2$ Expected genotypic and phenotypic ratios:

♀ \ ♂	7B	5b	
11 BB	77 BBB	55 BBb	Trisomic portion: Segregation ratio 283 : 5 or 56.6 : 1
12 Bb	84 BBb	60 Bbb	
1 bb	7 Bbb	5 bbb	
7 B	49 BB	35 Bb	Disomic portion: Segregation ratio 119 : 25 or 4.76 : 1
5 b	35 Bb	25 bb	

FIGURE 8.48 Diagrammatic sketch showing maximum equational chromatid segregation telotrisomic analysis. (Modified from Reeves, A. F., G. S. Khush, and C. M. Rick. 1968. *Can. J. Genet. Cytol.* 10: 937–940. With permission; Adapted from Koo, D-H. et al. 2015. *PLOS ONE*, DOI:10.1371/journal.pone.0137747.)

trisomic ratio with Triplo 2S. Likewise genes *v* and *gs5* exhibited a trisomic ratio with Triplo 2L but a disomic ratio with Triplo 2S. These results suggest that genes *v* and *gs5* are on the long arm and *f* and *gs6* on the short arm of chromosome 2 (Table 8.48).

The gene order can be reversed on the linkage map from telotrisomic analysis. In barley, the genes *gs5* and *e* were assigned to the short and long arm of chromosome 2, respectively (Robertson, 1971). From telotrisomic analysis, gene *gs5* showed the trisomic ratio and gene *e* the disomic ratio with Triplo 2L. This indicates that the gene order should be reversed.

With the use of multiple marker stocks in telotrisomic analysis, the precise gene order on a linkage map can be obtained (Rhoades, 1940; Khush and Rick, 1968b; Tsuchiya and Singh, 1982). Khush and Rick (1968b) studied four genes (*r*, *wf*, *rv*, and *sf*) with Triplo 3L of tomato. The genes *r*

TABLE 8.48

F$_2$ Segregation Results of Genes f (= lg), gs6, v, gs5 with Triplo 2L and Triplo 2S of Barley

Triplo	Marker Genes	2x			2x +1 telo			Total			χ²		
		A[a]	a	Total	A	a	Total	A	a	Total	3:1	5:1	7:1
2L	f (=lg)	62	6	68	30	11	41	92	17	109	5.13	0.09	
2L	gs6	69	26	95	30	14	44	99	40	139	1.05		
2L	v	122	50	172	124	0	124	246	50	296	10.37	0.00	5.21
2L	gs5	55	14	69	30	0	30	85	14	99	6.23	0.46	0.24
2S	f (=lg)	95	16	111	71	0	71	166	16	182	25.51	8.12	2.29
2S	gs6	50	3	53	41	0	41	91	3	94	23.84	12.29	7.44
2S	v	121	26	147	56	11	67	177	37	214	6.77	0.06	4.59
2S	gs5	56	8	64	19	5	24	75	13	88	4.91	0.23	0.41

Source: From Tsuchiya, T. and R. J. Singh. 1982. *Theor. Appl. Genet.* 61: 201–208. With permission.

[a] A, normal; a, mutant.

and *wf* showed disomic ratios while *rv* and *sf* were trisomic; *r* and *wf* should be located on 3S, *rv*, and *sf* on 3L (Table 8.49). The orientation of the linkage group, reading from the end of the short arm, was suggested as *r-wf-sy-ru*-centromere-*rv-sf* (Khush and Rick, 1968b). Tsuchiya and Singh (1982) used a multiple marker stock *cu2-uz* of barley for telotrisomic analysis. From conventional linkage analysis gene *cu2* was located farther from the centromere than gene *uz*. With Triplo 3L, both genes showed a trisomic ratio. However, more recessive homozygotes were obtained for gene *cu2* than for *uz* in the diploid portion of an F$_2$ population, indicating *cu2* is closer to the centromere than *uz* in the long arm of chromosome 3.

Sometimes, a smaller number of recessive homozygotes than the expected 25% for a disomic ratio are obtained in the F$_2$ population even if the gene is not located on the telocentric chromosome. A wrong conclusion can be drawn if simply a chi-square value is calculated for a total population without a chromosome count. The results obtained in barley for genes *als*, *yst2*, *f2*, and *x$_s$* on chromosome 3 are such examples. Chi-square values for these genes fit the trisomic ratio (Table 8.50). However, when chromosomes were studied in the F$_2$, a high proportion of trisomic plants was found to consist of recessive homozygotes. Their frequencies were too high to consider them as a trisomic

TABLE 8.49

F$_2$ Segregation Results of Genes r, wf, rv, and sf with Triplo 3L of Tomato

	2x				2x + 1 telo			
	A[a]	a	Total	% Recessive	A	a	% Recessive	χ² (3:1)
r	94	38	184	28.7	39	13	25.0	0.00
wf	99	33	184	25.0	40	12	23.1	0.10
rv	102	18	180	15.0	59	1	1.6	17.41
sf	104	16	180	13.3	59	1	1.6	17.41

Source: From Khush, G. S. and C. M. Rick. 1986b. *Cytologia* 33: 137–148. With permission.

[a] A, normal; a, mutant.

TABLE 8.50

F₂ Segregation Results of 15 Combinations between Triplo 3L (3HL) and Various Marker Stocks

Marker Genes	2x			2x + 1 telo			Total			χ^2		
	A[a]	a	Total	A	a	Total	A	a	Total	3:1	5:1	7:1
cu2	140	18	158	54	0	54	194	18	212	30.81	10.20	3.11
uz	148	10	158	54	0	54	202	10	212	46.52	21.79	11.74
wst	53	4	57	44	0	44	97	4	101	23.84	14.95	6.73
als	41	2	43	38	0	38	79	2	81	18.62	11.75	7.45
gs2	49	9	58	45	0	45	94	9	103	14.52	4.66	1.84
zb	99	11	110	43	0	43	142	11	153	25.88	9.90	3.94
cer-zn³⁴⁸	61	12	73	52	1	53	113	13	126	14.49	3.66	0.55
Yst	107	27	134	50	14	64	157	41	198	1.94		
x_c	38	21	59	25	7	32	63	28	91	1.61		
al	61	8	69	29	6	35	90	14	104	7.39	0.77	0.08
f2	163	25	188	114	3	117	277	28	305	40.71	12.27	3.07
yst2	46	8	54	29	4	33	75	12	87	5.82	0.52	0.14
x_s	65	9	74	42	3	45	107	12	119	14.12	3.71	0.63
a_n	37	9	46	17	7	24	54	16	70	0.17		

Source: From Tsuchiya, T. and R. J. Singh. 1982. *Theor. Appl. Genet.* 61: 201–208. With permission.

[a] A = normal; a = mutant.

ratio even if the possibilities of random chromatid crossing over were considered. From the results, it seems to be essential to count the chromosome number of every F₂ plant in telotrisomic analysis. There is an alternative to testing the trisomic or disomic ratios. The chromosome counts of recessive homozygotes only will provide fairly accurate results regarding the segregation ratios, disomic, or trisomic.

Singh et al. (1996) studied the segregation of 26 genes with seven telotrisomics (Triplo 1S, 2L, 3L, 5L, 8S, 9S, 10S) of rice. For example, of the five genes of chromosome 1 tested with Triplo 1S, marker gene *d-18* showed a trisomic ratio suggesting that this gene is on the short arm of chromosome 1 while genes *chl-6*, *spl-1*, *z8*, and *gf2* segregated in a disomic fashion suggesting that these genes are on the long arm of chromosome 1 (Table 8.51). Based on the segregation of markers in

TABLE 8.51

F₂ Segregation Results of Five Combinations between Triplo 1S and Various Marker Stocks

Marker Genes	2x			2x + 1 telo			Total			χ^2 (2x) 3:1
	A	a	Total	A	a	Total	A	a	Total	
d-18	325	122	447	102	0	102	427	122	549	1.25
chl-6	262	105	367	62	32	94	324	137	461	2.25
spl-6	91	29	120	42	11	53	133	40	173	0.04
z-8	186	56	242	69	29	98	255	85	340	0.45
gf-2	116	54	170	24	9	33	140	63	203	4.12

Source: From Singh, K., D. S. Multani, and G. S. Khush. 1996. *Genetics* 143: 517–529. With permission.

telotrisomics and secondary trisomics, centromere position on eight linkage groups and orientation of 10 linkage groups of rice were established (Singh et al., 1996).

Telotrisomic analysis by the backcross procedure is useless because clear cut segregation ratios are not obtained regardless of parental genotypes or segregation of chromosomes. Since mono-telotrisomics are fertile enough to generate F_2's, it is not advisable to conduct backcrossing.

8.2.2.5 Acrotrisomics

8.2.2.5.1 Introduction

Trisomic plants carrying an extra acrocentric chromosome are designated acrotrisomics. In barley, six acrotrisomics ($1L^{1S}$, $3L^{3S}$, $4L^{4S}$, $5S^{5L}$, $6S^{6L}$, and $7S^{7L}$) have been studied and identified morphologically and cytologically, and have been used in locating genes physically in chromosomes (Tsuchiya et al., 1986).

8.2.2.5.2 Sources of Acrotrisomics

Acrotrisomics in barley have originated from the progeny of related primary trisomics ($4L^{4S}$ = Triplo 4), telotrisomics ($1L^{1S}$ = Triplo 1S), triploid hybrids ($3L^{3S}$), and unrelated telotrisomic ($5S^{5L}$ = Triplo 1S). In sugar beets, Romagosa et al. (1985) identified an acrotrisomic $9S^{9L}$ in the progeny of a primary trisomic, Triplo 9.

Acrocentric chromosomes in barley may have originated by a breakage in chromosome arm(s), preferably adjacent to heterochromatic regions, and followed by the healing of the broken ends or fusion with a second broken end. If the telomere concept of Muller (1940) is accepted, all the breakage should have at least two breaks with an intact telomere. However, according to McClintock (1941c), broken ends of meiotic chromosomes will heal when chromosomes enter into sporophytic tissues. In barley and also in sugar beets, acrotrisomics originated in the progenies of primary trisomics, telotrisomics, and triploids. It is reasonable to assume that a single breakage occurred either at one site ($3L^{3S}$, $5S^{5L}$) or at two sites ($1L^{1S}$, $4L^{4S}$) in the extra chromosome, which went through a breakage-fusion-bridge cycle and the broken end(s) healed as suggested by McClintock (1941c).

8.2.2.5.3 Identification of Acrotrisomics

8.2.2.5.3.1 Morphological Identification Morphologically, acrotrisomics $1L^{1S}$, $3L^{3S}$, $4L^{4S}$, and $5S^{5L}$ were similar to their corresponding primary trisomics and telotrisomics for the long arm of chromosomes 1L, 3L, 4L, and 5L. Acrotrisomic $1L^{1S}$ plants possess long, narrow leaves, many tillers, rather small spikes, long awns, and narrow seeds. Acrotrisomic $3L^{3S}$ plants show pale green color, revoluted leaves with tip extremely twisted, and prominent hairs on the surface of leaf blades, compact spike, slightly shorter awns, and high pollen and ovule sterility. Acro $4L^{4S}$ plants carry short, thick culms, slightly shorter, dark-green revoluted leaves, and short and compact spikes. They resemble the Triplo 4 (Robust). Although acro $5S^{5L}$ is deficient of the 60% of distal portion of the long arm of chromosome 5 and carries a complete short arm, morphologically acro $5S^{5L}$ plants are similar to Triplo-5 and Triplo-5L. This suggests that the 40% proximal segment of the long arm has almost the same effect on plant morphology as the complete long arm.

8.2.2.5.3.2 Cytological Identification Precise identification of acrotrisomic plants in barley has been possible based on combining the aceto-carmine and Giemsa N-banding techniques (Singh and Tsuchiya, 1982a,b; Tsuchiya et al., 1984; Shahla and Tsuchiya, 1986, 1987). Shahla and Tsuchiya (1987) reported that in acro $1L^{1S}$, long and short arms were 37.5% and 73.0% deficient, respectively. Acro $3L^{3S}$ carried an intact long arm (3L) without an apparent deficiency while 3S had 77.8% deficiency (Figure 8.49a,b). The proximal band in the 3S was approximately 33% and the breakage occurred in the proximal heterochromatic band (Tsuchiya et al., 1984). Acrocentric $4L^{4S}$ was found to be deficient for both arms, 4L = 31.7%, 4S = 59.3% (Tsuchiya et al., 1984). In acro $5S^{5L}$, the short arm was intact while 60% of distal 5L was deficient (Shahla and Tsuchiya, 1986).

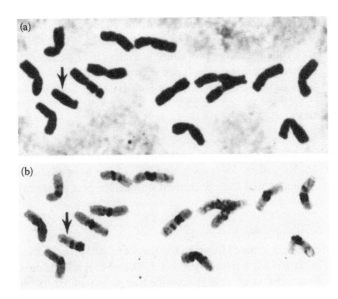

FIGURE 8.49 Aceto-carmine (a) and Giesma N-banded (b) acrotrisomics of a plant with $2n = 14 + 1$ acro $3L^{3S}$. Arrow shows an acro3L^{3S}.

The only acrotrisomic, acro 9S^{9L}, reported in sugar beets carried a complete short arm but 45% of the distal portion of 9L was missing (Romagosa et al., 1985).

Although Giemsa C- and N-banding techniques have facilitated the precise identification of individual chromosomes of barley and also various aneuploid stocks, location of correct breakage points has not been possible (Singh and Tsuchiya, 1982a,b). Barley pachytene chromosomes have not been utilized to identify deficiencies and translocations because they lack morphological landmarks such as euchromatin and heterochromatin differentiation (Singh and Tsuchiya, 1975b). Utilizing pachytene chromosomes, break points have been located precisely in the aneuploids of tomato (Khush and Rick, 1968a,b; Khush, 1973) and maize (Rhoades, 1955).

8.2.2.5.3.3 Meiotic Behavior of Acrocentric Chromosome The extra acrocentric chromosome pairs with its normal homologue at diakinesis and metaphase-I forming 1III + 6II in a majority of sporocytes (Table 8.52) or remains as a univalent giving the 7II + 1I configuration (Figure 8.50a). The trivalents may be V-shaped, rod-shaped, ring-and-rod shaped (Figure 8.50b), or Y-shaped. As is expected from metaphase-I configurations, 8-7 chromosome migration predominates in a large number of sporocytes (range 53.7–81.0%). Sometimes an acrocentric chromosome lags at the equatorial plate, while the remaining chromosomes have already reached their respective poles.

TABLE 8.52
Meiotic Chromosome Behavior of Acrotrisomics of Barley

Acrotrisomic Types	Metaphase-I (%) Configurations		Anaphase-I (%) Chromosome Migration			Author
	1III + 6II	1I + 7II	8-7	7-1-7	7 + 1-1 + 7	
1L^{1S}	71.0	29.0	73.7	22.2	4.1	Shahla and Tsuchiya (1987)
3L^{3S}	66.9	33.1	77.8	13.4	8.8	Singh, unpublished data
4L^{4S}	65.2	34.8	81.0	14.7	3.5	Hang (1981)
5S^{5L}	71.0	29.0	53.7	29.9	16.4	Shahla and Tsuchiya (1986)

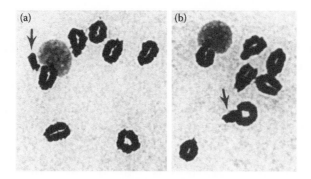

FIGURE 8.50 Chromosome configurations at diakinesis in plants with $2n = 14 + 1$acro3L^{3S}. (a) 7II + 1I (arrow shows one univalent). (b) 6II + 1III (ring- and rod-shaped trivalent, arrow).

Occasionally, acrocentric chromosomes divide longitudinally, giving 8-8 chromosome separation, though the frequency is low (Table 8.52). The lagging chromosomes will be eliminated before reaching the poles in the first division, and, if entered into dyads, will be eliminated in the second meiotic division.

8.2.2.5.4 *Transmission of the Extra Chromosome in Acrotrisomics*

Transmission of acro 1L^{1S}, 3L^{3S}, 4L^{4S}, and 5S^{5L} in selfed populations (F$_2$) was 29.4%, 33.3%, 30.5% and 37.6%, respectively (Tsuchiya et al., 1984; Shahla and Tsuchiya, 1986). The transmission rate is rather lower in acrotrisomics × diploid crosses. For example acro 1L^{1S} showed 25.4% transmission (Shahla and Tsuchiya, 1987), acro 4L^{4S} transmitted in 22.1% of plants (Hang, 1981) while 5S^{5L} was found in 35.8% of plants (Shahla and Tsuchiya, 1986). Acrocentric chromosomes may be transmitted through pollen in low frequencies, 1.3% (4L^{4S}), 9% (5S^{5L}).

8.2.2.5.5 *Genetic Segregation in Acrotrisomics*

The theoretical segregation ratios in acrotrisomic analysis are the same as those observed for telotrisomic analysis (Reeves et al., 1968). All the available four acrotrisomics (1L^{1S}, 3L^{3S}, 4L^{4S}, and 5L^{5S}) of barley have been utilized in the physical localization of genes (Tsuchiya et al., 1984; Shahla and Tsuchiya, 1987). Genetic segregation ratios in F$_2$ population between acro3L^{3S} and four genes (a_c, *yst2*, a_n, and x_s) are cited as an example. These genes are located on 3S. Two genes (a_c and *yst2*) showed a trisomic ratio suggesting that these genes are located within the proximal segment consisting of 22.2% of the heterochromatic segment of the short arm of the acro 3L^{3S} (Table 8.53).

TABLE 8.53

Genetic Segregation Ratios in F$_2$ Population of Acrotrisomics 3L^{3S} x Four Recessive Genetic Stocks

Genes	2x			2x + 1 acro			Total		
	+[a]	a	Total	+	a	Total	+	a	Total
Albino (a_c)	45	14	59	32	0	32	78	14	92
Yellow streak 2 (*yst2*)	84	12	96	34	0	34	121	12	133
Albino (a_n)	49	17	66	32	9	41	81	26	107
Xantha (x_s)	59	15	74	30	12	42	89	27	116

Source: From Tsuchiya, T. et al. 1984. *Theor. Appl. Genet.* 68: 433–439. With permission.
[a] +, normal; a, mutant.

TABLE 8.54

Genetic Segregation Ratios in F₂ Population of Acrotrisomics 5S⁵ᴸ x Six Recessive Genetic Markers

Genes	2x			2x + 1 acro			Total		
	+[a]	a	Total	+	a	Total	+	a	Total
Fragile stem 2 (fs2)	66	13	79	63	0	63	129	13	142
Golden (g)	89	25	114	30	0	30	119	25	144
Chlorina 3 (f3)	77	16	93	53	0	53	130	16	146
Chlorina 7 (f7)	59	19	78	51	11	62	110	30	140
Third outer glume (trd)	65	26	91	42	16	58	107	42	149
Intermediate spike (int-a¹)	68	23	91	43	15	58	111	38	149

Source: From Shahla, A. and T. Tsuchiya. 1986. *Can. J. Genet. Cytol.* 28: 1026–1033. With permission.

[a] +, normal; a, mutant.

Six morphological markers were analyzed with acro 5S⁵ᴸ of barley by Shahla and Tsuchiya (1986). The segregation results are summarized in Table 8.54. Genes *fs2*, *g*, and *f3* segregated in a trisomic ratio with acro 5S⁵ᴸ and no homozygous recessives were observed among acrotrisomic progeny. Genes *f7*, *trd*, and *int-a¹* segregated in a disomic ratio suggesting that these genes are located on the missing region of the long arm of chromosome 5.

Acrotrisomic linkage mapping is an excellent tool to locate genes precisely in a particular region of a chromosome. However, this method has a serious problem, since acrocentric chromosomes originate by breakage in chromosome arm (s) and healing of the broken ends. The mode of production of deficiencies is of two types, shown in Figure 8.51. A terminal deficiency needs only one break in the arm (Figure 8.51a), but an intercalary segmental deficiency needs at least two breaks. An acentric fragment gets eliminated and a telomere attaches to the broken end of the acrocentric chromosome (Figure 8.51b). Precise physical location of genes may not be possible if genes are located in the attached telomeric segment. Furthermore, if an intercalary deficiency (deletion) is small and cannot be detected by the Giemsa banding technique, examination of pachytene chromosomes should be attempted.

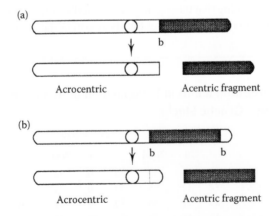

FIGURE 8.51 Diagrammatic sketch of the mode of origin of an acrocentric chromosome. (a) Single break at point b produces acrocentric chromosome with healed broken end and acentric fragment. (b) Two breaks at point b with the result of an acentric chromosome and the original telomere attached to from an acrocentric chromosome. (From Tsuchiya, T. et al. 1984. *Theor. Appl. Genet.* 68: 433–439. With permission.)

8.2.2.6 Compensating Trisomics

8.2.2.6.1 Introduction

In a compensating trisomic, a missing normal chromosome is replaced either by two tertiary chromosomes or by a secondary and a tertiary chromosome. Blakeslee (1927) was the first who reported a compensating trisomic, "Nubbin," where a missing chromosome 1.2 was compensated by tertiary chromosomes 1.9 and 2.5, and the trisomic was designated by the chromosome formula $2n - 1.2 + 1.9 + 2.5$. Compensating trisomics have been reported in *Datura stramonium* (Blakeslee, 1927; Avery et al., 1959), *Triticum monococcum* (Smith, 1947), *Lycopersicon esculentum* (Khush and Rick, 1967a), *Pennisetum glaucum* (Saini and Minocha, 1981), and *Secale cereale* (de Vries and Sybenga, 1989).

8.2.2.6.2 Sources of Compensating Trisomics

The compensating trisomic type Nubbin in *Datura stramonium* was isolated among the progeny of a plant that was exposed to radium (Blakeslee, 1927). Furthermore, Avery et al. (1959) summarized the results of seven compensating trisomics in *D. stramonium*; all were produced experimentally. Khush and Rick (1967a) isolated three compensating trisomics in tomato in the progenies of tertiary or translocated chromosomes. Similarly, Saini and Minocha (1981) reported a compensating trisomic in pearl millet that was isolated from the progeny of a multiple interchange trisomic.

According to Avery et al. (1959), compensating trisomics can be isolated in several ways. One way is by crossing a secondary trisomic, such as $2n + 3.3$ as a female with an interchange heterozygote of the formula $2n = 3.21$ and $2n + 4.22$ as a male. The chromosome end arrangement in F_1 would be 3.3-3.21-21.22-22.4-4.3. In the F_2 generation, they found 68 plants were normal diploid, 2 plants were $2n + 3.3$, 2 plants were compensating trisomic $2n - 3.4 + 3.3 + 4.22$, and 4 plants were not identified.

de Vries and Sybenga (1989) isolated four telocentric-tertiary compensating trisomics in rye from four different reciprocal translocations and three different telocentrics.

8.2.2.6.3 Identification of Compensating Trisomics

8.2.2.6.3.1 Morphological Identification Morphologically, depending on the chromosome involved, compensating trisomic types exhibit morphological features of the corresponding trisomics. The tomato compensating trisomics $2n - 3S.3L + 3L.3L + .3S$ and $2n - 3S.3L + 3S.3S + 3L.3L$ expressed several morphological traits of Triplo 3 (Khush and Rick, 1967a).

8.2.2.6.3.2 Cytological Identification The true identification of compensating trisomics is possible by observing the meiotic chromosome pairing pattern. Chromosome configurations during meiosis depend on the types of compensating trisomics. Khush and Rick (1967a) identified three compensating trisomics of tomato at pachynema. In diiso compensating trisomics, $(2n - 3S.3L + 3S.3S + 3L.3L$ and $2n - 7S.7L + 7S.7S + 7L.7L)$, isochromosomes paired internally, leaving normal chromosomes (3S.3L; 7S.7L) as univalents. In the monoteloiso compensating trisomic, $2n - 3S.3L + 3L.3L + 3S$, a trivalent configuration was more frequent. In a ditertiary compensating trisomic, $2n - S.1L. + 1S.2L + 3S.1L$ of pearl millet, Saini and Minocha (1981) recorded the maximum association of seven chromosomes (1VII + 4II) in 14.4% of sporocytes. The other chromosome configurations observed were 1V + 5II (31.8%), 1III + 6II (28.8%), 1IV + 1III + 4II (10.6%), 1IV + 5II + 1I (6.1%), and 7II + 1I (6.1%). At anaphase-I, an 8-7 chromosome migration predominated (91.7%).

8.2.2.6.4 Transmission of the Extra Chromosome in Compensating Trisomics

In seven compensating trisomics of *Datura stramonium,* one normal chromosome was replaced by two tertiary chromosomes. As expected, compensating trisomics threw diploids, parent types, related tertiaries and primaries, and also unrelated chromosomal types. A tomato compensating trisomic type $(2x - 3S.3L + 3L.3L + .3S)$ transmitted diploids (56.3%), compensating trisomic-parental types (40.5%), telotrisomics (2.7%), and diiso compensating trisomics (0.5%) (Khush and Rick, 1967a).

8.2.2.6.5   Genetic Segregation in Compensating Trisomics

Compensating trisomics have been used to locate marker genes to their respective chromosomes in *Datura stramonium* (Avery et al., 1959), *Triticum monococcum* (Smith, 1947), and *L. esculentum* (Khush and Rick, 1967a). Moreover, markers can be located in a particular arm of a chromosome because certain compensating types throw secondary, tertiary, and telotrisomics in their progeny.

A disomic ratio (3:1::3:1) is observed if a gene is not located on the compensated chromosome. In case a marker gene is located on the compensated chromosome, a compensating trisomic ratio (0:1::1:0) is expected that is a completely different ratio from ratios observed in other trisomics (Khush, 1973).

Generally, compensating trisomics are made heterozygous for marker genes to be studied and the F_1 heterozygous compensating trisomic is crossed onto the diploid recessive homozygote. It is expected that all the *n* chromosome male gametes will carry a recessive allele, with the exception of random chromatid crossing over, and all $2x$ progenies will be recessive homozygote. Utilizing this procedure, Avery et al. (1959) located a gene for albino-11 in chromosome 11.12. The procedure is as follows: A compensating trisomic for chromosome 1.2 ($2n - 1.2 + 1.9 + 2.5$) was hybridized by a recessive homozygote mutant albino-11 (albino-11 grafted on green stock produced flowers and capsules). The F_1 compensating trisomic plants were crossed onto albino-11 and the progeny segregated into 199 albino seedlings and 36 normal. The occurrence of 36 normal plants indicated either that the gene albino-11 is located away from the centromere and by double reduction *n* gametes with normal alleles were produced or that the compensating chromosomes were transmitted through male gametes at a low frequency.

A similar approach was used by Khush and Rick (1967a) to assign a marker gene *ru* to the short arm of chromosome 3 of tomato. A compensating trisomic, $2n - 3S.3L(+) + 3L.3L (-) + .3S(+)$ with normal + allele, was crossed by a recessive mutant gene *ru* [3S.3L(*ru*) + 3S.3L(*ru*)]. Two kinds of female gametes [$n = 12$, 3S.3L(+); $n = 11 + 3L.3L + 3S(+)$] and only one kind of male gamete [$n = 11 + 3S.3L(ru)$] are expected. In the F_1 diploid ($2n = 24$, + *ru*) and compensating trisomic [$2n - 3S.3L(ru) + 3L.3L + .3S(+)$] carrying *ru* gene in heterozygous state are expected; generally, diploids are discarded. The elimination of telo 3S (+) in somatic tissues resulted in chimeric condition and expression of the *ru* character. This indicated that gene *ru* is located in the short arm of chromosome 3.

Various types of aneuploids in the progeny of a test cross between a compensating trisomic for chromosome 3 and the diploid marker stock (*ru*) are expected:

1. Diploid: [3S.3L (*ru*) + 3S.3L (*ru*)].
2. Compensating trisomic (parental type): [3S.3L (*ru*) + 3L.3L (–) + 3S (+)].
3. Secondary trisomic: [3S.3L (*ru*) + 3S.3L (*ru*) + 3L.3L (–)].
4. Telotrisomic: [3S.3L (*ru*) + 3S.3L (*ru*) + .3S (–)].
5. Compensating trisomic (double isochromosome): [3S.3L (*ru*) + 3S.3S (+) + 3L.3L (–)].

8.2.2.7   Novel-Compensating Diploid

8.2.2.7.1   Introduction

In compensating diploids one normal chromosome is replaced by two telocentric chromosomes representing both arms, or four telocentric chromosomes compensate for a pair of normal homologue. The compensation results in genetically balanced plants.

8.2.2.7.2   Sources of Novel-Compensating Diploid

The occurrence of compensating diploid plants is very rare in diploid species. Such plants have been reported in *Nigella doerfleri* (Strid, 1968), *Hordeum vulgare* (Tsuchiya, 1973; Singh and Tsuchiya,

1981c, 1993), *Pennisetum glaucum* (Pantulu and Narasimha Rao, 1977; Manga et al., 1981), and *Secale cereale* (Zeller et al., 1982; Melz and Winkel, 1986). Tsunewaki (1963) identified a tetraploid strain of Emmer wheat from Tibet with $2n = 26 + 4$ telo (15 pairs). In *Nigella doerfleri*, Strid (1968) observed about half of the plants in a natural population with $2n = 14(10 + 4$ telocentrics) originating from the Island of Los, Kikladhes (Greece). The diploid chromosome constitution of *N. doerfleri* is $2n = 12$.

Tsuchiya (1973) found a plant with $2n = 13 + 2$telos (1L + 1S) in the progeny of a cross between diploid ($2n = 14$) and $2n = 14 + 1$telo1S of barley. Singh and Tsuchiya (1981c) isolated a plant with $2n = 13 + 1$acro3L^S + 1telo3S in an F_2 population of a cross $2n = 14 + 1$acro3L^S × yellow streaks (*yst2*). In $2n = 13 + 1$telo1L + 1telo1S plants, 1telo1L and 1telo1S compensate for one normal chromosome 1 and similarly in plants with $2n = 13 + 1$acro3L^{3S} + 1telo3S, one acro3L^{3S} and one telo3S replace a normal chromosome 3. Hang et al. (1998) isolated and identified barley compensating (partial) diploid ($2n = 13 + 1$acro6S^L) from the progenies of $2n = 14 + 1$acro6S^L.

The origin of plants with $2n = 13 + 1$acro3L^{3S} + 1telo3S has been assumed as follows. This particular plant originated in the F_2 population of $2n = 14 + 1$acro3L^{3S} × *yst2*. The misdivision of a normal chromosome 3 may have occurred at anaphase-I in a sporocyte in one of the spikelets of an F_1 acrotrisomic 3L^{3S} plant and by chance seven chromosomes moved to one pole and $6 + 1$acro3L^{3S} + 1telo3S (telo3S is the product of misdivision) moved to the other pole. Thus, one gamete of this spikelet contained $6 + 1$acro3L^S + 1telo3S chromosomes. The gamete with $6 + 1$acro3L^{3S} + 1telo3S is genetically balanced and may have survived, a seed with $2n = 13 + 1$acro3L^S + 1telo 3S chromosome was generated by fertilization with a normal 7 chromosome gamete. Compensating diploid plants in other plant species may have originated in a similar fashion.

8.2.2.7.3   Identification of Novel-Compensating Diploid

8.2.2.7.3.1   Morphological Identification   Morphologically, no visible differences are expected among compensating diploid plants. In barley, plants with the formulas $2n = 14$, $13 + 1$telo1L + 1telo1S, $12 + 2$telo1L + 2telo1S (Tsuchiya, 1973) and $2n = 14$, $13 + 1$acro3L^{3S} + 1telo3S, $2n = 12 + 2$acro3L^{3S} + 2telo3S (Singh and Tsuchiya, 1981c, 1993) were indistinguishable morphologically. In this situation, chromosome count is required to separate these cytotypes.

8.2.2.7.3.2   Cytological Identification   True identity of novel-compensating diploids is based on chromosome count and karyotype analysis (Figure 8.52a,b). Banding techniques can be used where arms are similar.

8.2.2.7.4   Meiotic Behavior in Novel: Compensating Diploid

In general, meiosis is normal in compensating diploids. Meiotic chromosome behavior in plants with $2n = 13 + 1$acro3L^{3S} + 1telo3S and $2n = 12 + 2$acro3L^{3S} + 2telo3S will be described in detail.

8.2.2.7.4.1   Plants with $2n = 13 + 1$ acro3L^{3S} + 1 telo 3S   The general trend of chromosome association in plants with $2n = 13 + 1$acro3L^{3S} + 1telo3S was .3S-3S.3L-3L^{3S} and this type of pairing behavior was readily recognized at late diplotene to diakinesis because of the differences in length of telo 3S and acro 3L^{3s} (Figure 8.53a–c). Of the 312 metaphase-I sporocytes analyzed, 262 sporocytes (84%) showed 1III + 6II and the remaining 50 sporocytes (16%) had 6II + 1 heteromorphic (het.) II + 1I configurations. A heteromorphic bivalent constituted a normal chromosome 3 and an acro 3L^{3s} and often telo3L remained as an univalent. The acro3L^{3S} was more tightly associated with the long arm of chromosome 3 by forming an interstitial chiasma than was telo3S (Figure 8.53c).

Chromosome migration at anaphase-I was influenced by the types of trivalent configuration. The highest frequency (78.2%) of sporocytes with V-shape trivalent resulted in the highest frequency (87.9%) of sporocytes with 7 (normal chromosomes) – 8 (6 normal + 1acro 3L^{3S} + 1telo3S) chromosome disjunction (Figure 8.53d). Moreover, a low frequencies of sporocytes showed the following

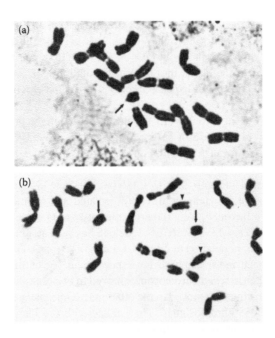

FIGURE 8.52 Somatic metaphase chromosomes of novel-compensating diploids of barley. (a) $2n =$ 13 + 1acro3L^{3S} + 1telo3S, 22.2% of proximal portion of the short arm of chromosome 3 is in trisomic condition; (b) $2n =$ 12 + 2acro3L^{3S} + 2telo3S, 22.2% of proximal portion of the short arm of chromosome 3 is in tetrasomic condition. Arrows show telo3S and arrow-heads acro3L^{3S}. (From Singh, R. J., and T. Tsuchiya. 1993. *Genome* 36: 343–349. With permission.)

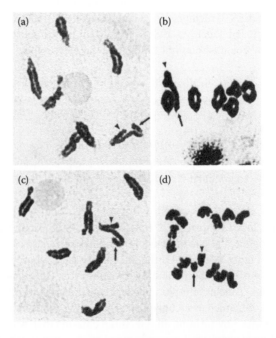

FIGURE 8.53 Meiotic chromosomes of plants with $2n =$ 13 + 1acro3L^{3S} + 1telo3S of barley. (a, b) Early diakinesis showing 1 III + 6 II chromosome configuration. (c) Metaphase-I showing 1 rod-shaped trivalent + 6 bivalents; arrow indicates an interstitial chiasma between long arm of chromosome 3 and acro3L^{3S}. (d) Anaphase-I showing 7-8 (6 + 1acro3L^{3S} + 1telo3S) chromosome separation. Arrows show telo3S and arrow-heads acro3L^{3S}. (From Singh, R. J. and T. Tsuchiya. 1993. *Genome* 36: 343–349. With permission.)

chromosome separation: 7 normal + 1acro3L^{3S} – 6 normal + 1telo3S = (9.3%); 7 normal + 1telo3S – 7 normal + 1acro3L^{3S} = (2.8%). The spores generated from 7–8 chromosome disjunction are genetically balanced and function normally. However, spores with 7 + 1acro3L^{3S} chromosomes and 7 + 1telo3S are functional only through the female and produced plants with $2n = 14 + 1$ acro 3L^{3S} and $2n = 14 + 1$telo3S after fertilization with normal 7-chromosome gametes. Such gametes are expected from a rod-shaped trivalent where 1 normal + 1telo3S or 1 normal + 1acro3L^{3S} moves to one pole leaving 1telo3S or 1acro 3L^{3S} for the other pole. The spores with $n = 6 + 1$acro3L^{3S} and $n = 6 + 1$telo3S will be aborted because of deficiency. This indicates that male and female spores with $n = 7$ and $n = 6 + 1$acro3L^{3S} + 1telo3S will be functional in equal frequencies. The almost normal anaphase-I resulted in almost completely normal telophase-I (97.3% cells).

8.2.2.7.4.2 Plants with $2n = 12 + 2$ acro $3L^{3S} + 2$ telo 3S Chromosome association in plants with $2n = 12 + 2$acro3L^{3S} + 2telo3S from diplotene to early diakinesis was exclusively 8II (Figure 8.54a, b). Six bivalents were from the normal 12 chromosomes and two bivalents were from 2acro3L^{3S} and 2telo3S. A bivalent from the telo3S was easily identified because it showed pointed centromeric structure toward the spindle pole and was shorter in length than the bivalent derived from acro 3L^{3S}. Although 22.2% of the short arm of chromosome 3 was in tetrasomic condition, a quadrivalent configuration was not recorded at diplotene or later meiotic stages.

The majority of the sporocytes (97.2%) at metaphase-I in plants with $2n = 12 + 2$acro 3L^{3S} + 2 telo3S showed 8II (Figure 8.53b). Only six sporocytes of the 216 studied had 7II + 2I in which two univalents were always telo 3S (Figure 8.53c). At anaphase-I, 90% of the sporocytes showed 8(6 + 1acro3L^{3S} + 1telo3S) – 8(6 + 1acro3L^{3S} + 1telo3S) chromosome separation (Figure 8.54d). Sometimes, bridges without fragments were recorded. At metaphase-II, all of the 200 dyads carried eight chromosomes. This indicates that meiosis is fairly normal in plants with $2n = 12 + 2$acro 3L^{3S} + 2telo3S.

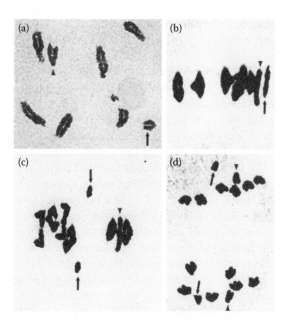

FIGURE 8.54 Meiotic chromosomes of plants with $2n = 12 + 2$acro3L^{3S} + 2telo3S of barley. (a) Diakinesis with 8II (6II + 1IIacro3L^{3S} + 1IItelo3S). (b) Metaphase-I showing 8II. (c) Metaphase-I showing 7II + 2I. (d) Anaphase-I showing 8-8 chromosome separation. Arrows show telo3S and arrow-heads acro3L^{3S}. (From Singh, R. J. and T. Tsuchiya. 1993. *Genome* 36: 343–349. With permission; Adapted from Hang, A., C. S. Burton, and K. Satterfield. 1998. *J. Genet. & Breed.* 52: 161–165.)

TABLE 8.55

Pollen and Seed Fertility (%) of Novel-Compensating Diploid with Chromosomes Constitution of $2n = 13 + 1$ acro $3L^{3S} + 1$ telo 3S and $2n = 12 + 2$ acro $3L^{3S} + 2$ telo 3S Compared with $2n = 14$ Siblings in Barley

Chromosome Constitution 2n	Number of Plants	Pollen Fertility (% of Pollen Grains)			Number of Plants	Seed Fertility (Range)		
		Good	(Range)	Degenerated		Average	Plant (Range)	Spike (Range)
14	5	98.2	(98.0–99.3)	1.8	5	96.61	(88.9–100)	(88.9–100)
$13 + 1$acro$3L^{3S} + 1$telo3S	5	97.3	(95.1–98.8)	2.7	8	89.52	(84.0–100)	(82.1–100)
$12 + 2$acro$3L^{3S} + 2$telo3S	5	98.3	(98.1–99.2)	1.2	5	90.48	(76.9–100)	(76.9–100)

Source: From Singh, R. J. and T. Tsuchiya. 1993. *Genome* 36: 343–349. With permission.

The normal meiosis in the above two novel-compensating diploids and also in other crops resulted in normal pollen and seed fertility. Fertility was not different from the diploid (Table 8.55).

Fedak (1976) studied meiosis and breeding behavior of an eight-paired barley produced by Wiebe et al. (1974). He observed 8II in 87.7% of the sporocytes. A low frequency (8.8%) of cells showed 1IV + 6II. This suggests that the material is still heterozygous for a small interchange involving a normal chromosome. Despite quadrivalent association, eight-paired barley bred true. However, seed fertility was only 65.7%.

8.2.2.7.5 Breeding Behavior of Novel-Compensating Diploid

Plants with various chromosome constitutions are expected in the progenies of compensating diploids. The results obtained in compensating diploid with $2n = 13 + 1$acro $3L^{3S} + 1$telo3S of barley will be cited here as an example. Somatic chromosome counts of 172 plants were obtained from the selfed progeny of a plant with $2n = 13 + 1$acro$3L^{3S} + 1$telo3S. Plants with chromosome constitutions 14, $13 + 1$acro$3L^{3S} + 1$telo3S (Figure 8.51a) and $12 + 2$acro$3L^{3S} + 2$telo3S (Figure 8.52b) were obtained in a ratio of 1:2:1 (Table 8.56). Several other aneuploid types were also obtained with low frequencies (Table 8.56). A similar 1:2:1 ratio was observed by Tsuchiya (1973) and Strid (1968).

As expected, a 1:1 ratio was observed for $2n = 14$ and $2n = 13 + 1$acro$3L^{3S} + 1$telo3S plants in diploid ($2n = 14$) × $2n = 13 + 1$acro$3L^{3S} + 1$telo3S and reciprocal crosses. However, the percentage range of plants with $2n = 13 + $acro$3L^{3S} + 1$telo3S was higher compared to diploid in the progenies where plants with $2n = 13 + $acro$3L^{3S} + 1$telo3S were used as a male parent. This suggests that male gametes with $n = 6 + 1$acro$3L^{3S} + 1$telo 3S are genetically balanced and in this particular case compete better than normal $n = 7$ chromosome gametes in fertilization, although 22.2% of the short arm was in a duplicate disomic condition in the gametes. The 22.2% segment, however, is heterochromatic.

Plants with $2n = 12 + 2$acro$3L^{3S} + 2$telo3S chromosomes were completely stable. Somatic chromosomes of 95 seedlings in the selfed progenies of plants with $2n = 12 + 2$acro$3L^{3S} + 2$telo 3S were counted. Ninety-four seedlings were parental type and one plant was triploid ($2n = 18 + 3$acro $3L^{3S} + 3$telo3S). This triploid plant may have originated from the fertilization of an unreduced female gamete ($12 + 2$acro$3L^{3S} + 2$telo3S) with a male gamete containing $n = 6 + 1$acro$3L^{3S} + 1$telo3S. All the plants were $2n = 13 + 1$acro$3L^{3S} + 1$telo3s in crosses between plants with $2n = 12 + 2$acro$3L^{3S} + 2$telo3S and $2x$. Similar results were reported by Manga et al. (1981) in pearl millet.

8.2.2.7.6 Usefulness of Novel-Compensating Diploid

The present case of an eight-paired barley is another example of establishing a barley strain with the basic chromosome number $x = 8$. Eight-paired barleys have been reported by Tsuchiya (1969, 1973) and Wiebe et al. (1974). Tsuchiya (1969) isolated an eight-paired barley from the progenies of

TABLE 8.56

Breeding Behavior in $2n = 13 + 1acro3L^{3S} + 1telo3S$ and $2n = 12 + 2acro3L^{3S} + 2telo3S$ Novel-Compensating Diploid Plants in Barley

	Frequency (%) of Plants with Somatic Chromosome Numbers				Total Number of Plants Studied	Chi Square	
Parental Material $2n$	14	Type I	Type II	Other Types		1:2:1	1:1
$13 + 1acro3L^{3S} + 1telo3S$ self	46(26.7)	80(46.6)	39(22.7)	7[a] (4.1)	172	0.41	
$13 + 1acro3L^{3S} + 1telo3S \times 2x$	48(52.2)	44(47.8)	–	–	92		0.17
Range (%)	(42.9–57.0)	(43.0–57.1)	–	–	–		
$2x \times 13 + 1acro3L^{3S} + 1telo3S$	56(44.1)	71(55.9)	–	–	127		1.77
Range (%)	(31.8–54.5)	(45.5–68.2)	–	–	–		
$12 + 2acro3L^{3S} + 2telo3S$ self	–	–	94(100.00)	1[b]	94		
$12 + 2acro3L^{3S} + 2telo3S \times 2x$	–	284	3[c]	–	287		
$2x \times 12 + 2acro3L^{3S} + 2telo3S$	3[c]	107	–	–	110		

Source: From Singh, R. J. and T. Tsuchiya. 1993. *Genome* 36: 343–349. With permission.

[a] Includes: 1($2n = 14 + 1acro3L^{3S}$), 3($2n = 14 + 1telo3S$), 1($2n = 13 + 2acro3L^{3S} + 1telo3S$), 1($2n = 13 + 1acro3L^{3S} + 2telo3S$), 1 (chimera, $2n = 14, 15, 16$).

[b] Triploid ($2n = 18 + 3acro3L^{3S} + 3telo3S$).

[c] Resulted from possible selfing.

Triplo 6, which had a paracentric inversion in the extra chromosome 6, and showed 61% seed set, although meiosis was normal. The eight-paired barley reported by Wiebe et al. (1974) was derived from the progeny of tertiary trisomics and showed reduced vigor and partial seed set. However, another eight-paired barley ($2n = 12 + 2telo1L + 2telo1S$), as expected, was indistinguishable from the diploid because it was genetically balanced (Tsuchiya, 1973). Recently, Hang and Tsuchiya (1992) produced nine-paired barley by crossing two eight-paired ($2n = 12 + 2acro3L^{3S} + 2telo3S$; $2n = 12 + 2telo1L + 2telo1S$) barleys. The nine-paired barley was qualitatively and quantitatively similar to diploid barley.

Hagberg (1962) successfully produced a duplication for barley breeding by crossing two translocation lines involving the same chromosomes by using the principle established by Gopinath and Burnham (1956). The novel-compensating plant with $2n = 12 + 2acro3L^{3S} + 2telo3S$ is tetrasomic for the 22.2% proximal segment of the short arm in the acro $3L^{3S}$ and may be useful in barley breeding programs. If a dominant gene (genes) for desired traits is located in the 22.2% segment, this particular region will carry four doses and will be superior to two doses carried by diploid barley. Furthermore, the four doses of a gene are assumed to be fixed in the population because a quadrivalent was not observed at meiosis. This assumption was confirmed by growing the selfed progeny of plants with $2n = 12 + 2acro3L^{S} + 2telo3S$; all the plants were green. The gene $yst2$ is in duplex condition ($Yst2, Yst2, yst2, yst2$); two telo 3S carry the recessive allele and two acro $3L^{3S}$ carry the dominant allele. If a quadrivalent formation occurs, the selfed progeny should segregate in a 35:1 ratio for green and yellow streak. All the plants were green, confirming the absence of quadrivalent formation.

8.2.2.8 Monosomics and Nullisomics

8.2.2.8.1 Introduction

The term monosome ($2n - 1$), first coined by Blakeslee (1921), designates a primary monosome (monosomic) where one of the chromosomes is missing from the normal diploid complement. Tertiary monosomics possess a translocated chromosome, that is, arms from two different chromosomes. The loss of a chromosome in a diploid species has a more drastic effect on plant morphology than when it occurs in a polyploid species. Monosomics are very useful for locating genes to specific chromosomes, in the assignment of linkage groups, in the study of the genetic control of chromosome pairing and in the manipulation of chromosomes. Monosomics have been described in great detail by Burnham (1962), Khush (1973), and Weber (1983, 1991). Therefore, only classical examples will be discussed in this chapter.

An individual lacking one chromosome pair is called nullisome (nullisomic) and is symbolized as $2n - 2$. For example, hexaploid wheat constitutes $2n = 6x = 42$ chromosomes. If a pair of chromosome such as 1A is deleted, the plant will carry $2n = 6x = 40$ and this particular plant is identified as a nulli-1A. A complete series of nullisomics in wheat cv. Chinese Spring has been established (Sears, 1944, 1954)

8.2.2.8.2 Sources of Monosomics and Nullisomics

Occasionally, monosomics originate spontaneously in the progenies of normal diploid and polyploid populations (Table 8.57). However, the frequency of occurrence of monosomic plants is relatively higher after treatment with various physical or chemical mutagens, in the progenies of haploids, polyploids, and aneuploids, in interspecific crosses and in the progenies of plants with certain genetic systems (Khush, 1973; Edwards et al., 1980; Weber, 1983, 1991).

8.2.2.8.2.1 Spontaneous Origin
Monosomic $(2n - 1)$ plants appear spontaneously, although rarely, in the progenies of diploids and polyploids. Blakeslee and Belling (1924a) observed sectoral chromosomal chimeras for monosomics $(2n - 1)$, trisomics $(2n + 1)$, and tetraploids $(4x)$ in *Datura stramonium*. The deficiency was for one of the largest chromosomes and offspring from the $2n - 1$ branch failed to show individuals of the parental types. In addition, Bergner et al. (1940) reported only one $2n - 1$ plant among 20,879 plants of diploid *D. stramonium*.

Monosomics can be isolated through screening of cultivated varieties. However, it requires considerable effort. Hacker and Riley (1963) identified 6 nullisomics, 40 monosomics, 4 trisomics, and 3 with telocentrics among 3453 plants in a hexaploid $(2n = 6x = 42)$ oat cultivar Sun II (1.53% aneuploids). McGinnis (1962) reported 0.6% aneuploids in the sampled population of oat (Table 8.58). Riley and Kimber (1961) screened four wheat cultivars and 0.69% of the population proved to be monosomic.

8.2.2.8.2.2 Chemical Treatments and X-Rays
Monosomics have been generated in populations after colchicine treatments and also from X-rayed pollen (Table 8.57). Bergner et al. (1940) found seven monosomic $(2n - 1 = 23)$ plants among 2135 grown plants after treating seeds of *Datura stramonium* with colchicine. They suggested that the occurrence of monosomics was 70 times as frequent as was observed in the control. Smith (1943) recorded a total of 26% off-type in colchicine-treated plants. One monosomic was in *Nicotiana langsdorffii* and three were in F$_1$ *Nicotiana langsdorffii* × *N. sanderae* colchicine-treated germinating seeds.

Bergner et al. (1940) assumed the chromosome loss was due to anaphase-lagging or nondisjunction in cells recovering from the effect of colchicine treatment. However, according to Smith (1943) the loss of chromosomes was due to an abnormal scattering of chromosomes in colchicine-arrested cells which was followed during recovery by multipolar spindle formation.

Irradiation of pollen often produces monosomics. The tube nucleus of a pollen grain mediates metabolism of the pollen grain, and the loss of one chromosome from the generative nucleus or one of the sperm nuclei, relatively inert metabolically, is compatible with survival and functioning of the pollen grain (Weber, 1983). Rick (1943) found two monosomic plants $(2n = 13)$ in *Petunia hybrida* after X-raying pollen; one was a primary monosome and the other was a tertiary monosome.

Rick and Khush (1961) X-rayed pollen of nonmutant tomato plants and pollinated plants containing recessive markers. Among 2312 plants that were analyzed, nine plants monosomic $(2n = 23)$ for chromosome 11 were identified. Subsequently, Khush and Rick (1966) identified cytologically

TABLE 8.57

Sources of Monosomics

Crop Species	n	Methods	References
Datura stramonium	12	Spontaneous	Blakeslee and Belling (1924a); Blakeslee and Avery (1938)
		Colchicine treatment	Bergner et al. (1940)
Lycopersicon esculentum	12	X-ray of pollen	Rick and Khush (1961); Khush and Rick (1966)
Nicotiana longsdorfii	10	Colchicine treatment	Smith (1943)
Zea mays	10	$3x \times 2x$	McClintock (1929a)
		X-ray of pollen	Morgan (1956); Baker and Morgan (1966)
		Genetic system	Weber (1983)
		Trisomics	Einset (1943)
		Knob and B chromosome interaction	Rhoades et al. (1967)
Petunia hybrid	7	X-ray of pollen	Rick (1943)
Nicotiana tabacum	24	Interspecific hybrids, asynapsis	Clausen and Cameron (1944)
Nicotiana alata	9	Spontaneous	Avery (1929)
Gossypium hirsutum	26	Spontaneous, trisomics, monosomics	Endrizzi (1963); Brown and Endrizzi (1964); Endrizzi and Ramsay (1979)
		X-ray, neutron, gamma ray, intervarietal hybrids, interspecific hybrids, cytological combinations, desynapsis	Edwards et al. (1980)
Avena byzantine	21	Haploids, aneuploids, autotriploids	Nishiyama (1970)
Avena sativa	21	Interspecific hybrids	Rajhathy and Dyck (1964)
		Spontaneous	McGinnis (1962); Hacker and Riley (1963); Ekingen (1969)
		X-ray	Costa-Rodrigues (1954); Rajhathy and Dyck (1964); Andrews and McGinnis (1964); Chang and Sadanaga (1964); Schulenburg (1965)
		Gamma ray	Singh and Wallace (1967b)
		Nitrous oxide	Dvořák and Harvey (1973)
Triticum aestivum	21	Haploids, nullisomics, trisomics	Sears (1954)
		Monosomics, translocations, X-ray pollen	Riley and Kimber (1961)
Clarkia amoena	7	Spontaneous	Snow (1964)
Glycine max	20	Desynapsis	Skorupska and Palmer (1987)
		Primary trisomics	Xu et al. (2000b)
Brassica napus	19	Intergeneric and interspecific hybrids	Fan and Tai (1985)
Oryza sativa	12	Interracial hybrid	Seshu and Venkataswamy (1958)
		Gamma ray	

a primary monosome, mono-11, and 18 tertiary monosomics. All the interchanges occurred in the kinetochore, and deficiencies were tolerated for only 15 of the 24 arms of the complement.

Baker and Morgan (1966) X-irradiated maize pollen containing the dominant asynaptic allele (*As*) and pollinated a recessive asynaptic (*as*) female. Of the 5593 X_1 kernels expressing markers flanking the kinetochore, three plants were monosomic ($2n = 19$) for chromosome 1. Two more plants were recovered which had lost two of the three markers, and they probably were tertiary monosomics. This suggests that the monosomic in maize for chromosome 1, the longest, is tolerated by the maize genome, but it is difficult to maintain.

TABLE 8.58

Aneuploids Resulting from Physical and Chemical Treatments of Hexaploid Oat

Methods	Oat Cultivars	No. Seedlings Screened	% Aneuploid	References
Haploidy	Garry	224,000	0.0	Rajhathy and Dyck (1964)
	Victory	224,000	0.0	Rajhathy and Dyck (1964)
	Rodney	224,000	0.0	Rajhathy and Dyck (1964)
Selection	Garry	4203	0.6	McGinnis (1962)
	Sun II	3453	1.5	Hacker and Riley (1963)
	Interspecific populations	2970	30.5	Rajhathy and Dyck (1964)
X-irradiation 300–500r	Garry	200	30.8	Rajhathy and Dyck (1964)
600r	Garry	293	9.9	Andrews and McGinnis (1964)
600r	Rodney	160	23.1	Andrews and McGinnis (1964)
30kr	Borreck	220	13.6	Schulenburg (1965)
700r	Cherokee	233	9.8	Maneephong and Sadanaga (1967)
300r	Missouri 04047	279	7.2	Costa-Rodrigues (1954)
Sonic Vibration 20 min.	Garry	24	4.2	Andrews and McGinnis (1964)
Myleran 10^3M	Garry	113	0.9	Andrews and McGinnis (1964)
	Rodney	90	0.0	Andrews and McGinnis (1964)
8-ethoxycaffeine 10^3M	Garry	72	1.4	Andrews and McGinnis (1964)
	Rodney	126	0.8	Andrews and McGinnis (1964)
Nitrous oxide	Garry		36.0	Dvořák and Harvey (1973)
Selection	Borreck	244	0.5	Schulenburg (1965)
Selection	Borreck	3500	2.4	Ekingen (1969)
	Zenshin	3000	2.5	Ekingen (1969)
EMS (3%)	Borreck	158	4.4	Schulenburg (1965)

Five monosomics ($2n = 23$) of rice were produced by treatment of pollen with gamma rays by Wang and Iwata (1996). One monosomic was tertiary and remaining four contained normal chromosome. Mono-9 and mono-10 were identified cytologically.

By utilizing X-rays, gamma rays, and chemical mutagens, several attempts were made to induce monosomics in hexaploid oat (Table 8.58). The frequencies of monosomic plants ranged from 0.0% to 36.0%. Rajhathy and Thomas (1974) suggested that the production of aneuploids by radiation or chemical mutagens is not desired in hexaploid oat because these mutagens also induce minor chromosome structural changes and these changes cannot be readily detected in polyploid species since polyploid species can tolerate considerable amounts of deficiency and duplication.

8.2.2.8.2.3 Haploids, Polyploids, and Aneuploids Haploids have been an excellent source for isolating aneuploids particularly monosomics, in hexaploid wheat ($2n = 6x = 42$) cv. Chinese Spring. Nineteen of the possible 21 monosomics were derived from the progenies of haploids (Sears, 1939, 1954). The remaining two monosomics were obtained from a nullisomic III (3B). Based on studies of meiosis in haploids published before 1939, Sears (1939) summarized four ways to obtain monosomics in wheat from the progenies of haploids.

Haploids, obtained from twin seedlings, also generated monosomics in hexaploid oat (Rajhathy and Dyck, 1964; Nishiyama et al., 1968; Nishiyama, 1970). Nishiyama (1970) isolated 7 (1–7) of the 21 possible monosomics in *Avena byzantina* cv. Kanota. The remaining monosomics were found from the progenies of aneuploids ($2n+$) and autotriploids ($3x$).

In maize, McClintock (1929a) found an individual in an F_2 generation of $3x \times 2x$ where microsporocytes showed a $2n - 1$ chromosome complement. The monosomic plant was very small with a poorly developed tassel and in all cases a 9II + 1I chromosome configuration was observed.

Monosomics have also been isolated from the progenies of trisomics (Einset, 1943; Brown and Endrizzi, 1964; Xu et al., 2000b). Einset (1943) obtained five monosomics and one monosomic and one fragment chromosome plants among 1916 plants in the crosses of $2x + 1 \times 2x$ maize plants. Xu et al. (2000b) discovered mon-3 and mono-6 of the soybean in the backcross progenies of triplo 3 (BC_3) and triplo 6 (BC_4), respectively. Triplo 3 (BC_2) was female and triplo 6 was male in backcross to cv. Clark 63. Mon-3 probably originated from an $n - 1$ sperm and mono-6 from an $n - 1$ egg. The derivation of monosomics from the primary trisomics is generally associated with nondisjunction and lagging of chromosomes during meiosis, leading to male and female spores lacking one complete chromosome. Mostly, the missing chromosome in a monosomic plant is the same as the extra chromosome in the primary trisomic parent.

8.2.2.8.2.4 Asynaptic and Desynaptic Lines Clausen and Cameron (1944) obtained monosomics in *Nicotiana* in asynaptic male sterile female × normal male crosses. Aside from an occasional triploid, the offspring consisted of unbalanced diploids; relatively few were simple monosomics or trisomics, and most of them were double or triple monosomics or monosomic-trisomic combinations. Sears (1954) isolated 17 of the possible 21 monosomics in the progenies of partially asynaptic Nulli-III (3B) of hexaploid wheat cv. Chinese Spring. In *Gossypium hirsutum* a monosomic M13 was obtained from an asynaptic strain (Brown and Endrizzi, 1964). Skorupska and Palmer (1987) identified a monosomic ($2n = 39$) in soybean from the progenies of a desynaptic mutant.

8.2.2.8.2.5 Intra- and Interspecific Hybrids Clausen and Cameron (1944) isolated monosomics in *N. tabacum* × *N. sylvestris* or *N. tomentosa* crosses. However, a considerable amount of heterozygosity was found that confused the recognition and establishment of monosomic series.

A high frequency of monosomics in oat was obtained by Rajhathy and Dyck (1964) among the progenies of backcrossed pentaploid hybrids (Table 8.58). However, they questioned the usefulness of this source of monosomic lines because of the heterogeneity of the progeny. Chromosome rearrangements that could arise from such wide crosses could result in disturbances in meiotic behavior and lead to seed sterility in hybrids between the isolated monosomic lines and established varieties when the latter are used in monosomic analysis. However, these lines could be backcrossed to a desired cultivar to establish the aneuploid series in a uniform genetic background.

8.2.2.8.2.6 Genetic System A unique genetic system, the *r-X1* deficiency in maize, has been discovered for generating primary monosomics (Weber, 1991). The *r-X1* deficiency includes the *R* locus on chromosome 10. The dominant *R* locus is necessary for anthocyanin production in the aleurone of the endosperm of maize kernels. In induction of monosomics, if *R/r-X1* plants (heterozygous for the deficiency) are test crossed as female parents by an *r/r* male parent, among the test cross progeny about 55%–66% of the kernels are colored (*Rr*), while the remainder are colorless (*r/r-X1*), deficiency-bearing kernels. The deficiency is only transmitted through the female and is not transmitted through pollen. The progeny of *Rr* are always diploid but the progeny of *r/r-X1* segregate in equal (10%–18%) proportions of monosomics and trisomics; the remaining plants are diploid or occasionally multiply aneuploid. Thus, aneuploids are produced only from ovules bearing the *r-X1* deficiency.

Weber (1983) suggested that a gene is located on chromosome 10 within the segment corresponding to the deficiency which is necessary for normal chromosomal disjunction at postmeiotic divisions during the mega gametophyte (embryo sac) mitotic divisions. The loss occurs post-meiotically in the *r-X1* deficiency-bearing embryo sacs. Thus, the full haploid chromosome complement is present in these embryo sacs and nondisjunction occurs at one or more of the three mitotic divisions during embryo sac development, generating some aneuploid nuclei in these embryo sacs; even though nullisomic nuclei are present in these embryo sacs, they do not abort because other nuclei in the same embryo sac are haploid or disomic for the monosomic chromosome. When $n - 1$ egg in such embryo sacs is fertilized by haploid pollen, a monosomic is produced.

Monosomics generated utilizing the r-$X1$ system are selected as follows. A male parent recessive for a sporophyte-expressed mutation is crossed onto an R/r-$X1$ female homozygous for the dominant allele of this mutation. Progeny expressing the recessive phenotype are usually monosomic for the chromosome carrying this mutation; however, some plants with deficiencies including the marker mutation are also recovered.

Weber (1983) has recovered primary monosomics for each of the 10 maize chromosomes from a cross between two inbred lines. He crossed an r/r male that also carried a recessive marker for each of the ten chromosomes (Mangelsdorf's multiple chromosome tester) onto an R/r-$X1$ female which bore the corresponding dominant alleles (Table 8.59). In the F_1, r/r-$X1$ plants were selected. Five of the markers in Mangelsdorf's tester are expressed in the sporophyte ($bm2$, lg, gl, j, and g on chromosomes 1, 2, 7, 8, and 10, respectively). Progeny of this cross expressing one of these markers are usually monosomic for the chromosome that carries the marker; however, partial chromosome losses are also occasionally observed.

The other five markers express in the endosperm of kernels. Thus, plants monosomic for the chromosomes bearing these markers cannot immediately be detected. Plants from the above cross that are of subnormal stature and have at least 50% aborted pollen are identified as putative monosomics. These are crossed with a stock that is a, su, pr, y, w x, and R. If a plant is monosomic for chromosomes 3, 4, 5, 6, or 9, the test cross will produce only kernels of the recessive phenotype for the chromosome bearing that gene. Diploid and all other monosomics will give a 1:1 ratio.

In hexaploid wheat monosomics, one expects to obtain an average of about 3% nullisomic plants from the selfed progeny (Sears, 1953b). This result is not always recorded experimentally, since the frequency of nullisomics depends on the genetic background of monosomics, chromosome type, and growing conditions. The selfed progenies of monosomics in Chinese Spring transmitted nullisomics with the range of 0.9% (5B) to 10.3% (3B) (Sears, 1944). In *Avena byzantina* cv. Kanota, Morikawa (1985) found nullisomics in the progeny of only four monosomic lines and the frequency was highly variable (mono-8, 42.1%; mono-19, 41.2%; mono-9, 2.7%, and mono-17, 2.8%). Seventeen monosomic lines did not transmit nullisomics.

TABLE 8.59
Mangelsdorf's Multiple Chromosome Markers

Female (*R/r*-X1)	Male (*r/r*)	Chromosome Number	Marker Gene	Frequency
Bm2	*bm2*	1	Brown midrib	0.03
Lg	*lg*	2	Liguleless	1.24
A	*a*	3	Anthocyaninless	0.21
Su	*su*	4	Sugary endosperm	0.29
Pr	*pr*	5	Red aleurone	0.06
Y	*y*	6	Yellow endosperm	1.89
Gl	*gl*	7	Glossy seedling	1.01
J	*j*	8	Japonica striping	3.46
Wx	*wx*	9	Waxy endosperm	0.49
G	*g*	10	Golden plant	1.57

Source: From D. F. Weber. 1991. In *Chromosome Engineering in Plants: Genetics, Breeding, Evolution. Pt. A.* Gupta, P. K. and T. Tsuchiya, Eds., Elsevier, Amsterdam, pp. 181–209. With permission.

Kramer and Reed (1988) pollinated mono-F, H, L, M, P, and R of *N. tabacum*-Purpurea by irradiated pollen of *N. glutinosa*. Two nulli haploid plants ($2n = 23$) were recovered only from a cross with mono-H. Subsequently, nullisomic ($2n = 46$) plants were obtained after chromosome doubling.

8.2.2.8.3 Identification of Monosomics and Nullisomics

8.2.2.8.3.1 Morphological Identification Monosomic plants in diploids are generally more drastically modified morphologically than in polyploids. Monosomics in *Petunia hybrida* (Rick, 1943), *Lycopersicon esculentum* (Rick and Khush, 1961; Khush and Rick, 1966), *Avena strigosa* (Andrews and McGinnis, 1964), *Zea mays* (McClintock, 1929a; Einset, 1943; Baker and Morgan, 1966), and *Oryza sativa* (Wang and Iwata, 1996) were weak and reduced in size.

Monosomics in maize generated by the *r-X1* deficiency procedure were relatively more vigorous than those monosomics in maize reported earlier. Weber (1983) emphasized that monosomic plants need "tender care." Monosomics in maize are smaller than their diploid sibs and as in primary trisomics, chromosome imbalance affects the plant in a distinctive manner.

Maize monosomics mature slower than diploids; however, they were not retarded at any stage of their growth (Weber, 1983, 1991). In contrast, the growth state of monosomic tomato plants was extremely slow in the seedling stage; after 2–3 months they showed vigorous growth, and some reached large size (Khush and Rick, 1966). Monosomics in rice were dwarf and completely sterile.

Two monosomics in the soybean expressed contrasting morphological features. Mono-3 was slow in vegetative growth (dwarf) with few branches but produced plumper pods and larger seeds than those in the disomic sibs. Mono-6 was indistinguishable from the disomic prior to flowering stage. Mono-6 set wrinkled pods, and smaller seeds than those in the disomic plants (Xu et al., 2000b). A definite conclusion regarding the loss of a chromosome from the diploid complement can be drawn only after isolating all possible 20 monosomics in the soybean.

At the tetraploid level, the loss of one chromosome is not quite as severe as it is in diploids. In *Nicotian tabacum*, the 24 monosomics produced by Clausen and Cameron (1944) differ from one another and from normal sibs in a specific ensemble of quantitative and qualitative morphological features that can be classified accurately on the basis of their morphology; some monosomics were quite distinctive, while others required more careful examination. Similarly, monosomics in *Gossypium hirsutum* can be recognized by distinct morphological characteristics, including modifications of both vegetative and reproductive structures, such as smaller or narrower leaves, smaller flowers or flower parts and smaller, longer or partially collapsed bolls (Brown and Endrizzi, 1964; Endrizzi and Brown, 1964).

Fan and Tai (1985) reported two monosomics in *Brassica napus* and both were morphologically indistinguishable from normal disomics.

In hexaploid wheat and oat, monosomics differ little from normal sibs. Monosomics in wheat cv. Chinese Spring, grown under very favorable conditions, were difficult or impossible to distinguish from normal, except for mono-IX (5A) (Sears, 1954).

Hacker and Riley (1965) compared the morphological features of monosomics and nullisomics with the euploids of oat cultivar Sun II and were able to identify 13 distinct monosomic lines; but only a few discernible differences were observed between monosomic plants. Ekingen (1969) observed no morphologically distinguishable features in monosomics of oat, while nullisomics were clearly distinguished from each other as well as from the corresponding mono- or disomic plants.

Nullisomic plants in Chinese Spring wheat can be distinguished from normal sibs as well as from the corresponding monosomics by morphological features at seedling and maturity stages (Table 8.60). However, nulli-1B, 7A, 7B, and 7D are difficult to distinguish from disomics; they are identified by intercrossing (Sears, 1954). Nullisomics from the homoeologous group 7 differed very little from disomics at the seedling and adult plant stages and were distinguishable only by a slight reduction in vigor and height, and by certain spike traits.

Four of the 21 possible nullisomics of hexaploid oats were reduced in vigor and plant height (Chang and Sadanaga, 1964; Ekingen, 1969; Morikawa, 1985). Morikawa (1985) described

TABLE 8.60
Morphological Features of Nullisomics of *Triticum aestivum* cv. Chinese Spring

Homoeologous Group	Morphological Features
1. (1A, 1B, 1D)	Three nullisomics of this group are reduced in plant height in varying degree. Spikes are a little less dense than normal, with slightly stiffer glumes; they are both female and male sterile.
2. (2A, 2B, 2D)	All three nullisomics are very dwarfish with greatly reduced tillers; all are male fertile but female sterile. The spikes have thin, papery glumes and are completely awnless.
3. (3A, 3B, 3D)	Nullisomics are identified at the seedling stage by their narrow, short, stiff leaves, and at maturity nullisomics are dwarfed with narrow leaves and short spikes.
4. (4A, 4B, 4D)	Nullisomics of all three have narrow leaves and slender culms. Mature plants are dwarfed and male sterile.
5. (5A, 5B, 5D)	Nullisomics have narrow leaves and slender culms, are late in maturity, spikes are reduced in size and have small glumes and seeds; they are female fertile and male sterile.
6, (6A, 6B, 6D)	Nullisomics have narrow leaves, slender culms and narrow, spreading outer glumes; they are straggly in appearance, and all are female fertile.
7. (7A, 7B, 7D)	Nullisomics differ very little from normal at the seedling and maturity stages, are distinguishable only by a slight reduction in vigor and height and by certain spike characters. The seed fertility of 7B and 7D is nearly normal, but is greatly reduced in 7A by pistilloidy.

Source: Adapted from Sears, E. R. 1954. *Mo. Agric. Exp. Stn. Res. Bull.* 572: 1–58.

morphological characteristics of four nullisomics of *Avena byzantina* cv. Kanota. Nulli-8 showed strong desynapsis, complete sterility, and fatuoid character. Nulli-9 was extremely weak, often dying before flowering, produced many tillers and was self-sterile. Nulli-17 was very weak and grass-like with comparatively large florets and was self-sterile. Nulli-19 was shorter in height, with virescent appearance in some environments and was sterile.

Nulli-H of *Nicotiana tabacum* carried smaller flowers than those of the disomic and mono-H (Kramer and Reed, 1988).

8.2.2.8.3.2 Cytological Identification Through observations of pachytene chromosomes, monosomics were identified in tomato (Khush and Rick, 1966) and maize (Weber, 1983). Khush and Rick (1966) cytologically identified a primary monosome (mono-11) and 18 tertiary monosomes in tomato. In tertiary monosomes, all interchanges occurred in the kinetochore. Wang and Iwata (1996) identified mono-9 and mono-11 in rice based on mitotic metaphase chromosome karyotype.

Like trisomics, monosomics are identified based on karyotype analysis of mitotic metaphase chromosomes, pachytene chromosomes, and univalents and lagging chromosomes. However, karyotype analysis alone could not facilitate precise identification of monosomics. Olmo (1936) was not able to distinguish cytologically the monosomics of *Nicotiana tabacum* despite differences between *N. sylvestris* and *N. tomentosa* chromosomes.

Hexaploid wheat ($2n = 6x = 42$) is an allopolyploid and contains chromosomes of A, B, and D genome diploid ($2n = 2x = 14$) species. Thus, it is necessary to know which chromosomes belong to A, which to B and which to D. Morrison (1953) and Sears (1954) measured univalents at meiotic metaphase-I and telophase-II of the monosomics of *Triticum aestivum* cv. Chinese Spring. It was, however, not possible to distinguish all of the chromosomes based on karyotype measurement.

To assign chromosomes to the D genome, Sears (1958) and Okamoto (1962) crossed $2n - 1$ plants with a tetraploid wheat (*Triticum dicoccum*, AABB) and examined the pattern of meiotic pairing in the F_1. When the pattern was $14II + 6I$, the monosome belonged to the D genome and when $13II + 8I$, the monosome belonged to either the A or the B genome. By utilizing monotelosomics ($2n - 1$ telo), they distinguished the chromosomes belonging to the A-genome from those belonging to the B-genome. Okamoto (1962) crossed all monotelosomics, except for IV (4A), with synthetic

TABLE 8.61
The Homoeologous Groups, Chromosomes Belonging to the A, B, and D Genomes and the Original Numbers (Parentheses) Assigned to These Chromosomes in Hexaploid Wheat cv. Chinese Spring

Homoeologous Group	Genome A	Genome B	Genome D
1	1A (XIV)	1B (I)	1D (XVII)
2	2A (XIII) [II][a]	2B (II) [XIII]	2D (XX)
3	3A (XII)	3B (III)	3D (XVI)
4	4A (IV) (4B)[b]	4B (VIII) (4A)	4D (XV)
5	5A (IX)	5B (V)	5D (XVIII)
6	6A (VI)	6B (X)	6D (XIX)
7	7A (XI)	7B (VII)	7D (XXI)

Source: From Okamoto, M. 1962. *Can. J. Genet. Cytol.* 4: 31–37. With permission.

[a] Chapman and Riley (1966).

[b] Dvořák (1983b).

tetraploid AADD (*T. aegilopoides* × *T. tauschii*). In the F_1, if a telocentric chromosome belonged to A-genome, chromosomes formed heteromorphic bivalent. Failure to form heteromorphic bivalent suggested that the telocentric belonged to the B-genome (Table 8.61). He assigned VIII to 4B and IV to 4A.

Subsequent reports (Chapman et al., 1976; Dvořák, 1976) disputed the designation of 4A because telosome 4A failed to pair with any *T. urartu* (A genome) chromosomes. Based on karyotype (Dvořák, 1983b), Giemsa C- and N-banding techniques (Chen and Gill, 1984), and *in situ* hybridization (Rayburn and Gill, 1985a), it was confirmed that 4A belonged to the B genome. Heterochromatic banding procedures did help to solve the controversy of 4A and also facilitated identification of all the 42 chromosomes of Chinese Spring (Gill and Kimber, 1974; Gerlach, 1977; Endo, 1986; Naranjo et al., 1988; Shang et al., 1988a; Dvořák et al., 1990). However, at the 7th International Wheat Genetics Symposium held at Cambridge, England, the following conclusion was reached: "The proposal was not fully accepted and it was finally agreed to recommend that 4 A be designated 4 B and that 4 B be designated 4" (Kimber and Tsunewaki, 1988).

Several attempts were made to identify monosomics of hexaploid oat by karyotype analysis of metaphase chromosomes (McGinnis and Taylor, 1961; McGinnis and Andrews, 1962; Hacker and Riley, 1965; McGinnis and Lin, 1966; Singh and Wallace, 1967a; Ekingen, 1969; Nishiyama, 1970; Hafiz and Thomas, 1978; Morikawa, 1985). It was not possible to identify all monosomic lines due to such factors such as pretreatment, fixation, staining procedure, and chromosome preparation technique. Morikawa (1985) could only divide oat karyotype into four groups: (i) Three pairs of satellite chromosomes (Sat); (ii) four pairs of metacentric chromosomes (M); (iii) seven pairs of submetacentric chromosomes (SM), and (iv) seven pairs of subterminal chromosomes (ST). The A- and B-genome chromosomes of oat do not express a diagnostic pattern of C-bands (Fominaya et al., 1988a,b). Thus assignment of chromosomes to their respective genomes has not yet been established clearly.

Another technique often used is to intercross the same or different monosomic lines. The chromosome pairing at metaphase-I is examined. The failure of two monosomes to pair with each other (double monosomics) will indicate the monosomic lines are different. Monosomics can also be identified by translocation tester stocks, if available (Endrizzi and Brown, 1964).

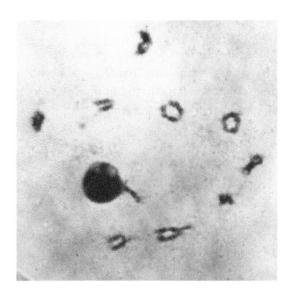

FIGURE 8.55 A photomicrograph of mono-6 in maize at diakinesis showing 9II + 1I configuration. Note the univalent chromosome 6 associated with the nucleolus. (From Plewa, M. J. and D. F. Weber. 1975. *Genetics* 81: 277–286. With permission.)

8.2.2.8.3.3 Meiosis in Monosomics and Nullisomics In monosomic plants $(2n - 1)$, generally chromosome associations of bivalents (II) and one univalent (I) are observed at diakinesis and metaphase-I. Nonhomologous chromosome association was frequently recorded at pachynema in the univalent. The frequency of trivalent association is substantially lower in monosomics of diploids than in monosomics of tetraploid and hexaploid plants. In maize monosomics $(2n - 1 = 19)$, Weber (1983) observed 9II + 1I in 99.3% of the cells and a trivalent configuration in 0.7% cells. He suggested that a trivalent configuration might be an artifact when a univalent chromosome is adjacent to or superimposed on a bivalent. Figure 8.55 shows a sporocyte of mono-6 at diakinesis, where a univalent chromosome 6 is attached with the nucleolus.

It is interesting to note that maize being a diploid tolerates a considerable degree of deficiencies as doubly and triply monosomic plants have been isolated and studied cytologically. Double and triple monosomics are identified by the simultaneous loss of marker genes on two and three nonhomologous chromosomes. Nonhomologous univalents rarely pair at diakinesis (Figure 8.55) or metaphase-I (Figure 8.56a,b) and segregate independently at anaphase-I (Weber, 1991).

In four rice monosomics $(2n = 23)$, 100% sporocytes contained 11II + 1I chromosome configuration (Wang and Iwata, 1996). Similarly, two monosomics of the soybean showed exclusively 19II + 1I (Xu et al., 2000a).

Nicotiana tabacum $(2n = 4x = 48)$ and *Brassica napus* $(2n = 38)$ are allotetraploids. Clausen and Cameron (1944) observed in monosomics of *N. tabacum*, in addition to usual configurations of 23II + 1I, two more types of associations: (1) the occurrence of a trivalent, 1III + 22II and (2) desynaptic cells with 22II + 3I, 21II + 5I, etc. Both abnormalities were associated with mono-D and mono-S and in other monosomes, trivalent configurations were rare.

Chromosomal affinities were also recorded in *B. napus*. Fan and Tai (1985) observed in mono-1, 18II + 1I in 85.3% cells and 1III + 17II in 14.7% cells. But in mono-2, the 1III + 17II configuration occurred in 61.9% of cells, a figure substantially higher than with mono-1.

The trivalent configuration was not recorded in monosomics of hexaploid $(2n = 6x = 42)$ wheat, but desynaptic cells were recorded. Morrison (1953) studied meiotic pairing in 8 of the 21 monosomes of wheat and observed 20II + 1I in 97.6% of the sporocytes. Low frequencies of 19II + 3I

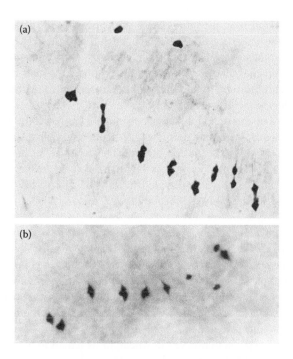

FIGURE 8.56 Photomicrographs of double (a) and triple (b) monosomics in maize. (a) A metaphase-I cell with 8II + 2I. (b) A metaphase-I cell with 7II + 3I. (Courtesy of D. F. Weber).

(2.1%) and 18II + 5I (0.3%) were also found. A definite conclusion cannot be drawn from these results because not all monosomics were studied.

Person (1956) reported the levels of meiotic irregularities were much higher in heterozygous wheat materials than when observed in stable, more or less homozygous varieties. The percentage of cells showing desynapsis was the highest (35.8%) in F_1 intervarietal hybrids but the extent of desynaptic cells showed progressive reduction during successive backcrosses: ($BC_1 = 20.2\%$, $BC_2 = 19.2\%$, $BC_3 = 15.7\%$, $BC_4 = 15.3\%$, $BC_5 = 12.7\%$, $BC_6 = 10.2\%$, and $BC_7 = 7.6\%$). Meiotic irregularities sometimes produce a monosomic plant deficient for a chromosome other than that deficient in the monosomic parent. Person (1956) termed this phenomenon "Univalent shift." Analogous results were reported in three monosomics of oat where early generations had a higher incidence of partial asynapsis and the recovery of homozygosity in later backcross generations resulted in a more stable meiosis (Rajhathy and Thomas, 1974).

One expects to observe $n - 1$ bivalents at metaphase-I in nullisomics, but it is not always observed. Nulli-8 of oat showed 5II + 30I while nulli-9, 17, and 19 exhibited 20II (Morikawa, 1985). Sears (1954) recorded desynapsis in nulli-2A and nulli-3B. Desynapsis occurred in nulli-5D at low temperature (12–15°C) but pairing was normal at 20°C and 28°C (Riley et al., 1966b). By contrast, in nulli-5 homoeologous chromosome pairing prevailed that resulted in multivalent association. It was demonstrated that a pairing control gene, *Ph*, is located on 5BL because there was no homoeologous synapsis in the presence of 5BL, but it occurs in the presence of 5BS (Riley and Chapman, 1967).

8.2.2.8.3.4 Behavior of Univalent Chromosomes at Meiosis Behavior of univalents at meiosis has been studied very thoroughly in monosomics of *N. tabacum* (Olmo, 1936) and *T. aestivum* cv. Chinese Spring (Sears, 1952a,b; Morrison, 1953; Person, 1956). During the disjunction of chromosomes at anaphase-I to telophase-I or anaphase-II to telophase-II, the univalent may be included in either daughter nucleus or may lag, divide, or misdivide. In maize, the univalent passes to one pole at telophase-I in over half of the cells because at prophase-II 58.3% of the cell pairs contain nine chromosomes in one cell and 10 in the other, 17.9% of the prophase-II cell pairs have 9 + 1 monod in each cell indicating the univalent divided at telophase-I, and the rest have nine chromosomes indicating the monod chromosome was lost (Weber, personal communication). This can be demonstrated by analyzing tetrads in mono-6. Of the four microspores, two left microspores contain one nucleolus (presence of chromosome 6) and two right microspores are without nucleolus indicating chromosome 6 is lost during meiosis (Figure 8.57).

During the early phase of anaphase-I, bivalents disjoin normally but the univalent remains more or less stationary at the equatorial plate and does not show any visible sign of longitudinal split; however, it undergoes division later while chromosomes of the bivalents have already reached their respective poles. Thus, the daughter univalents lagging between the poles are eliminated from the macronuclei. In wheat, the univalent divided in 96% of sporocytes at first meiotic division (Sears, 1952a). Similarly, Darlington (1939b) found 98% misdivision of univalents in *Fritillaria kamtschatkensis.*

The misdivision of univalent mono-5A in four different strains of hexaploid wheat has been elaborated extensively in classic papers of Sears (1952a,b). The misdivision observed in wheat can be divided into three classes based on the number and kinds of arms going to one pole as follows: (i) One normal chromatid passes to one pole and two arms of the other chromatid either pass separately to the other pole or one or both remain acentric on the plate; (ii) two identical arms move to one pole and the other two arms either go to the other pole or one or both remain acentric on the plate; (iii) three arms go to one pole and the fourth arm either moves to the other pole or remains on the plate. There is a fourth type, not observed in wheat, in which all four arms migrate to one pole.

The misdivision of mono-5A in Chinese Spring wheat was observed at telophase-I in 39.7% of cells and at telophase-II at least 29.3% of cells contained a laggard. The rate of misdivision of chromosome 5A was influenced by the genetic background because it ranged from 13.7% (Hope) to 39.7% (Chinese Spring) and also varied from chromosome to chromosome (Table 8.62). The misdivision of a univalent at telophase-I produced both telocentrics and isochromosomes. At telophase-II, some of the isochromosomes misdivide again to yield additional telocentrics. The telocentrics and isochromosomes generated during microsporogenesis are lost or not transmitted through pollen. Thus, all of the telocentrics and isochromosomes recovered are produced during megasporogenesis. However, it has been suggested in *Fritillaria* (Darlington, 1940) and also in maize (Rhoades, 1940) that isochromosomes are produced through misdivision of telocentrics during pollen mitosis and can be transmitted.

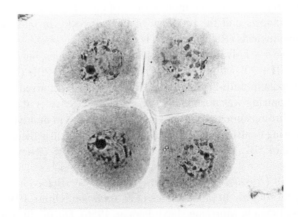

FIGURE 8.57 A quartet of microspores from a mono-6 plant. Nucleoli are present in the two haploid cells each containing chromosome 6, whereas nucleolar blebs are present in the two cells nullisomic for chromosome 6 because they do not contain chromosome 6, an NOR. (From Weber, D. F. 1978. *Can. J. Genet. Cytol.* 20: 97–100. With permission.)

TABLE 8.62

Frequency of Misdivision of Various Chromosomes at Telophase-I in Several Wheat Cultivars

Chromosome	Variety	Misdivision at TI	
		%	Total No. Cells
5A	Chinese Spring	39.7	126
5A	Hope	13.7	95
5A	Thatcher	17.5	63
5A	Red Egyptian	23.6	191
2B	Red Egyptian	11.4	44
5B	Chinese Spring	26.7	15
7A	Hope	29.4	17

Source: Adapted from Sears, E. R. 1952a. *Chromosoma (Berlin)* 4: 535–550.

8.2.2.8.4 Fertility of Monosomics and Nullisomics

Pollen and seed fertility in monosomic lines depends on the ploidy level, the chromosome involved, environmental conditions, and genetic background. Pollen fertility in mono-11 of tomato was 25% and in tertiary monosomes ranged from 5% to 40% around a mode of 25%. As is expected, the monosomics with higher pollen fertility generally set fruits with a higher number of seeds (Khush and Rick, 1966). Pollen fertility in maize monosomics ranged from 3.4% (mono-8) to 46.9% (mono-6). It is expected to observe 50% pollen fertility when the univalent chromosome in a monosomic plant is not lost during meiosis. However, observations on pollen fertility in maize monosomics revealed the following levels: mono-2 = 16.8%, mono-4 = 40.2%, mono-6 = 46.9%, mono-7 = 37.0%, mono-8 = 25.8%, mono-9 = 3.4%, and mono-10 = 43.9%. The univalent chromosome is frequently lost, producing a high frequency of microspores with nine chromosomes (Weber, 1983). Pollen fertility (seed set) in soybean monosomics was: mono-3, 8.8% (59 seeds); mono-6, 20% (176 seeds). Disomic plant had 99% pollen fertility and produced 342 seeds (Xu et al., 2000b).

In *Nicotiana tabacum*, Clausen and Cameron (1944) recorded that most of the monosomics produced fewer seeds per capsule than disomics with the exception of mono-A, which produced more seeds. Brown and Endrizzi (1964) found that seed and mote counts in *Gossypium* offered a reliable technique for recognition of monosomics, and that pollen fertility in monosomics proved to be inconclusive in distinguishing disomic and monosomic plants.

Seed set in monosomics of *Avena byzantina* cv. Kanota studied by Morikawa (1985) ranged from 69.1% (mono-21) to 99.4% (mono-19). The disomic plants showed a seed set of 99.4%.

8.2.2.8.5 Breeding Behavior of Monosomics and Nullisomics

The monsomic condition is not transmitted through female or male in diploid species. Examples are *Datura stramonium* (Blakeslee and Belling, 1924a), *Oryza sativa* (Seshu and Venkataswamy, 1958), and *Zea mays* (Weber, 1983). In diploid species, the $n-1$ spores abort and cannot compete with normal spores because they are physiologically and genetically unbalanced. In cotton, male transmission of $n-1$ gametes is rare (Endrizzi and Ramsay, 1979).

Unbalanced gametes are functional to a certain degree in polyploid species. The classical work of Sears (1953b, 1954) in hexaploid wheat cv. Chinese Spring demonstrates that after selfing monosomics, three types of offspring (disomic, monosomic, and nullisomic) occur in frequencies which vary. Some wheat plants lacking the chromosome concerned are obtained, and about 75% of the female gametes carry $n = 20$ chromosomes. The deviation from the expected 50% is due to the irregular behavior and resultant frequent loss of the unpaired chromosome during meiosis. Functioning male gametes predominantly contain $n = 21$ chromosomes, because pollen competition strongly favors pollen with the $n = 21$ chromosome number, and pollen of that constitution is involved in 90%–99% of fertilization.

The frequency of nullisomics in hexaploid wheat depends on the degree of functioning of 20-chromosome pollen. A selfed progeny of monosomic plants should segregate, chromosomally, 24% disomics ($2n$), 73% monosomics ($2n - 1$), and 3% nullisomics ($2n - 2$) (Table 8.63). The frequencies of nullisomics in Chinese Spring wheat ranged from 0.9% (nulli-5B, nulli-6B) to 7.6% (nulli-3B) (Sears, 1954).

Tsunewaki and Heyne (1960) examined the transmission rate of 21 monosomics cytologically in five monosomic series established in five wheat cultivars. The frequencies of the monosomics ranged from 52.2% (mono-2B) to 89.2% (mono-3B).

Avena sativa is also an allohexaploid, but it differs from hexaploid wheats in degree of aneuploid tolerance (Morikawa, 1985). Nullisomics were not found in the 17 monosomic lines, and monotelosomics ($2n = 40 + 1$telo.) with low frequencies were obtained in the progenies of four monosomics. The selfed progenies of monosomics threw monosomic plants at a high frequency ranging from 35.5% (mono-12) to 97.8% (mono-5).

The female transmission rate of monosomics in *N. tabacum* ($2n = 4x = 48$) ranged from 5.1% (mono-W) to 81.9% (mono-E). Monosomics which exhibited high ovular abortion showed a strongly depressed transmission rate (Clausen and Cameron, 1944).

The results discussed thus far suggest that the frequency of monosomics does not fall within the range of expectation. The actual frequency of monosomics is either too high or too low. It can be attributed to the elimination of the univalent. Greenleaf (1941) investigated cytologically mono-P of *N. tabacum* with the objective of determining the cause of aberrant transmission rates. Two main causes were established:

1. Slow rate of development of the $n - 1$ embryo sacs.
2. A high frequency of aborted embryo sacs prior to the attainment of the eight-nucleate condition.

According to Rajhathy and Thomas (1974), the genetic background of the monosomic lines and the particular chromosome involved have a significant effect on breeding behavior of monosomics.

Nullisomics are expected to be stable and to breed true. Sears (1954) examined selfed progenies of nine nullisomics of Chinese Spring wheat. With the exception of nulli-3B, the eight nullisomics were fairly stable (Table 8.64). Nulli-3 transmitted aberrant plants that helped in isolating monosomics.

TABLE 8.63

Expected Transmission of the Monosomics in *Triticum aestivum*

	n (21-Chromosome Pollen) 0.96	$n - 1$ (20-Chromosome Pollen) 0.04
n (21 chromosome eggs) 0.25	$2n$ (42) plants 0.24	$2n - 1$ (41) plants 0.01
$n - 1$ (20 chromosome eggs) 0.75	$2n - 1$ (41) plants 0.72	$2n - 2$ (40) plants 0.03

Source: Adapted from Sears, E. R. 1953b. *Am. Nat.* 87: 245–252.
Euploids ($2n = 42$) = 0.24.
Monosomics ($2n = 41$) = 0.73.
Nullisomics ($2n = 40$) = 0.03.

TABLE 8.64

Chromosome Constitution of Selfed Progenies of Nullisomic Plants in *Triticum aestivum*

Nullisome	No. Offspring Grown	$2n = 40$	$2n = 39$	$2n = 41$	% $2n - 2$
1B	68	65	1	2	95.6
7B	66	61	3	2	92.4
7A	15	10	3	2	66.7
3A	9	9	0	0	100.0
1A	12	11	0	1	91.7
1D	4	4	0	0	100.0
6D	2	2	0	0	100.0
7D	15	13	1	1	86.7

Source: Adapted from Sears, E. R. 1954. *Mo. Agric. Exp. Stn. Res. Bull.* 572: 1–58.

8.2.2.8.6 Genetic Studies

The concept of using monosomics for associating genetic markers with a particular chromosome was first discovered by Bridges (1921) from the studies of "Diminished" monosome (mono-4) in *Drosophila melanogaster*. Since then monosomics (primary) have effectively been utilized in genetic and linkage studies in tomato (Khush and Rick, 1966), maize (Weber, 1983), tetraploid tobacco (Clausen and Cameron, 1944), tetraploid cotton (Endrizzi, 1963; Endrizzi and Ramsay, 1979; Endrizzi and Ray, 1991), and in hexaploid wheat (Unrau, 1950; Sears, 1953b; Khush, 1973).

As in simple primary trisomics, genes are tested against whole chromosomes when monosomics and nullisomics are used. By the monosomic method not only are single genes assigned to a chromosome but multigenes are also located.

8.2.2.8.6.1 Locating Recessive Genes

All monosomic stocks carrying a dominant allele, such as *DD*, are crossed as a female by disomic stocks with the recessive trait (*dd*) as a male. Monosomic plants produce two kinds of gametes: n (*D*), $n - 1$(–) and disomic male plants are expected to produce only one type, n (*d*), of gametes. In critical crosses, all F_1 monosomic offspring are recessive [*d* (–)] and disomic plants are dominant (*Dd*). This suggests that a recessive mutant is located in the missing chromosome. Thus, monosomics give results like sex-linked characters in animals. A monosomic is hemizygous for all genes on the univalent (pseudo-dominance).

In noncritical combinations, all F_1 offspring (monosomics, disomics) should show the dominant character and the F_2 should segregate in a normal disomic 3 dominant: 1 recessive fashion. However, $n - 1$ gamete does not function in diploid species. Thus, this technique is applicable only for tetraploid and hexaploid crops.

Bridges (1921) crossed *Drosophila melanogaster* mono-4 (Diminished) as a female with the recessive mutant bent and eyeless as male. All F_1 flies that were mono-4 were bent or eyeless. However, crosses of mono-4 with recessive markers from chromosomes 1, 2, and 3 showed all F_1 flies with dominant character.

8.2.2.8.6.2 Locating Dominant Genes

All monosomic stocks carrying recessive characters (*d*–) are crossed by disomic stocks which carry the dominant allele (*DD*). All F_1 disomic and monosomic plants will show the dominant trait. Backcross or F_2 progeny will identify the critical cross. In F_2, all recessive homozyotes are expected to be nullisomics if a gene is located on the missing chromosome, otherwise nullisomic plants should segregate for dominant and recessive characters.

Unrau (1950) associated seven of nine genes with a particular chromosome in wheat by monosomic and nullisomic methods. For example, red glumes in Federation-41 wheat are dominant

TABLE 8.65

Summary of the Segregation for Glume Color in F_2 from Crosses between 17 Monosomics of Chinese Spring and Federation 41, Red Glume Color

F1 Plants	Red Glumes	White Glumes	% White Glumes[a]	Total
Mono-1B Total (excluding Mono-1B) of	528	38	6.7	566
16 monosomics	10,975	3462	24.0	14,437

Source: Adapted from Unrau, J. 1950. *Sci. Agric.* 30: 66–89.

[a] All white glumed plants were nullisomics.

and in crosses with Chinese Spring segregates in a 3(red):1 (white) ratio. Federation-41 wheat was crossed on to 17 monosomics of Chinese Spring wheat. In F_2, all except mono-1B showed a 3:1 ratio (Table 8.65). The proportion of white glume was very low and white glumed plants were nullisomic. This conclusively demonstrates that a gene for red glume color is associated with chromosome 1B.

8.2.2.8.6.3 Locating Duplicate Genes Duplicate genes can be located in the chromosomes by monosomic analysis in polyploid species. In *Nicotiana tabacum* ($2n = 4x = 48$), Clausen and Cameron (1944) assigned hairy-filament to chromosome A (subgenome *tomentosa*) and O (subgenome *sylvestris*) and yellow-burley to chromosome B (subgenome *tomentosa*) and chromosome O by means of monosomics.

Association of hairy-filaments (*hf1, hf2*) to their respective chromosomes was accomplished by examining the segregation of smooth versus hairy-filament in F_2. A ratio of 3 normal: 1 hairy-filament was observed in the mono-A fraction and a 15 normal: 1 hairy-filament ratio in the disomic fraction. If a gene is not associated, it should segregate in a 15:1 ratio. Thus, hairy-filament showed association with chromosomes A and O because a high proportion of hairy-filament segregants was observed in the progenies of the F_1 mono-types heterozygous for hairy-filament, and with the other monosomics, the segregation is in satisfactory agreement with a 15:1 ratio for duplicate genes (Table 8.66). Similarly, they associated another duplicate gene yellow-burley (*yb1, yb2*) with mono-B [F_2 = green, 66: yellow-burley, (*yb1*), 23] and with mono-O (BC$_1$ = green, 19: yellow-burley (*yb2*), 17]. Thus, *yb1* was located in B chromosome of subgenome *tomentosa* and *yb2* in the O chromosome of subgenome *sylvestris*.

Thus, modifications in ratios in F_2 and BC$_1$ generations, produced from heterozygous F_1 monosomics, help to associate a gene or genes with particular chromosomes.

8.2.2.9 Alien Addition Lines

8.2.2.9.1 Introduction

The transfer of agriculturally important traits from alien species to cultigens by interspecific and intergeneric crosses, known as wide hybridization, has been demonstrated in numerous crops (Goodman et al., 1987). Thus, wide hybridization is a valuable tool for creating genetic variability in plant breeding by broadening the germplasm base of the cultigens (Smith, 1971; Harlan, 1976; Hadley and Openshaw, 1980; Stalker, 1980; Zenkteler and Nitzsche, 1984; Tanksley and McCouch, 1997; Zamir, 2001). Exploitation of wild relatives of the crop plants is often hampered because of poor crossability, early embryo abortion, hybrid seed inviability, hybrid seedling lethality, and hybrid sterility due to low chromosome pairing. However, these barriers have been overcome by (1) the assemblage of diverse germplasm; (2) application of growth hormones to reduce embryo abortion; (3) improved culture conditions; (4) restoration of the seed fertility by doubling the chromosomes of sterile F_1 hybrids; and (5) utilization of bridge crosses where direct crosses are not possible. Introgression of useful genetic traits from wild relatives (donor parent) to cultigens (recipient parent) is achieved by producing alien addition, substitution, and translocation lines.

TABLE 8.66

Segregation of Hairy-Filament in F_2 after Selfing F_1 Monosomics Heterozygous for Hairy-Filament in *Nicotiana tabacum*

Monosomic Types	Smooth-Filament	Hairy-Filament	% Hairy-Filament
A	25	15	60.0
B	47	1	2.1
C	44	4	9.1
D	40	2	5.0
E	33	2	6.1
G	36	0	0.0
H	38	2	5.3
M	39	5	12.8
N	41	1	2.4
O	30	15	50.0
P	37	2	5.4
S	37	1	2.7
Totals (excluding A and O)	394	20	5.1
Expected 15:1	388	26	6.7

Source: From Clausen, R. E. and D. R. Cameron. 1944. *Genetics* 29: 447–477. With permission.

The objective of the following two chapters is not to report a comprehensive review on alien addition and alien substitution lines. The literature on wheat and its allied species and genera are so voluminous that it is beyond reach to review it all here.

8.2.2.9.2 Production of Alien Addition Lines (AALs)

Monosomic alien addition lines (MAALs) and disomic alien addition lines (DAALs) have been produced in polyploid and diploid species according to the procedure described by O'Mara (1940). Modifications in the technique depended on ploidy levels and the nature of the crops. In some species all the possible AALs are available, while in others a complete set is lacking (Table 8.67). The procedure involves the production of interspecific or intergeneric F_1 hybrids, induction of an amphiploid, and generation of BC_1, isolation of monosomics in BC_2 and BC_3 and selection of disomic additions after selfing of the MAALs. The term "alien addition races" was originally proposed by Clausen (Gerstel, 1945).

8.2.2.9.2.1 Hexaploid Species Hexaploid wheat ($2n = 6x = 42$, genome formula AABBDD) and diploid rye ($2n = 2x = 14$, RR) crosses result in F_1 polyhaploid ($2n = 4x = 28$, ABDR) sterile plants. Sometimes seeds are obtained because of unreduced gamete formation. Generally, octoploid ($2n = 8x = 56$, AABBDDRR) triticale is synthesized by doubling the chromosomes with colchicine. The synthesized octoploid triticale is fertile and is backcrossed to wheat (BC_1) resulting in heptaploid ($2n = 7x = 49$, AABBDDR) plants. The heptaploid plants mostly exhibit 21II + 7I chromosome pairing at metaphase-I. The seven rye univalents are randomly excluded or included in the gametes. MAALs ($2n = 42W + 1R–7R$) are isolated either after selfing the heptaploid or the heptaploid is again crossed by wheat (BC_2). DAALs are isolated in the selfed population of MAALs (Figure 8.58).

In general, the higher ploidy (female) × lower ploidy (male) crosses are more compatible than the reciprocals. However, there are a few exceptions (McFadden and Sears, 1944, 1946; Röbbelen and Smutkupt, 1968). Thomas (1968) isolated six of the possible seven MAALs of *Avena hirtula* ($2n = 2x = 14$, As) from crosses to *A. sativa* ($2n = 6x = 42$, AACCDD) by way of an *A. hirtula*

TABLE 8.67
A Partial List of Alien Addition Lines

Recipient Species	n	Donor Species	n	Alien Addition Lines		Author
				MAALs	DAALs	
Allium cepa	8	*Allium fistulosum*	8	4	–	Peffley et al. (1985)
Allium fistulosum	8	*Allium cepa*	8	8	–	Barthes and Ricroch (2001); Shigyo et al. (1996)
Avena sativa	21	*Avena barbata*	14	1	1	Thomas et al. (1975)
Avena sativa	21	*Avena hirtula*	7	6	4	Thomas (1968)
Avena sativa	21	*Avena strigosa*	7	1	–	Dyck and Rajhathy (1963)
Avena sativa	21	*Zea mays*	10	–	5	Ananiev et al. (1997)
Beta vulgaris	9	*Beta patellaris*	9	1	–	Heijbroek et al. (1983); Speckmann et al. (1985)
Beta vulgaris	9	*Beta patellaris*	9	9	–	Meshbah et al. (1997)
Beta vulgaris	9	*Beta procumbens*	9	1	–	Savitsky (1975)
Beta vulgaris	9	*Beta procumbens*	9	9	–	Lange et al. (1988)
Beta vulgaris	9	*Beta procumbens*	9	3	–	Speckmann et al. (1985)
Beta vulgaris	9	*Beta webbiana*	9	9	–	Reamon-Ramos and Wricke (1992)
Beta vulgaris	9	*Beta corolliflora*	19	9	–	Gao et al. (2001)
Brassica campestris	10	*Brassica alboglabra*	9	4	–	Cheng et al. (1997)
Brassica campestris	10	*Brassica oleracea*	9	8	8	Quiros et al. (1987)
Brassica napus	19	*Brassica campestris*	10	1	–	McGrath and Quiros (1990)
Brassica napus	19	*Brassica nigra*	8	?	–	Jahier et al. (1989)
Brassica napus	19	*Brassica nigra*	8	6	–	Chevre et al. (1991)
Brassica oleracea	9	*Brassica nigra*	8	5	–	Chèvre et al. (1997)
Raphanus sativus	9	*Brassica oleracea*	9	7	–	Kaneko et al. (1987)
Cucurbita moschata	20	*Cucurbita palmata*	20	6	–	Graham and Bemis (1979)
Diplotaxis erucoides	7	*Brassica nigra*	8	7	–	This et al. (1990)
Glycine max	20	*Glycine tomentella*	39	22	–	Singh et al. (1998)
Gossypium hirsutum	26	*Gossypium sturtianum*	13	4	–	Rooney et al. (1991)
Lycopersicon esculentum	12	*Solanum lycopersicoides*	12	6	–	DeVerna et al. (1987)
Lycopersicon esculentum	12	*Solanum lycopersicoides*	12	12	–	Chetelat et al. (1998)
Lolium multiflorum	14	*Festuca drymeja*	7	4	–	Morgan (1991)
Nicotiana tabacum	24	*Nicotiana glutinosa*	12	1	1	Gerstel (1945)
Nicotiana tabacum	24	*Nicotiana paniculata*	10	1	1	Lucov et al. (1970)
Nicotiana tabacum	24	*Nicotiana plumbaginifolia*	10	3	3	Cameron and Moav (1957)
Nicotiana plumbaginifolia	10	*Nicotiana sylvestris*	12	12	–	Suen et al. (1997)
Oryza sativa	12	*Oryza officinalis*	12	12	–	Shin and Katayama (1979)
Oryza sativa	12	*Oryza officinalis*	12	12	–	Jena and Khush (1989)
Oryza sativa	12	*Oryza australiensis*	12	8	–	Multani et al. (1994)
Triticum durum	14	*Agropyron elongatum*	7	7	6	Mochizuki (1962)
Triticum durum	14	*Aegilops umbellulata*	7	7	2	Makino (1976, 1981)
Triticum durum	14	*Dasypyrum villosum*	7	6	–	Blanco et al. (1987)

(Continued)

TABLE 8.67 (*Continued*)
A Partial List of Alien Addition Lines

Recipient Species	n	Donor Species	n	Alien Addition Lines		Author
				MAALs	DAALs	
Triticum durum	14	*Secale cereale*	7	2	–	Sadanaga (1957)
Triticum durum	14	*Triticum tauschii*	7	6	–	Makino (1981)
Triticum durum	14	*Triticum tauschii*	7	7	–	Dhaliwal et al. (1990)
Triticum turgidum	14	*Hordeum chilense*	7	7	–	Fernandez and Jouve (1988)
Triticum turgidum	14	*Triticum distichum*	14	8	–	Fominaya et al. (1997)
Triticum aestivum	21	*Aegilops caudata*	7	–	6	Friebe et al. (1992)
Triticum aestivum	21	*Aegilops geniculata*	14	1	13	Friebe et al. (1999)
Triticum aestivum	21	*Aegilops markgrafii*	7	–	5	Peil et al. (1998)
Triticum aestivum	21	*Aegilops ovata*	14	–	4	Landjeva and Ganeva (1999)
Hordeum vulgare	7	*Hordeum bulbosum*	7	2	–	Thomas and Pickering (1988)
Triticum aestivum	21	*Aegilops searsii*	7	7	7	Pietro et al. (1988)
Triticum aestivum	21	*Aegilops speltoides*	7	–	7	Friebe et al. (2000)
Triticum aestivum	21	*Aegilops uniaristata*	7	–	5	Miller et al. (1997)
Triticum aestivum	21	*Aegilops variabilis*	14	–	14	Friebe et al. (1996)
Triticum aestivum	21	*Aegilops sharonensis*	7	1	1	Miller et al. (1982a)
Triticum aestivum	21	*Aegilops comosa*	7	1	1	Riley et al. (1966a)
Triticum aestivum	21	*Agropyron ciliare*	14	1	6	Wang et al. (2001)
Triticum aestivum	21	*Agropyron elongatum*	7	7	–	Dvorak and Knott (1974)
Triticum aestivum	21	*Agropyron elongatum*	7	–	1	Konzak and Heiner (1959)
Triticum aestivum	21	*Haynaldia villosa*	7	6	5	Hyde (1953)
Triticum aestivum	21	*Hordeum chilense*	7	6	–	Miller et al. (1982a)
Triticum aestivum	21	*Hordeum vulgare*	7	5	6	Islam et al. (1981)
Triticum aestivum	21	*Roegneria ciliaris*	14	1	6	Wang et al. (2001)
Triticum aestivum	21	*Secale cereale*	7	3	3	O'Mara (1940)
Triticum aestivum	21	*Secale cereale*	7	3	4	Riley and Chapman (1958b)
Triticum aestivum	21	*Secale cereale*	7	–	7	Riley and Macer (1966)
Triticum aestivum	21	*Secale cereale*	7	7	5	Evans and Jenkins (1960)
Triticum aestivum	21	*Secale cereale*	7	–	7	Driscoll and Sears (1971)
Triticum aestivum	21	*Secale cereale*	7	6	6	Bernard (1976)
Triticum aestivum	21	*Thinopyrum intermedium*	21	1	1	Wienhues (1966)
Triticum aestivum	21	*Thinopyrum intermedium*	21	–	6	Forster et al. (1987)
Triticum aestivum	21	*Thinopyrum intermedium*	21	–	6	Larkin et al. (1995)
Triticum aestivum	21	*Thinopyrum bessarabicum*	21	–	6	William and Mujeeb-Kazi (1995)
Triticum aestivum	21	*Triticum longissimum*	7	7	7[a]	Feldman and Sears (1981)
Triticum aestivum	21	*Triticum ovatum*	14	4	–	Mettin et al. (1977)
Triticum aestivum	21	*Triticum umbellulatum*	7	1	1	Sears (1956)
Triticum aestivum	21	*Triticum umbellulatum*	7	6	6	Kimber (1967)
Triticum aestivum	21	*Triticum umbellulatum*	7	1	5	Friebe et al. (1995)
Triticum aestivum	21	*Triticum variabile*	14	9	–	Jewell and Driscoll (1983)

[a] Six distinct DAALs by C-banding technique (Hueros et al., 1991)

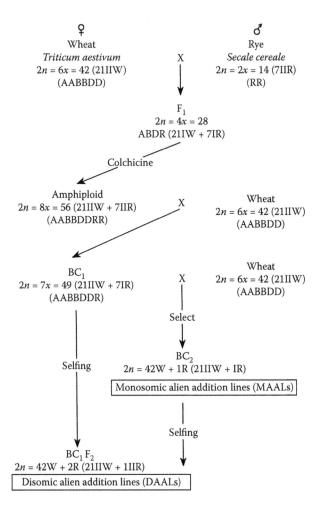

FIGURE 8.58 Method for obtaining monosomic and disomic addition lines of wheat ($2n = 6x = 42$) – rye ($2n = 2x = 14$).

(female) by *A. sativa* (male) cross (Table 8.68). Islam et al. (1981) observed 1.3% seed set in a *Triticum aestivum* cv. Chinese Spring ($2n = 42$) (female) × *Hordeum vulgare* cv. Betzes ($2n = 14$) (male) cross, while 15.4% seed set was obtained with the reciprocal cross.

Occasionally, direct interspecific or intergeneric crosses are not successful. In each case, bridge crosses may be utilized (Sears, 1953a; DeVerna et al., 1987). Sears (1953a) added seven chromosomes

---

TABLE 8.68

Frequency of *Avena* Plants with Various Chromosome Numbers in BC$_2$ Population and Progeny of Selfed BC$_1$ Plants

Generation	Number of Plants	Proportion of Plants with 2n Chromosome Numbers						
		42	42+ telo	43	44	45	46	47
BC$_2$	189	0.63	0.01	0.26	0.06	0.04	0.0	0.0
BC$_1$-selfed	39	0.18	0.02	0.33	0.28	0.13	0.03	0.03

Source: From Thomas, H. 1968. *Can. J. Genet. Cytol.* 10: 551–563. With permission.

---

of *Haynaldia villosa* to hexaploid wheat by using a tetraploid ($2n = 4x = 28$) wheat (*T. dicoccoides*) as a bridging species. The amphidiploid (AABBVV) was crossed to *T. aestivum* and backcrossed to *T. aestivum*. Two of the 53 backcross progenies were AABBDDV. Subsequently, Hyde (1953) isolated five of the possible seven DAALs and six of the possible seven MAALs. Friebe et al. (1987) examined *T. aestivum–Dasypyrum villosum* (formerly *H. villosa*) addition lines of E. R. Sears using C-banding technique and identified 6 DAALs.

Transfer of useful genes of economic importance to wheats from alien species has been very successful compared to other important crops. Shepherd and Islam (1988) list the following sequence of 24 species utilized for introgressing desired genes by producing alien additions, substitutions, and translocations: (1) *Secale cereale*, (2) *S. montanum*, (3) *Aegilops umbellulata*, (4) *Ae. variabilis*, (5) *Ae. comosa*, (6) *Ae. mutica*, (7) *Ae. longissima*, (8) *Ae. sharonensis*, (9) *Ae. bicornis*, (10) *Ae. searsii*, (11) *Ae. ventricosa*, (12) *Agropyron elongatum*, (13) *Ag. intermedium*, (14) *Ag. glaucum*, (15) *Ag. trichophorum*, (16) *Haynaldia villosa*, (17) *Elymus trachycaulus*, (18) *Hordeum vulgare*, (19) *H. chilense*, (20) *Triticum urartu*, (21) *T. thaoudar*, (22) *T. monococcum*, (23) *T. timopheevi*, (24) *T. aestivum* added to *S. cereale*.

8.2.2.9.2.2 Tetraploid Species

A complete set of the seven possible MAALs was obtained from crosses of tetraploid wheat, *T. durum* ($2n = 4x = 28$, AABB) with *Agropyron elongatum* ($2n = 2x = 14$) (Mochizuki, 1962) and from *T. durum* with *Ae. umbellulata* ($2n = 2x = 14$, $C^u C^u$) (Makino, 1976) crosses based on principles utilized at the hexaploid level. Makino (1976, 1981) obtained allotriploid F_1 hybrids ($2n = 3x = 21ABC^u$) between *T. durum* and *Ae. umbellulata*. The allotriploid hybrid plants were backcrossed with pollen from *T. durum* and the pentaploid plants ($2n = 5x = 35$, $AABBC^u$), obtained by the fertilization of unreduced female gametes (ABC^u) with normal male gametes (AB), were again backcrossed to *T. durum*. In the BC_2, seven possible MAALs ($2n = 29$) were isolated. In a similar way, Blanco et al. (1987) produced six of the possible seven MAALs from crosses of *Dasypyrum villosum* ($2n = 2x = 14$, VV) to *T. durum*.

8.2.2.9.2.3 Diploid Species

Monosomic alien addition lines (allotrisomics) have been produced in *Beta vulgaris* ($2n = 2x = 18$) (Savitsky, 1975; Heijbroek et al., 1983; Speckmann et al., 1985; Lange et al., 1988), *Oryza sativa* ($2n = 2x = 24$) (Shin and Katayama, 1979; Jena and Khush, 1989), and *Lycopersicon esculentum* (DeVerna et al., 1987). Savitsky (1975) added a single chromosome of *B. procumbens* ($2n = 2x = 18$) to *B. vulgaris*. The added *B. procumbens* chromosome carried a gene for nematode (*Heterodera schachtii*) resistance.

DeVerna et al. (1987) produced the 12 possible MAALs ($2n = 25$) from crosses of *Solanum lycopersicoides* ($2n = 2x = 24$, SS) to *Lycopersicon esculentum* ($2n = 2x = 24$, LL) by using *L. pennellii* ($2n = 2x = 24$, PP) as a bridging species.

Monosomic alien addition lines of *Oryza sativa* (AA) and *O. officinalis* (CC) were produced by Shin and Katayama (1979) and by Jena and Khush (1989). The initial cross made by Shin and Katayama (1979) was between a tetraploid ($2n = 4x = 48$) *O. sativa* (AAAA) and a diploid ($2n = 24$) strain of *O. officinalis* (CC). The allotriploid ($2n = 3x = 21$, AAC) F_1 plants were backcrossed to *O. sativa*. The progeny segregated plants with chromosome numbers ranging from $2n = 24$ to $2n = 54$ (Table 8.69).

Jena and Khush (1989) crossed diploid *O. sativa* and *O. officinalis*. The F_1 (AC) plants were obtained through embryo rescue technique. An amphidiploid (AACC) was not produced but F_1 plants were backcrossed to *O. sativa*. A growth hormone mixture, gibberellic acid (GA3) + naphthalene acetic acid (NAA) + kinetin (K) in the proportion of 100 to 25–5 mg/L, was sprayed on with an atomizer 24 h postpollination two times a day for 5 days. Of the 41,437 spikelets pollinated, only 539 (1.3%) set seeds. A total of 367 BC_1 plants were recovered through culture (1/4 strength MS medium). Of the 367 BC_1 plants, 357 were allotriploids and 10 were hypotriploids. This result indicates that some of the unreduced female gametes contained all the A and C genome chromosomes and produced allotriploid zygotes with AAC constitution when fertilized with gametes carrying an

TABLE 8.69

Chromosome Numbers in AAC x AA Crosses of *Oryza sativa* and *O. officinalis*

Chromosome Numbers	Number of Plants (%)	Number of Plants (%)
24	51 (12.6)	25 (26.6)
25	33 (8.1)	40 (42.5)
26	33 (8.1)	11 (11.7)
27	60 (14.8)	10 (10.6)
28	81 (20.0)	4 (4.3)
29	69 (17.0)	3 (3.2)
30	50 (12.3)	1 (1.1)
31	12 (3.0)	
32	11 (2.7)	
33	3 (0.7)	
36	1 (0.2)	
54	1 (0.2)	
Total	405	94

Source: Adapted from Shin, Y.-B. and T. Katayama. 1979. *Jpn. J. Genet.* 54: 1–10; Jena, K. K. and G. S. Khush. 1989. *Genome* 32: 449–455.

A chromosome. Allotriploid F_1 plants were backcrossed to *O. sativa* and 94 plants were recovered. Chromosome numbers ranged from $2n = 24$ to 30, much narrower than those recovered ($2n = 24$ to 54) by Shin and Katayama (1979) (Table 8.69). Twelve morphological types, each containing a complete chromosome complement of *O. sativa* and a single different chromosome of *O. officinalis,* were isolated and identified.

The wealth of untapped genetic diversity of wild perennial *Glycine* species has not been exploited in the soybean breeding programs because of their extremely low crossability with *G. max* and a need to employ *in vitro* embryo rescue methods to produce F_1, BC_1, and BC_2 plants (Singh et al., 1998). Wide hybrids produced in the genus *Glycine* are few compared with those in cereals. Most soybean researchers were unable to produce beyond amphidiploid stage. Singh et al. (1998) produced, for the first time, BC_1, BC_2, BC_3, and monosomic alien addition lines (MAALs) from *G. max* ($2n = 40$; GG) female × *G. tomentella* ($2n = 78$; DDEE) male → F_1 ($2n = 59$; GDE) → Colchicine treatment → amphidiploid ($2n = 118$; GGDDEE) female × soybean cv. Clark 63 male → BC_1 [$2n = 76$ (expected $2n = 79$; GGDE)] female × soybean cv. Clark 63 male → BC_2, BC_3, BC_4 → produced plants with $2n = 40$, $40 + 1$, $40 + 2$, $40 + 3$. Singh et al. (1998) distinguished morphologically 22 MAALs. The main hurdle was to produce BC_1 plants. It consumed 5 years of hybridization, perseverance, and motivation to break the unthinkable barrier. Similar observation was recorded by Jacobsen et al. (1994) to obtain BC_1 plant from potato and tomato somatic hybrids. Their experience is informative to readers. The first backcross is the most difficult step in starting a successful backcross program. The addition of an extra chromosome of *G. tomentella* to the $2n$ soybean complement modified several morphological traits including flowering habit, plant height, degree of pubescence, seed fertility, number of seed per pod and plant, pod and seed color, and seed yield. Disomic alien addition lines (DAALs) were isolated but plants died prematurely. The female transmission rate of an extra *G. tomentella* chromosome in MAALs averaged 36.5% and male transmission averaged 11.7%.

8.2.2.9.3 Identification of Alien Addition Lines

Alien addition lines can be identified by gross morphology, karyotype of the added chromosomes, meiotic chromosome pairing in F_1 plants obtained by intercrossing two DAALs, isozyme analysis, and by genetic tests.

8.2.2.9.3.1 Morphological Identification   Monosomic and disomic alien addition lines can be identified morphologically from the recipient parent and among themselves because the extra added chromosome modifies the specific vegetative and spike morphological features. The alterations may be both qualitative and quantitative. Most of the modifications are due to the interaction between genes of the recipient and donor parents and each chromosome has a specific effect on plant morphology and fertility.

Each disomic wheat–rye addition expresses diagnostic morphological traits such as stature of plants, color of leaves, compact and lax spikes, tapered head, hairy-neck, awned, and seed fertility. (O'Mara, 1940; Riley and Chapman, 1958b; Evans and Jenkins, 1960). Figure 8.59 shows spikes of seven "Chinese Spring" wheat—"Imperial" rye disomic alien addition lines together with parents and amphiploid. Based on morphological features of spikes, seven DAALs can be easily identified and distinguished. Similarly spikes of ditelosomic alien addition lines of "Chinese Spring" wheat–"Imperial" rye exhibit contrasting distinguishing features. It should be noted that long arms of rye chromosomes alter spike morphology more than that of short arms (Figure 8.60).

The wheat-barley disomic addition lines produced by Islam et al. (1981) differed morphologically from "Chinese Spring" and from each other. An extra pair of barley chromosomes expressed diagnostic characters in hexaploid wheat background identical to the effect of an extra chromosome in primary trisomic condition. Primaries Bush (Triplo 1) and Slender (Triplo 2) were similar

FIGURE 8.59   Spikes of "Chinese Spring" wheat (W), "Imperial" rye (R), amphiploid (A), and seven disomic (2n = 42W + 2R) addition lines. (Courtesy of Dr. B. Friebe).

FIGURE 8.60   Spikes of eleven ditelosomic "Chinese Spring" wheat–"Imperial" rye (2n = 42 + 2telo) addition lines. L = long arm; S = short arm (Courtesy of Dr. B. Friebe).

to DAALs 1 and 2, respectively. DAAL 5 was not isolated because of complete seed sterility. In contrast, primary Pseudo-normal (Triplo 5) was quite fertile (Tsuchiya, 1960a, 1967).

The addition of single chromosomes (monosomic = MAAL) from wild diploid wheat to tetraploid ($2n = 4x = 28$) wheat alters morphological features of plants more drastically than those recorded in MAALs of hexaploid wheat. Several sets have been produced in *T. durum* (Table 8.67).

Makino (1976) produced and identified the seven possible monosomic additions (C^u1 to C^u7) derived from *T. durum* and *Ae. umbellulata*. The C^u1 line was distinct from others in having a waxless spike. MAALs C^u2 and C^u3 carried the satellite chromosomes; C^u2 expressed reduced culm and top internode length, reduced ear density and increased rachis length, while C^u3 had only reduced culm and top internode lengths. MAAL C^u4 was characterized by red seed and carried a dense spike. The culm length of line C^u5 was short (84.7 cm), as was MAAL C^u2 (81.9 cm), suggesting that genes for short culm of *Ae. umbellulata* are probably located on chromosomes C^u2 and C^u5. MAAL C^u6 expressed thicker culm (3.2 mm) than *T. durum* (2.9 mm), but C^u7 did not show specific characteristics.

The plants of a DAAL ($2n = 4x = 50$) from a *Nicotiana tabacum* by *N. glutinosa* crosses were slightly reduced in overall length, were later in maturity, carried greater compactness of the inflorescence and shorter internodes than normal plants (Gerstel, 1945).

The addition of an alien chromosome to a diploid chromosome complement modifies the morphological features in the same way as observed in autotrisomics. Jena and Khush (1989) produced 12 possible MAALs from *Oryza sativa* by *O. officinalis* crosses. The addition lines differed from their diploid sibs by an array of morphological traits and resembled the 12 simple primary trisomics of *O. sativa*. Khush (2010) summarized available genetic stock of MAALs of *O. sativa* and seven distantly related species [*O. officinalis* (CC), *O. punctata* (BB), *O. australiensis* (EE), *O. bracyantha* (FF), *O. granulate* (GG), *O. minuta* (BBCC), and *O. latifolia*, CCDD)].

8.2.2.9.3.2 Cytological Identification   The alien chromosomes in addition lines can be identified from observations of chromosome morphology (total length, relative length, and arm ratio) and cytological chromosome markers (heterochromatic and euchromatic distribution, heterochromatic knobs, kinetochore position, and presence of satellite).

O'Mara (1940) and Riley and Chapman (1958b) designated wheat–rye addition lines arbitrarily. Evans and Jenkins (1960) designated DAALs derived from wheat–rye crosses based on karyotype analysis. For example, DAAL I carried the longest chromosome and DAAL VII the satellite chromosome. Gupta (1971, 1972) suggested designating wheat–rye addition lines on the basis of homoeologous relationships between rye and wheat chromosomes related mainly to substitution-compensation. The rye chromosomes in wheat–rye addition lines were numbered 1R to 7R. This nomenclature is far from being complete and is occasionally controversial due to the lack of correspondence at the gene level and in naturally occurring translocations (Sybenga, 1983). Thus, the ability of an alien chromosome to be matched against the wheat homoeologous grouping depends on the genetic equivalence of the alien chromosome to the wheat group.

By utilizing the Giemsa C-banding technique, Darvey and Gustafson (1975) identified rye chromosomes in four sets of wheat–rye disomic addition lines (Imperial, Dakold, King II, Petkus) and verified the homoeologous chromosome (1R to 7R) relationships established by genetic studies of seed shriveling in wheat and triticale (Darvey, 1973). Singh and Röbbelen (1976) examined Holdfast-King II wheat–rye disomic addition lines by the Giemsa C-banding technique. The addition line V was found to carry a pair of nucleolus organizer rye (1R) chromosomes (Figure 8.61) and addition lines II and VI each showed a deletion in its short arm.

Giemsa C- and N-banding and genomic *in situ* hybridization (GISH) techniques have helped identify wheat–rye addition lines precisely and it has been determined that Chinese Spring–Imperial additions are complete. Furthermore, Giemsa C-banding also identified ditelosomic additions of Chinese Spring–Imperial. Telocentric chromosomes showed characteristic telomeric and intercalary heterochromatic bands (Figure 8.62).

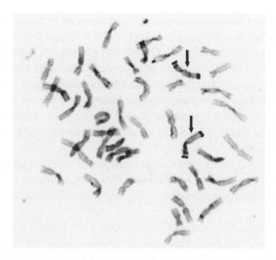

FIGURE 8.61 Giemsa C-banded somatic metaphase of a disomic wheat ("Holdfast")–rye ("King II") addition line showing $2n = 42$ wheat + 2 satellite rye (V = 1R) chromosomes (arrow). (From Singh, R. J. and G. Röbbelen. 1976. *Z. Pflanzenüchtg.* 76: 11–18. With permission.)

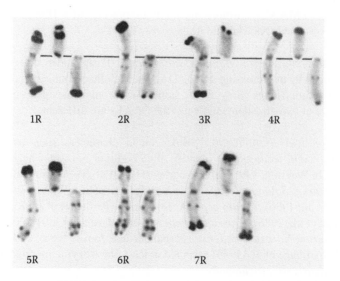

FIGURE 8.62 Giemsa C-banded somatic metaphase of eleven ditelosomic "Chinese Spring" wheat–"Imperial" rye addition lines showing characteristic telomeric and intercalary heterochromatic bands. (From Mukai, Y., B. Friebe, and B. S. Gill. 1992. *Jpn. J. Genet.* 67: 71–83. With permission.)

8.2.2.9.3.3 Meiosis in Alien Addition Lines It is expected that a pair of alien chromosomes in DAALs should form a bivalent at metaphase-I. However, it is not always recorded (Gerstel, 1945; Hyde, 1953; Riley and Chapman, 1958b; Kimber, 1967; Bernard, 1976; Islam et al., 1981; Orellana et al., 1984).

Orellana et al. (1984) examined meiotic pairing in "Chinese Spring"–"Imperial" and "Holdfast"–"King II" addition and substitution lines using the C-banding technique. They observed decrease in pairing of both wheat and rye homologous chromosomes and suggested that chromosome pairing in DAALs and substitution lines is a complex process influenced by factors such as genes controlling meiotic pairing, constitutive heterochromatin, and cryptic wheat–rye interactions.

In wheat-barley DAALs, chromosome pairing was influenced by the added barley chromosomes (Islam et al., 1981). Synapsis was reduced in DAAL A (76.9%) and D (84.8%), while it was almost normal (22 II) in other DAALs (B = 91.3%; C = 90.2%; E = 98.6%; F = 90%). The wheat cultivar Chinese Spring showed 21 II in 95.9% of the sporocytes. It should be noted that MAALs A and B showed a much higher frequency of univalents than Chinese Spring, and the difference was significant with MAAL A. This observation indicates that barley chromosome A (chromosome 4) causes significantly increased asynapsis of at least one pair of wheat chromosomes in monosomic addition, and presumably in disomic addition also. They found no evidence of any meiotic pairing between wheat and barley chromosomes.

The lagging alien chromosomes generally misdivide in the first meiosis and may generate telocentric chromosomes. Ditelosomic addition lines have been isolated and identified in the progeny of "Chinese Spring" wheat–"Imperial" rye (Figure 8.62) and wheat–barley DAALs (Islam and Shepherd, 1990).

An alien chromosome generally remains as a univalent in MAALs and the chromosomes from the recipient parent synapse regularly. However, a very low frequency of trivalent formation was occasionally seen (Savitsky, 1975; Shin and Katayama, 1979; Jena and Khush, 1989). Shin and Katayama (1979) recorded 0–1 trivalents in 10 of the 12 MAALs of *Oryza sativa* and *O. officinalis* while Jena and Khush (1989) observed a trivalent in three sporocytes of MAAL 3. Furthermore, desynapsis was found in MAALs G and H of Shin and Katayama, but it was not recorded by Jena and Khush (1989).

In *Raphanobrassica* MAALs, chromosome configurations of 9II + 1I predominated and no cell with 1III + 8II was observed (Kaneko and Matsuzawa, 1987).

8.2.2.9.3.4 Intercrossing Disomic Alien Addition Lines

Positive identification of addition lines can also be determined by intercrossing among DAALs. Meiotic chromosome pairing in F_1 hybrids is analyzed. Alien addition lines carrying the same two chromosomes generally form a bivalent, while the occurrence of two univalents indicates the DAALs are different.

8.2.2.9.3.5 Biochemical Identification

Biochemical characters such as isozyme banding patterns are a very useful technique to identify alien addition lines of wheat–rye (Miller, 1984; Benito et al., 1991a,b; Wehling, 1991), wheat–barley (Hart et al., 1980; Islam and Shepherd, 1990), *Lycopersicon esculentum-Solanum lycopersicoides* (Chetelat et al., 1989), *Cucurbita moschata–C. palmata* (Weeden et al., 1986), *Allium cepa–A. fistulosum* (Peffley et al., 1985), *Beta vulgaris–B. procumbens* (Lange et al., 1988), *Brassica campestris–B. oleracea* (Quiros et al., 1987), *Triticum turgidum* Con V. *durum-Hordeum chilense* (Fernandez and Jouve, 1988), and wheat–*H. chilense* (Miller et al., 1982b). Lange et al. (1988) assigned at least one isozyme marker to eight of the nine MAALs of *Beta vulgaris–B. procumbens*.

By using wheat–barley addition lines, Islam and Shepherd (1990) summarized the assigned genes controlling at least 58 isozymes to specific barley chromosomes or chromosome arms. Thus, wheat-barley DAALs are very useful material for determining the chromosome arm location of protein and isozyme structural genes in barley (Table 8.70). Furthermore, these lines are helping to construct a restriction fragment length polymorphism (RFLP) map for barley (Heun et al., 1991).

8.2.2.9.4 Breeding Behavior of Alien Addition Lines

8.2.2.9.4.1 Monosomic Alien Addition Lines (MAALs)

In MAALs, an alien chromosome remains as a univalent at meiotic metaphase-I, and during gametogenesis it can be either eliminated or included in one of the gametes. Thus, transmission of the extra chromosome is expected to be lower than that observed in autotrisomics. The female transmission rates of the extra chromosome in MAALs with an extra *Oryza officinalis* chromosome were not observed in six MAALs (chromosomes 2, 3, 5, 7, 8, and 11). The highest male transmission rate was 14.6% in MAAL 6.

TABLE 8.70

Association of Isozyme Markers and Seed Protein Characters of Barley Using Wheat–Barley Alien Addition Lines

Barley Chromosome		Gene Locus
1	*Amy 2* (α-Amylase)	*Est 3, 5* (Esterase)
	CM a,c (A hordeins)	*Pgd 1* (Phospho gluconate dehydrogenase)
	Enp 1 (Endopeptidase)	*Prx 4* (Peroxidase)
2	*Est 7* (Esterase)	**Isa 1** (Inhibitor subtilisin, amylase)
	Gpd 1 (Glucose-6-phosphate dehydrogenase)	Ndh 4 (NADH dehydrogenase)
	Idh 2 (Isocitrate dehydrogenase)	*Prx 2* (Peroxidase)
3	*Aat 3* (Aspartate aminotransferase)	*Est 1,2,4,10* (Esterase)
	CMEa [A hordeins (CM proteins)]	*Mdh 2* (Malate dehydrogenase)
	Itc 1a (Inhibitor trypsin, chymotrypsin)	*Tpi 1* (Triose phosphate isomerase)
4	*Acp 2* (Acid phosphatase)	**CM b,d** [A hordeins(CM proteins)]
	Adh 1 (Alcohol dehydrogenase)	*Ibf1* (Iodine-binding factor)
	Adh 2 (Alcohol dehydrogenase)	*Ndh 1* (NADH dehydrogenase)
	ß-Amy 1 (ß-Amylase)	**Paz 1** (Protein Z4 in endosperm = Antigen 1a in beer)
	Cat 1 (Catalase)	Pgm 1 (Phospho gluco mutase)
5	*Aco 2* (Aconitase hydratase)	**Hor 3** (D hordeins)
	Gdh 1 (Glutamate dehydrogenase)	**Ica 1,2** (Inhibitor chymotrypsin, *Aspergillus* protease)
	Gpi 1 (Glucose phosphate isomerase)	*Mdh 1* (Malate dehydrogenase)
	Hor 1 (C,B hordeins)	*Pgd 2* (Phospho gluconate dehydrogenase)
	Hor 2 (C,B hordeins)	*Adh 3* (Alcohol dehydrogenase)
6	*Aat 2* (Aspartate aminotransferase)	*Amp 1* (Amino peptidase)
	α-*Amy 1* (α-Amylase)	*Dip 1* (Dipeptidase)
	Aco 1 (Aconitate hydratase)	*Dip 2* (Dipeptidase)
	Acp 3 (Acid phosphatase)	
7	*Est 9* (Esterase)	**Paz 2** (Protein Z7 in endosperm = Antigen 1b in beer)
	Gpi 2 (Glucose phosphate isomerase)	*Sdh 1* (Shikimate dehydrogenase)
	Isozyme (plain type)	*Tpi 2* (Triose Phosphate isomerase)
	Seed protein (bold type)	

Source: From Islam, A. K. M. R. and K. W. Shepherd. 1990. *Biotechnology in Agriculture and Forestry, Wheat.* Springer-Verlag, Berlin. pp. 128–151. With permission.

In autotrisomics, male transmission was not recorded in Triplo 1, 2, 3, 4, and 7 and the highest male transmission was 27.3% in Triplo 9, a major nucleolus organizer chromosome (Table 8.71).

Blanco et al. (1987) examined the transmission rate of the extra chromosome after selfing in six MAALs of *Triticum durum–Dasypyrum villosum* (Table 8.72). The transmission of the extra chromosome through the female was variable, ranging from 5% (MAAL A) to 15.7% (MAAL B). Plants with $2n = 28 + 1$ telo and $2n = 29 + 1$ telo were also identified. The occurrence of telocentric chromosomes indicates misdivision of a univalent that occurred at a higher rate than in *Triticum durum–Ae. umbellulata* where monotelotrisomic plants were not found (Table 8.73).

The frequency of DAALs in the selfed progeny of MAALs in tetraploid wheat was very low because the extra alien chromosome causes more unbalance than those recorded at 6× level. Therefore, transmission of the extra alien chromosome through the male is too low (Makino, 1981; Blanco et al., 1987; Dhaliwal et al., 1990) and generally the plants are sterile. Joppa and McNeal (1972) reported six D-genome disomic addition lines of *Durum* wheat. Three DAALs (1D, 3D, 6D) were male sterile while 4D and 5D were stable and were partially male fertile.

TABLE 8.71

Female and Male Transmission Rates of the Extra Chromosome in the MAALs of *Oryza sativa–O. officinalis*

MAAL	Total	2n	2n+1	%(2n+1)	%Autotrisomics	Total	2n	2n+1	%(2n+1)	% Autotrisomic
			$(2n+1) \times 2n$					$2n \times (2n+1)$		
1	35	32	3	8.6	15.5	–	–	–	–	0.0
2	76	69	7	9.2	31.6	32	32	0	0.0	0.0
3	72	67	5	6.9	17.8	–	–	–	–	0.0
4	323	267	56	17.3	43.9	144	134	10	6.9	0.5
5	151	141	10	6.6	32.7	50	50	0	0.0	1.6
6	156	145	11	7.0	37.9	35	35	0	0.0	0.0
7	267	235	32	12.0	31.1	40	40	0	0.0	0.0
8	125	108	17	13.6	25.5	49	49	0	0.0	14.3
9	316	270	46	14.5	35.5	19	17	2	11.7	27.3
10	472	363	109	23.1	27.4	76	74	2	2.6	1.5
11	375	326	49	13.1	39.7	16	16	0	0.0	2.7
12	655	479	176	26.8	37.1	89	76	13	14.6	5.6

Source: Adapted from Jena, K. K. and G. S. Khush. 1989. *Genome* 32: 449–455; From Khush, G. S. et al. 1984. *Genetics* 107: 141–163. With permission; Khush, G. S. and R. J. Singh. 1991. *Chromosome Engineering in Plants: Genetics, Breeding, and Evolution.* Part A. P. K. Gupta and T. Tsuchiya, Eds., pp. 577–598. With permission.

TABLE 8.72

Breeding Behavior of the Selfed Monosomic Addition Lines of *Dasypyrum Villosum* in *Triticum Durum*

Lines	No. Plants Examined	28	28+telo	29	29+telo.	30
		% Frequency (2n)				
A	119	91.6	2.5	5.0	0.9	0.0
B	159	67.9	12.0	15.7	1.9	2.5
C	130	89.2	1.5	8.5	0.0	0.8
D	31	77.4	9.7	9.7	0.0	3.2
E	137	76.6	11.0	11.0	0.7	0.7
F	29	72.4	6.9	13.8	3.4	3.4
Total	605	(Mean) 79.8	7.3	10.6	1.0	1.3

Source: Adapted from Blanco, A., R. Simeone, and P. Resta. 1987. *Theor. Appl. Genet.* 74: 328–333.

8.2.2.9.4.2 Disomic Alien Addition Lines (DAALs) DAALs are expected to breed true. However, it is not always observed. Differences in meiotic cycles between recipient and donor chromosomes lead to the elimination of donor chromosomes. For example, in wheat–rye DAALs, the rye chromosomes fail to function normally in the wheat nucleus and the tendency for asynapsis may be due to the absolute homozygosity of the rye chromosomes achieved during backcrossing and selfing (O'Mara, 1940).

Miller (1984) summarized the breeding behavior of "Chinese Spring" wheat–"Imperial" rye DAALs grown over several years in Cambridge, England. An average of 86.8% plants carried

TABLE 8.73
Breeding Behavior of the Selfed Monosomic
Addition Lines of *Aegilops umbellulata*
Chromosomes in *Triticum durum*

Line	No. of Progenies Examined	No. of Plants Examined	% Frequency ($2n$)		
			28	29	30
C^u1	1	87	79.3	20.7	
C^u2	2	121	81.8	16.5	2.5
C^u3	2	99	85.7	13.1	1.0
C^u4	1	91	90.1	9.9	
C^u5	1	66	95.5	4.5	
C^u6	1	27	74.1	25.9	
C^u7	4	248	93.1	6.9	

Source: From Makino, T. 1976. *Can. J. Genet. Cytol.* 18: 455–462.
With permission.

expected $2n = 44$ chromosomes and the range was 86% (4R) to 95% (2R). Plants with $2n = 42, 43$, and 45 chromosomes in a low frequencies were also identified (Table 8.74).

Dvořák and Knott (1974) studied the stability of wheat–*Agropyron elongatum* DAALs. The majority of segregating plants contained the expected $2n = 44$ chromosome number, ranging 96.6% for DAAL VII to 85.5% for DAAL IV. The MAAL ($2n = 43$) plants were identified for all the lines, but the frequency was low. Plants with $2n = 43 + 1$ telo, and $2n = 45$ were also recorded. This result indicates that DAALs are not stable and throw aneuploids in segregating populations. In contrast, the only DAAL of *N. tabacum-N. glutinosa* reported was meiotically normal and chromosomally fairly stable (Gerstel, 1945). The majority of the self-pollinated progeny of DAALs of wheat–*T. umbellulatum* carried the expected $2n = 44$ chromosomes. However, plants with $2n = 43$ were also found in which the alien chromosomes failed to form bivalents.

TABLE 8.74
Breeding Behavior of the Selfed Disomic
Alien Addition Lines of 'Imperial' Rye
Chromosomes in Chinese Spring Wheat

Chromosome	Chromosome Numbers			
	42	43	44	45
1R	0.05	0.05	0.90	–
2R	–	0.05	0.95	–
3R	–	0.07	0.93	–
4R	–	0.14	0.86	–
5R	0.07	0.23	0.63	0.07
6R	0.04	0.07	0.89	–
7R	0.04	0.04	0.92	–
Mean	0.029	0.093	0.868	0.010

Source: From Miller, T. E. 1984. *Can. J. Genet. Cytol.*
26: 578–589. With permission.

8.2.2.9.5   Fertility in Alien Addition Lines

Seed fertility in both MAALs and DAALs is influenced by the ploidy level of the recipient parent, the genomic affinity between the recipient and donor species, and the genetic constitution of the added chromosomes.

Seed fertility in MAALs at $2x$ level is similar to the seed set observed in autotrisomics. MAAL 4 of *Oryza sativa–O. officinalis* was completely sterile (Jena and Khush, 1989); likewise Triplo 4 of autotrisomics of *O. sativa* was totally sterile (Khush et al., 1984). Analogous results were found in MAALs of *Beta vulgaris–B. procumbens* (Speckmann et al., 1985).

Seed set in MAALs of *T. durum–Ae. umbellulata* ranged from 32% (C^u) to 84% (C^u4), while the selfed seed set in the recipient *T. durum* was 100% (Makino, 1976). Similarly, seed set was lower in MAALs of *T. durum–Dasypyrum villosum* compared to the recipient parent (Blanco et al., 1987).

Seed set in MAALs and DAALs at the $6x$ level is expected to be better than at the $4x$ and $2x$ levels because the extras cause more imbalance at $2x$ and $4x$ than at $6x$. The extent of seed fertility depends on the degree of taxonomic closeness between wheat and the alien species. Both MAALs and DAALs derived from crosses of *T. aestivum* and *Secale cereale, Agropyron, Haynaldia,* or *Aegilops* are relatively more fertile than the wheat–barley additions. DAAL 5 of wheat–barley was not isolated because barley chromosome 5 causes total male and female sterility in the presence of wheat chromosomes (Islam et al., 1981). Despite complete homology and almost normal meiosis in wheat and alien species disomic additions, seed fertility varied from line to line.

8.2.2.10   Alien Substitution Lines

8.2.2.10.1   Introduction

Exotic germplasm (the alien species), relatives to cultigens, is a rich reservoir for economically valuable traits and has been proven to be useful source for increasing genetic variability in many crops either by conventional plant breeding methods or by molecular techniques. Exotic germplasm includes all germplasm that does not have immediate usefulness without hybridization and selection for adaptation for a given region (Hallauer and Miranda Filho, 1981). Alien genes are introgressed into cultigens by hybridization followed by several backcrossing to the recurrent parent or by genetic transformation. However, the majority of examples of alien substitution lines are available from common wheat. These lines are of two types:

1. Substitution of a total genome of a cultigen into the cytoplasm of an alien species (genome substitution)
2. Substitution of a single chromosome or a pair of chromosomes into the chromosome complement of a cultigen (chromosome substitution)

Sometimes, substitution of a segment (s) of alien chromosome into a cultigen occurs either spontaneously or is induced by a mutagen to produce chromosomal interchange-segmental substitution (Gupta, 1971).

Shepherd and Islam (1988) listed 176 entries of alien addition lines, 84 entries of alien substitution lines and 58 entries of translocation lines involving alien and wheat chromosomes. Such endeavor is lacking for other crop plants. During past decade, considerable efforts were taken on improving tomato, potato, Alliums, Brassicas, soybean, rice, and maize by using alien germplasm. The genetic transformation has revolutionized the production of pest resistant (Bt maize and cotton) and herbicide tolerant (roundup ready soybean) crops.

8.2.2.10.2   Types of Alien Substitution

8.2.2.10.2.1   Genomic Substitution   Cytoplasms from alien species are an excellent source of cytoplasmic male sterility (CMS) (Virmani and Edwards, 1983). CMS lines are produced by alien species (female) × cultigen (male) crosses, followed by a recurrent backcross method to eliminate

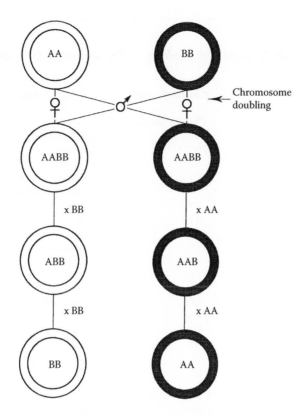

FIGURE 8.63 Substitution of genome complements after doubling of chromosomes through colchicine treatment. (From Kihara, H. 1951. *Cytologia* 16: 177–193. With permission.)

the undesirable traits donated by the alien species (Figure 8.63). Thus, a modified cultivar known as an alloplasmic line is derived.

In the tetraploid *Triticum durum* ($2n = 4x = 28$) and the hexaploids *T. aestivum* ($2n = 6x = 42$), alloplasmic lines expressing CMS are derived from related genera (Kihara, 1951; Fukasawa, 1953; Maan, 1973a, b; Sasakuma and Maan, 1978). The taxonomy of genera *Triticum* and *Aegilops* is changing frequently; however, a revised and accepted classification is summarized in Table 8.75 that contains nuclear genome symbols and synonyms (B. S. Gill, personal communication).

Alloplasmic lines are not easy to produce. Examples are numerous, but only one is cited here. Smutkupt (1968) observed 61% seed set in a cross of common wheat cv. Chinese Spring ($2x = 42$) female × rye cv. Petkuser ($2x = 14$) (male), while the reciprocal cross set only 1% seed. This difference may be attributed to relatively slow growth of wheat pollen tubes in the styles of rye. In rye (female) × wheat (male) crosses, 36% of the seeds were shriveled and required embryo culture to rescue alloplasmic F_1 plants. By contrast, the reciprocal cross produced 94% fully-developed seeds.

Occasionally, alloplasmic wheat lines show reduction in seed germination, alteration in growth habit (normal to reduced growth), variable seed fertility (full fertility to total sterility), maturity changes (early to delayed), and seed variability (normal plump to shriveled) (Tsunewaki et al., 1983; Virmani and Edwards, 1983).

Alloplasmic lines of common wheat and triticale exhibit multiple sporocytes (syncytes) in anthers. Tsuji et al. (1983) examined 11 alloplasmic lines of wheat cv. "Rosner." The ploidy level ranged from $3x$ to $40x$, and the modal level was $8x$. However, the frequency of syncytes was not high (range 0.2%–4.6%). The occurrence of syncytes was mostly correlated with male sterility.

TABLE 8.75

Taxonomy of the Genus *Triticum* and *Aegilops*, 2*n* Chromosome, Genomes, and Their Synonyms[a]

Species	Genome Symbol	Synonyms
Diploids (2*n* = 2*x* = 14)		
Triticum monococcum subsp. *aegilopoides*	A^{m}A^m	*T. monococcum* subsp. *boeoticum*, *T. aegilopoides*,
Triticum monococcum subsp. *monococcum*	A^{m}A^m	*T. monococcum*
Ttriticum urartu	AA	*T. monococcum*
Aegilops speltoides var. *speltoides*	SS	*T. speltoides*, *Ae. speltoides* subsp. *speltoides*
Aegilops speltoides var. *lingustica*	SS	*T. speltoides*, *Ae. speltoides* subsp. *lingustica*
Aegilops bicornis var. *bicornis*	S^{b}S^b	*T. bicorne*
Aegilops bicornis var. *anathera*	S^{b}S^b	*T. bircorne*
Aegilops longissima	S^{l}S^l	*T. longissimmum*
Aegilops sheronensis	S^{l}S^l	*T. sharonensis* (invalid)
Aegilops searsii	S^{s}S^s	*T. searsii* (invalid)
Aegilops mutica var. *mutica*	TT	*T. tripsacoides*
Aegilops mutica var. *loliacea*	TT	*T. tripsacoides*
Aegilops tauschii	DD	*Ae. squarrosa*, *T. tauschii*
Aegilops comosa var. *comosa*	MM	*T. comosum*
Aegilops comosa var. *subventricosa*	MM	*Ae. comosa* subsp. *heldreichii*
Aegilops uniaristata	NN	*T. uniaristatatum*
Aegilops caudata	CC	*T. dichasians* (invalid), *Ae. markgrafii*
Aegilops umbellulata	UU	*T. umbellulatum*
Tetraploids (2*n* = 4*x* = 28)		
Triticum turgidum subsp. *carthlicum*	BBAA	*T. turgidum*
Triticum turgidum subsp. *dicoccoides*	BBAA	*T. turgidum*
Triticum turgidum subsp. *dicoccum*	BBAA	*T. turgidum*
Triticum turgidum subsp. *durum*	BBAA	*T. turgidum*
Triticum timopheevii subsp. *timopheevii*	GGAA	*T. timopheevii*
Aegilops crassa	DD<u>M</u>M	*T. crassaum*
Aegilops cylindrica	DDCC	*T. cylindricum*
Aegilops geniculata	<u>M</u>MUU	*T. ovatum*
Aegilops neglecta	UU<u>M</u>M	
Aegilops biuncialis	UU<u>M</u>M	*T. macrochaetum*
Aegilops columnaris	UU<u>M</u>M	*T. columnare* (in valid)
Aegilops triuncialis var. *triuncialis*	UUCC; CCUU	*T. triunciale*
Aegilops ventricosa	DD<u>N</u>N	*T. ventricosum*
Hexaploids (2*n* = 6*x* = 42)		
Triticum aestivum subsp. *aestivum*	BBAADD	
Triticum aestivum subsp. *compactum*	BBAADD	
Triticum aestivum subsp. *compactum*	BBAADD	
Triticum aestivum subsp. *macha*	BBAADD	
Triticum aestivum subsp. *spelta*	BBAADD	
Triticum aestivum subsp. *sphaerococcum*	BBAADD	
Triticum zhukovskyi	GGAAA^{m}A^m	
Aegilops recta (= *Ae. triaristata*)	UU<u>MM</u>NN	*T. neglectum*
Aegilops crassa	DD<u>D</u>DMM	*T. crassum*
Aegilops vavilovii	DDMMSS	*T. syriacum*
Aegilops juvenalis	DD<u>M</u>MUU	*T. juvenale*

Source: Adapted from Feldman, M. and A. A. Levy. 2012. *Genetics* 192: 763–774.

[a] Based on van Slageren www.ksu.edu/wgrc/taxonomy. Kimber and Tsunewaki, 1988; Feldman and Levy, 2012 (underlined designation indicates a modified genome).

8.2.2.10.2.2   Chromosome Substitution and Translocation Lines   When alien chromosomes compensate completely for the absence of cultigen chromosomes in sporophytes and gametophytes, the phenomenon is known as chromosome substitution. Substitution is only possible when there is a close genetic relationship between the substituting alien chromosome and the replaced recipient (cultigen) chromosome. A considerable amount of research has been done in producing alien substitution and translocation lines in common wheat, with the objective to breed wheat carrying resistance to pathogens and pests (Friebe et al., 1996). Sometimes, substitution of a complete chromosome donates several undesirable traits from the wild species. In such cases, segmental substitutions are produced (Sears, 1956b).

1. Production of chromosome substitution and translocation lines

   Common wheat, being an allohexploid ($2n = 6x = 42$; AABBDD), tolerates a considerable degree of aneuploidy. It hybridizes rather easily with its wild relative genera such as *Secale*, *Aegilops*, *Agropyron,* and *Hordeum* and useful traits have been introgressed into tetraploid and hexaploid wheats (Sharma and Gill, 1983; Shepherd and Islam, 1988). Of all the substitution lines studied, available wheat–rye substitutions have been the most extensively examined.

   Wheat (W)–rye (R) substitutions of spontaneous origin are routinely found in nature; several widely grown European wheat cultivars contain mainly the 1RS (1BL) substitution or the 1R chromosome is involved in translocations (1RS.1BL; 1RS.1AL) with genomes of wheat (Mettin et al., 1973; Zeller, 1973; Zeller and Baier, 1973; Münzer, 1977; Zeller and Hsam, 1983). By using the Giemsa C-banding technique, Lukaszewski (1990) examined 207 entries in six major 1989 U.S. wheat nurseries and 30 entries in the 21st International Winter Wheat Performance nurseries. Among entries of the United States, 4.3% of lines contained 1RS.1AL and 7.1% of lines contained 1RS.1BL translocations. The international entries predominantly carried 1RS/1BL translocations and the frequency was higher (38.1%). Javornik et al. (1991) recorded 36 out of 59 Yugoslav wheat cultivars carrying 1BL.1RS translocations.

   Substitution lines are produced experimentally. A prerequisite to isolate substitution lines is to have available monosomic alien addition lines (MAALs) and disomic alien addition lines (DAALs). In producing wheat–rye substitution lines, the DAALs ($2n = 42W + 2R$) are used as a pollen parent and wheat monosomics ($2n = 41W$) as a female parent. The F_1 plants containing $2n = 20IIW + 1IW + 1IR$ are either selfed or pollinated onto DAALs. It is expected that the F_1 plants will produce four kinds of female and male spores: $20W + 0R$, $20W + 1R$, $21W + 0R$, $21W + 1R$ in the ratio of 9:3:3:1 assuming both rye chromosomes are excluded in 75% of gametes (Sears, 1944). Since nullisomic ($20W + 0R$) male spores will not function, and the missing wheat chromosome is fully compensated by a rye chromosome, the functional male spores will be in the ratio of 0:3:3:1. However, in a noncompensating combination, male spores will segregate in a ratio of 0:0:3:1. In the F_1, plants with $2n = 41W + 2R$ ($20IIW + 1IW + 1IIR$) are selected cytologically, and after selfing these plants disomic substitution lines ($2n = 40W + 2R$) are selected because 1W univalent is usually eliminated during gametogenesis (Figure 8.64). This technique can also be used to produce ditelocentric addition lines. The telocentric chromosomes are easy to identify cytologically and can be used for locating genes on a particular arm of a chromosome. The same procedure was used to produce wheat–*Agropyron* substitution lines (Dvořák, 1980).

   Merker (1979) felt that the above procedure for producing wheat–rye substitution lines requires a considerable amount of cytological work. His procedure is as follows: Pollinate monosomic wheats ($2n = 41W$) by octoploid triticales ($2n = 8x = 56$; AABBDDRR). The F_1 plants are expected to segregate two cytotypes: 1. $2n = 5x = 49$ ($42W + 7R$) and 2. $2n = 5x - 1 = 48$ ($41W + 7R$). Substitution lines are expected after selfing $2n = 48$ chromosome plants where a missing wheat chromosome (in question) has been substituted by a rye chromosome which possesses the best compensating ability. The deficiency of this method

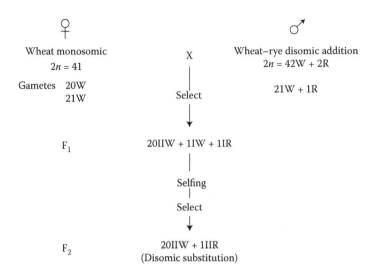

FIGURE 8.64 Production of wheat–rye substitution line.

is that another rye chromosome may substitute for the questionable wheat chromosome. By contrast, if substitution lines are produced from DAALs, a rye chromosome is forced to substitute for a specific wheat chromosome because the rye chromosome has no preference. However, both methods require cytological work.

Friebe and Larter (1988) produced a complete set of isogenic wheat–rye D-genome substitutions by crossing an inbred line of spring rye cv. "Prolific" to a tetraploid wheat. The A and B genomes were extracted from the hexaploid wheat cv. "Thatcher," and selection for wheat–rye substitution lines was carried out in BCF_3 to BCF_6 families.

Joppa and Williams (1977) isolated 14 possible substitution monosomics by crossing the appropriate Chinese Spring nullisomic-tetrasomic with *T. turgidum* L. cv. *durum* "Longdon" ($2n = 28$, AABB). The F_1 plants were allowed to self-pollinate. Selection for disomic substitution for a D-genome chromosome pair (homoeologous A or B genome chromosome pair) was conducted in F_2 and F_3 generations. In another study, Joppa and Maan (1982) isolated a substitution line in the F_2 population from a cross of *durum* wheat selection 56-1 and *T. boeticum*. One *T. boeticum* chromosome compensated for the loss of a *durum* chromosome 4B. Present day *durum* and *T. aestivum* wheats are structurally different from the originally synthesized tetraploid and hexaploid wheats because of chromosomal interchanges among A B D genomes of wheats (Okamoto and Sears, 1962). For example, *T. urartu* is considered to be A genome donor of the AA genome of cultivated wheat (Chapman et al., 1976). However, it has been established that only six A genome chromosome pairs are present in tetraploid emmer wheat and hexaploid bread wheat. It is now known that chromosome 4A is not from the A genome donor species but from an unknown species and appears to have multiple structural rearrangement in comparison to 4A chromosome of *T. monococcum* and *T. urartu* (Naranjo, 1990; Zeller et al., 1991). Based on isozyme and molecular homoeoloci, Liu et al. (1992) furnished evidence that present 4AL chromosome constitutes chromosome segments of 7BS and 5AL and telomere of 7BL came from 5AL. These observations suggest that wheat genomes underwent considerable structural changes during the course of evolution of present day wheats and still these changes are occurring.

2. Wheat–rye substitutions and homoeologous relationships
   Homoeologous relationships among the three genomes (ABD) of common wheat cv. "Chinese Spring" were established and classified into seven homoeologous groups by

Sears (1966b), based on the knowledge of nullisomic tetrasomic series where a pair of extra chromosomes of one genome could successfully compensate for the loss of two chromosomes from the other two genomes. The three genomes of common wheat are closely related and originated from a common progenitor.

Several nomenclature systems have been used to designate rye chromosome (Table 8.76; Sybenga, 1983). However, to avoid confusion and to adopt an universal nomenclature, rye chromosomes are being assigned to the respective homoeologous wheat chromosomes based on morphological and cytological markers such as characteristic C-bands (Figures 8.65), meiotic chromosome pairing, telocentrics, interchanges, substitution and compensation ability, and genetic tests (O'Mara, 1947; Sears, 1968; Gupta, 1971; Zeller and Hsam, 1983; Miller, 1984; Naranjo and Fernández-Rueda, 1991). The Giemsa C-karyogram of "Imperial" rye in "Chinese Spring" wheat is considered standard. The results are briefly summarized below:

a. Chromosome 1R

It is a nucleolus organizing chromosome, substitutes completely for wheat chromosomes 1A, 1B, and 1D, and carries genes for resistance to pests and diseases. The 1A (1R) and 1B(1R) substitutions and 1B.1R translocations are common in many wheat cultivars (Table 8.77).

TABLE 8.76

The Equivalence of the Original Nomenclature to the Homoeological Nomenclature of the Lines of the Major Wheat–Rye Addition Series

	Homoeologous Groups and Rye Chromosomes Designation						
	1, 1R	2, 2R	3, 3R	4, 4R[a]	5, 5R	6, 6R	7, 7R[a]
"Holdfast"–"King II"	V	III	–	IV	I	II	VII
"Kharkov"–"Dakold"	VII	II	–	V	VI	IV	III
"Chinese Spring"–"Imperial"	E	B	G	C	A	F	D
"Chinese Spring"–"King II"	1R	2R	3R	4R	5R	6R	7R
"Chinese Spring"–S. montanum	1R	2R	–	4R	5R	6R	–
"Fec 28"–"Petkus 10"	F	D	C	A	E	–	B

Source: From Miller, T. E. 1984. *Can. J. Genet. Cytol.* 26: 578–589. With permission.

[a] Partial homoeology to group 4 and group 7.

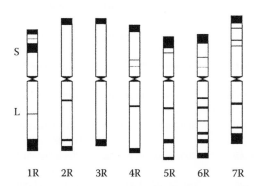

FIGURE 8.65 An idiogram of Imperial rye in Chinese Spring wheat addition line, represents from karyogram from Figure 8.60. (Redrawn from Mukai, Y., B. Friebe, and B. S. Gill. 1992. *Jpn. J. Genet.* 67: 71–83.)

TABLE 8.77
Alien Chromosome Substitutions and Transfer of Useful Traits to Cultigen from Their Allied Species and Genera

Recipient	Donor	Procedure	Gene Transferred	Authors
Beta vulgaris	*Beta patellaris*	Substitution	Nematode resistance	Salentijn et al. (1994)
Beta vulgaris	*Beta procumbens*	Substitution	Nematode resistance	Salentijn et al. (1994)
Brassica napus	*Brassica nigra*	Substitution	Blackleg	Chevre et al. (1996)
Brassica napus	*Sinapsis arvensis*	Substitution	Blackleg	Snowdon et al. (2000)
Cicer arietinum	*Cicer reticulatum*	Homoeologous pairing	Cyst nematode	Di Vito et al. (1996)
Hordeum vulgare	*Hordeum bulbosum*	Irradiation-induced	Powdery mildew	Pickering et al. (1995)
Lycopersicon esculentum	*Lycopersicon chilense*	Homoeologous pairing	Cucumber mosaic virus	Stamova and Chetelat (2000)
Lycopersicon esculentum	*Lycopersicon pennellii*	Homoeologous pairing	Total soluble solids	Ehed and Zamir (1994)
Lycopersicon esculentum	*Lycopersicon pennellii*	Homoeologous pairing	Potato aphids	Hartman and St. Clair (1999b)
Lycopersicon esculentum	*Lycopersicon pennellii*	Homoeologous pairing	Beat army worm	Hartman and St. Clair (1999a)
Lycopersicon esculentum	*Lycopersicon peruvianum*	Homoeologous pairing	Root-knot nematode	Doganlar et al. (1997)
Lycopersicon esculentum	*Solanum lycopersicoides*	Homoeologous pairing	Male-fertility	Chetelet et al. (1997)
Oryza sativa	*Oryza australeinsis*	Homoeologous pairing	Bacterial blight	Multani et al. (1994)
Oryza sativa	*Oryza australeinsis*	Homoeologous pairing	Brown plant hopper	Multani et al. (1994)
Oryza sativa	*Oryza brachyantha*	Homoeologous pairing	Bacterial blight	Brar and Khush (1997)
Oryza sativa	*Oryza latifolia*	Homoeologous pairing	Bacterial blight	Multani et al. (1994)
Oryza sativa	*Oryza latifolia*	Homoeologous pairing	Brown plant hopper	Multani et al. (1994)
Oryza sativa	*Oryza latifolia*	Homoeologous pairing	Whitebacked plant hopper	Multani et al. (1994)
Oryza sativa	*Oryza longistaminata*	Homoeologous pairing	Bacterial blight	Brar and Khush (1997)
Oryza sativa	*Oryza minuta*	Homoeologous pairing	Bacterial blight	Amante-Bordeos et al. (1997)
Oryza sativa	*Oryza minuta*	Homoeologous pairing	Rice blast	Amante-Bordeos et al. (1997)
Oryza sativa	*Oryza minuta*	Homoeologous pairing	Brown plant hopper	Amante-Bordeos et al. (1997)
Oryza sativa	*Oryza nivara*	Homoeologous pairing	Grassy stunt	Brar and Khush (1997)
Oryza sativa	*Oryza officinalis*	Homoeologous pairing	Bacterial blight	Brar and Khush (1997)
Oryza sativa	*Oryza officinalis*	Homoeologous pairing	Brown plant hopper	Brar and Khush (1997)
Oryza sativa	*Oryza officinalis*	Homoeologous pairing	Whitebacked plant hopper	Brar and Khush (1997)
Oryza sativa	*Oryza perennis*	Homoeologous pairing	Cytoplasmic male sterility	Dalmacio et al. (1995)
Triticum aestivum	*Secale cereale*	1B(1R) Substitutions and translocations	Powdery mildew resistance	Mettin et al. (1973); Zeller (1973)

(Continued)

TABLE 8.77 (*Continued*)
Alien Chromosome Substitutions and Transfer of Useful Traits to Cultigen from Their Allied Species and Genera

Recipient	Donor	Procedure	Gene Transferred	Authors
Triticum aestivum	*Secale cereale*	1B(1R) Substitutions and translocations	Stem and leaf rust resistance	Mettin et al. (1973); Zeller (1973)
Triticum aestivum	*Secale cereale*	1B(1R) Substitutions and translocations	Stripe rust resistance	Zeller and Hsam (1983)
Triticum aestivum	*Secale cereale*	1B(1R) Substitutions and translocations	Powdery mildew	Zeller and Hsam (1983)
Triticum aestivum	*Secale cereale*	1A.1R translocation	Green bug resistance	Zeller and Fuchs (1983)
Triticum aestivum	*Secale cereale*	1A.1R translocation	Powdery mildew resistance	Zeller and Fuchs (1983)
Triticum aestivum	*Secale cereale*	1A.1R translocation	Stem and leaf rust resistance	Zeller and Fuchs (1983)
Triticum aestivum	*Secale cereale*	4A.2R translocation[a]	Powdery mildew resistance	Driscoll and Jensen (1964)
Triticum aestivum	*Secale cereale*	4A.2R translocation[a]	Leaf rust resistance	Driscoll and Jensen (1964)
Triticum aestivum	*Secale cereale*	3A.3R translocation	Stem rust resistance	Stewart et al. (1968)
Triticum aestivum	*Secale cereale*	3A.3R translocation	Powdery mildew resistance	Lind (1982)
Triticum aestivum	*Secale cereale*	6BS.6RL translocation	Wheat mildew resistance	Lind (1982)
Triticum aestivum	*Secale cereale*	2BS.2RL translocation	Hessian fly resistance	Fribe et al. (1990a)
Triticum aestivum	*Secale cereale*	2BS.2RL translocation	Stripe rust resistance	Fribe et al. (1990a)
Triticum aestivum	*Secale cereale*	1BL.1RS translocations	Powdery mildew resistance	Friebe et al. (1989)
Triticum aestivum	*Secale cereale*	Irradiation-induced translocations	Hessian fly resistance	Friebe et al. (1991a)
Triticum aestivum	*Secale cereale*	4BL.5RL translocation	High copper tolerance	Schlegel et al. (1991)
Triticum aestivum	*Secale cereale*	2A.2R translocation	Green bug resistance	Friebe et al. (1995b)
Triticum aestivum	*Triticum monococcum*	Bridge cross	Stem rust resistance	McIntosh et al. (1984)
Triticum aestivum	*Agropyron elongatum*	Irradiation-induced translocation	Leaf rust resistance	Wienhues (1966); Sharma and Knott (1966)
Triticum aestivum	*Agropyron elongatum*	Irradiation-induced translocation	Leaf and stem resistance	Dvořák and Knott (1977)
Triticum aestivum	*Agropyron elongatum*	Homoeologous pairing	Leaf rust resistance	Sears (1973)
Triticum aestivum	*Agropyron elongatum*	Substitution	Wheat streak mosaic virus	Larson and Atkinson (1973)
Triticum aestivum	*Agropyron elongatum*	Substitution	Salt tolerance	Omielan et al. (1991)
Triticum aestivum	*Agropyron intermedium*	Translocations	Leaf rust resistance	Wienhues (1966)
Triticum aestivum	*Agropyron intermedium*	Irradiation-induced translocation	Leaf rust resistance	Friebe et al. (1993)

(Continued)

TABLE 8.77 (*Continued*)
Alien Chromosome Substitutions and Transfer of Useful Traits to Cultigen from Their Allied Species and Genera

Recipient	Donor	Procedure	Gene Transferred	Authors
Triticum aestivum	*Aegilops comosa*	Homoeologous pairing	Yellow rust resistance	Riley et al. (1966a); Chapman and Johnson (1968)
Triticum aestivum	*Aegilops longissimum*	Homoeologous pairing	Powdery mildew resistance	Ceoloni et al. (1988)
Triticum aestivum	*Aegilops sharonensis*	Irradiation-induced	Male fertility	King et al. (1991)
Triticum aestivum	*Aegilops speltoides*	Homoeologous pairing	Leaf rust resistance	Dvořák and Knott (1980); Dvořák (1977)
Triticum aestivum	*Aegilops umbellulata*	Irradiation-induced translocation	Leaf rust resistance	Sears (1956b)
Triticum aestivum	*Aegilops variabilis*	Homoeologous pairing	Powdery mildew resistance	Spetsov et al. (1997)
Triticum aestivum	*Aegilops ventricosa*	Homoeologous pairing	Cereal cyst nematode	Delibes et al. (1993)
Triticum aestivum	*Aegilops uniaristata*	Homoeologous pairing	Aluminum tolerance	Miller et al. (1997)
Triticum aestivum	*Thinopyrum distichum*	Translocation	Leaf rust resistance	Morais et al. (1988)
Triticum aestivum	*Thinopyrum intermedium*	Translocation	Wheat streak mosaic virus	Chen et al. (1998)
Triticum aestivum	*Thinopyrum intermedium*	Addition	Barley yellow dwarf	Larkin et al. (1995)
Triticum aestivum	*Thinopyrum intermedium*	Addition	Rust resistance	Larkin et al. (1995)
Triticum aestivum	*Thinopyrum intermedium*	Substitution	Barley yellow dwarf	Sharma et al. (1995)
Triticum aestivum	*Triticum araraticum*	Homoeologous pairing	Stem rust resistance	Dyck (1992)
Triticum aestivum	*Triticum monococcum*	Homoeologous pairing	Powdery mildew resistance	Shi et al. (1998)
Triticum aestivum	*Triticum turgidum*	Homoeologous pairing	Leaf rust resistance	Dyck (1984)
Triticum aestivum	*Haynaldia vilosa*	Substitution	Powdery mildew resistance	Liu et al. (1988)
Triticum aestivum	*Agropyron intermedium* × *Triticum monococcum*	Substitutions and translocations	Wheat streak mosaic virus / Green bug resistance	Friebe et al. (1991b) / Friebe et al. (1991b)
Triticum aestivum	*Aegilops speltoides* × *Triticum monococcum*	Homoeologous pairing	Leaf rust resistance	Kerber and Dyck (1990)
Avena sativa	*Avena barbata*	Irradiation-induced translocation	Mildew resistance	Aung and Thomas (1978)

(Continued)

TABLE 8.77 (*Continued*)
Alien Chromosome Substitutions and Transfer of Useful Traits to Cultigen from Their Allied Species and Genera

Recipient	Donor	Procedure	Gene Transferred	Authors
Avena sativa	*Avena abyssinica* ×			
	Avena strigosa	Irradiation-induced translocations	Crown rust resistance	Sharma and Forsberg (1977)
Trifolium repens	*Trifolium nigrescens*	Polyploidization	Clover cyst nematode	Hussain et al. (1997)

Note: *Agropyron elongatum*, *Elytrigia elongata*, *Thinopyrum elongata*, and *Lophopyrum elongatum* are one species.
[a] 4B.5R (B. Friebe, personal communication).

b. Chromosome 2R

It is the longest chromosome in the rye chromosome complement and both arms have prominent heterochromatic bands. Chromosome 2R substitutes reasonably well for wheat chromosomes 2B and 2D. Interchanges involving 2RL.2AS, 2BL.2RS, and 2BS.2RL (Figure 8.66) are rather common. However, interchange 4A.2R and location of homoeoloci *Gli-2* on 2RS, 6AS, 6BS, and 6DS suggest small structural differences in 2R. Based on the location of gene *Gli-2* on chromosome 6R^m (m = *S. montanum*), Shewry et al. (1985) postulated the presence of an interchange between 2R and 6R^m,

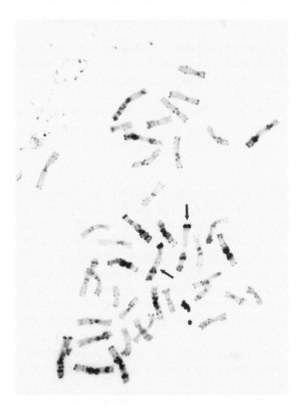

FIGURE 8.66 Giemsa C-banded mitotic metaphase of Hessian fly-resistant wheat–rye translocation line, 2BS.2RL (arrows). (From Friebe, B. et al. 1990a. *Theor. Appl. Genet.* 79: 385–389. With permission.)

probably present also in wheat. Chromosome 2R contains genes for resistance to powdery mildew, Hessian fly, and leaf rust.

c. Chromosome 3R

It contains a terminal heterochromatic prominent band on both arms, substitutes for wheat chromosomes 3A, 3D. The 3RS (short arm) shows homoeology with 3AS, 3BS, and 3DS, but 3RL (long arm) contains little homoeology with 3AL, 3BL, and 3DL. Furthermore, the distal region of 3RL has homoeology with 6AL, 6BL, and 6DL. Chromosome 3R contains genes for resistance to wheat mildew.

d. Chromosome 4R

Chromosome 4R carries a prominent telomeric heterochromatic band on the short arm and a medium size telomeric band on the long arm (Figure 8.63) and substitutes for wheat chromosomes 4A, 4B, and 7B (Miller, 1984). Chromosome pairing results suggest homoeology of 4RS to 4BS and 4DS and no affinity with 4A, but to some extent is homoeologous to 4AL; chromosome 4RL contains some homoeology to 6AS, 6BS, and 6DS and also wheat group 7 chromosomes, which is in agreement with the detection of the 4RL.7RS translocation (Koller and Zeller, 1976). Thus, chromosome 4R includes chromosome segments of homoeologous groups 4 and 7. According to Liu et al. (1992), present 4R is composed of 4RL and 7RS.

e. Chromosome 5R

This chromosome contains a large terminal heterochromatic band on the short arm. Genetically, chromosome 5R is identified by the "hairy neck" gene *HP* (controls peduncle pubescence) and substitutes for wheat chromosomes 5A, 5B, and 5D. As far as homoeologous relations of arms are concerned, 5RS is homoeologous to 5AS, 5BS, and 5DS (chromosome pairing and genetic data). However, substitution 5R (4A) and interchanges 4A/5R suggest that 5R has some homoeology with wheat groups 4.

f. Chromosome 6R

It possesses a prominent telomeric heterochromatic band on the short arm and substitutes for wheat chromosomes 6A, 6B and 6D. Chromosome 6RL shows homoeology with 6AL, 6BL, and 6DL and some homoeology to 3AL, 3BL, and 3DL (genetic data). This relation may be due to a reciprocal translocation involving 3RL and 6RL in rye relative to wheat. Pairing between 6RL and the long arm of wheat group 7 chromosomes suggests that an additional interchange involving 6RL and 7RL could have occurred during the evolution of rye (Naranjo and Fernández-Rueda, 1991).

g. Chromosome 7R

This chromosome carries telomeric heterochromatic bands on both arms and shows partial homoeology with wheat groups 4 and 7 chromosomes. The 7RS does not pair with 7AS and 7DS but pairs with 5BL, 5DL and 7BS (the terminal segment of 7S is homoeologous to the long arm of group 5 chromosomes and the 7RS.5RL translocation in rye relative to wheat). The 7RL shows homoeology to 7AL, 7BL, and 7DL as is demonstrated by genetic and RFLP data and shows affinity with 2AS, 2BS, and 2DS. Thus, chromosome 7 is a modified chromosome containing several rearrangements with respect to wheat chromosomes. Liu et al. (1992) recorded that 7R is very much modified due to interchanges; 7RS constitutes segments of 7RS (the region proximal to kinetochore), 4RL, and telomere from 5RL.

Incomplete homoeology of 4R, 5R, and 7R to the corresponding homoeologous D-genome chromosomes is due to the occurrence of reciprocal translocations in these chromosomes (Taketa et al., 1991).

In addition to wheat–rye addition and substitutions, attempts are being made to produce addition and substitution lines from more distantly-related wild species of wheats. Wheat–*Agropyron* substitution lines have been produced in order to transfer disease resistance genes from *Agropyron* to common wheat (Knott, 1964; Wienhues, 1966, 1971, 1973; Knott et al., 1977; Dvořák, 1980) and emmer wheat (Tsujimoto et al., 1984). The scientific botanical name of *A. elongatum* has a confused history. This particular species is also named as *Elytrigia elongata, Thinopyrum elongatum*, and *Lophopyrum elongatum*.

Examples of wheat–*Aegilops* substitutions are a few and no systematic effort has been made to produce all possible substitution lines. However, two classical examples of gene introgression from *Aegilops* species to common wheat need to be mentioned. Sears (1956b) transferred a small segment of *Ae. umbellulata* chromosome conferring resistance to leaf rust to wheat cv. Chinese Spring. The translocation was a substitution with 2III recovered following irradiation. Riley et al. (1968) transferred a gene for yellow rust resistance from *Ae. comosa* to wheat cv. Chinese Spring through genetically-induced homoeologous recombination.

Miller et al. (1982a) found an *Ae. sharonensis* chromosome, which substitutes for homoeologous group 4 (ABD) of Chinese Spring.

8.2.2.10.3   Practical Application of Substitution Lines

Knowledge of substitution lines has enhanced our understanding of homoeologous relationships among three wheat genomes and allied genera such as *Secale, Agropyron,* and *Aegilops*. Useful traits for economic importance have been introgressed into several major crops (Table 8.77).

Transfer of useful genetic traits from alien species to cultigens is not limited to only wheat but it is being explored for several diploid and polyploid species (Table 8.77). Jena and Khush (1990) transferred genes for resistance to insects and diseases from *Oryza officinalis* to *O. sativa*. Singh et al. (1990, 1993) produced for the first time intersubgeneric hybrids with fertile plants from a cross of soybean, *Glycine max*, and a wild perennial species *G. tomentella*. It took a decade to isolate monosomic alien addition lines (Singh et al., 1998) and a gene for resistance to soybean cyst nematode has been introgressed from *Glycine tomentella* to the soybean (Riggs et al., 1998).

The wider crosses not only take patience but determination, commitment, and, frequently, a good deal of skill and ingenuity. Not surprisingly, that the easy materials are more likely to be used is very valuable and encouraging for cytogeneticists involved in the wide hybridization (Harlan, 1976).

9 Genome Analysis

9.1 INTRODUCTION

Understanding the genomic relationships among species is important to systematists, evolutionary biologists, cytogeneticists, molecular biologists, and plant breeders. The taxonomic nomenclature of species and their evolutionary relationships can be refined by cytogenetic evidences such as chromosome morphology (karyotypes), crossability, hybrid viability, meiotic chromosome pairing, and molecular (isozymes, RFLP, RAPD, chloroplast and mitochondrial DNA, and other methods) approaches. Phylogenetic relationships among species can be understood more precisely by a multidisciplinary approach rather than through reliance on a single technique (Jauhar, 1990, 1996; Jauhar et al., 1991; Singh et al., 1992a). For example, van der Maesen (1986) combined the genus *Atylosia* with the genus *Cajanus* based on morphological, cytological, and chemo-taxonomic data. However, the chromosome pairing and molecular methods to establish genomic relationships among species will be the main theme of this chapter.

9.2 CLASSICAL TAXONOMY AND GENOME DESIGNATIONS

A taxonomic nomenclature of taxa based on morphological traits is the traditional and valuable starting point for cytogeneticists as well as for molecular geneticists. The genus *Glycine* Willd. has been selected as an example to elucidate the importance of taxonomy.

Based on classical taxonomy, the genus *Glycine* has been divided into two subgenera, *Glycine* and *Soja* (Moench) F. J. Hermann. Subgenus *Glycine* currently consists of 26 wild perennial species, and all the diploid ($2n = 40$) species are indigenous to Australia and associated areas (Table 9.1). *Glycine tabacina* (Labill.) Benth., with $2n = 40$, 80 chromosomes, has been found in Australia (New South Wales, Queensland, Australian Capital Territory), Taiwan, South Pacific Islands (New Caledonia, Fiji, Tonga, Vanuatu), and West Central Pacific Islands (Mariana, Ryukyu). All accessions of *G. tabacina* collected outside of Australia are tetraploid, and within Australia, tetraploid *G. tabacina* predominates with occasional diploid. *Glycine tomentella* Hayata has four cytotypes ($2n = 38, 40, 78, 80$). Aneuploidy ($2n = 38, 78$) is found in New South Wales and in the adjoining regions of Queensland. Diploid ($2n = 40$) and tetraploid ($2n = 80$) cytotypes have been collected from various islands off Queensland, Australia, and Papua New Guinea. *Glycine tomentella* accessions from the Philippines, Indonesia, and Taiwan are tetraploid. *Glycine dolichocarpa* contains $2n = 80$ chromosomes and is distributed in Taiwan and morphologically it looks similar to 80-chromosome *G. tomentella*. *Glycine hirticaulis* has extremely narrow distribution in Australia and contains $2n = 40$ and 80 diploids and tetraploids, respectively.

Subgenus *Soja* contains *G. max* (L.) Merr., a cultigen, and its wild annual progenitor *G. soja* Sieb. and Zucc., both of which carry $2n = 40$ chromosomes. *Glycine soja* is found in Peoples' Republic of China, Russia, Korea, Japan, and Taiwan. Table 9.1 shows species, $2n$ chromosome number, nuclear and plastome genomes, and geographical distribution of *Glycine* species.

Classical taxonomy has played a major role in the identification and nomenclature of new species in the genus *Glycine*. *Glycine clandestina* ($2n = 40$) has been observed to be morphologically a highly variable species (Hermann, 1962). The curved and short pod forms have been segregated from *G. clandestina snsu lato* as *G. cyrtoloba* (Tindale, 1984) and *Glycine microphylla* (Tindale, 1986), respectively. The separation of both morphological forms is logical because both forms differ genomically from other species (Singh and Hymowitz, 1985b,c; Singh et al., 1988). Recently, two

TABLE 9.1
Taxonomy of the Genus *Glycine* Willd.[a]

Species	2*n*	N	C	Distribution
		Genome Symbol[b]		
Subgenus Glycine				
G. albicans Tind. and Craven	40	I	A	Australia
G. aphyonota B. Pfeil	40	Unknown	Unknown	Australia
G. arenaria Tindale	40	H	A	Australia
G. argyrea Tindale	40	A_2	A	Australia
G. canescens F. J. Hermann.	40	A	A	Australia
G. clandestina Wendl.	40	A_1	A	Australia
G. curvata Tindale	40	C_1	C	Australia
G. cyrtoloba Tindale	40	C	C	Australia
G. falcata Benth.	40	F	A	Australia
G. hirticaulis Tindale and Craven	40	H_1	A	Australia
	80	Unknown	Unknown	Australia
G. lactovirens Tindale and Craven	40	I_1	A	Australia
G. latifolia (Benth.) Newell and Hymowitz	40	B_1	B	Australia
G. latrobeana (Meissn.) Benth.	40	A_3	A	Australia
G. microphylla (Benth.) Tindale	40	B	B	Australia
G. montis-douglas B.E. Pfeil and Craven	40	Unknown	Unknown	Australia
G. peratosa B. E. Pfeil and Tindale	40	A_5	A	Australia
G. pescadrensis Hayata	80	AB_1	A	Australia, Taiwan, Japan
G. pindanica Tindale and Craven	40	H_2	A	Australia
G. pullenii B. Pfeil, Tindale, and Craven	40	Unknown	Unknown	Australia
G. rubiginosa Tindale and B. E. Pfeil	40	Unknown	Unknown	Australia
G. stenophita B. Pfeil and Tindale	40	B_3	B′	Australia
G. dolichocarpa Tateishi and Ohashi	80	Unknown	Unknown	South East Coast of Taiwan
G. tabacina (Labill.) Benth.	40	B_2	B	Australia
	80	(Complex)	Unknown	Australia, West Central and South Pacific Islands
G. tomentella Hayata	38	E	A	Australia, Papua New Guinea
	40	D, D_1, D_2, D_3	A	Australia
	78	(Complex)	Unknown	Australia, Papua New Guinea
	80	(Complex)	Unknown	Australia, Papua New Guinea, Philippines, Taiwan
Subgenus Soja (Moench) F. J. Herm.				
G. soja Sieb. and Zucc.	40	G	G	China, Russia, Taiwan, Japan, Korea
G. max (L.) Merr.	40	G_1	G_1	Cultigen

[a] From Kollipara, K. P., Singh, R. J., and Hymowitz, T. 1997. *Genome* 40: 57–68. With permission; Doyle, J. J. et al. 2000. *Syst. Bot.* 25: 437–448. With permission.

[b] N, nuclear; C, chloroplast.

more morphotypes have been segregated from *G. clandestina* and have been identified taxonomically as *G. rubiginosa* and *G. peratosa* (Pfeil et al., 2001).

The genomes of diploid species are assigned capital letter symbols according to the degree of chromosome homology between species in F_1 hybrids (Kihara and Lilienfeld, 1932). Similar letter symbols are designated for species whose interspecific F_1 hybrids show normal chromosome pairing. Minor chromosome differentiation is indicated by placing subscripts after the letter. Highly differentiated species are designated by different letter symbols because their hybrids exhibit highly irregular chromosome pairing (see *Triticum*—Kihara and Nishiyama, 1930; *Brassica*—U, 1935; *Gossypium*—Beasley, 1940; *Oryza*—Nezu et al., 1960; *Glycine*—Singh and Hymowitz, 1985b).

Genome designations of the *Glycine* species have been made arbitrarily, assuming the diploid chromosome number to be $2n = 40$. Beginning with *G. canescens* (A genome), the genomes of the other species were designated based on chromosome pairing, hybrid seedling lethality, and hybrid seed inviability in interspecific hybrids. The same genome symbol designations were assigned to species possessing similar genomes but the genomes were differentiated by placing a subscript after the letter such as A_1 (*G. clandestina*), A_2 (*G. argyrea*), and A_3 (*G. latrobeana*); B (*G. microphylla*), B_1 (*G. latifolia*), B_2 (*G. tabacina*), and B_3 (*G. stenophita*). Genome symbols were assigned to species as soon as their relationships were established: C = *G. cyrtoloba*; C_1 = *G. curvata*; D = *G. tomentella*; E = *G. tomentella*, $2n = 38$; F = *G. falcata* (Singh et al., 1988, 1992a). It should be pointed out that cytogenetic studies, isozyme banding patterns, and molecular investigations have demonstrated at least three distinct genomic forms in diploid *G. tomentella* (Singh et al., 1988, 1992a; Brown, 1990; Doyle et al., 1990a). The genomic designations are broad, functional, and primarily based on chromosome pairing rather than evolutionary definition. The frequency of bivalent formation is the primary direct measure of chromosome homology, whereas crossability (generating fertile hybrids) is an indirect measure of chromosome homology. Genomic groups typically correlate with evolutionary groups; they do not reflect speciation *per se*.

9.3 GENOMIC RELATIONSHIPS AMONG DIPLOID SPECIES

9.3.1 CROSSING AFFINITY

Crossability rate is an excellent indirect measure for estimating the degree of genomic relationship between parental species. Interspecific crosses involving parental species with similar genomes usually set normal pods and seeds, while in crosses between genomically dissimilar species seed abortion is common or the hybrid is sterile. Certain hybrid combinations do not set seed and sometimes crosses are successful only in one direction and often are genotype-dependent (*Gossypium*—Skovsted, 1937; *Hibiscus*—Menzel and Martin, 1970; *Vigna*—Chen et al., 1983; *Cajanus*—Dundas et al., 1987; *Capsicum*—Pickersgill, 1988; *Glycine*—Singh et al., 1988; *Cuphea*—Ray et al., 1989; *Medicago*—McCoy and Bingham, 1988).

In the genus *Glycine*, intragenomic crosses set mature pods while pod abortion is common in intergenomic hybrids (Table 9.2). In *Avena*, Baum and Fedak (1985) reported no internal barriers to gene flow among the diploid species of Strigosa group (*A. hirtula*, A. *wiestii*, A. *strigosa*, and A. *atlantica*).

The failure of intragenomic crosses can be attributed to a wide range of phenomena (e.g., genetically determined species isolation mechanism, pollen tube dysfunction on alien stigmas, genetic incompatibilities that produce physiologically unfit embryos) but is often correlated with long periods of reproductive isolation. Menzel and Martin (1970) attempted to cross *Hibiscus cannabinus* ($2n = 36$) and *H. asper* ($2n = 36$) and were not successful (*H. cannabinus* × *H. asper* 158 crosses; reciprocal, 130 crosses). Both species possess a similar genome, but are isolated geographically. *Hibiscus asper* is a native of West Africa, whereas *H. cannabinus* is found in East Africa. However, reproductive isolation is not always a predictor of incompatibility. *Oryza sativa* ($2n = 24$), a native of Asia, and *O. glaberrima* ($2n = 24$), an inhabitant of Africa, are isolated geographically, but hybridize readily and produce vigorous hybrids with complete meiotic chromosome pairing (12II). However, the hybrids are sterile (Nezu et al., 1960; Bouharmont et al., 1985).

TABLE 9.2

Pod Set in F₁ Hybrids between Wild Perennial *Glycine* Species

Cross	Genome	Pod Set/Total Florets	% Pod Set
G. argyrea × *G. canescens*	$A_2 \times A$	6/26	23.3
G. microphylla × *G. latifolia*	$B \times B_1$	3/40	7.5
G. latifolia × *G. tabacina*	$B_1 \times B_2$	18/147	12.2
G. canescens × *G. clandestina*	$A_1 \times A_1$	9 + 1[a]/86	10.5 + 1.2[a]
G. canescens × *G. microphylla*	$A \times B$	1[a]/20	5.0[a]
G. canescens × *G. latifolia*	$A \times B_1$	14[a]/462	4.9[a]
G. canescens × *G. tabacina*	$A \times B_2$	0/395	0.0
G. canescens × *G. tomentella*	$A \times D$	21/150	14.0
G. latifolia × *G. tomentella*	$B_1 \times D$	8[a]/280	2.9[a]

Source: From Singh, R. J., K. P. Kollipara, and T. Hymowitz. 1988. *Genome* 30: 166–176. With permission.

[a] Aborted pods.

9.3.2 Chromosome Pairing

The chromosome pairing (chiasmata frequency) method "Analysartoren-Methode" established by Kihara and his associates (Kihara 1924, 1930; Kihara and Nishiyama, 1930; Kihara and Lilienfeld, 1932; Lilienfeld, 1951; Kihara, 1963) to ascertain the genomes of hexaploid wheats ($2n = 6x = 42$) and their relatives, originally founded by Rosenberg (1909), is still a reliable technique (Kimber and Feldman, 1987; Jauhar and Crane, 1989; Wang, 1989), despite a few objections (de Wet and Harlan, 1972; Baum et al., 1987).

A prerequisite for Kihara's method is to ascertain chromosome pairing relationships among all available diploid species in order to establish diploid analyzers. Based on this principle, genomic relationships among diploid species have been assessed for several important genera such as *Arachis* (Smartt et al., 1978; Wynne and Halward, 1989), *Avena* (Rajhathy and Thomas, 1974; Ladizinsky, 1974; Nishiyama et al., 1989), *Brassica* (U, 1935; Röbbelen, 1960; Prakash and Hinata, 1980), *Bromus* (Armstrong, 1981), *Cajanus* (Subrahmanyam et al., 1986), *Capsicum* (Egawa and Tanaka, 1984; Pickersgill, 1988), *Dactylis* (Lumaret, 1988), *Eleusine* (Chennaveeraiah and Hiremath, 1974), *Glycine* (Singh and Hymowitz, 1985b, c; Singh et al., 1988, 1992a), *Gossypium* (Skovsted, 1937; Beasley, 1940; Stephens, 1947, 1950; Phillips, 1966; Endrizzi et al., 1985; Percival and Kohel, 1991), *Helianthus* (Chandler et al., 1986), *Hibiscus* (Menzel and Martin, 1970), *Hordeum* (Bothmer et al., 1986), *Nicotiana* (Goodspeed, 1954; Smith, 1968), *Oryza* (Nezu et al., 1960; Nayar, 1973), *Paspalum* (Burson, 1981a,b), *Pennisetum* (Jauhar, 1981; Pantulu and Rao, 1982), *Secale* (Khush, 1962; Stutz, 1972; Singh and Röbbelen, 1977) and *Triticum* and its allied genera (Lilienfeld, 1951; Dewey, 1984; Kimber and Feldman, 1987).

The degree of chromosome pairing in interspecific hybrids provides an important cytogenetic context for inferring phylogenetic relationships among species, enhances our understanding of evolution of the genus, and provides information about the ancestral species. Smith (1968) recognized five categories of chromosome pairing in F₁ hybrids:

- Complete or almost complete pairing
- "Drosera scheme" pairing
- High, variable pairing
- Low, variable pairing
- Minimal pairing

In category one, intragenomic hybrids, the pairing of chromosomes is by definition complete in 75%–100% of the sporocytes, although, occasionally, species will differ by chromosomal interchanges. The classical examples of chromosomal interchanges differentiating species are the species of *Oenothera* (Cleland, 1962), *Datura* (Avery et al., 1959), and *Secale* (Schiemann and Nürnberg-Krüger, 1952; Riley, 1955). Sometimes, genomically similar species differ by inversions (Singh et al., 1988). "*Drosera scheme*" pairing ($2x$II + $1x$I) occurs in tetraploid ($4x$) × diploid ($2x$) crosses and has been named after chromosome pairing in *Drosera longifolia* ($2n = 4x = 40$) × *Dosera rotundifolia* ($2n = 2x = 20$) (Rosenberg, 1909). High and low variable chromosome pairing occurs in intergenomic hybrids where some but not all chromosomes retain some level of genetic and structural homology. The degree of pairing between chromosomes depends on genomic affinity of the parental species. Minimal chromosome pairing (all chromosomes remain univalent in over 50% of the sporocytes) occurs in F_1 hybrids where two species have diverged sufficiently to obscure any ancestral chromosome homology.

Several numerical methods have been developed to interpret chromosome pairing data (Gaul, 1959; Driscoll et al., 1979, 1980; Kimber et al., 1981; Alonso and Kimber, 1981; Kimber and Alonso, 1981). Menzel and Martin (1970) used a genome affinity index (GAI) for comparing the degree of homology among parental genomes. The mean number of "bivalent-equivalents" (II-equiv.) is divided by the base chromosome number. Since base chromosome numbers have been debated for a majority of the crop species, it is suggested to use, for diploid species, gametic chromosome numbers instead of a base chromosome number of parents to compute GAI. For example, a base chromosome $x = 10$ has been suggested for the genus *Glycine* ($2n = 40$) (Darlington and Wylie, 1955). However, *Glycine* species with $2n = 2x = 20$ have not been identified thus far (Singh and Hymowitz, 1985b; Kumar and Hymowitz, 1989). A bivalent and trivalent is each equal to one II-equiv., but a quadrivalent with chiasmata in at least three arms is counted as two II-equivalent.

9.3.2.1 Intragenomic Chromosome Pairing

Generally, species with similar genomes exhibit complete or almost complete chromosome pairing in their hybrids (Figure 9.1a). Sometimes, species differ by chromosomal interchanges or by paracentric inversions (Figure 9.1b).

In the genus *Triticum*, A-genome species ($2n = 2x = 14$), *T. monococcum*, and *T. urartu* are found in the wild in the same geographical area. They exhibit seven bivalents at metaphase-I and mainly the Drosera scheme (7II + 7I) of pairing in the *T. turgidum* ($2n = 4x = 28$) × *T. urartu* cross (Johnson, 1975) suggesting that *T. turgidum* is an allotetraploid carrying an A-genome derived either from *T. monococcum* or *T. urartu*. Furthermore, *T. monococcum* or *T. urartu* possess identical karyotypes (Giorgi and Bozzini, 1969), and insignificantly different nuclear DNA content (Furuta et al., 1986), suggesting that they carry similar genomes.

Kerby and Kuspira (1987) identified five diploid ($2n = 2x = 14$) species in the section *Sitopsis*: *Aegilops speltoides* (S), *Aegilops sheronensis* (S^{sh}), *Aegilops longissima* ($S^l S^l$), *Aegilops bicornis* (S^b), and *Aegilops searsii* (S^s). Hybrids between *Aegilops longissima* and *Aegilops sheronensis* display univalents (0–2), bivalents (5–7), trivalents (0–1), and quadrivalents (0–1) and almost normal pollen (81.3%–87.1%) and seed (81.0%–95.0%) fertility. This suggests that both species carry similar genomes and differ only by a reciprocal translocation (depends on the type of quadrivalents-alternate, ring). Although *Aegilops bicornis* is given a different superscript genome symbol (S^b), it shows almost the same meiotic behavior, pollen, and seed set in crosses with *Aegilops sheronensis* and *Aegilops longissima* (Feldman et al., 1979).

Hybrids *Aegilops speltoides* (S) × *Aegilops bicornis* (S^b) and *Aegilops sheronensis* (S^{sh}) × *Aegilops speltoides* (S) showed 3–7 bivalents and set no seed. Furthermore, crosses *Aegilops speltoides* × *Aegilops longissima* and *Aegilops searsii* × *Aegilops longissima* showed a quadrivalent configuration together with univalents, bivalents, and trivalents (Feldman et al., 1979). This indicates a considerable amount of diversity among the B-genome species of the genus *Aegilops*.

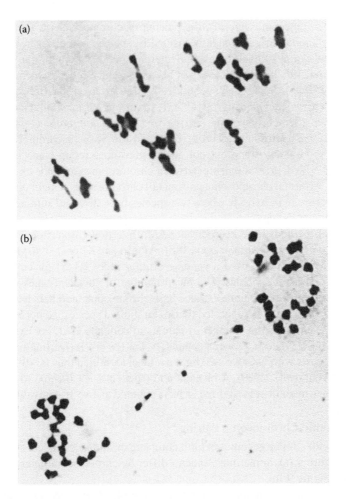

FIGURE 9.1 Meiosis in *Glycine latifolia* (B_1B_1) × *Glycine microphylla* (BB): (a) metaphase-I showing 20 bivalents. (From Singh, unpublished results). (b) Anaphase-I showing a chromatin bridge and an acentric fragment in interspecific hybrid of *Glycine clandestina* × *Glycine canescens*. (Figure 8b from Singh, R. J., K. P. Kollipara, and T. Hymowitz. 1988. *Genome* 30: 166–176. With permission.)

In the genus *Glycine*, all F_1 plants from crosses among A and B genome species display 20 bivalents in a majority of sporocytes. However, *G. microphylla* (B) × *G. tabacina* (B_2) F_1's showed a 0.83 GAI value, suggesting some differences between genomes of B and B_2 (Table 9.3). A chromatin bridge and an acentric fragment at anaphase-I was clearly seen in A × A_1, B × B_1, B_1 × B_2, and B × B_2 hybrids.

Based on classical taxonomy, *G. soja* and *G. max* are different species (Hermann, 1962). Both species carry $2n = 40$ chromosomes, hybridize readily, produce viable, vigorous, and fertile hybrids, and sometimes differ by a reciprocal translocation (Palmer et al., 1987; Singh and Hymowitz, 1988) or by a paracentric inversion (Ahmad et al., 1977). Therefore, *G. soja* and *G. max* possess similar genomes (G).

Another approach for determining the chromosome homology between species is to analyze the chromosome pairing of F_1 hybrids at pachynema. This technique, however, is feasible only for species with analyzable pachynema chromosomes. By utilizing pachynema chromosomes, Singh and Hymowitz (1988) identified 20 pachynema bivalents of *G. max* × *G. soja* F_1 hybrids. Heteromorphic regions were observed only in chromosomes 6 and 11, while pairing was completely normal all along the length of the long and short arms of the remaining chromosome pairs.

TABLE 9.3

Interspecific Hybrids with Complete or Almost Complete Chromosome Pairing at Diakinesis and Metaphase-I

Hybrids (Genomes)	Average Chromosome Pairing (Range)			GAI	Authority
	I	II	III + IV		
1. *Glycine* (n = 20)					
Argyrea × *canescens* (A$_2$ × A)	0.8(0–4)	19.6(18–20)		0.99	Singh et al. (1988)
Argyrea × *clandestina* (A$_2$ × A$_1$)	0.1(0–2)	19.9(19–20)		1.0	Singh et al. (1988)
Microphylla × *tabacina* (B × B$_2$)	6.8(0–12)	16.6(14–20)		0.83	Singh et al. (1988)
Latifolia × *microphylla* (B$_1$ × B)	2.0(0–4)	19.0(18–20)		0.95	Singh and Hymowitz (1985c)
Latifolia × *tabacina* (B$_1$ × B$_2$)	0.6(0–4)	19.7(18–20)		0.99	Singh and Hymowitz (1985c)
2. *Gossypium* (n = 13)					
Arboreum × *herbaceum* (A$_2$ × A$_1$)	0.2	11.1	0.9	0.99	Gerstel (1953)
Anomalum × *triphyllum* (B$_1$ × B$_2$)	0.10	12.90		1.0	Douwes (1953)
Raimondii × *gossypioides* (D$_5$ × D$_6$)	1.14	12.43		0.96	Menzel and Brown (1955)
Stocksii × *areysianum* (E$_1$ × E$_2$)	0.16	12.20		0.97	Douwes (1953)
Somalense × *areysianum* (E$_2$ × E$_3$)	0.20	12.90		0.99	Douwes (1953)
3. *Oryza* (n = 12)					
Sativa × *perennis* (A × A)	0.0	12.0		1.0	Nezu et al. (1960)
Sativa × *glaberrima* (A × A^g)	0.0	12.0		1.0	Nezu et al. (1960)
Glaberrima × *perennis* (A^g × A)	0.0	12.0		1.0	Nezu et al. (1960)
Officinalis × *eichingeri* (C × C)	0.0	12.0		1.0	Ogawa and Katayama (1973)
4. *Hordeum* (n = 7)					
Publiflorum × *comosum* (H × H)	0.24(0–2)	6.81(5–7)	0.05(0–1)	0.98	Bothmer et al. (1986)
Publiflorum × *patagonicum* (H × H)	0.19(0–4)	6.90(5–7)		0.99	Bothmer et al. (1986)
Comosum × *euclaston* (H × H)	0.40(0–6)	6.76(4–7)		0.97	Bothmer et al. (1986)
Patagonicum × *flexuosum* (H × H)	0.79(0–4)	6.60(5–7)		0.94	Bothmer et al. (1986)
Patagonicum × *stenostachys* (H × H)	0.43(0–4)	6.79(5–7)		0.97	Bothmer et al. (1986)

A close genomic relationship between a cultigen and its wild progenitor suggests that they belong to one biological species or primary gene pool, the GP-1 of Harlan and de Wet (1971). Diploid (2n = 26) cultivated cotton (*G. herbaceum*-A$_1$, *G. arboreum*-A$_1$) are closely related to the B genome (*G. anomalum*-B$_1$, *G. typhyllum*-B$_2$, and *G. capitis-viridis*-B$_3$) wild species (Endrizzi et al., 1985). Hybrids A$_1$ × B$_1$ and A$_2$ × B$_1$ exhibit almost complete chromosome pairing (Skovsted, 1937). However, more recent phylogenetic studies suggest that the two *Gossypium* A genome species are more closely related to the F genome species, *Gossypium longicalyx* (Cronn et al., 2002). Cultivated rice (*O. sativa*, 2n = 24) and its wild annual progenitor *O. nivara* (2n = 24) are cross-compatible and show essentially normal chromosome pairing in F$_1$, indicating that *O. sativa* and *O. nivara* have the same genomic (A) constitution (Dolores et al., 1979). Dundas et al. (1987) recorded almost complete pairing involving *Cajanus cajan* (2n = 22) with 2n = 22 chromosome *C. acutifolius* and *C. confertiflorus*. In the classical study of *Nicotiana*, 90% of the intrasectional hybrid combinations displayed complete or almost complete chromosome pairing (Goodspeed, 1954). In *Hordeum*, North and South American species have the same (H) genome designation, although complete genome homology (GAI, range = 0.52–0.82) is lacking (Bothmer et al., 1986). For reason of clarity, I suggest to designate genome symbol H$_1$ to North American species and H$_2$ to South American species.

In general chloroplast DNA variation results support isoenzyme, molecular, cytological, and crossing data (Doebley et al., 1992). This suggests that evolutionary divergence within a group occurred because of geographical isolation and accumulation of chromosomal structural changes during the time (Stebbins, 1950; Grant, 1971).

A wrong conclusion can be reached if genomic relationship is based on fertility or sterility of F_1 hybrids without analyzing the chromosome pairing. One might expect to obtain fertile F_1 hybrids between two genomically similar species, but find the hybrids to be sterile. This could be due to cryptic structural hybridity, complementary lethal genes, or differentiation in genes and chromosomal structures (Stebbins, 1950). A classic example of cryptic structural hybridity is the *Primula verticillata* (2n = 18) × *Primula floribunda* (2n = 18) hybrid. Chromosome pairing was almost normal (nine loosely associated bivalents), but the hybrid was completely sterile (Newton and Pellew, 1929). In *Oryza*, intragenomic hybrids between the A (Nezu et al., 1960) and C genome species (Ogawa and Katayama, 1973) exhibit cryptic structural hybridity. Hybrids of *Avena canariensis* (AcAc) × *A. damascena* (AdAd) showed seven normal bivalents but almost complete sterility after selfing (Leggett, 1984; Nishiyama et al., 1989). Likewise F_1's from a *Triticum monococcum* × *Triticum urartu* cross-exhibited complete chromosome pairing (6.97II), yet the hybrids were completely self-sterile (Dhaliwal and Johnson, 1982). In *Dactylis*, meiosis in the diploid hybrids was so regular that it was assumed that reduction in hybrid fertility was caused by genic or cryptic structural hybridity (Lumaret, 1988).

One expects to observe normal chromosome pairing in morphologically similar species. However, there are some exceptions. For example, aneudiploid (2n = 38) and diploid (2n = 40) *G. tomentella* are morphologically indistinguishable, yet at metaphase-I, a majority of the chromosomes remain as univalents (GAI 0.31), while bivalents are loosely connected and rod shaped (Singh et al., 1988). Similarly, *Paspalum notatum* (2n = 20) × *P. vaginatum* (2n = 20) F_1's exhibited 20 I in 57% of the sporocytes, 1II + 18I (weakly synapsed) in 30% of the sporocytes. These species belong to different taxonomic groups but have similar morphological features (Burson, 1981a). Likewise, *Cuphea procumbens* (2n = 18) and *Cuphea crassiflora* (2n = 24) are morphologically similar, but their hybrids exhibit univalents predominantly (Ray et al., 1989).

9.3.2.2   Intergenomic Chromosome Pairing

The genomic relationships among genomically dissimilar species can be determined by the degree of chromosome pairing in their hybrids. The extent of chromosome association in the hybrids elucidates structural homology in the parental chromosomes, hence furnishes evidence regarding the progenitor species (Phillips, 1966; Smith, 1968).

Usually the F_1 hybrids generated from genomically unlike parents (different biological species) are germinated through *in vitro* techniques. In general, hybrids are weak, slow in vegetative and reproductive growth and sterile. The lack of seed set is attributed to the reduced amount of chromosome pairing that results in the formation of chromosomally unbalanced gametes. In the subgenus *Glycine*, A and B genome species show intergenomic chromosome association (Figure 9.2). Hybrid seed inviability, seedling lethality, and vegetative lethality are a common occurrence in distant intergenomic crosses (Stebbins, 1958; Hadley and Openshaw, 1980). The reduced chromosome pairing between distantly-related species is attributed to structural differences between chromosomes (chromosomal) perpetuated during speciation rather than to asynaptic or desynaptic genes found in certain genetic stocks. According to Stebbins (1958), doubling chromosomes of chromosomally sterile F_1 hybrids restores fertility by providing each genome a set of perfect homologs with which to pair at meiosis. Chromosome pairing is regular and often fertility is restored. In contrast, desynapsis persists in giving genic sterility after doubling the chromosomes of asynaptic or desynaptic genetic stocks.

9.3.2.3   Variable and Minimum Chromosome Pairing

Variable (semi-homologous–homoeologous) and minimum chromosome pairing are common in intergenomic F_1 hybrids (Tables 9.4 and 9.5). The classical examples reported are from the genera

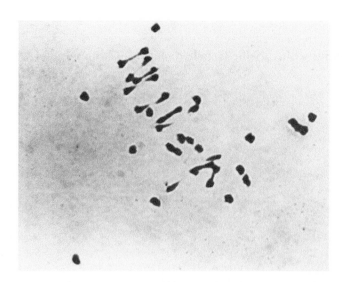

FIGURE 9.2 Meiosis in *Glycine latifolia* ($2n = 40$) × *Glycine canescens* ($2n = 40$) hybrid; metaphase-I showing 20 univalents + 10 bivalents. (Adapted from Singh, R. J. and Hymowitz, T. 1985b. *Theor. Appl. Genet.* 71: 221–230.)

Brassica (U, 1935; Mizushima, 1950a,b; Attia and Röbbelen, 1986a,b), *Glycine* (Grant et al., 1984; Singh et al., 1988, 1992a), *Gossypium* (Skovsted,, 1937; Endrizzi and Phillips, 1960; Phillips, 1966), *Hordeum* (Bothmer et al., 1986), *Nicotiana* (Goodspeed, 1954), *Oryza* (Nezu et al., 1960; Shastry and Ranga Rao, 1961; Katayama, 1982), *Paspalum* (Burson, 1981a,b), and *Triticum* (Kihara and Lilienfeld, 1932; Dewey, 1984).

In intergenomic hybrids, most of the bivalents are rod-shaped and are weakly associated, suggesting a weak chromosomal (genomic) affinity. Thus, the higher the number of univalents, the lesser the genomic affinity. In *Nicotiana*, approximately 90% of the intersectional diploid hybrids showed semihomologous chromosome pairing (Goodspeed, 1954). A considerable amount of chromosome homology exists among A, B, and D genomes of *Triticum aestivum* ($2n = 6x = 42$) (Dewey, 1984). Chromosome association at metaphase-I ranged from 1III + 12I to 7II in *Triticum monococcum* ($2n = 14$, A) × *Triticum* [*Aegilops speltoides* ($2n = 14$, B)] (Kihara and Lilienfeld, 1932).

Chromosome pairing among three elementary species of *Brassica*, namely *Brassica campestris* (AA, $2n = 20$), *Brassica oleracea* (CC, $2n = 18$), and *Brassica nigra* (BB, $2n = 16$), showed a high degree of chromosome association in A × C (mean 0.02 V,VI + 0.29IV + 0.30III + 7.3II + 2.3I) and a low degree of affinity in B × C (mean 0.02III + 1.9II + 13.6I) and A × B (0.01IV + 0.7III + 4.36II + 8.75I) crosses (Attia and Röbbelen, 1986a). This suggests that *Brassica nigra* is distantly related to *Brassica oleracea* and *Brassica campestris*.

Minimum chromosome pairing is usually observed in intersubgeneric and intergeneric hybrids. For example, an intersubgeneric hybrid of *Glycine max* ($2n = 40$, G) and *Glycine clandestina* ($2n = 40$, A$_1$) exhibits 40 univalents in 40.9% of the sporocytes and 1–6 are loosely paired rod-shaped bivalents (Singh et al., 1987a). Burson (1981a,b) observed very little genome homologies among five diploid ($2n = 20$) *Paspalum* species (*Paspalum intermedium*, I; *Paspalum jurgensii*, J; *Paspalum vaginatum*, D; *Paspalum setaceum*, S; and *Paspalum notatum*, N) as a majority of the sporocytes carried 20 univalents. In a *Nicotiana glauca* ($2n = 24$) × *Nicotiana plumbaginifolia* cross, over 75% of the sporocytes showed zero bivalents (Goodspeed, 1954). Williams and Pandey (1975) recorded only 0–2 rod-shaped bivalents in a *Nicotiana glutinosa* ($2n = 24$) × *Nicotiana glauca* ($2n = 24$) cross. In certain Eurasian and South American *Hordeum* species hybrids, Bothmer et al. (1986) observed negligible chromosome pairing (Table 9.5). Diploid ($2n = 2x = 14$) species hybrids of *Hordeum* and *Secale* exhibited 14 univalents in a majority of the sporocytes (Gupta and Fedak, 1985).

TABLE 9.4
Interspecific Hybrids with Variable (Semihomologous) Chromosome Pairing at Diakinesis and Metaphase-I

Hybrids (Genomes)	Average Chromosome Pairing (Range)			GAI	Authority
	I	II	III + IV		
1. *Glycine* (*n* = 20)					
Clandestina × *microphylla* ($A_1 \times B$)	23.6(18–30)	8.2(5–11)		0.41	Singh et al. (1988)
Latifolia × *argyrea* ($B_1 \times A_2$)	19.7(12–30)	10.2(5–14)		0.51	Singh et al. (1988)
Canescens × *microphylla* ($A \times B$)	22.2(14–26)	8.9(7–13)		0.45	Singh and Hymowitz (1985b)
Latifolia × *canescens* ($B_1 \times A$)	20.9(12–32)	9.5(4–14)		0.48	Singh and Hymowitz (1985b)
Tomentella × *clandestina* ($D \times A_1$)	13.6(4–22)	13.3(9–18)		0.67	Singh and Hymowitz (1985b)
Argyrea × *tomentella* ($D \times A_2$)	18.6(8–24)	10.7(8–16)		0.54	Singh et al. (1993)
Argyrea × *cyrtoloba* ($A_2 \times C$)	30.8(20–38)	4.6(1–10)		0.23	Singh et al. (1988)
Latifolia × *cyrtoloba* ($B_1 \times C$)	29.9(26–36)	5.05(2–7)		0.25	Singh et al. (1988)
Tomentella × *canescens* ($E \times A$)	30.0(25–37)	4.5(1–7)		0.23	Singh et al. (1988)
Tomentella × *tomentella* ($E \times D$)	26.6(19–33)	6.2(3–10)		0.31	Singh et al. (1988)
2. *Gossypium* (*n* = 13)					
Herbaceum × *sturtianum* ($A_1 \times C_1$-n)	10.53	7.46	0.17	0.59	Phillips (1966)
Arboreum × *raimondii* ($A_2 \times D_5$)	13.79	5.90	0.13	0.47	Endrizzi and Phillips (1960)
Arboreum × *stocksii* ($A_2 \times E_1$)	14.90	5.50		0.42	Skovsted (1937)
Stocksii × *arboreum* ($E_1 \times A_2$)	18.40	3.80		0.29	Beasley (1942)
Anomalum × *sturtianum* ($B_1 \times C_1$-n)	9.23	8.09	0.16	0.64	Phillips (1966)
Anomalum × *davidsonii* ($B_1 \times D_3$-d)	19.90	3.05		0.23	Skovsted (1937)
Anomalum × *klotzschianum* ($B_1 \times D_3$-k)	17.45	4.26	0.01	0.33	Phillips (1966)
Anomalum × *stocksii* ($B_1 \times E_1$)	20.57	2.70		0.21	Douwes (1951)
Sturtianum × *armourianum* (C_1-n × D_2-1)	8.45	8.2	0.35	0.66	Skovsted (1937)
3. *Oryza* (*n* = 12)					
Sativa × *officinalis* ($A \times C$)	14.31(16–24)	4.8(0–4)	0.022 (0–1)	0.40	Shastry et al. (1961)
Punctata × *eichingeri* ($B \times C$)	18.52	2.74		0.23	Katayama and Ogawa (1974)
Punctata × *officinalis* ($B \times C$)	17.02	3.49		0.29	Katayama and Ogawa (1974)
4. *Hordeum* (*n* = 7)					
Bogdani × *patagonicum* ($I \times H$)	7.38(0–13)	3.33(1–7)		0.48	Bothmer et al. (1986)
Brevisubulatum × *brachyantherum* ($I \times H$)	10.24(6–14)	1.88(0–4)		0.27	Bothmer et al. (1986)
Brevisubulatum × *brachyantherum* ($I \times H$)	3.20(0–8)	5.24(3–7)		0.75	Bothmer et al. (1986)
5. *Brassica* (*n* = 8, 9, 10)					
Oleracea × *campestris* ($C \times A$)	2.3	7.3	0.61[a]		Attia and Röbbelen (1986a)
Oleracea × *nigra* ($C \times B$)	13.6	1.9	0.02		Attia and Röbbelen (1986a)
Campestris × *nigra* ($A \times B$)	8.75	4.36	0.71		Attia and Röbbelen (1986a)

[a] Includes V, VI (0.02).

TABLE 9.5

Interspecific Hybrids with Minimum Chromosome Pairing at Diakinesis and Metaphase-I

Hybrids (Genomes)	Average Chromosome Pairing (Range)			GAI	Authority
	I	II	III + IV		
1. *Glycine* (*n* = 20)					
Microphylla × *falcata* (B × F)	38.6(32–40)	0.7(0–4)		0.04	Singh et al. (1992a)
Latifolia × *falcata* (B₁ × F)	37.8(28–40)	1.12(0–6)		0.06	Singh et al. (1988)
Tomentella × *microphylla* (E × B)	38.2(35–39)	0.40(0–2)		0.02	Singh et al. (1992a)
Max × *clandestina* (G × A₁)	37.8(28–40)	1.15(0–6)		0.06	Singh et al. (1987a)
2. *Gossypium* (*n* = 13)					
Somalense × *australe* (E₂ × C)	23.78	1.11		0.09	Phillips (1966)
Somalense × *bickii* (E₂ × G₁)	25.58	0.21		0.02	Phillips (1966)
3. *Oryza* (*n* = 12)					
Sativa × *australiensis* (A × E)	23.6	0.2(0–1)		0.02	Nezu et al. (1960)
Sativa × *officinalis* (A × C)	24.0	0.0(0–1)		0.00	Nezu et al. (1960)
Officinalis × *australiensis* (C × E)	23.0	0.5(0–3)		0.04	Nezu et al. (1960)
4. *Hordeum* (*n* = 7)					
Marinum × *bogdani* (X × H)	13.71(12–14)	0.14(0–1)		0.02	Bothmer et al. (1986)
Marinum × *brachyantherum* (X × H)	13.10(10–14)	0.45(0–2)		0.06	Bothmer et al. (1986)
Bulbosum × *patagonicum* (I × H)	13.90(12–14)	0.06(0–2)		0.01	Bothmer et al. (1986)
Marinum × *muticum* (X × H)	13.00(10–14)	0.50(0–2)		0.07	Bothmer et al. (1986)
Roshevitzii × *patagonicum* (I × H)	13.42(11–14)	0.18(0–1)	0.03(0–1)	0.03	Bothmer et al. (1986)
5. *Paspalum* (*n* = 10)					
Jurgensii × *intermedium* (J × I)	16.48(10–20)	1.70(0–5)		0.17	Burson (1981a)
Jurgensii × *vaginatum* (J × D)	18.22(8–20)	0.88(0–6)		0.09	Burson (1981a)
Jurgensii × *setaceum* (J × S)	17.73(10–20)	1.14(0–5)		0.11	Burson (1981a)
Intermedium × *notatum* (I × N)	18.43(12–20)	0.78(0–4)		0.08	Burson (1981b)
Notatum × *vaginatum* (N × D)	18.83(12–20)	0.59(0–4)		0.06	Burson (1981b)

9.4 GENOMES OF *GLYCINE*

In the genus *Glycine*, genome A shows partial genome homology with genome B and genome D but a stronger genomic affinity with D than with B. Genome B shows no similarity with D because hybrids are seedling lethal (Singh et al., 1988). Genome C from *Glycine cyrtoloba* shows slightly stronger (may be insignificant) affinity with B (B × C = 6.9II + 26.2I; B₁ × C = 5.05II + 29.9I) than with A (A₂ × C = 4.6II + 30.8I). Moreover, A × C and A₁ × C are vegetative lethal. Likewise, genome E from aneudiploid *Glycine tomentella* (2*n* = 38) exhibits limited chromosome pairing with genomes A and D (Table 9.4). In A × E at metaphase-I, the average chromosome associations (range) were 4.5II (1–7) + 30.0I (25–37) and in D × E they were 6.2II (3–10) + 26.6I (19–33). In these crosses, bivalents are usually rod-shaped and are loosely connected at metaphase-I; univalents are scattered in the cytoplasm (Figure 9.3a). In the subgenus *Glycine*, *Glycine falcata* is a unique species because it differs from the other species in several morphological traits (Hermann, 1962; Hymowitz and Newell, 1975), seed protein composition (Mies and Hymowitz, 1973), oil and fatty acid contents (Chavan et al., 1982), ribosomal gene variation (Doyle and Beachy, 1985), and phytoalexin production (Keen et al., 1986). Chromosome pairing results support the uniqueness of genome (F) of *Glycine falcata* because they show minimum chromosome synapsis with A and B

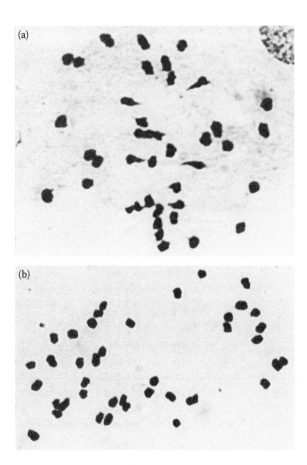

FIGURE 9.3 Meiosis in interspecific *Glycine* hybrids: (a) metaphase-I showing 31 univalents + 4 bivalents in *Glycine tomentella* (2*n* = 38, EE) × *Glycine canescens* (2*n* = 40, AA). (b) Metaphase-I showing 40 univalents in (*Glycine clandestina* × *Glycine canescens*) × *Glycine falcata*. (From Singh, R. J., K. P. Kollipara, and T. Hymowitz. 1988. *Genome* 30: 166–176. With permission.)

genomes (Figure 9.3b). Metaphase-I in F_1's A × F and B × F showed 40 univalents in a majority of the sporocytes. An average chromosome association (range) of 1.12II (0–6) + 37.8I (28–40) was found in B_1 × F (Table 9.5).

Information on chromosome pairing (translated into genomic affinity index) in intergenomic hybrids may shed some light regarding which genome is most likely to be the progenitor genome of the genus. In the genus *Glycine*, it is proposed that *Glycine falcata* is the ancestral species. The absence of chromosome homology of F with A and B genomes suggests that A and B genome species diverged into two evolutionary pathways from a common gene pool; which one differentiated first is uncertain. Divergence within A and B genomes occurred because of geographical isolation. Chromosomal structural changes and morphological variants were maintained due to the autogamous nature of pollination. In further divergence from A genome species, *Glycine tomentella* evolved and it is likely that *G. clandestina* (A_1A_1) may have played a role because D has closer affinity (16.1II + 7.9I) with *G. clandestina* than those observed with *G. canescens* (AA) (13.0II + 14.0I) or with *G. argyrea* (A_2A_2) (12.6II + 14.8I). Aneudiploid (2*n* = 38) *Glycine tomentella* (E) originated from D because both cytotypes are morphologically indistinguishable and E has a higher genomic affinity with D (0.31) than that recorded with A (0.23). Genome E could not be derived from B, C and F genomes because of no genomic affinity. It is suggested that *Glycine cyrtoloba* (C) probably originated from B genome species and *Glycine curvata* (C_1C_1) is a variant of *Glycine cyrtoloba*.

Darlington and Wylie (1955) proposed that $x = 10$ is the basic chromosome number for the cultivated soybean. Based on this proposal, Singh et al. (2001) hypothesized a putative ancestor with $2n = 20$ chromosomes for the genus *Glycine* and carrying at least a pair of nucleolus organizer regions (NORs). Although such a progenitor is currently unknown, it would be most likely found in Southeast Asia (Cambodia, Laos, and Vietnam) and has not been collected and identified thus far. Whether tetraploidization ($2n = 4x = 40$) involved auto- (spontaneous chromosome doubling) or allo- (interspecific hybridization followed by chromosome doubling) polyploidy of the progenitor species and whether it occurred prior to dissemination or after cannot be substantiated experimentally because we do not know where the progenitor of the genus *Glycine* originated. The progenitor of the wild perennial species of the subgenus *Glycine* radiated out into several morphotypes depending on the growing conditions in the Australian subcontinent. These species have never been domesticated and remain as wild perennials. By contrast, the pathway of migration from a common progenitor to China is assumed as: wild perennial ($2n = 4x = 40$; unknown or extinct) → wild annual ($2n = 4x = 40$; *Glycine soja*) → Soybean ($2n = 4x = 40$; cultigen). All known species of the genus *Glycine* exhibit diploid-like meiosis and are inbreeders (Singh and Hymowitz, 1985a).

Alloploylpoidization probably played a key role in the speciation of the genus *Glycine*. This implies that the 40-chromosome *Glycine* species and the 80-chromosome *Glycine tabacina* and *Glycine tomentella* are tetraploid and octoploid, respectively. The expression of four rDNA loci in *Glycine curvata* and *Glycine cyrtoloba* (Singh et al., 2001) strongly supports the hypothesized allotetraploid origin suggested by cytogenetic (Singh and Hymowitz, 1985a,b; Xu et al., 2000a) and molecular studies (Shoemaker et al., 1996).

9.5 GENOMES OF *GOSSYPIUM*

The genus *Gossypium* consists of 45 diploid ($2n = 2x = 26$) and 5 allotetraploid ($2n = 4x = 52$) species (Fryxell, 1992). Diploid species constitute eight genomic groups (A, B, C, D, E, F, G, and K) and have a wide geographical distribution. Allotetraploids are limited to the New World (Table 9.6; Figure 9.4).

Genomic relationships among *Gossypium* species ($2n = 26$) and their genome designations have been based on chromosome pairing in intra- and interspecific hybrids and correlate strongly with geographical distribution, morphological features, karyotype analysis, seed protein banding patterns, and phylogenetic analysis of gene sequences (Stephens, 1947, 1950; Endrizzi et al., 1985; Cronn et al., 2002).

The A genome species (*Gossypium herbaceum* = A_1; *Gossypium arboreum* = A_2) are the cultigens of Africa and Asia and differ by a reciprocal translocation (Gerstel, 1953). Of the three B genome species, *Gossypium anomalum* (B_1) has been studied extensively, has disjunct distribution in northern and southern Africa and shows almost complete pairing with B_2 (*Gossypium triphyllum*—Southern Africa) and B_3 (*Gossypium capitis-viridis*—Cape Verde Islands). The C, G, and K genomes are confined to Australia. Edwards and Mirza (1979) designated *Gossypium bickii*, an Australian wild species, genome symbol G based on karyotype comparison with *Gossypium herbaceum* (A_1) and *Gossypium sturtianum* (C_1). Total chromosome length of *Gossypium bickii* was found to be shorter (55.54 µm) than C_1 (74.04 µm) and A_1 (66.54 µm) genome species. Furthermore, they did not observe nucleolus organizer regions in *Gossypium bickii*. However, this study needs to be repeated.

The D-genome species are found in the New World (Arizona, Mexico, Galapagos Islands, and Peru). It is interesting to note that despite a wide geographical distribution, D genome species are morphologically variable, but carry similar genomes. The E genome species, *Gossypium stocksii* (E_1), *Gossypium somalense* (E_2), *Gossypium areysianum* (E_3), and *Gossypium incanum* (E_4), are distributed from eastern Africa, north and east through the southern tip of the Arabian peninsula to Pakistan (Phillips, 1966). Three species, *G. benadirense*, *G. briccettii*, and *G. vollesenii*, have also been assigned to the E genome but are poorly known (Percival et al., 1999). All the E genome species possess similar genomes.

TABLE 9.6
Species, Genomic Symbols, and Distribution of the Genus *Gossypium* L.

Species	Genome	Distribution-Gene Pool
Diploid ($2n = 26$)		
A. Subgenus *Sturtia* (R. Brown) Todaro [All the indigenous Australian species]		
1. Section *Sturtia* [Species do not deposit terpenoid aldehydes in the seeds (gossypol)]		
i. *G. sturtianum* J. H. Willis	C_1	Australia-GP-3
ii. *G. robinsonii* F. Mueller	C_2	Australia-GP-3
2. Section *Grandicalyx* Fryxell [Unusual perennial, thick underground root-stock, fat bodies on seeds]		
i. *G. costulatum* Todaro	K_1	Australia-GP-3
ii. *G. cunninghamii* Todaro	K_3	Australia-GP-3
iii. *G. exiguum* Fryxell, Craven, and Stewart	K_8	Australia-GP-3
iv. *G. rotundifolium* Fryxell, Craven, and Stewart	K_{12}	Australia-GP-3
v. *G. enthyle* Fryxell, Craven, and Stewart	K_7	Australia-GP-3
vi. *G. nobile* Fryxell, Craven, and Stewart	K_{11}	Australia-GP-3
vii. *G. pilosum* Fryxell	K_5	Australia-GP-3
viii. *G. pulchellum* (C. A. Gardner) Fryxell	K_4	Australia-GP-3
ix. *G. londonderriense* Fryxell, Craven, and Stewart	K_9	Australia-GP-3
x. *G. marchantii* Fryxell, Craven, and Stewart	K_{10}	Australia-GP-3
xi. *G. populifolium* (Bentham) F. Mueller ex Todaro	K_2	Australia-GP-3
xii. *G. napoides* Stewart, Wendel, and Craven	K	Australia-GP-3
xiii. Other wild species -3		Australia-GP-3
3. Section → *Hibiscoidea* Todaro [Species do not deposit terpenoid aldehydes in the seeds]		
i. *G. australe* F. Mueller	G	Australia-GP-3
ii. *G. nelsonii* Fryxell	G_3	Australia-GP-3
iii. *G. bickii* Prokhanov	G_1	Australia-GP-3
B. Subgenus *Houzingenia* (Fryxell) Fryxell [New world → primarily Mexico; large shrubs or small trees]		
1. Section → *Houzingenia*		
a. Subsection: *Houzingenia*		
i. *G. thurberi* Todaro	D_1	Mexico, Arizona (United States)-GP-2
ii. *G. trilobium* (DC) Skovsted	D_8	Mexico-GP-2
b. Subsection: *Integrifolia* (Todaro) Todaro		
i. *G. davidsonii* Kellog	D_{3-d}	California (United States)-GP-2
ii. *G. klotzschianum* Anderson	D_{3-k}	California (United States)-GP-2
c. Subsection: *Caducibracteolata* Mauer		
i. *G. armourianum* Kearney	D_{2-1}	Mexico-GP-2
ii. *G. harknessi* Brandegee	D_{2-2}	Mexico-GP-2
iii. *G. turneri* Fryxell	D_{10}	Mexico-GP-2
2. Section → *Erioxylum* (Rose and Standley) Prokhanov		
a. Subsection: *Erioxylum*		
i. *G. aridum* (Rose and Standley ex Rose) Skovsted	D_4	Mexico-GP-2
ii. *G. lobatum* H. Gentry	D_7	Mexico-GP-2
iii. *G. laxum* Phillips	D_9	Mexico-GP-2
iv. *G. schwendimanii* Fryxell and S. Koch	D_{11}	Mexico-GP-2

(Continued)

TABLE 9.6 (*Continued*)

Species, Genomic Symbols, and Distribution of the Genus *Gossypium* L.

Species	Genome	Distribution-Gene Pool
b. Subsection: *Selera* (Ulbrich) Standley		
i. *G. gossypioides* (Ulbrich) Fryxell	D_6	Mexico-GP-2
c. Subsection: *Austroamericana* Fryxell		
i. *G. raimondii* Ulbrich	D_5	Peru-GP-2
C. Subgenus *Gossypium* L.		
1. Section → *Gossypium*		
a. Subsection: *Gossypium*		
i. *G. herbaceum* L.	A_1	Africa, Asia-GP-2
ii. *G. arboreum* L.		
b. Subsection: *Anomala* Todaro		
i. *G. anomalum* Wawra and Peyritsch	B_1	Africa-GP-2
ii. *G. triphyllum* (Harvey and Sonder) Hochreutiner	B_2	Africa-GP-2
iii. *G. capitis-viridis* Mauer		
c. Subsection: *Longiloba* Fryxell		
i. *G. longicalyx* J.B. Hutchinson and Lee	B_3	Cape Verde Islands-GP2
d. Subsection: *Pseudopambak* (Prokhanov) Fryxell		
i. *G. benadirense* Mattei	E	African-Arabian-GP-3
ii. *G. bricchettii* (Ulbrich) Vollesen	E	African-Arabian-GP-3
iii. *G. vollesenii* Fryxell	E	African-Arabian-GP-3
iv. *G. stocksii* Masters ex. Hooker	E_1	African-Arabian-GP-3
v. *G. somalense* (Gürke) J.B. Hutchinson	E_2	African-Arabian-GP-3
vi. *G. areysianum* Deflers	E_3	African-Arabian-GP-3
vii. *G. incanum* (Schwartz) Hillcoat	E_4	African-Arabian-GP-3
Allotetraploid ($2n = 4x = 52$)		
D. Subgenus *Karpas* Rafinesque [The allotetraploid cottons]		
i. *G. hirsutum* L.	AD_1	Central America-GP1
ii. *G. barbadens* L.	AD_2	South America-GP1
iii. *G. tomentosum* Nuttall ex Seemann	AD_3	Hawaiian Islands-GP-1
iv. *G. mustelinum* Miers ex Watt	AD_4	NE Brazil-GP1
v. *G. darwinii* Watt	AD_5	Galapagos Islands-GP1

Source: From Percival, A. E., J. F. Wendel, and J. M. Stewart. 1999. *Cotton: Origin, History, Technology, and Production.* W. C. Smith, Ed., John Wiley & Sons, Inc., pp. 33–66. With permission. Campbell, B. T. et al. 2010. *Crop Sci.* 50: 1161–1179.

Phillips and Strickland (1966) suggested that *Gossypium longicalyx* from Africa carries the F genome as it differs from the D genome (21.60I in D × F cross). Schwendiman et al. (1980) confirmed the F genome designation to *Gossypium longicalyx* because it differs from *Gossypium stocksii* karyologically. According to Schwendiman et al. (1980), *Gossypium longicalyx* appears to be a "mixed" genome related to all the other genomes (except D). Recent phylogenetic analysis suggests a close relationship between *G. longicalyx* and the two A genome species (Cronn et al., 2002).

Considering karyotype reduction during the course of speciation (Stebbins, 1950), the relative chromosome size of *Gossypium* species reveals the following karyotypic (in decreasing order) relationships. C (largest) → EF → B → A → G → D (smallest) (Stephens, 1947; Katterman and Ergle, 1970; Edwards and Mirza, 1979). Thus, species with the C genome should be considered ancestral. However, by utilizing the frequency of univalents observed in the intergenomic hybrids,

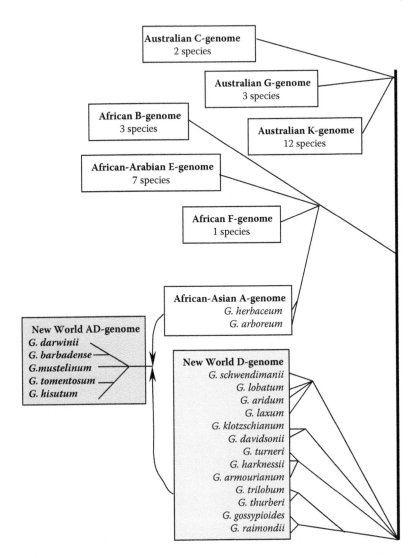

FIGURE 9.4 Evolutionary history of the *Gossypium*. Genomic relationships among diploid species. The origin of allotetraploids following intergenomic hybridization. (From Percival, A. E., J. F. Wendel, and J. M. Stewart. 1999. *Cotton: Origin, History, Technology, and Production*. W. C. Smith, Ed., John Wiley & Sons, Inc., pp. 33–66. With permission.)

Phillips (1966) postulated that the E genome is ancestral in the genus because it has negligible genomic affinity with A, B, C, D, and G genomes. However, the E genome has a greater affinity with B and A than with C and G genomes. The evolutionary pathways of the genus *Gossypium* shown in Figure 9.4 are based on relationships among diploid species. The origin of the allotetraploids (New World AD-genome) following intergenomic hybridization is described. More recent data (Cronn et al., 2002) highlight the difficulty in pinpointing the basal divergence in the genus *Gossypium*. The early stages of evolution are characterized by a rapid radiation and thus no genome can be regarded as "ancestral." The traditional conception of the genus being replaced by a rapid radiation of the proto-*Gossypium* milieu into the D, ABF, E, CGK lineages, with the divergence between the New World (D genome) and Old World/Oceanic lineages (ABCEGK), is most likely to have occurred first.

9.6 ANCESTORS OF ALLOPOLYPLOID SPECIES

Interspecific hybridization between morphologically and genomically distinct diploid species followed by chromosome doubling (amphidiploidy-allopolyploidization) has played a prominent role in the speciation of plants. A large number of our agriculturally important crops such as *Brassicas*, coffee, cotton, oat, sugarcane, tobacco, and wheat are natural allopolyploids. The main characteristic features of allopolyploids are vigorous and aggressive growth habit, diploid-like meiosis (allosyndesis), normal fertility, and true breeding.

Traditionally ancestors of allopolyploids were inferred based on geographical distribution, morphological features, chromosome count, karyotype analysis, and isozyme banding patterns. Molecular phylogenetic studies have confirmed many of these early hypotheses, supplemented by data from interspecific hybridizations, resynthesis of allopolyploids, and comparison of the morphological traits, meiotic pairing and fertility in the allopolyploids and their hybrids. These procedures have been used to explain the ancestors of natural allopolyploid species of genera *Brassica* (Prakash and Hinata, 1980), *Glycine* (Singh et al., 1989; Singh et al., 1992b), *Gossypium* (Beasley, 1940; Harland, 1940), *Nicotiana* (Goodspeed and Clausen, 1928; Clausen, 1932; Greenleaf, 1941), and *Triticum* (McFadden and Sears, 1946). These examples are briefly outlined as follows.

9.6.1 Brassica (*B. carinata* $2n = 4x = 34$; *B. juncea* $2n = 4x = 36$; *B. napus* $2n = 4x = 38$)

The genus *Brassica* consists of three elementary or basic genome species (*B. campestris* $n = 10 = $ A; *B. oleracea* $n = 9 = $ C; *B. nigra* $n = 8 = $ B) and is secondary polyploid from an extinct species with a base chromosome number $x = 6$. This was elegantly demonstrated by Röbbelen (1960) based on pachynema chromosome analysis.

It was observed that *Brassica nigra* (B) is a double tetrasomic (genome A B C DD E FF), *Brassica oleracea* (C) is triple tetrasomic (genome A BB CC D EE F), and *Brassica campestris* is doubly tetrasomic and a hexasomic (AA B C DD E FFF). Each letter designates a particular chromosome. Thus, three basic genomes are derived from a basic chromosome number $x = 6$. Natural allotetraploids *Brassica napus* ($n = 19$, AC), *Brassica juncea* ($n = 18$, AB), and *Brassica carinata* ($n = 17$, BC) exhibit diploid-like meiosis and disomic inheritance.

U (1935) determined the ancestral species of *Brassica napus*, *Brassica junicea*, and *Brassica carinata* by studying the meiotic chromosome pairing in triploid hybrids of *Brassica napus* × *Brassica campestris*, *Brassica napus* × *Brassica oleracea*, *Brassica carinata* × *Brassica oleracea*, and *Brassica carinata* × *Brassica nigra* (Figure 9.5). The chromosome association of $2x$ bivalent + x univalent predominated in triploid hybrids. Further confirmation of the ancestral genomes of allotetraploids is provided by synthesizing amphidiploids. Mizushima (1950b) synthesized *Brassica carinata* ($2n = 34$, BBCC) by hybridizing *Brassica nigra* ($2n = 16$, BB) and *Brassica oleracea* ($2n = 18$, CC), and subsequently doubled the chromosomes by colchicine. Amphidiploid ($2n = 4x = 34$, BBCC) exhibited 0–4IV and 9–17II at metaphase-I. Anaphase-I and second meiosis was regular in 80% of the cells and pollen was nearly fertile. In contrast, *B. carinata* synthesized by Pearson (1972) showed normal meiosis (17II) and high seed fertility. Similarly, *Brassica juncea* ($2n = 36$, AABB) and *Brassica napus* ($2n = 38$, AACC) have been generated artificially by crossing *Brassica campestris* ($2n = 20$, AA) × *Brassica nigra* ($2n = 16$, BB) and *Brassica campestris* ($2n = 20$, AA) × *Brassica oleracea* ($2n = 18$, CC), respectively. Synthesized *Brassica juncea* and *Brassica napus* exhibit normal meiosis, resemble morphologically the respective natural allotetraploid species, and are fertile (Prakash and Hinata, 1980).

It has been observed that the established allotetraploid has a more stable meiotic behavior than the newly synthesized amphidiploids of the same species. Such stabilization is based on genetic control (Attia and Röblelen, 1986b).

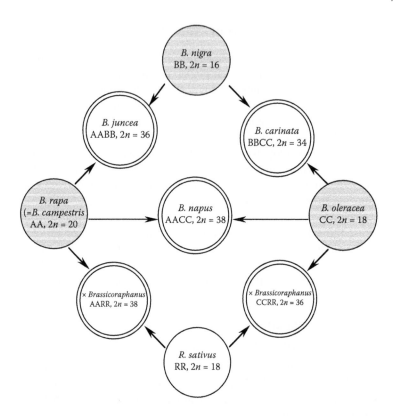

FIGURE 9.5 The *Brassica-Raphanus* diamond (extension of U's triangle), showing the species relationships of the major cruciferous crops. [*Brassica nigra* (black mustard); *Brassica juncea* (Indian or brown mustard); *Brassica rapa* (Oriental vegetables); *Brassica carinata* (Ethiopian mustard); *Brassica oleracea* (various types of cole crops); *Brassica napus* (Oil-seed rape, fodder rape); × *Brassicoraphanus*, AARR (Cultivar group Raparadish); CCRR, Radicole; *Raphanus sativus*. (Fodder radish, oil-seed radish). (From Lange, W. et al. 1989. *Euphytica* 40: 1–14. With permission.)

Studies on seed protein electrophoresis (Uchimiya and Wildman, 1978), chloroplast DNA variation (Palmer et al., 1983; Yanagino et al., 1987), mitochondrial DNA variation (Palmer, 1988), and nuclear restriction fragment length polymorphisms (RFLPs) (Song et al., 1988) have confirmed the established cytogenetic studies that three allotetraploid species are derived from the interspecific hybridization of three elementary diploid species. Yanagino et al. (1987) observed significant correlation between the species relationships as seen with the chloroplast DNA and cytogenetic studies, but not with those based on numerical morphological taxonomy. Another example is for the genus *Oryza* where molecular approaches simply verified the phylogenetic relationships among species established based on taxonomy, hybridization, isozymes, and cytogenetics (Wang et al., 1992).

Lange et al. (1989) extended the *Brassica* triangle of U (1935) to the *Brassica-Raphanus* diamond (Figure 9.5), representing the genomic relationships among species of the majority of the Cruciferous crops. The introduction of *Raphanus sativus* in interspecific hybridization in *Brassica* created two new hybrids:

1. Radicole = *Raphanus sativus* (female) × *Brassica oleracea* (male)→ amphidiploid → × *Brassicoraphanus (2n = 36; CCRR)*.
2. Raparadish = *Raphanus sativus* (female) × *Brassica rapa* (male) → amphidiploid → × *Brassicoraphanus (2n = 38; AARR)*. Both radicole and raparadish resemble fodder rape and are resistance to the beet cyst nematode. The poor seed production is the major disadvantage in radicole and raparadish.

9.6.2   *Glycine* (*G. tomentella* $2n = 4x = 80$)

In the genus *Glycine*, 80-chromosome *G. tomentella* has wide geographical distribution, aggressive and vigorous growth habit, and diploid-like meiosis in its accessions. Meiotic pairing in intra- and interspecific hybrids suggested that the 80-chromosome *Glycine tomentella* evolved through allopolyploidization (Singh and Hymowitz, 1985a; Singh et al., 1987b). Morphologically, 40- and 80-chromosome *Glycine tomentella* are indistinguishable, indicating that one of the ancestors of 80-chromosome tomentella is a 40-chromosome *Glycine tomentella*.

Triploid F_1 hybrids of *Glycine tomentella* ($2n = 80$-Queensland = Qld.) × *Glycine canescens* ($2n = 40$, AA) and *Glycine tomentella* ($2n = 80$-Qld) × *Glycine tomentella* ($2n = 40$, DD) exhibit the "Drosera Scheme" (20 bivalents + 20 univalents) of pairing in a majority of the cells (Table 9.7; Figure 9.6a). Therefore, *Glycine canescens* and *Glycine tomentella* ($2n = 40$) are the probable ancestors of 80-chromosome *Glycine tomentella* from Qld, Australia (Putievsky and Broué, 1979; Singh and Hymowitz, 1985a,b; Singh et al., 1989).

The artificially synthesized amphidiploid (AADD) of *Glycine tomentella* ($2n = 40$, DD) × *Glycine canescens* ($2n = 40$, AA) is morphologically almost identical to the natural 80-chromosome *Glycine tomentella*. Average chromosome associations (ranges) in AADD was 0.13 IV (0–1) + 39.6 II (38–40) + 0.3 I (0–2). A total of 75% of the sporocytes showed 40 bivalents (Figure 9.6b) and 90% of the sporocytes exhibited 40–40 chromosome migration at anaphase-I. Pollen fertility was 84.8% and the plants set normal pods and seeds (Table 9.8).

Singh et al. (1989) hybridized the synthesized amphidiploid (AADD) with four *Glycine tomentella*-Qld ($2n = 80$) accessions (Table 9.8). Hybrid seed germinated normally. Seedlings grew vigorously, carried the expected chromosome numbers of $2n = 80$, and were highly fertile. A majority of the sporocytes showed complete pairing. The conclusion drawn from these results is that *Glycine canescens* (AA) and *Glycine tomentella* (DD) are the ancestors of *Glycine tomentella*-Qld ($2n = 80$). A similar approach was used to determine the putative diploid ancestors of tetraploid ($2n = 80$) *Glycine tabacina* (Singh et al., 1992b). By using the chloroplast DNA polymorphism technique, Doyle et al. (1990b) reached the same conclusion.

9.6.3   *Gossypium* (*G. hirsutum*; *G. barbadense*; *G. mustelinum*;
G. tomentosum; *G. darwinii*, $2n = 4x = 52$)

The genus *Gossypium* carries five allotetraploid ($2n = 4x = 52$) species, and all are distributed in the Americas (New World). It has been demonstrated that allotetraploid cottons are of an amphidiploid origin between diploid ($2n = 2x = 26$) cultivated Asiatic (Old World) cottons and diploid wild American cottons (Skovsted, 1934, 1937; Wendel, 1989).

Skovsted (1934, 1937) reported that the American allotetraploid species (*Gossypium hirsutum*; *Gossypium barbadense*) contain a set of 13 large and 13 small chromosomes. The small chromosomes are contributed by the American diploid (D-genomes) species while large chromosomes are donated by cultivated Asiatic (A-genome) cotton. Triploid hybrids between American tetraploids × Asiatic diploids (Skovsted, 1934; Beasley, 1942) and American tetraploids × American diploids exhibit the "Drosera Scheme" (13II + 13I) of chromosome association (Table 9.9). This suggests that the New World tetraploid cottons are allotetraploid with only the A and D genome species being the donor parents because triploids F_1's from *Gossypium barbadense* (AD_2) × *Gossypium anomalum* (B_1) and *Gossypium barbadense* × *Gossypium stocksii* (E_1) exhibit mainly univalents (Table 9.9).

Beasley (1940) and Harland (1940) independently synthesized amphidiploid ($A_2A_2D_1D_1$) by doubling the chromosome number in *Gossypium arboreum* (cultivated Asiatic cotton) × *Gossypium thurberi* (wild American cotton). Synthesized allotetraploid plants were female fertile but were male sterile because boll and seed set were normal when pollinated with the American $n = 26$ chromosome pollen. These studies demonstrate convincingly that the cultivated Asiatic cotton-A genome and wild American cotton-D genome are the ancestors of the tetraploid American cottons. The

TABLE 9.7

Parental Accessions, Origin of F$_1$, Number of F$_1$ Plants Studied, Meiotic Pairing at Diakinesis/Metaphase-I, and Seed Set in Interspecific F$_1$ Hybrids in the Genus *Glycine*

Hybrid	Origin of F$_1$[a]	No. F$_1$ Plants	2n	Chromosome Association[b] Univalent	Bivalent	Trivalent	Total PMC	Seed Set
CAN (2n = 40) × TOM (2n = 80)								
440928 × 441005	SC	3	60	24.9(18–34)	16.9(13–21)	0.44(0–1)	25	ST[c,d]
TOM (2n = 80) × CAN (2n = 40)								
441005 × 440932	S	3	60	23.5(20–30)	18.3(15–20)	–	12	ST[e]
446958 × 440928	SC	2	60	23.0(19–30)	18.2(15–20)	0.2(0–1)	10	ST[f]
TOM (2n = 40) × TOM (2n = 80)								
505267 × 505214	SC	2	60	20.4(18–24)	19.8(18–21)	–	18	ST
505222 × 505256	SC	3	60	20.8(16–26)	19.6(17–22)	–	51	ST
505267 × 441005	SC	1	60	20.0(18–24)	20.0(18–21)	–	30	ST

Note: CAN, *Glycine canescens*; TOM, *Glycine tomentella*.

[a] S, seed; SC, seed culture.

[b] Means with range in parentheses.

[c] ST, sterile.

[d] From Singh, R. J. and Hymowitz, T. 1985b. *Theor. Appl. Genet.* 71: 221–230. With permission.

[e] From Singh, R. J. and Hymowitz, T. 1985c. *Z. Pflanzenzüchtg.* 95: 289–310. With permission.

[f] From Singh, R. J., K. P. Kollipara, and T. Hymowitz. 1987b; *Genome* 29: 490–497. With permission.

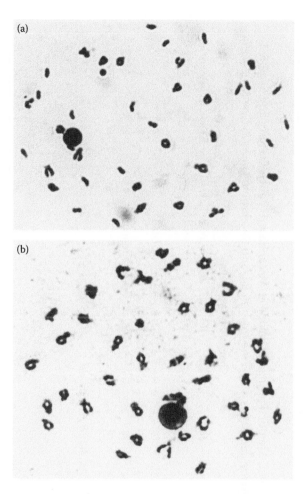

FIGURE 9.6 Meiosis in: (a) an intraspecific hybrid of *Glycine tomentella* (2*n* = 40, DD) × *Glycine tomentella* (2*n* = 80), diakinesis showing 20 univalents + 20 bivalents; (b) Synthesized amphidiploid (*Glycine tomentella* (2*n* = 40, DD) × Glycine *canescens* (2*n* = 40, AA) F₁ 2*n* = 40 AD = colchicine treatment = 2*n* = 80, AADD = H321CT) diakinesis showing 40 bivalents. (From Singh, R. J., K. P. Kollipara, and T. Hymowitz. 1989. *Genome* 32: 796–801. With permission.)

question of where hybridization and amphidiploidy occurred are still being debated (Valiček, 1978; Endrizzi et al., 1985; Wendel, 1989). Seed protein electrophoresis also revealed that the genomes similar to A₁ (*Gossypium herbaceum*) and D₅ (*Gossypium raimondii*) are the probable ancestors of the natural allotetraploid genomes (Cherry et al., 1970).

Another approach toward elucidating the relationship between the A and D genome species and the tetraploid American cotton species is to synthesize amphidiploids involving genomes other than A and D, to hybridize the newly synthesized allotetraploids with natural tetraploid American cottons and to analyze crossability rate, hybrid seed viability, meiotic chromosome pairing, and pollen and seed fertility of hybrids. Brown (1951) identified a spontaneously induced amphidiploid of *Gossypium davidsonii* (D₃-d) and *Gossypium anomalum* (B₁). Meiotic chromosome pairing in the amphidiploid was 26 bivalents in 9 out of the 23 sporocytes studied. Univalents ranged from 2 to 4, and some sporocytes showed a trivalent or a quadrivalent. Most bivalents were normally synapsed with two chiasmata and the number of bivalents averaged 25.0. Amphidiploid *davidsonii–anomalum* set aborted bolls when crossed with *G. hirsutum* and seeds failed to germinate. This shows that *Gossypium anomalum* (B₁) is not a genome donor species to allotetraploid American cottons.

TABLE 9.8

Parental Accessions, Origin of F_1, Number of F_1 Plants Studied, Meiotic Pairing at Diakinesis/Metaphase-I, Pollen Fertility and Seed Set in F_1 Hybrids of Synthesized Amphidiploid ($2n = 80$) of *Glycine canescens* × *Glycine tomentella*, and 80-chromosome *Glycine tomentella*

Hybrid	Origin of F_1[a]	F_1 Plants	$2n$	Chromosome Association[b] I	II	III	IV	Total PMC	Pollen Fertility (%)	Seed Set F[c]
TOM ($2n = 40$) × [CAN × CAN ($2n = 40$)] CT, $2n = 80$ 446993 × [440936 × 440928] = H321CT	S	1	80	0.3(0–2)	39.6(38–40)	0.0	0.13(0–1)	40	84.8	F
H321CT × 441005 ($2n = 80$)	S	2	80	4.7(0–14)	37.2(33–40)	0.0	0.20(0–1)	42	89.5	F
H321CT × 505214 ($2n = 80$)	S	2	80	0.4(0–2)	39.8(39–40)	0.0	0.0	30	99.4	F
H321CT × 505215 ($2n = 80$)	S	2	80	2.0(0–6)	38.6(35–40)	0.0	0.2(0–2)	33	81.3	F
H321CT × 505256 ($2n = 80$)	S	4	80	3.6(0–10)	37.6(34–40)	0.0	0.3(0–2)	85	86.0	F

Source: From Singh, R. J., K. P. Kollipara, and T. Hymowitz. 1989. *Genome* 32: 796–801. With permission.

Note: TOM, *G. tomentella*; CAN, *G. canescens*; CT, colchicine treatment; H, hybrid.

[a] S, seed; SC, seed culture.

[b] Means with ranges in parentheses.

[c] F, fertile; ST, sterile.

TABLE 9.9

Chromosome Pairing in Triploid ($2n = 39$) Interspecific Hybrids of Cotton

Hybrids ($4x \times 2x$)	Average Chromosome Pairing		
	I	II	III
G. barbadense (AD)$_2$ × *G. aridum* (D$_4$)	13.99	12.28	0.15
G. hirsutum (AD)$_1$ × *G. aridum* (D$_4$)	13.15	12.40	0.35
G. barbadense (AD)$_2$ × *G. armourianum* (D$_{2-1}$)	14.75	12.05	0.05
G. darwinii (AD)$_5$ × *G. armourianum* (D$_{2-1}$)	13.20	12.00	0.60
G. barbadense (AD)$_2$ × *G. trilobum* (D$_8$)	12.70	12.40	0.50
G. hirsutum (AD)$_1$ × *G. trilobum* (D$_8$)	13.50	12.45	0.20
G. barbadense (AD)$_2$ × *G. anomalum* (B$_1$)	33.80	2.60	
G. barbadense (AD)$_2$ × *G. stocksii* (E$_1$)	37.90	0.55	

Source: From Skovsted, A. 1937. *J. Genet.* 34: 97–134. With permission.

9.6.4 *NICOTIANA* (*N. tabacum* $2n = 4x = 48$)

The genus *Nicotiana* contains species with $n = 9$, 10, 12, 16, 19, 20, 21, 22, and 24 chromosomes from a base $x = 6$ chromosome species that probably is now extinct. *Nicotiana tabacum* is an allo-tetraploid ($2n = 4x = 48$) cultivated species, belonging to the genus *Nicotiana*, subgenus *tabacum*, and section *Genuinae*. Because of its commercial value, *Nicotiana tabacum* has been studied extensively cytogenetically (Goodspeed, 1954; Smith, 1968).

The allotetraploid nature of *Nicotiana tabacum* was demonstrated by Goodspeed and Clausen (1928) by studying a series of triploid F$_1$ hybrids involving *Nicotiana sylvestris* ($2n = 2x = 24$; S^1 genome), *Nicotiana tomentosa* ($2n = 2x = 24$; T^1 genome), and *Nicotiana tabacum* (SSTT) triangle. The F$_1$ hybrids of *Nicotiana sylvestris* × *Nicotiana tabacum* and *Nicotiana tomentosa* × *Nicotiana tabacum* exhibited classical "Drosera Scheme" (12II + 12I) chromosome pairing suggesting that *Nicotiana sylvestris* and *Nicotiana tomentosa* are the probable ancestors of *N. tabacum,* which most likely originated in Central South America. Both diploid species are found there today, and haploid *Nicotiana tabacum* ($2n = 2x = 24$) displays complete lack of meiotic chromosome pairing similar to that in *Nicotiana sylvestris* × *Nicotiana tomentosa* F$_1$ hybrids (Goodspeed, 1954).

Clausen (1932), on the basis of morphological features of *Ncotiana sylvestris* × *Nicotiana tomentosiformis* and *Nicotiana sylvestris* × *Nicotiana tomentosa*, suggested that *Nicotiana tomentosiformis* satisfies more nearly the requirements of a progenitor of *Nicotiana tabacum* than does *Nicotiana tomentosa*. However, *Nicotiana tomentosa* and *Nicotiana tomentosiformis* are closely related and are differentiated by a reciprocal translocation (Goodspeed, 1954).

Greenleaf (1941) synthesized amphidiploids *Nicotiana sylvestris* × *Nicotiana tomentosa* and *Nicotiana sylvestris* × *Nicotiana tomentosiformis*. Both amphidiploids showed normal meiosis and over 90% pollen fertility but were completely female sterile, although megasporogenesis was quite normal. The sterility was due to genetic factors. Morphologically, the *Nicotiana sylvestris* × *Nicotiana tomentosiformis* amphidiploid closely resembled *Nicotiana tabacum*.

Gerstel (1960) used a genetic approach to learn whether the chromosomes of *Nicotiana tomentosiformis* or those of *Nicotiana otophora* are more nearly homologous with one genome of *Nicotiana tabacum*. Several recessive marker stocks of *Nicotiana tabacum* (SSTT) were crossed to both diploid species carrying the corresponding dominant allele T^{1}T^1. The chromosome number of interspecific triploid (STT^1) F$_1$ hybrids was doubled by treating emerging seedlings with 0.1% aqueous colchicine solution for 3 h at room temperature. The segregation ratios were obtained by back crossing the amphidiploids (SSTTT^{1}T^1) to recessive tester stocks of *Nicotiana tabacum*. The

TABLE 9.10

Segregation of Morphological Traits in 6x Amphidiploids of *Nicotiana*

Morphological Traits	(*N. tabacum* × *N. tomentosiformis*)				(*N. tabacum* × *N. otophora*)			
	++	−	Ratio	6x as	++	−	Ratio	6x as
fs (Fasciated)	87	29	3.0:1	−	195	56	3.5:1	−
Pb (Purple bud)	147	44	3.3:1	−	329	39	8.4:1	−
	381	114	3.3:1	−	342	77	4.4:1	−
ws (White seedling)	762	95	3.5:1	−	1887	98	9.1:1	−
	2481	309	3.5:1	−	644	44	6.8:1	−
yb (Early yellow leaves)	314	100	3.1:1	−	159	19	8.4:1	−
yg (Yellow green)	107	23	4.7:1	−	785	53	14.8:1	−

Source: From Gerstel, D. U. 1960. *Genetics* 45: 1723–1734. With permission.

Note: ++ = dominant; − = recessive.

amphidiploids (TTT^{1}T^1) had the duplex genotype (ZZzz). The observed segregation ratios reflect directly the genetic output of the amphidiploid parents. A ratio of 5:1 (random chromosome segregation) or 3.7:1 (chromatid segregation) is expected if chromosomes of T and T^1 genomes form multivalents. However, the observed ratios were smaller even than the 3.7:1 ratio expected with random chromatid segregation but might fit the maximum equational (3.5:1) segregation (Table 9.10). Segregation ratios were higher in hexaploid (*Nicotiana tabacum* × *Nicotiana otophora*) crosses. This indicates that *Nicotiana tomentosiformis* is more closely related than to *Nicotiana otophora* to *Nicotiana tabacum* in chromosome homology.

A similar approach was used by Gerstel (1963) to show that *N. sylvestris* has a close genomic affinity with a genome of *Nicotiana tabacum*. However, it was concluded "some chromosomes of the two species have remained completely homologous while others have become differentiated to some degree during evolution."

According to Gerstel (1960) "It is quite likely that none of the now living species of that taxonomic group is identical with the parent that once went into ancestral *Nicotiana tabacum*, but perhaps it is not entirely futile to attempt to designate the modern species closest to that ancestor."

9.6.5   *Triticum* (*T. aestivum*, 2n = 6x = 42)

On the basis of plant morphology and nuclear genomic constitution, species of the genus *Triticum* can be classified into three main groups: einkorn (2n = 2x = 14), emmer (2n = 4x = 28) and dinkel (2n = 6x = 42) (Sears, 1975). The emmer and dinkel wheat are allopolyploids. The mode of chromosome pairing in triploid (2n = 3x = 21) and pentaploid (2n = 5x = 35) hybrids helped Kihara to uncover the ancestral species of the allopolyploid wheat. Triploid (*Triticum turgidum*, 2n = 4x = 28 × *Triticum monococcum*, 2n = 2x = 14) and pentaploid (*Triticum turgidum* × *Triticum aestivum*) hybrids exhibit 7II + 7I and 14II + 7I chromosome configurations, respectively. This indicates that diploid (*Triticum monococcum*), tetraploid (*Triticum turgidum*), and hexaploid (*Triticum aestivum*) have one (AA), two (AABB), and three (AABBDD) sets of seven chromosomes each and that the A genome is present in tetraploid and hexaploid wheat (Kihara and Nishiyama, 1930). This suggests that hexaploid wheat are allohexaploids and originated from the hybridization of one species of the emmer group (AABB) and *T. tauschii* (DD) (Figure 9.7). Feldman and Levy (2012) proposed that allopolyploid wheat can achieve genomic pasticity through the induction of a series of cardinal nonadditive genomic changes. Some of them, genetic and epigenetic, are rapid and non-Mendelian, occurring during or immediately after the formation of the allopolyploid (revolutionary

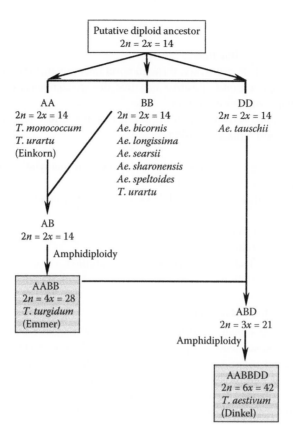

FIGURE 9.7 Diagrammatic figure of the origin of hexaploid wheat.

changes). Other changes occur sporadically over a long period of time for diploidization during the evolution of allopolyploid wheat (evolutionary changes) shown below:

1. Structural changes:
   a. Revolutionary changes occur during or soon after allopolyploidization, lead to diploidization, and are often reproducible.
      Genetic:
      – Elimination of low-copy DNA sequences
      – Elimination, reduction, or amplification of high-copy sequences
      – Intergenomic invasion of DNA sequences
      – Elimination of rRNA and 5S RNA genes
      Epigenetic
      – Chromatin remodeling
      – Chromatin modification
      – Heterochromatinization
      – Small RNA activation or repression
   b. Evolutionary changes facilitated by allopolyploidy in wheat occur during the evolution of species by promoting species biodiversity.
      – Chromosomal repatterning (intra- and intergenomic translocations)
      – Introgression of chromosomal segments from alien genomes and production of recombinant genomes
      – Intergenomic transposons invasion

2. Functional changes:

Genetic changes
- Gene loss/loss of function
- Rewriting of gene expression through novel intergenomic interactions
- Novel dosage response (positive, negative, dosage compensation)
- Gene suppression or activation
- Transcriptional activation of transposons that may affect nearby genes
- New transposon insert/excision

Epigenetic
- Gene silencing or activation through changes in methylation or small RNAs or through chromatin modifications
- Transposon silencing in small RNAs or through chromatin modifications

Evolution changes:
- Subfunctionalizations
- Neofunctionlization
- New dosage effects through copy number variation
- New allelic variations

9.6.5.1 The Source of the A Genome

It has been demonstrated, as shown above, and also verified by Giemsa C-banding (Gill and Kimber, 1974; Gill et al., 1991) and sequences of purothionins—small basic proteins (Jones et al., 1982)—that *Triticum monococcum* is the A genome donor. Another diploid species, *Triticum urartu*, which was suggested as the B genome donor (Johnson, 1975), is similar genomically to *Triticum monococcum* (Dvořák, 1976; Chapman et al., 1976). Indications are that *Triticum urartu* rather than *Triticum monococcum* is the source of the A genome (Kerby and Kuspira, 1987).

9.6.5.2 The Source of the B Genome

The B genome donor species is a puzzle for wheat cytogeneticists and it is still an unsettled issue (Kerby and Kuspira, 1988). Based on geographical distribution, morphological features, karyotype analysis, Giemsa C- and N-banding patterns, meiotic chromosome pairing, nuclear DNA content, *in situ* hybridization studies, restriction fragment patterns of rRNA genes, DNA:DNA hybridization, and seed protein banding pattern and cytoplasmic compatibility and incompatibility studies, six diploid wild species have been proposed as the possible sources of the B genome (Table 9.11). All of the species except *Triticum urartu* belong to the *Sitopsis* section, are cross-compatible, and exhibit almost normal meiosis and various degrees of pollen and seed sterility (Feldman et al., 1979). By comparing all the possible evidences reported thus far, Kerby and Kuspira (1987, 1988) suggest "*Aegilops searsii* is the source of the B-genome, if this set of chromosomes is monophyletic in origin." However, ctDNA revealed that the cytoplasm of emmer and common wheat originated from *Aegilops longissima* and not from *Aegilops speltoides*, *Aegilops sharonensis*, *Aegilops bicorne*, *Aegilops searsii*, or *Triticum urartu* (Tsunewaki and Ogihara, 1983). Meiotic chromosome pairing of *Triticum aestivum* involving *Aegilops speltoides*, *Aegilops sharonensis*, and *Aegilops longissima* revealed that none of these three *Aegilops* are B genome donor of the wheat (Fernández-Calvín and Orellana, 1994).

9.6.5.3 The Source of the D Genome

Pathak (1940), on the basis of morphological traits, geographical distribution, karyotype analysis, and susceptibility to rust, proposed that *Aegilops squarrosa* (*Triticum tauschii*), a diploid wild species, was the donor of the D genome of hexaploid wheat. This was verified experimentally by Kihara (1944, 1947) and McFadden and Sears (1944, 1946) independently. McFadden and Sears (1944, 1946) synthesized allohexaploids by doubling the chromosomes of F_1 triploid hybrids of *T. turgidum*

TABLE 9.11

Species, Genome Symbols, and Authority Reported Genomes of Hexaploid Wheat

Species	Genome $n = 7$	Authority
Triticum monococcum	A	Kihara (1924)
Aegilops speltoides	B?	Pathak (1940); Sarkar and Stebbins (1956); Daud and Gustafson (1996)
Aegilops longissima	B?	Tanaka (1956)
Aegilops sheronensis	B?	Kushnir and Halloran (1981)
Aegilpos bicornis	B?	Sears (1956a)
Aegilops searsii	B?	Feldman and Kislev (1977)
T. urartu	B?[a]	Johnson (1975)
Aegilops tauschii	D	Pathak (1940); Kihara (1944); McFadden and Sears (1944)

[a]   *T. urartu = T. monococcum.*

(*Aegilops dicoccoides*) × *Triticum tauschii*. The synthesized hexaploid plants resembled *Triticum aestivum* for several morphological traits, formed 21 bivalents, and were fertile. Furthermore, hybrids between synthetic hexaploid wheat and cultivated hexaploid wheat also showed 21 bivalents, and showed 96.3% seed set. This suggests that synthesized and cultivated wheat have similar genomes, hence *Triticum tauschii* is the D genome donor species.

The genomic affinity among species established on the basis of the chromosome pairing method is valid, as has been shown from several classical examples, and is being verified by biochemical and molecular approaches. It is suggested that classical taxonomy is a foundation for genome analysis and should not be disturbed.

9.7   GENOME ANALYSIS BY MOLECULAR METHODS

During the past decade, literature on genomic relationships (plant phylogenetic relationships) has been dominated by molecular data, including nuclear (RFLP, AFLP, RAPD, SSR, sequence variation in the gene such as ITS region of rDNA), extra nuclear (chloroplast and mitochondrial DNA) DNA variation, and genomic *in situ* hybridization (GISH) by multicolor FISH. This latter approach is extremely powerful where production of interspecific or intergeneric hybrids is not feasible by conventional method.

Molecular methods thus far have verified the genomic relationships established by cytogenetics in the genus *Arachis, Avena, Glycine, Gossypium, Hordeum, Lycopersicon, Oryza, Phaseolus, Pisum, Secale, Vigna, Zea, Triticum,* and many more. The intention here is to use *Glycine* as an example where classical taxonomy is clear; genomic relationships based on cytogenetics, molecular methods, and chloroplast DNA have been exceedingly examined.

9.7.1   GENOME ANALYSIS BY NUCLEOTIDE SEQUENCE VARIATION

Kollipara et al. (1997) determined phylogenetic relationships among 18 species of the genus *Glycine* from nucleotide sequence variation in the internal transcribed spacer (ITS) region of nuclear ribosomal DNA (Figure 9.8). Of a total of 648 characters used in the analyses, 16, 215, 168, 199, and 50 belonged to the 5S, ITS1, 5.8S, ITS2, and LS sequence, respectively. The mean length of ITS1 was 16.4 nucleotides longer than that of the ITS2 sequence. The alignment assumed a total of 18 indels, excluding the ambiguous regions. Only 9 of these 18 indels were phylogenetically informative. The length of the gaps ranged from one (in most cases) to five (indel 5) nucleotides.

The sequence divergences obtained from the pairwise comparisons of unambiguous characters ranged from 0.2% between *Glycine max* and *Glycine soja* to 8.6% between *Glycine hirticaulis* and

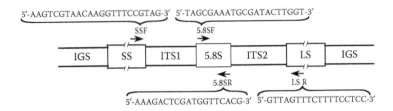

FIGURE 9.8 The general structure of the nrDNA showing the PCR and sequencing strategy. The relative positions of the small subunit (SS), large subunit (LS), 5.8S, internal transcribed spacers (ITS1 and ITS2), the intergenic spacer (IGS), and the primers (with sequences) are indicated. (Redrawn from Kollipara, K. P., R. J. Singh, and T. Hymowitz. 1997. *Genome* 40: 57–68. With permission.)

Glycine falcata. The highest divergence was recorded between *Glycine hirticaulis* and *Glycine falcata*, both in the ITS1 (24.9%) and ITS2 (10.3%) sequences. *Glycine falcata* showed maximum divergence (2.4%) from *Glycine lactovirens*, *Glycine cyrtoloba*, and *Glycine curvata* in its 5.8S sequence. *Glycine soja* and *Glycine max* differed by one nucleotide in the entire sequence.

Analysis of the entire sequence, excluding the ambiguous sequence, resulted in 16 maximally parsimoneous trees with equal length of 176 steps. The strict consensus trees with bootstrap and decay values are shown in Figure 9.9. The CI (consistency index) of these trees was 0.632 when uninformative characters were excluded. The RI (retention index) of these most parsimonious trees was 0.735.

The *Glycine* species with the same letter genome symbols, assigned based on cytogenetics, were resolved as monophyletic groups on the strict consensus tree (Figure 9.9). The newly described species, *Glycine albicans, Glycine arenaria, Glycine hirticaulis, Glycine lactovirens, and Glycine pindanica,* formed two distinct clades. The clade containing *Glycine arenaria, Glycine hirticaulis,* and *Glycine pindanica* supported by a bootstrap value of 86 and decay value of +3 was assigned H genome. The other clade consisting of *Glycine albicans* and *Glycine lactovirens* supported by a bootstrap value of 100 and decay value of +≥5 was assigned I genome. The evolutionary tree estimated by the maximum likelihood method was essentially similar in its topology to the maximally parsimonius trees. The only difference was with respect to the placement of *Glycine falcata* in relation to the outgroup taxa and the C genome clade.

The ITS region (nrDNA) is a multigene family. However, in the soybean, the nrDNA is mapped to a single locus on the short arm of chromosome 13 based on the location of the nucleolus organizer region by pachytene chromosome analysis (Singh and Hymowitz, 1988) and also by fluorescent *in situ* hybridization using ITS as a probe (Singh et al., 2001). The wild perennial *Glycine* species also contain one pair of NOR chromosome, like those in the soybean, except for *Glycine curvata* and *Glycine cyrtoloba,* which have two NOR chromosomes (Singh et al., 2001). The ITS region appeared highly homogenous from this study, since direct sequencing of the PCR fragments did not show polymorphisms in the nucleotide sequences.

Comparison of strict consensus trees based on sequences revealed that the ITS1 and ITS2 sequences complemented each other in resolving the major clades, the A and B genome species, *Glycine microphylla, Glycine latifolia,* and *Glycine tabacina* as a sister group to the clade with A, D, H, and I genome species, while the ITS2 resolved the A genome species, *Glycine argyrea, Glycine canescens, Glycine clandestina,* and *Glycine latrobeana* from the clade containing B, D, E, H, and I genome species. The phylogenetic tree was found to be more robust owing to the complementary nature of both regions.

The robustness of a hypothesis can be tested by assessing its congruence with a phylogenetic hypothesis developed from different methods, like morphological, geographical, biochemical, molecular, etc., which yield the identical trees. In the genus *Glycine,* extensive cytogenetic and limited biochemical investigations supported by classical taxonomic data contributed to the

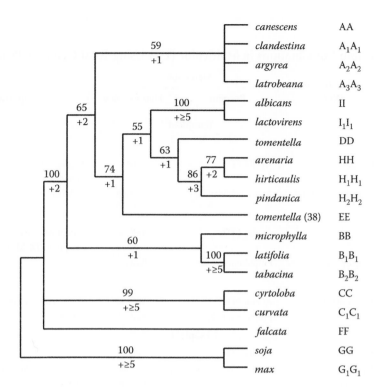

FIGURE 9.9 Strict consensus of the 16 most parsimonious trees with equal length of 176 steps obtained from equally weighted parsimony analysis, using the Branch-and-Bound option of the PAUP program, of the entire unambiguously aligned ITS region shown in Figure 9.8 (CI excluding the uninformative characters = 0.632; RI = 0.735). The numbers above the branches indicate the number of times a monophyletic group occurred in 100 bootstrap replicates. The numbers shown below the branches indicate the additional number of steps over 176 (the total number of steps in the shortest tree) needed to collapse that branch (decay values). Decay analysis with tree length ≥5 steps longer than the most parsimonious trees could not be performed owing to the computational constraints. The taxa and the genome symbols are indicated; tomentella (38) indicates *Glycine tomentella* with $2n = 38$. (From Kollipara, K. P., R. J. Singh, and T. Hymowitz. 1997. *Genome* 40: 57–68. With permission.)

establishment of intergenomic relationships among 13 of the 20 species. The phylogeny established from molecular data helped assign genome symbols: H to *Glycine arenaria*, H_1 to *Glycine hirticaulis*, H_2 to *Glycine pindanica*, I to *Glycine albicans*, and I_2 to *Glycine lactovirens*. The H and I genome groups were strongly supported by the higher bootstrap and decay values, suggesting the close relationship among the species within these clades (Figure 9.9). However, cpDNA phylogeny was found to be congruent with the inference based on cytogenetic studies with respect to B, C, and G genome species but remaining species were classified as an A-plastome group (Doyle et al., 2000). It was concluded that cpDNA is not the perfect phylogenetic tool for establishing genomic relationships among species of the subgenus *Glycine* as a whole, as plastomes of species within groups are poorly differentiated. Its utility is often limited at lower taxonomic levels among closely related taxa (Doyle, 1991; Doyle et al., 1998).

A multidisciplinary approach (cytogenetics and ITS of rDNA) was applied by Singh et al. (1998) to uncover the genomic diversity in *Glycine canescens* and diploid *Glycine tomentella* of western Australia. Cytogenetic results demonstrated that *Glycine canescens* from western Australia are genomically similar; however, they differ by a paracentric inversion from the standard *Glycine canescens* from South Australia. On the other hand, 40-chromosome *Glycine tomentella* accessions, morphologically similar, from western Australia are highly diverse genomically. Cytogenetics and

TABLE 9.12

Labeling-Detection Combinations in Multicolor FISH Using Total Genomic Probes

	Labeling of Probe DNA	Detection Reagent	Fluorescence	Application
		Detection		
	Labeling of Probe DNA	**Reagent**	**Fluorescence**	**Application**
		Tetraploid		
Method 1				
1st genome	Biotin-dUTP	Avidin-FITC	Yellow	*Aegilops* (CCDD, CCUU, DDNN),
				synthetic amphidiploid
2nd genome	Digoxigenin-dUTP	Antidig-rhodamine	Orange	(AASS)
Method 2				
1st genome	Biotin-dUTP	Avidin-FITC	Yellow	Tetraploid wheat (AABB, AAGG),
				Aegilops (DDMM, UUMM, UUSS)
2nd genome	Cold (blocking)	Propidium iodide	Red	
Method 3				
1st genome	Biotin-dUTP	Avidin-FITC	Yellow	*Millium montianum*
2nd genome	None	Propidium iodide	Red	Tobacco
		Hexaploid		
1st genome	Biotin-dUTP	Avidin-FITC	Yellow	Common wheat
2nd genome	Digixigenin-dUTP	Antidig-rhodamine	Orange	Synthetic amphidiploid
3rd genome	Cold (blocking)		Brown	(AABBNN, AAGGUU)
		Octoploid		
1st genome	Biotin-dUTP	Avidin-FITC	Yellow	Synthetic amphidiploid
2nd genome	Digixigenin-dUTP	Antidig-rhodamine	Orange	(AABBDDEE, AABBDDRR)
3rd genome	Cold (blocking)		Brown	
4th genome	2/3 bio. + 1/3 dig.	Avidin-FITC + Antidig-rhodamine	Yellow	

Source: From Mukai, Y. 1996. *Methods of Genome Analysis in Plants.* P. P. Jauhar, Ed., CRC Press, Boca Raton, FL, pp. 181–192.

ITS investigations supported the four isozyme groupings of Doyle and Brown (1985). This demonstrates that diploid *Glycine tomentella* of western Australia is complex, and from an evolutionary viewpoint, it is actively radiating out into several genomic variants.

9.7.2 GENOME ANALYSIS BY MULTICOLOR FISH

Genome analysis in ployploid plants can be determined by multicolor genomic *in situ* hybridization (Heslop-Harrison and Schwarzcher, 1996; Mukai, 1996; Shishido et al., 1998; Sánchez-Morán et al., 1999). Mukai (1996) described labeling-detection combinations to discriminate genomes of polyploid *Aegilops* and *Triticum* (Table 9.12). Multicolor FISH expresses genomes much clearer in synthetic amphidiploids than that observed in the natural amphidiploids. This may be attributed to chromosomal interchanges among different genomes during evolution. Although multicolor FISH technique is a powerful cytogenetic tool to determine the genome of polyploid crops, this technique has some limitations. If more than three probes, FITC (green), rhodamine (red), and AMCA (blue), are applied, chromosomes cannot be counterstained by propidium iodide (red) or DAPI (blue).

10 Cytogenetic Aberrations in Tissue Culture

10.1 INTRODUCTION

Chromosomal aberrations are a common occurrence in cell and tissue culture-derived calluses and their regenerates. In general, numerical and structural changes in chromosomes are attributed to spindle failure that causes endoreduplication, c-mitosis, nuclear fragmentation (amitosis), multipolar configurations, and lagging chromosomes. These changes are induced by (1) media composition, (2) age of the callus, (3) nature of the callus, morphogenic versus nonmorphogenic, (4) genetic background of the explants, and (5) kinds of media, solid versus liquid (Bayliss, 1973, 1980; Evans and Reed, 1981; Wersuhn and Dathe, 1983; D'Amato, 1985; Ogura, 1990; Greier, 1991).

10.2 CHROMOSOMAL ABERRATIONS IN CALLUS

10.2.1 MEDIA COMPOSITION

Chromosomal aberrations induced in cultures are the result of direct influence of chemical substances present in the medium. Medium compositions that can produce more organized and meristematic state cause less somaclonal variation (Choi et al., 2000a,b; Bregitzer et al., 2002). However, the majority of reports agree that chromosome aberrations are generated during culture by growth hormones such as 2,4-D (Torrey, 1959, 1967; Venketeswaran, 1963; Shimada and Tabata, 1967; Mohandas and Grant, 1972; Nuti Ronchi et al., 1976; Kar and Sen, 1985; Ziauddin and Kasha, 1990), IAA (Huskins and Steinitz, 1948; Naylor et al., 1954; Nishiyama and Taira, 1966), NAA (Dermen, 1941; Berger and Witkus, 1948; Zhang et al., 1987), and kinetin (Torrey, 1959; Van't Hof and McMillan, 1969; Nuti Ronchi et al., 1976). The medium containing yeast extract and coconut water is also responsible for chromosomal aberrations (Straus, 1954; Inomata, 1982; Kar and Sen, 1985). Yeast extract contains many substances including nucleic acid that cause chromosome stickiness and breakage (Kodani, 1948).

Venketeswaran (1963) examined cytologically the cell aggregates of *Vicia faba* ($2n = 2x = 12$) grown in a medium containing 2, 4-D and observed different ploidy levels with a predominance of aneuploids. Naylor et al. (1954) reported that 2 mg/L IAA induced mitosis without cytokinesis in tobacco pith tissues grown in a modified White's agar medium that resulted in multinucleate cells and an abnormally lobed nucleus. Dermen (1941) observed $4x$, $8x$, and $16x$ cells in bean ($2n = 2x = 22$) by NAA treatment and suggested that a mixture of diploid and polyploid tissues may be induced in some parts of treated plants.

Van't Hof and McMillan (1969) studied the pea root ($2n = 2x = 14$) callus and observed that the callus was smooth and spherical in shape when kinetin was added to the medium while it was rough and irregular in medium without kinetin. Chromosome counts indicated that in medium with kinetin the mitotic index in polyploid cells increased rapidly during 0–7 days in culture (mixed cell populations) and a slow phase occurred from 7 to 14 days (consisting of approximately 50% $2x$ and 50% $4x$ cells). In contrast, they observed only diploid cells in callus grown in medium without kinetin. Matthysse and Torrey (1967a,b) recorded that diploid cells do not require kinetin or kinetin-like substances. Zhang et al. (1987) recorded variable frequencies of haploids ($2n = x = 6$), tetraploids ($2n = 4x = 24$), and breakage in chromosomes in the cotyledon attachment region in induced calluses of *Vicia faba* ($2n = 2x = 12$), cultured in a medium containing various concentrations of NAA

335

and kinetin. In *Haplopappus gracilis*, the increase in frequency of polyploid cells was associated with decrease in supply of nutrient (Singh and Harvey, 1975b; Singh et al., 1975).

Chromosomal aberrations induced in culture are dependent on the kind of growth hormone used. Published results suggest that 2, 4-D induces greater amounts of polyploidy than NAA (Jacobsen, 1981; Jha and Roy, 1982; Kar and Sen, 1985). Libbenga and Torrey (1973) recorded in pea cortical nuclei that both auxin and cytokinin were needed to induce chromosome doubling.

Investigations by Singh and Harvey (1975a) and Bayliss (1975) contradict the above results. Their results showed that the higher the concentration of 2, 4-D in the culture medium, the lower the frequency of mitotic irregularities.

Preferential selection of haploid cells from a mixed population of cells of varying ploidy levels has been demonstrated cytologically by Gupta and Carlson (1972), following parafluorophenylalanine (PFP) treatment. They used pith calluses of tobacco ($2n = 48$) that consisted of haploid and diploid cells. The growth of diploid cells was inhibited at the level of 9.0 g/mL PFP and diploid cells turned black and died at 15.0 g/mL of PFP. In contrast, growth of haploid cells was not affected.

Do chromosome aberrations arise during culture or do they pre-exist in the cultured tissues? To answer this question, Singh (1986) analyzed chromosomes of five calluses each after 1, 2, 3, 6, 7, 8, 10, and 15 days of inoculation of barley immature embryos, although embryos were not callused in the beginning. It is evident from Table 10.1 that chromosome aberrations, mainly tetraploid, initiated in cultured embryos after 6 days. After 8 days, all five cultured embryos showed low mitotic index. The interphase nuclei were enlarged, with condensed chromatin (endoreduplication). Tetraploid cells began to increase gradually after 10 days. These observations suggest that chromosome aberrations are induced during the culture (Cooper et al., 1964; Shimada and Tabata, 1967; Singh and Harvey, 1975b; Balzan, 1978; Armstrong et al., 1983; Sree Ramulu et al., 1984b; Natali and Cavallini, 1987; Pijnacker et al., 1989). However, variable chromosome numbers in root tips of *Allium cepa* $2n = 2x = 16$ (Partanen, 1963), *Pisum sativum* $2n = 2x = 14$ (Torrey, 1959), pith tissues of *Nicotiana tabacum* $2n = 4x = 48$ (Murashige and Nakano, 1967; Shimada and Tabata, 1967) (Table 10.2), and hypocotyl tissues of *Haplopappus gracilis* $2n = 2x = 4$ (Singh and Harvey, 1975b) suggest that chromosome aberrations pre-exist. If these tissues are grown *in vitro*, polyploid and aneuploid cells are expected to be generated more frequently than tissues lacking initial chromosome abnormalities.

Enhanced tolerance to NaCl in tobacco is attributed to chromosome number. Kononowicz et al. (1990a) found cells adapted to 428 mM NaCl were predominantly hexaploid ($2n = 6x = 72$) while unadapted cells contained normal ($2n = 4x = 48$) chromosome number. Enrichment of the cell

---

TABLE 10.1

Chromosome Count in Immature Embryo-Derived Callus of Barley cv. Himalaya after 1–15 days in Culture

Days after Embryo Cultured	2n Chromosome Number						Cells Studied	Total % Diploid
	14	14 + 1 telo.	15	28	13 + 2 telo.	Fragments		
1	182	–	–	–	–	–	182	100.0
2	130	–	–	–	–	–	130	100.0
3	248	–	–	–	–	–	248	100.0
6	202	–	1	1	–	–	204	99.0
7	255	1	–	2	–	–	258	98.8
8	50	–	–	3	1	–	54[a]	92.6
10	372	–	–	9	2	2	385	96.6
15	149	–	–	11	–	–	160	93.1

Source: Adapted from Singh, R. J. 1986. *Theor. Appl. Genet.* 72: 710–716.

[a] Low mitotic index, condensation of chromatin in the interphase cells.

TABLE 10.2

Chromosome Numbers in Tobacco Pith and Calluses

Period (Year)	No. Cells Examined	Chromosome Numbers[a]
0[b]	53	$2n = 48(25), 96(28)$
0[c]	44	$2n = 48(4), 96(31), 192(7)$, aneuploids (2); 182, 184
1	15	$2n = 96(4), 192(3)$, aneuploids (8): 54, 74, 86, 88, 92, 156, 158, 304
6	12	$2n =$ aneuploids (12): 108, 122, 124, 140, 146, 148(2), 152(2), 154, 160, 174

Source: From Murashige, T. and R. Nakano. 1967. *Am. J. Bot.* 54: 963–970. With permission.

[a] 0 denotes number of cells. Normal chromosome ($2n = 48$).

[b] Pith freshly excised from region of the stem 3.5–10.5 cm below apex.

[c] Pith freshly excised from region of the stem 15.5–22.5 cm below apex.

population for hexaploid cells occurred only after exposure to higher NaCl (428 mM) not at a lower level of NaCl (171 mM). Hexaploidy induced by higher NaCl bred true for at least 25 cell generations after removed from NaCl exposure. The hexaploidy could not be correlated with several phenotypic changes associated with plants regenerated from adapted cells, including male sterility and increased salt tolerance (Kononowicz et al., 1990b).

10.2.2 AGE OF CALLUS

Generally, the older calluses lose the capacity to regenerate plants because they carry either high ploidy in the tissues or accumulate an increasing number of aneuploid cells that lead to the loss of balanced chromosomal constitution of the cells (Murashige and Nakano, 1965, 1967; Shimada and Tabata, 1967; Torrey, 1967; Yazawa, 1967; Heinz et al., 1969; Shimada, 1971; Novák, 1974; Ogihara and Tsunewaki, 1979; Roy, 1980; Kar and Sen, 1985; Orton, 1985; Lavania and Srivastava, 1988).

Murashige and Nakano (1965, 1967) observed the complete loss of morphogenic potentialities in 1- to 6-year-old calluses of tobacco (*Nicotiana tabacum* cv. Wisconsin) and that was associated with increased polyploidy ($2x \rightarrow 4x \rightarrow 8x$) and highly variable aneuploidy (Table 10.2). Similarly, Shimada (1971) found only aneuploid cells in 2-year-old calluses of tobacco, though only 10 cells were studied, while 64% of the cells carried expected $2n = 48$ chromosomes in 3- to 7-month-old calluses.

Novák (1974) studied changes of karyotype in 92-, 157-, 254-, and 339-day-old leaf calluses of *Allium sativum* ($2n = 2x = 16$). Older calluses showed high frequencies of polyploid and aneuploid cells and chromosomes ranged from haploid to hypertetraploid (Table 10.3). Chromatin bridges, laggards, and chromosome fragments at anaphase were also recorded.

Twelve-month-old calluses derived from immature petioles of celery carried normal chromosome complements in 2.5% of cells and lost completely the plant regenerability trait. However, 6-month-old calluses showed the normal karyotype of $2n = 2x = 22$ in 84% cells and plant regenerability was normal (Orton, 1985). In contrast, aging did not influence chromosomal aberrations in maize (Gresshoff and Doy, 1973; Edallo et al., 1981; McCoy and Phillips 1982). McCoy and Phillips (1982) observed a high frequency of diploid ($2n = 2x = 20$) cells in 4-month (96%–97%) and 8-month (94%–95%) old immature embryo-derived calluses of *Zea mays*. As expected, these calluses were highly morphogenic.

10.2.3 NATURE OF CALLUS (MORPHOGENIC VS. NONMORPHOGENIC)

The morphogenic ability of a callus depends on the chromosome constitution of the cells. Balanced chromosome number in a callus is a prerequisite for regenerating plants. This is in agreement with the

TABLE 10.3

Frequency (%) of Chromosomal Aberrations in AS-1 Strain of *Allium sativum* (2*n* = 16) Callus as Related to the Length in Culture

Chromosomal Type	Age of Culture (Days)			
	92	157	254	339
Haploid (2*n* = *x* = 8)	0.75	1.51	1.59	2.03
Hypodiploid (> 8 < 16)	1.49	3.79	8.73	4.73
Diploid (2*n* = 2*x* = 16)	53.73	46.21	42.86	26.35
Hyperdiploid (> 16 < 32)	20.89	6.82	9.52	12.16
Tetraploid (2*n* = 4*x* = 32)	14.18	34.85	33.33	45.27
Hypertetraploid (>32)	8.96	6.82	3.97	9.46

Source: From Novák, F. J. 1974. *Caryologia* 27: 45–54. With permission.

results reported in *Daucus carota* (Smith and Street, 1974), *Hordeum vulgare* (Scheunert et al., 1978; Singh, 1986; Gaponenko et al., 1988), *Haworthia setata* (Ogihara and Tsunewaki, 1979), *Nicotiana tabacum* (Murashige and Nakano, 1965, 1967; Mahfouz et al., 1983; Wersuhn and Sell, 1988), *Pisum sativum* (Torrey, 1959, 1967), *Triticum aestivum* (Ahloowalia, 1982), and *Zea mays* (Balzan, 1978).

Singh (1986) examined cytologically 10 morphogenic calluses of barley (Table 10.4). The expected diploid (2*n* = 2*x* = 14) chromosome cells predominated in all morphogenic calluses (Figure 10.1a). The percentage of diploid cells ranged from 74.1% to 100% (Table 10.4); however, haploid (2*n* = *x* = 7)

TABLE 10.4

Chromosome Analysis in Immature Embryo-Derived Calluses of Barley cv. Himalaya

Callus Types	Callus Nos.	2*n* Chromosomes Number							Total Cells	Diploid (%)
		7	14	15	21	28	56	Others[a]		
Morphogenic	1	–	190	–	–	–	–	–	190	100.0
	2	–	170	–	–	–	–	–	170	100.0
	3	2	198	–	–	–	–	–	200	99.0
	4	–	200	–	–	2	–	–	202	99.0
	5	–	100	–	–	1	–	–	101	99.0
	6	–	118	–	–	3	–	–	121	97.5
	7	–	147	–	–	17	–	–	164	89.6
	8	–	100	–	2	10	2	–	114	87.7
	9	–	60	–	–	16	2	–	78	76.9
	10	–	117	–	3	33	5	–	158	74.1
Nonmorphogenic	1	–	49	2	2	49	2	44	148	33.1
	2	–	10	1	3	9	–	59	82	12.2
	3–7[b]									

Source: Adapted from Singh, R. J. 1986. *Theor. Appl. Genet.* 72: 710–716.

[a] Others (no. of cells observed are in parenthesis): 13 + 2 telocentrics (5), 14 + 1 telocentrics (7), 14 + 2 metacentrics (2), 18 (2), 19 (5), 19 + 1 acrocentric (1), 21 + 1 telocentric (8), 21 + 2 telocentrics (5), 21 + 3 telocentrics (2), 21 + 1 ring (3), 21 + 1 dicentric (1), 22 (6), 22 + 1 ring (1), 22 + 2 dicentrics (4), 23 (1), 24 (6), 24 + 1 fragment (5), 24 + 1 acrocentric (1), 25 (2), 25 + 1 telocentric (3), 27 (4), 27 + 1 telocentric (1), 7 + 29 telocentrics (1), fragment chromosomes (3), clumped chromosomes (14), uneven cell division (5), and high uncountable ploidy chromosome number (5).

[b] Lacked mitotic cell division.

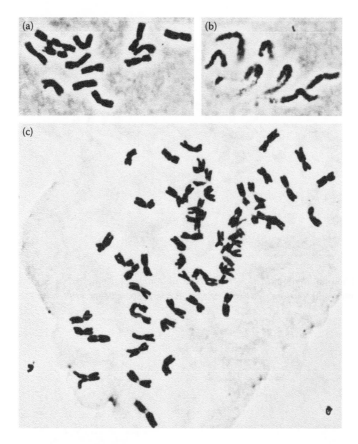

FIGURE 10.1 Photomicrographs of cells observed in morphogenic callus of barley. (a) $2n = 14$ (From R. J. Singh, unpublished results). (b) $2n = 7$, (c) $2n = 56$. (From Singh, R. J. 1986. *Theor. Appl. Genet.* 72: 710–716. With permission.)

(Figure 10.1b), tetraploid $(2n = 4x = 28)$, octoploid $(2n = 8x = 56)$ (Figure 10.1c), and triploid $(2n = 3x = 21)$ cells were also recorded in low frequencies. The occurrence of these chromosomal types varied from callus to callus. The calluses with the higher number of tetraploid cells were not as morphogenic as those that showed largely diploid cells. In contrast, diploid cells were comparatively lower in nonmorphogenic calluses (Table 10.4). The majority of aneuploid cells also carried chromosome structural changes such as ring, acrocentric, dicentric, telocentric, and fragment chromosomes (Figure 10.2). Similar results have been recorded in several plant species (Torrey, 1967; Sacristán and Wendt-Gallitelli, 1971; Novák, 1974; Orton, 1980; Singh, 1981; Murata and Orton, 1984).

It is interesting to note that aneuploid cell chromosome counts were around haploid, diploid, triploid, and tetraploid numbers. The uneven chromosome separation during cell division (Figure 10.3a) may have contributed to the origin of cells with haploid, triploid, and uncountable microchromosomes (Figure 10.3b). This supports the observations of Chen and Chen (1980), who suggested that triploid cells might have originated from $4x$ cells through reductional grouping of chromosomes, accompanied by multipolar formation. Thus, the loss of regenerability of a callus is attributed to an increased frequency of cells with polyploid and aneuploid chromosome numbers.

Tissue culture may play an important role in restructuring the chromosome after interspecific and intergeneric hybridization. Once a desired chromosome combination is accomplished, effort should be directed to bring the chromosome number to a genetic balance level by media modifications (Torrey, 1967; Evans et al., 1984). Multiplication of tetraploid cells in pea root callus was favored by the medium supplements yeast extract and kinetin, but by nutrient modifications Torrey (1967) obtained uniformly

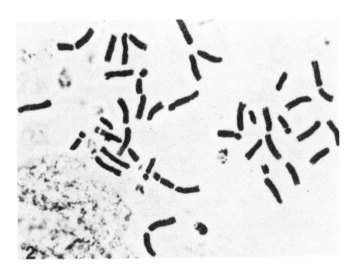

FIGURE 10.2 Photomicrograph of a cell observed in nonmorphogenic callus of barley showing 7 complete + 29 telocentric chromosomes. (From Singh, R. J. 1986. *Theor. Appl. Genet.* 72: 710–716. With permission.)

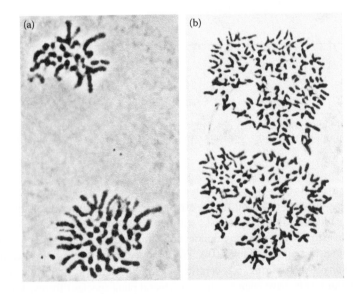

FIGURE 10.3 Photomicrographs of cells observed in nonmorphogenic callus of barley. (a) Telophase with unequal chromosome on each pole; (b) Two daughter nuclei with numerous microchromosomes. (From Singh, R. J. 1986. *Theor. Appl. Genet.* 72: 710–716. With permission.)

diploid cells. Similarly, Evans et al. (1984) established liquid cell cultures of the SU/SU genotype of *N. tabacum* ($2n = 48$) that maintained stable chromosome number for 6 years in culture.

10.2.4   GENETIC BACKGROUND OF THE EXPLANTS

Chromosomal instability induced in the culture and plant regenerability are often influenced by the genotype of the explants (Murashige and Nakano, 1967; Heinz et al., 1969; Kao et al., 1970; Sacristán, 1971; Okamoto et al., 1973; Asami et al., 1975; Novák et al., 1978; Scheunert et al., 1978; Jacobsen, 1981; Browers and Orton, 1982a,b; Wersuhn and Dathe, 1983; Bajaj and Gill, 1985;

Sree Ramulu et al., 1985; Rhodes et al., 1986; Lee and Phillips, 1987; Pijnacker and Sree Ramulu, 1990; Ohkoshi et al., 1991; Ruíz et al., 1992). Kao et al. (1970) analyzed the chromosomes of the suspension cultures of *Triticum monococcum* ($2n = 2x = 14$), *T. aestivum* ($2n = 6x = 42$), *Glycine max* ($2n = 2x = 40$), *Melilotus alba* ($2n = 2x = 16$), and *Haplopappus gracilis* ($2n = 2x = 4$). All the cultures except *H. gracilis* carried chromosome aberrations.

Asami et al. (1975) examined chromosomes in 6-month-old calluses of four aneuploid stocks (nulli-5 B tetra-5 D, ditelo 5 A^L, ditelo 5 B^L, and ditelo 5 D^L) and a disomic control ($2n = 6x = 42$) of *Triticum aestivum* cv. Chinese Spring. The disomic control showed normal chromosome complements of $2n = 42$ in 97% of the cells and only one cell carried $2n = 40$. By contrast, aneuploid stocks showed a wide range of chromosome distribution as shown below:

Stocks	Chromosome	Frequency of (Range) $2n = 42$ (%)
Disomic	40–42	97
Nulli-5 B tetra-5 D	24–86	22
Ditelo 5 A^L	27–126	52
Ditelo 5 B^L	28–85	58
Ditelo 5 D^L	35–84	68

Browers and Orton (1982a) recorded a clear-cut genotypic response on chromosome stability in a 9-week-old suspension culture of celery ($2n = 2x = 22$) *Apium graveolens* var. *rapeceum*. Keeping the culture conditions unchanged, PI 169001 contained 80% diploid and 20% nondiploid cells and was highly morphogenic. In contrast, suspension culture of PI 171500 was 31% diploid and 69% nondiploid and failed to regenerate plants.

Sree Ramulu et al. (1985) studied the chromosomes in calluses and suspension cultures of mono-haploid ($2n = x = 12$), dihaploid ($2n = 2x = 24$), and tetraploid ($2n = 4x = 48$) potatoes (*Solanum tuberosum*) after 7 and 14 days of culture. Polyploidization ($1x \rightarrow 2x \rightarrow 4x \rightarrow 8x$) was more rapid in monohaploid than in dihaploid and tetraploid potatoes. After 14 days of culture, chromosome number ranged from 12 to 96 for monohaploid, 24 to 48 for dihaploid, and 45 to 200 for tetraploid. The occurrence of aneuploid cells between two polyploid levels ($2x$ and $4x$; $4x$ and $8x$) could be attributed to chromosome elimination or addition to the nearest ploidy level due to unequal chromosome separation and chromosome fragmentations because aneuploid cells are physiologically and genetically more unbalanced than the cells with the exact multiple of basic chromosomes.

10.2.5 KIND OF MEDIA (SOLID VS. LIQUID)

The extent of chromosomal aberrations generated in solid (agar) and liquid (suspension) media is a controversial issue. Mitra et al. (1960) examined chromosomes of carrot ($2n = 2x = 18$) cells cultured in liquid and solid media. Liquid culture showed haploid ($2n = x = 9$), diploid ($2n = 2x = 18$), tetraploid ($2n = 4x = 36$), and octoploid ($2n = 8x = 72$) cells while diploid cells were recorded in the solid medium. Demoise and Partanen (1969), working with garden peony (*Paeonia suffruticosa*), $2n = 2x = 10$, suggested that liquid medium favors higher polyploids than solid medium, though results were not very conclusive.

A failure to regenerate plants from suspension and protoplasts is attributed to the extensive chromosome damage generated in culture. Karp et al. (1987a) examined the chromosomes of suspension cells and protoplasts derived from embryogenic callus of two wheat cultivars. Chromosome numbers ranged from 20 to 100. A majority of cells carried numbers from 30 to 39. Chromosome aberrations such as deletions, dicentrics, fragments, telocentrics, and megachromosomes were recorded. These chromosome aberrations created cells with unbalanced chromosomes, making them physiologically and genetically unfit to regenerate normal plants.

However, Singh and Harvey (1975a) contradict the above notion. They studied the chromosome constituents of solid and liquid cultures of *H. gracilis* ($2n = 2x = 4$). Polyploid cells decreased (28 days = 88% → 46 days = 75.9% → 122 days = 90% → 147 days = 66.7% → 220 days = 61.7%) in solid medium with time whereas the frequency of polyploid cells increased (14 days = 56.9% → 94 days = 59.2% → 175 days = 84.7% → 258 days = 92.9%) in liquid culture. Stable karyotype was observed in bulk of *Brachycome dichromosomatica* ($2n = 2x = 4$) suspension culture and only one cell line carried $2n = 5$ chromosomes after 3 years (Nagl and Pfeifer, 1988).

10.3 MECHANISM OF CHROMOSOMAL ABERRATIONS IN CULTURE

It is expected that all the cells of somatic tissues of plants are genetically identical and their chromosome constitutions are similar. However, it is not universally accepted. In pea roots and pith cells of tobacco, diploid and polyploid cells (polysomaty) have been observed in various frequencies and it occurs during tissue maturation.

Cell division in somatic tissues follows regularly the path of DNA synthesis → mitosis → cytokinesis. When cytokinesis is delayed, cells fail to divide equationally, resulting in doubling somatic chromosomes. This is called endomitosis (endoreduplication). The majority of reports suggest that polyploidization in cultures is generated through endoreduplication (D'Amato, 1985).

The occurrence of tetraploid ($2n = 4x = 28$) and octoploid ($2n = 8x = 56$) cells in both morphogenic and nonmorphogenic calluses in the barley suggests that polyploidization followed in a progressive fashion ($2x \rightarrow 4x \rightarrow 8x$). This suggests that chromosome doubling occurred during the culture because initially cultured embryos carried only diploid ($2n = 2x = 14$) cells. The observation of anaphase cells revealed that chromosomes separated during mitotic anaphase but failed to reach their respective poles. This may have been due to disturbed spindle formation (C-mitosis). Therefore, such cells will have a doubled chromosome number in the next mitotic division and the cycle will keep on repeating (Singh, 1986). It is assumed that some media components act like colchicine (Levan, 1938) and inhibit formation of the spindle during cell division causing chromosome doubling.

Unequal chromosome separation at mitotic telophase may be the cause of the occurrence of haploid cells. Haploid cells in appreciably low frequency have been recorded in *Daucus carota* ($2n = 2x = 18$) suspension cultures (Mitra et al., 1960), callus cultures of *Allium sativum* ($2n = 2x = 16$) (Novák, 1974), *Hordeum vulgare* ($2n = 2x = 14$) (Singh, 1986), *Triticum aestivum* ($2n = 6x = 42$) (Novák et al., 1978), and protoplast-derived calluses of *Solanum tuberosum* ($2n = 4x = 48$) (Sree Ramulu et al., 1984b). However, haploid cells may not compete with diploid or balanced polyploid cells during mitosis and they will be eliminated eventually.

The quantitative trait loci (QTLs) controlling callus growth and subsequent shoot regeneration have been identified by 222 markers in doubled haploid (DH) lines derived from the cross between Steptoe × Morex barleys (Mano et al., 1996). Morex contributed two QTLs for callus growth (*Qcg 1, Qcg 2*) and Steptoe contributed three QTLs for shoot regeneration (*Qsr 1, Qsr 2, Qsr 3*) while Morex contributed *Qsr 4*. In the multiple genic model, two QTLs for callus growth and four QTLs for shoot regeneration accounted for 47.8% and 49.8% of the total variation in the barley population, respectively. Chromosomal regions associated with green plant regeneration in anther culture-derived wheat plants have been identified (Torp et al., 2001). Four QTLs for 80% green plant were on 2AL, and 2BL and one QTL explaining 31.5% of the genetic variation for green plant formation were detected on 5BL.

Chromosome breakage in culture is a nonrandom phenomenon that occurs preferentially in the heterochromatic region or at euchromatin/heterochromatin junctions (Michaelis and Rieger, 1958; Döbel et al., 1978; Jørgensen and Anderson, 1989). Barley calluses produced high frequencies of telocentric, acrocentric, and fragment chromosomes (Singh, 1986). All the chromosomes in barley carry centromeric heterochromatin and a few of them have intercalary regions (Singh and Tsuchiya, 1982b). The occurrence of acrocentric chromosome suggests that breaks occur at interband regions and also occur at heterochromatic regions. That has been observed in the cultures of *Crepis capillaris* (Sacristán, 1971; Ashmore and Gould, 1981), maize (McCoy et al., 1982; Lee and Phillips, 1987; Benzion and Phillips, 1988; Fluminhan et al., 1996), oat (Johnson et al., 1987a,b), wheat (Fedak et al., 1987), and wheat–rye hybrids (Armstrong et al., 1983; Lapitan et al., 1984).

Sacristán (1971) studied the extent of preferential chromosome aberrations in 3-month-old calluses of *Crepis capillaris* ($2n = 2x = 6$). The nucleolus organizer chromosome showed the highest (82.3%) chromosomal rearrangement followed by the shortest chromosome (64.6%). The longest chromosome showed the least (47.0%). It was reasoned that in the case of the SAT-chromosome the break region corresponds to a region of late DNA synthesis.

In this respect, the mode of breakage in chromosomes induced in culture is not different from that produced by X-ray treatment or any other mutagen (Khush and Rick, 1968a; Natarajan and Ahnström, 1969; Jancey and Walden, 1972; Gill et al., 1980; Inomata, 1982; Nuti Ronchi, 1986a,b). Khush and Rick (1968a), working with tomato, found that about 60% of the breaks induced by X-ray treatment occurred in heterochromatin. Thus, late-replicating DNA in heterochromatic regions is more vulnerable to breakage than the euchromatic regions (Sacristán, 1971; Döbel et al., 1978; McCoy et al., 1982; Armstrong et al., 1983; Murata and Orton, 1984).

The decrease in the frequency of heritable variation with increasing exposure to 2, 4-D in regenerated soybean plants expressed partial sterility, complete sterility, curled and wrinkled leaves, dwarfism, chlorophyll deficiency, chlorophyll chimera, indeterminate to determinate flowering, lack of unifoliate, yellow-edged cotyledons, and isozyme variants (Shoemaker et al., 1991). These mutants were not examined cytologically and an inheritance study was not conducted.

10.4 CHROMOSOMAL ABERRATIONS IN REGENERATED PLANTS

Phenotypic variability among cell and tissue culture-derived regenerants may be attributed to epigenetic, genetic, and chromosomal changes induced by the culture conditions (Evans and Reed, 1981; Evans et al., 1984; Sibi, 1984; D'Amato, 1985; Karp, 1986; Vasil, 1988; Stelly et al., 1989; Wersuhn, 1989; Oono, 1991; Skirvin et al., 2000). The culture-induced variants have been called "calliclones" (Skirvin and Janick, 1976; Skirvin, 1978), "protoclones" (Shepard et al., 1980), and a widely used term "somaclones" (Larkin and Scowcroft, 1981). The frequency of somaclonal variation is at a higher rate (up to 10% per cycle of regeneration) than chemical- or radiation-induced mutation. This makes somaclonal variation a viable alternative to mutagenesis and a valuable tool for a plant breeder to introduce variation into breeding programs (Skirvin et al., 2000).

Epigenetic variations are due to the results of culture stress and these variations are not transmitted from generation to generation. Thus, these changes are acquired traits and are not genetically controlled.

The genetic variations are induced during culture due to single nuclear gene mutations. The mutants exhibit Mendelian inheritance. A large number of plant species have been regenerated from cell and tissue cultures carrying somaclonal variation; the nature of mutation has been elucidated in only a few cases (Table 10.5). van den Bulk (1991) summarizes application of cell and tissue culture for disease resistance breeding. A total of 14 crops were listed. Inheritance was not determined in 19 studies. In banana, peach, potato, and sugarcane, disease resistance was transmitted after vegetative reproduction and resistance was transmitted to the progeny in tomato, sugarcane, rice, rape seed, maize, and celery. Moreover, heritable isozyme mutants have been recorded in the regenerants of broccoli (Orton and Browers, 1985), wheat (Ryan and Scowcroft, 1987), wild barley (Breiman et al., 1987), and blackberry (McPheeters and Skirvin, 2000). Both dominant and recessive mutants isolated from culture are similar to those mutants obtained spontaneously in nature or induced by mutagens (Neuffer and Sheridan, 1980; Evans and Sharp, 1983, 1986; Gavazzi et al., 1987; Ullrich et al., 1991). Genetic variations for quantitative traits (maturity, height, lodging, seed yield, seed weight, protein, and oil contents) were derived from tissue culture of the soybean; however, the magnitude of the genetic variation was relatively small (Hawbaker et al., 1993).

TABLE 10.5

Recessive and Dominant Gene Mutations Observed in Regenerated Plants

Crops	Mutants	Authority
Recessive mutants		
Oryza sativa	Early heading, albino, short culm, sterility	Fukui (1983)
	Dwarf mutants	Sun et al. (1983)
Lactuca sativa	Seedling mutants	Engler and Grogan (1984)
Lycopersicon esculentum	Fruit color, male sterility, seedling mutants	Evans and Sharp (1983)
	Disease resistance	Evans et al. (1984)
	Chlorophyll mutants	Buiatti et al. (1985)
Brassica napus	Yellow seed color	George and Rao (1983)
Medicago sativa	Flower color	Groose and Bingham (1984)
Zea mays	Shrunken kernel	McCoy and Phillips (1982)
Nicotiana sylvestris	Streptomycine sulphate resistant	Maliga (1981)
Glycine max	Male sterility	Graybosch et al. (1987)
Sorghum biclor	Male sterility	Elkonin et al. (1994)
Dominant mutants		
Nicotiana tabacum	Herbicide resistance	Chaleff and Parsons (1978)

A majority of morphological variants observed among the regenerated plants is due to numerical (aneuploidy, polyploidy) and structural (deletions, duplications, interchanges, inversions) chromosome changes induced during the culture. Generally, a high frequency of regenerants from diploid species carries normal chromosome complements. On the other hand, regenerants from polyploid species such as sugarcane, wheat, oat, triticale, potato, and tobacco have a comparatively higher frequency of plants with aberrant chromosome numbers. This is due to the fact that polyploid species can tolerate to a greater extent aneuploidy because of the buffering capacity of the polyploid condition than true diploid species.

Despite many potential uses claimed for somaclonal variation, and substantial efforts by scores of individuals, the fact remains that thus far there is not a single example of any significantly important new variety of any major crop species developed as a result of somaclonal variation (Vasil, 1990). It may not be true for the horticultural crops where tissue culture-derived somaclones were released as cultivars (Skirvin et al., 1994).

10.4.1 Diploid versus Polyploid Species

Chromosomal variations among regenerated plants of several diploid species are shown in Table 10.6. It is clearly seen that a very high proportion of regenerated plants carried normal chromosome complements. Gould (1979) observed no chromosomal and phenotypically aberrant *Brachycome dichromosomatica* ($2n = 4$) regenerated plants from a year-old culture; this suggests again that diploid species are highly stable in culture. The exceptions are *Pisum sativum* and *Brassica oleracea*. In *Pisum sativum*, 9 of the 20 regenerants were mixoploid and in *Brassica oleracea* ($2n = 18$), of the 71 regenerants from stem pith explants, 6 plants were diploid, 54 plants tetraploid, and 11 octoploid. The occurrence of polyploid plants may be attributed to the source of explants. Pith cells carry diploid and polyploid cells (polysomaty) (Murashige and Nakano, 1967). This suggests that during shoot and root morphogenesis in diploid species, cells with unbalanced chromosome numbers cannot compete with balanced diploid cells in cell division. This enhances the regeneration of a large number of plants with normal diploid chromosome complements.

TABLE 10.6
Chromosomal Variations among Regenerated Plants in the Diploid Species

Species	2n		Frequency of Regenerants with						Authority
		2n = 2x	2n = 4x	2n = 8x	Mixoploid	Aneuploids	% 2n = 2x	
Apium graveolens	22	44	0	0	0	2	95.7	Orton (1985)
Hordeum vulgare	14	42	0	0	0	0	100.0	Karp et al. (1987b)
Lolium multiflorum	14	52	1	0	0	0	98.1	Jackson et al. (1986)
Lolium multiflorum	14	52	0	0	0	0	100.0	Jackson and Dale (1988)
Lotus corniculatus	24	91	0	3	6	0	91.0	Damiani et al. (1985)
Pennisetum americanum	14	30	0	0	0	0	100.0	Swedlund and Vasil (1985)
Pisum sativum	14	11	0	0	9	0	55.0	Natali and Cavallini (1987)
Secale cereale	14	14	0	0	0	0	100.0	Lu et al. (1984)
Sorghum arundinaceum	20	10	0	0	0	0	100.0	Boyes and Vasil (1984)
Zea mays	20	108	1	0	0	1	98.2	Edallo et al. (1981)
Zea mays	20	119	0	0	3	2	96.0	McCoy and Phillips (1982)

By contrast, morphogenic capability in polyploid species is not influenced by aneuploidy, polyploidy, and chromosome structural changes. A few polyploid species such as potato ($2n = 4x = 48$), wheat ($2n = 6x = 42$), and triticale ($2n = 6x = 42$) will be cited here as examples:

1. Potato (*Solanum tuberosum*)

   Cultivated potato is vegetatively propagated and autotetraploid crop. Several authors have regenerated plants by mesophyll protoplasts and shoot tip cultures of potato and have recorded a considerable amount of genetic and chromosomal variability (Table 10.7).

   Shepard et al. (1980) isolated several morphological variants among regenerants from leaf mesophyll protoplasts of the potato cultivar Russet Burbank. The variants showed compact growth habit, an earliness for tuber set, smooth and white skin, high yield, a requirement for less light (13 h) to flower than the parent (16 h), and resistance to early and late blight. The inheritance of these important traits was not determined. Gill et al. (1987) studied these variants cytologically at pachynema of meiosis. All the variants carried the expected $2n = 4x = 48$ chromosomes. The high yielding variant did not carry chromosomal aberrations, but in the remaining mutants, the phenotypic alterations were due to interchanges, deletions, inversions, and duplications.

   It is evident from Table 10.7 that tetraploid potato regenerants can tolerate a high number of chromosomes; for example, Gill et al. (1986) identified 27.8% of the regenerated plants with $2n = 72$–96 chromosomes; Creissen and Karp (1985) found 21.5% plants with $2n = 73$–96 chromosomes. Furthermore, addition and deletion of one or two chromosomes around a mode of $2n = 48$ showed little morphological change or sometimes plants were indistinguishable from plants carrying normal $2n = 48$ chromosomes. Similar results have been reported in tobacco (Sacristán and Melchers, 1969; Nuti Ronchi et al., 1981).

   Rietveld et al. (1993) produced about 1000 tuber disc-derived potato plants from cultivars Kennebec, Russet Burbank, and Superior. Plant regenerability trait was genotype-dependent. Of the three cultivars, Russet Burbank somaclones expressed the greatest variability for most traits; however, only from 1.0% to 1.3% somaclonal population exhibited morphological aberrations.

2. Wheat (*Triticum aestivum*)

   Hexaploid wheat ($2n = 6x = 42$) is an allohexaploid and contains three (A B D) genomes. Karp and Maddock (1984) studied chromosomes of 192 regenerated plants derived from immature embryo callus of four hexaploid wheat cultivars. A total of 71% of the regenerants carried the expected $2n = 42$ chromosomes and 29% of the plants were aneuploid

---

TABLE 10.7

Chromosomal Variation (%) among Regenerated Plants in Potato (*Solanum tuberosum*, $2n = 4x = 48$)

Species	48	48[a]	48[b]	Authority
S. tuberosum cv. Majestic	57.0	21.5	21.5	Creissen and Karp (1985)
S. tuberosum cv. Maris Bard	63.6	8.4	28.0	Fish and Karp (1986)
S. tuberosum cv. Bintje	32.6	55.8	11.6	Sree Ramulu et al. (1983)
S. tuberosum cv. Bintje	71.0	21.0	7.9	Sree Ramulu et al. (1984a)
S. tuberosum cv. Bintje	39.3	46.1	14.6[c]	Sree Ramulu et al. (1986)
S. tuberosum cv. Russet Burbank	61.1	11.2	27.8	Gill et al. (1986)

[a] One or two chromosomes.
[b] Higher than 49 chromosomes.
[c] Mixoploid (included).

TABLE 10.8

Chromosome Constitution in Somatic Cells of Regenerated Plants from Immature Embryo-Derived Calluses of Triticale cv. Welsh

	Number of Plants with 2n Chromosome				
Age of Culture	42	41	Others	% 2n = 42	Total no. Plants
1 month	33	7	14	61.1	54
6 months	10	12	31	19.6	51
Control 1	43	5	1	87.8	49
Control 2	12	5	3	60.0	20

Source: From Armstrong, K. C., C. Nakamura, and W. A. Keller. 1983. *Z. Pflanzenzüchtg.* 91: 233–245. With permission.

$(2n = 38-45)$. The most frequently observed numbers were $2n = 41$ or 43. Chromosome interchanges were recorded; however, no plants with chimaerism, polyploidy, or inversion were found. In contrast, all the regenerated plants carried normal chromosome complements of $2n = 42$ in the studies of Shimada and Yamada (1979), Ahloowalia (1982), and Ozias-Akins and Vasil (1982). However, they studied only a few plants cytologically.

It is essential to study the chromosomes of regenerated plants at both mitosis and meiosis. A wrong conclusion can be drawn unless both stages are analyzed. Fedak et al. (1987) regenerated four plants from suspension cultures of wheat cultivar Chinese Spring. All four plants had $2n = 42$ chromosomes, but they carried chromosomal interchanges, while meiosis was normal in the parent.

Chromosomal aberrations were also higher among the regenerated plants of tetraploid wheat ($2n = 4x = 28$). Bennici and D'Amato (1978) recorded a wide range of chromosomes in root tips ($2n = 6-30$) and shoot tips ($2n = 6-756$). However, a majority of root and shoot somatic tissues carried $2n = 28$ chromosomes.

3. Triticale ($2n = 6x = 42$)

Like hexaploid wheat, triticale also is an allohexaploid (A B R) genome and regenerates plants from mature and immature embryos by organogenesis and embryogenesis (Sharma et al., 1981; Nakamura and Keller, 1982; Lazar et al., 1987; Bebeli et al., 1988). Armstrong et al. (1983) studied the chromosomes of regenerants derived from 1- to 6-month-old calluses of triticale cultivar Welsh. Plants regenerated from 1-month-old callus carried $2n = 42$ chromosomes in 61.1% plants while regenerants from 6-month-old callus showed 42 chromosomes in 19.6% of the plants (Table 10.8). Chromosome aberrations such as telocentrics, duplications, deletions, and interchanges were recorded. It is evident that chromosome breakage occurred near or adjacent to heterochromatic regions near centromeres and chromosome aberrations occurred during culture. Chromosome breakage at heterochromatic regions was also demonstrated by Lapitan et al. (1984) in a study of 10 regenerated amphidiploid ($2n = 56$) plants of wheat × rye (AABBDDRR) by utilizing Giemsa C-banding technique. They recorded three wheat/rye and one wheat/wheat chromosome translocation, seven deletions and five amplifications of heterochromatic bands of rye chromosomes. Twelve of the 13 break points in chromosomes involved in translocations and deletions occurred in heterochromatin.

10.4.2 AGE OF CALLUS

Based on chromosome analysis of the calluses, one should expect to obtain a higher frequency of regenerants with altered chromosomes from the older cultures (Choi et al., 2000b, 2001). This has been demonstrated in maize and oat. Lee and Phillips (1987) determined the chromosomes of 78 regenerants from 3- to 4-month-old cultures of maize ($2n = 2x = 20$). All the plants carried normal chromosome constitution. In contrast, 189 plants regenerated from 8- to 9-month-old cultures showed 91 plants (48%) cytologically abnormal. Interchanges were the most frequent alterations (38/45); a high frequency (35%) of deficient chromosomes was also detected. A low frequency of tetraploids and a trisomic plant were also identified. In a similar study, McCoy and Phillips (1982) observed fewer cytologically abnormal plants (4/59) in an 8-month-old culture of maize than those reported by Lee and Phillips (1987).

McCoy et al. (1982) conducted an extensive study on chromosome analysis, at meiosis, of plants regenerated from 4-, 8-, 12-, 16-, and 20-month-old cultures of the oat cultivars Lodi and Tippecanoe. Oat is an allohexaploid ($2n = 6x = 42$) with A, B, and C genomes. They demonstrated that the frequency of chromosomal aberrations was increased with the culture age and was also genotype dependent. After 4 months in culture, 49% of the Lodi regenerants carried altered chromosomes and 12% of the Tippecanoe regenerates were abnormal. But after 20 months in culture, 88% of the Lodi and 48% of the Tippecanoe regenerated plants were cytogenetically abnormal. Chromosomal aberrations such as interchanges, monosomics, trisomics, and deletions were frequently observed. Long-term (4–11 years) culture-derived somaclones of rice contained deletion in plastid genome and that was associated with the accumulation of starch granules (Kawata et al., 1995).

10.4.3 Source of Explant Materials

It has been demonstrated that regenerants from protoplast culture generally carry a higher degree of aneuploidy and phenotypic variability than those plants regenerated from immature embryo or shoot tip culture. Newell et al. (1984) reported in *Brassica napus* ($2n = 38$) that regenerated plants carried 87% $2n = 38$, 7.8% monosomics, 2.6% trisomics, and 1.3% tetraploids. In contrast, protoplast-derived regenerates were 44% $2n = 38$, 20% hypodiploid, and 36% tetraploid or hypotetraploid.

Kanda et al. (1988) studied chromosomes in seven plants regenerated from 7- to 10-month-old protoplast callus of rice. They observed one diploid ($2n = 24$), four tetraploids ($2n = 48$), and two aneuploids ($2n = 46$) plants. Contrarily, Kobayashi (1987) regenerated 25 plants from orange (*Citrus sinensis*) protoplasts and recorded no significant variations among somaclones in leaf and flower morphology, leaf oil, isozyme-banding pattern, and chromosome numbers.

10.4.4 Culture Conditions

Chromosomal aberrations among regenerated plants are influenced by the media components and also by the way cultures are maintained. Fish and Karp (1986) obtained a higher frequency of euploid potato plants by media modifications. They screened cytologically 178 protoplast-derived and observed 63.6% of the regenerants carried $2n = 48$ chromosomes, but in their earlier studies (Karp et al., 1982) only 4% of the regenerants were euploids.

Ogihara (1981), working with *Haworthia setata*, observed an average of 89.2% diploid ($2n = 14$), 6.2% tetraploid ($2n = 28$), 0.4% monosomic ($2n = 13$), and 4.1% modified chromosomes among regenerants from culture maintained by subculturing. Chromosome aberrations were higher in plants from the culture maintained by cloning: diploid = 42.2%; tetraploid = 25.8%; and modified chromosomes = 34.0%.

10.5 CHROMOSOMAL ABERRATIONS IN HAPLOID CALLUS AND THEIR REGENERANTS

Chromosomal variability is comparatively higher in anther-derived calluses and their regenerated plants than from plants regenerated from somatic tissues. Chromosome numbers of anther-derived plants are often doubled (Sacristán, 1971, 1982; Bennici, 1974; Mix et al., 1978; Chen and Chen, 1980; Wenzel and Uhrig, 1981; Keathley and Scholl, 1983; Santos et al., 1983; Toriyama et al., 1986; Pohler et al., 1988; Wersuhn and Sell, 1988; Kudirka et al., 1989).

Sacristán (1971) studied karyotypic changes in callus cultures from haploid ($2n = 3$) and diploids ($2n = 6$) *Crepis capillaris*. The degree of polyploidization and chromosomal interchanges in haploid culture was considerably higher than in diploid culture.

In anther culture, if plants regenerate through embryogenesis, the majority of them are haploid while plants regenerated from callus are diploid. It has been suggested that usually many cells take part in the formation of a bud, while single cells give rise to somatic embryos (Constantin, 1981). In *Brassica napus* anther culture, Sacristán (1982) observed that plants derived through embryogenesis were haploid, while most of the plants regenerated from callus were diploid, and no haploid plants were obtained. Toriyama et al. (1986) regenerated 15 mature plants from protoplasts that were isolated from cell suspensions of anther callus in rice. Four plants were haploid ($2n = 12$) and 11 plants were diploid ($2n = 24$). Diploids were uniform morphologically but seed set varied from 0% to 95%. Pohler et al. (1988) studied chromosomes of callus tissues and regenerated plants from anther culture of hexaploid triticale. Chromosome variation was the highest ($2n = 18–43$) in the calluses and it was the least pronounced in plants ($2n = 20–43$) derived from embryoids. In anther calluses, 24.8% cells carried $2n = 21$ and 15.6% cells $2n = 42$ chromosomes. A total of 184 plants were analyzed by root tip count. Plants derived from embryoids showed a higher frequency (83.3%) of euhaploids ($2n = 3x = 21$) than plant regenerants from calluses (31.6%).

Utilizing liquid culture medium, Uhrig (1985) regenerated 313 plants from anther culture of diploid ($2n = 2x = 24$) potato. The majority of plants (80.2%) retained diploid chromosome numbers; however, low frequencies of haploid (0.6%) and triploid (0.3%) plants were obtained. Chromosome doubling occurred in 13.7% of the plants and mixoploidy was recorded in 5.2% of plants. A large number of embryoids can be generated with fewer chromosomal aberrations by suitable genotype and media modifications.

Tempelaar et al. (1985) recorded that monohaploid potato ($2n = 12$) polyploidized at a faster rate than dihaploid ($2n = 24$) and tetraploid ($2n = 48$). Of 36 plants studied cytologically from monohaploid culture, 14 were dihaploid, tetraploid or mixoploid and only two plants were monohaploids. Karp et al. (1984) scored nearly 100% doubled monohaploids ($2n = 2x = 24$) from a single leaf regeneration cycle of monoploid ($2n = x = 12$) potatoes and 50% of the regenerants from doubled monohaploid leaves were tetraploids ($2n = 4x = 48$). Very few mixoploids and aneuploids were found. Wenzel and Uhrig (1981) regenerated 6000 androgenic clones from microspores ($n = 12$) of dihaploid potato ($2n = 24$). The chromosome numbers of about 90% of these plants doubled spontaneously and yielded fertile plants, and only 10% of the plants were monohaploid.

Pijnacker and Ferwerda (1987) studied karyotypic changes in cell suspension, calluses, and their regenerants of an S-(2-amino ethyl) cystein resistant cell line of a dihaploid ($2n = 24$) potato. Chromosome numbers in cell suspension varied between $2n = 33$ and 151, with peaks near 40 and 75 chromosomes. In a comparison of a 2-year-old callus culture and a morphogenic callus, chromosome numbers ranged from 36 to 217 and 33 to 130, respectively. However, the regenerates carried $2n = 34$ or 35 chromosomes.

Bennici (1974) studied chromosomes in roots from old and young calluses of haploid ($2n = x = 9$) *Pelargonium*. In roots from young calluses, 71.42% of cells were diploid ($2n = 18$) while roots from old calluses had only 13.81% diploid cells, and the remaining cells possessed polyploid and aneuploid chromosome numbers.

Wersuhn and Sell (1988) recorded mainly tetraploid plants from anther culture of *N. tabacum* cv. Samsun ($2n = 48$). Plants regenerated from anther culture were more highly aberrant cytologically than plants regenerated from seed.

Qiren et al. (1985) regenerated 1715 plants from anther culture of rice. Chromosome counts revealed: diploid ($2n = 24$), 35.7%; haploid ($2n = 12$), 49.1%; polyploids, 5.1%; aneuploids, 10.2%.

A high frequency of regenerants from anther culture is albino in cereals such as wheat (Marsolais et al., 1984), barley (Mix et al., 1978), and rye (Wenzel et al., 1977). Of the 390 anther-derived plants in rye, 97 plants were green and 293 were albino. A chromosome count of 100 plants showed 31 plants with $2n = x = 7$, 63 plants with $2n = 2x = 14$, 5 plants with $2n = 1x + 2x$, and 1 plant with $2n = 4x = 28$ (Wenzel et al., 1977). Similarly, in barley, Mix et al. (1978) obtained 600 green plants among 4000 regenerants and only 20% of the green plants were haploid. The remaining plants were diploid, triploid, tetraploid, and aneuploid. Diploid plants were completely fertile.

10.6   CHROMOSOMAL ABERRATIONS IN SOMATIC HYBRIDS

Somatic hybridization by protoplast fusion is an alternative approach to transfer alien genes where sexual crosses are not successful in distantly-related and incompatible species (Bhojwani et al., 1977; McComb, 1978; Kumar and Cocking, 1987; Wolters et al., 1994a; Jacobsen et al., 1995; Waara and Glimelius, 1995). Several interspecific somatic hybrids involving species of the genera *Brassica*, *Datura*, *Daucus*, *Medicago*, *Nicotiana*, and *Petunia* and have been reported (Table 10.9) and only a few intergeneric somatic hybrids have been produced (see D'Amato, 1985). Somatic hybridization technique helped the resynthesis of alloployploid crops to create their genetic variability and restore ploidy level and heterozygosity after breeding at reduced ploidy level in polyploid crops (Waara and Glimelius, 1995). It is essential that one of the fusants should regenerate complete plants through protoplast.

Although considerable efforts have been made to identify somatic hybrids based on isozyme-banding patterns and molecular approaches, the information on systematic chromosome analysis is lacking. Somatic hybrid plants are expected to carry the somatic chromosome ($2n$) constitutions of both parents (Table 10.9). For example, the somatic hybrid of tetraploid potato ($2n = 4x = 48$) and tomato ($2n = 2x = 24$) should possess $2n = 6x = 72$ chromosomes.

The chromosomal stability of a somatic hybrid depends on the degree of closeness between the genomes of the fusion parents. Somatic hybrids can be divided into three categories: (A) stable, (B) partially stable, and (C) unstable.

10.6.1   STABLE SOMATIC HYBRIDS

Carlson et al. (1972) generated stable somatic hybrid plants ($2n = 42$) by protoplast fusion of *Nicotiana glauca* ($2n = 24$) and *N. langsdorffii* ($2n = 18$). All the regenerated plants were similar to the sexually synthesized amphidiploids. Evans et al. (1982) recovered stable somatic hybrids ($2n = 96$) of *N. tabacum* ($2n = 48$) and *N. nesophila* ($2n = 48$). Variation in clones was attributed to cytoplasmic segregation and mitotic recombination. Schenck and Röbbelen (1982) synthesized for the first time rapeseed (*Brassica napus*, $2n = 38$) by fusion of protoplast from *B. oleracea* ($2n = 18$) and *B. campestris* ($2n = 20$). Hybrid plants flowered and set seed. Schnabelrauch et al. (1985) obtained stable and unstable somatic hybrids in *Petunia parodii* and *P. inflata*. Fish et al. (1988) generated somatic hybrids of dihaploid *Solanum tuberosum* ($2n = 24$) and *S. brevidens* ($2n = 24$). Tetraploid ($2n = 48$) plants were completely stable (Table 10.9).

Heath and Earle (1995) produced fertile 51 amphiploids (aacc) and one putative hexaploid (aacccc) *Brassica napus* by somatic hybrids through protoplast fusion of *Brassica oleracea* var. *botrytis* and *Brassica rapa* var. *oleifera* (high erucic content in seed oil; 22:1). An erucic acid content as high as 57.4% was found in the seed oil of one regenerated plant. Hybrids recovered carried large seed size, lodging resistance, and nonshattering seed pods.

10.6.2   PARTIALLY STABLE SOMATIC HYBRIDS

Chromosome elimination during callus growth and also in regenerants occurs in some somatic hybrids (Dudits et al., 1977; Krumbiegel and Schieder, 1979; Hoffmann and Adachi, 1981; Lazar

TABLE 10.9
Chromosome Variations in Regenerated Plants by Protoplast Fusion

Somatic Hybrids	Expected 2n	Observed 2n	Authority
Arabidopsis thaliana (2n = 8x = 40) + Brassica campestris (2n = 20)	60	35–45	Hoffmann and Adachi (1981)
Brassica oleracea (2n = 18) + B. campestris (2n = 20)	38	8–54	Schenck and Röbbelen (1982)
Brassica napus (2n = 38) + Arabidopsis thaliana (2n = 10)	48	38–86	Forsberg et al. (1994)
Medicago sativa (2n = 32) + M. coerulea (2n = 16)	48	48	Pupilli et al. (1995)
Lycopersicon esculentum (2n = 24) + L. peruvianum (2n = 48)	72	72	Parokonny et al. (1997)
Lycopersicon esculentum (2n = 24) + Solanum lycopersicoides (2n = 24)	48	48–68	Handley et al. (1986)
Nicotiana glauca (2n = 24) + N. langsdorffii (2n = 18)	42	42	Carlson et al. (1972)
Nicotiana glauca (2n = 24) + N. langsdorffii (2n = 18)	42	60–66	Morikawa et al. (1987)
Nicotiana tabacum (2n = 48) + Nicotiana nesophila (2n = 48)	96	96	Evans et al. (1982)
Nicotiana sylvestris (2n = 24) + N. plumbaginifolia (2n = 20)	44	33–54	Parokonny et al. (1994)
Oryza sativa (2n = 24) + Porteresia coarctata (2n = 48)	72	72	Jelodar et al. (1999)
Petunia parodii (2n = 14) + P. inflata (2n = 14)	28	14–36	Schnabelrauch et al. (1985)
Solanum melongena (2n = 24) + S. sisymbriifolium (2n = 24)	48	38–48	Gleddi et al. (1986)
Solanum tuberosum (2n = 24) + S. brevidens (2n = 24)	48	45–89	Fish et al. (1988)
Solanum tuberosum (2n = 24) + Lycopersicon esculentum (2n = 24)	48	50–72	Melchers et al. (1978)
Solanum tuberosum (2n = 24) + Solanum brevidens (2n = 24)	48	48–72	Gravrilenko et al. (2002)

et al., 1981; de Vries et al., 1987). Krumbiegel and Schieder (1979) selected 13 somatic hybrids of *Datura inoxia* (2*n* = 24) and *Atropa belladona* (2*n* = 48). Chromosome numbers ranged from 84 to 175. Chromosome instability was recorded, but no evidence of chromosome elimination was observed. Hoffmann and Adachi (1981) studied somatic hybrids of *Arabidopsis thaliana* (2*n* = 40) and *Brassica campestris* (2*n* = 20) and observed elimination of chromosomes from both species. Chromosome analysis of 10 plants showed 2*n* = 35–45 chromosomes; the somatic hybrid should have consisted of 2*n* = 60 (Table 10.9). Chromosome numbers in different vegetative organs were stable. Chromosome structural changes were observed. This suggests that chromosome instability arose later in the stage of the plant growth because initial somatic hybrid plants were stable (Gleba and Hoffmann, 1978). de Vries et al. (1987) recorded no preferential loss of species-specific chromosomes in the *Nicotiana plumbaginifolia* (2*n* = 48) × *Solanum tuberosum* (2*n* = 12) somatic hybrid.

10.6.3 UNSTABLE SOMATIC HYBRIDS

Preferential loss of chromosomes occurs in those somatic hybrids where parents are genomically incompatible and distally related (Binding, 1976; Binding and Nehls, 1978; Maliga et al., 1978; Zenkteler and Melchers, 1978; Wetter and Kao, 1980; Pental et al., 1986). This may be attributed to the asynchronous mitotic cycle (Bennett et al., 1976). Uniparental chromosome elimination has been found in sexual hybrids of *Hordeum vulgare* (2*n* = 14) and *H. bulbosum* (2*n* = 14, 28), where *H. bulbosum* chromosomes are eliminated during early embryo development and only haploid (2*n* = *x* = 7) embryos with *H. vulgare* chromosomes are recovered (Kasha and Kao, 1970; Subrahmanyam and Kasha, 1973). A similar phenomenon was observed in wheat × maize (Laurie and Bennett, 1989), wheat × pearl millet (Laurie, 1989), and oat (2*n* = 6*x* = 42) × maize (2*n* = 2*x* = 20) (Rines and Dahleen, 1990) crosses. In these crosses, maize and pearl millet chromosomes were eliminated during the first few cell-division cycles in most of the embryos. Isolated oat-maize addition lines are an excellent genetic stock to study the expression of maize genes in oat (Muehlbauer et al., 2000)

Maliga et al. (1978) analyzed chromosome numbers in somatic hybrids of *Nicotiana tabacum* (2*n* = 48) and *N. knightiana* (2*n* = 24) and recorded variation in chromosome numbers within individual plants. Variegation in leaf and flower color and segregation for morphological traits in vegetatively multiplied plants were attributed to segregation of chromosomes in the somatic cells, a result of numerical chromosomal instability.

Wetter and Kao (1980) recorded preferential loss of *Nicotiana glauca* chromosomes in the *Glycine max* (2*n* = 40) + *N. glauca* (2*n* = 18) somatic hybrid. Similarly, Pental et al. (1986) observed the elimination of tobacco chromosomes in *N. tabacum* (2*n* = 48) + *Petunia hybrida* (2*n* = 14) somatic hybrids.

Binding and Nehls (1978) investigated cytologically somatic hybrid calluses of *Vicia faba* (2*n* = 12) and *Petunia hybrida* (2*n* = 14). The putative fusants carried predominantly nuclei or chromosomes of one or the other species and a few chromosomes of the other species. If a cell carried predominantly *Vicia faba* chromosomes, *Petunia hybrida* chromosomes were eliminated and *vice versa*.

Melchers et al. (1978) identified somatic hybrids of potato (2*n* = 24) and tomato (2*n* = 24) based on morphological features, isozyme-banding patterns, and chromosome counts. Somatic hybrid plants are expected to carry 2*n* = 48 chromosomes. Chromosome numbers in three plants ranged from 50 to 56 and a fourth plant carried 2*n* = 72 (Table 10.9). The fourth plant was developed by triple fusion (two potato + one tomato protoplast) because this particular regenerant contained significantly fewer staining bands of the tomato small subunit. Triple fusion is rather common in protoplast fusion (Schenck and Röbbelen, 1982; Morikawa et al., 1987). Ninnemann and Jüttner (1981) analyzed volatile patterns of potato + tomato hybrids of Melchers and found no evidence of preferential elimination of chromosomes. However, they did not study the chromosomes. This

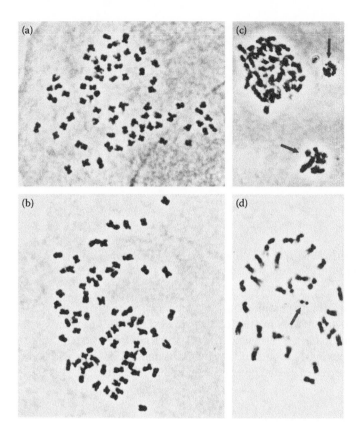

FIGURE 10.4 Photomicrographs of cells observed in regenerants of potato ($2n = 48$) and tomato ($2n = 24$) somatic hybrids. (a) $2n = 72$; (b) $2n = 68$; (c) A cell showing the chromosome elimination (arrows); (d) $2n = 29$ (note a minute chromosome—arrow).

contradicts the statement of Melchers et al. (1978) that some of the plants are chimeras containing tissues with different chromosome numbers.

A detailed chromosome analysis of potato ($2n = 48$) and tomato ($2n = 24$) somatic hybrids developed by Shepard et al. (1983) was conducted by the author. Somatic hybrid plants are expected to carry $2n = 72$ and the initial hybrid did carry $2n = 72$ chromosome (Figure 10.4a). Chromosome numbers in vegetative cuttings of potato + tomato somatic hybrids differed from cell to cell, root to root, cutting to cutting, plant to plant, and hybrid to hybrid. Chromosome numbers ranged from $2n = 24$ and $2n = 120$. It appears that each organ is a mixture of cells composed of various chromosome numbers. The deviation from the expected 72 chromosomes in all the cuttings of somatic hybrid is due to chromosome elimination (Figure 10.4b–d). The occurrence of 48(+4) (Figure 10.5a) and 24 (Figure 10.5b), chromosome cells suggests that these cells probably carry tomato and potato (plus a few tomato) genome chromosomes, respectively. It is evident that chromosome elimination is gradual and occurs in each mitotic cycle. In contrast to this observation, Shepard et al. (1983) reported that somatic hybrid cuttings displayed chromosome numbers ranging from 62 to 72 depending on the cutting. The most frequent encountered chromosome number for the "Rutgers" hybrid was 70, with greater variability observed for the "Nova" hybrids. These data indicate a degree of mitotic instability and some chromosome segregation in vegetative cuttings but not wholesale chromosome elimination. This statement is without evidence. Wolters et al. (1994b) cytologically examined a total of 107 somatic hybrids of tomato + potato. Most (79%) hybrids were aneuploid, some were hyperploid, and others were hypoploid. Chromosome number ranged from 34 to 72. Preferential tomato chromosome elimination in the tomato + potato fusion hybrids and

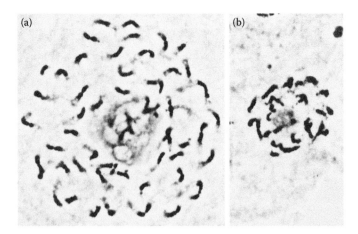

FIGURE 10.5 Photomicrographs of cells observed in regenerants of potato ($2n = 48$) and tomato ($2n = 24$) somatic hybrids. (a) A prophase cell with $2n = 52$; (b) A prophase cell with $2n = 24$.

variable number of tomato chromosomes (6–11) in BC_1 were recorded by Garriga-Calderé et al. (1997). This contradicts the assumption of Shepard et al. (1983). An extensive chromosome analysis of the potato + tomato ($2n = 72$) somatic hybrid will identify derived potato and tomato lines. After BC_1 and BC_2 generations, genomic *in situ* hybridization (GISH) revealed homoeologous chromosome pairing between tomato and potato chromosomes (Wolters et al., 1994b).

11 Transgenic Crops

11.1 INTRODUCTION

The progress made in creating genetically modified organism (GMO) during 1997–2016 is substantial (http://www.isb.vt.edu/release-summary-data.aspx). For successful genetic engineering of crops, an alien gene (donor species) linked to a reporter gene and other DNA sequences essential for insertion and expression is inserted into single and totipotent cells of the recipient species. The transformed cells proliferate *in vitro* during selection and regeneration creating a novel cultivar. Genetic transformation has improved many economically important crops. One such example is the soybean, a leguminous crop widely used for oil, feed, and soyfood products (Singh and Hymowitz, 1999).

11.2 PRESENT STATUS OF TRANSGENIC CROPS

To date, 10 transgenic crops (Figure 11.1) consisting of 10 phenotypes (Figure 11.2) have been evaluated and released in the United States (http://www.isb.vt.edu/release-summary-data.aspx).

Many transformants carry valuable traits: tolerance to herbicides, agronomic performance, resistance to insects, virus, bacterial, nematode and pathogens, quality traits (protein, oil, carbohydrate and fatty acids including amino acids composition), modified reproductive capacity, photosynthetic, enhancement and yield increase, delayed senescence, enhanced flavor and texture, longer shelf life, and more healthful produce (Dunwell, 2000). The number of approved permits from 1997 to 2016 is shown in Figure 11.3. The production of GMO has been revolutionized for the last decade by private and public institutions. Monsanto Company, for example, has played a major role in applying for the maximum (7053) number of permits followed by Dupont-Pioneer, Syngenta, and others (Figure 11.4). The number of states that issued permits to grow transgenic crops is shown in Figure 11.5. Hawaii is the leader followed by Puerto Rico, Illinois, Iowa, California, and others.

Monsanto has joined and acquired the following seed companies:

Parent company
- Pharmacia Corporation, The United States

Subsidiaries
- Sementes Agroceres SA
- Plant Breeding International Cambridge, Ltd., The United Kingdom
- HybriTech Seed International, Inc., The United States
- Holden's Foundation Seeds, Inc., The United States
- Jacob Hartz Seed Co., Inc., The United States
- First Line Seeds, Ltd., Canada
- Deklb Genetics Corporation, The United States
- Cargill International Seeds, The United States
- Agracetus, Inc., The United States
- Calgene, Inc., The United States
- Asgrow Seed Company, The United States

Joint venture
- Renessen, The United States
- Monsoy, Brazil
- Mahyco, India

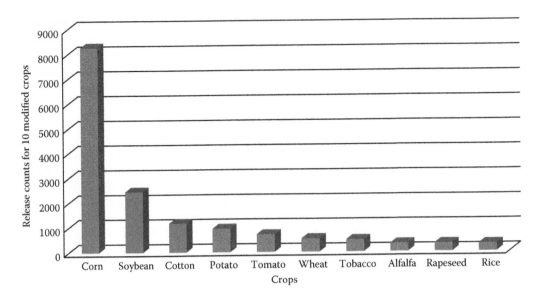

FIGURE 11.1   Number of released counts for 10 transgenic crops.

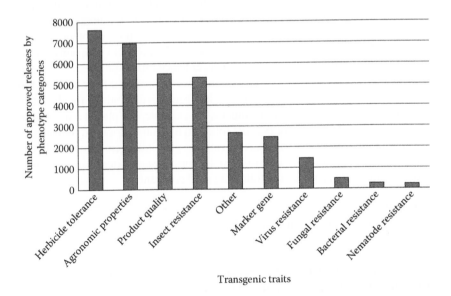

FIGURE 11.2   Number of approved releases by phenotypes; herbicide tolerance predominates.

11.3   PRODUCTION OF TRANSGENIC CROPS

Several procedures are available for producing transgenic crops (Table 11.1). Foreign DNA is delivered into recipient species through *Agrobacterium*, particle bombardment, electroporation of protoplast, polyethylene glycol (PEG), and microinjection of protoplasts (Songstad et al., 1995). Recently, Joersbo (2001) proposed three selection systems using glucuronic acid, mannose, and xylose that resulted in the higher transformation frequencies compared with the frequently used kanamycin selection.

The development of gene transfer methodology is crop dependent. For example, alien genes of economic importance can be delivered into soybean by *Agrobacterium*-mediated transformation of cotyledonary explants (Hinchee et al., 1988; Clemente et al., 2000), and by particle bombardment,

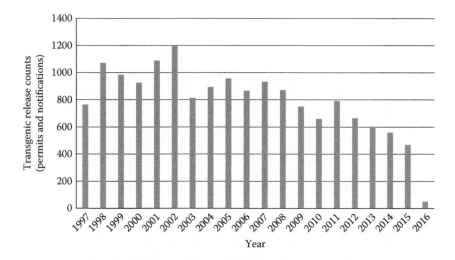

FIGURE 11.3 Number of transgenic crops released from 1997 to 2016.

invented by Klein et al. (1987) (Christou, 1997). Hinchee et al. (1988) successfully isolated glyphosate-tolerant transformants using the *Agrobacterium*-mediated gene transfer method and Padgette et al. (1995) reported the stability of the transformants. The soybean is marketed as "Roundup Ready® soybean" and is widely grown in the United States and is spreading into other soybean growing countries of the world. Genetic transformation created Bollgard® cotton, YieldGard® maize (www.biotechbasics.com), FLAVR SAVR™ tomato (Kramer and Redenbaugh, 1994) and golden rice (Potrykus, 2001). Bollgard® cotton is commercially grown in the United States, Australia, China, Mexico, South Africa, and Argentina and was recently approved for India. The FLAVR SAVR™ tomato containing superior consumer quality, shelf life, and flavor failed to attract consumers as it was more expensive than nontransgenic tomato. Figure 11.5 shows the top 10 most frequent locations for crop release in the United States.

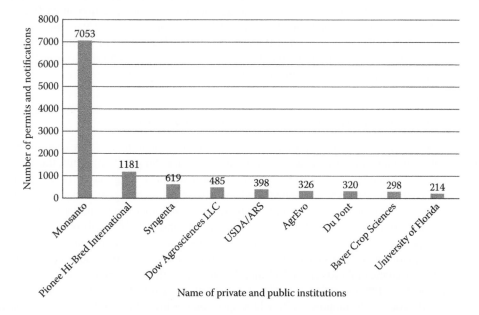

FIGURE 11.4 Number of permits and notifications to private and public institutions; Monsanto is on the top with 7053.

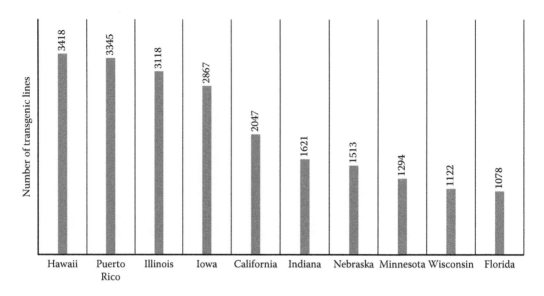

FIGURE 11.5 Number of transgenic lines being grown in 10 states of the United States.

11.4 CHROMOSOMAL ABERRATIONS IN TRANSGENIC CROPS

Chromosomal aberrations (numerical and structural changes) are a common occurrence in cell and tissue culture-derived calluses and their regenerants (Singh, 1993). Selection and osmotic treatment exacerbate cytological aberrations in transformed barley cultures, but bombardment itself did not appear to affect the frequency of cytological aberrations in cells of callus cultures (Choi et al., 2001). Chromosomally abnormal soybean cultures used in transformation experiments regenerate mostly sterile plants (Singh et al., 1998). Thus, soybean is one of the most difficult large seeded legumes to regenerate plants from cell and tissue cultures. The earlier reports suggest that reproducible methods of regeneration depend on type and age of explants, media composition (basic salts and growth hormone combinations), and genotypes (Christianson et al., 1983; Barwale et al., 1986; Lazzeri et al., 1987a,b; Wright et al., 1987; Bailey et al., 1993). The degree of sterility in transgenic soybean plants is related to the period the cultures are nurtured on 2, 4-D prior to transformation, post-transformation stresses, and the genetic background of the explant.

11.4.1 TIME IN CULTURE WITH 2, 4-D PRIOR TO TRANSFORMATION

Generally, older cultures lose the capacity to regenerate plants due to high ploidy or an increasing frequency of aneuploidy that leads to a chromosomal imbalance (Singh, 1993). Thus, these cultures will generate a low frequency of morphologically deformed plants. Singh et al. (1998) examined chromosome counts (mitotic pro-metaphase and metaphase) in embryogenic suspension cultures (nontransgenic) and in roots from developing somatic embryos (either transgenic or nontransgenic) from T_o, T_1, and T_2 transgenic two Asgrow genotypes A 2242 and A 2872. Seeds of A2242 (control) carried the expected $2n = 40$ chromosomes. The chromosome counts of suspension cultures and germinating somatic embryos from culture 817 are useful (Table 11.2). Tetraploidy occurred even in the germinating somatic embryos from 7.83 months on 2, 4-D; three samples showed $2n = 80$ normal chromosomes and two samples, in addition to tetraploidy, carried $2n = 79 + 1$ dicentric chromosomes. Cells after 11.26 months on 2, 4-D suspension cultures showed $2n = 80$ chromosomes and transgenic plants recovered from these cultures were tetraploid. Embryo suspensions of culture 826, which were on 2, 4-D for 4.20 months, displayed cells with $2n = 80$ chromosomes. Suspension

TABLE 11.1

List of Genetically Modified Crops Released for Commercial Cultivation

Crops	Traits	Modification	% Modified in the United States	% Modified in World
Alfalfa	Tolerance to glyphosate or glufosinate	Genes added	Planted in the US from 2005–2010 Court injunction; 2011 approved for sale	–
Apples	Delayed browning	Genes added for reduced polyphenol oxidase (PPO) production from other apples	2015 approved for sale	
Canola/rapeseed	Tolerance to glyphosate or glufosinate; high laurate, oleic acid canola	Genes added	87% (2005)	21%
Maize	Tolerance to herbicides glyphosate, glufosinate, and 2, 4-D: insect resistance-added enzymes, alpha amylase that converts starch into sugar to facilitate ethanol production, resistance to virus	Genes, some from Bt, added	Herbicide resistance (2013), 85%; Bt, (2013), 76%; stacked, (2013); 71%	26%
Cotton	Insect resistance	Genes some from Bt added	Herbicide resistance (2013), 82%; Bt, (2013), 75%; Stacked (2013), 71%	49%
Eggplant	Insect resistance	Genes from Bt	Negligible	Negligible
Papaya (Hawaiian)	Resistance to the papaya ringspot virus	Genes added	80%	
Potato (food)	Resistance to Colorado beetle, potato leaf roll virus and potato virus; reduced acrylamide when fried and reduced bruising	Bt cry3A, coat protein from PVY; innate potatoes added genetic material coding for mRNA for RNA interference	0%	0%
Potato (starch)	Antibiotic resistance gene, used for selection better starch production	Antibiotic resistance gene from bacteria, modifications to endogenous starch producing enzymes	0%	0%
Rice	Enriched with beta-carotene (a source of vitamin A)	Genes from maize and a common soil microorganism	Forecast to be on the market in 2015 or 2016	

(Continued)

TABLE 11.1 (*Continued*)
List of Genetically Modified Crops Released for Commercial Cultivation

Crops	Traits	Modification	% Modified in the United States	% Modified in World
Soybean	Tolerance to glyphosate or glufosinate; reduced saturated fats (high oleic acid); kills susceptible insects and resistance to virus	Herbicide resistant gene taken from bacteria added knocked out native genes that catalyze saturation, gene for one or more Bt crystal proteins added	94% (2014)	77%
Squash	Resistance to watermelon, cucumber and zucchini/courgette yellow mosaic virus	Viral coat protein genes	13% (2005)	
Sugar beet	Tolerance to glyphosate, glufosinate	Genes added	95% (2010); regulated 2011; deregulated 2012	9%
Sugarcane	Pesticide tolerance High sucrose content	Genes added		
Sweet peppers	Resistance to cucumber mosaic virus	Viral coat proteins		Small quantity grown in China
Tomato	Suppression of the enzyme polygalacturonase (PG), retarding fruit softening after harvesting while at the same time retaining both color and flavor of the fruit	Antisense gene of the gene responsible for PG enzyme production added	Taken off the market due to commercial failure	Small quantities grown in China

Source: Data from https://en.wikipedia.org/wiki/Genetically_modified_crops.

TABLE 11.2

Chromosome Analysis at Somatic Metaphase in the Transgenic Asgrow Soybean Genotype A2242

Culture ID	R0 Phenotype	Origin of Roots	Months on 2, 4-D	Number Samples	2n	Karyotype
A2242	Control			5	40	Normal
22-1	Diploid	T1	6.43	7	40	Normal
22-1	Tetraploid	T1	6.90	5	40; 41	39 + (1); 38 + (3)[a]
22-1	Diploid	T1	6.90	2	40	Normal
22-1	Diploid	T1	6.96	9	40	Normal
22-1	Diploid	T1	6.96	2	80	Normal
22-1	Diploid	T1	7.00	8	40	Normal
22-1	Diploid	T1	7.30	9	80	Normal
22-1	Diploid	T1	7.96	8	80	Normal
22-1	Diploid	T1	9.00	1	80	Normal
22-1	Tetraploid	T0	11.47	3	80	Normal
22-1	Tetraploid	T0	15.36	1	80	Normal
22-1	–	Embryo	15.36	5	80	3,80;2,79 + 1[b]
22-1	–	Embryo	16.73	4	80	3,80;1,40 + 80[c]
817	–	Embryo	7.17	4	40	3,40;1,39 + 1[d]
817	–	Embryo	7.33	5	40	4,40;1,39 + 1[d]
817	–	Embryo	7.83	5	80	3,80;2,79 + 1[e]
817	–	Embryo	8.70	1	40	Normal
817	–	T0	9.13	1	40	Normal
817	–	Suspension	11.26	1	80	Normal
817	–	Embryo	12.43	2	80	Normal
825	–	Suspension	6.86	1	40	Normal
828	–	Embryo	2.86	1	40	Normal
826	–	Suspension	4.20	1	80	Normal

Source: From Singh, R. J. et al. 1998. *Theor. Appl. Genet.* 96: 319–324. With permission.

[a] Three small metacentric chromosomes.

[b] One megachromosome.

[c] Chimaera 40 + 80 chromosomes.

[d] Long chromosome.

[e] One sample with 79 + 1 dicentric chromosomes and other sample with 79 + 1 fused centromeric chromosomes.

culture 825 (6.86 months on 2, 4-D) and somatic embryo culture 828 (only 2.86 months on 2, 4-D) showed diploid cells. The germinating embryos derived from cultures ranging from 7.17 to 16.73 months on 2, 4-D possessed imbalanced chromosomes (Table 11.2). The older cell lines, therefore, are not desirable for transformation because chromosomal and genetic abnormalities occur with age. This may lead to difficulty in recovery of regenerable and fertile plants.

In order to ensure morphologically normal transformants with complete fertility, Stewart et al. (1996) bombarded 3-month-old globular-stage embryos of soybean cv. "Jack" and used post-bombardment selection for transgenic line on a solidified medium instead of liquid medium.

Transgenic plants were not analyzed cytologically but all plants were fertile and morphologically normal. Maughan et al. (1999) isolated four normal and fertile β-casein transgenic soybean cv. "Jack" plants via particle bombardment of 3- to 4-month-old somatic embryos nurtured on solid medium. Hazel et al. (1998) examined growth characteristics and transformability of embryogenic cultures of soybean cv. "Jack" and Asgrow A2872. The most transformable cultures constituted of tightly packed globular structures and cytoplasm-rich cells in the outermost layers of the tissues with the highest mitotic index. On the other hand, the outer layers of the less transformable culture had more lobed cells with prominent vacuoles. Santarém and Finer (1999) used 4-week-old proliferative embryogenic tissue of soybean cv. "Jack" maintained on semisolid medium for transformation by particle bombardment. They produced fertile transgenic soybean 11–12 months following culture initiation. They claim a significant improvement over bombardment of embryogenic liquid suspension culture tissue of soybean where transformability was recorded at least 6 months after culture initiation (Hazel et al., 1998). Clemente et al. (2000) produced 156 primary transformants by *Agrobacterium*-mediated transformation. Glyphosate-tolerant shoots were identified after 2–3 months of selection. The R_0 (T_0) plants were fertile (262 seed per plant) and did not express gross phenotypic abnormalities. This suggests that transformation in the soybean must be conducted using young (1–3 months) cultures.

The plants regenerated from cultures also inherited chromosomal abnormalities (Table 11.2). Chromosomes of nine T_1 populations from culture 22-1 were examined. The parental cultures were in contact with 2, 4-D for 6.43 to 9.00 months prior to transformation. Although T_0 plants from eight cultures expressed normal, diploid morphological features, T_1 progenies from four populations carried $2n = 80$ chromosomes and four populations had $2n = 40$ chromosomes (Figure 11.6a). The T_0 phenotype of one T_1 population of culture 22-1 was similar to that of tetraploid. These plants showed dark-green leathery leaves and produced mostly one-seeded pods. Chromosome counts from 5 T_1 seedlings showed $2n = 39 + 1$ small metacentric chromosome (Figure 11.6b) in three plants, and one plant each contained $2n = 38 + 3$ small metacentric chromosomes (Figure 11.6c) and $2n = 40$ chromosomes. The 40-chromosome plant may have had a small deletion, which could not be detected cytologically, or may have carried desynaptic or asynaptic genes. Four T_0-derived

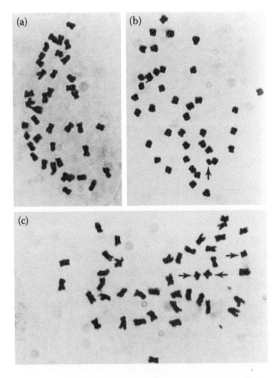

FIGURE 11.6 Mitotic chromosomes in root tips of seedlings from Asgrow soybean line A2242 R_1 generation. (a) $2n = 40$ showing normal karyotype; (b) $2n = 39 + 1$ small metacentric chromosomes (arrow); (c) $2n = 38 + 3$ small metacentric chromosomes (arrows). (From Singh, R. J. et al. 1998. *Theor. Appl. Genet.* 96: 319–324. With permission.)

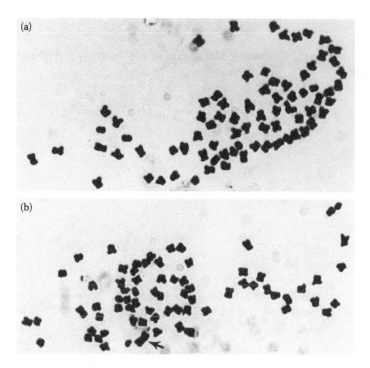

FIGURE 11.7 Mitotic metaphase chromosomes in root tips of seedlings from Asgrow soybean line A2242 grown in agar. (a) $2n = 80$ showing normal karyotype; (b) $2n = 79 + 1$ mega (monocentric) chromosome (arrow). (From Singh, R. J. et al. 1998. *Theor. Appl. Genet.* 96: 319–324. With permission.)

plants from culture 22-1 (three plants from cultures after 11.47 months on 2, 4-D and one plant after 15.36 months on 2, 4-D) were morphologically tetraploid and were cytologically confirmed to have $2n = 80$ chromosomes. Three transformants had $2n = 80$ (Figure 11.7a) and two contained $79 + 1$ megachromosome (Figure 11.7b). Most reports agree that structural and numerical changes in chromosomes induced during culture are caused by 2, 4-D (Singh, 1993). Thus, it is advised to examine cytologically the chromosomes in pretransformed material to ensure that it is devoid of structural and numerical changes (Poulsen, 1996).

It has been established that various stresses such as medium composition, age of culture, and nature of culture—morphogenic versus nonmorphogenic, genotype of the explant, media, and *in vitro* culture during the passage of plant regeneration—induce chromosomal instability (Singh, 1993; D'Amato, 1995). Chromosomal constitution in transgenic versus nontransgenic barley ($2n = 14$) demonstrated that the extent of ploidy changes in transgenic plants was intensified, presumably due to additional stresses that occurred during transformation. The delivery of foreign DNA into plant cells involves several stressful events, such as vacuuming, cellular damage by microprojectile bombardment, selection process, and growth of dying transformed cells for prolonged time during recovery, that cause chromosomal aberrations (Choi et al., 2000a, 2001). Chromosomal aberrations are attributed to spindle failure by growth hormones that causes endoreduplication, c-mitosis, nuclear fragmentation (amitosis), multipolar configurations, and lagging chromosomes (Singh, 1993).

11.4.2 GENETIC BACKGROUND OF THE EXPLANT

Chromosomal aberrations induced during culture are genotype-dependent (Singh, 1993). Singh et al. (1998) observed tissues and primary transformants from soybean genotype A2872 did not display chromosomal abnormalities (Table 11.2) while soybean genotype A2242 was cytologically highly unstable (Table 11.3). Nine selfed seeds from A2872 (control) showed $2n = 40$ chromosomes

TABLE 11.3

Chromosome Analysis at the Somatic Metaphase in the Transgenic Asgrow Soybean Genotype A2872

Culture ID	R0 Phenotype	Origin of Roots	Months on 2, 4-D	No. Samples	$2n$	Karyotype
A2872 Control				9	40	Normal
821 Suspension			8.26	1	40	Normal
821 Embryo			9.23	5	40	Normal
2-1	Diploid	T1	22.76	1	40	Normal
2-2	Diploid	T1	11.20	1	40	Normal
2-2	Diploid	T1	18.03	1	40	Normal
2.4	Diploid	T1	13.16	1	40	Normal
2-4	Diploid	T1	20.00	1	40	Normal
5-1	Diploid	T1	20.50	2	40	Normal
5-1	Diploid	T1	21.50	2	40	Normal
5-2	Diploid	T1	19.06	4	40	Normal
5-2	Diploid	T1	20.20	1	40	Normal
5-2	Diploid	T1	20.73	1	40	Normal
5-2	Diploid	T1	21.20	2	40	Normal
5-2	Diploid	T1	29.00	3	40	Normal
5-2	Diploid	T1	29.43	3	40	Normal
5-2	Tetraploid	T1	29.43	5	40	Normal
90-2	Diploid	T1	19.06	1	40	Normal
90-2	Diploid	T1	32.30	4	40	Normal
801	–	Embryo	9.23	2	40	Normal

Source: From Singh, R. J. et al. 1998. *Theor. Appl. Genet.* 96: 319–324. With permission.

with normal karyotype. Culture age on 2, 4-D ranged from 8.26 to 32.30 months. Embryo suspensions, germinating embryos, and T_0 and T_1 plants expressed normal diploid-like phenotypes and, as expected, all plants carried $2n = 40$ chromosomes. However, five T_1 plants from a 29.43-month-old culture of 5-2 had tetraploid morphological features with one-seeded pods, but showed $2n = 40$ chromosomes (Table 11.4). These abnormalities may be genetic (desynaptic or asynaptic, male sterile and female fertile). These morphotypes often express tetraploid phenotypes (slow growth, dark-green leathery leaves, clustered flowers, empty pods) with partial to complete sterility carrying one seed per pod (Palmer and Kilen, 1987). Interestingly, the primary transformants from A2872 contained $2n = 40$ chromosomes though the culture was on 2, 4-D for 32.30 months. This shows that a genotype may be highly responsive to culture conditions but may be prone to chromosomal aberrations (Hermsen, 1994).

Aragão et al. (2000) developed a soybean transformation protocol that produced a high frequency of fertile transgenic soybeans that was genotype-independent. Their technique includes microparticle bombardment (*ahas* gene: a selectable marker gene isolated from *Arabidopsis thaliana* that contains a mutation at position 653 bp) of the soybean meristematic region and culturing in selection medium with imazapyr herbicides followed by multiple shooting induction. They claimed a 200-fold increase in the recovery of transgenic soybean plants over the methods of Christou (1997).

11.4.3 SEED FERTILITY IN THE TRANSGENIC CROPS

Morphological variants, particularly seed sterility in transgenic crops, are attributed to chromosomal aberrations (Austin et al., 1995; El-Kharbotly et al., 1995; Fütterer and Potrykus, 1995; Lynch

TABLE 11.4

Culture Identification (ID), Plant ID, Months in Culture with 2, 4-D before Transformation, Total Number of Transformants Recovered, Seed Set Range,[a] and Chromosome Number in Selected Primary Transformants in Soybean Cultivar Asgrow 2242

Culture ID	Plant ID	Month 2, 4-D	Total Number Plants	Seed set Range	2n 40	80	41
828-1	671-3-4	3.20	6	0–363	14	–	–
828-1	668-8-1	5.50	4	5–50	9	–	–
828-1	668-1-3	5.60	9	14–219	10	–	–
828-1	668-1-6	5.60	5	6–65	9	–	1
828-1	668-1-5	5.60	2	15–24	7	–	3
828-1	668-1-8	5.60	2	85–101	7	–	–
828-1	668-1-12	5.60	9	75–185	18	–	–
828-1	668-1-13	5.60	5	59–178	10	–	–
828-1	668-2-3	5.60	10	6–53	–	20	–
828-1	668-2-7	5.60	7	6–111	–	10	–
828-1	668-4-3	5.60	6	2–12	4	–	–
828-1	668-4-15	5.60	5	8–51	1	–	–
828-1	668-1-1	6.00	4	0–225	4	–	–
22-1	549-2-2	6.43	1	25	3	–	1
22-1	549-4-6	6.43	2	63–206	7	–	3
22-1	549-4-10	6.43	5	90–321	20	–	–
22-1	557-1-6	7.30	3	9–65	–	10	–
22-1	557-2-3	7.30	7	90–321	–	15	–
817	610-5-1	7.33	10	10–495	54	–	–
817	610-6-1	7.83	5	56–190	23	–	14 + 1[a]
817	610-7-1	7.83	10	52–329	–	4	–
22-1	557-2-7	7.96	10	33–350	–	30	–
22-1	557-5-2	7.96	9	65–210	–	9	–
22-1	566-5-1	9.50	1	27	–	10	–
828-2	677-3-1	11.00	12	50–268	5	5	–
817	647-1-1	12.43	5	0–19	5	–	–
817	647-2-1	12.43	5	25–66	–	3	–
817	647-15-7	12.43	5	30–133	–	10	–
817	647-15-4	12.43	2	7–36	10	–	–
817	652-6-1	13.30	12	0–235	–	9	–
828-1	668-7-2	13.80	2	28–38	4	–	2
22-1	587-5-1	15.36	2	7–18	–	2	–
22-1	587-3-1	15.36	3	4–8	–	6	–
22-1	609-6-1	16.00	1	9	–	3	–

[a] T.M. Klein, personal communication.

et al., 1995; Schulze et al., 1995; Shewry et al., 1995; Hadi et al., 1996; Liu et al.,1996; Choi et al., 2000a, 2002). *Agrobacterium*-mediated transformation produced diploid ($2n = 24$) (El-Kharbotly et al., 1995) and tetraploid ($2n = 2x = 48$) potato (Conner et al., 1994). The diploid transgenic potato that conferred resistance to *Phytophthora infestans* inherited reduced-male fertility while female fertility was not so markedly influenced. All transformed tetraploid potatoes contained normal chromosome complement but morphological changes including low yield and small tubers were observed in the field grown plants compared to the control.

The *Agrobacterium*-mediated transformation has been extremely effective to dicotyledonous crops. However, Ishida et al. (1996) developed an *Agrobacterium*-mediated transformation protocol to produce transgenic maize in high frequency. They produced 120 morphologically normal transformants and 70% of them set normal seeds. Plants were not examined cytologically.

Christey and Sinclair (1992) obtained kale, rape, and turnip transformants through *Agrobacterium rhizogenes*-mediated transformation. Transgenic plants were successfully produced from hairy roots. Morphological changes with an increase in leaf edge serration, leaf wrinkling, and plagiotropic roots were observed in some plants while in other lines phenotypic alterations were barely noticeable. Transgenic plants were not examined cytologically.

Toriyama et al. (1988) produced five transgenic rice plants after direct gene transfer into protoplasts through electroporation-mediated transformation. One plant was diploid ($2n = 24$), three plants were triploid ($2n = 3x = 36$), and one plant was unidentified. Ghosh Biswas et al. (1994) produced 73 transgenic rice cv. IR 43 plants by direct transfer of genes to protoplasts and 29 plants reached maturity in the greenhouse. Eleven plants flowered but did not produce seed. However, two protoplast-derived nontransgenic plants set seeds. Protoplast-derived plants (transgenic and nontransgenic) had fewer tillers, narrower leaves and were shorter in height than seed-derived plants. Fertile transgenic rice plants through electroporation-mediated transformation have been produced (Shimamoto et al., 1989; Xu and Li, 1994).

Another method used to insert a foreign DNA into the protoplast is by using a polyethylene glycol-mediated DNA transformation. Hall et al. (1996) reported high efficiency procedure for the generation of transgenic sugar beets from stomatal guard cells. They examined ploidy level by flow cytometry and recorded 75% diploid transformants. Seed production was normal and the average frequency of germination was 96%. Lin et al. (1995) transferred 61 rice plants (via polyethylene glycol-mediated transformation) to the greenhouse and 28 of them were fertile. Seed set per plant ranged from 10 to 260. The cause of seed sterility was not determined.

Microprojectile bombardment-mediated primary transformants (T_0) in soybean genotype A2242 exhibited a range of fertility that depended on culture medium and duration of time on 2, 4-D (Table 11.3). Chromosome counts of 10 seeds from randomly selected T_0 plants revealed normal chromosomes ($2n = 40$) from the young cultures regardless of seed set. Tetraploidy ($2n = 80$) and aneuploidy (near diploidy) were predominant in plants obtained from the older cultures (Table 11.4). Culture 828-2 after 11 months on 2, 4-D produced 12 T_0 plants. Of the 10 plants examined, 5 plants were diploid and 5 plants were tetraploid. Seed fertility in 9 plants was more than 101 seeds per plant while seed set in 3 plants ranged from 21 to 100. Occasionally, seed set in T_0 diploid plants was low (2–12; 0–19; 7–36). Thus, seed sterility may be attributed to culture conditions (epigenetic), desynaptic and asynaptic gene mutations, or minor chromosomal deletions which cannot be detected cytologically. Schulze et al. (1995) observed fruit development after selfing transgenic cucumber; however, none of the harvested fruits contained seeds. Simmonds and Donaldson (2000) produced fertile transgenic soybeans from young proliferative cultures via particle bombardment while sterile plants were recovered from 12- to 14-month-old cultures. The cause of sterility was not established. Choi et al. (2000a) recorded 75% (15/20 normal lines) fertile lines from microprojectile bombardment-mediated transgenic oat while 36% (10/28) lines had low seed set. Thus, transformation procedure and age of culture play an important role in producing completely fertile, stable, and normal transformants.

11.5   CYTOLOGICAL BASIS OF GENE SILENCING

The loss or low expression and unexpected segregation (gene silencing) of foreign genes are routinely observed in transformed crops (Fromm et al., 1990; Fütterer and Potrykus, 1995; Meyer, 1995; Senior, 1998; Kooter et al., 1999). Foreign DNA in transgenic common bean plants, produced from particle-bombardment, was not expressed in 12 (44%) plants and two plants showed a poor transmission of the inserted gene (1:10) in R_1 (T_1) generation although all plants had normal phenotype and were fertile (Aragão et al. 1996). This may be due to chimaerism in T_0 plants.

An extensive review on genetic transformation and gene expression in the Poaceae by Fütterer and Potrykus (1995) reveals that the expressions of transgenes in the progeny of transgenic plants are quite unpredictable. The departure from the Mendelian inheritance (unstable transgenes) in transformants occurs when inserts are located in highly repetitive sequences (heterochromatic regions of chromosomes) or in the extra chromosomal DNA (mitochondrial or chloroplast genomes). Several genomic factors such as aberrant crossing-over during meiosis, spontaneous or induced mutations, ploidy, aneuploidy, sex chromosomes and transposable elements (Maessen, 1997), and nucleolar dominance (Pikaard, 1999) cause deviations in the Mendelian inheritance of the transgenes.

In transformation experiments, genes may be physically present but gene activity may be poorly expressed or totally lost in subsequent generations. This phenomenon is known as cosuppression (Matzke and Matzke, 1995; Stam et al., 1997; Matzke et al., 2000). In cosuppression, foreign genes (transgenes) cause the silencing of endogenous plant genes if they are sufficiently homologous (Stam et al., 1997). An excellent example of cosuppression was shown in tobacco by Brandle et al. (1995). They produced a transgenic tobacco line carrying the mutant *A. thaliana* acetohydroxyacid synthase gene *csrl-1* that expressed a high level of resistance to the sulfonylurea herbicide chlorsulfuron. The instability of herbicide resistance was observed during subsequent field trials that was not anticipated from the initial greenhouse screening. Hemizygous plants from this line were resistant but homozygous plants (59%) were damaged by the herbicide. Damage was correlated with cosuppression of the *csrl-1* transgene and the endogenous tobacco AHAS genes, *sur A* and *sur B*. The difference in the performance of glyphosate® tolerance in sugar beet transformants between greenhouse and the field was also reported by Mannerlöf et al. (1997). The disparity was attributed to differences between the environments, variation in gene expression caused by copy number, position, or methylation effects. Based on this information, they suggested that cosuppression was triggered by agro-climatic conditions and the initial greenhouse study was not predictable.

The stability of transformants is associated with the insertion of transgenes in the regions of chromosomes. Stable inserts have been located in the vicinity of telomere as they have a tendency to integrate toward the distal end of chromosome arms which are gene-rich regions (Matzke and Matzke, 1998; Figure 11.8). According to Matzke et al. (2000), different regions of the genomes vary in their ability to tolerate foreign DNA resulting in erratic expression. Iglesias et al. (1997) demonstrated by fluorescence *in situ* hybridization (FISH) that two stably expressed inserts in tobacco were present in the vicinity of telomeres while two unstably expressed inserts were located at intercalary and paracentromeric regions. By contrast, by using FISH in transgenic barley (Choi et al., 2002) recorded integration of insert was preferred distal positions on barley chromosomes:

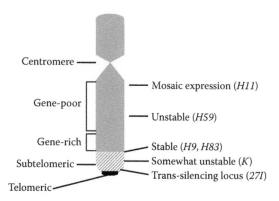

FIGURE 11.8 Influence of chromosomal location on transgene expression. The chromosomal locations of genetically well-characterized transgene loci (in italics, right) in tobacco as determined by FISH are shown relative to landmarks (left) on a model chromosome. (From Matzke, A. J. M. and M. A. Matzke. 1998. *Current Opinion Plant Biol.* 1: 142–148. With permission.)

7 (37%) out of 19 lines examined had integration sites in distal positions of the chromosomes; 4 (21%) in telomeric regions, 3 (16%) in centromeric regions, 3 (16%) in satellite regions, and 2 (10%) had subtelomeric regions.

It has been demonstrated that gene silencing in transgenic crops is homology-dependent (Matzke et al., 2000). Expression of transgenes driven by barley endosperm-specific promoters was much more stable in its inheritance pattern than driven by the constitutive promoters such as rice actin promoter and maize ubiquitin promoter (Cho et al., 1999, 2002; Choi et al., 2002). The recipient plants often have effective defense mechanisms against the uptake, integration, and faithful maintenance of the foreign genes or their transcribed products. DNA methylation is widely associated with gene activation (Kumpatla et al., 1998).

Singh et al. (1998) presented a cytological clue that may help explain aberrant segregation ratios or loss of transgene sequences which may be applicable in some cases. For example, the selfed population of a plant with $2n = 39 + 1$ metacentric chromosomes identified in soybean genotype A2242 was expected to segregate plants in a ratio of 1 ($2n = 40$):2 ($2n = 39 + 1$ metacentric):1 ($2n = 38 + 2$ metacentrics). Diploid plants will be normal and fertile and may not express the introgressed genes if this gene is in the deleted chromosomes. Matzke et al. (1994) attributed an erratic inheritance in a transgenic tobacco line to aneuploidy ($2n = 49$ or 50). Transgne silencing is common in diploid barley (M.-J. Cho, personal communication).

An early chromosome count prior to pretransformation of young cells, callus, and embryo suspension should be used as transformation targets to reduce the possibility of somaclonal variation and regenerability problems. Stability of inserts in a crop should be field tested in wide agro-eco-geo-climatic conditions (day length, drought, heat, irregular weather) for several years before releasing as a commercial cultivar.

12 Relationship between Cytogenetics and Plant Breeding

12.1 INTRODUCTION

Modern crop improvement programs develop elite lines by multidisciplinary approaches. Development of these lines is dependent on crop breeding objectives, the type of crops, and end-use products (long term vs. short-term, reproductive cycle, molecular-aided vs. conventional, and loss vs. benefit). Other important qualities include: inheritance patterns and heritability of the selected characters (sex expression, flowering date, disease and insect resistance, yield heterosis, modifying plant architecture and quality); horticultural and agronomic traits; and nutritional quality, including processing and consumer preferences from available germplasm (Singh and Lebeda, 2007). Thus, crop improvement requires a multidisciplinary approach such as agronomy, physiology, microbiology, plant pathology and entomology, genetics, cytogenetics, plant breeding, genetic resources, distant hybridization, and molecular biology (Figure 12.1).

Kölreuter (1761; https://archive.org/details/djosephgottliebk01klre) was the first to propose that F_1 hybrids in *Nicotiana*, *Dianthus*, *Mathiola*, and *Hyoscyamus* inherit equal heredity materials of both parents. However, in the nineteenth and twentieth century, humans tried to understand in more detail the rules and processes of evolution, domestication, genetics, and breeding of plants. In the field of evolutionary biology, the background was built by Darwin (1859), domestication by Vavilov (see Löve, 1992) and Harlan (1992), and genetics and breeding by Mendel (Orel, 1996). These germinal ideas and their combination created the background of modern biology and genetics. Practical application and utilization of this knowledge applied to studies of plant genetic resources has led to advances in plant breeding.

Modern cultivated crops are ancient cultivated plants that have played an important role in human development. *In situ* cultivation of plants began more than 10,000 years ago as result of changes in human lifestyle from nomadic and seminomadic to sedentary life (Hancock, 2004). Sauer (1969) outlined the requirement for plant domestication and early cultivation. As a crucial factor he considered the status of land, that is, the first cultivated lands would have been those that required little preparation. Environmental conditions (rainfall and temperatures favorable for growing plants) were considered important. These conditions played a major role for developing diverse food plants. Vavilov postulated at least eight centers of origin and diversity of cultivated plants (Figure 12.2; Vavilov, 1950).

Our currently diverse group of modern crops originated from ancestral species found on all habitable continents some cultivated for more than 10,000 years, and others only recently developed. From their original centers of origin crops have been spread by humans throughout the world, often to climates far from their natural range. For example, tomato originated in the tropics of the New World but is now a major vegetable crop in temperate climates. Developing novel traits that facilitate adaptation to alien environments such as frost-hardiness in potato is a goal of modern vegetable breeding (Phillips and Rix, 1993). For future continuation of domestication and improvement of

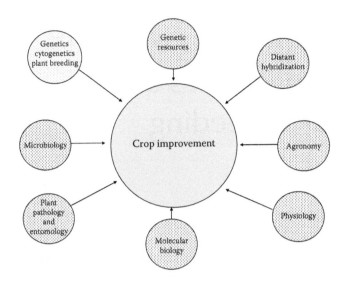

FIGURE 12.1 Diagrammatic sketch of a multidisciplinary approach from crop improvement. (Modified from Pohelman, J. M. and Sleeper, D. A. 1995. *Breeding Field Crops*, 4th ed., Iowa State University Press, Ames, IA.)

useful plants, it is necessary that regional as well as global genetic resources be appropriately collected, maintained, and preserved before being extinct. The potential of existing biological diversity will be properly directed for the benefit of all peoples (Rubatzky and Yamaguchi, 1997).

12.2   GENE POOL CONCEPT

A prerequisite for the exploitation of wild species to improve cultivars is to have complete comprehension and understanding of the taxonomic and evolutionary relationships between cultigen and its wild allied species (Hawkes, 1977). Harlan and de Wet (1971) proposed the concept of three gene pools based in hybridization among species. These are primary (GP-1), secondary (GP-2), and tertiary (GP-3) (Figure 12.3).

12.2.1   PRIMARY GENE POOL (GP-1)

The primary gene pool consists of biological species, and crossing within this gene pool is easy. Hybrids are vigorous, exhibit normal meiotic chromosome pairing, and possess total fertility. The gene segregation is normal and gene exchange is generally easy. GP-1 is further subdivided into subspecies A, which includes cultivated races and subspecies B, which includes spontaneous races. Soybean cultivars, and land races and their wild annual progenitor, *Glycine soja*, are included in GP-1. The GP-1 in allotetraploid cultivated cotton (*Gossypium hirsutum*, *Gossypium barbadense)* includes the wild tetraploid species *Gossypium tomentosum*, *Gossypium darwinii,* and *Gossypium mustelinum*. Interspecific hybrids are fertile. Introgression between *Gossypium hirsutum* and *Gossypium barbadense* occurs (Percival et al., 1999).

A race is a biological unit with some integrity, originated in some geographical region at some time in the history of the crop; it is not clearly separable as a species but has a distinct cohesion of morphology, geographical distribution, and ecological adaptation and frequently of breeding behavior (Harlan and de Wet, 1971).

Zeven (1998) defined race as "an autochthonous landrace is a variety with a high capacity to tolerate biotic and abiotic stress, resulting in high yield stability and an intermediate yield level under a low input agricultural system."

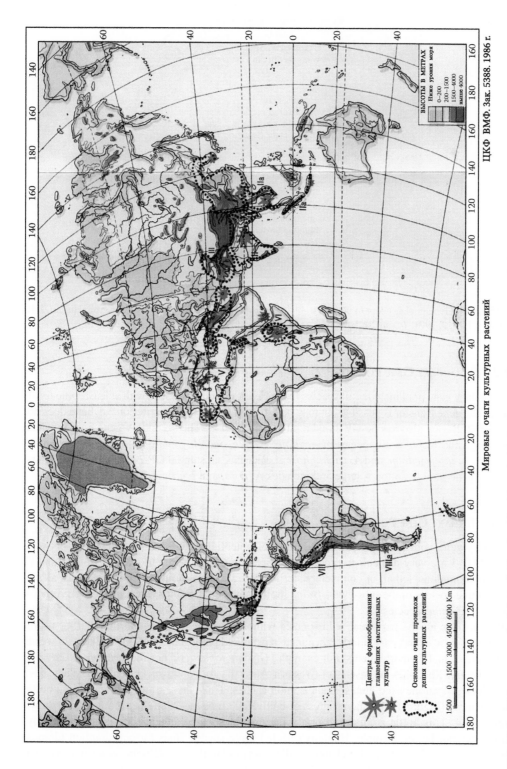

FIGURE 12.2 Eight centers of origin and diversity of cultivated plants. (Adapted from Vavilov, N. I. 1987. *Origin and Geography of Cultivated Plants*, Nakuka, Leningrad, the USSR.)

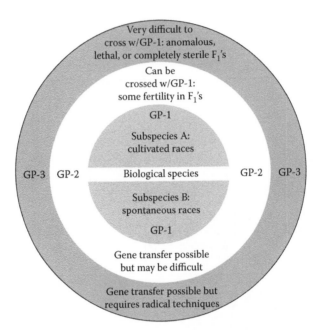

FIGURE 12.3 Gene pool concept in plants established based on hybridization. (Redrawn from Harlan, J. R. and de Wet, J. M. J. 1971. *Taxon* 20: 509–517.)

12.2.2 SECONDARY GENE POOL (GP-2)

The secondary gene pool includes all species that can be crossed with GP-1 with at least some fertility in F_1. The gene transfer is possible but may be difficult. Based on this criterion, soybean does not have a GP-2. Barley also has no GP-2 and a rather small GP-3. The secondary gene pool in wheat is very large and includes all species of *Aegilops*, *Secale*, and *Haynaldia* plus at least *Agropyron elongatum*, *A. intermedium*, and *A. trichophorum*. Rice has substantial GP-2 and a small GP-3. The secondary gene pool in cotton is large and includes A, D, B, and F genome diploid species.

12.2.3 TERTIARY GENE POOL (GP-3)

The tertiary gene pool is the extreme outer limit of potential genetic resource. Hybrids between GP-1 and GP-3 are very difficult to produce, require *in vitro* technique to rescue F_1 plant, and F_1 plants are anomalous, lethal, or completely sterile. Based on this definition, GP-3 includes all the wild perennial species of the subgenus *Glycine* for the soybean. GP-3 for wheat includes *Agropyron* and several *Elymus* species.

GP-3 includes all the perennial wild species without economic values but they harbor useful genes of economic values particularly resistance to pests and pathogens. They grow like wild weeds in wide agro climatic environments. For example, *Glycine tomentella* has four cytotypes ($2n = 38$, 40, 78, 80) and flourishes in deserts of Western Australia to tropics of Queensland. They have been collected from the sand dunes, islands, and beaches. Thus, these traits are locked in a bottle of the "Genies" (Figure 12.4). The progress of cultivated crops has been diagrammatically shown in Figure 12.4. This bottle is unlocked producing undomesticated annual species spontaneously; *Glycine soja* can be an example. Humans selected large seeded lines for consumption (domestication); a wild undomesticated and intermediate form classified as *Glycine gracilis* is known. This species is morphologically intermediate between *G. soja* and *G. max*. This illustrates that domestication from wild annual species to domesticated lines is a long-term process and during this

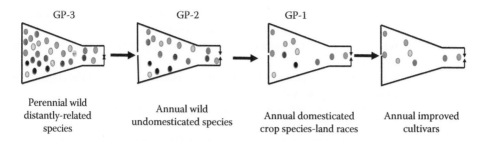

GP-3 GP-2 GP-1

Perennial wild
distantly-related
species

Annual wild
undomesticated species

Annual domesticated
crop species-land races

Annual improved
cultivars

FIGURE 12.4 Diagrammatic progression of the domestication of cultivated crops from the wild perennial species. (Modified and redrawn from Tanksley, S. D. and McCouch, S. R. 1997. *Science* 277: 1063–1066.)

progression, the genetic treasure from GP-3 is unlocked. There are many barriers (crossability, seed abortion, hybrid seedling lethality, and inviability; also described below) to overcome for unlocking the bottle filled with the genetic treasure.

12.3 CYTOGENETIC TECHNIQUES AND PLANT BREEDING

Cell division has played a major role in plant breeding. Male sterility (genetic, cytoplasmic, and cytoplasmic genetic) produced hybrid maize, sorghum, pearl millet, and rice. Meiotic mutants usually cause male sterility. These are known as premeiotic genes (premeiotic interphase), asynaptic genes (from leptonema to Zygonema), desynaptic genes (from pachynema to metaphase-I), and chromosomal disjunction (spindle) genes (from anaphase-I to tetrad). Many male sterile genes have been found during pollen mitosis. Male sterility genes cause complete breakdown of microsporogenesis but macrosporogenesis is typically uninfluenced. Genetic male sterility is due to failure to produce and shed viable pollen, controlled by monogenic recessive genes. These genes are used in population development for abiotic and biotic stresses. Cytoplasmic male sterile plants are completely sterile regardless of the nuclear constitution. Cytoplasmic genetic male sterility is due to interaction of cytoplasm and nuclear genes; fertile when fertility restorer genetic stocks are used as a pollinator. This system is used for hybrid maize, sorghum, pearl millet, rice, and pigeon pea.

The creation of haploidy by interspecific hybridization between cultivated crops and allied species enhanced crop breeding. In zygote development of *H. vulgare* (2n = 14) × *H. bulbosum* (2n = 14), the selective elimination of *H. bulbosum* chromosomes during mitosis leads to the haploid genome of vulgare chromosome producing haploid barley. However, embryo rescue technique is required.

Diploid-like meiosis is observed in allopolyploid crops. In tetraploid (2n = 28) and hexaploid (2n = 42) wheat, chromosome pairing within genome is controlled by a gene, *Ph*, located on 5BL chromosome (Okamoto, 1957). The *Ph* gene is known as a pairing suppressor gene that permits chromosome pairing between homologous genomes. The removal of the *Ph* gene causes pairing among three (AABBDD) wheat genomes leading to sterility.

Plant breeders modify breeding programs depending on mode of reproduction which depends on the floral modification (Elliott, 1958). These are

1. Unisexual flowers
   a. Male and female flowers on the same plant (monoecious)
   b. Male and female flowers on different plants (dioecious)
2. Complete or hermaphroditic flowers
   a. Maturation of stigmas and anthers of the same flower at different times (dichogamous)
      i   Maturation of anthers ahead of stigmas (proterandrous or protandrous)
      ii Maturation of stigmas ahead of anthers (proterogynous or protogynous)

b. Maturation of stigmas and anthers of the same flower simultaneously (homogamous).
   i. At pollination time, the flowers are open (chasmogamous).
      A. Self-pollination is impossible because of the relative positions of the stigmas and anthers.
      B. Self-pollination is possible because of the relative positions of the stigmas and anthers (herkogamous).
         • The stamens and styles are of the same length in all flowers (homomorphous or homostylous).
         • The stamens and styles are of different lengths on separate plants (heteromorphous).
            • Both the stamens and styles are of different length (heterostylous).
            • Only the stamens differ in length (heteroantherous).
   ii. At pollination time, the flowers are closed (cleistogamous).
      The alternation of generation in flowering plants includes sporophytic and gametophytic stages that complete the life cycle of plants. When sperm nuclei conjugate with the egg and antipodal nuclei in the same flower to produce viable zygotes, this is known as self-fertilization (autogamy). By contrast, when pollen of one plant pollinates the stigmas of another plant, this is known as cross-pollination (allogamy). Several floral modifications enhance cross-pollinations. These are
      1. The separation of plants by sexes
      2. The separation of staminate and pistillate inflorescences on the same plant
      3. A flower structure that prevents self-pollination
         a. Unequal maturation of stamens and pistils
         b. Flower shape and the arrangement of stigma and anthers
         c. Protective film over stigma surfaces
      4. Self-incompatibility mechanism
      5. Genetic and cytoplasmic mechanism

The following are crossability barriers in plants:

1. Geographical isolation.
2. Photoperiod differences.
3. Genetic differences.
4. Self-and cross incompatibility.
5. Separation of sexes.
6. Failure of pollen to germinate onto stigma.
7. Abnormalities such as bursting of pollen tubes in style.
8. Growth of pollen tube is so slow to reach to the ovary.
9. Pollen tube reaches to the ovary but fails to fertilize the egg and antipodal.
10. Embryo development is arrested after fertilization.
11. Seeds do not develop beyond 19–21 days postpollination commonly recorded in wide hybridization and immature seeds are rescued through tissue culture media.
12. F_1 plants are sterile due to lack of meiotic chromosome affinity between parental chromosomes.

However, sexual reproduction is prevented in certain plants. Such changes occur during disturbances in normal embryo sac development and viable embryos are formed without union of male and female gametes. This phenomenon is known as apomixis (asexual reproduction). Plants are usually uniform but variants are isolated in facultative apomixis. Apomixis is common in forage grasses, citrus, mango, blackberries, guayule, and ornamental shrubs (Chapter 6). Vegetative reproduction is routinely used for micropropagation such as in orchids. Onion, garlic, potato, sweet potato, flower bulbs, and sugarcane are vegetatively propagated. Grafting is used for producing disease-free roses and grapes.

Chromosome number of most of the economically important crops is known and chromosome maps have been established based on karyotype analysis. Physical mapping of genes for abiotic and biotic stresses by molecular cytogenetics helps breeders to grow the number of plants in segregating generations; the closer the genes, the larger the number of population required to recover recombinant types. The knowledge of chromosome number of parental species is extremely valuable to plant breeders when they are looking for desirable genes from the secondary and tertiary gene pools.

Physical mapping of genes on specific regions of chromosomes in maize, barley, tomato, and bread wheat has been established by the deficiency method. Balanced tertiary trisomics have been proposed to produce hybrid barley. Chromosomeal aberrations such as numerical changes (auto-polyploidy and allopolyploidy) either induced or identified in nature have created many economic crops. Autotriploid watermelon and banana and autotetraploid rye, potato, alfalfa, and tobacco are classic examples. Emmer and bread wheat and various *Brassica* species were induced by nature. Aneuploidy such as trisomics (primary, secondary, tertiary, telotrisomics, and acrotrisomics) have been exploited to associate qualitative genes with specific chromosomes and physical mapping of genes on the chromosomes in diploid crops such as maize, barley, rye, tomato, rice, and others. Likewise, monosomics and nullisomics have been used for association of genes with specific chromosomes in tetraploid and hexaploid wheat.

12.4   RECOMBINATION IN PLANT BREEDING

Genetic recombination has played a major role in crop improvement and is the most important element of plant breeding. The recombination can be chromosomal due to exchange due to crossing over. The closer the genes on a chromosome, the frequency of recombinant individuals is going to be fewer. Based on Mendelian inheritance 50% of the progenies from a heterozygote parent will be of recombinant type. The exception to the Mendelian laws was discovered shortly after its rediscovery (1900). The law does not apply if genes are arranged in coupling (*AABB* × *aabb*) and repulsion (*AAbb* × *aaBB*) and located on the same chromosome. This is due to linkage, coined by Morgan, from his studies in *Drosophila*. This suggests that genes are arranged on chromosomes in a linear fashion that became the basis for developing genetic linkage maps. Fewer recombinants are expected if two loci are closer than those with the greater distance. Furthermore, crossing over occurs more often in certain regions of chromosomes than in others. This suggests that linkage holds both parent characteristics together.

For plant breeding, recombination is suppressed by asynaptic and desynaptic meiotic mutants, chromosome translocations, inversions, and lack of chiasma formation that suppresses the crossing-over. In autotetraploids, tetrasomic inheritance occurs by contrast in disomic inheritance in diploids. It is considered that several economic useful traits such as yield, quality, and other abiotic stresses are controlled by quantitative traits. These traits are complex and controlled by many major and minor genes.

The genetic recombination between two parents is associated with a genetic system and the nature of crops: self-pollinated or cross-pollinated. This is an integral part of plant speciation that is the result of natural selection under different environments. Highly self-pollinated crops maintain recombination if parents are heterozygous but homozygosity is achieved rapidly after few generation of selfing. By contrast, cross-pollinated crops are usually highly heterozygous and common in forage crops. Genotypic stability is achieved by recombination of genetic variability; the greater the variability, the greater are the possibilities of responses to selection. In crops once the maximum heterozygosity is achieved, asexual reproduction like apomixis and asexual propagation like vegetative propagation-micro propagation occur (Chapter 6). Thus, breeding programs are planned according to the nature of crops and degree of recombination.

12.5 MUTATION IN PLANT BREEDING

Artificial mutation creates genetic variations in plants not available in the germplasm collections. It is sudden heritable desired change in plants induced by radiation and chemical mutagens. These plants are called mutagenic plants. Mutation breeding has been used in varietal improvement since the 1930s and more than 3200 mutant varieties were officially released for commercial use in more than 210 plant species from more than 70 countries (http://www-naweb.iaea.org/nafa/pbg/mutation-breeding.html). Breeders have used mutagenesis to broaden the genetic base of germplasm and use the mutant lines directly as new varieties or as sources of new germplasm in varietal improvement programs. Mutation breeding should be supplemented with the conventional methods of selection and hybridization as an aid in the creation of genetic variability (Elliott, 1958).

12.6 POLYPLOIDY IN PLANT BREEDING

Polyploidization, auto- and allopolyploidy, has played a major role in crop improvement. Autopolyploid plants originate through the union of unreduced gametes (AAA) while allopolyploids originate when two species (genomes A and B) are hybridized producing an F_1 hybrid with AB genome. F_1 plants are sterile because of a lack of homology between parental chromosomes. Chromosomes are doubled (AABB) through spontaneous somatic chromosome doubling or by colchicine treatment. These plants are known as allotetraploid. Both auto- and allopolyploidy have been described in Chapter 8. Autotriploid watermelon and banana and autotetraploid rye, potato, coffee, alfalfa, groundnut (peanut), and sweet potato are classic examples of the role of autopolyploidy. Similarly, allopolyploidy contributed the evolution of tobacco, sugarcane, plum, loganberry, strawberry, emmer wheat, bread wheat, oats, cotton, and various Brassicas. Triticale is a manufactured crop.

12.7 PLANT INTRODUCTION IN PLANT BREEDING

Crop improvement depends on the objectives of the plant breeder, available germplasm resources, nature of the plants (autogamous to allogamous), agro-climatic conditions, and inheritance of abiotic and biotic stress traits. The understanding of the center of origin of crops and allied species and their diversity is necessary. Vavilov postulated at least eight centers of origin and diversity of cultivated plants (Figure 12.2).

1. *Chinese Center of Origin:* The crops of this region are cereals and legumes (broomcorn millet, Italian millet, Japanese barnyard millet, Koaliang, buckwheat, hull-less barley, soybean, adzuki bean, velvet bean), roots, tuber, and vegetables (Chinese yam, radish, Chinese cabbage, onion, cucumber), fruits and nuts (pears, Chinese apple, peach, apricot, cherry, walnut, litchi), sugar, drug, and fiber plants (sugarcane, opium poppy, ginseng, camphor, hemp).
2. *Indian Center of Origin:* This center has been divided into:
   a. *Indo-Burma:* Main center (India) includes Assam and Burma, but not Northwest India, Punjab, nor Northwest Frontier Provinces. The crops of this region are cereals and legumes (rice, chickpea, pigeon pea, Urd bean, mung bean, rice bean, and cowpea), vegetables and tubers (eggplant, cucumber, radish, taro, yam), fruits (mango, orange, tangerine, citron, tamarind), sugarcane, coconut palm, sesame, safflower, tree cotton, oriental cotton, jute, crotalaria, kenaf) and spices, stimulants, dyes (hemp, Black pepper, gum Arabic, sandal wood, indigo, cinnamon tree, croton, bamboo).
   b. *Siam-Malaya-Java:* Includes slat Indo Malayan Center: Includes Indo China and the Malay Archipelago. This center includes cereals and legumes (Job's tears, velvet bean), fruits (pummelo, banana, breadfruit, and mangosteen), and oil, sugar, spice, and fiber plants (candlenut, coconut palm, sugarcane, clove, nutmeg, black pepper, Manila hemp).

3. *Central Asiatic Center of Origin:* This center includes Northwest India (Punjab, Northwest Frontier Provinces and Kashmir), Afghanistan, Tajikistan, Uzbekistan, and Western Tian-Shan. The crops are grains, legumes, and others (common wheat, club wheat, shot wheat, peas, lentil, horse bean, chickpea, mung bean, mustard, flax, sesame), fiber plants (hemp, cotton), vegetables (onion, garlic, spinach, carrot), and fruits (pistachio, pear, almond, grape, apple).

4. *Middle East Center of Origin:* This region includes Asia Minor, all of Transcaucasia, Iran, and the highlands of Turkmenistan. The crops are grains and legumes (einkorn wheat, durum wheat, poulard wheat (*Triticum turgidum*), oriental wheat, Persian wheat, two-rowed barley, rye, Mediterranean oats, common oats, lentil and lupin), forage crops (alfalfa, Persian clover, fenugreek, vetch, hairy vetch), and fruits (fig, pomegranate, apple, pear, quince, cherry, hawthorn).

5. *Mediterranean center of origin:* This includes the borders of the Mediterranean Sea. The crops are cereals and legumes (durum wheat, emmer wheat, Polish wheat, spelt, Mediterranean oats, sand oats, canarygrass, grass pea, lupin), forage crops (Egyptian clover, white clover, crimson clover, serradella), oil and fiber plants (flax, rape, black mustard, olive), vegetables (garden beat, cabbage, turnip, lettuce, asparagus, celery, chicory, parsnip, rhubarb), ethereal oil, and spice plants (caraway, anise, thyme, peppermint, sage, hop).

6. *Ethiopia:* Abyssinian hard wheat, poulard wheat, emmer, Polish wheat, barley, grain sorghum, pearlmillet, African millet, cowpea, flax, teff, sesame, castor bean, garden cress, coffee, okra, myrrh, and indigo.

7. *South Mexican and Central American Center:* This region includes the southern section of Mexico, Guatemala, Honduras, and Costa Rica. The crops are grains and legumes (maize, common bean, lima bean, tepary bean, jack bean, and grain amaranth), melon plants (Malabar gourd, winter pumpkin, chayote), fiber plants (upland cotton, bourbon cotton, and henequen), sweet potato, arrow root, pepper, papaya, guava, cashew, wild black cherry, chochenial, cherry tomato, and cacao.

8. *South American Center:* This region includes Peruvian, Ecuadorean, and Bolivian Center. The crops are root tubers (Andean potato, other endemic cultivated species), grains and legumes (starchy maize, lima bean, and common bean), root tubers (edible canna and potato), vegetable crops (pepino, tomato, ground cherry, pumpkin, pepper), fiber plants (Egyptian cotton), and fruits (cocoa, passion flower, guava, heilborn, quinine tree, tobacco, cherimoya).

   a. Isles of Chile for potato.

Since the observation of the crop diversity in the world and establishment of the center of diversity proposed by Vavilov, major effort has been directed to collect, maintain, and preserve crops from the centers of diversity. This became mandatory because adaptation of highly yielding cultivars and genetically modified crops are taking over the niches of wild annual and perennial species. Fifteen international centers have been established, below, and these are under CGIAR (the Consultative Group for International Agricultural Research):

1. Africa Research Center (http://www.cgiar.org/cgiar-consortium/research-centers/africarice/)
2. Biodiversity International (http://www.cgiar.org/cgiar-consortium/research-centers/bioversity-international/)
3. Center for International Forestry Research (CIFOR) (http://www.cgiar.org/cgiar-consortium/research-centers/center-for-international-forestry-research-cifor/)
4. International Center for Agricultural Research in the Dry Areas (ICARDA), http://www.cgiar.org/cgiar-consortium/research-centers/international-center-for-agricultural-research-in-the-dry-areas-icarda/)

5. International Center for Tropical Agriculture (CIAT) (http://www.cgiar.org/cgiar-consortium/research-centers/international-center-for-tropical-agriculture-ciat/)
6. International Crop Research Institute for the Semi-Arid Tropics (ICRISAT) (http://www.cgiar.org/cgiar-consortium/research-centers/international-crops-research-institute-for-the-semi-arid-tropics-icrisat/)
7. International Food Policy Research Institute (IFPRI) (http://www.cgiar.org/cgiar-consortium/research-centers/international-food-policy-research-institute-ifpri/)
8. International Institute of Tropical Agriculture (IITA) http://www.cgiar.org/cgiar-consortium/research-centers/international-institute-of-tropical-agriculture-iita/)
9. International Livestock Research Institute (ILRI) (http://www.cgiar.org/cgiar-consortium/research-centers/international-livestock-research-institute-ilri/)
10. International Maize and Wheat Improvement Center (CIMMYT) (http://www.cgiar.org/cgiar-consortium/research-centers/international-maize-and-wheat-improvement-center-cimmyt/)
11. International Rice Research Institute (IRRI) (http://www.cgiar.org/cgiar-consortium/research-centers/international-rice-research-institute-irri/)
12. International Potato Center (CIP) (http://www.cgiar.org/cgiar-consortium/research-centers/international-potato-center-cip/)
13. International Water Management Institute (IWMI) (http://www.cgiar.org/cgiar-consortium/research-centers/international-water-management-institute-iwmi/)
14. World Agroforestry Centre (ICRAF) (http://www.cgiar.org/cgiar-consortium/research-centers/world-agroforestry-centre/)
15. World Fish (http://www.cgiar.org/cgiar-consortium/research-centers/worldfish/)

All international centers are unique and have different mandates. However, CIMMYT and IRRI have modified wheat and rice, respectively, that produced high yielding wheat and rice varieties bringing a green revolution.

12.8 WIDE HYBRIDIZATION AND PLANT BREEDING

Based on reviewing literature on hybridization, Harlan and de Wet (1971) proposed three gene pool concepts based on cytogenetics: primary (GP-1), secondary (GP-2), and tertiary (GP-3) for utilization of germplasm resources for crop improvement and this has played a key role in improving crops. The primary gene pool consists of landraces and biological species. The secondary gene pool includes species that can yield at least partially fertile F_1 on hybridization with GP-1. The tertiary gene pool is the outer limit of the potential genetic resources for breeding. Prezygotic and postzygotic barriers can cause partial or complete hybridization failure, inhibiting introgression between GP-1 and GP-2. Genetic resources are developed with integrated, multidisciplinary approaches through plant exploration, taxonomy, genetics, cytogenetics, plant breeding, microbiology, plant pathology, entomology, agronomy, physiology, hybridization, and molecular biology, including cell and tissue culture and genetic transformation. These efforts have produced superior crops, rich in nutrients, with resistance to abiotic and biotic stress, high yield, and extended shelf-life (Chapter 8). Breeders use the following method to overcome the hybrid barrier, and wide hybridization in soybean is cited as an example:

1. The assemblage of large germplasm
2. Application of growth hormones to reduce the embryo abortion
3. Improve culture conditions
4. Restoration of the seed fertility by doubling the chromosomes of sterile F_1 hybrids
5. Utilization of bridge crosses where direct crosses are not possible

The genetic resources of approximately 26 wild perennial species of the *genus Glycine* Willd. subgenus *Glycine* have not been exploited to broaden the genetic base of soybean (*G. max*, $2n = 40$). These species are extremely morphologically diverse and grow in diverse climatic and soil conditions and have a wide geographical distribution in Australia and surrounding islands (Chung and Singh, 2008). Considering the gene pool concept of Harlan and de Wet (1975), *G. max* and *G. soja* Sieb. and Zucc. ($2n = 40$) are included in the primary gene pool because both species are cross-compatible and produce viable fertile F_1 plants and are taxonomically included in the subgenus *Soja* (Moench) F. J. Hermann. *Glycine max* does not have a secondary gene pool. If we follow the definition of the tertiary gene pool (hybrids between the primary gene pool and the tertiary gene pool are anomalous, lethal or completely sterile and gene transfer is either not possible with known techniques or else rather extreme or radical measures are required), all wild perennial *Glycine* species belong to tertiary gene pool.

Since publication of Palmer and Hadley (1968) and Ladizinsky et al. (1979), there were few attempts to produce intersubgeneric F_1 hybrids and all failed either at F_1 or amphidiploid stage (Singh and Nelson, 2015). The rate of intersubgeneric crossability rate between soybean and wild perennial *Glycine* species can be enhanced by growth hormone (100 mg GA_3, 25 mg NAA, and 5 mg kinetin/L) spraying pollinated gynoecia 24 h postpollination one time a day for 19–21 days. The methodology is described in Chapter 2. Figures 12.5 and 12.6 show flow charts of isolating fertile plants of a *G. max* cv. Dwight ($2n = 40$) and *G. tomentella* ($2n = 78$) PI 441001. It is interesting to note that *G. tomentella* PI 441001 cytoplasm preferentially eliminates PI 441001 chromosome (Figure 12.7). In BC_2 carrying Dwight cytoplasm chromosome segregation is $2n = 55$–59 (Figure 12.7a) while in PI 441001 cytoplasm chromosome number in BC_2 plants ranged from $2n = 41$–50 (Singh and Nelson, 2014; Figure 12.7b). We need to exploit cytoplasm of other accessions of wild perennial species to discover producing haploidy in soybean. Such a

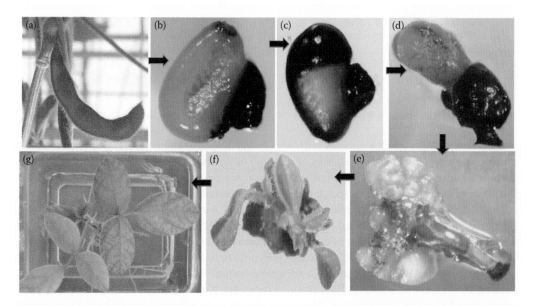

FIGURE 12.5 Immature seed rescue of *Glycine max* cv. "Dwight" ($2n = 40$) × *G. tomentella* PI 441001 ($2n = 78$). (a) One 21-day-old hybrid pod showing three developing small seeds; (b) one immature seed in seed maturation (SM) medium after 1 week; (c) seed coat turns black but cotyledons are green and healthy; (d) cotyledons emerging from a seed with black seed coat; (e) multiple embryos; (f) one developing shoot with leaves; (g) a small seedling with roots and shoot in a rotting medium. (From Singh, R. J. and Nelson, R. L. 2015. *Theor. Appl. Genet.* 128: 1117–1136.)

FIGURE 12.6 Morphological and cytological identification of F_1 from *Glycine max* cv. "Dwight" $(2n = 40) \times G.$ *tomentella* PI 441001 $(2n = 78)$. (a) Two seedlings growing in soil in a pot in a greenhouse, growth of one seedling is slower (arrow) than the other seedling; (b) mitotic metaphase showing $2n = 59$ chromosomes; (c) meiotic metaphase-I showing one ring bivalent, three rod bivalents (associated loosely; arrow) and univalents (most univalents are located at the poles, but a few are scattered in the cytoplasm); (d) one amphidiploid plant growing in a pot in a greenhouse showing stunted growth with one pod (insert), a mature pod with one seed; (e) mitotic metaphase showing $2n = 118$ chromosomes, all chromosomes are similar morphologically; (f) BC_1 plants growing in a pot in a greenhouse showing twining trait of PI 441001; (g) plant 07H1–26 (BC_2 plant) showing droopy branches, narrow trifoliolate leaves, and black seed (insert right bottom); (h) mitotic metaphase cell of 07H1–26 with $2n = 56$ chromosomes showing three (two from "Dwight" + one from PI 441001) satellite chromosomes (arrows); (i) morphology of one BC_3F_1 plant (07H6–3; $2n = 41$) showing morphological traits like "Dwight."

system is routinely used in wide crosses of cereals. This has laid the foundation to exploit GP-3 in soybean.

In cotton, GP-3 includes species with C, G, and K (Australian) and E (African-Arabian) genome species. The GP-3 species may or may not be hybridized easily with upland cotton and introgression of traits presents difficulties that are not easily resolved (Percival et al., 1999). Brubaker et al. (1999) modified the proposed two procedures for the utilization of cotton GP-3 germplasm to broaden the germplasm base of cultivated tetraploid $(2n = 4x = 52)$ cotton. A diagrammatic flow scheme, modified from Brubaker et al. (1999), is shown in Figure 12.8. One is hexaploid pathway and the other is tetraploid pathway. Intergeneric hybrids were totally sterile in both types of crosses. In hexaploid strategy, hybridize the diploid $(2n = 2x = 26)$ wild species carrying economically useful traits to a tetraploid cotton. Produce hexaploid $(2n = 6x = 78)$ by treating F_1 with the colchicine. Backcross to recurrent parent and isolate modified tetraploid cotton after BC_2 to BC_4 generations. They were able to isolate derived fertile lines by hexaploid pathway. In the tetraploid pathway, hybridize wild diploid $(2n = 2x = 26)$ species to an A or D genome diploid-bridging species. Produce tetraploid $(2n = 4x = 52)$ by treating F_1 with the colchicine. Backcross to recurrent parent and isolate two populations of modified tetraploid cotton or intercross two populations. However, they have not been able to isolate derived fertile lines from the tetraploid pathway. Thus, difficult materials are not easily approachable and require considerable efforts.

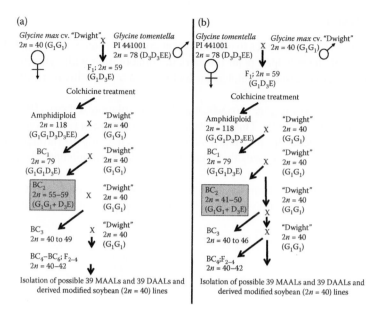

FIGURE 12.7 Diagrammatic sketch (redrawn) for isolating MAALs and DAALs from *Glycine max* cv. "Dwight" × *G. tomentella* PI 441001 (a) and *G. tomentella* PI 441001 × *G. max* cv. "Dwight" (b). Note, chromosome number of plants in BC$_2$ generation; shown in a box.

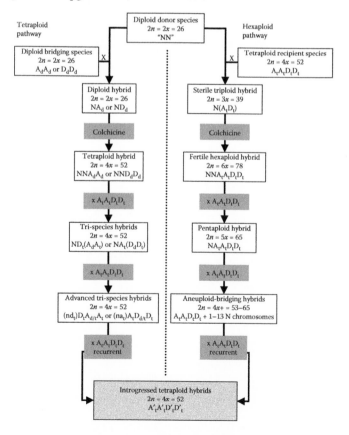

FIGURE 12.8 Diagrammatic flow pathways for transferring genes from GP-3 to cultivated cotton. (Courtesy of C. L. Brubaker.)

12.9   CONCLUSION

1. The contribution of cytogenetics to our important crop plants on planet earth is boundless.
2. Cytogenetic tools have helped plant breeders in collaboration with cytogeneticists, plant pathologists, entomologists, and molecular biologists to breed cultivars resistant to abiotic and biotic stresses.
3. The discovery of dwarfing genes for wheat and rice-revolutionized rice (IRRI) and wheat (CIMMYT) during the 1960s. This achievement led to the green revolution.
4. Transfer of desired genes from distantly-related species to cultigens through wide hybridization helped create new hybrids that were not feasible from incompatible and unrelated species or genera.
5. Transfer of desirable traits from distantly-related organisms through genetic transformation has developed crops not possible by conventional breeding methodology.

Glossary[*]

ABA: Abscisic acid.

Abaxial: The surface of a leaf facing away from the axis or stem of a plant.

Aberration: Variation of chromosome structure either induced by mutagen or natural mutation.

Aberration rate: The portion of chromosomal changes as compared to normal chromosomes.

Abiotic stress: Tolerance of plants to problem soils, heat and cold stress, and other adverse climatic conditions.

Abjection: The separating of a spore from a sporophore or sterigma by a fungus.

Ablastous: Without germ or bud.

Abort: The collapse of seeds prior to maturation.

Abortive: Defective or barren; development arrested while incomplete or imperfect.

Abscise: Separation of leaves from stem or flowers from raceme.

Abscission: Rejection of plant organs.

Abscission layer: Cells of abscission zone disintegrate causing the separation of plant organs (leaf, fruit, pod, and branch).

Abscission zone: Layer of cells which brings about the abscission of plant organs.

Abscisic acid: Plant growth hormone involved in fruit growth and rejection of plant organs.

Accessions: A sample of plant germplasm maintained in a gene bank for preservation and utilization by scientists.

Accession number: Each accession is assigned a particular number; such as PI (plant introduction) number in the USDA gene bank.

Accessory bud: Buds located at or near the nodes but not in the axils of leaves.

Accessory chromosome: Supernumerary or B-chromosome.

Acentric fragment: A chromosome fragment without a kinetochore.

Aceto-carmine stain: A dye (prepared in 45% acetic acid + 1%–2% carmine powder) used for staining plant chromosomes.

Aceto-orcein stain: In this stain, carmine is replaced by orcein.

Achene: Seed and pericarp attached only at the funicules, the seed usually tightly enclosed by the fruit wall, as in the sunflower and buckwheat.

Achiasmate: Meiosis without chiasma. Chromosome 4 in *Drosophila* which is mostly heterochromatic is without chiasma. Heterochromatin has been implicated in chromosome pairing for achiasmatic disjunction.

Achromatic: Chromosome regions that do not stain with specific dyes.

Achromatin: The part of the nucleus that does not stain with basic dyes.

A-chromosome: One of the standard chromosomes of an organism.

Acorn: One seeded fruit—oak tree.

Acquired characters: Changes in phenotype of an organism arise purely by environmental influences during development and not the result of gene action (Lamarckian inheritance).

Acquired immunity: Resistance developed in an organism during pre-immunization.

Acquired mutation: A noninheritable genetic change in an organism-somatic mutation.

Acridine orange: A dye that functions as a fluorochrome and a mutagen.

Acrocarpic: Fruits and/or seeds are formed on the top of a stem of a plant.

[*] Allard, 1960; Chapman and Peat, 1992; Darlington and Mather, 1949; Fahn, 1990; Fehr, 1987; Lewin, 1990; Nelson et al. 1970; Rieger et al. 1976; http://linkage.rockefeller.edu; Schlegel, R.H.J. 2010. *Dictionary of Plant Breeding.* 2nd ed. CRC Press, Boca Raton, FL.

Acrocentric chromosome: A chromosome consisting of one complete arm and a small piece of the other arm.

Acrosyndesis: Incomplete end-to-end chromosome pairing.

Acrotrisomics: An individual with a normal chromosome complement plus an extra acrocentric chromosome.

Actinomorphic: Sepals and petals radiating from the center of the flower; the sepals are all similar to one another, as are the petals to one another.

Active collection: A collection of germplasm used for regeneration, multiplication, distribution, characterization, and evaluation; maintained in sufficient quantity to distribute on request to other scientists.

Adaptation: The process by which an organism undergoes modifications (developmental, behavioral, anatomical, physiological) in such a way to survive perfectly in a given environment.

Adapted race: Physiological race.

Addition line: A cell line of individual carrying chromosomes or chromosome arms in addition to the normal standard chromosome set.

Additive: Quantitative effects of genes added together to produce phenotypes that is the sum total of the negative and positive effects of the individual polygenes.

Additive genes: Gene interaction without dominance (if allele), or without epistasis (if nonallele).

Adelphogamy: Sib pollination.

Adjacent-1 distribution: In an interchange, heterozygote *nonhomologous* kinetochores move to opposite poles.

Adjacent-2 distribution: In an interchange, heterozygote *homologous* chromosomes move to opposite poles.

Adjacent segregation: In reciprocal translocation heterozygote, the segregation of translocated and a normal chromosome giving unbalanced spores with duplications and deficiencies.

Adnate: Fusion of unlike parts (fusion of palea to the caryopsis in *Bromus* grass).

Adosculation: The fertilization of plants by pollen falling on the pistils.

Adspersed: To have a wide distribution.

Adventitious: Growing of a shoot from an unusual place (roots from a leaf and stem).

Adventitious embryony (Adventive embryony): Embryo develops from the nucellus or integument cells of the ovule.

Adventitious plant: An individual plant developed from somatic tissues.

Adventitious roots: Formation of the roots from stem nodes or any part of a plant.

Adventive: Not native or not fully established in a new habitat or environment.

AFLP: Amplified fragment length polymorphisms.

Agamogenesis: Any reproduction without the male gamete.

Agamogony: A type of apomixis in which cells undergo abnormal meiosis during megasporogenesis, resulting in a diploid embryo sac.

Agamospecies: Populations morphologically differentiated from one another and reproducing apomictically.

Agamospermy: All types of apomixis in which seeds are formed by asexual means; it does not include vegetative reproduction.

Agrobiodiversity: The variety and variability of animals, plants, and microorganisms used directly or indirectly for food and agriculture; it comprises the diversity of genetic resources and species used for food, fuel, fodder, fiber, and pharmaceuticals.

Agrobiology: The study of plant life in relation to agriculture, especially with regard to plant genetics, cultivation, and crop yield.

Agrobiotechnology: Modern biological knowledge and methods that can be applied to human goals in agriculture.

Albuminous seed: A mature seed which contains endosperm.

Aleurone: One or more outermost layers of endosperm that stores substantial quantities of protein, for example, Bran.

Alien addition line: A line containing an unaltered chromosome complement of one species (generally a cultigen) plus a single chromosome (monosomic alien addition lines, MAALs) or a pair of chromosomes (disomic alien addition lines, DAALs) from an alien species.

Alien chromosome: A chromosome from a related species.

Alien chromosome transfer: Transfer of individual alien chromosomes from alien species to cultigen through cytogenetic methods.

Alien gene transfer: The transfer of genes from exotic species to cultigen.

Alien germplasm: Genes from distantly wild species.

Alien species: An organism that has invaded or been introduced by humans and is growing in a new region; a species that serves as a donor of chromosomes, or chromosome segments to be transferred to a recipient species or genotype.

Alien substitution line: A line in which cultivated one or more chromosomes are substituted from one or more chromosomes of an alien species.

Allele: An allele is merely a particular form of the same gene, can be dominant or recessive and can be in homozygous and heterozygous conditions.

Allele mining: An approach to access new and useful genetic variation in crop plant collections.

Allelic: Two genes similar—same gene.

Allelism: The genes occupying the same relative position (locus) on homologous chromosomes.

Allelopathic: The populations are not in physical contact and live in different geographic areas.

Allodiploid: Cells or individuals in which one or more chromosome pairs are exchanged for one or more pairs from another species.

Allogamy: Cross fertilization.

Allogene: Recessive allele.

Allogenetic: Cells or tissues related but sufficiently dissimilar in genotype to interact antigenically.

Allohaploid: A haploid cell or individual derived from an allopolyploid and composed of two or more different chromosome sets.

Alloheteroploid: Heteroploid individuals whose chromosomes are derived from various genomes.

Allopathy: Different populations are not geographically in contact.

Alloplasm: Cytoplasm from an alien species that has been transferred by backcrossing into a cultivated species.

Alloplasmic: An individual having cytoplasm of an alien species and nucleus of a cultivated species.

Allopolyploid: An organism containing different sets of chromosomes derived from two or more genomically diverse species. For example, allotetraploid, allohexaploid, and allooctoploid.

Allosyndesis: Meiotic chromosome pairing of completely or partially homologous (homoeologous) chromosomes in allopolyploids.

Allotetraploid: A tetraploid individual having two different diploid genomes.

Alternate leaf: A plant with one leaf at a node.

Alternate segregation: In reciprocal translocation heterozygote, during meiotic anaphase I, the movement of both normal chromosomes one pole and both translocated chromosomes to the other pole, resulting in genetically balanced spores.

Alternation of generation: In the life cycle of many plants, gametophyte phase alternates with a sporophyte phase.

Alternative disjunction: The distribution of interchange chromosomes at anaphase I is determined by their centromere orientation; chromosomes located alternatively in the pairing configuration are distributed to the same spindle pole.

Ameiosis: The failure of meiosis and its replacement by nuclear division without reduction of the chromosome number.

Ameiotic mutants: Meiosis does not ensue and pollen mother cells degenerate.

Amitosis: Absence of mitosis.

Amixis: Reproduction in which the essential events of sexual reproduction are absent.

Amphicarpous: A plant producing two classes of fruit that differ either in form or in time of ripening.

Amphidiploid: An individual which contains entire somatic chromosome sets of two species produced after chromosome doubling and exhibits diploid-like meiosis.

Amphikaryon: The nucleus of zygote produced after fertilization.

Amphimixis: Seed develops by a fusion of a female and a male gamete—normal sexual reproduction.

Amphitene: Zygonema or zygotene.

Amplification: The production of additional copies of a chromosome sequence, found as intra-chromosomal or extra-chromosomal DNA.

Anandrous: Free from anthers.

Anaphase (mitosis): Anaphase is the shortest stage of mitosis, mutual attachment between homologous chromosomes is ceased, sister chromatid at the kinetochore region at metaphase repulse and two chromatids, equal distribution of chromosomes, move toward the opposite poles at the spindle.

Anaphase-I (meiosis): Migration of homologous chromosomes to the opposite poles.

Anatropus ovule: Ovule is completely inverted.

Androdioecious: Male and hermaphroditic flowers on separate plants.

Androecium: Refers to all stamens of a flower; a male reproductive organ of a plant.

Androgenesis: Synonymous: male parthenogenesis. The progeny inherits the genotype of the male gamete nucleus.

Androgynophore: A stalk like elongation of the floral axis between the perianth and stamens, which elevates the androecium and gynoecium.

Androsome: Any chromosome exclusively present in the male nucleus.

Anemophily: Wind pollination.

Aneucentric: Chromosome possessing more than one centromere.

Aneuhaploid: Chromosome number in aneuhaploid is not exact multiple of basic set but an individual is deficient for one or more chromosomes.

Aneuploid: An organism or species containing chromosomes other than an exact multiple of the basic (x) chromosome number.

Aneuploid reduction: Reduction of the genetic variability by decreasing the chromosome number.

Aneusomatic: An organism having both euploid and aneuploid cells.

Angiosperm: Seeds formed within an ovary.

Anisoploid: An individual with an odd number of chromosome sets in somatic cells.

Annealing: The pairing of complementary single strands of DNA to form a double helix.

Annual: A plant completing the life cycle within a year.

Anther: Part of the stamen bearing pollen grains.

Anthesis: The time of flowering in a plant.

Antimitotic: Substance that leads to the cessation of mitosis.

Antipodal cells: A group of three cells each with a haploid nucleus formed during postmegasporogenesis, located at the chalazal end of the mature embryo sac.

Anucleolate: Without nucleus.

Aperture: This term is for the pollen grain. A depressed area of characteristic shape in the wall; the pollen tube emerges via such an area.

Apetalous: A plant without flowers.

Apex: The terminal portion of the shoot or root in which the apical meristem is located.

Apocarpy: The carpels are free in a flower and ovary.

Apogamety: Embryo is formed from synergids and antipodals other than the egg without fertilization.

Apomeiosis: A type of gametophytic apomixis where failure of both chromosome reduction and recombination occurs.

Apomict: An organism produced by apomixis.

Apomixis: A mode of asexual reproduction in plants where a seed is produced without sexual fusion of female and male gametes. Synonymous: Agamospermy.

Apospory: Embryo sac is formed from somatic cells in the nucellus or chalaza of the ovule.

Asexual reproduction: A type of reproduction that does not involve the union of male and female gametes; vegetative propagation.

Astral rays: Distinctive lines forming a network between the two centrioles. It is common in animal cells.

Asymmetric: Nonadditivity and interactions of various kinds will cause normal distribution curve to appear skewed or asymmetrical.

Asymmetric cell division: Division of a cell into two cells of different chromosome numbers.

Asymmetrical fusion: A cell formed by the fusion of dissimilar cells; heterokaryon.

Asymmetrical karyotype: Characterized by the set containing some large and some small chromosomes.

Asynapsis: A complete failure of homologous chromosome pairing during meiosis.

Autoallopolyploid: It is confined to hexaploidy or higher levels of polyploidy.

Autobivalent: A bivalent at meiotic metaphase I that is formed from two structurally and genomically completely identical chromosomes.

Autofertility: Self-fertility.

Autogamy: Self-pollination predominates, is a common and widespread condition in angiosperms, particularly in annual herbs.

Autogenomatic: Genomes that are completely homologous and pair normally during meiosis.

Automixis: Fusion of two haploid nuclei in a meiotic embryo sac without fertilization of the two egg nucleus producing maternal-like progeny.

Autonomous apomixis: No dependence on pollination, for example, *Texraxacum* and *Hieracium*.

Autopolyploid: An individual containing three or more genomically identical sets of chromosomes.

Autoployploidization: The doubling of chromosome number by failure of chromosomes to divide equationally in mitosis following chromosome replication.

Autoradiography: This method detects radioactivity-labeled molecules by their effects in creating an image on photographic film.

Autosegregation: Spontaneous changes in genotype in the egg cells that develop without fertilization; may occur in obligate as well as facultative apomixis.

Autosome: Any chromosome other than sex (X or Y) chromosome.

Autosyndesis: The pairing of complete or partial homologues of chromosome.

Autotetraploid ($4x$, AAAA): An individual possessing four identical sets of homologous chromosomes.

Autotriploid ($3x$, AAA): An individual possessing three identical sets of homologous chromosomes.

Autotroph: An organism that synthesizes its own food materials.

Autozygote: A diploid individual in which the two genes of a locus are identical by descent.

Auxin: Hormones that promote cell elongation, callus formation, fruit production, and other functions in plants.

Auxotroph: An organism that synthesizes most of its cellular constituents but ingests others.

Avirulent: Parasite unable to infect and cause disease in a host plant.

Backcross: The crossing of an F_1 heterozygote to one of its parents.

Backcross breeding: Recurrent backcross to one of the parents of a hybrid.

Back mutation: Restores wild type.

Bacteria: Bacteria are single cell, microscopically small organism with simple structure usually reproduce asexually with very short generation time (20 min). A single bacterium produced clone genetically identical. Accidental changes (mutation) lead to new type of clone.

They may be autotrophs (lost the ability to produce usually amino acids) or prototrophs (can produce these nutrients).

Bacteriophage: Any virus that infects bacteria.

Bacterium: A microscopic, single-celled prokaryotic organism.

Balanced lethal: A situation in which only certain combinations of gametes survive in zygotes. For example, heterozygous velans/gaudens survive and homozygotes die in the *Oenothera lamarkiana*.

Balbiani ring: An extremely large puff at a band of a polytene chromosome.

Banding pattern: The linear pattern of deeply and weakly stained bands on the chromosomes after C-N-G-banding techniques.

B-A translocation: These interchanges are obtained by breakage in the A chromosomes and in the B chromosomes and after reciprocal exchanges A^B and B^A chromosomes are produced.

Base collection: A collection of germplasm kept for a long time in cold storage.

Base pairs: A partnership of A (adenine) with T (thymine) or of C (cytosine) with G (guanine) in a DNA double helix. In RNA, thymine is replaced by uracil (A = U; C = G).

Base sequence: The order of nucleotide bases in a DNA molecule.

Base sequence analysis: A technique, often automated, for determining the base sequence.

Base substitution: A form of mutation in which one of the bases in DNA replaces another.

Basic chromosome number: The haploid chromosome number of the ancestral species represented as x (basic genome). For example, hexaploid wheat is identified as $2n = 6x = 42$ (AABBDD).

B-chromosome: These are also called supernumerary chromosomes, derived from the A chromosome complement; usually heterochromatic but genetically not completely inert.

Bidirectional replication: When two replication forks move away from the same origin in different directions.

Biennial: An individual living for two years; typically flowering and fruiting the second year.

Bimitosis: The simultaneous occurrence of two mitoses in binucleate cells.

Binucleate: Cells with two nuclei.

Biodiversity: The existence of a wide amount of diversity present in nature.

Biological species concept: A system in which organisms are classified in the same species if they are potentially capable of interbreeding and producing fertile offspring.

Biopiracy: The patenting of plants and other biological material, formerly held in common, and their exploitation for profit.

Biotechnology: The set of biotechnology methods developed by basic research and now applied to research (recombinant DNA technology) and product development.

Biotype: A group of individuals having the same genotype; may be homozygous or heterozygous.

Bipartitioning: Normal meiosis.

Bisexual (=perfect): A flower containing both stamens and carpels.

Bivalent: A pair of homologous chromosomes synapsed during meiosis.

Bivalent formation: The association of two homologous chromosomes as a ring or rod configuration depending on chiasma formation.

Botany: The science of biology that deals with plant life.

bp: An abbreviation of base pairs; distance along DNA is measured in bp.

Bract: A modified leaf-like structure occurring in the inflorescence.

Bridge-breakage-fusion-bridge cycle: A process that can arise from the formation of dicentric chromosomes; daughter cells are formed that differ in their content of genetic material due to duplications and/or deletions in the chromosomes.

Bridge cross: A method of overcoming the incompatibility between two species, the F_1 of one cross is used as a bridge to hybridize third species.

Bulb: An underground storage organ, containing a short, flattened stem with roots on its lower surface, and above its fleshy leaf base and leaves, surrounded by protective scale leaves.

Caducous: Plant parts which fall off early or prematurely.

Callose: A polysaccharide (β-1, 3 glucan) present in sieve areas, walls of the pollen tubes.

Callus: A disorganized mass of undifferentiated plant cells obtained through culture from small pieces of plant tissues.

Calyx: All the sepals (outer part of a flower) of a flower.

Carmine: A red dye used to stain the chromosomes; usually it is dissolved in 45% acetic acid known as aceto-carmine.

Carnoy's fixative: A fixative used for cereal chromosome study; consists of 6 parts 95% ethanol: 3 parts chloroform: 1 part acetic acid.

Carpel: Pistil; a female reproductive organ includes ovary, style, and stigma.

Caryopsis: A characteristic feature of plants of the Poaceae (Gramineae) in which the pericarp and testa are fused.

C-banding: Constitutive heterochromatin, exhibits bands after staining with Giemsa, is redundant and usually present in the proximity of kinetochore, telomere, and in the nucleolus organizer region.

Cell: The smallest, membrane-bound, living unit of biological structure capable of autoreduplication.

Cell culture: A process of culturing and growing cells through *in vitro* and regenerating plants by several subcultures.

Cell cycle: The period from one division to the next consists of G_1, S, G_2, and M.

Cell division: The reproduction of a cell by karyokinesis and cytogenesis.

Cell furrow: The wall formed between the two daughter nuclei at the end of telophase in animals.

Cell plate: The wall formed between the two daughter nuclei after karyokinesis.

Center of diversity: A local region where a particular taxon exhibits greater genetic diversity than it does anywhere else; developed by N. I. Vavilov.

Center of domestication: The area believed to be that in which a particular drop species was first cultivated.

Centimorgan (cM): A unit of measure of recombination frequency.

Centric: A chromosome segment having kinetochore.

Centric fission: A chromosomal structural change that results in two acrocentric or telocentric chromosomes; as opposed to centric fusion.

Centric fusion: The whole-arm fusion of chromosomes by the joining together of two telocentric chromosomes to form one chromosome.

Centriole: Small hollow cylinders consisting of microtubules that become located near the poles during mitosis.

Centromere: See kinetochore.

Centromere interference: An inhibitory influence by the centromere on crossing over and the distribution of chiasmata in its vicinity.

Centromere misdivision: A transverse instead of lengthwise division of the centromere resulting in telocentric chromosomes.

Centromeric index: (CI) = short arm length/chromosome length $\times$ 100.

Chalaza: The region on an ovule where the nucellus and integuments connect with the funiculus; the opposite end to the micropyle.

Character: Phenotypic trait of an organism controlled by genes and may be dominant or recessive and may be governed by a single gene or multigenes.

Chasmogamy: Fertilization after opening of flower.

Chiasma: An exchange of chromatid segment during first division of meiosis that results in the genetic crossing over; plural: chiasmata.

Chiasma interference: The occurrence, less frequent or more frequent than expected by chance, of two or more crossing over and chiasmata in a given segment of a chromosomal pairing configuration and/or chromosome.

Chiasma localization: The physical position of a chiasma in a pairing configuration and/or chromosome.

Chiasmate: Meiosis with normal chiasma formation.

Chiasma terminalization: The progressive shift of chiasmata along the arms of paired chromosomes from their points of origin toward terminal position.

Chimaera: A plant composed of two or more genetically distinct types of tissues due to mutation, segregation, irregularity of mitosis (mosaic).

***Chlamydomonas*:** It is a unicellular haploid green alga (*Chlamydomonas reinhardi*). It contains two motile flagella, a small distinct nucleus with haploid chromosomes and numerous mitochondria. Chloroplast constitutes 50% of the cell that helps to nurture photrophically in light utilizing CO_2 as its carbon source. However, *Chlamydomonas* can grow heterotrophically on an acetate culture without light. Asexual cycle takes about 6 h and sexual cycles less than 2 weeks.

Chlorenchyma: A chloroplast-containing parenchyma tissue such as the mesophyll and other green tissues.

Chlorophyll: The major photosynthetic green pigment present in the chloroplast.

Chloroplast: A cell organelle characterized by containing chlorophyll and enzymes required for photosynthesis. Chloroplast contains DNA and is capable of self-replication and protein synthesis.

C-value: The total amount of DNA in a haploid genome.

Chromatid: A half-chromosome (sister chromatids) that originates from longitudinal splitting of a chromosome during mitotic anaphase or meiotic anaphase-II.

Chromatid aberration: Chromosomal changes produced in one chromatid due to result of spontaneous or induced mutation.

Chromatid bridge: Dicentric chromatid with kinetochores passing to opposite poles at anaphase.

Chromatid segregation: Double reduction, used in autopolyploidy.

Chromatin: It is the complex of DNA and protein in the nucleus of the interphase cell; originally recognized by reaction with stains specific for DNA.

Chromatin reconstitution: The reconstitution of chromatin with chromosomal constituents previously removed by chromatin dissociation.

Chromocenters: Darkly stained bodies in the interphase nuclei consisting of fused heterochromatic telomeres.

Chromomeres: They are knob-like regions along the entire length of pachytene chromosomes, occupying specific positions on the chromosomes, larger heterochromatic chromomeres stain relatively darker than the euchromatic chromomeres.

Chromomere pattern: The linear order and distribution of chromomeres along a chromosome.

Chromonema: The chromosome threads at leptonema.

Chromosomal: Refers to the structure, constituents, and function of chromosomes.

Chromosomal aberrations: An abnormal chromosomal complement resulting from the loss, duplication, and rearrangement of genetic materials.

Chromosomal inheritance: The inheritance of genetic traits located in the chromosomes (Mendelian inheritance).

Chromosomal sterility: Sterility in F_1 plants is caused by the lack of homology between the parental chromosomes.

Chromosome: An important, complex organelle of the nucleus, containing genes. The number is maintained constant from generation to generation by mitotic and meiotic divisions; it consists of protein, nucleic acid, and genes.

Chromosome arm: One of the two parts of a chromosome divided by an intercalary centromere.

Chromosome complement: The total chromosome constitution of an organism.

Chromosome configurations: An association of chromosomes during meiosis in the form of univalent, bivalent, trivalent, quadrivalent (tetravalent), and so on.

Chromosome congression: Chromosomes move to the equatorial plate half way between two poles of the spindle by the help of spindle fibers.

Chromosome drift: The shift in the chromosome number.

Chromosome elimination: The loss of chromosomes from nuclei due to genic, chromosomal, or genomic causes.

Chromosome fragmentation: The breakage of chromosomes by radiation.

Chromosome map: A linear representation of a chromosome in which genetic markers belonging to particular linkage groups are plotted based on their relative distances—physical mapping. Chromosome maps are: (1) cytological maps—according to karyotype of chromosomes; (2) genetic maps—according to crossing over, recombination, and map units; (3) cytogenetic maps—gene located in a particular chromosome based on cytologically identifiable chromosome aneuploids such as deletions, duplications, translocations, inversions, trisomics, and other chromosomal aberrations; (4) restriction fragment length polymorphism (RFLP) map—a linear graph of sites on DNA cleaved by various restriction enzymes.

Chromosome mosaicism: The presence of cell populations of various karyotypes in the same individual.

Chromosome movement: The movement of chromosomes during mitosis and meiosis as a prerequisite for the anaphase separation of chromatids and/or chromosomes.

Chromosome mutation: Any structural change involving the gain, loss, or translocation of chromosome parts; it can arise spontaneously or be induced experimentally by physical and chemical mutagens.

Chromosome number: The specific somatic chromosome number ($2n$) of a given species.

Chromosome pairing: The highly specific side-by-side association of homologous chromosomes during meiotic prophase.

Chromosome segregation: The separation of the members of a pair of homologous chromosomes in a manner that only one member is present in any postmeiotic nucleus.

Chromosome set: A minimum complement of chromosomes derived from the gametic complement of a putative ancestor (basic chromosome-genome).

Chromosome size: The physical shape and size of a chromosome.

Chromosome substitution: The replacement of one or more chromosomes by others from another source by spontaneous events or crossing over.

Chromosome walking: It describes the sequential isolation of clones carrying overlapping sequences of DNA, allowing regions of the chromosome to be spanned. Walking is often conducted in order to reach a particular locus.

Class intervals: Collection of observations into groups. Class intervals may be either large or small depending on the need of the investigators.

Cleavage: Division of cytoplasmic portion in animals.

Cleistogamous flower: Pollination and fertilization occur before flowers open insuring complete self-pollination.

Climbing: Ascending on other plants.

Clonal propagation: Asexual propagation from a single cell or plant (vegetative propagation).

Clone: An organism propagated asexually, vegetatively, from the same individual; genetically identical.

Cloning vector: A plasmid or phage that is used to carry inserted foreign DNA for the purpose of producing more material or a protein product.

C-mitosis: Formation of two daughter cells in one cell induced by colchicine resulting in double chromosome number.

CMS: Cytoplasmic male sterility.

Coding strand: DNA has the same sequence as mRNA.

Codominance: Genes when both alleles of a pair are fully expressed in the heterozygote.

Codon: A triplet of nucleotides which represents an amino acid.

Coefficient of variation (C.V.): Standard deviation expressed as percentage of the mean.

Coenocyte: A multinucleate cell resulting from the repeated division of nuclei without cytokinesis.

Cohort: The group of individuals that enter the breeding program in the same season.

Coiling: When chromosome cores first become visible in late prophase of mitosis.

Colchicine: An alkaloid chemical induces polyploidy by inhibiting the formation of spindles, delaying the division of kinetochore.

Colchiploidy: Polyploidy induced by colchicine.

Coleoptile: A sheath surrounding the apical meristem and leaf primordia of the grass family.

Coleorhiza: A sheath of tissues surrounding the radicle of the grass embryo.

Companion species: Commonly a weed species that grows in close proximity to a crop species and which may be ancestral to crop; the two species may exchange genes such as *Glycine soja* and *Glycine max* growing side by side.

Compatible: A congenial interaction between a parasite and its host in plant disease.

Compensating diploids: One normal chromosome that is replaced by two telocentric chromosomes representing either arms or four telocentric chromosomes that compensate for a pair of normal homologues.

Compensating trisomics: A missing normal chromosome is replaced either by two tertiary chromosomes or by a secondary and a tertiary chromosome.

Compilospecies: A genetically aggressive plundering-related species of their heredities, and in some cases it may completely assimilate a species causing extinction.

Complementary genes: Two nonallelic gene pairs complement each other in their effect on the same trait. For example, purple-flowered F_1 plants are produced by crossing two white-flowered parents in sweet pea. Selfing of F_1 produces nine purple and seven white; purple arose as the complementary effect of dominant alleles at two different gene pairs.

Complete flower: A flower has sepals, petals, stamens, and pistils.

Complex heterozygous: Special type of genetic system based on the heterozygosity for multiple reciprocal translocations.

Complex locus: A cluster of two or more genes closely linked and functionally related.

Composite: A mixture of several genotypes maintained in bulk from one generation to the next.

Concerted evolution: The ability of two related genes to evolve together as though constituting a single locus.

Configuration: An association of two or more chromosomes at meiosis and segregating independently of other associations at anaphase.

Constitutive heterochromatin: It is differentiated by darker staining during interphase and prophase. In most of the mammals, constitutive heterochromatin is located in proximity of the kinetochore. It is an inert state of permanently nonexpressed sequences, usually highly repetitive DNA.

Constitutive mutation: The gene that is usually regulated to be expressed without regulation.

Continuous spindle fibers: Connect the two polar regions with each other and remain from early prometaphase to early prophase.

Controlling elements: Originally found and identified in maize by their genetic property, they are transposable units. They may be autonomous (transpose independently) or nonautonomous (transpose only in the presence of an autonomous element).

Convergent coorientation: In a trivalent, two chromosomes face one pole while a third one orients to the other pole.

Coorientation: Orientation of kinetochores at metaphase-I.

Core collection: The basic sample of a germplasm collection; designated to represent the wide range of diversity in terms of morphology, geography range, or genes.

Cotyledon: One of the two halves of a dicot seed; the cotyledons develop into the first two leaves of the young dicot plants.

Coupling: When linked recessive alleles (ab) are present in one homologous chromosome and the other homologous chromosome carries the dominant allele (AB). In the case of repulsion, each homologue contains a dominant and a recessive gene (Ab, aB).

Crossability: The ability of two individuals, species, or populations to hybridize.

Crossbreeding: Outbreeding.

Crossing-over: The exchange of chromatid segments of homologous chromosomes by breakage and reunion following synapsis during prophase of meiosis.

Crossing-over map: A genetic map constructed by utilizing crossing-over frequencies as a measure of the relative distance between genes in one linkage group.

Crossing-over value: The frequency of crossing-over between two genes.

Cryptic structural hybridity: Hybrids with chromosomes structurally different but not visible by metaphase-I chromosome pairing.

ctDNA: Chloroplast DNA.

Culm: Aerial stems of grasses, usually vertical but can be prostrate or spreading.

Cultivar: A category within a species of crop plants; cultivated variety.

Cuticle: A fatty substance which is almost impermeable to water is on the outer walls of the epidermal cells.

C value: The total amount of DNA in a haploid genome.

Cytochimera: Tissues contain different chromosome numbers in a single plant organ.

Cytodifferentiation: The sum of processes by which, during the development of the individual, the zygote, specialized cells, tissue, and organs are formed.

Cytogamy: The fusion of cells.

Cytogenetic map: A map showing the locations of genes on a chromosome.

Cytogenetics: A hybrid science that deals with the study of cytology (study of cells) and genetics (study of inheritance). It includes the study of the number, structure, function, and movement of chromosomes and numerous aberrations of these properties as they relate to recombination, transmission, and expression (phenotype) of the genes.

Cytogony: The reproduction by a single cell.

Cytokinesis: The division of cytoplasm and organelles in equal proportion. In plants, it occurs by the formation of a cell plate, while in animals it begins by furrowing. In lower organisms, cytokinesis does not follow after telophase, resulting in multinucleate cells.

Cytokinin: Plant hormones that stimulate shoot in tissue culture.

Cytological hybridization: See *in situ* hybridization.

Cytological maps: The genes located on the maps based on cytological findings such as deficiencies, duplications, inversions, and translocations.

Cytology: The study of cells.

Cytolysis: Breaking up of the cell wall.

Cytomixis: The extrusion of chromatin from one cell into the cytoplasm of an adjoining cell.

Cytoplasm: A portion of the cell apart from the nucleus.

Cytoplasmic inheritance: Inheritance of traits whose determinants are located in the cytoplasmic organelles, also known as non-Mendelian inheritance.

Cytosine: A nitrogenous base, one member of the base pair G-C (guanine and cytosine).

Cytoskeleton: Networks of fibers in the cytoplasm of the eukaryotic cells.

Cytosol: The volume of cytoplasm in which organelles are located.

Cytotaxonomy: The study of natural relationships of organisms by a combination of cytology and taxonomy.

Cytotype: Chromosome numbers other than the standard chromosome complement of that species. For example, *Glycine tomentella* contains accessions with four chromosome numbers ($2n = 38, 40, 78, 80$).

Daughter cell: The cells resulting from the division of a single cell.

Daughter chromosome: One of the chromatids of mitotic anaphase-I or meiotic anaphase-II.

Deciduous: Plants that shed all their leaves at the end of each growing season.

Decondensation stage: A stage between interphase and prophase of mitosis in which heterochromatin is decondensed for a short time.

Deficiency: The loss of a segment from a normal chromosome that may be terminal or intercalary. The term deletion refers to an internal deficiency.

Dehydration: Loss of water during reactions in which large molecules are synthesized from smaller ones; the reverse of hydrolysis.

Deletion: The loss of a chromosomal segment from a chromosome set.

Denaturation of DNA or RNA: Conversion from the double-stranded to the single-stranded state accomplished by heating.

Dendrogram: An evolutionary tree diagram may order objects, individual genes on the basis of similarity.

Deoxyribose: The four-carbon sugar that characterizes DNA.

Desynapsis: The premature separation of homologous chromosomes during diakinesis/metaphase-I; often genetically controlled.

Detassel: Removal of male inflorescence (tassel) from maize.

Developmental genetics: The study of the operation of genes during development.

Diakinesis (meiosis): Chromosomes continue to shorten and thicken. This is a favorable stage to count chromosomes. Chromosomes lay well-spaced out uniformly in the nucleus and often form a row near the nuclear membrane.

Diallel cross: Production of all possible hybrid combinations from a set of parents.

Diaspore: Any part of a plant dispersed from the parent plant and functions in reproduction.

Dicentric chromosome: A chromosome with two kinetochores.

Dichogamous: Protandry; male and female organs mature at different times allowing cross-pollination.

Didiploid: Two different diploid chromosome sets present in one cell or organism (allotetraploid).

Differentiation: It occurs in cell, tissue, and organ cultures where a plant develops from meristematic cells.

Diffuse kinetochore: Centromere activity is spread over the surface of the entire chromosome in certain insects of the order Hemiptera and in plant Luzula.

Diffuse stage: Prophase stage (meiosis) in which the chromosomes become reorganized; they even disappear.

Dihybrid cross: A genetic cross involving two genes.

Diisosomic: An individual that has a pair of homologous isochromosomes from one arm of a particular chromosome.

Diisotrisomic: An individual that lacks one chromosome but carries two homologous isochromosomes of one arm of a particular chromosome.

Dikaryon: A dinucleate cell.

Dioecious: A plant bears either staminate or pistillate flowers.

Diplo-haplontic: This phenomenon occurs in higher plants and in many algae and fungi. Life cycle of a typical plant contains spore-bearing ($2n$) and gamete-bearing (n) stages.

Diploid: An organism with two sets of chromosomes (two complete genomes). For example, barley is a diploid with $2n = 2x = 14$.

Diploidization: In polyploidy, chromosomes exhibit diploid-like meiosis; genetically controlled.

Diploidy: The presence of two homologous sets of chromosomes in somatic cells.

Diploid-like meiosis: An allopolyploid showing bivalent chromosome pairing that, sometimes, is under genetic control.

Diplonema (meiosis): A stage at which sister chromatids begin to repulse at one or more points in a bivalent.

Diplontic: Mature individuals are diploid by mitotic divisions and differentiations. Meiosis produces only haploid gametes. Diplophase is more prominent than the haplophase in humans, higher animals, and some algae.

Diplontic selection: Diploid cells overgrow the aneuploid cells in culture.

Diplospory: The embryo sac is formed from a megasporocyte as a result of abnormal meiosis.

Direct tandem inversions: It is due to two successive inversions involving chromosome segments directly adjacent to each other.

Discontinuous variation: Phenotypic distinct variation caused by genes.

Disjunct: A genus or species whose representatives are at separate locations; once distribution may have been continuous.

Disjunction: Migration (separation) of chromosomes to opposite poles at meiotic anaphase.

Disomic inheritance: It occurs by the association of chromosomes in bivalents at meiosis.

Dispermic: An egg fertilized by two male gametes.

Distal chromosome: A part of a chromosome, which is relative to another part, is farther from the kinetochore (opposed to proximal).

Distant hybridization: The crossing of species of different genera.

Ditelomonotelosomic: An individual that has a pair of telocentric chromosomes for one arm and a single telocentric chromosome for the other arm.

Ditelosomic: An individual that has two telocentrics of one chromosome arm.

Ditelotrisomic: An individual that has two telocentric chromosomes from one arm plus one complete chromosome of the homologue.

Ditertiary compensating trisomics: An individual with a compensating trisomics chromosome in which a missing chromosome is compensated by two tertiary chromosomes.

Divergence: A pattern of evolution in which distinction among groups occurs through natural selection and environments—two groups are not alike.

Dizygous: It is also known as fraternal twins. When two different eggs are each fertilized by a different sperm; they have different genotypes.

DMSO: Dimethyl sulfoxide.

DNA: Deoxyribose nucleic acid. All hereditary information is encoded.

DNase: An enzyme that attacks bond in DNA.

Domestication: The selective breeding by humans of species in order to accommodate human needs.

Dominance: When one member of a pair of alleles manifests itself more or less, when heterozygous, than its alternative allele.

Donor parent: A parent from which one or more genes controlling economically important traits are transferred to recurrent parent by backcrossing.

Double crossing-over: The production of a chromatid in which crossing-over has occurred twice.

Double cross: Crossing two single cross F_1 hybrids.

Double fertilization: The union of one sperm nucleus with the egg nucleus and one with polar nuclei (secondary nucleus) to form an embryo and endosperm in flowering plants.

Double heterozygote: Heterozygote in respect of two genes.

Double haploid: A diploid plant obtained from spontaneous or induced chromosome doubling of a haploid cell or plant, usually after anther or microspore culture.

Double reduction: Chromatid segregation.

Drosera scheme pairing: A triploid hybrid showing $2x$ bivalent + $1x$ univalent configurations at diakinesis or metaphase-I.

Duplex genotype: AAaa.

Duplication: When an extra piece of chromosome segment is attached to either the same chromosome (homologous) or transposed to another member (nonhomologous) of the genome.

Dyad: Two cell product of first meiosis.

Dysploid: A species with differing basic chromosome numbers (e.g., $x = 5$; 6; 7; 8; etc.). Basic chromosome numbers in dysploid are stable.

Dysploidy: Abnormal polyploidy.

Ear: In maize, the female is an ear from which protrudes the stigma (silks).

Ecology: A branch of science that deals with the structure and function of nature; a study of the relationship between organisms and their environments.

Ectopic pairing: Chromosome pairing between an allele and nonallelic (ectopic) homologous sequence due to some homology-searching mechanism.

Egg: The female gamete as opposite to sperm.

Egg apparatus: It includes egg cell and two synergid cells at the micropyle end of the embryo sac.

Electron microscope: A microscope that uses an electron beam as a light source.

Electrophoresis: A biochemical technique that utilizes an electrical current to separate mixtures of similar substances.

Emasculation: Removal of anthers from a flower.

Embryo culture: The excision of young embryos under aseptic conditions and transferring them on suitable artificial nutrient media for germination or callusing.

Embryogenic cells: Cells that have completed the transition from somatic state to one in which no further exogenously applied stimuli are necessary to produce the somatic embryos.

Embryology: To study the development of reproductive organs in an organism.

Embryo: Rudimentary plant in a seed.

Embryoid: An embryo-like structure observed in cell and tissue cultures.

Embryo sac: It is a mega gametophyte of flowering plants, includes egg, synergids, polar nuclei, and antipodals.

Endemic: A native and confined to a certain geographical area.

Endogamy: Sexual reproduction in which the mating partners are more or less closely related.

Endomitosis: Separation of chromosomes but two sister chromatids lay side by side and retention within a nucleus.

Endoplasmic reticulum: The outer membrane of the nuclear envelope.

Endopolyploidy: Polyploidy that arises from repeated chromosome replication without cytokinesis.

Endoreduplication: Chromosomes replicate time after time without condensation and mitosis and appear as bundles of multiple chromatids, for example, polytene chromosomes.

Endosome: A nucleolus-like organelle that does not disappear during mitosis.

Endosperm: Triploid tissue that originates from the fusion of a sperm nucleus with the secondary nucleus and nourishes the embryo during an early growth of the seedling.

Endothelium: The inner epidermis of the integument, next to the nucellus, where cells become densely cytoplasmatic and secretory. It is also known as integumentary tapetum.

Endothesium: A layer of cells situated in the pollen-sac wall situated below the epidermis.

Enneaploid: A polyploid with nine chromosome sets ($9x$).

Entomophilous: Pollination by insects.

Environment: The external conditions that affect growth and development of an organism.

Ephemeral: A plant completing its life cycle in considerably less than 12 months (very short-lived).

Epidermis: The outermost layer of cells.

Epigeal: Seed germination in which the cotyledons are raised above the ground by elongation of the hypocotyl (beans).

Epigenetic: A change in phenotype of an individual without altering the genotype.

Epipetalous stamen: A stamen that is congenitally attached to a petal.

Epiphytotic: An uncontrollable spread of a destructive plant disease.

Epistasis: Dominance of one gene over a nonallelic gene.

Equational division: A division of each chromosome into exact longitudinal halves which are incorporated equally into two daughter nuclei, also known as mitosis.

Equational plate: An arrangement of the chromosomes in which they lie approximately in one plane, at the equator of the spindle; seen during mitotic and meiotic metaphase.

Erectoides: A mutant plant with short erect stature and compact appearance.

EST: Expressed sequence tag.

Ethnobotany: The science that deals with the folklore knowledge of plants.

Euchromatin: The decondensed region of a chromosome containing a majority of the genes and stains lightly.

Euchromatization: The spontaneous or induced change of heterochromatin into euchromatin.

Euchromosome: A chromosome showing the typical features of the standard complement of a given species.

Euhaploid: A haploid genome showing no deviating number of chromosomes compared with the standard genome of the species.

Eukaryotes: An organism with true nucleus.

Eumeiosis: Normal occurring meiosis.

Euploid: An organism possessing chromosomes in exact multiples of the basic (x) chromosome number such as $2x$, $3x$, $4x$, and so on.

Eusom: A plant showing each member of the chromosome complement with the same copy number.

Euspory: The typical sporogenesis with normal flow of reduction division—reduced apomixis.

Eutelomere: A weakly staining subterminal segment adjacent to the protelomere.

Exalbuminous seed: A seed devoid of endosperm after maturity.

Exine: The outer wall of a mature pollen grain.

Exocarp: The outer most layer of the fruit wall; it may be the skin of the fruit, a leathery rind, or quite hard.

Exogamy: Outbreeding.

Exotic species: A species that is not native to a region.

Explant: An excised piece of tissue used to initiate tissue culture through *in vitro*.

Expressivity: The degree of phenotypic effect of a gene in terms of deviation from the normal phenotype.

Extrachromosomal inheritance: Inheritance that is not controlled by nuclear genes but by cytoplasmic organelles.

Extra chromosome: B-chromosome.

Extranuclear genes: Genes located in the cytoplasmic organelles such as mitochondria and chloroplasts.

F_1, F_2, etc.: Filial generations. The offspring resulting from experimental crossing (F_1) and selfing (F_2).

Facultative apomixis: The plant produces seed both apomictically and sexually.

Facultative heterochromatic: Chromatin which is heterochromatinized, may or may not be condensed in interphase.

Female ("Venus looking glass"): Female person, animal, or plant. In plants, having pistil and no stamens; capable of being fertilized and bearing fruits.

Feral: An uncultivated crop.

Fertility: Ability of a plant to produce offspring.

Fertilization: Fusion of a male sperm with an egg cell (formation of zygote).

Feulgen staining: A cytological colorless stain, utilizes Feulgen or Schiff's reagent. Chromosomes are colored, cytoplasm and nucleolus are colorless and are specific for DNA.

Filament: The stalk of a stamen.

Filiform apparatus: A complex wall outgrowth in a synergid cell.

FISH: Fluorescence *in situ* hybridization.

Fission: The division of one cell by cleavage into two daughter cells, or a chromosome into two arms.

Fixation: The killing of cells or tissues by chemicals in such a way that the internal structure of a cell is preserved.

Flag leaf: The last leaf before inflorescence emergence.

Fluorescence *in situ* hybridization (FISH): A cytological technique for the detection of repetitive DNA sequences such as ribosomal DNA sequences (NOR sequences).

Fluorescence microscopy: The principle of fluorescence microscopy is based on the light transmitted or reflected by the specimen. The specimen is stained with a fluorochrome (a dye that emits a longer wavelength when exposed to shorter wavelength of light). The fluorescent region of a specimen becomes bright against a dark background.

Frequency distribution: Frequencies of class intervals provide frequency distribution.

Funiculus: The stalk by which the ovules are attached to the placenta.

G_1: The period of the eukaryotic cell cycle between the last mitosis and the start of DNA replication (synthesis).

G_2: The period of the eukaryotic cell cycle between the end of DNA replication (synthesis) and the start of the next mitosis.

Gamete: A sex cell that contains half the chromosome complement of the zygote formed after meiosis that participates in fertilization.

Gametic chromosome number: n.

Gametocyte: A cell that produces gametes through division; a spermocyte or oocyte-germ cell.

Gametogenesis: The formation of male and female gametes.

Gametophyte: A part of life cycle of plant which produces the gametes where there is an alternation of generations.

Gametophytic apomixis: Agamic seed formation in which the embryo sac develops from an unreduced initial; includes both apospory and diplospory.

Gametophytic incompatibility: A series of alleles at one locus determines whether the pollen will be functional or nonfunctional. Similar gene in pollen and stylar tissue will cause inhibition of pollen tube growth. Thus, the ability of a pollen to function is determined by its own genotype and not by the plants that produce it.

G-banding: A cytological technique which exhibits a striated pattern in chromosomes.

G-Bands: Bands observed in the chromosomes after Giemsa staining. The chromosomes are pretreated with a dilute trypsin solution, urea or protease. Bands have been used to identify human and plant chromosomes.

Gene: The unit of inheritance located mainly in chromosomes at fixed loci and contains genetic information that determines characteristics of an organism.

Genebank: An establishment that preserves and maintains seeds, plants, and DNA of cultivated, wild and danger of extinction species.

Gene center: Center of origin of a given crop plant.

Gene dosage: It refers to the number of copies of a particular gene in the genome.

Gene flow: The contribution of a particular gene or genes to a population through interbreeding.

Gene-for-gene theory: Plant-pathogen interaction.

Gene frequency: Proportion in which alternative alleles of a gene occur in a population.

Gene interaction: Modification of gene action by a nonallelic gene or genes.

Gene location: Determination of physical or relative distances of a gene on a particular chromosome.

Gene pool: The reservoir of different genes of a certain plant species (lower and higher taxa) available for crossing and selection; divided based on hybridization into primary, secondary, and tertiary gene pools.

Generative nucleus: A product (two daughter nuclei) of division of primary nucleus in the pollen grain. One usually divides into two sperm nuclei in the pollen tube while vegetative nucleus does not divide again.

Genetic: A property obtained by an organism by virtue of its heredity.

Genetic code: The correspondence between triplets in DNA (or RNA) and amino acids in proteins.

Genetic combination: For n pairs of homologous chromosomes, only $1/2n - 1$ gametes will contain parental combination.

Genetic drift: Random change in gene frequency in a population due to sampling error.

Genetic erosion: The loss of genetic information that occurs when highly adaptable cultivars are developed and threaten the survival of their more locally adapted ancestors, which form the genetic base of the crop.

Genetic male sterility: Failure to produce and shed viable pollen, controlled by genes.

Genetic marker: An allele used to identify a gene or a chromosome or a chromosome segment.

Genetic map: The linear arrangement of gene loci on a chromosome determined from genetic recombination.

Genetic mapping: The assignment of genes to a specific linkage group and determination of the relative distance of these genes to other known genes in that particular linkage group.

Genetics: The science that deals with heredity and variation.

Genome: A complete haploid set of chromosomes of a species.

Genome affinity index (GAI): It is used for comparing the degree of homology among parental genomes. The mean number of bivalent-equivalents is divided by the basic chromosome number.

Genome doubling: Chromosome doubling of F_1 hybrid for producing auto- or allo- polyploidy.

Genome segregation: The segregation of sets of chromosomes (genomes) during mitosis in the somatic tissues that occurs in polyploid species and also in some wide crosses.

Genome symbol: Assignment of symbol to species, usually by capital letter, by cytogenetic and molecular methods.

Genomic *in situ* hybridization (GISH): A cytological technique by which genome specific probe is used. DNA sequence specific to a particular genome serves as a marker to identify genome, translocations, and alien addition and substitution lines.

Genotype: The description of the genes contained by a particular organism for the character or characters—the entire genetic constitution.

Genotypic ratio: Proportion of different genotypes in a population.

Genus: The first element of the scientific name. It includes a taxonomic group of related species.

Germ cell: Gamete.

Germination: The emergence of root and shoot from a seed; may be epigeal—elongation of hypocotyl or hypogeal—very little or not at all elongation of hypocotyl.

Germplasm: The total of the hereditary materials available into a species and genus.

Germplam enhancement: Gene transfer through sexual or asexual methods from germplasm accessions to transfer useful genes of economic importance to cultivated crops.

GMO: Genetically modified organism.

Golgie apparatus: All golgie bodies in a given cell are not like staining bodies usually engaged in secretion.

Golgie bodies: A cell organelle made up of closely packed broad cisternae and small vesicles.

Guanine: A nitrogenous base, one member of the base pair G-C (guanine and cytosine).

Gynoecium: The female portion of the flower—stigma, style, and ovary.

Gynomonoecy: A plant having both female and hermaphrodite flowers.

Gynophore: An elongation of the floral axis between the stamens and the carpels, forming a stalk that elevates the gynoecium.

Haploid: An organism containing gametic (*n*) chromosome number.

Haploid-initiator gene: A partially dominant haploid-initiator gene (*hap*) in barley; controls the abortion or the survival of abnormal embryos and endosperms. Plants homozygous for the *hap* gene produce progenies that include 10%–14% haploid.

Haplontic: In most primitive unicellular or filamentous algae and protozoa, haplophase is more prominent than the diplophase. Haplophase occupies from meiosis to fertilization.

Haplotype: The particular combination of alleles in a defined region of some chromosome, in effect the genotype in miniature.

Haustorium: A specialized organ that draws nutrients from another organ.

Helobial endosperm: A type of endosperm that develops in a way intermediate between that by which a cellular and a nuclear endosperm develops.

Hemialloploid: Segmental allopolyploid.

Hemigamy: Sometimes the sperm nucleus does not unite with the egg nucleus and divides independently and simultaneously. The embryo sac nuclei are unreduced. Chimeric embryos (somatic + half the somatic chromosome) develop.

Hemiploid: An individual with haploid chromosome number.

Hemizygous: A gene allele opposite from a deficiency.

Hermaphroditic: Flowers include male and female sexes.

Hereditary: Transmissible from parent to offspring.

Hereditary determinant: Gene.

Heredity: The process that brings about the biological similarity and dissimilarity between parents and progeny.

Heterocarpy: The production of two or more types of fruits by a plant.

Heterocaryosis: The presence of two or more genetically different nuclei within single cells of a mycelium.

Heterochromatin: The region of a chromosome that takes differential stain. These regions contain condensed chromatin.

Heterochromosome: Any chromosome that differs from the autosomes in size, shape, and behavior.

Heterofertilization: Fertilization of the nuclei of endosperm and embryo-forming cells by genetically different gametes.

Heterogametic sex: An organism has different sex chromosomes (X and Y, Z and W) and autosome chromosomes.

Heterogenetic: Chromosome pairing between more or less different genomes in amphidiploids.

Heterogenetic chromosome pairing: The F_1 hybrids of genomically divergent species lack chromosome affinity and form only a small number of loosely synapsed bivalents.

Heterokaryotype: A chromosome complement that is heterozygous for any sort of chromosome mutations.

Heteromorphic bivalent: A bivalent in which one of the chromosomes differs structurally. The chromosomes may differ in size or kinetochore positions.

Heteroplasma: Extranuclear genes—not alike (analogous to heterozygote for nuclear genes).

Heteroploid: An individual containing a chromosome number other than the true monoploid or diploid number.

Heteropycnosis: Differential spiralized regions of chromosomes. Positive at interphase or early prophase, negative at late prophase or metaphase.

Heterosis: Heterozygote superiority of an F_1 hybrid over parents with respect to some character or characters.

Heterozygous: A zygote derived from the union of gametes unlike in respect of the constitution of their chromosomes or from genes in a heterozygote (e.g., Aa). Also, used in case of hybrid for: interchanges, duplications, deficiencies, inversions—opposite of homozygotes.

Hexaploid: A polyploid with six sets of basic chromosomes.

Highly repetitive DNA: It is equated with the satellite DNA which is composed of many tandem repeats (identical or related) of a short basic repeating unit. (= simple sequence DNA).

Hilum: The scar on the surface of a seed revealed by abscission of the funicle; the point of attachment of the seed to the pod.

Histology: Study of tissues.

Holocentric: Diffused kinetochore. Every point along the length of chromosome exhibits centromeric activity.

Holandric genes: The genes located on Y chromosomes.

Homoeobox: A short stretch of nucleotides whose base sequence is virtually identical in all the genes that contain it. It determines when particular groups of genes are expressed during development.

Homoeologous chromosomes: Partially homologous chromosomes; for example, in wheat ABD genome chromosomes are known as homoeologous.

Homogametic sex: An organism has 2X chromosomes and autosome.

Homogamy: The male and female parts of a flower mature simultaneously.

Homogenetic chromosome pairing: Synapsis of homologous chromosomes of the same genome.

Homologous chromosomes: Like chromosomes showing synapsis all along their length during pachynema. For n pairs of homologous chromosomes, only $1/2n - 1$ gametes will contain parental combination.

Homomorphic incompatibility: It is due to relative length of the stamens and the style.

Homoplasmon: Extranuclear gene—alike (analogous to homozygote for nuclear genes).

Homozygote: An organism having like allelles at corresponding loci on homologous chromosomes (AA, aa).

Hotspot: A site in the chromosome at which the frequency of mutation (or recombination) is very much increased.

Hybrid: The progeny from mating parents of different genotypes, stains, species, and genera.

Hybrid inviability: The death of F_1 hybrid plants due to aberrant development process.

Hybrid lethality: The failure of hybrids to produce viable offspring.

Hybridization: The production of a hybrid by crossing two individuals of unlike genetic constitution.

Hybrid vigor: See heterosis.

Hyperploid: A diploid organism with an extra chromosome.

Hypertriploid: An individual containing chromosome numbers higher than $3x$.

Hypogeal emergence: The elongation of the epicotyl and the cotyledons remaining below ground; common in monocots.

Hypoploid: A diploid organism lacking an extra chromosome.

Ideotype: An ideal plant form formulated to assist in reaching selection goals.

Idiogram: A diagrammatic representation of chromosome morphology of an organism.

Illegitimate crossing-over: Unequal crossing-over.

Immunity: A plant that does not permit infection by a parasitic organism.

Inbred: A line originated by continued self-fertilization accompanied by selection.

Inbreeding: Mating of closely related plants or animals.

Included inversions: A segment that is part of an inverted segment is inverted once again.

Incompatibility: Inability of a male gamete to fertilize female gamete (egg), thus preventing the zygote formation.

Indehiscent: A fruit that does not open by sutures, pores, or caps, the seeds being released by the rotting away of the fruit wall.

Independent assortment: The random distribution to the meiotic products (gametes) of genes located on different chromosomes. A dihybrid individual with RrYy genotype will produce four gametic genotypes (RY, Ry, rY, ry) in equal proportion.

Independent inversions: Such inversions occur in independent parts of the chromosomes and the two resultant inverted segments are separated from each other by an uninverted (normal) chromosome segment.

Indigenous: An organism originated naturally in a particular area.

Induced mutation: Induction of mutation by mutagens.

Infertile: Not able to produce viable spores.

Inflorescence: Spikelets arranged on a common branching system (panicle, raceme, spike).

Inheritance: The transmission of genetic information from parents to progenies.

***In situ* hybridization:** A cytological technique conducted by denaturing the DNA of cells squashed on a microscope slide so that reaction is feasible with an added single-stranded DNA or RNA, which labeled and following hybridization gene is located on certain sites of a chromosome.

Integument: Layers of cells surrounding the ovule.

Intercalary: Chromosomal segments located besides terminal region of chromosomes.

Interchange (reciprocal translocation): The reciprocal exchange of terminal segments of non-homologous chromosomes.

Interchange trisomics: An additional chromosome to diploid set, which is composed of two different chromosomes (translocation).

Intergeneric cross: A cross between two species from two different genera.

Interkinesis: The time gap between cell divisions I and II. It is short or may not occur at all.

Internode: The interval on a stem or rachis between two nodes.

Interphase: The period between mitotic cell division (or intermitotic period) which is divided into G1, S, and G2.

Interspecific cross: The cross between two parents from different species.

Interstitial segment: A chromosome region between the centromere and a site of rearrangement.

Intine: The inner wall of a mature pollen grain.

Intrachromosomal: Within a chromosome.

Intragenomic pairing: Pairing of chromosomes in monohaploid.

Introgressive hybridization: The incorporation of gene or genes of a species into gene pool of another species.

Inversion: Reversal of gene order within a chromosome, may be paracentric which does not include kinetochore or pericentric which involves kinetochore in the inverted region.

Inversion heterozygote: One chromosome carries normal gene sequence while the other chromosome contains an inverted gene sequence; may be paracentric or pericentric.

In vitro: Outside a living organism.

Irradiation: Exposure of an organism to X-rays or other radiation to enhance mutation rates.

Irregular flower: The perianth constituted in such a way that it is not possible to divide it into two equal halves, as in the canna flower.

Isochromocentric: Nuclei showing as many chromocenters as chromosomes.

Isochromosome: A chromosome with two identical arms; it is usually derived from telocentric chromosomes.

Isodicentric chromosome: An abnormal chromosome containing a duplication of part of the chromosome including the centromere; the resulting chromosome contains two centromeres and a point of symmetry that depends on the position of the breakpoint.

Isogenic lines: Genetically uniform, homozygous lines.

Isolation: The separation of one group from another preventing mating between and among groups.

Isosomic: Individuals showing isochromosomes.

Isotelocompensating trisomic: A compensating trisomic; a missing chromosome is compensated for one telocentric and one tertiary chromosome.

Isotertiary compensating trisomic: A compensating trisomic; a missing chromosome is compensated for by one isochromosome and one tertiary chromosome.

Isotrisomic: An individual containing one isochromosome to the normal chromosome complement.

Kappa particles: Cytoplasmic kappa particles cause a killer trait in *Paramecium* that is genotypic dependent. Kappa particles reduce in cells containing the genes k, s1, and s2. Some races of killer paramecium produce paramicin, which is lethal to other races (sensitive). Certain portions of kappa particles contain proteinaceous R-bodies. These are responsible and required for the toxic action of kappa particles on sensitive type. The behavior and structure of R-bodies are similar to any other bacterial structure that has been established by chemical composition and electron microscopic studies.

Karyoevolution: Evolutionary change in the chromosome set, expressed as number, structure of chromosomes.

Karyogamy: The fusion of two gametic nuclei.

Karyogram: Ideogram.

Karyokinesis: The process of nuclear division.

Karyology: Study of the nucleus.

Karyostatasis: The stage of cell cycle in which there is no visible dividing activity of the nucleus, but metabolic and synthetic activity.

Karyotype: The usual chromosome number, size, and morphology of the chromosome set of a cell of an organism. The complete set of chromosomes. For example, barley contains $2n = 2x = 14$.

kb: An abbreviation for 1000 base pairs of DNA or 1000 bases of RNA.

Kinetochore: The region of a chromosome attached to a spindle fiber during movement of the chromosomes at mitosis and meiosis.

Kinetochore-misdivision: Division of kinetochore crosswise instead of lengthwise particularly during meiosis.

Kinetochore-orientation: The process of orientation of kinetochore during prometaphase of mitosis and meiosis.

Knob: A heterochromatic darkly stained region that may be terminal or intercalary. It is often used as a landmark to identify specific chromosomes.

Lagging: Delayed migration of chromosome from the equatorial plate to the poles at mitotic or meiotic anaphase and consequently excluded from the daughter nuclei producing aneuploid gametes.

Lampbrush chromosome: A special type of chromosome found in the primary oocyte nuclei of vertebrate and invertebrate. Such chromosomes are due to prolonged diplonema. Chromosomes exhibit fuzzy appearance due to paired loops. These loops vary in size and shape from chromosome to chromosome and have been used to construct cytological maps.

Landrace: A primitive cultivar evolved from a natural population and selected for cropping.

Leaky mutation: It allows some residual expression of gene.

Leptonema (meiosis): Chromonemata become distinct from one another and appear as very long and slender threads. Nuclei contain a large nucleolus and a distinct nuclear membrane.

Lethal gene: Any gene in which a lethal mutation can be obtained (deletion of the gene).

Leucoplast: A plastid devoid of pigment.

Linkage: Association of qualitative and quantitative traits in inheritance due to location of marker genes in proximity on the same chromosome.

Linkage drag: The inheritance of undesirable genes along with a beneficial gene due to their close linkage.

Linkage group: It includes all loci that can be connected (directly or indirectly) by linkage relationships; equivalent to haploid chromosome. For example, barley has seven linkage groups corresponding to seven haploid chromosome numbers. All genes located in one chromosome form one linkage group.

Linkage map: A graphical representation of the positions of genes along a chromosome.

Linkage value: Recombination fraction: the proportion of crossovers versus parental types in a progeny. It may vary from 0% to 50%.

Localized kinetochore: A chromosome carrying normal permanently kinetochore localized to which the spindle fiber is attached during chromosome separation.

Locus: The position of a gene in a chromosome.

Lysenkoism: The doctrines of proclaimed in the former Soviet Union by T. D. Lysenko, which does not accept the gene concept and is based on the inheritance of acquired traits.

M_1, M_2: Symbols used to designate first and second generations after treatment with a mutagenic agent.

Maceration: The artificial separation of the individual cell from a tissue by enzyme treatment.

Macroevolution: Evolution above the species level.

Male: ("Mars shield and spear") Male person, animal, or plant; generally used in contrast to female.

Male sterility: The male sterile genes (*ms*) cause complete breakdown of microsporogenesis but macrosporogenesis (megasporogenesis) remains completely uninfluenced.

Map distance: It is measured at cM (centi Morgan) = percent recombination.

Marker: An identifiable physical location on a chromosome, determined by inheritance.

Maternal inheritance: Inheritance of a trait through female.

Maturity group: Division of soybean cultivars in North America into 12 groups based on how flowering responds to dark periods; maturity groups are selected for the latitude so that flowering is initiated for maximum yield.

Mean: Central value of a set of sample observations.

Median: The value for which 50% of the observations are on either side.

Median centromere: A centromere that is located midway on chromosomes resulting in two equal arm lengths.

Megabase (Mb): Unit length for DNA fragments equal to 1 million nucleotides and roughly equal to 1 cM.

Megagametophyte: Female gametophyte; embryo sac within the ovule of angiosperms.

Megaspore: A gametophyte bearing only female gametes.

Megaspore mother cells (MMCs): Produce eight nuclei (megaspores) after one of four haploid macrospores becomes functional in the embryo sac. Two mitotic divisions produce eight nuclei (one egg, two synergids, three antipodals, and two polar nuclei).

Meiosis: A phenomenon that occurs in reproductive organs of male and female organisms, producing gametes with the haploid (n) chromosome constitution. The first division is reductional and the second division is equational.

Meiotic duration: The time required to complete the cell cycle from prophase to telophase under certain conditions.

Meiotic segregation: The random orientation of bivalents at equatorial plate at metaphase-I regulates the segregation of paternal and maternal chromosomes to the daughter nuclei.

Meristem: An actively growing tissue that produces cells that undergo differentiation to form mature tissues.

Mesocarp: The middle layer of the fruit wall; often the fleshy edible portion.

Mesophyll: The photosynthetic parenchymatus tissue situated between the two epidermal layers of the leaf.

Mesopolyploid: Individuals have undergone moderate aneuploid-induced base number shift. Apomicts may contain complete or nearly complete duplicate sets of genes.

Metacentric chromosome: A chromosome with a median kinetochore (both arms are equal in length).

Metakinesis: Prometaphase.

Metaphase (mitosis): Chromosomes are arranged at the equatorial plate. Complete breakdown of the nuclear membrane and disappearance of the nucleolus occur.

Metaphase arrest: The arrest of cell division at mitotic metaphase, usually by application of specific agents.

Metaphase-I (meiosis): Bivalents are arranged at the equatorial plate with their kinetochores facing its poles. Chromosomes reach their maximum contractions, nuclear membrane and nucleolus disappear, and spindle fibers appear.

Metaphase plate: The movement and arrangement of all the chromosomes midway between the two poles of the spindle.

Metastatis: The ability of tumor cells to leave their site of origin and migrate to other locations in the body where a new colony is established.

Metaxenia: Tissues outside the embryo sac are influenced by the pollen source.

Maternal inheritance: Phenotypic differences found between individuals of identical genotype due to an effect of maternal inheritance gene (uniparental inheritance).

Microchromosomes: Tiny dot-like chromosomes that are too small for their centromere and individual arms to be resolved under the light microscope.

Microgametogenesis: The development of pollen grains.

Microgametophyte: The male gametophyte of a heterosporous plant.

Micronucleus: A nucleus separate from the main nucleus formed at telophase by one or more lagging chromosomes or fragments.

Micropyle: An opening of the ovule.

Microsomes: The fragmented pieces of endoplasmic reticulum associated with ribosomes.

Microsporangium: A structure that produces the microspores in a heterosporous plant.

Microspore: A gametophyte producing only male gametes—a pollen grain.

Microsporocyte: A cell that differentiates into a microspore.

Microspore Mother Cells: Diploid cells in the stamens (pollen mother cells, PMCs) give rise to haploid microspores after meiosis.

Microtubules: Manipulation of chromosomes in eucaryotes is due to microtubules. They are organized into spindle-shaped bodies, and disappear at the end of cell division.

Minichromosomes: Small chromosomes with kinetochores, totally heterochromatic.

Minute chromosomes: Usually very tiny chromosome segments as a result of chromosome aberrations.

Misdivision: Aberrant chromosome division in which transversal but no longitudinal separation of the centromere occurs.

Mitochondrion: A cell organelle found in the cytoplasm. Mitochondrial DNA is responsible for extra nuclear inheritance. It is the site of most cellular respiration.

Mitosis: A process of cell division known as equational division because the exact longitudinal division of the chromosomes into two chromatids occurs. The precise distribution into two identical daughter nuclei leads to the generation of two cells identical to the original cell from which they were derived.

Mitotic anaphase: Daughter chromosomes are pulled to opposite poles by the spindle fibers.

Mitotic apparatus: An organelle consisting of three components: (1) the asters which form the centrosome; (2) the gelatinous spindle; and (3) the traction fibers, which connect the centromeres of the various chromosomes to either centrosome.

Mitotic crossing over: Somatic crossing-over.

Mitotic cycle: The sequence of steps by which the genetic material is equally divided before the division of a cell into two daughter cells opens.

Mitotic index (MI): The fraction of cells undergoing mitosis in a given sample; usually the fraction of a total of 1000 cells that are undergoing division at one time.

Mitotic inhibition: Induced or spontaneous inhibition of mitotic division.

Mitotic prophase: Chromosomes begin to reappear, are uniformly distributed in the nucleus, and are more or less spirally coiled and seem longitudinally double.

Mitotic metaphase: Chromosomes are shrunk to the maximum limit, kinetochores move onto the equatorial plate.

Mitotic poison: Any substance that hampers proper mitosis.

Mitotic recombination: The recombination of genetic material during mitosis and the process of asexual.

Mitotic telophase: By the end of anaphase, spindle fibers disappear and chromosomes at each pole form a dense boll. Nucleolus, nuclear membrane, and chromocenters re-emerge. Chromosomes lose their stainability.

Mixoploidy: A failure in chromosome migration results in the formation of multiploid microsporocytes.

Mode: The values that occur most frequently.

Modifying genes: The genes that by interaction affect the phenotypic expression of nonallelic genes or gene. Gene may be an enhancer (intensifies the expression) or reducer (decreases the expression of other gene).

Monocentric: One centromere per chromosome.

Monoecious: Both staminate and pistillate flowers on the same plant.

Monogenic character: A character determined by a single gene.

Monohybrid cross: A genetic cross that involves only one pair of genes.

Monoploid: An organism having basic (x) chromosome number.

Monosomics: Where one of the chromosomes is missing from the normal chromosome complement ($2n - 1$).

Monotelotrisomics: An individual with $2n + 1$ telocentric chromosomes.

Monozygous twins: It is also known as identical twins. Twins arise from a single zygote formed by the fertilization of a single egg by a single sperm. In this case, the developing egg cleaves at an early division forming two separate embryos instead of the usual one.

Morphogenesis: The differentiation of structures during development.

Mother cell: A diploid nucleus which by meiosis yields four haploid nuclei—the microspore or pollen mother-cell (P.M.C.) and the megaspore or embryo-sac mother-cell in flowering plants.

mt DNA: Mitochondrial DNA.

Multiform aneuploidy: An organism contains cells with different aneuploid chromosome numbers producing the tissues with chromosome mosaicism.

Multiple factors: Many genes involved in the expression of any one quantitative character.

Multivalent: An association of more than two completely or partially homologous chromosomes during meiosis, generally observed in chromosomal interchanges and in autopolyploidy.

Mutagen: A physical or chemical agent that significantly raises the frequency of mutation above spontaneous rate. Mutations are called induced mutations.

Mutagenic: Capable of causing mutation.

Mutant: An organism carrying a gene resulted from mutation.

Mutation: Any heritable change in the genetic material (sequence of genomic DNA) not caused by segregation or genetic recombination, may occur spontaneously or induced experimentally. A detectable and heritable change in the genetic material, transmitted to succeeding generations.

n: haploid chromosome number symbol.

Natural crossing: It occurs between individuals within a population.

N-bands: This procedure was originally developed to stain nucleolus organizer chromosomes. Chromosomes are treated at low pH (4.2 ± 0.2) in 1 M NaH_2Po_4 for 3 min at high temperature $(94 \pm 1°C)$. Chromosomes are stained with Giemsa after incubation and washing.

Necrotic: Dead plant tissues usually caused by disease, insect activity, or nutrient deficiency.

Neocentromeres: Centromere region is replaced by a secondary center of movement; chromosome ends move first during anaphase-I of meiosis.

Neopolyploid: Polyploid individuals of the recent origin.

Neospecies: New species.

Neurospora: The common *Neurospora crassa* is a bread mold. The normal vegetative phase is a branching haploid individual with hyphae which forms a spongy pad, the mycelium. Mycelium fuses for each other but nuclei do not (heterokaryon), although mixed and remain in the same cytoplasm. Asexual occurs by repeated mitosis of nuclei or by the mitotic formation of special haploid spores (conidia). Conidia can form a new mycelium.

Node: The point at which a leaf is inserted or attached with the stem.

Nondisjunction: The failure of separation of paired chromosomes at mitotic and meiotic anaphase resulting in one daughter cell receiving both and another daughter cell none of the chromosomes.

Nonchromosomal: It shows a non-Mendelian inheritance; the genes are extra-chromosomally located in chloroplasts and mitochondria.

Noncoorientation: In a quadrivalent, two nonhomologous centromeres are on opposite sites and centromeres of two translocated chromosomes are stretched between the other two and not attached to the opposite poles. Noncooriented quadrivalent configuration results in a 3:1 chromosome segregation, always produces unbalanced gametes, and generates trisomics in the progenies.

Nondisjunction: It can occur during mitosis and meiosis. Chromosomes fail to separate at metaphase resulting in their passage to the same pole. This may occur due to failure of chromosome synapsis, or from multivalent formation.

Nonhomologous: Chromosomes carrying dissimilar chromosome segments and genes that do not pair during meiosis.

Nonhomologous chromosome association: Pairing of nonhomologous chromosomes at meiotic metaphase-I; usually observed in haploid.

Nonlocalized centromeres: The spindle attachment is not confined to a strictly localized chromosome area.

Nonrecurrent parent: The parent of a hybrid not used again as a parent is backcrossing.

Northern blotting: A molecular biology technique for transferring RNA from an agarose gel to a nitrocellulose filter on which it can be hybridized to a complementary DNA.

Nucellus: Ovule tissues internal to the integuments and surrounding the embryo sac.

Nuclear division: The division of the cell nucleus by mitosis, meiosis, and amitosis.

Nuclear envelope: A doubled-layer outer membrane of the nucleus which separates nucleus and the cytoplasm of the cell in eukaryotes.

Nuclear fragmentation: The degeneration of the nucleus by partition of the nucleus into more or less different parts.

Nuclear pore: Holes in the nuclear envelope and are assumed to be used for transport of macromolecules.

Nucleic acid: A large molecule composed of nucleotide subunits.

Nucleolar chromosome: The chromosome that carries the nucleolus organizer region.

Nucleolar dominance: Amphiplasty.

Nucleotype: The gross physical characterization of the nucleus; its mass and particularly amount of DNA content.

Nucleolus: A spherical body found inside the nucleus is associated with a specific chromosomal segment (the nucleolus organizer), is active from telophase to the following prophases, and is the site of ribosomal RNA (rRNA) synthesis.

Nucleolus organizer region: The chromosome region that is associated with the nucleolus, responsible for the formation of the nucleolus and contains ribosomes, referred to as the NOR.

Nucleosome: The basic structural subunit of chromatin, consisting of ~200 bp of DNA and an octamer of histone of proteins.

Nucleus: A spheroidal most important body within the cells of eukaryotes which contain chromosomes.

Nulliplex: In polyploidy, one recessive allele is carried by all chromosomes (*aaaa*).

Nullisomics: An individual lacking one chromosome pair from the normal chromosome complement ($2n - 2$).

Nullitetrasomics: In allohexaploid wheat, one pair of homologous chromosomes is missing and is substituted by two pairs of homologous chromosomes.

Null mutation: Physical deletion of a gene.

Obligate apomixis: In plants where no sexual reproduction occurs.

Ontogeny: The development of an organism from fertilization to maturity.

Oöcyte: The cell that becomes the egg by meiosis.

Oögonium: A mitotically active cell which gives rise to oocytes.

Organelle: Any structure of characteristic morphology and function in a cell, such as nucleus, mitochondrion, chloroplast, etc.

Organogensis: Formation of organs (structural and functional units—flowers).

Outbreeding: A mating of genetically unrelated plants.

Overdominance: A heterorygote (Aa) is superior in phenotype than either homozygotes (AA, aa).

Overlapping inversions: A part of an inverted segment being inverted a second time together with an adjacent segment that was not included in the first inversion segment.

Ovum: An unfertilized egg—ovule, egg.

p: Human chromosome nomenclature for the short arm (petit, French for short).

P_1: Denotes parents of a first filial (F_1) generation.

Pachynema: Synapsis of homologous chromosomes is completed. Chromosomes are noticeably thicker and shorter than those in leptonema. Nucleolus is clearly visible and certain chromosomes may be attached to it.

Pairing: Chromosome pairing—synapsis.

Pairing segment: The segment of chromosomes (X and Y chromosome) which pairs and cross-over (pairing segment) and the remaining segments which do not pair are known as differential segment.

Paleopolyploid: Ancient polyploids; all their $2x$—progenitors apparently have become extinct.

Pangenesis: Proposed by Charles Darwin, all structures and organs throughout the body contribute copies of themselves to a sex cell.

Panicle: A much branched raceme in which flowers are borne on the ultimate branchlets, for example, rice and oat.

Panmixis: Random mating of population.

Paracentric inversion: A type of intrachromosomal structural change that does not include the kinetochore of the chromosome.

Paramecia: They are unicellular, slipper-shaped animal, maintain three individual nuclei; two are small (micronuclei) with diploid chromosome number and the third nucleus is large (macronucleus) and is polyploid. Micronuclei divide mitotically by binary fusion. Macronucleus constricts in the middle and separates into two halves.

Parameters: Values of the larger population from which the sample has been taken are known as parameters.

Paranemic coiling: Chromatids are easily separated laterally.

Parasexuality: Crossing over and genetic recombination occurs in viruses and fungi by means other than the regular meiosis. The diploid nuclei are borne from rare fusion of two genetically unlike nuclei in a heterokaryon. Haploid nuclei are produced from the diploid with new combination of genes.

Parthenogenesis: Embryo is produced from a haploid or diploid unfertilized egg cell with or without the stimulus of pollination and fertilization.

Parthenocarpy: The development of fruits without seeds as a result from lack of pollination, a failure of fertilization, or death of embryo at an early stage of development.

Pathogen: An organism causing disease in plants.

PCR: (Polymerase chain reaction); a molecular technique in which cycles of denaturation, annealing with primer, and extension with DNA polymerase are used to amplify the number of copies of a target DNA sequences by $>10^6$ times.

Pedicel: The stalk that supports the paired of spikelet or flowers.

Pedigree: The record of the ancestry of an individual or a cultivar.

Peduncle: The stalk that supports an inflorescence including the stalk that supports a flower.

Pentaploid: $5x$.

Penetrance: The proportion of genotypes that exhibit an expected phenotype, depends both on genotype and the environment.

Perennial: The plants continue to grow through the years (2 or more), flower and fruit repeatedly.

Perfect flower: A flower possessing both stamen and pistil.

Pericentric inversion: A type of inversion that includes the kinetochore within the inverted segments.

Permanent heterozygote: Individuals in which chromosomal heterozygosity is fixed either due to balanced lethal factors, translocations, or inversions and that always breed true; for example, *Oenothera*.

Phenocopy: The phenotype of a given genotype changed by external conditions to resemble the phenotype of a different genotype.

Phenotype: The observable external appearance of an organism produced by the interaction of genotype and the environment.

Phylogeny: The relationships of groups of organisms (taxonomic groups) as expressed by their evolutionary relationships.

Pistil: The female reproductive organ (gynoecium) of a flower consisting of ovary, style, and stigma.

Pistillate: A female flower containing a functional gynoecium, but lacks well-developed functional stamens.

Placenta: The region of attachment of the ovules to the carpel.

Plasmagene: An extranuclear hereditary gene that shows non-Mendelian inheritance.

Plasma membrane: A membrane that surrounds the boundary of every cell.

Plasmotype: All plasmagenes (total hereditary complement of a cell) constitute the plasmon or plasmotype.

Plastid: A self-cytoplasmic organelle of plant cells. Plastid includes chloroplasts (chlorophyll), chromoplasts (carotenoides), and leucoplasts (no visible pigments).

Plastogene: Genes located in the plastids showing non-Mendelian inheritance.

Plectonemic coiling: Chromatids cannot be separated without unwinding the coil.

Pleiotropy: The multiple distinct phenotypic effect of a single gene.

Ploidy: The number of chromosome sets per cell of an organism as haploid, diploid, and polyploid.

Plumule: The embryonic shoot—coleoptile.

Point mutation: A change in single base pairs (gene mutation).

Polarity: The effect of a mutation in one gene in influencing the expression of subsequent genes in the same transcription unit.

Pollen grain: Haploid male spores produced after meiosis—male gametophyte.

Pollen mother cell: Microsporocyte, abbreviated as PMC, produces pollen grains as a result of microsporogenesis.

Pollen tube: A germinating pollen grain that carries male gametes to the ovum.

Pollination: Natural or artificial placement of pollen on the receptive stigma of a female flower.

Pollinium: A pollen mass, as in the milk weeds and orchids.

Polycentromeres: Each chromosome is attached by many spindle fibers.

Polyembryony: The production of several embryos in a seed.

Polygamous: The plants produce both perfect and imperfect flowers.

Polygene: A gene that individually exerts two slight effects but together with several or many genes controls quantitative character.

Polyhaploid: Haploid (*n*) individuals arising from polyploid species.

Polymitosis: Occurrence of a series of mistoses in rapid succession with or without division of chromosomes.

Polymorphism: The occurrence of two or more genetically different types in the population—allelic variation. In a molecular term, changes in DNA influence the restriction pattern.

Polyploid: An organism containing more than two sets of chromosome.

Polyploidization: Induction of polyploids artificially or naturally from haploid and diploid plants.

Polyspory: It is characterized by numerous cytological abnormalities because complete sets of asynchronously expressed genes are not present. It is associated with the frequent occurrence of aneuploid series formation and stabilization (Poaceae).

Polytene chromosome: A giant chromosome consisting of many identical chromosomes closely associated along their entire length (endomitosis). In *Drosophila melanogaster* salivary gland chromosomes about 5000 bands can be recognized. Homologous polytene chromosomes pair along the length in the somatic cells (somatic pairing). Any chromosomal structural changes could be determined by differences in the pairing pattern. Changes in appearance of bands in certain tissues occur at different times, either condensing or expanding to form "puffs".

Population: A group of organisms inhabiting an area, which share a common gene pool.

Position effect: A change in the expression of a gene brought about by the translocation to a new site in the genome. (A previously active gene may become inactive if placed near heterochromatin.)

P-particles in *Paramecium*: A group of about 10 different particles (kappa, gamma, delta, pi, mu, lambda, alpha, and tau) are found in the cytoplasm of *Paramecium aurelia*. These are called P-particles. These particles consist of DNA, RNA, and proteins.

Primary constriction: Synonymous with centromere or kinetochore.

Primary trisomics: An individual with normal chromosome complement plus an extra complete standard chromosome is designated as a simple primary trisomic.

Primer: A short sequence, often, RNA that is paired with one strand of DNA and provides a free 3′-OH end at which a DNA polymerase starts.

Primordium: A cell or organized group of cells in the earliest stage of differentiation.

Probe: Single-stranded DNA or RNA molecules of specific base sequence, labeled either radioactively or immunologically, used to detect the complementary base sequence by hybridization.

Procumbent: A plant stem that lives on the ground for all or most of its length.

Proembryo: An earliest stage of embryo development.

Prokaryote: An organism lacking nuclei—Bacteria.

Prophase (mitosis): It is the first stage of mitosis, cells prepare for cell division. Coiling of chromosomes (condensation) occurs and chromosomes appear as a thread-like structure. Prophase includes splitting of chromosomes longitudinally into two duplicates (chromatids) and the centrosome and dissociation of each half to opposite sides of the nucleus, the synthesis of mitotic apparatus, the disappearance of nucleolus, and the beginning of the breakdown of the nuclear membrane.

Protandry: Maturation of anthers before pistils.

Protelomere: A terminal deep-staining structure with sharp limits, normally composed of one to three dark staining large chromomeres.

Protoclone: A clone derived from a protoplast.

Protogyny: Maturation of pistils before anthers.

Protoplast: The plant cell exclusive of the cell wall.

Pseudogamy: In plants where an embryo develops from a diploid or haploid egg cell with a sperm penetrated but without fertilization and egg cell nucleus remains functional.

Pseudogenes: These are inactive genes but are stable components of the genome derived by mutation of an ancestral active gene.

Pseudoisochromosome: Chromosome ends are homologoues due to reciprocal translocation, chromosomes pair at meiotic metaphase-I like an isochromosome but the interstitial segments (proximal to kinetochore) are nonhomologous.

Pubescence: A general term used for hairiness.

Punnett square: A checkerboard used in finding the all possible zygotes produced by a fusion of parental gametes. This also results in determining the genotypic and phenotypic ratios.

Pure line: The descendent obtained from self-fertilization of a single homozygous parent.

Purine: A nitrogen-containing single-ring, basic compound that occurs in nucleic acids. The purines in DNA or RNA are adenine and guanine.

Pyrimidine: A nitrogen-containing, double-ring, basic compound that occurs in nucleic acids. The pyrimidines in DNA are cytosine and thymine; in RNA, cytosine and uracil.

Q: Human chromosome nomenclature for the long arm.

Q-bands: Caspersson et al. (1968) showed that chromosomes of many plant species stained with quinacrin mustard exhibit bright and dark zones under ultraviolet light. This method was later used for the identification of individual human chromosomes (A-T rich region of a chromosome). .

Quadripartitioning: During meiosis, cytokinesis does not occur after meiosis I and is postponed until after the second division.

Quadrivalent: An association of four chromosomes generally observed in a reciprocal translocation or in autotetraploids.

Qualitative inheritance: Inheritance of qualitative traits which is controlled by a single gene (discontinuous inheritance).

Quantitative inheritance: Inheritance involving the simultaneous action of many genes in a cumulative or additive fashion, each of which produces a small effect. (Height, weight, and size are few traits.)

Quartet: The four nuclei produced at the end of meiosis (=tetrad).

Quasibivalent: Pseudo bivalent—Bivalent-like association during meiosis due to stickiness rather than by chiasmata.

Race: A genetically distinct mating group within a species.

Raceme: An unbranched, elongate, indeterminate inflorescence on padicellate flowers.

Rachilla: The central axis of a spikelet is in the grass family.

Radicle: Embryonic root.

Radioautography: A photographic technique in which a radioisotope is taken in by cell and can be traced with a sensitive photographic film.

Radioisotope: A radioactive isotope of an element.

Radiomimetic: A chemical that mimics ionizing radiation in terms of damage to genes and chromosomes resulting in gene and chromosome mutations.

Random mating: In a population where an individual has equal probability to mate with any individual of the opposite sex.

Range: The distance between two extreme values.

R-bands: The lightly stained G-bands become darkly stained after R-banding staining. R-bands are obtained after treating chromosomes at low pH (4–4.5) at 88°C incubation in NaH_2PO_4 (1 M). R-bands are reversible if pH is adjusted to 5.5–5.6.

Reassociation of DNA: The pairing of complementary single strands of DNA to form a double helix.

Recessive: A member of an allelic pair of contrasting traits which is not expressed.

Recessive lethal: Recessive homozygous is lethal.

Reciprocal cross: Male and female parents are reversed.

Reciprocal recombination: Reciprocal exchange of one part of chromosome with part of another chromosome.

Recombination: The occurrence of new combinations of genes in hybrids, not present in the parents, due to independent assortment and crossing over in crosses between genetically different parents.

Recurrent parent: In a backcrossing program, a parent that is used in successive backcrosses.

Reduction: A process that halves the somatic chromosome number at meiosis; segregation of homologous includes a member of each set but these chromosomes are no longer the unaltered parental chromosomes.

Regeneration: The production of plants from cells and tissues through *in vitro* culture.

Regulatory gene: It codes for an RNA or protein product whose function is to control the expression of other genes.

Relational coiling: Chromatids are twisted about each other during early prophase.

Renaturation: The return of a denatured nucleic acid or protein to its native configuration.

Renner complexes: In *Oenothera*, several genetic factor complexes are found and these complexes segregate as a whole in meiosis, each gamete carrying one or the other. Certain complexes are lethal in gametes, (gametophytic lethal) while some combinations are lethal zygotes (zygotic lethal). Specific names were assigned to these complexes by Renner and they are known as Renner complexes.

Repetitive DNA: Nucleotide sequences that are present in the genome in many copies.

Reproductive isolation: It is an isolating mechanism of evolution where a population is isolated by several genetically controlled mechanisms which prevent gene exchange between two populations.

Resting stage: Interphase; any nucleus not undergoing division. However, the nucleus is very active metabolically.

Restitution: The spontaneous rejoining of broken ends of chromosomes to produce the original chromosome configuration.

Restitution nucleus: A single nucleus with unreduced chromosomes is produced through failure of one of the divisions of meiosis.

Restorer genes: In cytoplasmic-genetic male sterility, certain male parents carry genes that restore the pollen producing ability of male-sterility cytoplasm.

Restriction enzyme: It recognizes specific short sequences of (usually) unmethylated DNA and cleaves the duplex randomly.

Restriction fragment length polymorphism (RFLP): Inherited differences in sites for restriction enzymes that result in differences in the lengths of the fragments produced by cleavage with the relevant restriction enzyme. RFLPs are used to genetic mapping to link the genome directly to a conventional genetic marker.

Restriction map: A linear array of sites on DNA cleaved by various restriction enzymes.

Reverse tandem inversions: Two inverted segments are adjacent to each other but mutually interchanged.

Rhizome: An underground horizontal stem with scale leaves and auxiliary buds that serves as a means of vegetative propagation.

Ribosome: Small cellular components composed of specialized ribosomal RNA and protein; site of protein synthesis.

Ribosomes: Responsible for the synthesis of long-chained molecules known as proteins.

Ring bivalent: A ring-shaped chromosome association at diakinesis or metaphase-I of meiosis with a chiasma in both arms.

Ring chromosome: A physically circular chromosome, produced as a result of chromosomal structural changes and usually unstable mitotically and meiotically.

RNA: Ribonucleic acid; long chain of nucleotides connected by phosphate-to-sugar bonds.

S^1 nuclease: An enzyme that specifically degrades unpaired (single stranded) sequence of DNA.

Salivary gland chromosomes: In certain tissues of insects (flies, mosquitoes, midges) of Diptera, polytene chromosomes are found in the interphase nuclei. Many extra replications of each chromosome occur in a single nucleus (endopolyploidy), and all lined up together in parallel fashion (polytene), resulting in thick chromosomes. Discovered and identified first in *Drosophila melanogaster.*

S-allele: An allele of a gene controlling incompatibility in many allogamous plants.

SAT-chromosome: A chromosome with a secondary constriction, usually associated with a nucleolus.

Satellite: A small-terminal segment of chromosome, mostly associated with the nucleolus and is known as nucleolus organizer chromosome.

Satellite DNA: It is highly repetitive, mostly found in heterochromatin and is usually not transcribed. Many satellite DNA sequences are localized to the kinetochore.

Scale leaf: A dry rudimentary version of layer greener leaves observed on the same plant.

Scarification: A method for mechanically making slit on the seed coat usually for wild species seeds that facilitates imbibition of water and germination. In a very hard seed coat, sulfuric acid is used for a few seconds.

Scion: A portion of actively growing shoot of a plant is grafted onto a stock of another.

Scutellum: A part of the embryo of the Poaceae (Gramineae), considered similar to a cotyledon. It serves as an organ that transfers nutrients from the endosperm to the other parts of the germinating embryos.

Secondary chromosomes: A chromosome consists of both arms homologous (an isochromosome).

Secondary constriction: It is usually associated with the regions when the nucleolus is formed or associated (nucleolus organizer).

Secondary gene pool: Genes from secondary gene pool can be transferred to cultivated crops of primary gene pool with some difficulty compared to species of the primary gene pool.

Secondary trisomic: An individual carrying an extra isochromosome in addition to its normal somatic chromosome complement.

Second division restitution (SDR): Due to premature cytokinesis before the second meiotic division takes place; sister chromatids end up in the same nucleus.

Second division segregation: First meiotic division is equational and second meiotic division is reducational.

Seed: The fertilized and ripened ovule with an embryo and generally with a food reserve (endosperm or cotyledons).

Seedling: A young plant grown from a seed.

Segmental allopolyploidy: Characterized by genomes intermediate in degree of similarity and generally exhibiting preferential pairing ($B_1B_1B_2B_2$).

Segregation: The migration of homologous chromosomes into daughter nuclei during meiosis resulting in separation of the genes due to recombination in the offspring.

Self-compatible: A plant that can be self-fertilized.

Self-fertilization: An individual produces seeds by the fusion of male and female gametes from the same plant.

Self-incompatibility: An individual is self-sterile although both male and female gametes are produced at the same time and are functional.

Self-pollination: The transfer of pollen from anther to stigma of the same plant.

Self-production: The transfer of pollen to the stigma of the same plant.

Semigamy: Embryo develops from an egg cell with a sperm penetration but without fertilization.

Semisterility: An organism heterozygous for a gene or chromosome (reciprocal translocation) exhibits approximately 50% male and female abortion.

Senescence: The phase of plant growth that extends from flowering to full maturity.

Sequence-tagged sites (STSs): It is a short stretch of genomic sequence that can be detected by the PCR.

Sequencing: Determination of the order of nucleotides (base sequences) in a DNA or RNA molecule or the order of amino acids in a protein.

Sex chromosomes: The chromosomes that determine the sex of an organism (X and Y chromosomes).

Sex linkage: The inheritance of a sex chromosome is coupled with that of a given gene located on the X chromosome.

Sexual reproduction: It requires meiosis and fertilization alternating in a cycle.

Short-day plant: A plant in which flowering is induced by daily exposure to less than 12 h of light.

Siblings: Progenies from the same plant (parents).

Sigma virus in *Drosophila*: Sigma virus is found in the cytoplasm of *Drosophila* which is cytoplasmically inherited for CO_2 sensitivity. However, it recovers when CO_2 is removed. Sensitivity is caused by a small virus-like particle named "sigma." Maintenance of sigma is genetically controlled.

Silent mutation: It does not change the product of a gene.

Simple sequence DNA: It is equal to satellite DNA.

Simplex: Aaaa.

Single cross: Crossing between two genetically different parents.

Sister chromatids: The copies of a chromosome produced by its replication during interphase.

SMC: Sperm or spore mother-cell.

Somaclonal variation: Vegetative nonsexual plant cells can be propagated *in vitro* in an appropriate nutrient medium; the cells proliferate in an undifferentiated (disorganized) pattern to form callus, or in a differentiated (organized) manner to form a plant with roots and shoots. Regenerated plants express morphological variations including genetic changes not present in the parents. These aberrant mutants are called somaclonal variation.

Somatic cell: The cells other than the germ cells.

Somatic chromosome doubling: Doubling the chromosomes of vegetative tissues of an organism experimentally such as by colchicine treatment.

Somatic crossing over: Crossing over during mitosis of somatic cells.

Somatic mutation: A mutation that occurs in the somatic tissues and is not inherited.

Somatic pairing: Synapsis of homologous chromosomes in the somatic cells.

Somatoplasm: The tissues that are essential for the functioning of the organism but lack the property of entering into sexual reproduction.

Somato plastic sterility: Embryos derived from suffering survivor less frequently than embryos produced by outcrossing.

Southern blotting: A molecular biology procedure which transfers denatured DNA from an agarose gel to a nitrocellulose filter where it can be hybridized with a complementary nucleic acid.

Spatial order: Chromosomes or genomes may occupy certain domains within the nucleus and/or dividing cells.

Speciation: The differentiation of two or more species from originally one species. It may be due to sympatric (without geographical isolation), allelopathic (with geographical isolation), or instantaneous (allopolyploidy) speciation.

Species: A kind of plant or animal, its distinctness seen in morphological, anatomical, and cytological and chemical discontinuities presumably brought about reproductive isolation in nature from all other organisms. Populations capable of interbreeding and producing viable fertile offsprings.

S period: The period of DNA (replication) synthesis during interphase.

Spermatogonia: These cells are in testes, divide mitotically and produce a group of diploid primary sporocyte. Primary sporocytes divide meiotically generating two secondary spermatocytes and produce four spermatids after second mitotic division.

S-phase: Synthesis of DNA occurs in eukaryotic cell cycle.

Spike: An elongate, unbranched, intermediate inflorescence in which flowers are sessile (without a stalk).

Spikelet: In the grass family, each flower is subtended by two bracts which are called spikelets. The central axis of a spikelet is the Rachilla.

Spindle: A bipolar oriented structure made of protein fibers, organized by centrosomes or kinetochores that functions in orientation, coorientation, and separation of chromosomes during metaphase and anaphase.

Spindle attachment region: Kinetochore.

Spontaneous mutation: A normally occurring mutation.

Spore: A reproductive cell in plant, product of meiosis and produces gametes (microspore, mega or macrospore) after three mitoses.

Sporocyte: A spore mother cell.

Sporogenesis: Formation of haploid spores in the higher plants.

Sporophyte: A part of life cycle of plants where spores are produced.

Sporophytic incompatibility: The incompatibility relationship of pollen is determined by plant producing it (heterostyle plants).

Sporulation: Formation of spores.

Stamen: The pollen-bearing organ of the flower, consisting of filament and anther. It is a basic unit of androecium.

Staminate: A flower with stamens; the carpels being rudimentary or suppressed.

Statistics: All values computed from the sample data are known as statistics.

Sterility: Failure of plants to produce function gametes or viable zygotes.

Stigma: The region of a gynoecium usually at the apex of the style on which compatible pollen grains germinate.

Stolon: Horizontal stems produced above ground and roots at nodes.

Stomata: Pores on the underside of leaves that control gas exchange and water loss.

Structural homozygosity: Homozygosity for chromosome mutations.

Style: A portion of gynoecium between ovary and stigma through which pollen tubes grow.

Subculture: Transfer of cultured tissues of an organism to a fresh medium.

Submedian: Kinetochore located nearer one end than another.

Subspecies: A taxonomic subdivision of species; genomically similar and capable of interbreeding.

Substitution line: A line in which one or more chromosomes are replaced by one or more chromosomes of a donor species.

Supernumerary chromosomes: Chromosomes also known as B-chromosomes or accessory chromosomes. They are largely heterochromatic and believed to be genetically inert or contain few coding genes.

Symbiosis: Two or more species living together; parasitic or mutualistic.

Symmetric Hybrid: Stable allopolyploid hybrid.

Symmetrical karyotype: All chromosomes are of about the same size.

Sympatric: Two or more species occupy the same geographical area or overlap in their distribution.

Synapsis: Pairing of homologous chromosomes during meiosis.

Synaptonemal complex: The morphological structure of synapsed chromosomes under the electron microscope.

Syncyte: A multinucleate cell produced by inhibition of cytokinesis in mitosis, migration of nucleus from one cell to another and fusion of two cells.

Syndesis: Synapsis.

Synergids: The two haploid cells that lie besides the egg cell in a mature embryo sac.

Syngamy: The union of male and female gametes during fertilization to produce zygote—sexual reproduction.

Synizesis: Clumping of chromosomes in one side of nucleus (= synizetic knot).

Syntenic: Gene loci located on the same chromosome.

Synthetic amphidiploid: An artificially produced amphidiploid.

Systematics: The classification of an organism on the reconstruction of evolutionary relatedness among living organisms.

Tandem duplication: A fragment chromosome is inserted into the partner chromosome.

Tandem repeats: Multiple copies of the same sequence.

Tandem satellites: Separate constrictions in the larger satellites.

Tapetum: Refers to innermost layer of pollen-sac wall.

Tassel: The staminate inflorescence of maize.

Taxon: A term refers to taxonomic group of an organism (family, genus, and species).

Taxonomy: Systematics; the study of the classification of living organisms (family, genus, species, variety).

T-bands: (T = terminal bands) Terminal bands are observed after treating chromosomes at high temperature (87°C) at pH 6.7, and after Giemsa staining show telomere of some chromosomes preferential staining.

Telocentric chromosome: A chromosome having a terminal kinetochore and one complete arm of a normal chromosome.

Telomere: The end of a chromosome arm.

Telophase I (meiosis): Chromosomes reach to their respective poles, nuclear membrane and nucleolus start to develop and eventually two daughter nuclei each the haploid chromosomes are generated.

Telotrisomic: An individual with a normal chromosome complement plus an extra telocentric chromosome is called telotrisomic.

Terminal chromosomal association: The end-to-end chromosome pairing.

Terminalization: The movement of chiasma toward the ends of the chromosomes.

Tertiary chromosome: A chromosome formed by interchange between nonhomologous chromosomes.

Tertiary gene pool: Gene transfer from species of tertiary gene pool to the cultivated crop of primary gene pool usually requires special crossing and embryo rescue techniques in order to get viable derived lines.

Tertiary trisomics: An individual containing an interchanged nonhomologous chromosome in addition to the normal somatic chromosome complement.

Test cross: A cross of a heterozygote to the homozygous recessive to determine the heterozygosity or linkage.

Tetrad (meiosis): Four uninucleate cells with the haploid chromosome number produced at the end of meiosis.

Tetraploid: An organism having four basic chromosome sets (4x) in the nucleus.

Tetrasome: An organism with one chromosome in the complement represented four times.

Tetraspory: Meiotic karyokinesis occurs but cytokinesis does not occur.

Tissue culture: The maintenance or growth of tissues *in vitro* in a way that may allow further differentiation and preservation of architecture or function or both.

Totipotency: The capacity of a cell cultured *in vitro* to regenerate into a plant.

Transgenic: A plant or animal modified by genetic engineering to contain DNA from an external source.

Translocation: A chromosomal structural aberration involving the reciprocal exchange of terminal segments of nonhomologous chromosomes.

Translocation tester set: A series of more or less defined homozygous reciprocal translocations.

Transposable elements: Gene loci capable of being transposed from one spot to another within and among the chromosomes of the complement.

Transposition: The transfer of a chromosome segment to another position due to intra- and interchromosomal structural changes.

Tribe: A rank between family and genus.

Trigeneric hybrid: A spontaneous or experimental hybrid among three genomes of different genera.

Triploid: An organism with three basic sets (3x) of chromosomes in the nucleus may be auto- or alloploid.

Trisomic: An organism containing a normal chromosome complement and one extra chromosome.

Trisomic analysis: A method to locate a gene on the chromosome by using trisomics; if a gene is associated with the chromosome, disomic ratio is modified.

Triplex: AAAa.

Trivalent: An association of three homologous chromosomes.

True-breeding: An organism homozygous for a trait.

Tube nucleus: Vegetative nucleus.

Tuber: A swollen underground stem tip that contains stored food material and serves as a source for vegetative propagation.

Twining: A plant coiling around objects as a measure of support.

Twins: Two embryos developed simultaneously within an ovary. Fraternal twins; originate from two fertilized eggs. Identical twins; arise from only one fertilized egg.

Umbel: A flat-topped inflorescence, like sunflower, in which all of the flowers are borne or pedicel of approximately equal length and all arise from the apex of the main axis.

Unbalanced translocation: A type of chromosome translocation in which a loss of chromosomal segments results in a deleterious effect.

Uninemy: Single strandedness of DNA in the chromosome.

Uniparental inheritance: A kind of inheritance in which only one parent provides genes to the progeny; a phenomenon, usually exhibited by extranuclear genes, in which all progenies have the phenotype of only one parent.

Unipolar: A cell spindle with one pole.

Unisexual (= imperfect): A flower with either stamens or carpel but not both.

Univalent: An unpaired chromosome during meiosis.

Univalent shift: A meiotic irregularity in monosomics that produces a monosomic plant deficient for a chromosome other than that originally deficient in the monosomic parent.

Unreduced gametes: Gametes not resulting from common meiosis, and so showing the number of chromosomes per cell that is characteristic of a sporophyte; they spontaneously arise as a consequence of irregular division in anaphase-I of meiosis resulting in spontaneously (meiotic) polyploidization.

Valance crosses: Crossing of individuals of different ploidy levels.

Variance: The extent to which values within frequency distribution depart from the mean is called variance. It is large when values are spread out and small when values are close together.

Variegation: Mosaic phenotype is a widespread phenomenon attributed to plastid variation, genetic instability, instability of the phenotypic expression of genes, somatic crossing over, and somatic instability of chromosomes (breakage-fusion-bridge cycle), and transposable elements.

Variety: In classical taxonomy, a subdivision of a species, also termed cultivars. A group of individuals distinct in form and function and genetically uniform.

Vegetative nucleus: The tube nucleus of a pollen grain.

Vegetative reproduction: Reproduction of the higher plants either naturally (runners, bulbs, rhizomes, etc.) or artificially (cutting and grafting, etc.).

Viruses: They are minuscule agents, through electron microscope several kinds of viruses have been identified, Shape ranges from round to rod-like to polyhedral with attached tail (T-even bacteriophage). Size ranges from vaccinia virus to as large as a small bacterium. The structure of virus is like a protein coat containing a single chromosome. Bacteriophage lambda does not kill all the infected bacteria. They are called lysogenic bacteria. The inactive form virus is known as the prophage (temperate phage). By contrast, virulent phages always cause destruction to the bacterium.

Vivipary: Bearing seeds that germinate within a flower or inflorescence.

Weismannism: A concept proposed by August Weismann that acquired traits are not inherited and only changes in the germplasm are transmitted from generation to generation.

Wide cross: Wide hybridization.

Wide hybridization: Cross combination between taxonomically distantly-related species or genera.

Wild type: Normal phenotype of an organism.

*x***:** A symbol for basic number of chromosomes in a polyploid species.

X- chromosome: A chromosome concerned with the determination of sex.

Xenia: Effect of genotype of pollen on embryo and endosperm.

Xenogamy: Cross-pollination

Xerophyte: A type of plant of arid habitats.

Xylem: A specialized vascular tissue that conducts water and nutrients in plants and provides mechanical support in vascular plants.

Yeast: Yeast is included in fungi that produce asci and ascospores (ascomycetes). Yeast may be haploid or diploid, unicellular organism and multiply by budding (*Saccharomyces cerevisiae*).

Y-chromosome: The sex chromosome found only in males in heterozygous condition.

Zygomorphic: The perianth constituted in such a way that only a median plane will yield two equal halves, as in most orchids or the sweet pea flower.

Zygonema (meiosis): The stage at which homologues begin to pair.

Zygote: A fertilized egg.

Zygotic embryo: An embryo derived from fusion of male and female gametes.

Zygotic or somatic chromosome number: $2n$.

Appendix A[*]

Johann Gregor Mendel (20 July 1822–6 January 1984)
The Founder of the Modern Science of Genetics

A.1 INTRODUCTORY REMARKS

Experience of artificial fertilization, such as is effected with ornamental plants in order to obtain new variations in color, has led to the experiments which will here be discussed. The striking regularity with which the same hybrid forms always reappeared whenever fertilization took place between the same species induced further experiments to be undertaken, the object of which was to follow up the developments of the hybrids in their progeny.

To this object numerous careful observers, such as Kölreuter, Gärtner, Herbert, Lecoq, Wichura and others, have devoted a part of their lives with inexhaustible perseverance. Gärtner, especially in his work *Die Bastarderzeugung im Pflanzenreiche*, has recorded very valuable observations; and quite recently Wichura published the results of some profound investigations into the hybrids of the Willow. That, so far, no generally applicable law governing the formation and development of hybrids has been successfully formulated can hardly be questioned by anyone who is acquainted with the extent of the task, and can appreciate the difficulties with which experiments of this class have to contend. A final decision can only be arrived at when we shall have before us the results of *detailed experiments* made on plants belonging to the most diverse orders.

Those who survey the work done in this department will arrive at the conviction that among all the numerous experiments made, not one has been carried out to such an extent and in such a way as to make it possible to determine the number of different forms under which the offspring of the hybrids appear, or to arrange these forms with certainty according to their separate generations, or definitely to ascertain their statistical relations.

It requires indeed some courage to undertake a labor of such far-reaching extent; this appears, however, to be the only right way by which we can finally reach the solution of a question the importance of which cannot be overestimated in connection with the history of the evolution of organic forms.

Mendel's paper, "Experiments in Plant Hybridization," records the results of such a detailed experiment. This translation was made by the Royal Horticultural Society of London. The original paper was published in the which appeared in 1986. *Versuche über Pflanzen-hybriden. Verhandlungen des naturforschenden Ver-eines in Brünn*, Bd. IV für das Jahr 1865, Abhand-lungen, 3–47; cross

---

[*] From Sinnott, E. W., L. C. Dunn, and Th. Dobzhansky 1960. Principles of Genetics. McGraw-Hill Book Company, Inc. New York, Toronto, London.

reference from Sinnott et al. (1950). This experiment was practically confined to a small plant group, and is now, after eight years' pursuit, concluded in all essentials. Whether the plan upon which the separate experiments were conducted and carried out was the best suited to attain the desired end is left to the friendly decision of the reader.

A.2 SELECTION OF THE EXPERIMENTAL PLANTS

The value and utility of any experiment are determined by the fitness of the material to the purpose for which it is used, and thus in the case before us it cannot be immaterial what plants are subjected to experiment and in what manner such experiment is conducted.

The selection of the plant group which shall serve for experiments of this kind must be made with all possible care if it is desired to avoid from the outset every risk of questionable results.

The experimental plants must necessarily:

1. Possess constant differentiating characteristics.
2. The hybrids of such plants must, during the flowering period, be protected from the influence of all foreign pollen, or be easily capable of such protection.
3. The hybrids and their offspring should suffer no marked disturbance in their fertility in the successive generations.

Accidental impregnation by foreign pollen, if it occurred during the experiments and were not recognized, would lead to entirely erroneous conclusions. Reduced fertility or entire sterility of certain forms, such as occurs in the offspring of many hybrids, would render the experiments very difficult or entirely frustrate them. In order to discover the relations in which the hybrid forms stand toward each other and also toward their progenitors it appears to be necessary that all member of the series developed in each successive generations should be, *without exception*, subjected to observation.

At the very outset special attention was devoted to the Leguminosae on account of their peculiar floral structure. Experiments which were made with several members of this family led to the result that the genus *Pisum* was found to possess the necessary qualifications.

Some thoroughly distinct forms of this genus possess characters which are constant, and easily and certainly recognizable, and when their hybrids are mutually crossed they yield perfectly fertile progeny. Furthermore, a disturbance through foreign pollen cannot easily occur, since the fertilizing organs are closely packed inside the keel and the anthers burst within the bud, so that the stigma becomes covered with pollen even before the flower opens. This circumstance is especially important. As additional advantages worth mentioning, there may be cited the easy culture of these plants in the open ground and in pots, and also their relatively short period of growth. Artificial fertilization is certainly a somewhat elaborate process, but nearly always succeeds. For this purpose the bud is opened before it is perfectly developed, the keel is removed, and each stamen carefully extracted by means of forceps, after which the stigma can at once be dusted over with the foreign pollen.

In all, 34 more or less distinct varieties of Peas were obtained from several seedsmen and subjected to a two year's trial. In the case of one variety that was noticed, among a larger number of plants all alike, a few forms which were markedly different. These, however, did not vary in the following year, and agreed entirely with another variety obtained from the same seedsman; the seeds were therefore doubtless merely accidentally mixed. All the other varieties yielded perfectly constant and similar offspring; at any rate, no essential difference was observed during two trial years. For fertilization 22 of these were selected and cultivated during the whole period of the experiments. They remained constant without any exception.

Their systematic classification is difficult and uncertain. If we adopt the strictest definition of a species, according to which only those individuals belong to a species which under precisely the same circumstances display precisely similar characters, no two of these varieties could be referred to one species. According to the opinion of experts, however, the majority belong to the species *Pisum sativum*; while the rest are regarded and classed, some as sub-species of *P. sativum*, and some as independent species, such as *P. quadratum*, *P. saccharatum*, and *P. umbellatum*. The positions, however, which may be assigned to them in a classificatory system are quite immaterial for the purposes of the experiments in question. It has so far been found to be just as impossible to draw a sharp line between the hybrids of species and varieties as between species and varieties themselves.

A.3   DIVISION AND ARRANGEMENT OF THE EXPERIMENTS

If two plants which differ constantly in one or several characters be crossed, numerous experiments have demonstrated that the common characters are transmitted unchanged to the hybrids and their progeny; but each pair of differentiating characters, on the other hand, unite in the hybrid to form a new character, which in the progeny of the hybrid is usually variable. The object of the experiment was to observe these variations in the case of each pair of differentiating characters, and to deduce the law according to which they appear in successive generations. The experiment resolves itself therefore into just as many separate experiments are there are constantly differentiating characters presented in the experimental plants.

The various forms of Peas selected for crossing showed differences in length and color of the stem; in the size and form of the leaves; in the position, color, size of the flowers; in the length of the flower stalk; in the color, form, and size of the pods; in the form and size of the seeds; and in the color of the seed-coats and of the albumen (endosperm). Some of the characters noted do not permit of a sharp and certain separation, since the difference is of a "more or less" nature, which is often difficult to define. Such characters could not be utilized for the separate experiments; these could only be applied to characters which stand out clearly and definitely in the plants. Lastly, the result must show whether they, in their entirety, observe a regular behavior in their hybrid unions, and whether from these facts any conclusion can be reached regarding those characters which possess a subordinate significance in the type.

The characters which were selected for experiment relate:

1. To the *difference in the form of the ripe seeds*. These are either round or roundish, the depressions, if any, occur on the surface, being always only shallow; or they are irregularly angular and deeply wrinkled (*P. quadratum*).
2. To the *difference in the color of the seed albumen* (endosperm). The albumen of the ripe seeds is either pale yellow, bright yellow and orange colored, or it possesses a more or less intense green tint. This difference of color is easily seen in the seeds as their coats are transparent.
3. To the *difference in the color of the seed-coat*. This is either white, with which character white flowers are constantly correlated; or it is gray, gray-brown, leather-brown, with or without violet spotting, in which case the color of the standards is violet, that of the wings purple, and the stem in the axils of the leaves is of a reddish tint. The gray seed-coats become dark brown in boiling water.
4. To the *difference in the form of the ripe pods*. These are either simply inflated, not contracted in places; or they are deeply constricted between the seeds and more or less wrinkled (*P. saccharatum*).
5. To the *difference in the color of the unripe pods*. They are either light to dark green, or vividly yellow, in which coloring the stalks, leaf-veins, and calyx participate.*
6. To the *difference in the position of the flowers*. They are either axial, that is, distributed along the main stem; or they are terminal, that is, bunched at the top of the stem and arranged almost in a false umbel; in this case the upper part of the stem is more or less widened in section (*P. umbellatum*).
7. To the *difference in the length of the stem*. The length of the stem is very various in some forms; it is, however, a constant character for each, in so far that healthy plants, grown in the same soil, are only subject to unimportant variations in this character. In experiments

---

* One species possesses a beautifully brownish-red colored pod, which when ripening turns to violet and blue. Trials with this character were only begun last year.

with this character, in order to be able to discriminate with certainty, the long axis of 6–7 ft. was always crossed with the short one of 3/4 ft. to 1 [and] 1/2 ft.

Each two of the differentiating characters enumerated above were united by cross-fertilization. There were made for the

1st experiment	60 fertilizations on 15 plants.
2nd experiment	58 fertilizations on 10 plants.
3rd experiment	35 fertilizations on 10 plants.
4th experiment	40 fertilizations on 10 plants.
5th experiment	23 fertilizations on 5 plants.
6th experiment	34 fertilizations on 10 plants.
7th experiment	37 fertilizations on 10 plants.

From a larger number of plants of the same variety, only the most vigorous were chosen for fertilization. Weakly plants always afford uncertain results, because even in the first generation of hybrids, and still more so in the subsequent ones, many of the offspring either entirely fail to flower or only form a few and inferior seeds.

Furthermore, in all the experiments reciprocal crossings were effected in such a way that each of the two varieties which in one set of fertilizations served as seed-bearer in the other set was used as the pollen plant.

The plants were grown in garden beds, a few also in pots, and were maintained in their natural upright position by means of sticks, branches of trees, and strings stretched between. For each experiment a number of pot plants were placed during the blooming period in a greenhouse, to serve as control plants for the main experiment in the open as regards possible disturbance by insects. Among the insects which visit Peas the beetle *Bruchus pisi* might be detrimental to the experiments should it appear in numbers. The female of this species is known to lay the eggs in the flower, and in so doing opens the keel; upon the tarsi of one specimen, which was caught in a flower, some pollen grains could clearly be seen under a lens. Mention must also be made of a circumstance which possibly might lead to the introduction of foreign pollen. It occurs, for instance, in some rare cases that certain parts of an otherwise normally developed flower wither, resulting in a partial exposure of the fertilizing organs. A defective development of the keel has also been observed, owing to which the stigma and anthers remained partially covered. It also sometimes happens that the pollen does not reach full perfection. In this event there occurs a gradual lengthening of the pistil during the blooming period, until the stigmatic tip protrudes at the point of the keel. This remarkable appearance has also been observed in hybrids of *Phaseolus* and *Lathyrus*.

The risk of false impregnation by foreign pollen is, however, a very slight one with *Pisum*, and is quite incapable of disturbing the general result. Among more than 10,000 plants which were carefully examined there were only a very few cases where an indubitable false impregnation had occurred. Since in the greenhouse such a case was never remarked, it may well be supposed that *Bruchus pisi*, and possibly also the described abnormalities in the floral structure, were to blame.

A.4   [F₁] THE FORMS OF THE HYBRIDS*

Experiments which in previous years were made with ornamental plants have already affording evidence that the hybrids, as a rule, are not exactly intermediate between the parental species. With some of the more striking characters, those, for instance, which relate to the form and size of the leaves, the pubescence of the several parts, etc., the intermediate, indeed, is nearly always to be

---

* Mendel throughout speaks of his cross-bred peas as "hybrids," a term which many restrict to the offspring of two distinct *species*. He, as he explains, held this to be only a question of degree.

seen; in other cases, however, one of the two parental characters is so preponderant that it is difficult, or quite impossible, to detect the other in the hybrid.

This is precisely the case with the Pea hybrids. In the case of each of the seven crosses, the hybrid-character resembles* that of one of the parental forms so closely that the other either escapes observation completely or cannot be detected with certainty. This circumstance is of great importance in the determination and classification of the forms under which the offspring of the hybrids appear. Henceforth in this paper those characters which are transmitted entire, or almost unchanged in the hybridization, and therefore in themselves constitute the characters of the hybrid, are termed the *dominant*, and those which become latent in the process *recessive*. The expression "recessive" has been chosen because the characters thereby designated withdraw or entirely disappear in the hybrids, but nevertheless reappear unchanged in their progeny, as will be demonstrated later on.

It was furthermore shown by the whole of the experiments that it is perfectly immaterial whether the dominant character belongs to the seed plant or to the pollen plant; the form of the hybrid remains identical in both cases. This interesting fact was also emphasized by Gärtner, with the remark that even the most practiced expert is not in a position to determine in a hybrid which of the two parental species was the seed or the pollen plant.

Of the differentiating characters which were used in the experiments the following are dominant:

1. The round or roundish form of the seed with or without shallow depressions.
2. The yellow coloring of the seed albumen.
3. The gray, gray-brown, or leather brown color of the seed-coat, in association with violet-red blossoms and reddish spots in the leaf axils.
4. The simply inflated form of the pod.
5. The green coloring of the unripe pod in association with the same color of the stems, the leaf-veins, and the calyx.
6. The distribution of the flowers along the stem.
7. The greater length of stem.

With regard to this last character it must be stated that the longer of the two parental stems is usually exceeded by the hybrid, a fact which is possibly only attributable to the greater luxuriance which appears in all parts of plants when stems of very different lengths are crossed. Thus, for instance, in repeated experiments, stems of 1 ft. and 6 ft. in length yielded without exception hybrids which varied in length between 6 ft. and 7 [and] 1/2 ft.

The *hybrid seeds* in the experiments with seed-coat are often more spotted, and the spots sometimes coalesce into small bluish-violet patches. The spotting also frequently appears even when it is absent as a parental character.

The hybrid forms of the *seed-shape* and of the *[color of the] albumen* are developed immediately after the artificial fertilization by the mere influence of the foreign pollen. They can, therefore, be observed even in the first year of experiment, whilst all the other characters naturally only appear in the following year in such plants as have been raised from the crossed seed.

A.5 THE FIRST GENERATION [BRED] FROM THE HYBRIDS

In this generation there reappear, *together with the dominant* characters, also the *recessive* ones with their peculiarities fully developed, and this occurs in the definitely expressed average proportion of 3:1, so that among each four plants of this generation three display the dominant character and one the recessive. This relates without exception to all the characters which were investigated in the experiments. The angular wrinkled form of the seed, the green color of the albumen, the white

---

* Note that Mendel, with true penetration, avoids speaking of the hybrid-character as "transmitted" by either parent, thus escaping the error pervading the older views of heredity.

color of the seed-coats and the flowers, the constriction of the pods, the yellow color of the unripe pod, of the stalk, of the calyx, and of the leaf venation, the umbel-like form of the inflorescence, and the dwarfed stem, all reappear in the numerical proportion given, without any essential alteration. *Transitional forms were not observed in any experiment.*

Since the hybrids resulting from reciprocal crosses are formed alike and present no appreciable difference in their subsequent development, consequently these results can be reckoned together in each experiment. The relative numbers which were obtained for each pair of differentiating characters are as follows:

- Expt. 1: Form of seed. From 253 hybrids 7324 seeds were obtained in the second trial year. Among them were 5474 round or roundish ones and 1850 angular wrinkled ones. Therefrom the ratio 2.96:1 is deduced.
- Expt. 2: Color of albumen. 258 plants yielded 8023 seeds, 6022 yellow, and 2001 green; their ratio, therefore, is as 3.01:1.

In these two experiments each pod yielded usually both kinds of seed. In well-developed pods which contained on the average 6–9 seeds, it often happened that all the seeds were round (Expt. 1) or all yellow (Expt. 2); on the other hand there were never observed more than five wrinkled or five green ones on one pod. It appears to make no difference whether the pods are developed early or later in the hybrid or whether they spring from the main axis or from a lateral one. In some few plants only a few seeds developed in the first formed pods, and these possessed exclusively one of the two characters, but in the subsequently developed pods the normal proportions were maintained nevertheless.

As in separate pods, so did the distribution of the characters vary in separate plants. By way of illustration, the first 10 individuals from both series of experiments may serve.

	Experiment 1 Form of Seed		Experiment 2 Color of Albumen	
Plants	Round	Angular	Yellow	Green
1	45	12	25	11
2	27	8	32	7
3	24	7	14	5
4	19	10	70	27
5	32	11	24	13
6	26	6	20	6
7	88	24	32	13
8	22	10	44	9
9	28	6	50	14
10	25	7	44	18

As extremes in the distribution of the two seed characters in one plant, there were observed in Expt. 1 an instance of 43 round and only 2 angular, and another of 14 round and 15 angular seeds. In Expt. 2 there was a case of 32 yellow and only 1 green seed, but also one of 20 yellow and 19 green.

These two experiments are important for the determination of the average ratios, because with a smaller number of experimental plants they show that very considerable fluctuations may occur. In counting the seeds, also, especially in Expt. 2, some care is requisite, since in some of the seeds of many plants the green color of the albumen is less developed, and at first may be easily overlooked. The cause of this partial disappearance of the green coloring has no connection with the hybrid-character of the plants, as it likewise occurs in the parental variety. This peculiarity is also confined to the individual and is not inherited by the offspring. In luxuriant plants, this appearance was frequently noted. Seeds which are damaged by insects during their development often vary in color and form, but with a little practice in sorting, errors are easily avoided. It is almost superfluous to mention that the pods must remain on the plants until they are thoroughly ripened and have become dried, since it is only then that the shape and color of the seed are fully developed.

Expt. 3: Color of the seed-coats. Among 929 plants, 705 bore violet-red flowers and gray-brown seed-coats; 224 had white flowers and white seed-coats, giving the proportion 3.15 to1.

Expt. 4: Form of pods. Of 1181 plants, 882 had them simply inflated, and in 299 they were constricted. Resulting ratio, 2.95 to1.

Expt. 5: Color of the unripe pods. The number of trial plants was 580, of which 428 had green pods and 152 yellow ones. Consequently these stand in the ratio of 2.82 to1.

Expt. 6: Position of flowers. Among 858 cases 651 had inflorescences axial and 207 terminal. Ratio, 3.14 to1.

Expt. 7: Length of stem. Out of 1064 plants, in 787 cases the stem was long, and in 277 short. Hence a mutual ratio of 2.84 to1. In this experiment the dwarfed plants were carefully lifted and transferred to a special bed. This precaution was necessary, as otherwise they would have perished through being overgrown by their tall relatives. Even in their quite young state, they can be easily picked out by their compact growth and thick dark-green foliage.[*]

If now the results of the whole of the experiments be brought together, there is found, as between the number of forms with the dominant and recessive characters, an average ratio of 2.98–1, or 3–1.

The dominant character can have here a *double signification*—viz. that of either a parental character or a hybrid-character.[†] In which of the two significations it appears in each separate case can only be decided in the following generation. As a parental character it must pass over unchanged to the whole of the offspring; as a hybrid-character, on the other hand, it must maintain the same behavior as in the first generation [F_1].

A.6 [F_2] THE SECOND GENERATION [BRED] FROM THE HYBRIDS

Those forms which in the first generation [F_2] exhibit the recessive character do not further vary in the second generation [F_3] as regards this character; they remain *constant* in their offspring.

It is otherwise with those which possess the dominant character in the first generation. Of these *two*-thirds yield offspring which display the dominant and recessive characters in the proportion of 3–1, and thereby show exactly the same ratio as the hybrid forms, while only *one*-third remains with the dominant character constant.

The separate experiments yielded the following results:

Expt. 1: Among 565 plants which were raised from round seeds of the first generation [F_1], 193 yielded round seeds only, and remained therefore constant in this character; 372, however, gave both round and wrinkled seeds, in the proportion of 3–1. The number of the hybrids, therefore, as compared with the constants is 1.93–1.

Expt. 2: Of 519 plants which were raised from seeds whose albumen was of yellow color in the first generation, 166 yielded exclusively yellow, while 353 yielded yellow and green seeds in the proportion of 3–1. There resulted, therefore, a division into hybrid and constant forms in the proportion of 2.13–1.

For each separate trial in the following experiments, 100 plants were selected which displayed the dominant character in the first generation, and in order to ascertain the significance of this, ten seeds of each were cultivated.

---

[*] This is true also of dwarf or "cupid" sweet peas.

[†] This paragraph presents the view of the hybrid-character as something incidental to the hybrid, and not "transmitted" to it—a true conception here expressed probably for the first time.

Expt. 3: The offspring of 36 plants yielded exclusively gray-brown seed-coats, while of the offspring of 64 plants some had gray-brown and some had white.

Expt. 4: The offspring of 29 plants had only simply inflated pods; of the offspring of 71, on the other hand, some had inflated and some constricted.

Expt. 5: The offspring of 40 plants had only green pods; of the offspring of 60 plants some had green, some yellow ones.

Expt. 6: The offspring of 33 plants had only axial flowers; of the offspring of 67, on the other hand, some had axial and some terminal flowers.

Expt. 7: The offspring of 28 plants inherited the long axis, of those of 72 plants some the long and some the short axis.

In each of these experiments, a certain number of the plants came constant with the dominant character. For the determination of the proportion in which the separation of the forms with the constantly persistent character results, the two first experiments are especially important, since in these a larger number of plants can be compared. The ratios 1.93–1 and 2.13–1 gave together almost exactly the average ratio of 2–1. Experiment 6 gave a quite concordant result; in the others the ratio varies more or less, as was only to be expected in view of the smaller number of 100 trial plants. Experiment 5, which shows the greatest departure, was repeated, and then in lieu of the ratio of 60:40, that of 65–35 resulted. *The average ratio of 2:1 appears, therefore, as fixed with certainty.* It is therefore demonstrated that, of those forms which possess the dominant character in the first generation, two-thirds have the hybrid-character, while one-third remains constant with the dominant character.

The ratio 3–1, in accordance with which the distribution of the dominant and recessive characters results in the first generation, resolves itself therefore in all experiments into the ratio of 2:1:1, if the dominant character be differentiated according to its significance as a hybrid-character or as a parental one. Since the members of the first generation [F$_2$] spring directly from the seed of the hybrids [F$_1$], *it is now clear that the hybrids form seeds having one or other of the two differentiating characters, and of these one-half develop again the hybrid form, while the other half yield plants which remain constant and receive the dominant or the recessive characters in equal numbers.*

A.7   THE SUBSEQUENT GENERATION [BRED] FROM THE HYBRIDS

The proportions in which the descendants of the hybrids develop and split up in the first and second generations presumably hold good for all subsequent progeny. Experiments 1 and 2 have already been carried through 6 generations; 3 and 7 through 5; and 4, 5, and 6 through 4; these experiments being continued from the third generation with a small number of plants, and no departure from the rule has been perceptible. The offspring of the hybrids separated in each generation in the ratio of 2:1:1 into hybrids and constant forms.

If A be taken as denoting one of the two constant characters, for instance the dominant, a the recessive, and Aa the hybrid form in which both are conjoined, the expression

$$A + 2Aa + a$$

shows the terms in the series for the progeny of the hybrids of two differentiating characters.

The observation made by Gärtner, Kölreuter, and others, that hybrids are inclined to revert to the parental forms, is also confirmed by the experiments described. It is seen that the number of the hybrids which arise from one fertilization, as compared with the number of forms which become constant, and their progeny from generation to generation, is continually diminishing, but that nevertheless they could not entirely disappear. If an average equality of fertility in all plants in all generations be assumed, and if, furthermore, each hybrid forms seed of which one-half yields hybrids again, while the other half is constant to both characters in equal proportions, the ratio of numbers for the offspring in each generation is seen by the following summary, in which **A** and **a** denote again the two parental characters, and **Aa** the hybrid forms. For brevity's sake it may be assumed that each plant in each generation furnishes only four seeds.

				Ratios		
Generation	A	Aa	a	A:	Aa:	a
1	1	2	1	1:	2:	1
2	6	4	6	3:	2:	3
3	28	8	28	7:	2:	7
4	120	16	120	15:	2:	15
5	496	32	496	31:	2:	31
n				$2^n - 1$:	2:	$2^n - 1$

In the tenth generation, for instance, $2^n - 1 = 1023$. There result, therefore, in each 2048 plants which arise in this generation 1023 with the constant dominant character, 1023 with the recessive character, and only two hybrids.

A.8 THE OFFSPRING OF HYBRIDS IN WHICH SEVERAL DIFFERENTIATING CHARACTERS ARE ASSOCIATED

In the experiments above described, plants were used which differed only on one essential character. [This statement of Mendel in light of present knowledge is open to some misconception. Though his work makes it evident that such varieties may exists, it is very unlikely that Mendel could have had seven pairs of varieties such that the members of each paired differed from each other in *one* considerable character (*wesentliches Merkmal*). The point is little theoretical or practical consequence, but a rather heavy stress is thrown in "*wesentlich*."]. The next task consisted in ascertaining whether the law of development discovered in these applied to each pair of differentiating characters when several diverse characters are united in the hybrid by crossing.

As regards the form of the hybrids in these cases, the experiments showed throughout that this invariably more nearly approaches to that one of the two parental plants which possesses the greater number of dominant characters. If, for instance, the seed plant has a short stem, terminal white flowers, and simply inflated pods; the pollen plant, on the other hand, a long stem, violet-red flowers distributed along the stem, and constricted pods; the hybrid resembles the seed parent only in the form of the pod; in the other characters it agrees with the pollen parent. Should one of the two parental types possess only dominant characters, then the hybrid is scarcely or not at all distinguishable from it.

Two experiments were made with a considerable number of plants. In the first experiment the parental plants differed in the form of the seed and in the color of the albumen; in the second in the form of the seed, in the color of the albumen, and in the color of the seed-coats. Experiments with seed characters give the result in the simplest and most certain way.

In order to facilitate study of the data in these experiments, the different characters of the seed plant will be indicated by A, B, C, those of the pollen plant by a, b, c, and the hybrid forms of the characters by Aa, Bb, and Cc.

Experiment 1

AB, Seed parents;	ab, Pollen parents;
A, form round;	a, form wrinkled;
B, albumen yellow.	b, albumen green.

The fertilized seeds appeared round and yellow like those of the seed parents. The plants raised therefrom yielded seeds of four sorts, which frequently presented themselves in one pod. In all, 556 seeds were yielded by 15 plants, and of these there were:

315 round and yellow,
101 wrinkled and yellow,
108 round and green,
32 wrinkled and green.

All were sown the following year. Eleven of the round yellow seeds did not yield plants, and three plants did not form seeds. Among the rest:

38 had round yellow seeds	AB
65 round yellow and green seeds	ABb
60 round yellow and wrinkled yellow seeds	AaB
138 round yellow and green, wrinkled yellow and green seeds	AaBb

From the wrinkled yellow seeds 96 resulting plants bore seed, of which:

28 had only wrinkled yellow seeds	aB
68 wrinkled yellow and green seeds	aBb

From 108 round green seeds 102 resulting plants fruited, of which:

35 had only round green seeds	Ab
67 round and wrinkled green seeds	Aab

The wrinkled green seeds yielded 30 plants which bore seeds all of like character; they remained constant *ab*.

The offspring of the hybrids appeared therefore under nine different forms, some of them in very unequal numbers. When these are collected and coordinated, we find:

38 plants with the sign	AB
35 " " " "	Ab
28 " " " "	aB
30 " " " "	ab
65 " " " "	ABb
68 " " " "	aBb
60 " " " "	AaB
67 " " " "	Aab
138 " " " "	AaBb

The whole of the forms may be classed into three essentially different groups. The first includes those with the signs *AB*, *Ab*, *aB*, and *ab*: they possess only constant characters and do not vary again in the next generation. Each of these forms is represented on the average 33 times. The second group includes the signs *ABb*, *aBb*, *AaB*, *Aab*: these are constant in one character and hybrid in another, and vary in the next generation only as regards the hybrid-character. Each of these appears on any average 65 times. The form *AaBb* occurs 138 times: it is hybrid in both characters, and behaves exactly as do the hybrids from which it is derived.

If the numbers in which the forms belonging to these classes appear be compared, the ratios of 1, 2, 4 are unmistakably evident. The numbers 33, 65, 138 present very fair approximations to the ratio numbers of 33, 66, 132.

The development series consists, therefore, of nine classes, of which four appear therein always once and are constant in both characters; the forms AB, ab resemble the parental forms, the two others present combinations between the conjoined characters A, a, B, b, which combinations are likewise possibly constant. Four classes appear always twice, and are constant in one character and hybrid in the other. One class appears four times, and is hybrid in both characters. Consequently, the offspring of the hybrids, if two kinds of differentiating characters are combined therein, are represented by the expression

$$AB + Ab + aB + ab + 2ABb + 2aBb + 2AaB + 2Aab + 4AaBb.$$

This expression is indisputably a combination series in which the two expressions for the characters A and a, B and b are combined. We arrive at the full number of the classes of the series by the combination of the expressions:

$$A + 2Aa + a$$
$$B + 2Bb + b.$$

Experiment 2

ABC, seed parents;	abc, pollen parents;
A, form round;	a, form wrinkled;
B, albumen yellow;	b, albumen green;
C, seed-coat gray-brown.	seed-coat white.

This experiment was made in precisely the same way as the previous one. Among all the experiments it demanded the most time and trouble. From 24 hybrids 687 seeds were obtained in all: these were all either spotted, gray-brown or gray-green, round or wrinkled. [Mendel does not state the cotyledon-color of the first crosses in this case; for as the coats were thick, it could not have been seen without opening or peeling the seeds.] From these in the following year 639 plants fruited, and as further investigation showed, there were among them:

8 plants ABC		22 plants $ABCc$		45 plants $ABbCc$
14 " ABc	17 " $AbCc$	36 " $aBbCc$
9 " AbC	25 " $aBCc$	38 " $AaBCc$
11 " Abc	20 " $abCc$	40 " $AabCc$
8 " aBC	15 " $ABbC$	49 " $AaBbC$
10 " aBc	18 " $ABbc$	48 " $AaBbc$
10 " abC	19 " $aBbC$	
7 " abc	24 " $aBbc$	
	14 " $AaBC$	78 " $AaBbCc$
	18 " $AaBc$	
	20 " $AabC$	
	16 " $Aabc$	

The whole expression contains 27 terms. Of these eight are constant in all characters, and each appears on the average 10 times; 12 are constant in two characters, and hybrid in the third; each appears on the average 19 times; six are constant in one character and hybrid in the other two; each appears on the average 43 times. One form appears 78 times and is hybrid in all of the characters. The ratios 10:19:43:78 agree so closely with the ratios 10, 20, 40, 80, or 1, 2, 4, 8, that this last undoubtedly represents the true value.

The development of the hybrids when the original parents differ in three characters results therefore according to the following expression:

$$ABC + ABc + AbC + Abc + aBC + aBc + abC + abc +$$
$$2ABCc + 2AbCc + 2aBCc + 2abCc + 2ABbC + 2ABbc +$$
$$2aBbc + 2aBbc + 2AaBC + 2AaBc + 2AabC + 2Aabc +$$
$$4ABbCc + 4aBbCc + 4AaBCc + 4AabCc + 4AaBbC +$$
$$4AaBbc + 8AaBbCc.$$

Here also is involved a combination series in which the expressions for the characters A and a, B and b, C and c are united. The expressions

$$A + 2Aa + a$$
$$B + 2Bb + b$$
$$C + 2Cc + c$$

give all the classes of the series. The constant combinations which occur therein agree with all combinations which are possible between the characters A, B, C, a, b, c; two thereof, ABC and abc, resemble the two original parental stocks.

In addition, further experiments were made with a smaller number of experimental plants in which the remaining characters by twos and threes were united as hybrids: all yielded approximately the same results. There is therefore no doubt that for the whole of the characters involved in the experiments the principle applies that *the offspring of the hybrids in which several essentially different characters are combined exhibit the terms of a series of combinations, in which the developmental series for each pair of differentiating characters are united.* It is demonstrated at the same time that *the relation of each pair of different characters in hybrid union is independent of the other differences in the two original parental stocks.* This is the principle of Independent Assortment.

If n represents the number of the differentiating characters in the two original stocks, 3^n gives the number of terms of the combination series, 4^n the number of individuals which belong to the series, and 2^n the number of unions which remain constant. The series therefore contains, if the original stocks differ in four characters, $3^4 = 81$ classes, $4^4 = 256$ individuals, and $2^4 = 16$ constant forms: or, which is the same, among each 256 offspring of the hybrids are 81 different combinations, 16 of which are constant.

All constant combinations which in Peas are possible by the combination of the said seven differentiating characters were actually obtained by repeated crossing. Their number is given by $2^7 = 128$. Thereby is simultaneously given the practical proof *that the constant characters which appear in the several varieties of a group of plants may be obtained in all the associations which are possible according to the laws of combination, by means of repeated artificial fertilization.*

As regards the flowering time of the hybrids, the experiments are not yet concluded. It can, however, already be stated that the time stands almost exactly between those of the seed and pollen parents, and that the constitution of the hybrids with respect to this character probably follows the rule ascertained in the case of the other characters. The forms which are selected for experiments of this class must have a difference of at least 20 days from the middle flowering period of one to that of the other; furthermore, the seeds when sown must all be placed at the same depth in the earth, so that they may germinate simultaneously. Also, during the whole flowering period, the more important variations in temperature must be taken into account, and the partial hastening or delaying of the flowering which may result therefrom. It is clear that this experiment presents many difficulties to be overcome and necessitates great attention.

If we endeavor to collate in a brief form the results arrived at, we find that those differentiating characters, which admit of easy and certain recognition in the experimental plants, all behave exactly alike in their hybrid associations. The offspring of the hybrids of each pair of differentiating characters are, one-half, hybrid again, while the other half are constant in equal proportions having the characters of the seed and pollen parents, respectively. If several differentiating characters are combined by cross-fertilization in a hybrid, the resulting offspring form the terms of a combination series in which the combination series for each pair of differentiating characters are united.

The uniformity of behavior shown by the whole of the characters submitted to experiment permits, and fully justifies, the acceptance of the principle that a similar relation exists in the other characters which appear less sharply defined in plants, and therefore could not be included in the separate experiments. An experiment with peduncles of different lengths gave on the whole fairly satisfactory results, although the differentiation and serial arrangement of the forms could not be effected with that certainty which is indispensable for correct experiment.

A.9   THE REPRODUCTIVE CELLS OF THE HYBRIDS

The results of the previously described experiments led to further experiments, the results of which appear fitted to afford some conclusions as regards the composition of the egg and pollen cells of hybrids. An important clue is afforded in *Pisum* by the circumstance that among the progeny of the hybrids constant forms appear, and that this occurs, too, in respect of all combinations of the associated characters. So far as experience goes, we find it in every case confirmed that constant progeny can only be formed when the egg cells and the fertilizing pollen are of like character, so that both are provided with the material for creating quite similar individuals, as is the case with the normal fertilization of pure species. We must therefore regard it as certain that exactly similar factors must be at work also in the production of the constant forms in the hybrid plants. Since the various constant forms are produced in *one* plant, or even in *one* flower of a plant, the conclusion appears logical that in the ovaries of the hybrids there are formed as many sorts of egg cells, and in the anthers as many sorts of pollen cells, as there are possible constant combination forms, and that these egg and pollen cells agree in their internal compositions with those of the separate forms.

In point of fact it is possible to demonstrate theoretically that this hypothesis would fully suffice to account for the development of the hybrids in the separate generations, if we might at the same time assume that the various kinds of egg and pollen cells were formed in the hybrids on the average in equal numbers. [This and the preceding paragraph contain the essence of the Mendelian principles of heredity.]

In order to bring these assumptions to an experimental proof, the following experiments were designed. Two forms which were constantly different in the form of the seed and the color of the albumen were united by fertilization.

If the differentiating characters are again indicated as *A*, *B*, *a*, *b*, we have:

AB, Seed parents;	*ab*, Pollen parents;
A, form round;	*a*, form wrinkled;
B, albumen yellow.	*B*, albumen green.

The artificially fertilized seeds were sown together with several seeds of both original stocks, and the most vigorous examples were chosen for the reciprocal crossing. There were fertilized:

1. The hybrids with the pollen of *AB*.
2. The hybrids with the pollen of *ab*.
3. *AB* with the pollen of the hybrids.
4. *ab* with the pollen of the hybrids.

For each of these four experiments, the whole of the flowers on three plants were fertilized. If the above theory be correct, there must be developed on the hybrids egg and pollen cells of the forms *AB, Ab, aB, ab,* and there would be combined:

1. The egg cells *AB, Ab, aB, ab* with the pollen cells *AB.*
2. The egg cells *AB, Ab, aB, ab* with the pollen cells *ab.*
3. The egg cells *AB* with the pollen cells *AB, Ab, aB,* and *ab.*
4. The egg cells *ab* with the pollen cells *AB, Ab, aB,* and *ab.*

From each of these experiments, there could then result only the following forms:

1. *AB, ABb, AaB, AaBb.*
2. *AaBb, Aab, aBb, ab.*
3. *AB, ABb, AaB, AaBb.*
4. *AaBb, Aab, aBb, ab.*

If, furthermore, the several forms of the egg and pollen cells of the hybrids were produced on an average in equal numbers, then in each experiment the said four combinations should stand in the same ratio to each other. A perfect agreement in the numerical relations was, however, not to be expected since in each fertilization, even in normal cases, some egg cells remain undeveloped or subsequently die, and many even of the well-formed seeds fail to germinate when sown. The above assumption is also limited in so far that while it demands the formation of an equal number of the various sorts of egg and pollen cells, it does not require that this should apply to each separate hybrid with mathematical exactness.

The first and second experiments had primarily the object of proving the composition of the hybrid egg cells, while the third and fourth experiments were to decide that of the pollen cells. [To prove, namely, that both were similarly differentiated, and not one or other only]. As is shown by the above demonstration the first and third experiments and the second and fourth experiments should produce precisely the same combinations, and even in the second year the result should be partially visible in the form and color of the artificially fertilized seed. In the first and third experiments the dominant characters of form and color, *A* and *B*, appear in each union, and are also partly constant and partly in hybrid union with the recessive characters *a* and *b*, for which reason they must impress their peculiarity upon the whole of the seeds. All seeds should therefore appear round and yellow, if the theory be justified. In the second and fourth experiments, on the other hand, one union is hybrid in form and in color, and consequently the seeds are round and yellow; another is hybrid in form, but constant in the recessive character of color, whence the seeds are round and green; the third is constant in the recessive character of form but hybrid in color, consequently the seeds are wrinkled and yellow; the fourth is constant in both recessive characters, so that the seeds are wrinkled and green. In both these experiments there were consequently four sorts of seed to be expected; namely, round and yellow, round and green, wrinkled and yellow, wrinkled and green.

The crop fulfilled these expectations perfectly. There were obtained in the

1st Experiment, 98 exclusively round yellow seeds;
3rd Experiment, 94 exclusively round yellow seeds.

In the 2nd Experiment, 31 round and yellow, 26 round and green, 27 wrinkled and yellow, 26 wrinkled and green seeds.

In the 4th Experiment, 24 round and yellow, 25 round and green, 22 wrinkled and yellow, 27 wrinkled and green seeds.

There could scarcely be now any doubt of the success of the experiment; the next generation must afford the final proof. From the seed sown there resulted for the first experiment 90 plants, and for the third 87 plants which fruited: these yielded for the

1st Exp.	3rd Exp.		
20	25	Round yellow seeds	*AB*
23	19	Round yellow and green seeds	*ABb*
25	22	Round and wrinkled yellow seeds	*AaB*
22	21	Round and wrinkled green and yellow seeds	*AaBb*

In the second and fourth experiments, the round and yellow seeds yielded plants with round and wrinkled yellow and green seeds, *AaBb.*

From the round green seeds, plants resulted with round and wrinkled green seeds, *Aab.*

The wrinkled yellow seeds gave plants with wrinkled yellow and green seeds, *aBb.*

From the wrinkled green seeds, plants were raised which yielded again only wrinkled and green seeds, *ab.*

Although in these two experiments likewise some seeds did not germinate, the figures arrived at already in the previous year were not affected thereby, since each kind of seed gave plants which, as regards their seed, were like each other and different from the others. There resulted therefore from the

2nd. Exp.	4th Exp.	
31	24	Plants of the form *AaBb*
26	25	Plants of the form *Aab*
27	22	Plants of the form *aBb*
26	27	Plants of the form *ab*

In all the experiments, therefore, there appeared all the forms which the proposed theory demands, and they came in nearly equal numbers.

In a further experiment the characters of *flower-color and length of stem* were experimented upon, and selection was so made that in the third year of the experiment each character ought to appear in *half* of all the plants if the above theory were correct. *A, B, a, b* serve again as indicating the various characters.

A, violet-red flowers.
a, white flowers.
B, axis long.
A, axis short.

The form *Ab* was fertilized with *ab*, which produced the hybrid *Aab*. Furthermore, *aB* was also fertilized with *ab*, whence the hybrid *aBb*. In the second year, for further fertilization, the hybrid *Aab* was used as seed parent, and hybrid *aBb* as pollen parent.

Seed parent, *Aab.*
Pollen parent, *aBb.*
Possible egg cells, *Ab,ab.*
Pollen cells, *aB, ab.*

From the fertilization between the possible egg and pollen cells, four combinations should result, viz.,

$$AaBb + aBb + Aab + ab.$$

From this it is perceived that, according to the above theory, in the third year of the experiment out of all the plants

Half should have	Violet-red flowers (*Aa*)	Classes 1, 3
" " "	White flowers (*a*)	" 2, 4
" " "	A long axis (*Bb*)	" 1, 2
" " "	A short axis (*b*)	" 3, 4

From 45 fertilizations of the second year 187 seeds resulted, of which only 166 reached the flowering stage in the third year. Among these the separate classes appeared in the numbers following.

Class	Color of Flower	Stem	
1	Violet-red	Long	47 times
2	White	Long	40 "
3	Violet-red	Short	38 "
4	White	Short	41 "

There subsequently appeared

The violet-red flower color (*Aa*)	in 85 plants
The white flower-color (*a*)	in 81 plants
The long stem (*Bb*)	in 87 plants
The short stem (*b*)	in 79 plants

The theory adduced is therefore satisfactorily confirmed in this experiment also.

For the characters of form of pod, color of pod, and position of flowers, experiments were also made on a small scale and results obtained in perfect agreement. All combinations which were possible through the union of the differentiating characters duly appeared, and in nearly equal numbers.

Experimentally, therefore, the theory is confirmed that *the pea hybrids form egg and pollen cells which, in their constitution, represent in equal numbers all constant forms which result from the combination of the characters united in fertilization.*

The difference of the forms among the progeny of the hybrids, as well as the respective ratios of the numbers in which they are observed, finds a sufficient explanation in the principle above deduced. The simplest case is afforded by the developmental series of *each pair of differentiating characters.* This series is represented by the expression $A + 2Aa + a$, in which A and a signify the forms with constant differentiating characters, and Aa the hybrid form of both. It includes in three different classes four individuals. In the formation of these, pollen and egg cells of the form A and a take part on the average equally in the fertilization; hence each form [occurs] twice, since four individuals are formed. There participate consequently in the fertilization

The pollen cells	$A + A + a + a$
The egg cells	$A + A + a + a$

It remains, therefore, purely a matter of chance which of the two sorts of pollen will become united with each separate egg cell. According, however, to the law of probability, it will always happen, on the average of many cases, that each pollen form A and a will unite equally often with each egg cell form A and a, consequently one of the two pollen cells A in the fertilization will meet with the egg cell A and the other with the egg cell a, and so likewise one pollen cell a will unite with an egg cell A, and the other with the egg cell a.

Pollen cells	A	$A\,a$	a
	↓	\ /	↓
	↓	X	↓
	↓	╱╲	↓
Egg cells	A	$A\quad a$	a

The result of the fertilization may be made clear by putting the signs for the conjoined egg and pollen cells in the form of fractions, those for the pollen cells above and those for the egg cells below the line. We then have

$$\frac{A}{A} + \frac{A}{a} + \frac{a}{A} + \frac{a}{a}$$

In the first and fourth term the egg and pollen cells are of like kind, consequently the product of their union must be constant, namely A and a; in the second and third, on the other hand, there again results a union of the two differentiating characters of the stocks, consequently the forms resulting from these fertilizations are identical with those of the hybrid from which they sprang. *There occurs accordingly a repeated hybridization.* This explains the striking fact that the hybrids are able to produce, besides the two parental forms, offspring which are like themselves A/a and a/A both give the same union Aa, since, as already remarked above, it makes no difference in the result of fertilization to which of the two characters the pollen or egg cells belong. We may write then $A/A + A/a + a/A + a/a = A + 2\,Aa + a$.

This represents the *average* result of the self-fertilization of the hybrids when two differentiating characters are united in them. In individual flowers and in individual plants, however, the ratios in which the forms of the series are produced may suffer not inconsiderable fluctuations. [Whether segregation by such units is more than purely fortuitous may perhaps be determined by seriation.]. Apart from the fact that the numbers in which both sorts of egg cells occur in the seed vessels can only be regarded as equal on the average, it remains purely a matter of chance which of the two sorts of pollen may fertilize each separate egg cell. For this reason the separate values must necessarily be subject to fluctuations, and there are even extreme cases possible, as were described earlier in connection with the experiments on the forms of the seed and the color of the albumen. The true ratios of the numbers can only be ascertained by an average deduced from the sum of as many single values as possible; the greater the number, the more are merely chance effects eliminated.

The developmental series for hybrids in which two kinds of differentiating characters are united contains among sixteen individuals nine different forms, viz.,

$$AB + Ab + aB + ab + 2ABb + 2aBb + 2AaB + 2Aab + 4AaBb.$$

Between the differentiating characters of the original stocks A, a, and B, four constant combinations are possible, and consequently the hybrids produce the corresponding four forms of egg and pollen cells: AB, Ab, aB, ab, and each of these will on the average figure four times in the fertilization, since sixteen individuals are included in the series. Therefore, the participators in the fertilization are

Pollen   cells: $AB + AB + AB + AB + Ab + Ab + Ab + Ab + aB + aB + aB + aB + ab + ab + ab + ab$.

Egg cells: $AB + AB + AB + AB + Ab + Ab + Ab + Ab + aB + aB + aB + aB + ab + ab + ab + ab$.

In the process of fertilization each pollen form unites on an average equally often with each egg cell form, so that each of the four pollen cells *AB* unites once with one of the forms of egg cell *AB*, *Ab*, *aB*, *ab*. In precisely the same way the rest of the pollen cells of the forms *Ab*, *aB*, *ab* unite with all the other egg cells. We obtain therefore:

$$\frac{AB}{AB} + \frac{AB}{Ab} + \frac{AB}{aB} + \frac{AB}{ab} + \frac{Ab}{AB} + \frac{Ab}{Ab} + \frac{Ab}{aB} + \frac{Ab}{ab}$$

$$+ \frac{aB}{AB} + \frac{aB}{Ab} + \frac{aB}{aB} + \frac{aB}{ab} + \frac{ab}{AB} + \frac{ab}{Ab} + \frac{ab}{aB} + \frac{ab}{ab}$$

or

$$AB + ABb + AaB + AaBb + ABb + Ab + AaBb + Aab + AaB + AaBb + aB + aBb + AaBb$$
$$+ Aab + aBb + ab = AB + Ab + aB + ab + 2ABb + 2aBb + 2AaB + 2Aab + 4AaBb$$

In precisely similar fashion is the developmental series of hybrids exhibited when three kinds of differentiating characters are conjoined in them. The hybrids form eight various kinds of egg and pollen cells: *ABC*, *ABc*, *AbC*, *Abc*, *aBC*, *aBc*, *abC*, *abc*, and each pollen form unites itself again on the average once with each form of egg cell.

The law of combination of different characters which governs the development of the hybrids finds therefore its foundation and explanation in the principle enunciated that the hybrids produce egg cells and pollen cells which in equal numbers represent all constant forms which result from the combinations of the characters brought together in fertilization.

A.10 EXPERIMENTS WITH HYBRIDS OF OTHER SPECIES OF PLANTS

It must be the object of further experiments to ascertain whether the law of development discovered for *Pisum* applies also to the hybrids of other plants. To this end several experiments were recently commenced. Two minor experiments with species of *Phaseolus* have been completed, and may be here mentioned.

An experiment with *Phaseolus vulgaris* and *Phaseolus nanus* gave results in perfect agreement. *Ph. nanus* had, together with the dwarf axis, simply inflated, green pods. *Ph. vulgaris* had, on the other hand, an axis 10 ft. to 12 ft. high, and yellow colored pods, constricted when ripe. The ratios of the numbers in which the different forms appeared in the separate generations were the same as with *Pisum*. Also the development of the constant combinations resulted according to the law of simple combination of characters, exactly as in the case of *Pisum*. There were obtained:

Constant Combinations	Axis	Color of the Unripe Pods	Form of the Ripe Pods
1	Long	Green	Inflated
2	"	"	Constricted
3	"	Yellow	Inflated
4	"	"	Constricted
5	Short	Green	Inflated
6	"	"	Constricted
7	"	Yellow	Inflated
8	"	"	Constricted

The green color of the pod, the inflated forms, and the long axis were, as in *Pisum*, dominant characters.

Another experiment with two very different species of *Phaseolus* had only a partial result. *Phaseolus nanus* L., served as seed parent, a perfectly constant species, with white flowers in short recemes and small white seeds in straight, inflated, smooth pods; as pollen parent was used *Ph. multiflorus*, W., with tall winding stem, purple-red flowers in very long recemes, rough, sickle-shaped crooked pods, and large seeds which bore black flecks and splashes on a peach-blood-red ground.

The hybrids had the greatest similarity to the pollen parent, but the flowers appeared less intensely colored. Their fertility was very limited; from 17 plants, which together developed many hundreds of flowers, only 49 seeds in all were obtained. These were of medium size, and were flecked and splashed similarly to those of *Ph. multiflorus*, while the ground color was not materially different. The next year 44 plants were raised from these seeds, of which only 31 reached the flowering stage. The characters of *Ph. nanus*, which had been altogether latent in the hybrids, reappeared in various combinations; their ratio, however, with relation to the dominant plants was necessarily very fluctuating owing to the small number of trial plants. With certain characters, as in those of the axis and the form of pod, it was, however, as in the case of *Pisum*, almost exactly 1:3.

Insignificant as the results of this experiment may be as regards the determination of the relative numbers in which the various forms appeared, it presents, on the other hand, the phenomenon of a remarkable change of color in the flowers and seed of the hybrids. In *Pisum* it is known that the characters of the flower- and seed-color present themselves unchanged in the first and further generations, and that the offspring of the hybrids display exclusively the one or the other of the characters of the original stocks. It is otherwise in the experiment we are considering. The white flowers and the seed-color of *Ph. nanus* appeared, it is true, at once in the first generation in one fairly fertile example, but the remaining 30 plants developed flower-colors which were of various grades of purple-red to pale violet. The coloring of the seed-coat was no less varied than that of the flowers. No plant could rank as fully fertile; many produced no fruit at all; others only yielded fruits from the flowers last produced, which did not ripen. From 15 plants only were well-developed seeds obtained. The greatest disposition to infertility was seen in the forms with preponderantly red flowers, since out of 16 of these only 4 yielded ripe seed. Three of these had a similar seed pattern to *Ph. multiflorus*, but with a more or less pale ground color; the fourth plant yielded only one seed of plain brown tint. The forms with preponderantly violet-colored flowers had dark brown, black-brown, and quite black seeds.

The experiment was continued through two more generations under similar unfavorable circumstances, since even among the offspring of fairly fertile plants there came again some which were less fertile and even quite sterile. Other flower- and seed-colors than those cited did not subsequently present themselves. The forms which in the first generation contained one or more of the recessive characters remained, as regards these, constant without exception. Also of those plants which possessed violet flowers and brown or black seed, some did not vary again in these respects in the next generation; the majority, however, yielded together with offspring exactly like themselves, some which displayed white flowers and white seed-coats. The red flowering plants remained so slightly fertile that nothing can be said with certainty as regards their further development.

Despite the many disturbing factors with which the observations had to contend, it is nevertheless seen by this experiment that the development of the hybrids, with regard to those characters which concern the form of the plants, follows the same laws as in *Pisum*. With regard to the color characters, it certainly appears difficult to perceive a substantial agreement. Apart from the fact that from the union of a white and a purple-red coloring a whole series of colors results, from purple to pale violet and white, the circumstance is a striking one that among 31 flowering plants only one received the recessive character of the white color, while in *Pisum* this occurs on the average in every fourth plant.

Even these enigmatical results, however, might probably be explained by the law governing *Pisum* if we might assume that the color of the flowers and seeds of *Ph. multiflorus* is a combination of two or more entirely independent colors, which individually act like any other constant character in the plant. If the flower-color A were a combination of the individual characters $A_1 + A_2 + \cdots$ which produce the total impression of a purple coloration, then by fertilization with the differentiating character, white color, a, there would be produced the hybrid unions $A_1a + A_2a + \cdots$ and so would it be with the corresponding coloring of the seed-coats.[*] According to the above assumptions, each of these hybrid color unions would be independent, and would consequently develop quite independently from the others. It is then easily seen that from the combination of the separate developmental series a complete color-series must result. If, for instance, $A = A_1 + A_2$, then the hybrids A_1a and A_2a form the developmental series—

$$A_1 + 2A_1a + a, \qquad A_2 + 2A_2a + a.$$

The members of this series can enter into nine different combinations, and each of these denotes another color:

$1\ A_1A_2$	$2\ A_1aA_2$	$1\ A_2a$
$2\ A_1A_2a$	$4\ A_1aA_2a$	$2\ A_2aa$
$1\ A_1a$	$2\ A_1aa$	$1\ aa$

The figures prescribed for the separate combinations also indicate how many plants with the corresponding coloring belong to the series. Since the total is 16, the whole of the colors are on the average distributed over each 16 plants, but, as the series itself indicated, in unequal proportions.

Should the color development really happen in this way, we could offer an explanation of the case above described, namely that of the white flowers and seed-coat color only appeared once among 31 plants of the first generation. This coloring appears only once in the series, and could therefore also only be developed once in the average in each 16, and with three color characters only once even in 64 plants.

It must, nevertheless, not be forgotten that the explanation here attempted is based on a mere hypothesis, only supported by the very imperfect result of the experiment just described. It would, however, be well worthwhile to follow up the development of color in hybrids by similar experiments, since it is probable that in this way we might learn the significance of the extraordinary variety in the coloring of our ornamental flowers.

So far, little at present is known with certainty beyond the fact that the color of the flowers in most ornamental plants is an extremely variable character. The opinion has often been expressed that the stability of the species is greatly disturbed or entirely upset by cultivation, and consequently there is an inclination to regard the development of cultivated forms as a matter of chance devoid of rules; the coloring of ornamental plants is indeed usually cited as an example of great instability. It is, however, not clear why the simple transference into garden soil should result in such a thorough and persistent revolution in the plant organism. No one will seriously maintain that in the open country the development of plants is ruled by other laws than in the garden bed. Here, as there, changes of type must take place if the conditions of life be altered, and the species possesses the capacity of fitting itself to its new environment. It is willingly granted that by cultivation the origination of new varieties is favored, and that by man's labor many varieties are acquired which, under natural conditions, would be lost; but nothing justifies the assumption that the tendency to formation of varieties is so extraordinarily increased that the species speedily lose all stability, and their offspring diverge into an endless series of extremely variable forms. Were the change in the conditions the sole cause of variability we might expect that those cultivated plants which are grown for centuries under almost identical conditions would again attain constancy. This, as is well known, is not the case since it is precisely under such circumstances that not only the most varied but also

---

[*] As it fails to take account of factors introduced by the albino, this representation is imperfect. It is however interesting to know that Mendel realized the fact of the existence of compound characters, and that the rarity of the white recessives was a consequence of this resolution.

the most variable forms are found. It is only the Leguminosae, like *Pisum, Phaseolus* [*1 Phaseolus* nevertheless is insect-fertilized], *Lens*, whose organs of fertilization are protected by the keel, which constitute a noteworthy exception. Even here there have arisen numerous varieties during a cultural period of more than 1000 years under most various conditions; these maintain, however, under unchanging environments a stability as great as that of species growing wild.

It is more than probable that as regards the variability of cultivated plants there exists a factor which so far has received little attention. Various experiments force us to the conclusion that our cultivated plants, with few exceptions, are *members of various hybrid series*, whose further development in conformity with law is varied and interrupted by frequent crossings *inter se*. The circumstance must not be overlooked that cultivated plants are mostly grown in great numbers and close together, affording the most favorable conditions for reciprocal fertilization between the varieties present and species itself. The probability of this is supported by the fact that among the great array of variable forms solitary examples are always found, which in one character or another remain constant, if only foreign influence be carefully excluded. These forms behave precisely as do those which are known to be members of the compound hybrid series. Also with the most susceptible of all characters, that of color, it cannot escape the careful observer that in the separate forms the inclination to vary is displayed in very different degrees. Among plants which arise from *one* spontaneous fertilization there are often some whose offspring vary widely in the constitution and arrangement of the colors, while that of others show little deviation, and among a greater number solitary examples occur which transmit the color of the flowers unchanged to their offspring. The cultivated species of *Dianthus* afford an instructive example of this. A white-flowered example of *Dianthus caryophyllus*, which itself was derived from a white-flowered variety, was shut up during its blooming period in a greenhouse; the numerous seeds obtained therefrom yielded plants entirely white-flowered like itself. A similar result was obtained from a subspecies, with red flowers somewhat flushed with violet, and one with flowers white, striped with red. Many others, on the other hand, which were similarly protected, yielded progeny which were more or less variously colored and marked.

Whoever studies the coloration which results in ornamental plants from similar fertilization can hardly escape the conviction that here also the development follows a definite law which possibly finds its expression *in the combination of several independent color characters*.

A.11   CONCLUDING REMARKS

It can hardly fail to be of interest to compare the observations made regarding *Pisum* with the results arrived at by the two authorities in this branch of knowledge, Köreuter and Gärtner, in their investigations. According to the opinion of both, the hybrids in outward appearance present either a form intermediate between the original species, or they closely resemble either the one or the other type, and sometimes can hardly be discriminated from it. From their seeds usually arise, if the fertilization was effected by their own pollen, various forms which differ from the normal type. As a rule, the majority of individuals obtained by one fertilization maintain the hybrid form, while some few others come more like the seed parent, and one or other individual approaches the pollen parent. This, however, is not the case with hybrids without exception. Sometimes the offspring have more nearly approached, some the one and some the other of the two original stocks, or they all incline more to one or the other side; while in other cases *they remain perfectly like the hybrid* and continue constant in their offspring. The hybrids of varieties behave like hybrids of species, but they possess greater variability of form and more pronounced tendency to revert to the original types.

With regard to the *form* of the hybrids and their *development*, as a rule an agreement with the observations made in *Pisum* is unmistakable. It is otherwise with the exceptional cases cited. Gärtner confesses even that the exact determination whether a form bears a greater resemblance to one or to the other of the two original species often involved great difficulty, so much depending upon the subjective point of view of the observer. Another circumstance could, however, contribute to render the results fluctuating and uncertain, despite the most careful observation and differentiation. For the experiments, plants were mostly used which rank as good species and are differentiated by a large number of characters. In addition to the sharply defined characters, where it is a question of greatly or less similarity, those characters must also be taken into account which are often difficult to define in words, but yet suffice, as every plant specialist knows, to give the forms a peculiar appearance. If it be accepted that the development of hybrids follows the law which is valid for *Pisum*, the series in each separate experiment must contain very many forms, since the number of terms, as is known, increases with the number of the differentiating characters as the powers of three. With a relatively small number of experimental plants the results therefore could only be approximately right, and in single cases might fluctuate considerably. If, for instance, the two original stocks differ in seven characters, and 100–200 plants were raised from the seeds of their hybrids to determine the grade of relationship of the offspring, we can easily see how uncertain the decision must become since for seven differentiating characters the combination series contains 16,384 individuals under 2187 various forms; now one and then another relationship could assert its predominance, just according as chance presented this or that form to the observer in a majority of cases.

If, furthermore, there appear among the differentiating characters at the same time *dominant* characters, which are transmitted entire or nearly unchanged to the hybrids, then in the terms of the developmental series that one of the two original parents which possesses the majority of dominant characters must always be predominant. In the experiment described relative to *Pisum*, in which three kinds of differentiating characters were concerned, all the dominant characters belonged to the seed parent. Although the terms of the series in their internal composition approach both original parents equally, yet in this experiment the type of the seed parent obtained so great a preponderance that out of each 64 plants of the first generation 54 exactly resembled it, or only differed in one character. It is seen how rash it must be under such circumstances to draw from the external resemblances of hybrids conclusions as to their internal nature.

Gärtner mentions that in those cases where the development was regular among the offspring of the hybrids the two original species were not reproduced, but only a few individuals which approached them. With very extended developmental series, it could not in fact be otherwise. For seven differentiating characters, for instance, among more than 16,000 individuals—offspring of the hybrids—each of the two original species would occur only once. It is therefore hardly possible that these should appear at all among a small number of experimental plants; with some probability, however, we might reckon upon the appearance in the series of a few forms which approach them.

We meet with an *essential difference* in those hybrids which remain constant in their progeny and propagate themselves as truly as the pure species. According to Gärtner, to this class belong the *remarkably fertile* hybrids *Aquilegia atropurpurea canadensis*, *Lavatera pseudolbia thuringiaca*, *Geum urbanorivale*, and some *Dianthus* hybrids; and, according to Wichura, the hybrids of the Willow family. For the history of the evolution of plants this circumstance is of special importance, since constant hybrids acquire the status of new species. The correctness of the facts is guaranteed by eminent observers, and cannot be doubted. Gärtner had an opportunity of following up *Dianthus Armeria deltoides* to the tenth generation, since it regularly propagated itself in the garden.

With *Pisum* it was shown by experiment that the hybrids form egg and pollen cells of *different* kinds, and that herein lies the reason of the variability of their offspring. In other hybrids, likewise, whose offspring behave similarly we may assume a like cause; for those, on the other hand, which remain constant the assumption appears justifiable that their reproductive cells are all alike and agree with the foundation-cell of the hybrid. In the opinion of renowned physiologists, for the purpose of propagation one pollen cell and one egg cells unite in Phanerogams [In *Pisum* it is placed beyond doubt that for the formation of the new embryo a perfect union of the elements of both reproductive cells must take place. How could we otherwise explain that among the offspring of the hybrids both original types reappear in equal numbers and with all their peculiarities? If the influence of the egg cell upon the pollen cell were only external, if it fulfilled the role of a nurse only, then the result of each fertilization could be no other than that the developed hybrid should exactly resemble the pollen parent, or at any rate do so very closely. The experiments so far have in no wise confirmed. An evident proof of the complete union of the contents of both cells is afforded by the experience gained on all sides that it is immaterial, as regards the form of the hybrid, which of the original species is the seed parent or which the pollen parent.] into a single cell, which is capable by assimilation and formation of new cells to become an independent organism. This development follows a constant law, which is founded on the material composition and arrangement of the elements which meet in the cell in a vivifying union. If the reproductive cells be of the same kind and agree with the foundation cell of the mother plant, then the development of the new individual will follow the same law which rules the mother plant. If by chance that an egg cell unites with a *dissimilar* pollen cell, we must then assume that between those elements of both cells, which determine opposite characters some sort of compromise is effected. The resulting compound cell becomes the foundation of the hybrid organism the development of which necessarily follows a different scheme from that obtaining in each of the two original species. If the compromise be taken to be a complete one, in the sense, namely, that the hybrid embryo is formed from two similar cells, in which the differences are *entirely and permanently accommodated* together, the further result follows that the hybrids, like any other stable plant species, reproduce themselves truly in their offspring. The reproductive cells which are formed in their seed vessels and anthers are of one kind, and agree with the fundamental compound cell [fertilized ovum].

With regard to those hybrids whose progeny is *variable*, we may perhaps assume that between the differentiating elements of the egg and pollen cells there also occurs a compromise, in so far that the formation of a cell as the foundation of the hybrid becomes possible; but, nevertheless, the arrangement between the conflicting elements is only temporary and does not endure throughout the life of the hybrid plant. Since in the habit of the plant no changes are perceptible during the whole period of vegetation, we must further assume that it is only possible for the differentiating elements to liberate themselves from the enforced union when the fertilizing cells are developed. In the formation of these cells all existing elements participate in an entirely free and equal arrangement, by which it is only the differentiating ones which mutually separate themselves. In this way the production would be rendered possible of as many sorts of egg and pollen cells as there are combinations possible of the formative elements.

The attribution attempted here of the essential difference in the development of hybrids to *a permanent or temporary union* of the differing cell elements can, of course, only claim the value of an hypothesis for which the lack of definite data offers a wide scope. Some justification of the opinion expressed lies in the evidence afforded by *Pisum* that the behavior of each pair of differentiating characters in hybrid union is independent of the other differences between the two original plants, and, further, that the hybrid produces just so many kinds of egg and pollen cells as there are possible constant combination forms. The differentiating characters of two plants can finally, however, only depend upon differences in the composition and grouping of the elements which exist in the foundation-cells of the same in vital interaction.

Even the validity of the law formulated for *Pisum* requires still to be confirmed, and a repetition of the more important experiments is consequently must to be desired, that, for instance, relating to the composition of the hybrid fertilizing cells. A differential may easily escape the single observer ["Dem einzelnen Beobachter kann leicht ein Differenziale entgehen], which although at the outset may appear to be unimportant, yet accumulate to such an extent that it must not be ignored in the total result. Whether the variable hybrids of other plant species observe an entire agreement must also be first decided experimentally. In the meantime we may assume that in material points an essential difference can scarcely occur, since the *unity* in the developmental plan of organic life is beyond question.

In conclusion, the experiments carried out by Kölreuter, Gärtner, and others with respect to *the transformation of one species into another by artificial fertilization* merit special mention. Particular importance has been attached to these experiments and Gärtner reckons them "among the most difficult of all in hybridisation."

If a species *A* is to be transformed into a species *B*, both must be united by fertilization and the resulting hybrids then be fertilized with the pollen of *B*; then, out of the various offspring resulting, that form would be selected which stood in nearest relation to *B* and once more be fertilized with *B* pollen, and so continuously until finally a form is arrived at which is like *B* and constant in its progeny. By this process the species *A* would change into the species *B*. Gärtner alone has effected 30 such experiments with plants of genera *Aquilegia, Dianthus, Geum, Lavatera, Lynchnis, Malva, Nicotiana,* and *Oenothera*. The period of transformation was not alike for all species. While with some a triple fertilization sufficed, with others this had to be repeated five or six times, and even in the same species fluctuations were observed in various experiments. Gärtner ascribes this difference to the circumstance that "the specific power by which a species, during reproduction, effects the change and transformation of the maternal type varies considerably in different plants, and that, consequently, the periods with which the one species is changed into the other must also vary, as also the number of generations, so that the transformation in some species is perfected in more, and in others in fewer generations." Further, the same observer remarks "that in these transformation experiments a good deal depends upon which type and which individual be chosen for further transformation."

If it may be assumed that in these experiments the constitution of the forms resulted in a similar way to that of *Pisum*, the entire process of transformation would find a fairly simple explanation. The hybrid forms as many kinds of egg cells as there are constant combinations possible of the characters conjoined therein, and one of these is always of the same kind as that of the fertilizing pollen cells. Consequently there always exists the possibility with all such experiments that even from the second fertilization there may result a constant form identical with that of the pollen parent. Whether this really be obtained depends in each separate case upon the number of the experimental plants, as well as upon the number of differentiating characters which are united by the fertilization. Let us, for instance, assume that the plants selected for experiment differed in three characters, and the species *ABC* is to be transformed into the other species *abc* by repeated fertilization with the pollen of the latter; the hybrids resulting from the first cross form eight different kinds of egg cells, viz., *ABC, ABc, AbC, aBC, Abc, aBc, abC,* and *abc*. These in the second year of experiment are united again with the pollen cells *abc*, and we obtain the series

$$AaBbCc + AaBbc + AabCc + aBbCc + Aabc + aBbc + abCc + abc.$$

Since the form *abc* occurs once in the series of eight terms, it is consequently little likely that it would be missing among the experimental plants, even were these raised in a smaller number, and the transformation would be perfected already by a second fertilization. If by chance it did not appear, then the fertilization must be repeated with one of those forms nearest akin, Aabc, aBbc, abCc. It is perceived that such an experiment must extend the farther *the smaller the number of experimental plants and the larger the number of differentiating characters* in the two original species; and that, furthermore, in the same species there can easily occur a delay of one or even of two generations such as Gärtner observed. The transformation of widely divergent species could generally only be completed in 5 or 6 years of experiment, since the number of different egg cells which are formed in the hybrid increases as the powers of 2 with the number of differentiating characters.

Gärtner found by repeated experiments that the *respective period of transformation* varies in many species, so that frequently a species *A* can be transformed into a species *B* a generation sooner than can species *B* into species *A*. He deduces therefrom that Kölreuter's opinion can hardly be maintained that "the two natures in hybrids are perfectly in equilibrium." It appears, however, that Kölreuter does not merit this criticism, but that Gätrner rather has overlooked a material point, to which he himself elsewhere draws attention, viz. that "it depends which individual is chosen for further transformation." Experiments which in this connection were carried out with two species of *Pisum* demonstrated that as regards the choice of the fittest individuals for the purpose of further fertilization it may make a great difference which of two species is transformed into the other. The two experimental plants differed in five characters, while at the same time those of species *A* were all dominant and those of species *B* all recessive. For mutual transformation *A* was fertilized with pollen of *B*, and *B* with pollen of *A*, and this was repeated with both hybrids the following year. With the first experiment, *B/A*, there were 87 plants available in the third year of experiment for selection of the individuals for further crossing, and these were of the possible 32 forms; with the second experiment, *A/B*, 73 plants resulted, which *agreed throughout perfectly in habit with the pollen parent*; in their internal composition, however, they must have been just as varied as the forms in the other experiment. A definite selection was consequently only possible with the first experiment; with the second the selection had to be made at random, merely. Of the latter only a portion of the flowers were crossed with the *A* pollen, the others were left to fertilize themselves. Among each five plants which were selected in both experiments for fertilization there agreed, as the following year's culture showed, with the pollen parent:

1st experiment	2nd experiment	
2 plants	-------	In all characters
3 plants	-------	In 4 characters
-------	2 plants	" 3 "
-------	2 "	" 2 "
-------	1 "	" 1 character

In the first experiment, therefore, the transformation was completed; in the second, which was not continued further, two or more fertilizations would probably have been required.

Although the case may not frequently occur in which the dominant characters belong exclusively to one or the other of the original parent plants, it will always make a difference which of the two possesses the majority of dominants. If the pollen parent has the majority, then the selection of forms for further crossing will afford a less degree of certainty than in the reverse case, which must imply a delay in the period of transformation, provided that the experiment is only considered as completed when a form is arrived at which not only exactly resembles the pollen parent in form, but also remains as constant in its progeny.

Gärtner, by the results of these transformation experiments, was led to oppose the opinion of those naturalists who dispute the stability of plant species and believe in a continuous evolution of vegetation. He perceives ["Es sicht" in the original is clearly a misprint for "Er sicht."] in the complete transformation of one species into another an indubitable proof that species are fixed with limits beyond which they cannot change. Although this opinion cannot be unconditionally accepted, we find on the other hand in Gärtner's experiments a noteworthy confirmation of that supposition regarding variability of cultivated plants which has already been expressed.

Among the experimental species there were cultivated plants, such as *Aquilegia atropurpurea* and *canadensis, Dianthus caryophyllus, chinensis,* and *japonicus, Nicotiana rustica* and *paniculata,* and hybrids between these species lost none of their stability after four or five generations.

Source: http://www.mendelweb.org/Mendel.html (Sinnott et al., 1950).

Mendel Web was conceived and constructed by Roger B. Blumberg, *rblum@netspace.org*

Appendix B

B.I SOURCES OF CHEMICALS

Two Color Fiber-FISH

Product	Company	Phone No.	Catalog No.	Quantity	Price
		Slide			
Poly-prep slides	Sigma	(800) 325-3010	P 0425	72 slides	31.00
		dUTPs			
Biotin-16-dUTP	Roche	(800) 262-1640	1 093 070	50 nmol	263.00
Dig-11-dUTP	Roche	(800) 262-1640	1 558 706	125 nmol	573.00
		Antibodies			
Fibers Green Detection (for Dig)					
FITC antidig (from sheep)	Roche	(800) 262-1640	1 207 741	200 µg	$133.20
FITC antisheep (rabbit)	Roche	(800) 262-1640	605 340	1 mL	$88.00
Fibers Red Detection (for Biotin)					
Texas red streptavidin	Vector Labs	(800) 227-6666	SA-5006	1 mg	$65.00
Biotinylated antistraeptavidin	Vector Labs	(800) 227-6666	BA-0500	500 µg	$70.00
		Antifade			
Prolong antifade	Molecular probes	(800) 438-2209	P-7481	Kit	$135.00
		Staining			
YoYo stain	Molecular probes	(800) 438-2209	Y-3601	200 µL (1 mM)	225.00

One Color Fiber-FISH

Product	Company	Phone No.	Catalog No.	Quantity	Price
		Antibodies			
Fibers Green Detection (for Biotin)					
FITC-Avidin	Vector Labs	(800) 227-6666	A-2001	5 mg	70.00
Biotin-antiavidin	Vector Labs	(800) 227-6666	BA-0300	500 µg	60.00

Source: From S. A. Jackson, R. M. Stupar, and J. Jiang. Personal communication.

Note: dUTPs come from Roche at 50 µL of 1 mM. The working concentration should be 0.5 mM, with dTTP:dUTP ratio of 2:1. Therefore, dilute dUTPs with 50 µL water, then mix with 200 µL 0.5 mM dTTP.

B.2 MATERIALS FOR FLOW CYTOMETRY

  i. Cell Trics (Partec): For filtration of the sample after chopping.
 ii. Eye protection against UV.
iii. Gilson-pipettes and tips: 1000 µL Gilson-pipette (blue tips) exclusively for PI. For non-toxic: 1000 µL Gilson (blue tips) 0 and a 100 µL Gilson (yellow tips).
 iv. Flow tubes: Provided by Partec or from MERCK.
  v. Wash bottles: For distilled water.
 vi. Distilled water: Glass distilled water.
vii. Water bath or dry incubator: Need to maintain 37°C for RNAse digestion.
viii. Sample tubes: Partec.

ix. Mercury high pressure lamp: Partec.

x. Hypochlorite: Normal bleach.

xi. Immersion gel: Optical gel; R. P. Cargille Labs, Cedar Grove, NJ 07009–1289.

xii. Glycerine: Mix it with the immersion gel, when the gel becomes dry.

xiii. Grease for laboratories: Very small amount needed to maintain the rubber fitting between flow cytometer and sample tube.

xiv. Normal razor blade: Gillette super silver.

xv. Pincers: For preparation of leaves.

xvi. Small Petri dishes.

xvii. Eppendorf-tubes: 2 mL to store the frozen RNase.

Source: R. Obermayer.

B.3   REAGENTS AND SOLUTIONS (www.ueb.cas.cz/olomouc1)

a. *Agarose Gel, 1.5% (w/v)*
   i.   3 g electrophoresis-grade agarose.
   ii. 200 mL 1 × TAE electrophoresis buffer.
   iii. Soak for 20 min at RT. Boil in a microwave oven to dissolve completely. Adjust volume to 200 mL with dH_2O. Pour gel into a casting form and let solidify.

b. *Amiprophos-Methyl Treatment Solution*
   i.   Prepare a 20 mM stock solution by dissolving 60.86 mg amiprophos-methyl in 10 mL cold acetone.
   ii. Store up to 1 year at –20°C in 1 mL aliquots.
   iii. Prepare the treatment solution immediately before use by combining amiprophos-methyl stock solution and 0.1× or 1× Hoagland's nutrient solution specified below:

Preparation of Amiprophos-Methyl Treatment Solution

Species	Hoagland's Solution		20 mM Amiprophos-Methyl	
	Concentration	Volume (mL)	Amount (µL)	Final Concentration (µM)
Vicia faba	1 ×	750	94.8	2.50
Pisum sativum	1 ×	750	380.0	10.00
Hordeum vulgare	0.1 ×	750	94.8	2.50
Secale cereale	0.1 ×	750	19.0	0.50

c. *4',6-Diamidino-2-Phenylindole (DAPI) Stock Solution, (0.1 mg/mL)*
   i.   Dissolve 5 mg DAPI in 50 mL dH_2O by stirring for 60 min.
   ii. Pass through a 0.22 µm filter to remove small particles.
   iii. Store up to 1 year at –20°C in 0.5 mL aliquots.
   Caution! *DAPI is a possible carcinogenic. It may be harmful if inhaled, swallowed, or absorbed through the skin, and may also cause irritation. Use gloves when handling. Be careful of particulate dust when weighing out the dye. Consult local institutional safety officer for specific handling and disposal procedures.*

d. *Ethidium Bromide Solution, 0.5 mg/mL*
   i.   Dissolve 5 mg EB in 10 mL dH_2O by stirring for 60 min.
   ii. Store up to 1 year at 4°C in the dark.
   iii. For working solution (0.5 µg/mL); dilute 1: 1000 in dH_2O. The working solution can be used several times.
   Caution! *Ethidium bromide is a powerful mutagen and is moderately toxic. It may be harmful if inhaled, swallowed, or absorbed through the skin. Use gloves when handling. Be careful of particulate dust when weighing out the dye. Consult local institutional safety officer for specific handling and disposal procedures.*

e. Formaldehyde Fixative
  i.   0.303 g Tris base (10 mM final).
  ii. 0.931 g Na_2 EDTA (10 mM final).
  iii. 1.461 g NaCl (100 mM final).
  iv. 250 µL Triton X-100 [0.1% (v/v) final].

Adjust volume to 200 mL with dH_2O, and adjust pH to 7.5 using 1 N NaOH. Add 37% (v/v) formaldehyde stock solution. Adjust final volume to 250 mL with dH_2O. Prepare fixative before use as concentrations shown in the following table for several species.

Preparation of Formaldehyde Fixative

Species	Formaldehyde Volume of Stock (mL)	Final (%) Concentration (v/v)
Vicia faba	27.0	4
Pisum sativum	20.0	3
Hordeum vulgare	13.5	2
Secale cereale	13.5	2

Caution! *Formaldehyde is toxic and is also carcinogenic. It is readily absorbed through the skin and is irritating or destructive to the skin, eyes, mucous membranes, and upper respiratory tract. Wear gloves and safety glasses. Always work in a chemical fume hood. Consult local institutional safety officer for specific handling and disposal procedures.*

f. Fructose Syrup
  i.   30 g fructose.
  ii. 20 mL dH_2O.
  iii. Incubate at 37°C overnight.
  iv. Add one crystal of thymol.
  v.   Store up to 1 year at 4°C.

g. Gamborg's B-5
  i.   Basal medium with minimal organics (macro- and micronutrients, and vitamins); G 5893, Sigma Chemical Co.

h. 5 N HCl
  i.   419.8 mL of 35% (v/v) HCl.
  ii. Adjust volume to 1000 mL with dH_2O.

i. Hoagland's Solution
  i.   Basal salt mixture (with macro- and micronutrients); H 2395, Sigma Chemical Co.
  ii. Prepare 500 mL according to manufacturer's direction.
  iii. Autoclave, store at RT.

j. Hydroxyurea Treatment Solution
Prepare treatment solution immediately before use by combining hydroxyurea and 0.1× or 1× Hoagland nutrient solution shown below:

Preparation of Hydroxyurea Treatment Solution

Species	Hoagland's Nutrient Solution Concentration	Volume (mL)	Hydroxyurea Amount (mg)	Final Concentration (mM)
Vicia faba	1×	750	71.3	1.25
Pisum sativum	1×	750	71.3	1.25
Hordeum vulgare	0.1×	750	114.0	2.00
Secale cereale	0.1×	750	142.6	2.50

k. LB01 Lysis Buffer
  i.   0.363 g Tris base (15 mM).
  ii. 0.149 g Na$_2$ EDTA (2 mM).
  iii. 0.035 g spermine 4HCl (0.5 mM).
  iv. 1.193 g KCl (80 mM).
  v.   0.239 NaCl (20 mM).
  vi. 200 µL Triton X-100 (0.1%).

Adjust volume to 200 mL with dH$_2$O, and adjust final pH to 7.5 using 1 N HCl. Filter through a 0.22 µm filter to remove small particles. Add 220 µL ß-mercaptoethanol and mix well. Store up to 1 year at –20°C in 10 mL aliquots.

  Caution! *ß-mercaptoethanol may be fatal if inhaled or absorbed through the skin and is harmful if swallowed. High concentrations are extremely destructive to the skin, eye, mucous membranes, and upper respiratory tract. Wear gloves and safety glasses and work in a chemical fume hood.*

l. LB01 Lysis Buffer (for Collecting Sorted Chromosomes)
  i.   0.545 g Tris (22.5 mM).
  ii. 0.223 g Na$_2$ EDTA (3 mM).
  iii. 0.052 g Spermine. 0.4HCl (0.75 mM).
  iv. 1.790 g KCl (120 mM).
  v.   0.351 g NaCl (30 mM).
  vi. 300 µL Triton X-100 (0.15%).

Adjust to 200 mL with dH$_2$O, adjust final pH to 7.5 using 1HCl. Filter through 0.22 µm filter to remove small particles, add 330 µL ß-mercaptoethanol and mix well, store at –20°C in 5 mL aliquots

m. Loading Buffer
  i.   2 mL 0.5 M EDTA, pH 8.0.
  ii. 100 mg SDS (1% w/v final).
  iii. 5 mg bromphenol blue (0.05% w/v final).
  iv. 5 mg xylene cyanol (0.05% w/v final).
  v.   5 mL 85% glycerol (42.5 v/v final).

Adjust volume to 10 mL with dH$_2$O, store up to 1 year at RT.

n. Magnesium Sulfate Stock Solution
  i.   Dissolve 1.23 g MgSO$_4$ · 7H$_2$O in 50 mL dH$_2$O.
  ii. Filter through a 0.22 µm filter to remove small particles.
  iii. Store at 4°C.

o. Mithramycine Stock Solution
  i.   Dissolve 50 mg mithramycine A in 50 mL dH$_2$O by stirring for 60 min.
  ii. Filter through a 0.22 µm filter to remove small particles.
  iii. Store up to 1 year at –20°C in 0.5 mL aliquots.

  Caution! *Mithramycine is a possible carcinogen. It may be harmful if inhaled, swallowed, or absorbed through the skin. Use gloves when handling. Be careful of particulate dust when weighing out dye. Consult local institutional safety officer for specific handling and disposal procedures.*

p. PCR Premix
  i.   5 µL 10× *Taq* DNA polymerase buffer (1× final; buffer does not contain MgCl$_2$).
  ii. 3 µL 25 mM MgCl$_2$ (1.5 mM final).
  iii. 1 µL 10 mM 4 dNTP mix (final 0.2 mM each dATP, dCTP, dGTP, dTTP).
  iv. 1 µL 50 µM forward primer (1 µM final).
  v.   1 µL 50 µM reverse primer (1 µM final).
  vi. 0.5 µL 5 U/µL *Taq* DNA polymerase (2.5 U/50 µL final).
  vii. 18.5 µL sterile dH$_2$O.

Mix well and microcentrifuge briefly (5–10 sec at 2000× g at RT). Prepare on ice shortly before use.

q. PRINS Buffer
   i. 0.605 g Tris base (10 mM final).
   ii. 1.864 g KCl (50 mM final).
   iii. 0.203 g $MgCl_2 \cdot 6H_2O$ (2 mM final).

Adjust volume to 500 mL with dH_2O, adjust pH to 8.0 using 1N HCl, and store up to 6 months at 4°C.

r. PRINS Reaction Mix
   i. 5 µL 10× *Taq* DNA polymerase buffer (1× final, containing 1.5 mM $MgCl_2$).
   ii. 5 µL 25 mM $MgCl_2$ (4 mM final).
   iii. 2.5 µL 2 mM dCTP/dGTP (0.1 mM each final).
   iv. 2 µL 0.2 mM fluorescein-12-dUTP (8 µM final).
   v. 2 µL 0.2 mM fluorescein-15-dATP (8 µM final).
   vi. 4.25 µL 0.2 mM dTTP (17 µM final).
   vii. 4.25 µL 0.2 mM dATP (17 µM final).
   viii. 5 µL 20 µM forward primer (2 µM final).
   ix. 5 µL 20 µM reverse primer (2 µM final).
   x. 0.6 µL 5 U/µL *Taq* DNA polymerase (3 U/50 µL final).

Adjust to 55 µL with sterile dH_2O (includes 5 µL for evaporation), prepare shortly before use. Actual composition of the mix should be optimized for a given primer pair and species.

s. Propidium Iodide (PI) Stock Solution
   i. Dissolve 50 mg (1 mg/mL) PI in 50 mL dH_2O.
   ii. Filter through a 0.22 µm filter to remove small particles.
   iii. Store at −20°C in 0.5 mL aliquots.

t. RNase Stock Solution
   i. Dissolve 25 mg (1 mg/mL) RNase in 25 mL dH_2O.
   ii. Filter through a 0.22 µm filter to remove small particles, heat to 90°C for 15 min to inactivate DNAse.
   iii. Store at −20°C in 0.5 mL aliquots.

u. Sheath Fluid SF 50
   i. 7.31 g NaCl (50 mM final).
   ii. Adjust volume to 2500 mL with dH_2O.
   iii. Sterilize by autoclaving.

v. Stop Buffer for PRINS
   i. 2.923 g NaCl (0.5 M final).
   ii. 1.861 g Na_2 EDTA (0.05 M final).

Adjust volume to 100 mL with dH_2O, adjust pH to 8.0 using concentrated NaOH, sterilize by autoclaving and store up to 6 months at 4°C.

w. TAE Electrophoresis Buffer, 50× and 1×
   i. 242 g Tris base (2 M).
   ii. 57.1 mL glacial acetic acid (1 M acetate final).
   iii. 100 mL 0.5 M EDTA, pH 8.0 (50 mM final).

Adjust volume to 1 L with dH_2O, store up to 1 year at RT, dilute 1:50 for working solution in dH_2O before use. Final 1× concentrations: 40 mM Tris, 20 mM acetate, and 1 mM EDTA.

   Caution! *Glacial acetic acid is volatile. Concentrated acids must be handled with great care. Wear gloves and safety glasses and work in a chemical fume hood.*

x. Tris Buffer
   i. 0.606 g Tris base (10 mM final).
   ii. 1.871 g Na_2 EDTA (10 mM final).
   iii. 2.922 g NaCl (100 mM final).

Adjust volume to 500 mL dH$_2$O, adjust pH to 7.5 using 1N NaOH, store up to 6 months at 4°C.

y. **Wash Buffer for PRINS**
   i.   1.161 g maleic acid (0.1 M final).
   ii. 0.876 g NaCl (150 mM final).
   iii. 0.5 μL Tween 20 (0.05% final).

Adjust volume to 100 mL with dH$_2$O, adjust pH to 7.5 using concentrated NaOH, sterilize by autoclaving, store up to 4 months at 4°C.

B.4 PREPARATION OF MEDIA

Constituent	Medium A	Medium B	Medium C	Medium D
Macronutrients (g/L)				
KNO$_3$	30.00	25.00	20.00	0.80
(NH$_4$)$_2$ SO$_4$	1.34	1.34	–	–
NH$_4$ NO$_3$	16.50	–	10.00	–
NaH$_2$ PO$_4$ · H$_2$O	1.50	1.50	1.00	0.165
KH$_2$ PO$_4$	1.70	–	3.00	–
MgSO$_4$ · 7H$_2$O	2.50	2.50	3.75	–
MgSO$_4$ · H$_2$O	–	–	–	5.76
Na$_2$ SO$_4$	–	–	–	2.00
CaCl$_2$ · 2H$_2$O	1.50	1.50	6.00	–
Ca (NO$_3$)$_2$	–	–	–	4.32
Micronutrients (mg/L)				
H$_3$BO$_3$	3.00	3.00	3.00	1.50
MnSO$_4$ · H$_2$O	10.00	10.00	10.00	7.00
ZnSO$_4$ · 7H$_2$O	2.00	2.00	2.00	3.00
KI	0.75	0.75	0.75	0.75
CuSO$_4$ · 5H$_2$O	0.025	0.025	0.025	0.10
Na$_2$ MoO$_4$ · 2H$_2$O	0.25	0.25	0.25	0.01
COCl$_2$ · 6H$_2$O	0.025	0.025	0.025	–
KCl	–	–	–	65.00
EDTA	–	–	26.10	–
Na$_2$ · EDTA	37.30	–	–	–
Fe · NaEDTA	–	73.40	–	73.40
FeSO$_4$ · 7 H$_2$O	27.80	–	24.90	–
KOH	–	–	15.035	–
Vitamins and Hormones (mg/L)				
Inositol	100.00	100.00	250.00	100.00
Nicotinic acid	1.00	1.00	1.00	1.00
Pyridoxine · HCl	1.00	1.00	1.00	1.00
Thiamine · HCl	10.00	10.00	10.00	1.00
Glycine	2.00	2.00	2.00	4.00
Ascorbic acid	100.00	50.00	–	–
L-Glutamine	730.00	–	–	–
Coumarin	–	–	–	9.00
IAA	0.20	0.30	–	1.00
NAA	2.00	–	–	–
Kinetin	1.28	0.75	–	–

Continued

Constituent	Medium A	Medium B	Medium C	Medium D
BAP	0.50	–	0.25	–
Sucrose (g/L)	100.00	30.00	25.00	10.00
Bacto-agar (g/L)	8.00	8.00	6.00	8.00
pH	5.8	5.8	5.6	6.0

Source: From Singh, R. J., K. P. Kollipara, and T. Hymowitz. 1987a. *Theor. Appl. Genet.* 74: 391–396. With permission.

Note: IAA, indol-3-acetic acid; NAA, 1-naphthalene acetic acid; BAP, 6-benzylaminopurine.

B.5 COMMON NAME, BOTANICAL NAME, AND 2n CHROMOSOME NUMBER(S) OF SOME IMPORTANT PLANTS

Common Name	Scientific Name	2n
Alfalfa	*Medicago sativa*	16, 32, 64
Almond	*Prunus amygdalus*	16
Anise	*Pimpinella anisum*	18, 20
Apple	*Malus domestica*	34
Apricot	*Prunus armeniaca*	16
Artichoke	*Cynara scolymus*	34
Ash (white)	*Fraxinus americana*	46, 92, 138
Ashoka	*Saraca indica*	24
Asparagus	*Asparagus officinalis*	20
Aster	*Aster novaeangliae*	10
Autumn crocus	*Colchicum autumnale*	38
Avocado	*Persea americana*	24
Bahiagrass	*Paspalum notatum*	20, 40
Bamboo	*Bambusa vulgaris*	72
Babul	*Acacia arabica*	52
Banana	*Musa sapientum*	22, 33, 44
Banyan tree	*Ficus benghalensis*	26
Barley	*Hordeum vulgare*	14
Bean-broad	*Vicia faba*	12
Castor oil	*Ricinus communis*	20
Lima	*Phaseolus lunatus*	22
Mung	*Vigna radiata*	22
String (French)	*Phaseolus vulgaris*	22
Beet	*Beta vulgaris*	18
Begonia	*Begonia rex*	32, 33, 34, 42, 43, 44
Bitter gourd	*Momordica charantia*	22
Blackberry	*Rubus alleghaniensis*	14
Black pepper	*Piper nigrum*	36, 48, 52, 54, 60, 65, 104, 128
Blueberry	*Vaccinium corymbosum*	48
Bluegrass (Kentucky)	*Poa pratensis*	28–98
Bottle gourd	*Lagenaria siceraria*	22
Broccoli	*Brassica oleracea* var. *Italica*	18
Brussels sprouts	*Brassica oleracea*	18
Buckwheat	*Fagopyrum esculentum*	16
Cabbage	*Brassica oleracea* var. *Capitata*	18
Cantaloupe	*Cucumis melo*	24

Continued

Common Name	Scientific Name	2*n*
Cardamom	*Elettaria cardamomum*	48, 52
Carnation	*Dianthus caryophyllus*	30, 90
Carrot	*Daucus carota*	18
Cashew nut	*Anacardium occidentale*	24, 30, 40, 42
Cassava	*Manihot esculenta*	36, 72
Cauliflower	*Brassica oleracea* var. *Botrytis*	18
Cedar (eastern red)	*Juniperus virginiana*	22, 33
Celery	*Apium graveolens*	22
Cherry (sour)	*Prunus cerasus*	32
Cherry (sweet)	*Prunus avium*	16
Chest nut (European)	*Castanea sativa*	24
Chickpea	*Cicer arietinum*	16
Chive	*Allium schoenoprasum*	16, 24, 32
Chrysanthemum	*Chrysanthemum morifolium*	54
Cinnamon	*Cinnamomum zeylanicum*	24
Clove	*Syzygium aromaticum*	22
Clover (red)	*Trifolium pratense*	14
Clover (Ladino)	*Trifolium repens*	32
Cocaine plant	*Erythroxylon coca*	24
Cocoa	*Theobroma cacao*	20
Coconut	*Cocos nucifera*	32
Coffee	*Coffea arabica*	44
Coriander	*Coriandrum sativum*	22
Cotton (upland)	*Gossypium hirsutum*	52
Cotton wood	*Populus deltoides*	38
Cowpea	*Vigna unguiculata*	22
Crabapple	*Pyrus ioensis* (or *Malus*)	34
Cranberry	*Vaccinium macrocarpon*	24
Crocus	*Crocus susianus*	12
Cucumber	*Cucumis sativus*	14
Cumin	*Cuminum cyminum*	14
Daffodil	*Narcissus pseudo-narcissus*	14
Daisy	*Bellis perennis*	18
Date Palm	*Phoenix dactilifera*	36
Dill	*Anethum graveolens*	22
Douglas fir	*Pseudotsuga taxifolia*	26
Eggplant	*Solanum melongena*	24
Fennel	*Foeniculum vulgare*	22
Fig	*Ficus carica*	26
Flax	*Linum usitatissimum*	30
Foxtail millet	*Setaria italica*	18
Garlic	*Allium sativum*	16
Geranium	*Pelargonium graveolens*	90
Ginger	*Zingiber officinale*	22
Gladiolus	*Gladiolus communis*	90, 180
Grape	*Vitis vinifera*	38
Grapefruit	*Citrus paradisi*	18, 27, 36
Grasspea	*Lathyrus sativus*	14
Guar	*Cyamopsis tetragonoloba*	14
Guava	*Psidium guajava*	22

Continued

Common Name	Scientific Name	2n
Guayule	*Parthenium argentatum*	54, 72
Hazelnut	*Corylus americana*	28
Hemp	*Cannabis sativa*	20
Hops	*Humulus lupulus*	20
Hyacinth-brown	*Hyacinthus orientalis*	16
Indian Mustard (red)	*Brassica juncea*	36
Iris (blue flag)	*Iris versicolor*	72, 84, 105
Ivy	*Hedera helix*	48
Jasmin	*Jasminum officinale*	26
Jojoba	*Simmondsia chinensis*	52
Juniper	*Juniperus communis*	22
Jute	*Corchorus olitoris*	14
Kiwi fruit	*Actinidia chinensis*	58
Lemon	*Citrus limon*	18, 36
Lentil	*Lens culinaris*	14
Lettuce	*Lactuca sativa*	18
Lily	*Lilium longiflorum*	24
Lime (acid)	*Citrus aurantifolia*	18
Lingonberry	*Vaccinium vitis-idaea*	24
Lotus (yellow)	*Nelumbo lutea*	16
Luffa	*Luffa cylindrica*	26
Lupin	*Lupinus luteus*	48
Macaroni wheat	*Triticum turgidum* var. *durum*	28
Mahogany	*Swietenia mahagoni*	46–48
Maize	*Zea mays*	20
Mango	*Mangifera indica*	40
Mahwah	*Bassia latifolia*	24
Maple (sugar)	*Acer saccharum*	26
Marigold	*Tagets erecta*	24
Millet-Foxtail	*Setaria italica*	18
Pearl	*Pennisetum glaucum*	14
Broomcorn	*Panicum miliaceum*	36
Sawa	*Echinochloa colona*	54
Milkweed	*Asclepias incarnata*	22
Morning-glory	*Ipomoea purpurea*	30
Muskmelon	*Cucumis melo*	24
Nutmeg	*Myristica fragrans*	44
Oak (pedunculate)	*Quercus robur*	24
(White)	*Quercus alba*	24
Oat	*Avena sativa*	42
Oil palm	*Elaeis guineensis*	32
Okra	*Hibiscus esculentus*	72, 144
Olive	*Olea europaea*	46
Onion	*Allium cepa*	16
Orange (bitter)	*Citrus aurantium*	18
(sweet)	*Citrus sinensis*	18, 27, 36, 45
Papaya	*Carica papaya*	18
Parsley	*Petroselinum crispum*	22
Parsnip	*Pastinaca sativa*	22
Pea	*Pisum sativum*	14

Continued

Common Name	Scientific Name	2n
Peach	*Prunus persica*	16
Peanut (groundnut)	*Arachis hypogaea*	40
Pear	*Pyrus communis*	34
Peepal tree	*Ficus religiosa*	26
Pecan	*Carya illinoensis*	32
Pepper	*Capsicum annuum*	24
Peppermint	*Mentha piperita*	36, 64, 66, 68, 70
Petunia	*Petunia hybrida*	14
Pine (red)	*Pinus* spp.	24
Pigeon pea	*Cajanus cajan*	22
Pineapple	*Ananas comosus*	50
Pistachio nut	*Pistacia vera*	30
Plum (European)	*Prunus domestica*	48
Poinsettia	*Euphorbia pulcherrima*	28
Poppy (opium)	*Papaver somniferum*	22
Potato	*Solanum tuberosum*	48
Pumpkin (summer squash)	*Cucurbita pepo*	40
Radish	*Raphanus sativus*	18
Ragweed	*Ambrosia trifida*	24
Raspberry (European)	*Rubus idaeus*	14, 21, 28
Ragi, Finger millet	*Eleusine coracana*	36
Redbud	*Cercis canadensis*	12
Red stem wormwood	*Artemisia scoparia*	16
Rhubarb	*Rheum officinale*	22, 44
Rice	*Oryza sativa*	24
Rose	*Rosa centifilia*	28
Rose	*Rosa damascena*	28
Rose	*Rosa multiflora*	14
Rubber	*Hevea brasiliensis*	36
Rye	*Secale cereale*	14
Safflower	*Carthamus tinctorius*	24
Sandalwood (Indian)	*Santalum album*	20
Sequoia (big tree)	*Sequoia gigantia*	22(44)
Sesame	*Sesamum indicum*	26
Snapdragon	*Antirrhinum majus*	16
Sorghum	*Sorghum bicolor*	20
Soybean	*Glycine max*	40
Spearmint	*Mentha spicata*	36, 48
Spinach	*Spinacia oleracea*	12
Squash (winter)	*Cucurbita maxima*	40
Sugarcane	*Saccharum officinarum*	80
Sunflower	*Helianthus annuus*	34
Sweet potato	*Ipomoea batatas*	60, 90
Sweet wormwood	*Artemisia annua*	18
Sycamore	*Acer pseudoplatanus*	52
Tamarind	*Tamarindus indica*	24
Tangerine	*Citrus nobilis*	18
Tea	*Camellia sinensis*	30

Continued

Common Name	Scientific Name	2n
Teak	*Tectona grandis*	24
Thyme	*Thymus vulgaris*	30
Timothy	*Phleum pratense*	42
Tobacco	*Nicotiana tabacum*	48
Tomato	*Lycopersicon esculentum*	24
Tulip	*Tulipa gesneriana*	24
Turmeric	*Curcuma longa*	64
Turnip	*Brassica rapa*	20
Vanilla	*Vanilla planifolia*	32
Violet (African)	*Saintpaulia ionantha*	28
Walnut (English)	*Juglans regia*	32
Watermelon	*Citrullus vulgaris*	22
Wax gourd	*Benincasa hispida*	24
Wheat	*Triticum aestivum*	42
Wild rice	*Zizania aquatica*	30
Winged bean	*Psophocarpus tetragonolobus*	18
Yam	*Dioscorea alata*	20, 30, 40, 50, 60, 70, 80
Zinnia	*Zinnia elegans*	24

Source: Adapted from Darlington, C. D. and A. P. Wylie. 1955. *Chromosome Atlas of Flowering Plants.* George Allen & Unwin Ltd., London.

References

Ahloowalia, B. S. 1969. Effect of temperature and barbiturates on a desynaptic mutant of rye grass. *Mut. Res.* 7: 205–213.

Ahloowalia, B. S. 1982. Plant regeneration from callus culture in wheat. *Crop Sci.* 22: 405–410.

Ahmad, F., S. N. Acharya, Z. Mir, and P. S. Mir. 1999. Localization and activity of rDNA genes on fenugreek (*Trigonella foenum-graecum* L.) chromosomes by fluorescent *in situ* hybridization and silver staining. *Theor. Appl. Genet.* 98: 179–185.

Ahmad, Q. N., E. J. Britten, and D. E. Byth. 1977. Inversion bridges and meiotic behavior in species hybrids of soybeans. *J. Hered.* 68: 360–364.

Ahokas, H. 1977. A mutant in barley: Triploid inducer. *Barley Genet. Newsl.* 7: 4–6.

Albertsen, M. C., and R. L. Phillips. 1981. Developmental cytology of 13 genetic male sterile loci in maize. *Can. J. Genet. Cytol.* 23: 195–208.

Alexander, P. 1969. Differential staining of aborted and nonaborted pollen. *Stain Tech.* 44: 117–122.

Allard, R. W. 1960. *Principles of Plant Breeding.* John Wiley & Sons, Inc. New York.

Allers, T., and M. Lichten. 2001a. Differential timing and control of noncrossover and crossover recombination during meiosis. *Cell* 106: 47–57.

Allers, T., and M. Lichten. 2001b. Intermediates of yeast meiotic recombination contain heteroduplex DNA. *Mol. Cell.* 8: 225–231.

Alonso, L. C., and G. Kimber. 1981. The analysis of meiosis in hybrids. II. Triploid hybrids. *Can. J. Genet. Cytol.* 23: 221–234.

Ambros, P. F., M. A. Matzke, and A. J. M. Matzke. 1986. Detection of a 17 kb unique sequence (T-DNA) in plant chromosomes by *in situ* hybridization. *Chromosoma (Berl.)* 94: 11–18.

Ananiev, E. V., O. Riera-Lizarazu, H. W. Rines, and R. L. Phillips. 1997. Oat-maize chromosome addition lines: A new system for mapping the maize genome. *Proc. Natl. Acad. Sci. U.S.A.* 94: 3524–3529.

Andrews, G. Y., and R. C. McGinnis. 1964. The artificial induction of aneuploids in *Avena. Can. J. Genet. Cytol.* 6: 349–356.

Aragão, F. J. L., L. M. G. Barros, A. C. M. Brasileiro, S. G. Ribeiro, F. D. Smith, J. C. Sanford, J. C. Faria, and E. L. Rech. 1996. Inheritance of foreign genes in transgenic bean (*Phaseolus vulgaris* L.) co-transformed via particle bombardment. *Theor. Appl. Genet.* 93: 142–150.

Aragão, F. J. L., L. Sarokin, G. R. Vianna, and E. L. Rech. 2000. Selection of transgenic meristematic cells utilizing a herbicidal molecule results in the recovery of fertile transgenic soybean [*Glycine max* (L.) Merril] plants at a high frequency. *Theor. Appl. Genet.* 101: 1–6.

Armstrong, K. C. 1981. The evolution of *Bromus inermis* and related species of *Bromus* Sect. *Pnigma. Bot. Jahrb. Syst.* 102: 427–443.

Armstrong, K. C., C. Nakamura, and W. A. Keller. 1983. Karyotype instability in tissue culture regenerants of triticale (x *Triticosecale* Wittmack) cv. 'Welsh' from 6-month-old callus cultures. *Z. Pflanzenzüchtg.* 91: 233–245.

Arora, O. P., and T. N. Khoshoo. 1969. Primary trisomics in Moss Verbena. *Euphytica* 18: 237–248.

Arthur, L., P. Ozias-Akins, and W. W. Hanna. 1993. Female sterile mutant in pearl millet: Evidence for initiation of apospory. *J. Hered.* 84: 112–115.

Arumuganathan, K., and E. D. Earle. 1991. Estimation of nuclear DNA content of plants by flow cytometry. *Plant Mol. Biol. Rep.* 9: 229–233.

Asami, H., T. Shimada, N. Inomata, and M. Okamoto. 1975. Chromosome constitution in cultured callus cells from four aneuploid lines of the homoeologous group 5 of *Triticum aestivum. Jpn. J. Genet.* 50: 283–289.

Ashmore, S. E., and A. R. Gould. 1981. Karyotypic evolution in a tumour derived plant tissue culture analysed by Giemsa C-banding. *Protoplasma* 106: 297–308.

Ashraf, M., and M. J. Bassett. 1987. Five primary trisomics from translocation heterozygote progenies in common bean, *Phaseolus vulgaris* L. *Theor. Appl. Genet.* 74: 346–360.

Asker, S. E., and L. Jerling. 1992. *Apomixis in Plants.* CRC Press, Inc., Boca Raton, FL.

Attia, T., and G. Röbbelen. 1986a. Cytogenetic relationship within cultivated *Brassica* analyzed in amphihaploids from the three diploid ancestors. *Can. J. Genet. Cytol.* 28: 323–329.

Attia, T., and G. Röbbelen. 1986b. Meiotic pairing in haploids and amphidiploids of spontaneous versus synthetic origin in rape, *Brassica napus* L. *Can. J. Genet. Cytol.* 28: 330–334.

Aung, T., and H. Thomas. 1978. The structure and breeding behaviour of a translocation involving the transfer of mildew resistance from *Avena barbata* Pott. into the cultivated oat. *Euphytica* 27: 731–739.

Austin, S., E. T. Bingham, D. E. Mathews, M. N. Shahan, J. Will, and R. R. Burgess. 1995. Production and field performance of transgenic alfalfa (*Medicago sativa* L.) expressing alpha-amylase and manganese-dependent lignin peroxidase. *Euphytica* 85: 381–393.

Avanzi, S., and P. L. Deri. 1969. Duration of the mitotic cycle in two cultivars of *Triticum durum*, as measured by ³H-thymidine labelling. *Caryologia* 22: 187–194.

Avery, A. G., S. Satina, and J. Rietsema. 1959. *Blakeslee: The genus Datura*. Ronald Press, New York, p. 289.

Avery, P. 1929. Chromosome number and morphology in *Nicotiana*. IV. The nature and effects of chromosomal irregularities in *N. alata* var. *grandiflora*. *Univ. of Calif. Publ. Bot.* 11: 265–284.

Avivi, L., and M. Feldman. 1980. Arrangement of chromosomes in the interphase nucleus of plants. *Hum. Genet.* 55: 281–295.

Awoleye, F., M. van Duren, J. Doležel, and F. J. Novak. 1994. Nuclear DNA content and *in vitro* induced somatic polyploidization cassava (*Manihot esculenta* Crantz) breeding. *Euphytica* 76: 195–202.

Ayonoadu, U. W., and H. Rees. 1968. The regulation of mitosis by B-chromosomes in rye. *Exp. Cell Res.* 52: 284–290.

Babcock, E. B., and M. Navashin. 1930. The genus *Crepis*. *Bibliogr. Genet.* 6: 1–90.

Bailey, M. A., H. R. Boerma, and W. A. Parrott. 1993. Genotype—Specific optimization of plant regeneration from somatic embryos of soybean. *Plant Sci.* 93: 117–120.

Bajaj, Y. P. S., and M. S. Gill. 1985. In vitro induction of genetic variability in cotton (*Gossypium* spp.). *Theor. Appl. Genet.* 70: 363–368.

Baker, B. S., A. T. C. Carpenter, M. S. Esposito, R. E. Esposito, and L. Sandler. 1976. The genetic control of meiosis. *Ann. Rev. Genet.* 10: 53–134.

Baker, E. P., and R. A. McIintosh. 1966. Chromosome translocations identified in varieties of common wheat. *Can. J. Genet. Cytol.* 8: 592–599.

Baker, R. L., and D. T. Morgan, Jr. 1966. Monosomics in maize induced by X-irradiation of the pollen. *Cytologia* 31: 172–175.

Balog, C. 1979. Studies on triploid *Allium triquetrum*. II. Metaphase I univalents and their influence on Anaphase I distribution. *Chromosoma (Berlin)* 73: 191–204.

Balog, C. 1984. Features of meiosis in triploid *Allium triquetrum*. *Cytologia* 49: 95–103.

Balzan, R. 1978. Karyotype instability in tissue cultures derived from the mesocotyl of *Zea mays* seedlings. *Caryologia* 31: 75–87.

Baranyi, M., and J. Greilhuber. 1995. Flow cytometric analysis of genome size variation in cultivated and wild *Pisum sativum* (Fabaceae). *Pl. Syst. Evol.* 194: 231–239.

Baranyi, M., and J. Greilhuber. 1996. Flow cytometric and Feulgen densitometric analysis of genome size variation in *Pisum*. *Theor. Appl. Genet.* 92: 297–307.

Barceló, P., and A. Martin. 1990. The identification of faba bean (*Vicia faba* L.) trisomics by translocation tester set. *Euphytica* 47: 45–48.

Barthes, L., and A. Ricroch. 2001. Interspecific chromosomal rearrangements in monosomic addition lines of *Allium*. Genome 44: 929–935.

Barton, D. W. 1950. Pachytene morphology of the tomato chromosome complement. *Amer. J. Bot.* 37: 639–643.

Barwale, U. B., M. M. Meyer Jr., and J. M. Widholm. 1986. Screening of *Glycine max* and *Glycine soja* genotypes for multiple shoot formation at the cotyledonary node. *Theor. Appl. Genet.* 72: 423–428.

Bashaw, E. C. 1980. Apomixis and its application in crop improvement. In *Hybridization of Crop Plants*. W. H. Fehr and H. H. Hadley, Eds., American Society of Agronomy/Crop Science of America, Maddison, WI. pp. 45–63.

Battaglia, E. 1955. Chromosome morphology and terminology. *Caryologia* 8: 179–187.

Battaglia, E. 1994. Nucleosome and nucleotype: A terminological criticism. *Caryologia* 47: 193–197.

Baum, B. R., J. R. Estes, and P. K. Gupta. 1987. Assessment of the genomic system of classification in the Triticeae. *Amer. J. Bot.* 74: 1388–1395.

Baum, B. R., and G. Fedak. 1985. *Avena atlantica*, new diploid species of the oat genus from Morocco. *Can. J. Bot.* 63: 1057–1060.

Bayliss, M. W. 1973. Origin of chromosome number variation in cultured plant cells. *Nature (London)* 246: 529–530.

Bayliss, M. W. 1975. The effects of growth *in vitro* on the chromosome complement of *Daucus carota* (L.) suspension cultures. *Chromosoma (Berl.)* 51: 401–411.

Bayliss, M. W. 1980. Chromosomal variation in plant tissues in culture. *Int. Rev. Cytol. (Suppl.)* 11A: 113–144.

Beadle, G. W. 1929. A gene for supernumerary mitosis during spore development in *Zea mays*. *Science* 70: 406–407.

Beadle, G. W. 1930. Genetical and cytological studies of Mendelian asynapsis in *Zea mays*. *Cornell Univ. Agric. Exp. Stn. Mem.* 129: 1–23.

Beadle, G. W. 1931. A gene in maize for supernumerary cell divisions following meiosis. *Cornell Univ. Agric. Exp. Stn. Mem.* 129: 3–12.

Beadle, G. W. 1932. A gene in *Zea mays* for the failure of cytokinesis during meiosis. *Cytologia* 3: 142–155.

Beadle, G. W. 1933. Further studies for asynaptic maize. *Cytologia* 4: 269–287.

Beadle, G. W. 1937. Chromosome aberration and gene mutation in *sticky* chromosome plants of *Zea mays*. *Cytologia* (Fujii Jubilee vol.) 1937: 43–56.

Beadle, G. W., and B. McClintock. 1928. A genic disturbance of meiosis in *Zea mays*. *Science* 68: 433.

Beasley, J. O. 1940. The production of polyploids in *Gossypium*. *J. Hered.* 31: 39–48.

Beasley, J. O. 1942. Meiotic chromosome behavior in species, species hybrids, haploids, and induced polyploids of *Gossypium*. *Genetics* 27: 25–54.

Beasley, J. O., and M. S. Brown. 1942. Asynaptic *Gossypium* plants and their polyploids. *J. Agric. Res.* 65: 421–427.

Bebeli, P., A. Karp, and P. J. Kaltsikes. 1988. Plant regeneration and somaclonal variation from cultured immature embryos of sister lines of rye and triticale differing in their content of heterochromatin. 1. Morphogenetic response. *Theor. Appl. Genet.* 75: 929–936.

Beckett, J. B. 1978. B-A translocations in maize. I. Use in locating genes by chromosome arms. *J. Hered.* 69: 27–36.

Bedbrook, J. R., J. Jones, M. O'Dell, R. J. Thompson, and R. B. Flavell 1980. A molecular description of telomeric heterochromatin in *Secale* species. *Cell* 19: 545–560.

Belling, J. 1921. The behavior of homologous chromosomes in a triploid *Canna*. *Proc. Natl. Acad. Sci. U.S.A.* 7: 197–201.

Belling, J. 1925. Homologous and similar chromosomes in diploid and triploid *Hyacinths*. *Genetics* 10: 59–71.

Belling, J., and A. F. Blakeslee. 1922. The assortment of chromosomes in triploid in *Daturas*. *Amer. Nat.* 56: 339–346.

Belling, J., and A. F. Blakeslee. 1924. The configurations and sizes of the chromosomes in the trivalents of 25-chromosome *Daturas*. *Proc. Nat. Acad. Sci. U.S.A.* 10: 116–120.

Belling, J., and A. F. Blakeslee. 1926. On the attachment of non-homologous chromosomes at the reduction division in certain 25-chromosome *Daturas*. *Proc. Natl. Acad. Sci. U.S.A.* 12: 7–11.

Benavente, E., and J. Orellana. 1984. Meiotic pairing of specific chromosome arms in triploid rye. *Can. J. Genet. Cytol.* 26: 717–722.

Bender, K., and H. Gaul. 1966. Zur frage der diploidisierung autotetraploider gerste. *Z. Pflanzenzüchtg.* 56: 164–183 (in German with English summary).

Benito, C., F. J. Gallego, C. Zaragoza, J. M. Frade, and A. M. Figueiras. 1991a. Biochemical evidence of a translocation between 6RL/7RL chromosome arms in rye (*Secale cereale* L.). A genetic map of 6R chromosome. *Theor. Appl. Genet.* 82: 27–32.

Benito, C., C. Zaragoza, F. J. Gallego, A. de la Peña, and A. M. Figueiras. 1991b. A map of rye chromosome 2R using isozyme and morphological markers. *Theor. Appl. Genet.* 82: 112–116.

Bennett, M. D. 1971. The duration of meiosis. *Proc. Roy. Soc. Lond., Ser. B.* 178: 277–299.

Bennett, M. D., P. Bhandol, and I. J. Leitch. 2000. Nuclear DNA amounts in angiosperms and their modern uses-807 new estimates. *Ann. Bot.* 86: 859–909.

Bennett, M. D., R. A. Finch, and I. R. Barclay. 1976. The time rate and mechanism of chromosome elimination in *Hordeum* hybrids. *Chromosoma* (Berlin) 54: 175–200.

Bennett, M. D., J. P. Gustafson, and J. B. Smith. 1977. Variation in nuclear DNA in the genus *Secale*. *Chromosoma* (Berlin) 61: 149–176.

Bennett, M. D., S. Johnston, G. L. Hodnett, and H. J. Price. 2000. *Allium cepa* L. cultivars from four continents compared by flow cytometry show nuclear DNA constancy. *Ann. Bot.* 85: 351–357.

Bennett, M. D., and I. J. Leitch. 1995. Nuclear DNA amounts in angiosperms. *Ann. Bot.* 76: 113–176.

Bennett, M. D., and I. J. Leitch. 1997. Nuclear DNA amounts in angiosperms–583 new estimates. *Ann. Bot.* 80: 169–196.

Bennett, M. D., and I. J. Leitch. 2001. Nuclear DNA amounts in pteridophytes. *Ann. Bot.* 87: 335–345.

Bennett, M. D., I. J. Leitch, and L. Hanson. 1998. DNA amounts in two samples of angiosperm weeds. *Ann. Bot.* 82 (Suppl A): 121–134.

Bennett, M. D., and H. Rees. 1970. Induced variation in chiasma frequency in rye in response to phosphate treatments. *Genet. Res.* 16: 325–331.

Bennici, A. 1974. Cytological analysis of roots, shoots and plants regenerated from suspension and solid *in vitro* cultures of haploid *Pelargonium*. *Z. Pflanzenzüchtg.* 72: 199–205.

Bennici, A., and F. D'Amato. 1978. *In vitro* regeneration of *Durum* wheat plants. 1. Chromosome numbers of regenerated plantlets. *Z. Pflanzenzüchtg.* 81: 305–311.

Benzion, G., and R. L. Phillips. 1988. Cytogenetic stability of maize tissue cultures: A cell line pedigree analysis. *Genome* 30: 318–325.

Berger, C. A., and E. R. Witkus. 1948. Cytological effects of alphanaphthalene acetic acid. *J. Hered.* 39: 117–120.

Berger, X. 1953. Untersuchungen über die embryologie partiell apomoktischer *Rubus* bastarde. *Ber Schweiz Bot. Ges.* 63: 224–266.

Bergey, D. R., D. M. Stelly, H. J. Price, and T. D. McKnight. 1989. *In situ* hybridization of biotinylated DNA probes to cotton meiotic chromosomes. *Stain Tech.* 64: 25–37.

Bergner, A. D., A. G. Avery, and A. F. Blakeslee. 1940. Chromosomal deficiencies in *Datura stramonium* induced by colchicine treatment. *Amer. J. Bot.* 27: 676–683.

Bergner, A. D., J. L. Cartledge, and A. F. Blakeslee. 1934. Chromosome behavior due to a gene which prevents metaphase pairing in *Datura*. *Cytologia* 6: 19–37.

Bernard, M. 1976. Étude des caractéristiques cytologiques, morphologiques et agronomiques de six lignées D' addition Blé-seigle. *Ann. Amélior. Plantes* 26: 67–91 (in French with English summary).

Bharathan, G., G. Lambert, and D. W. Galbraith. 1994. Nuclear DNA content of monocotyledons and related taxa. *Amer. J. Bot.* 81: 381–386.

Bhojwani, S. S., P. K. Evans, and E. C. Cocking. 1977. Protoplast technology in relation to crop plants: Progress and problems. *Euphytica* 26: 343–360.

Bicknell, R. A. 1999. Who will benefit from apomixis? *Biotechnol. Development. Monitor* 37: 17–21.

Bicknell, R. A., N. K. Borst, and A. M. Koltunow. 2000. Monogenic inheritance of apomixis in two *Hieracium* species with distinct developmental mechanisms. *Heredity* 84: 228–237.

Binding, H. 1976. Somatic hybridization experiments in *Solanaceous* species. *Molec. Gen. Genet.* 144: 171–175.

Binding, H., and R. Nehls. 1978. Somatic cell hybridization of *Vicia faba* + *Petunia hybrida*. *Molec. Gen. Genet.* 164: 137–143.

Blakeslee, A. F. 1921. Types of mutations and their possible significance in evolution. *Amer. Nat.* 55: 254–267.

Blakeslee, A. F. 1924. Distinction between primary and secondary chromosomal mutants in *Datura*. *Proc. Nat. Acad. Sci. U.S.A.* 10: 109–116.

Blakeslee, A. F. 1927. The chromosomal constitution of Nubbin, a compound (2n + 1) type in *Datura*. *Proc. Nat. Acad. Sci. U.S.A.* 13: 79–85.

Blakeslee, A. F. 1939. The present and potential service of chemistry to plant breeding. *Amer. J. Bot.* 26: 163–172.

Blakeslee, A. F., and A. G. Avery. 1937. Methods of inducing doubling of chromosomes in plants by treatment with colchicine. *J. Hered.* 28: 393–411.

Blakeslee, A. F., and A. G. Avery. 1938. Fifteen-year breeding records of 2n + 1 types in *Datura stramonium*. Cooperation in Research, Carnegie Inst. Washington Pub. No. 501, pp. 315–351.

Blakeslee, A. F., and J. Belling. 1924a. Chromosomal chimeras in the Jimson weed. *Science* 60: 19–20.

Blakeslee, A. F., and J. Belling. 1924b. Chromosomal mutations in the Jimson weed, *Datura stramonium*. *J. Hered.* 15: 195–206.

Blakey, C. A., S. L. Goldman, and C. L. Dewald. 2001. Apomixis in *Tripsacum*: Comparative mapping of a multigene phenomenon. *Genome* 44: 222–230.

Blanco, A., R. Simeone, and P. Resta. 1987. The addition of *Dasypyrum villosum* (L.) candargy chromosomes to *durum* wheat (*Triticum durum* Desf.). *Theor. Appl. Genet.* 74: 328–333.

Blixt, S. 1975. Why didn't Gregor Mendel find linkage? *Nature* 256: 206.

Bothmer, R. von., J. Flink, and T. Landström. 1986. Meiosis in interspecific *Hordeum* hybrids. I. Diploid combinations. *Can. J. Genet. Cytol.* 28: 525–535.

Bouharmont, J., M. Olivier, and M. Dumonte de Chassart. 1985. Cytological observations in some hybrids between the rice species *Oryza sativa* L. and *O. glaberrima* Steud. *Euphytica* 34: 75–81.

Boveri, T. 1904. *Ergebnisse uber die Konstitution der chromatischen Substanz des Zellkerns*. Fischer, Jena.

Boyes, C. J., and I. K. Vasil. 1984. Plant regeneration by somatic embryogenesis from cultured young inflorescences of *Sorghum arundinaceum* (DESV.) Stapf. Var. *Sudanense* (Sudan grass). *Plant Sci. Lett.* 35: 153–157.

Brandle, J. E., S. G. McHugh, L. James, H. Labbé, and B. L. Miki. 1995. Instability of transgene expression in field grown tobacco carrying the *csr1-1* gene for sulfonylurea herbicide resistance. *Bio/Technol.* 13: 994–998.

Brar, D. S., and G. S. Khush. 1997. Alien introgression in rice. *Plant Molec. Biol.* 35: 35–47.

Braver, G. and J. L. Blount. 1950. Somatic elimination of ring chromosomes in *Drosophila melanogaster*. *Genetics* 35: 98.

Bregitzer, P., S. Zhang, M.-J. Cho, and P. G. Lemaux. 2002. Reduced somaclonal variation in barley is associated with culturing highly differentiated, meristematic tissues. *Crop Sci.* 42: 1303–1308.

Breiman, A., D. Rotem-Abarbanell, A. Karp, and H. Shaskin. 1987. Heritable somaclonal variation in wild barley (*Hordeum spontaneum*). *Theor. Appl. Genet.* 74: 104–112.

Bridges, C. B. 1921. Genetical and cytological proof of non-disjunction of the fourth chromosome of *Drosophila melanogaster*. *Proc. Nat. Acad. Sci. U. S A.* 7: 186–192.

Bridges, C. B. 1922. The origin and variation in sexual and sex-limited characters. *Amer. Nat.* 56: 51–63.

Browers, M. A., and T. J. Orton. 1982a. A factorial study of chromosomal variability in callus cultures of celery (*Apium graveolens*). *Plant Sci. Lett.* 26: 65–73.

Browers, M. A., and T. J. Orton. 1982b. Transmission of gross chromosomal variability from suspension cultures into regenerated celery plants. *J. Hered.* 73: 159–162.

Brown, A. H. D. 1990. The role of isozyme studies in molecular systematics. *Aust. Syst. Bot.* 3: 39–46.

Brown, M. S. 1951. The spontaneous occurrence of amphiploidy in species hybrids of *Gossypium*. *Evolution* 5: 25–41.

Brown, M. S. 1958. The division of univalent chromosomes in *Gossypium*. *Amer. J. Bot.* 45: 24–32.

Brown, M. S. 1980. Identification of the chromosomes of *Gossypium hirsutum* L. by means of tanslocations. *J. Hered.* 71: 266–274.

Brown, M. S., and J. E. Endrizzi. 1964. The origin, fertility and transmission of monosomics in *Gossypium*. *Amer. J. Bot.* 51: 108–115.

Brown, S. W., and D. Zohary. 1955. The relationship of chiasmata and crossing over in *Lilium formosanum*. *Genetics* 40: 850–873.

Brown, W. V., and S. M. Stack. 1968. Somatic pairing as a regular preliminary to meiosis. *Bull. Torrey Bot. Club.* 95: 369–378.

Brthes, L., and A. Ricroch. 2001. Interspecific chromosomal rearrangements in monosomic addition lines of *Allium. Genome* 44: 929–935

Brubaker, C. L., A. H. D. Brown, A. McD. Stewart, M. J. Kilby, and J. P. Grace. 1999. Production of fertile hybrid germplasm with diploid Australian *Gossypium* species for cotton improvement. *Euphytica* 108: 199–213.

Buchholz, J. T., and A. F. Blakeslee. 1922. Studies of the pollen tubes and abortive ovules of the globe mutant of *Datura. Science* 55: 597–599.

Buiatti, M., G. Marcheschi, F. Tognoni, M. Lipuccidipaola, F. Collina Grenci, and G. Martini. 1985. Genetic variability induced by tissue culture in the tomato (*Lycopersicon esculentum*). *Z. Pflanzenzüchtg.* 94: 162–165.

Burnham, C. R. 1930. Genetical and cytological studies of semisterility and related phenomena in maize. *Proc. Nat. Acad. Sci. U.S.A.* 16: 269–277.

Burnham, C. R. 1934. Cytogenetic studies of an interchange between chromosomes 8 and 9 in maize. *Genetics* 19: 430–447.

Burnham, C. R. 1946. A gene for "long" chromosomes in barley. *Genetics* 31: 212–213.

Burnham, C. R. 1954. Tester set of translocations. *Maize Genet. Coop. Newsl.* 28: 59–60.

Burnham, C. R. 1956. Chromosomal interchanges in plants. *Bot. Rev.* 22: 419–552.

Burnham, C. R. 1962. *Discussion in Cytogenetics*. Burgess, Minneapolis, MN.

Burnham, C. R., and A. Hagberg. 1956. Cytogenetic notes on chromosomal interchanges in barley. *Hereditas* 42: 467–482.

Burnham, C. R., J. T. Stout, W. H. Weinheimer, R. V. Kowles, and R. L. Phillips. 1972. Chromosome pairing in maize. *Genetics* 71: 111–126.

Burnham, C. R., F. H. White, and R. W. Livers. 1954. Chromosomal interchanges in barley. *Cytologia* 19: 191–202.

Burns, G. W. 1969. *The Science of Genetics. An Introduction to Heredity*. The Macmillan Company, Collier-Macmillan Limited, London.

Burson, B. L. 1981a. Cytogenetic relationships between *Paspalum jurgensii* and *P. intermedium*, *P. vaginatum*, and *P. setaceum* var. *ciliatifolium*. *Crop Sci.* 21: 515–519.

Burson, B. L. 1981b. Genome relations among four diploid *Paspalum* species. *Bot. Gaz.* 142: 592–596.

Cabrera, A., M. C. Remirez, and A. Martin. 1999. Application of C-banding and fluorescence *in situ* hybridization for the identification of the trisomics of *Hordeum chilense*. *Euphytica* 109: 123–129.

Cabrero, J., and J. P. M. Camacho. 1982. Pericentric inversion polymorphism in *Aiolopus strepens* (Orthoptera: Acridiae): Effects of chiasma formation. *Caryologia* 35: 411–424.

Cameron, D. R., and R. (Milstein) Moav. 1957. Inheritance in *Nicotiana tabacum* XXVII. Pollen killer, an alien genetic locus inducing abortion of microspores not carrying it. *Genetics* 42: 326–335.

Carlson, P. S., H. H. Smith, and R. D. Dearing. 1972. Parasexual interspecific plant hybridization. *Proc. Nat. Acad. Sci. U.S.A.* 69: 2292–2294.

Carlson, W. R. 1983. Duplication of non-terminal A chromosome segments using B-A translocations. *Maydica* 28: 317–326.

Carlson, W. R. 1986. The B chromosome of maize. *CRC Crit. Rev. Plant Sci.* 3: 201–226.

Carlson, W. R. 1988. The cytogenetics of corn. In *Corn and Corn Improvement*. Agronomy monograph no. 18, 3rd ed., G. F. Sprague and J. W. Dudley, Eds., American Society of Agronomy/Crop Science Society of America/Soil Science Society of America, Madison, WI, pp. 259–343.

Carlson, W. R., and C. Curtis. 1986. A new method for producing homozygous duplications in maize. *Can. J. Genet. Cytol.* 28: 1034–1040.

Carlson, W. R., and R. R. Roseman. 1991. Segmental duplication of distal chromosomal regions in maize. *Genome* 34: 537–542.

Carvalho, C. R. D., and L. S. Saraiva. 1993. A new heterochromatin banding pattern revealed by modified HKG banding technique in maize chromosomes. *Heredity* 70: 515–519.

Casperson, T., G. Lomakka, and A. Moller. 1971. Computerized chromosome identification by aid of the quinacrin mustard fluorescence technique. *Heretitas* 67: 103–110.

Castleman, K. R., and J. H. Melnyk. 1976. An automated system for chromosome analysis: Final report. JPL document 5040-30. JET Propulsion Laboratory, California Inst. Tech., Pasadena, CA.

Catcheside, D. G. 1954. The genetics of *Brevistylis* in *Oenothera*. *Heredity* 8: 125–137.

Ceoloni, C., G. Del Signore, M. Pasquini, and A. Testa. 1988. Transfer of mildew resistance from *Triticum longissimum* in to wheat by *ph1* induced homoeologous recombination. In *Proc. 7th Int. Wheat Genet. Symp.* T. E. Miller and R. M. D. Koebner, Eds., Bath Press, Avon, U.K. pp. 221–226.

Chaleff, R. S., and M. F. Parsons. 1978. Direct selection *in vitro* for herbicide-resistant mutants of *Nicotiana tabacum*. *Proc. Natl. Acad. Sci. U.S.A.* 75: 5104–5107.

Chandler, J. M., C. C. Jan, and B. H. Beard. 1986. Chromosomal differentiation among the annual *Helianthus* species. *Syst. Bot.* 11: 354–371.

Chang, T. D., and K. Sadanaga. 1964. Breeding behavior, morphology, Karyotype and interesting results of six monosomes in *Avena sativa* L. *Crop Sci.* 4: 609–615.

Chao, S., J. M. Gardiner, S. Melia-Hancock, and E. H. Coe, Jr. 1996. Physical and genetic mapping of chromosome *9S* in maize using mutations with terminal deficiencies. *Genetics* 143: 1785–1794.

Chapman, G. P. 1992. Apomixis and evolution. In *Grass Evolution and Domestication*. G. P. Chapman, Ed., Cambridge University Press, London, pp. 138–155.

Chapman, G. P., and N. Busri. 1994. Apomixis in *Pennisetum*: An ultrastructural study. *Int. J. Plant Sci.* 155: 492–497.

Chapman, G. P., and W. E. Peat. 1992. *An Introduction to the Grasses (Including Bamboos and Cereals)*. C. A. B. International. Redwood Press Ltd., Melksham, U.K., p. 110.

Chapman, V., T. E. Miller, and R. Riley. 1976. Equivalence of the 4A genome of bread wheat with that of *Triticum urartu*. *Genet. Res. Camb.* 27: 69–76.

Chapman, V., and R. Riley. 1966. The allocation of the chromosomes of *Triticum aestivum* to the A and B genomes and evidence on genome structure. *Can. J. Genet. Cytol.* 8: 57–63.

Chaudhury, A. M., L. Ming, C. Miller, S. Craig, E. S. Dennis, and J. W. Peacock. 1997. Fertilization-independent seed development in *Arabidopsis thaliana*. *Proc. Natl. Acad. Sci. U.S.A.* 94: 4223–4228.

Chauhan, K. P. S., and W. O. Abel. 1968. Evidence for the association of homologous chromosomes during premeiotic stages in *Impatiens* and *Salvia*. *Chromosoma* (Berlin) 25: 297–302.

Chauhan, S. V. S., and T. Kinoshita. 1979. Histochemical localization of histones, DNA and proteins in the anthers of male-fertile and male-sterile plants. *Jpn. J. Breed.* 29: 287–293.

Chaven, C., T. Hymowitz, and C. A. Newell. 1982. Chromosome number, oil and fatty acid content of species in the genus *Glycine* subgenus *Glycine*. *J. Am. Oil Chem. Soc.* 59: 23–25.

Chen, B. Y., B. F. Cheng, R. B. Jørgensen, and W. K. Heneen. 1997. Production and cytogenetics of *Brassica campestris-alboglabra* chromosome addition lines. *Theor. Appl. Genet.* 94: 633–640.

Chen, C.-C., and C.-M. Chen. 1980. Changes in chromosome number in microspore callus of rice during successive subcultures. *Can. J. Genet. Cytol.* 22: 607–617.

Chen, C. C., and W. F. Grant. 1968a. Morphological and cytological identification of the primary trisomics of *Lotus pedunculatus* (Leguminosae). *Can. J. Genet. Cytol.* 10: 161–179.

Chen, C. C., and W. F. Grant. 1968b. Trisomic transmission in *Lotus pedunculatus*. *Can. J. Genet. Cytol.* 10: 648–654.

Chen, L. F., and R. G. Palmer. 1985. Cytological studies of tripoloids and their progeny from male-sterile *ms1* soybean. *Theor. Appl. Genet.* 71: 400–407.

Chen, N. C., L. R. Baker, and S. Honma. 1983. Interspecific crossability among four species of *Vigna* food legumes. *Euphytica* 32: 925–937.

Chen, P. D., and B. S. Gill. 1984. The origin of 4A and genomes B and G of tetraploid wheats. In *Proc. 6th Int. Wheat Genet. Symp.* S. Sakamoto, Ed., Kyoto University, Kyoto, Japan, pp. 39–48.

Chen, Q., B. Friebe, R. L. Conner, L. Laroche, J. B. Thomas, and B. S. Gill. 1998. Molecular cytogenetic characterization of *Thinopyrum intermedium*-derived wheat germplasm specifying resistance to wheat streak mosaic virus. *Theor. Appl. Genet.* 96: 1–7.

Cheng, Z., H. Yan, H. Xu, S. Tang, J. Jiang, M. Gu, and L. Zhu. 2001. Development and application of a complete set of rice telotrisomics. *Genetics* 157: 361–368.

Chennaveeraiah, M. S., and S. C. Hiremath. 1974. Genome analysis of *Eleusine coracana* (L.) Gaertn. *Euphytica* 23: 489–495.

Cherry, J. P., F. R. H. Kotterman, and J. E. Endrizzi. 1970. Comparative studies of seed proteins of species of *Gossypium* by gel electrophoresis. *Evolution* 24: 431–447.

Chetelat, R. T., P. Cisneros, L. Stamova, and C. M. Rick. 1997. A male-fertile *Lycopersicon esculantum* x *Solanum lycopersicoides* hybrids enables direct backcrossing to tomato at the diploid level. *Euphytica* 95: 99–108.

Chetelat, R. T., C. M. Rick, P. Cisneros, K. B. Alpert, and J. W. DeVerna. 1998. Identification, transmission, and cytological behavior of *Solanum lycopersicodes* Dun. monosomic alien addition lines in tomato (*Lycopersicon esculentum* Mill.). *Genome* 41: 40–50.

Chetelat, R. T., C. M. Rick, and J. W. DeVerna. 1989. Isozyme analysis, chromosome pairing, and fertility of *Lycopersicon esculentum* x *Solanum lycopersicoides* diploid backcross hybrids. *Genome* 32: 783–790.

Chévre, A. M., F. Eber, P. Barret, P. Dupuy, and J. Brace. 1997. Identification of the different *Brassica nigra* chromosomes from both sets of *B. oleracea–B. nigra* and *B. napus–B. nigra* addition lines with a special emphasis on chromosome transmission and self-incompatibility. *Theor. Appl. Genet.* 94: 603–611.

Chevre, A. M., F., Eber, P. This, P. Barret, X. Tangey, H. Brun, M. Delseny, and M. Renard. 1996. Characterization of *Brassica nigra* chromosomes and of blackleg resistance in *B. napus–B. nigra* addition lines. *Plant Breed.* 115: 113–118.

Chevre, A. M., P. This, F. Eber, M. Deschamps, M. Renard, M. Delseny, and C. F. Quiros. 1991. Characterization of disomic addition lines *Brassica napus—Brassica nigra* by isozyme, fatty acid, and RFLP markers. *Theor. Appl. Genet.* 81: 43–49.

Chirsten, H. R. 1952. Die embryologie von *Rubus idaeus* und von bastarden zwischen *Rubus caesius* und *Rubus idaes? Z. Indukt. Abstammungs Vererbungslehre* 84: 454–461.

Cho, M.-J., H. W. Choi, B. B. Buchanan, and P. G. Lemaux. 1999. Inheritance of tissue-specific expression of barley hordein promoter-*uidA* fusions in transgenic barley plants. *Theor. Appl. Genet.* 98: 1253–1262.

Cho, M.-J., H. W. Choi, W. Jiang, C. D. Ha, and P. G. Lemaux. 2002. Endosperm-specific expression of green fluorescent protein driven by the hordein promoter is stably inherited in transgenic barley (*Hordeum vulgare*) plants. *Physiologia Plant.* 115: 144–154.

Chochran, D. G. 1983. Alternate-2 disjunction in the German Cockroach. *Genetics* 104: 215–217.

Choi, H. W., P. G. Lemaux, and M.-J. Cho. 2000a. High frequency of cytogenetics aberration in transgenic oat (*Avena sativa* L.) plants. *Plant Sci.* 156: 85–94.

Choi, H. W., P. G. Lemaux, and M.-J. Cho. 2000b. Increased chromosomal variation in transgenic versus nontransgenic barley (*Hordeum vulgare* L.) plants. *Crop Sci.* 40: 524–533.

Choi, H. W., P. G. Lemaux, and M.-J. Cho. 2001. Selection and osmotic treatment exacerbate cytological aberrations in transformed barley (*Hordeum vulgare*). *J. Plant Phsiol.* 158: 935–943.

Choi, H. W., P. G. Lemaux, and M.-J. Cho. 2002. Use of fluorescence *in situ* hybridization for gross mapping of transgenes and screening for homozygous plants in transgenic barley (*Hordeum vulgare* L.). *Theor. Appl. Genet.* 106: 92–100.

Choo, T. M., E. Reinbergs, and K. J. Kasha. 1985. Use of haploids in breeding barley. In *Plant Breed. Rev. 3.* J. Janick, Ed., AVI Publishing, Westport, CT, pp. 219–252.

Choudhuri, H. C. 1969. Late DNA replication pattern in sex chromosomes of *Melandrium. Can. J. Genet. Cytol.* 11: 192–198.

Chow, C. and E. N. Larter. 1981. Centromeric banding in maize. *Can. J. Genet. Cytol.* 23: 255–258.

Christey, M. C., and B. K. Sinclair. 1992. Regeneration of transgenic kale (*Brassica oleracea* var. *acephala*), rape (*B. napus*) and turnip (*B. campestris* var. *rapifera*) plants via *Agrobacterium rhizogenes* mediated transformation. *Plant Sci.* 87: 161–169.

Christianson, M. L., D. A. Warnick, and P. S. Carlson. 1983. A morphogenetically competent soybean suspension culture. *Science* 222: 632–634.

Christou, P. 1997. Biotechnology applied to grain legumes. *Field Crops Res.* 53: 83–97.

Church, K. 1974. The effect of supernumerary heterochromatic chromosome segments on non-homologous chromosome associations in the grasshopper *Camula pellucida*. *Heredity* 33: 151–158.

Clark, F. J. 1940. Cytogenetic studies of divergent meiotic spindle formation in *Zea mays*. *Amer. J. Bot.* 27: 547–559.

Clausen, R. E. 1932. Interspecific hybridization in *Nicotiana* XIII. Further data as to the origin and constitution of *Nicotiana tabacum*. *Svensk Bot. Tidskr.* 26: 123–136.

Clausen, R. E., and D. R. Cameron. 1944. Inheritance in *Nicotiana tabacum*. XVIII. Monosomic analysis. *Genetics* 29: 447–477.

Clayberg, C. D. 1959. Cytogenetic studies of precocious meiotic centromere division in *Lycopersicon esculentum* Mill. *Genetics* 44: 1335–1346.

Cleland, R. E. 1922. The reduction divisions in the pollen mother cells of *Oenothera franciscana*. *Am. J. Bot.* 9: 391–413.

Cleland, R. E. 1962. The cytogenetics of *Oenothera*. *Adv. Genet.* 11: 147–237.

Clemente, T. E., B. J. LaVallee, A. R. Howe, D. Conner-Ward, R. J. Rozman, P. E. Hunter, D. L. Broyles, D. S. Kasten, and M. A. Hinchee. 2000. Progeny analysis of glyphosate selected transgenic soybeans derived from *Agrobacterium*-mediated transformation. *Crop Sci.* 40: 797–803.

Coleman, A. W., and L. J. Goff. 1985. Application of fluorochromes to pollen biology. I. Mithramycin and 4′, 6-diamidino-2 phenylindole (DAPI) as vital stains and for quantitation of nuclear DNA. *Stain Tech.* 60: 145–154.

Comings, D. E. 1980. Arrangement of chromatin in the nucleus. *Human Genet.* 53: 131–143.

Conger, A. G., and L. M. Fairchild. 1953. A quick-freeze method for making smear slides permanent. *Stain Tech.* 28: 281–283.

Conner, A. J., M. K. Williams, D. J. Abernethy, P. J. Fletcher, and R. A. Genet. 1994. Field performance of transgenic potatoes. *New Zealand J. Crop Hort. Sci.* 22: 361–371.

Constantin, M. J. 1981. Chromosome instability in cell and tissue cultures and regenerated plants. *Envirn. Exp. Bot.* 21: 359–368.

Contolini, C. S., and M. Y. Menzel. 1987. Early development of duplication-deficiency ovules in upland cotton. *Crop Sci.* 27: 345–348.

Cooper, D. C. 1952. The transfer of deoxyribose nucleic acid from the tapetum to the microsporocytes at the onset of meiosis. *Amer. Nat.* 86: 219–229.

Cooper, L. S., D. C. Cooper, A. C. Hildebrandt, and A. J. Riker. 1964. Chromosome numbers in single cell clones of tobacco tissue. *Am. J. Bot.* 51: 284–290.

Copenhaver, G. P., K. C. Keith, and D. Preuss. 2000. Tetrad analysis in higher plants. A budding technology. *Plant Physiol.* 124: 7–15.

Costa, J. M., A. Corey, P. M. Hayes, C. Jobet, A. Kleinhofs, A. Kopisch-Obusch, S. F. Kramer et al. 2001. Molecular mapping of the Oregon Wolfe barleys: A phenotypically polymorphic doubled-haploid population. *Theor. Appl. Genet.* 103: 415–424.

Costa-Rodrigues, L. 1954. Chromosomal aberrations in oats, *Avena sativa* L. *Agron. Lusitana* 16: 49–79.

Crane, C. F., and J. G. Carman. 1987. Mechanisms of apomixis *in Elymus rectisetus* from eastern Australia and New Zealand. *Amer. J. Bot.* 74: 477–496.

Cregan, P. B., K. P. Kollipara, S. J. Xu, R. J. Singh, S. E. Fogarty, and T. Hymowitz. 2001. Primary trisomics and SSR markers as tools to associate chromosomes with linkage groups in soybean. *Crop Sci.* 41: 1262–1267.

Creissen, G. P., and A. Karp. 1985. Karyotypic changes in potato plants regenerated from protoplasts. *Plant Cell Tiss. Org. Cult.* 4: 171–182.

Cronn, R. C., R. L. Small, T. Haselkorn, and J. F. Wendel. 2002. Rapid diversification of the cotton genus (*Gossypium*; Malvaceae) revealed by analysis of sixteen nuclear and chloroplast genes. *Amer. J. Bot.* 89: 707–725.

Curtis, C. A., and G. C. Doyle. 1992. Production of aneuploid and diploid eggs by meiotic mutants of maize. *J. Hered.* 83: 335–341.

Curtis, C. A., and A. J. Lukaszewaski. 1991. Metaphase I pairing of deficient chromosomes and genetic mapping of deficiency breakpoints in common wheat. *Genome* 34: 553–560.

Dalmacio, R., D. S. Brar, T. Ishii, L. A. Sitch, S. S. Virmani, and G. S. Khush. 1995. Identification and transfer of a new cytoplasmic male sterility source from *Oryza perennis* into indica rice (*O. sativa*). *Euphytica* 82: 221–225.

D'Amato, F. 1995. Aneusomaty *in vivo* and *in vitro* in higher plants. *Caryologia* 48: 85–103.

D'Amato, F. 1985. Cytogenetics of plant cell and tissue cultures and their regenerates. *CRC Crit. Rev. Plant Sci.* 3: 73–112.

Damiani, F., D. Mariotti, M. Pezzotti, and S. Arcioni. 1985. Variation among plants regenerated from tissue culture of *Lotus corniculatus* L. *Z. Pflanzenzüchtg.* 94: 332–339.

Darlington, C. D. 1929. Chromosome behaviour and structural hybridity in the *Tradescantiae*. *J. Genet.* 21: 207–286.

Darlington, C. D. 1932. The control of the chromosome by the genotype and its bearing on some evolutionary problems. *Amer. Nat.* 66: 25–51.

Darlington, C. D. 1939a. *The Evolution of Genetic Systems*. Basic Books, New York.

Darlington, C. D. 1939b. Misdivision and the genetics of the centromere. *J. Genet.* 37: 341–364.

Darlington, C. D. 1940. The origin of iso-chromosomes. *J. Genet.* 39: 351–361.

Darlington, C. D. 1958. *The Evolution of Genetic Systems. Revised and Enlarged Edition*. Basic Books, New York.

Darlington, C. D., and A. Hague. 1955. The timing of mitosis and meiosis in *Allium ascalonicum*. A problem of differentiation. *Heredity* 9: 117–127.

Darlington, C. D., and L. F. La Cour. 1969. *The Handling of Chromosomes*, 5th edition. George Allen & Unwin Ltd., London.

Darlington, C. D., and K. Mather. 1932. The origin and behaviour of chiasmata III. Triploid *Tulipa*. *Cytologia* 4: 1–15.

Darlington, C. D., and A. P. Wylie. 1955. *Chromosome Atlas of Flowering Plants*. George Allen & Unwin Ltd., London.

Darvey, N. L. 1973. Genetics of seed shrivelling in wheat and triticale. In *Proc. 4th Int. Wheat Genet. Symp.* E. R. Sears and L. M. S. Sears, Eds., Missouri Agric. Exp. Stn., Columbia, MO, pp. 155–159.

Darvey, N. L., and C. J. Driscoll. 1972. Evidence against somatic association in hexaploid Wheat. *Chromosoma* (Berlin) 36: 140–149.

Darvey, N. L., and J. P. Gustafson. 1975. Identification of rye chromosomes in wheat-rye addition lines and triticale by heterochromatin bands. *Crop Sci.* 15: 239–243.

Das, K. 1955. Cytogenetic studies of partial sterility in X-ray irradiated barley. *Ind. J. Genet. Plant Breed.* 15: 99–111.

Datta, A. K., and A. K. Biswas. 1985. An EMS-induced bushy mutant of *Nigella sativa* L. with desynaptic behaviour of chromosomes. *Cytologia* 50: 535–543.

Daud, H. M., and J. P. Gustafson. 1996. Molecular evidence for *Triticum speltoides* as a B-genome progenitor of wheat (*Triticum aestivum*). *Genome* 39: 543–548.

Davies, D. R. 1974. Chromosome elimination in inter-specific hybrids. *Heredity* 32: 267–270.

Delibes, A., D. Romero, S. Aguaded, A. Duce, M. Mena, I. Lopez-Braña, M.-F. Andrés, J.-A. Martin-Sanchez, and F. Garcia-Olmedo. 1993. Resistance to the cereal cyst nematode (*Heterodera avenae* Woll.) transferred from the wild grass *Aegilops ventricosa* to hexaploid wheat by a "stepping-stone" procedure. *Theor. Appl. Genet.* 87: 402–408.

Dellaporta, S. L., J. Wood, and J. B. Hicks. 1983. A plant DNA minipreparation: Version II. *Plant Mol. Biol. Rep.* 1: 19–21.

Delos Reyes, B. G., G. S. Khush, and D. S. Brar. 1998. Chromosomal location of eight isozyme loci in rice using primary trisomics and monosomic alien addition lines. *J. Hered.* 89: 164–168.

Demoise, C. F., and C. R. Partanen. 1969. Effects of subculturing and physical condition of medium on the nuclear behavior of a plant tissue culture. *Amer. J. Bot.* 56: 147–152.

Dempsey, E. 1959. Analysis of crossing-over in haploid genetics of asynaptic plants. *Maize Genet. Coop. Newsl.* 33: 54–55.

den Nijs, A. P. M., and G. E. van Dijk. 1993. Apomixis. In *Plant Breeding: Principles and Prospects*. M. D. Hayward, N. O. Bosemark, and I. Romagosa, Eds., Chapman & Hall, London, pp. 229–245.

Dermen, H. 1941. Intranuclear polyploidy in bean induced by naphthalene acetic acid. *J. Hered.* 32: 133–138.

De Robertis, E. D. P., W. W. Nowinski, and F. A. Saez. 1954. *General Cytology*, 2nd edition. W. B. Saunders, Philadelphia.

DeVerna, J. W., R. T. Chetelat, and C. M. Rick. 1987. Cytogenetic, electrophoretic, and morphological analysis of progeny of sesquidiploid *Lycopersicon esculentum-Solanum lycopersicoides* hybrids x *L. pennellii*. *Biol. Zent. Bl.* 106: 417–428.

de Vries, H. 1901. *Die Mutationstheorie*. Veit & Co., Leipzig.

de Vries, H., and K. Boedijn. 1923. On the distribution of mutant chracters among the chromosomes of *Oenothera lamarckiana*. *Genetics* 8: 233–238.

de Vries, J. N., and J. Sybenga. 1989. Meiotic behaviour of telo-tertiary compensating trisomic trisomics of rye: Evaluation for use in hybrid varieties. *Theor. Appl. Genet.* 78: 889–896.

de Vries, S. E., M. A. Ferwerda, A. E. H. M. Loonen, L. P. Pijnacker, and W. J. Feenstra. 1987. Chromosomes in somatic hybrids between *Nicotiana plumbaginifolia* and a monoploid potato. *Theor. Appl. Genet.* 75: 170–176.

Dewald, C. L., and B. Kindiger. 1994. Genetic transfer of gynomonoecy from diploid to triploid eastern gamagrass. *Crop Sci.* 34: 1259–1262.

de Wet, J. M. J. 1979. *Tripsacum* introgression and agronomic fitness in maize (*Zea mays* L.) *Proc. Conf. Broadening Genet. Base Crops, Wagenningen, 1978. Pudoc,* pp. 203–210.

de Wet, J. M. J. 1980. Origins of polyploids. In *Polyploidy: Biological Relevance.* W. H. Lewis, Ed., Plenum Press, New York, pp. 3–15.

de Wet, J. M. J., and J. R. Harlan. 1970. Apomixis, polyploidy, and speciation in *Dichanthium. Evolution* 24: 270–277.

de Wet, J. M. J., and J. R. Harlan. 1972. Chromosome pairing and phylogenetic affinities. *Taxon* 21: 67–70.

Dewey, D. R. 1984. The genomic system of classification as a guide to intergeneric hybridization with the perennial Triticeae. In *Gene Manipulation in Plant Improvement.* J. P. Gustafson, Ed., Plenum Press, New York, pp. 209–279.

Dhaliwal, H. S., B. Friebe, K. S. Gill, and B. S. Gill. 1990. Cytogenetic identification of *Aegilops squarrosa* chromosome additions in durum wheat. *Theor. Appl. Genet.* 79: 769–774.

Dhaliwal, H. S., and B. L. Johnson. 1982. Diploidization and chromosomal pairing affinities in the tetraploid wheats and their putative amphiploid progenitor. *Theor. Appl. Genet.* 61: 117–123.

Dhaliwal, H. S., T. J. Martin, and B. S. Gill. 1991. Identification of alien chromatin specifying resistance to wheat streak mosaic and greenbug in wheat germ plasm by C-banding and *in situ* hybridization. *Theor. Appl. Genet.* 81: 381–389.

Dhar, M. K., and S. Kaul. 2004. FISH reveals somatic association of nucleolus organizing regions in *Plantago ovata. Curr. Sci.* 87: 1336–1337.

Dhillon, T. S., and E. D. Garber. 1960. The genus *Collinsia.* X. Aneuploidy in *C. heterophylla. Bot. Gaz.* 121: 125–133.

Divito, M., K. B. Singh, N. Greco, and M. C. Saxena. 1996. Sources of resistance to cyst nematode in cultivated and wild *Cicer* species. *Genet. Res. Crop. Evol.* 43: 103–107.

Döbel, P., I. Schubert, and R. Rieger. 1978. Distribution of heterochromatin in a reconstructed karyotype of *Vicia faba* as identified by banding- and DNA-late replication patterns. *Chromosoma* (Berlin) 69: 193–209.

Dobzhansky, T., and M. M. Rhoades. 1938. A possible method for locating favorable genes in maize. *J. Am. Soc. Agron.* 30: 668–675.

Doebley, J., R. von Bothmer, and S. Larson. 1992. Chloroplast DNA variation and the phylogeny of *Hordeum* (Poaceae). *Am. J. Bot.* 79: 576–584.

Doganlar, S., A. Frary, and S. D. Tanksley. 1997. Production of interspecific F1 hybrids, BC_1, BC_2, and BC_3 populations between *Lycopersicon esculentum* and two accessions of *Lycopersicon peruvianum* carrying new root-knot nematode resistance genes. *Euphytica* 95: 203–207.

Doggett, H. 1964. Fertility improvement in autotetraploid sorghum I. cultivated autotetraploids. *Heredity* 19: 403–417.

Doležel, J. 1991. Flow cytometric analysis of nuclear DNA content in higher plants. *Phytochem. Anal.* 2: 143–154.

Doležel, J., P. Binarová, and S. Lucretti. 1989. Analysis of nuclear DNA content in plant cells by flow cytometry. *Biol. Plant.* 31: 113–120.

Doležel, J., J. Číhalíková, and S. Lucretti. 1992. A high-yield procedure for isolation of metaphase chromosomes from root tips of *Vicia faba* L. *Planta* 188: 93–98.

Doležel, J., and W. Göhde. 1995. Sex determination in dioecious plants *Melandrium album* and *M. rubrum* using high-resolution flow cytometry. *Cytometry* 19: 103–106.

Doležel, J., J. Greilhuber, S. Lucretti, A. Meister, M. A. Lysak, L. Nardi, and R. Obermayer. 1998. Plant genome size estimation by flow cytometry: Inter-laboratory comparison. *Ann. Bot.* 82 (Suppl A): 17–26.

Doležel, J., S. Lucretti, and I. Schubert. 1994. Plant chromosome analysis and sorting by flow cytometry. *Critical Rev. Plant Sci.* 13: 275–309.

Doležel, J., J. Macas, and S. Lucretti. 1999. Flow analysis and sorting of plant chromosomes. In *Current Protocols in Cytometry.* J. P. Robinson, Z. Darzynkiewicz, P. N. Dean, L. G. Dressler, A. Orfao, P. S. Rabinovitch, C. C. Stewart, H. J. Tanke, and L. L. Wheeless, Eds., John Wiley & Sons, New York, pp. 5.3.1.–5.3.33.

Dolores, R. C., T. T. Chang, and D. A. Ramirez. 1979. The cytogenetics of F_1 hybrids from *Oryza nivara* Sharma et Shastry x *O. sativa* L. *Cytologia* 44: 527–540.

Douwes, H. 1951. The cytological relationships of *Gossypium somalense* Gurke. *J. Genet.* 50: 179–191.

Douwes, H. 1953. The cytological relationships of *Gossypium areysianum* Deflers. *J. Genet.* 51: 611–624.

Dover, G. A. 1973. The genetics and interactions of "A" and "B" chromosomes controlling meiotic chromosome pairing in the Triticinae. In *Proc. 4th Int. Wheat Genet. Symp.* E. R. Sears and L. M. S. Sears, Eds., Missouri Agric. Exp. Stn., Columbia, MO, pp. 653–666.

Dover, G. A., and R. Riley. 1972. Prevention of pairing of homoeologous meiotic chromosomes of wheat by an activity of supernumerary chromosomes of *Aegilops*. *Nature* (London) 240: 159–161.

Doyle, G. G. 1986. Aneuploidy and inbreeding depression in random mating and self-fertilizing autotetraploid populations. *Theor. Appl. Genet.* 72: 799–806.

Doyle, J. J. 1991. The pros and cons of DNA systematic data: Studies of the wild perennial relatives of soybean. *Evolutionary Trends Plants* 5: 99–104.

Doyle, M. J., and A.H.D. Brown. 1985. Numerical analysis of isozyme variation in *Glycine tomentella*. *Biochem. Syst. Ecol.* 13: 413–419.

Doyle, J. J., and R. N. Beachy. 1985. Ribosomal gene variation in soybean (*Glycine*) and its relatives. *Theor. Appl. Genet.* 70: 369–376.

Doyle, J. J., J. L. Doyle, and A. H. D. Brown. 1990a. A chloroplast-DNA phylogeny of the wild perennial relatives of soybean (*Glycine* subgenus *Glycine*): Congruence with morphological and crossing groups. *Evolution* 44: 371–389.

Doyle, J. J., J. L. Doyle, A. H. D. Brown, and J. P. Grace. 1990b. Multiple origins of polyploids in the *Glycine tabacina* complex inferred from chloroplast DNA polymorphism. *Proc. Natl. Acad. Sci. U.S.A.* 87: 712–717.

Doyle, J. J., J. L. Doyle, A. H. D. Brown, and B. E. Pfeil. 2000. Confirmation of shared and divergent genomes in the *Glycine tabacina* polyploid complex (leguminosae) using histone H3-D sequences. *Syst. Bot.* 25: 437–448.

Doyle, J. J., M. Morgante, S. V. Tingey, and W. Powell. 1998. Size homoplasy in chloroplast microsatellites of wild perennial relatives of soybean (*Glycine* subgenus *Glycine*). *Mol. Biol. Evol.* 15: 215–218.

Driscoll, C. J. 1973. Minor genes affecting homoeologous pairing in hybrids between wheat and related genera. *Genetics* 74: s66.

Driscoll, C. J. 1986. Nuclear male sterility systems in seed production of hybrid varieties. *Critical Rev. Plant Sci.* 3: 227–256.

Driscoll, C. J., L. M. Bielig, and N. L. Darvey. 1979. An analysis of frequencies of chromosome configurations in wheat and wheat hybrids. *Genetics* 91: 755–767.

Driscoll, C. J., G. H. Gordon, and G. Kimber. 1980. Mathematics of chromosome pairing. *Genetics* 95: 159–169.

Driscoll, C. J., and N. F. Jensen. 1964. Chromosomes associated with waxlessness, awnedness and time of maturity of common wheat. *Can. J. Genet. Cytol.* 6: 324–333.

Dudits, D., Gy. Hadlaczky, E. Lévi, O. Fejér, Zs. Haydu, and G. Lázár. 1977. Somatic hybridisation of *Daucus carota* and *D. capillifolius* by protoplast fusion. *Theor. Appl. Genet.* 51: 127–132.

Dujardin, M., and W. W. Hanna. 1983. Apomictic and sexual pearlmillet x *Penniestum squamulatum* hybrids. *J. Hered.* 74: 277–279.

Dujardin, M., and W. W. Hanna. 1984. Cytogenetics of double cross hybrids between *Pennisetum americanum–P. purpureum* amphiploids and *P. americanum* × *P. squamulatum* interspecific hybrids. *Theor. Appl. Genet.* 69: 97–100.

Dujardin, M. and W. W. Hanna. 1987. Cytotaxonomy and evolutionary significance of two offtype millet plants derived from a pearlmillet × (Pearlmillet × *Pennisetum squamulatum*) apomictic hybrid. *J. Hered.* 78: 21–23.

Dujardin, M., and W. W. Hanna. 1988. Cytology and breeding behavior of a partially fertile triploid pearl millet. *J. Hered.* 79: 216–218.

Dundas, I. S., E. J. Britten, D. E. Byth, and G. H. Gordon. 1987. Meiotic behavior of hybrids of pigeonpea and two Australian native *Atylosia* species. *J. Hered.* 78: 261–265.

Dundas, I. S., K. B. Saxena, and D. E. Byth. 1982. Pollen mother cell and anther wall development in a photoperiod-insensitive male-sterile mutant of pigeon pea (*Cajanus cajan* (L.) Millsp.). *Euphytica* 31: 371–375.

Dunwell, J. M. 2000. Transgenic approaches to crop improvement. *J. Exp. Bot.* 51: 487–496.

Dvořák, J. 1976. The relationship between the genome of *Triticum urartu* and the A and B genomes of *Triticum aestivum*. *Can. J. Genet. Cytol.* 18: 371–377.

Dvořák, J. 1977. Transfer of leaf rust resistance from *Aegilops speltoides* to *Triticum aestivum*. *Can. J. Genet. Cytol.* 19: 133–141.

Dvořák, J. 1980. Homoeology between *Agropyrum elongatum* chromosomes and *Triticum aestivum* chromosomes. *Can. J. Genet. Cytol.* 22: 237–259.

Dvořák, J. 1983a. Evidence for genetic suppression of heterogenetic chromosome pairing in polyploid species of *Solanum* sect. *Petota*. *Can. J. Genet. Cytol.* 25: 530–539.

Dvořák, J. 1983b. The origin of wheat chromosomes 4A and 4B and their genome reallocation. *Can. J. Genet. Cytol.* 25: 210–214.

Dvořák, J., and B. L. Harvey. 1973. Production of aneuploids in *Avena sativa* L. by nitrous oxide. *Can. J. Genet. Cytol.* 15: 649–651.

Dvořák, J., and D. R. Knott. 1973. A study of somatic association of wheat chromosomes. *Can. J. Genet. Cytol.* 15: 411–416.

Dvořák, J., and D. R. Knott. 1974. Disomic and ditelosomic additions of diploid *Agropyron elongatum* chromosomes to *Triticum aestivum*. *Can. J. Genet. Cytol.* 16: 399–417.

Dvořák, J., and D. R. Knott. 1977. Homoeologous chromatin exchange in a radiation-induced gene transfer. *Can. J. Genet. Cytol.* 19: 125–131.

Dvořák, J., and D. R. Knott. 1980. Chromosome location of two leaf rust resistance genes transferred from *Triticum speltoides* to *T. aestivum*. *Can. J. Genet. Cytol.* 22: 381–389.

Dvořák, J., P. Resta, and R. S. Kota. 1990. Molecular evidence on the origin of wheat chromosomes 4A and 4B. *Genome* 33: 30–39.

Dyck, P. L. 1992. Transfer of a gene for stem rust resistance from *Triticum araraticum* to hexaploid wheat. *Can. J. Plant Sci.* 74: 671–673.

Dyck, P. L. 1994. The transfer of leaf rust resistance from *Triticum turgidum* ssp. *dicoccoides* to hexaploid wheat. *Genome* 35: 788–792.

Dyck, P. L., and T. Rajhathy. 1963. Cytogenetics of a hexaploid oat with an extra pair of chromosomes. *Can. J. Genet. Cytol.* 5: 408–413.

Dyck, P. L., and T. Rajhathy. 1965. A desynaptic mutant in *Avena strigosa*. *Can. J. Genet. Cytol.* 7: 418–421.

Dyer, A. F. 1965. The use of lacto-propionic orcein in rapid squash methods for chromosome preparations. *Stain Tech.* 38: 85–90.

Edallo, S., C. Zucchinali, M. Perenzin, and F. Salamini. 1981. Chromosomal variation and frequency of spontaneous mutation associated with *in vitro* culture and plant regeneration in maize. *Maydica* 26: 39–56.

Edwards, G. A., M. S. Brown, G. A. Niles, and S. A. Naqi. 1980. Monosomics of cotton. *Crop Sci.* 20: 527–528.

Edwards, G. A., and M. A. Mirza. 1979. Genomes of the Australian wild species of cotton. II. The designation of a new G genome for *Gossypium bickii*. *Can. J. Genet. Cytol.* 21: 367–372.

Edwardson, J. R. 1970. Cytoplasmic male sterility. *Bot. Rev.* 36: 341–420.

Egawa, Y., and M. Tanaka. 1984. Cytological relationships among three species of chilli peppers, *Capsicum chinense*, *C. frutescens* and *C. baccatum*. *Jpn. J. Breed.* 34: 50–56.

Eigsti, O. J., and A. P. Dustin. 1955. *Colchicine in Agriculture, Medicine, Biology and Chemistry*. Iowa State University Press, Ames.

Einset, J. 1943. Chromosome length in relation to transmission frequency of maize trisomes. *Genetics* 28: 349–364.

Ekberg, I. 1974. Cytogenetic studies of three paracentric inversions in barley. *Hereditas* 76: 1–30.

Ekingen, H. R. 1969. Über unterschiede in der häufigkeit verschiedener monosomen-typen in natürlichen und röntgenbestrahlten populationen des hafers, *Avena sativa*. *Z. Pflanzenzüchtg.* 61: 73–90.

Ekingen, H. R., T. Attia, and G. Röbbelen. 1977. Suppressor of homoeologous pairing in diploid *Aegilops squarrosa*. *Z. Pflanzenzüchtg.* 79: 72–73.

El-Kharbotly, A., E. Jacobsen, W. J. Stiekema, and A. Pereira. 1995. Genetic localisation of transformation competence in diploid potato. *Theor. Appl. Genet.* 91: 557–562.

Elkonin, L. A., T. N. Gudova, and A. G. Ishin. 1994. Inheritance of male sterility mutations induced in haploid tissue culture. *Euphytica* 80: 111–118.

Elliott, F. C. 1958. *Plant Breeding and Cytogenetics*. McGraw-Hill Book Company, New York.

Emerson, S. 1936. The trisomic derivatives of *Oenothera lamarckiana*. *Genetics* 21: 200–224.

Endo, T. R. 1986. Complete identification of common wheat chromosomes by means of the C-banding technique. *Jpn. J. Genet.* 61: 89–93.

Endo, T. R. 1988. Induction of chromosomal structural changes by a chromosome of *Aegilops cylindrica* L. in common wheat. *J. Hered.* 79: 366–370.

Endo, T. R. 1990. Gametocidal chromosomes and their induction of chromosome mutations in wheat. *Jpn. J. Genet.* 65: 135–152.

Endo, T. R. 2011. Cytological dissection of the Triticeae chromosomes by the gametocidal system. Chapter 12. In *Plant Chromosome Engineering: Methods and Protocols, Methods in Molecular Biology*, vol. 701. J. A. Birchler Ed., Springer Science + Business Media, LLC, pp. 247–257.

Endo, T. R., and B. S. Gill. 1984. Somatic karyotype, heterochromatin distribution and nature of chromosome differentiation in common wheat, *Triticum aestivum* L. em Thell. *Chromosoma* (Berlin) 89: 361–369.

Endo, T. R., and B. S. Gill. 1996. The deletion stocks of common wheat. *J. Hered.* 87: 295–307.

Endo, T. R. 2015. Advances in wheat Genetics: From Genome to Field, Y. Ogihara et al. (eds.); DOI: 10.1007/978-4-431-55675-6_8.

Endo, T. R., and Y. Mukai. 1988. Chromosome mapping of a speltoid suppression gene of *Triticum aestivum* L. based on partial deletions in the long arm of chromosome 5A. *Jpn. J. Genet.* 63: 501–505.

Endrizzi, J. E. 1963. Genetic analysis of six primary monosomes and one tertiary monosome in *Gossypium hirsutum*. *Genetics* 48: 1625–1633.

Endrizzi, J. E. 1974. Alternate-1 and alternate-2 disjunctions in heterozygous reciprocal translocations. *Genetics* 77: 55–60.

Endrizzi, J. E., and M. S. Brown. 1964. Identification of monosomes for six chromosomes in *Gossypium hirsutum*. *Am. J. Bot.* 51: 117–120.

Endrizzi, J. E., and R. J. Kohel. 1966. Use of telosomes in mapping three chromosomes in cotton. *Genetics* 54: 535–550.

Endrizzi, J. E., and L. L. Phillips. 1960. A hybrid between *Gossypium arboreum* L. and *G. raimondii* Ulb. *Can. J. Genet. Cytol.* 2: 311–319.

Endrizzi, J. E., and G. Ramsay. 1979. Monosomes and telosomes for 18 of the 26 chromosomes of *Gossypium hirsutum*. *Can. J. Genet. Cytol.* 21: 531–536.

Endrizzi, J. E., and D. T. Ray. 1991. Monosomic and monotelodisomic analysis of 34 mutant loci in cotton. *J. Hered.* 82: 53–57.

Endrizzi, J. E., E. L. Turcotte, and R. J. Kohel. 1985. Genetics, cytology, and evolution of *Gossypium*. *Adv. Genet.* 23: 271–375.

Engler, D. E., and R. G. Grogan. 1984. Variation in lettuce plants regenerated from protoplasts. *J. Hered.* 75: 426–430.

Enns, H., and E. N. Larter. 1962. Linkage relations of *ds*: A gene governing chromosome behaviour in barley and its effect on genetic recombination. *Can. J. Genet. Cytol.* 4: 263–266.

Esen, A., and R. K. Soost. 1973. Seed development in citrus with special reference to $2x$ x $4x$ crosses. *Am. J. Bot.* 60: 448–462.

Eshed, Y., and D. Zamir. 1994. Introgressions from *Lycopersicon pennellii* can improve the soluble-solids yield of tomato hybrids. *Theor. Appl. Genet.* 88: 891–897.

Evans, D. A., C. E. Flick, S. A. Kut, and S. M. Reed. 1982. Comparison of *Nicotiana tabacum* and *Nicotiana nesophila* hybrids produced by ovule culture and protoplast fusion. *Theor. Appl. Genet.* 62: 193–198.

Evans, D. A., and S. M. Reed. 1981. Cytogenetic techniques. In *Plant Tissue Culture Methods and Application in Agriculture*. T. A. Thorpe, Ed., Academic Press, NY, pp. 213–240.

Evans, D. A., and W. R. Sharp. 1983. Single gene mutations in tomato plants regenerated from tissue culture. *Science* 221: 949–951.

Evans, D. A., and W. R. Sharp. 1986. Applications of somaclonal variation. *Bio/Tech.* 4: 528–532.

Evans, D. A., W. R. Sharp, and H. P. Medina-Filho. 1984. Somaclonal and gametoclonal variation. *Am. J. Bot.* 71: 759–774.

Evans, G. M., and A. J. Macefield. 1973. The effect of B chromosomes on homoeologous pairing in species hybrids. 1. *Lolium temulentum* x *Lolium perenne*. *Chromosoma* (Berlin) 41: 63–73.

Evans, G. M., and H. Rees. 1971. Mitotic cycles in dicotyledons and monocotyledons. *Nature* (London) 233: 350–351.

Evans, L. E., and B. C. Jenkins. 1960. Individual *Secale cereale* chromosome additions to *Triticum aestivum*. I. The addition of individual "Dakold" fall rye chromosomes to "Kharkhov" winter wheat and their subsequent identification. *Can. J. Genet. Cytol.* 2: 205–215.

Fabergé, A. C. 1958. Relation between chromatid-type and chromosome-type breakage-fusion-bridge cycles in maize endosperm. *Genetics* 43: 737–749.

Fan, Z., and W. Tai. 1985. A cytogenetic study of monosomics in *Brassica napus* L. *Can. J. Genet. Cytol.* 27: 683–688.

Farquharson, L. I. 1955. Apomixis and polyembryony in *Tripsacum dactyloides*. *Am. J. Bot.* 42: 737–743.

Fedak, G. 1973. Increased chiasma frequency in desynaptic barley in response to phosphate treatments. *Can. J. Genet. Cytol.* 15: 647–649.

Fedak, G. 1976. Cytogenetics of Wiebe's 16 chromosome barley. *Can. J. Genet. Cytol.* 18: 763–768.

Fedak, G., K. C. Armstrong, and R. J. Handyside. 1987. Chromosome instability in wheat plants regenerated from suspension culture. *Genome* 29: 627–629.

Fedak, G., and S. B. Helgason. 1970a. The cytogenetics of a ditelotetrasomic line in barley. *Can. J. Genet. Cytol.* 12: 553–559.

Fedak, G., and S. B. Helgason. 1970b. Somatic association of chromosomes in barley. *Can. J. Genet. Cytol.* 12: 496–500.

Fedak, G., and T. Tsuchiya. 1975. Progress in the study of aneuploids in barley. *Genetica* 45: 177–190.

Fedak, G., T. Tsuchiya, and S. B. Helgason. 1971. Cytogenetics of some monotelotrisomic in barley. *Can. J. Genet. Cytol.* 13: 760–770.

Fedak, G., T. Tsuchiya, and S. B. Helgason. 1972. Use of monotelotrisomics for linkage mapping in barley. *Can. J. Genet. Cytol.* 14: 949–957.

Fedotova, Y. S., Y. F. Bogdanov, S. A. Gadzhiyeva, S. A. Sosnikhina, V. G. Smirnov, and E. I. Mikhailova. 1994. Meiotic mutants of rye *Secale cereale* L. 2. The nonhomologous synapsis in desynaptic mutants *sy7* and *sy 10. Theor. Appl. Genet.* 88: 1029–1036.

Feldman, M. 1966. The effect of chromosomes 5B, 5D, and 5A on chromosomal pairing in *Triticum aestivum. Proc. Natl. Acad. Sci. U.S.A.* 55: 1447–1453.

Feldman, M. 1968. Regulation of somatic association and meiotic pairing in common wheat. In *Proc. 3rd Int. Wheat Genet. Symp.* K. W. Finlay and K. W. Shepherd, Eds., Butterworth & Co. Ltd., Australia, pp. 169–178.

Feldman, M., and L. Avivi. 1973. The pattern of chromosomal arrangement in nuclei of common wheat and its genetic control. In *Proc. 4th Int. Wheat Genet. Symp.* E. R. Sears and L. M. S. Sears, Eds., Missouri Agric. Exp. Stn., Columbia, MO, pp. 675–684.

Feldman, M., and M. Kislev. 1977. *Aegilops searsii*, a new species of section *Sitopsis (Platystachy). Isr. J. Bot.* 26: 190–201.

Feldman, M., and A. A. Levy. 2012. Genome evolution due to alloploidization in wheat. *Genetics* 192: 763–774.

Feldman, M., and T. Mello-Sampayo. 1967. Suppression of homoeologous pairing in hybrids of polyploid wheats x *Triticum speltoides. Can. J. Genet. Cytol.* 9: 307–313.

Feldman, M., T. Mello-Sampayo, and E. R. Sears. 1966. Somatic association in *Triticum. Proc. Natl. Acad. Sci. U.S.A.* 56: 1192–1199.

Feldman, M., and E. R. Sears. 1981. The wild gene resources of wheat. *Sci. Am.* 244: 102–112.

Feldman, M., I. Strauss, and A. Vardi. 1979. Chromosome pairing and fertility of F_1 hybrids of *Aegilops longissima* and *Ae. searsii. Can. J. Genet. Cytol.* 21: 261–272.

Fernandez, J. A., and N. Jouve. 1988. The addition of *Hordeum chilense* chromosomes to *Triticum turgidum* conv. *durum.* biochemical, karyological and morphological characterizations. *Euphytica* 37: 247–259.

Fernández-Calvín, B. and J. Orellana. 1994. Metaphase I-bound arms frequency and genome analysis in wheat-*Aegilops* hybrids. 3. Similar relationships between the B genome of wheat and S or S = genomes of *Ae. speltoides, Ae. longissima,* and *Ae. sharonensis. Theor. Appl. Genet.* 88: 1043–1049.

Filion, W. G., and B. R. Christie. 1966. The mechanism of male sterility in a clone of Orchardgrass (*Dactylis glomerata* L.). *Crop Sci.* 6: 345–347.

Finch, R. A., and M. D. Bennett. 1979. Action of triploid inducer (*tri*) on meiosis in barley (*Hordeum vulgare* L.). *Heredity* 43: 87–93.

Fish, N., and A. Karp. 1986. Improvements in regeneration from protoplasts of potato and studies on chromosome stability. 1. The effect of initial culture media. *Theor. Appl. Genet.* 72: 405–412.

Fish, N., A. Karp, and M. G. K. Jones. 1988. Production of somatic hybrids by electrofusion in *Solanum. Theor. Appl. Genet.* 76: 260–266.

Fluminhan, Jr., A., M. L. R. de Aguiar-Perecin, and J. A. Dos santos. 1996. Evidence for heterochromatin involvement in chromosome breakage in maize callus culture. *Ann. Bot.* 78: 73–81.

Fominaya, A., S. Molnar, N.-S. Kim, Q. Chen, G. Fedak, and K. C. Armstrong. 1997. Characterization of *Thinopyrum distichum* chromosome using double fluorescence *in situ* hybridization, RFLP analysis of 5S and 26S rDNA, and C-banding by parents and addition lines. *Genome* 40: 689–696.

Fominaya, A., C. Vega, and E. Ferrer. 1988a. Giemsa C-banded karyotypes of *Avena* species. *Genome* 30: 627–632.

Fominaya, A., C. Vega, and E. Ferrer. 1988b. C-banding and nucleolar activity of tetraploid *Avena* species. *Genome* 30: 633–638.

Forer, A. and P. J. Wilson. 1994. A model for chromosome movement during mitosis. *Protoplasma* 179: 95–105.

Forsberg, J., M. Landgren, and K. Glimelius. 1994. Fertile somatic hybrids between *Brassica napus* and *Arabidopsis thaliana. Plant Sci.* 95: 213–223.

Forster, B. P., S. M. Reader, S. A. Forsyth, R. M. D. Koebner, T. E. Miller, M. D. Gale, and Y. Cauderon. 1987. An assessment of the homoeology of six *Agropyron intermedium* chromosomes added to wheat. *Genet. Res. Camb.* 50: 91–97.

Friebe, B., M. C. Cermeño, and F. J. Zeller. 1987. C-banding polymorphism and the analysis of nucleolar activity in *Dasypyrum villosum* (L.) Candary, its added chromosomes to hexaploid wheat and the amphiploid *Triticum dicoccum-D. villosum. Theor. Appl. Genet.* 73: 337–342.

Friebe, B., J. H. Hatchett, B. S. Gill, Y. Mukai, and E. E. Sebesta. 1991a. Transfer of Hessian fly resistance from rye to wheat via radiation-induced terminal and intercalary chromosomal translocations. *Theor. Appl. Genet.* 83: 33–40.

Friebe, B., J. H. Hatchett, R. G. Sears, and B. S. Gill. 1990a. Transfer of Hessian fly resistance from 'Chaupon' rye to hexaploid wheat via a 2BS/2RL wheat-rye chromosome translocation. *Theor. Appl. Genet.* 79: 385–389.

Friebe, B., M. Heun, and W. Bushuk. 1989. Cytological characterization, powdery mildew resistance and storage protein composition of tetraploid and hexaploid 1BL/IRS wheat-rye translocation lines. *Theor Appl. Genet.* 78: 425–432

Friebe, B., J. Jiang, B. S. Gill, and P. L. Dyck. 1993. Radiation-induced nonhomoeologous set of *Triticum aestivum-Aegilops geniculata* chromosome addition lines. *Genome* 42: 374–380.

Friebe, B., J. Jiang, D. R. Knott, and B. S. Gill. 1994. Compensation indices of radiation-induced wheat-Agropyron elongatum translocations conferring resistance to leaf rust and stem rust. *Crop Sci.* 34: 400–404.

Friebe, B., J. Jiang, W. J. Raupp, R. A. McIntosh, and B. S. Gill. 1996. Characterization of wheat-alien translocations conferring resistance to diseases and pests: Current status. *Euphytica* 91: 59–87.

Friebe, B., J. Jiang, N. Tuleen, and B. S. Gill. 1995. Standard karyotype of *Triticum umbellulatum* and characterization of derived chromosome addition and translocation lines in common wheat. *Theor. Appl. Genet.* 90: 150–156.

Friebe, B., N.-S. Kim, J. Kuspira, and B. S. Gill. 1990b. Genetic and cytogenetic analyses of the A genome of *Triticum monococcum*. VI. Production and identification of primary trisomics using the C-banding technique. *Genome* 33: 542–555.

Friebe, B., R. G. Kynast, and B. S. Gill. 2000. Gametocidal factor-induced structural rearrangements in rye chromosomes added to common wheat. *Chromosome Res.* 8: 501–511.

Friebe, B., R. G. Kynast, P. Zhang, L. Qi, M. Dhar, and B. S. Gill. 2001. Chromosome healing by addition of telomeric repeats in wheat occurs during the first mitotic divisions of the sporophyte and is a gradual process. *Chromosome Res.* 9: 137–146.

Friebe, B., and E. N. Larter. 1988. Identification of a complete set of isogenic wheat/rye D-genome substitution lines by means of Giemsa C-banding. *Theor. Appl. Genet.* 76: 473–479.

Friebe, B., Y. Mukai, H. S. Dhaliwal, T. J. Martin, and B. S. Gill. 1991b. Identification of alien chromatin specifying resistance to wheat streak mosaic and greenbug in wheat germ plasm by C-banding and *in situ* hybridization. *Theor. Appl. Genet.* 81: 381–389.

Friebe, B., L. L. Qi, S. Nasuda, P. Zhang, N. A. Tuleen, and B. S. Gill. 2000. Development of a complete set of *Triticum aestivum–Aegilops speltoides* chromosome addition lines. *Theor. Appl. Genet.* 101: 51–58.

Friebe, B., V. Schubert, W. D. Blüthner, and K. Hammer. 1992. C-banding pattern and polymorphism of *Aegilops caudata* and chromosomal constitutions of the amphiploid *T. aestivum–Ae. caudata* and six derived chromosome addition lines. *Theor. Appl. Genet.* 83: 589–596.

Friebe, B., N. A. Tuleen, E. D. Badaeva, and B. S. Gill. 1996. Cytogenetic identification of *Triticum peregrinum* chromosomes added to common wheat. *Genome* 39: 272–276.

Friebe, B. R., N. A. Tuleen, and B. S. Gill. 1999. Development and identification of a complete wheat-rye chromosomal translocations conferring resistance to greenbug. *Euphytica* 84: 121–125.

Friebe, B., W. Zhang, J. W. Raupp, B. S. Gill. and D. R. Porter. 1995. Non-homoeologous wheat—*Agropyron intermedium* chromosomal translocations conferring resistance to leaf rust. *Theor. Appl. Genet.* 86: 141–149.

Friedt, W. 1978. Untersuchungen an autotetraploiden gersten unter besonderer berücksichtigung der diploidisierung. I. Fertilität, vitalität und kornertrag. *Z. Pflanzenzüchtg.* 81: 118–139 (in German with English summary).

Fromm, M. E., F. Morrish, C. Armstrong, R. Williams, J. Thomas, and T. M. Klein. 1990. Inheritance and expression of chimeric genes in the progeny of transgenic maize plants. *Bio/Technol.* 8: 833–839.

Frost, H. B., M. M. Lesley, and W. F. Locke. 1959. Cytogenetics of a trisomic of *Matthiola incana*, involving a ring chromosome and somatic instability of singleness (versus doubleness) of flowers and shape of leaves. *Genetics* 44: 1083–1099.

Frost, H. B., and M. C. Mann. 1924. Mutant forms of *Matthiola* resulting from non-disjunction. *Am. Nat.* 58: 569–572.

Fryxell, P. A. 1957. Mode of reproduction of higher plants. *Bot. Rev.* 23: 135–233.

Fryxell, P. A. 1992. A revised taxonomic interpretation of *Gossypium* L. (Malvaceae). *Rheedia* 2: 108–165.

Fujigaki, J., and T. Tsuchiya. 1985. Karyotype analysis in a haploid plant of an inbred rye, *Secale cereale* L., by acetocarmine—Giemsa staining technique. *Z. Pflanzenzüchtg.* 94: 234–243.

Fujigaki, J., and T. Tsuchiya. 1990. Chromosome identification of seven primary trisomics in inbred rye (*Secale cereale* L.) by acetocarmine Giemsa C-banding. *Jpn. J. Genet.* 65: 209–219.

Fukasawa, H. 1953. Studies on restoration and substitution of nucleus of *Aegilotricum*. I. Appearance of male-sterile *durum* in substitution crosses. *Cytologia* 18: 167–175.

Fukui, K. 1983. Sequential occurrence of mutations in a growing rice callus. *Theor. Appl. Genet.* 65: 225–230.

Fukui, K. 1986. Standardization of karyotyping plant chromosomes by a newly developed chromosome image analyzing system (CHIAS). *Theor Appl. Genet.* 72: 27–32.

Fukui, K., and K. Kakeda. 1990. Quantitative karyotyping of barley chromosomes by image analysis methods. *Genome* 33: 450–458.

Fukui, K., and K. Kakeda. 1994. A critical assessment of karyotype analysis by imaging methods. *Jpn. J. Genet.* 69: 537–544.

Fukui, K., and K. Iijima. 1991. Somatic chromosome map of rice by imaging methods. *Theor. Appl. Genet.* 81: 589–596.

Fukui, K. and Y. Yamisugi. 1995. Mapping of C-banded *Crepis* chromosomes by imaging methods. *Chromo. Res.* 3: 79–86.

Fukui, K., and N. Ohmido. 2000. Visual detection of useful genes on plant chromosomes. *JARQ* 34: 153–158.

Fukui, K., N. Ohmido, and G. S. Khush. 1994. Variability in rDNA loci in genus *Oryza* detected through fluorescence *in situ* hybridisation. *Theor. Appl. Genet.* 87: 893–899.

Funaki, K., S. Matsui, and M. Sasaki. 1975. Location of nucleolar organizers in animal and plant chromosomes by means of an improved N-banding technique. *Chromosoma* (Berlin) 49: 357–370.

Furst, E., and T. Tsuchiya. 1983. New telosomic and acrosomic trisomics in barley. *Barley Genet. Newsl.* 13: 47–48.

Furuta, Y., K. Nishikawa, and S. Yamaguchi. 1986. Nuclear DNA content in diploid wheat and its relatives in relation to the phylogeny of tetraploid wheat. *Jpn. J. Genet.* 61: 97–105.

Fütterer, J., and I. Potrykus. 1995. Transformation of Poaceae and gene expression in transgenic plants. *Agronomie* 15: 309–319.

Gall, J. G., and M. L. Pardue. 1969. Formation and detection of RNA-DNA hybrid molecules in cytological preparations. *Proc. Natl. Acad. Sci. U.S.A.* 63: 378–383.

Gao, D., D. Guo, and C. Jung. 2001. Monosomic addition lines of *Beta corolliflora* Zoss in sugarbeet: Cytological and molecular-marker analysis. *Theor. Appl. Genet.* 103: 240–247.

Gaponenko, A. K., T. F. Petrova, A. R. Iskakov, and A. A. Sozinov. 1988. Cytogenetics of in vitro cultured somatic cells and regenerated plants of barley (*Hordeum vulgare* L.). *Theor. Appl. Genet.* 75: 905–911.

Garber, E. D. 1964. The genus *Collinsia* XXII. Trisomy in *C. heterophylla*. *Bot. Gaz.* 125: 46–50.

Garriga-Calderé, F., D. J. Fuigen, F. Filotico, E. Jacobsen, and M. S. Ramanna. 1997. Identification of alien chromosome through GISH and RFLP analysis and potential for establishing potato lines with monosomic additions of tomato chromosomes. *Genome* 40: 666–673.

Gates, R. R. 1908. A study of reduction in *Oenothera rubrinervis*. *Bot. Gaz.* 46: 1–34.

Gaul, H. 1959. A critical survey of genome analysis. In *Proc. Ist. Int. Wheat Genet. Symp.* B. C. Jenkins, Ed., Public Press, Winnipeg, Canada, pp. 194–206.

Gavazzi, G., C. Tonelli, G. Todesco, E. Arreghini, F. Raffaldi, F. Vecchio, G. Barbuzzi, M. G. Biasini, and F. Sala. 1987. Somaclonal variation versus chemically induced mutagenesis in tomato (*Lycopersicon esculentum* L.). *Theor. Appl. Genet.* 74: 733–738.

Geber, G., and D. Schweizer. 1988. Cytochemical heterochromatin differentiation in *Sinapsis alba* (*Cruciferae*) using a simple air-drying technique for producing chromosome spreads. *Pl. Syst. Evol.* 158: 97–106.

George, L., and P. S. Rao. 1983. Yellow-seeded variants in *in vitro* regenerants of mustard (*Brassica juncea* Coss var. RAI-5). *Plant Sci. Lett.* 30: 327–330.

Gerlach, W. L. 1977. N-banded karyotypes of wheat species. *Chromosoma (Berlin)* 62: 49–56.

Gerstel, D. U. 1945. Inheritance in *Nicotiana tabacum* XX. The addition of *Nicotiana glutinosa* chromosomes to tobacco. *J. Hered.* 36: 197–206.

Gerstel, D. U. 1953. Chromosomal translocations in interspecific hybrids of the genus *Gossypium*. *Evolution* 7: 234–244.

Gerstel, D. U. 1960. Segregation in new allopolyploids of *Nicotiana*. I. Comparison of 6x (*N. tabacum* x *tomentosiformis*) and 6x (*N. tabacum* x *otophora*). *Genetics* 45: 1723–1734.

Gerstel, D. U. 1963. Segregation in new allopolyploids of *Nicotiana* II. Discordant ratios from individual loci in 6x (*N. tabacum* x *N. sylvestris*). *Genetics* 48: 677–689.

Gerstel, D. U., B. L. Hammond, and C. Kidd. 1955. An additional note on the inheritance of apomixis in guayule. *Bot. Gaz.* 115: 89–93.

Ghosh Biswas, G. C., V. A. Iglesias, S. K. Datta, and I. Potrykus. 1994. Transgenic Indica rice (*Oryza sativa* L.) plants obtained by direct gene transfer to protoplasts. *J. Biotechnol.* 32: 1–10.

Gill, B. S. 1978. Cytogenetics of an unusual tertiary trisomic of tomato. *Caryologia* 31: 257–269.

Gill, B. S., C. R. Burnham, G. R. Stringam, J. T. Stout, and W. H. Weinheimer. 1980. Cytogenetic analysis of chromosomal translocations in the tomato: Preferential breakage in heterochromatin. *Can. J. Genet. Cytol.* 22: 333–341.

Gill, B. S., B. Friebe, and T. R. Endo. 1991. Standard karyotype and nomenclature system for description of chromosome bands and structural aberrations in wheat (*Triticum aestivum*). *Genome* 34: 830–839.

Gill, B. S., and G. Kimber. 1974. Giemsa C-banding and the evolution of wheat. *Proc. Nat. Acad. Sci. U.S.A.* 71: 4086–4090.

Gill, B. S., L. N. W. Kam-Morgan, and J. F. Shepard. 1986. Origin of chromosomal and phenotypic variation in potato protoclones. *J. Hered.* 77: 13–16.

Gill, B. S., L. N. W. Kam-Morgan, and J. F. Shepard. 1987. Cytogenetic and phenotypic variation in mesophyll cell-derived tetraploid potatoes. *J. Hered.* 78: 15–20.

Gill, B. S., J. L. Minocha, D. Gupta, and D. Kumar. 1969. Chromosome behaviour and seed setting in autotetraploid pearl millet. *Ind J. Genet. Plant Breed.* 29: 462–467.

Gill, B. S., S. S. Virmani, and J. L. Minocha. 1970. Primary simple trisomics in pearl millet. *Can. J. Genet. Cytol.* 12: 474–483.

Gill, K. S., B. S. Gill, T. R. Endo, and E. V. Boyko. 1996. Identification and high-density mapping of gene-rich regions in chromosome group 5 of wheat. *Genetics* 143: 1001–1012.

Gilles, A., and L. F. Randolph. 1951. Reduction of quadrivalent frequency in autotetraploid maize during a period of ten years. *Am. J. Bot.* 38: 12–17.

Gillies, C. B. 1975. Syneptonemal complex and chromosome structure. *Annu. Rev. Genet.* 9: 91–109.

Giorgi, B., and A. Bozzini. 1969. Karyotype analysis in *Triticum*. III. Analysis of the presumed diploid progenitors of polyploid wheats. *Caryologia* 22: 279–287.

Giraldez, R., M. C. Cermeño, and J. Orellana. 1979. Comparison of C-banding pattern in the chromosomes of inbred lines and open pollinated varieties of rye. *Z. Pflanzenzüchtg.* 83: 40–48.

Gleba, Y. Y., and F. Hoffmann. 1978. Hybrid cell lines *Arabidopsis thaliana* + *Brassica campestris*: No evidence for specific chromosome elimination. *Mol. Gen. Genet.* 165: 257–264.

Gleddi, S., W. A. Keller, and G. Setterfield. 1986. Production and characterization of somatic hybrids between *Solanum melongena* L. and *S. sisymbriifolium* Lam. *Theor. Appl. Genet.* 71: 613–621.

Godsen, J. R. 1997. *PRINS and In Situ PCR Protocols.* Humana Press, Totowa, NJ.

Golubovskaya, I. N. 1979. Genetic control of meiosis. *Int. Rev. Cytol.* 58: 247–290.

Golubovskaya, I. N. 1989. Meiosis in maize: *Mei* genes and conception of genetic control of meiosis. *Adv. Genet.* 26: 149–192.

Golubovskaya, I. N., and A. S. Mashnenkov. 1975. Genetic control of meiosis I. A meiotic mutation in maize (*Zea mays* L.) causing the elimination of the first meiotic division. *Genetika* 11(7): 11–17 (in Russian with English summary).

Golubovskaya, I. N., and A. S. Mashnenkov. 1976. Genetic control of meiosis II. A desynaptic mutant in maize induced by N-Nitroso-N Methylurea. *Genetika* 12(2): 7–14 (in Russian with English summary).

Golubovskaya, I. N., and D. V. Sitnikova. 1980. Three meiotic mutations of maize causing irregular segregation of chromosomes in the first division of meiosis. *Genetika* 16(4): 657–666 (in Russian with English summary).

Goodman, R. M., H. Hauptli, A. Crossway, and V. C. Knauf. 1987. Gene transfer in crop improvement. *Science* 236: 48–54.

Goodspeed, T. H. 1954. The genus *Nicotiana*. *Chronica Bot.* 16: 1–536.

Goodspeed, T. H. and P. Avery. 1941. The twelfth primary trisomic type in *Nicotiana sylvestris*. *Proc. Nat. Acad. Sci. U.S.A.* 27: 13–14.

Goodspeed, T. H., and P. Avery. 1939. Trisomic and other types in *Nicotiana sylvestris*. *J. Genet.* 38: 381–458.

Goodspeed, T. H., and R. E. Clausen. 1928. Interspecific hybridization in *Nicotiana*. VIII. The *Sylvestris - tomentosa-tabacum* hybrid triangle and its bearing on the origin of *tabacum*. *Univ. Calif. Publ. Bot.* 11: 245–256.

Gopinath, D. M., and C. R. Burnham. 1956. A cytogenetic study in maize of deficiency-duplication produced by crossing interchanges involving the same chromosomes. *Genetics* 41: 382–395.

Gottschalk, W. 1951. Untersuchungen am pachytän normaler und röntgenbestrahlter pollenmutterzellen von *Solanum lycopersicum. Chromosoma* (Berlin) 4: 298–341 (in German).

Gottschalk, W. 1987. Different intensity of the action of desynaptic genes on micro- and macrosporogenesis. *Cytologia* 52: 653–656.

Gottschalk, W., and S. R. Baquar. 1971. Desynapsis in *Pisum sativum* induced through gene mutation. *Can. J. Genet. Cytol.* 13: 138–143.

Gottschalk, W., and M. L. H. Kaul. 1974. The genetic control of microsporogenesis in higher plants. *The Nucleus* (Calcutta) 17: 133–166.

Gottschalk, W., and M. L. H. Kaul. 1980a. Asynapsis and desynapsis in flowering plants. I. Asynapsis. *The Nucleus* (Calcutta) 23: 1–15.

Gottschalk, W., and M. L. H. Kaul. 1980b. Asynapsis and desynapsis in flowering plants: II. Desynapsis. *The Nucleus* (Calcutta) 23: 97–120.

Gottschalk, W., and H. D. Klein. 1976. The influence of mutated genes in sporogenesis. A survey on the genetic control of meiosis in *Pisum sativum. Theor. Appl. Genet.* 48: 23–34.

Gould, A. R. 1979. Chromosomal and phenotypic stability during regeneration of whole plants from tissue culture of *Brachycome dichromosomatica* (2n = 4). *Austr. J. Bot.* 27: 117–121.

Graham, J. D., and W. P. Bemis. 1979. Six interspecific trisomics (2n *C. moschata* + 1 *C. palmata* chromosome) and one primary trisomic of *Cucurbita moschata. Cucurbit Genet. Coop. Rep.* 2: 37.

Graham, M. J., C. D. Nickell, and A. L. Rayburn. 1994. Relationship between genome size and maturity group in soybean. *Theor. Appl. Genet.* 88: 429–432.

Grant, J. E., J. P. Grace, A. H. D. Brown, and E. Putievsky. 1984. Interspecific hybridization in *Glycine* Willd. subgenus *Glycine* (Leguminosae). *Austr. J. Bot.* 32: 655–663.

Grant, V. 1971. *Plant Speciation.* Columbia University Press, New York.

Grant, V. 1981. *Plant Speciation.* Columbia University Press, New York.

Gravrilenko, T., J. Larkka, E. Pehu, and V.-M. Rokka. 2002. Identification of mitotic chromosomes of tuberous and non-tuberous *Solanum* species (*Solanum tuberosum* and *Solanum brevidens*) by GISH in their interspecific hybrids. *Genome* 45: 442–449.

Graybosch, R. A., M. E. Edge, and X. Delannay. 1987. Somaclonal variation in soybean plants regenerated from the cotyledonary node tissue culture system. *Crop Sci.* 27: 803–806.

Graybosch, R. A., and R. G. Palmer. 1988. Male sterility in soybean—an overview. *Am. J. Bot.* 75: 144–156.

Grazi, F., M. Umaerus, and E. Åkerberg. 1961. Observations on the mode of reproduction and the embryology of *Poa pratensis. Hereditas* 47: 489–541.

Greenleaf, W. H. 1941. The probable explanation of low transmission ratios of certain monosomic types of *Nicotiana tabacum. Proc. Natl. Acad. Sci. U.S.A.* 27: 427–430.

Greier, T. 1991. Chromosome variability in callus produced plants. In *Genetics and Breeding of Ornamental Species.* J. Harding, F. Singh and J. N. M. Mol, Eds., Kluwer Publishers, The Netherlands, pp. 79–106.

Greilhuber, J. 1986. Severely distorted Feulgen-DNA amounts in *Pinus* (Coniferophytina) after non-additive fixations as a result of meristematic self-tanning with vacuole contents. *Can. J. Genet. Cytol.* 28: 409–415.

Greilhuber, J. 1988. "Self-tanning"—A new and important source of stoichiometric error in cytometric determination of nuclear DNA content in plants. *Pl. Syst. Evol.* 158: 87–96.

Greilhuber, J. 1998. Intraspecific variation in genome size: A critical reassessment. *Ann. Bot.* 82 (Suppl A): 27–35.

Greilhuber, J., and I. Ebert. 1994. Genome size variation in *Pisum sativum. Genome* 37: 646–655.

Greilhuber, J., and R. Obermayer. 1997. Genome size and maturity group in *Glycine max* (Soybean). *Heredity* 78: 547–551.

Gresshoff, P. M., and C. H. Doy. 1973. *Zea mays*: Methods for diploid callus culture and the subsequent differentiation of various plant structures. *Austr. J. Biol. Sci.* 26: 505–508.

Griffor, M. C., L. O. Vodkin, R. J. Singh, and T. Hymowitz. 1991. Fluorescent *in situ* hybridization to soybean metaphase chromosomes. *Plant Mol. Biol.* 17: 101–109.

Grimanelli, D., O. Leblanc, E. Espinosa, E. Perotti, D. Gonzalez de Leon, and Y. Savidan. 1998a. Mapping diplosporous apomixis in tetraploid *Tripsacum*: One gene or several genes? *Heredity* 80: 33–39.

Grimanelli, D., O. Leblanc, E. Espinosa, E. Perotti, D. Gonzalez de Leon, and Y. Savidan. 1998b. Non-Mendelian transmission of apomixis in maize-*Tripsacum* hybrids caused by a transmission ratio distortion. *Heredity* 80: 40–47.

Grishaeva, T. M., and Y. F. Bogdanova. 2000. Genetic control of meiosis in *Drosophila. Russian J. Genet.* 36: 1089–1106.

Groose, R. W., and E. T. Bingham. 1984. Variation in plants regenerated from tissue culture of tetraploid alfalfa heterozygous for several traits. *Crop Sci.* 24: 655–658.

Gunthardt, H., L. Smith, M. E. Haferkamp, and R. A. Nilan. 1953. Studies on aged seeds. II. Relation of age of seeds to cytogenetic effects. *Agron. J.* 45: 438–441.

Gupta, N., and P. S. Carlson. 1972. Preferential growth of haploid plant cells *in vitro*. *Nature New Biol.* 239: 86.

Gupta, P. K. 1971. Homoeologous relationship between wheat and rye chromosomes. Present status. *Genetica* 42: 199–213.

Gupta, P. K. 1972. Cytogenetic evolution in the *Triticinae*. Homoeologous relationships. *Genetica* 43: 504–530.

Gupta, P. K., and G. Fedak. 1985. Meiosis in seven intergeneric hybrids between *Hordeum* and *Secale*. *Z. Pflanzenzüchtg.* 95: 262–273.

Gupta, P. K., and P. M. Priyadarshan. 1985. Triticale: Present status and future prospects. *Adv. Genet.* 21: 255–345.

Gupta, S. B., and P. Gupta. 1973. Selective somatic elimination of *Nicotiana glutinosa* chromosomes in the F_1 hybrids of *N. suaveolens* and *N. glutinosa*. *Genetics* 73: 605–612.

Gustafson, J. P., E. Butler, and C. L. McIntyre. 1990. Physical mapping of a low-copy DNA sequence in rye (*Secale cereale* L.). *Proc. Natl. Acad. Sci. U.S.A.* 87: 1899–1902.

Hacker, J. B., and R. Riley. 1963. Aneuploids in oat varietal populations. *Nature* (London) 197: 924–925.

Hacker, J. B., and R. Riley. 1965. Morphological and cytological effects of chromosome deficiency in *Avena sativa*. *Can. J. Genet. Cytol.* 7: 304–315.

Hadi, M. Z., M. D. McMullen, and J. J. Finer. 1996. Transformation of 12 different plasmids into soybean via particle bombardment. *Plant Cell Rep.* 15: 500–505.

Hadlaczky, Gy., Gy. Bisztray, T. Praznovszky, and D. Dudits. 1983. Mass isolation of plant chromosomes and nuclei. *Planta* 157: 278–285.

Hadlaczky, Gy., and L. Kalmán. 1975. Discrimination of homologous chromosomes of maize with Giemsa staining. *Heredity* 35: 371–374.

Hadley, H. H., and S. J. Openshaw. 1980. Interspecific and intergeneric hybridization. In *Hybridization of Crop Plants*. W. R. Fehr and H. H. Hadley, Eds., American Society of Agronomy and Crop Science Society of America, Madison, WI, pp. 133–159.

Hafiz, H. M. I., and H. Thomas. 1978. Genetic background and the breeding behaviour of monosomic lines of *Avena sativa* L. *Z. Pflanzenzuchtg.* 81: 32–39.

Hagberg, A. 1962. Production of duplications in barley breeding. *Hereditas* 48: 243–246.

Hagberg, A., and G. Hagberg. 1980. High frequency of spontaneous haploids in the progeny of an induced mutation in barley. *Hereditas* 93: 341–343.

Hagberg, A., and G. Hagberg. 1987. Production of spontaneously doubled haploids in barley using a breeding system with marker genes and the "hap"-gene. *Biol. Zenttralbl.* 106: 53–58.

Haig, D., and M. Westoby. 1991. Genomic imprinting in endosperm—Its effect on seed development in crosses between different ploidies of the same species, and its implications for the evolution of apomixis. *Phil. Trans. Royal. Soc. London. Series B-Biol. Sci.* 333: 1–13.

Hair, J. B. 1956. Subsexual reproduction in *Agropyron*. *Heredity* 10: 129–160.

Häkansson, A., and E. Ellerström. 1950. Seed development after reciprocal crosses between diploid and tetraploid rye. *Hereditas* 36: 256–296.

Hall, R. D., T. Riksen-Bruinsma, G. J. Weyens, I. J. Rosquin, P. N. Denys, I. J. Evans, J. E. Lathouwers et al. 1996. A high frequency technique for the generation of transgenic sugar beets from stomatal guard cells. *Nature Biotechnol.* 14: 1133–1138.

Hallauer, A. R. and J. B. Miranda Filho. 1981. *Quantitative Genetics in Maize Breeding*. Iowa State University Press, Ames, IA.

Hancock, J. F. 2004. *Plant Evolution and the Origin of Crop Species*, 2nd edition. CABI Publishing, Wallingford, UK.

Handley, L. W., R. L. Nickels, M. W. Cameron, P. P. Moore, and K. C. Sink. 1986. Somatic hybrid plants between *Lycopersicon esculentum* and *Solanum lycopersicoides*. *Theor. Appl. Genet.* 71: 691–697.

Hang, A. 1981. Cytogenetics of the acrotrisomic 4L4S in barley (*Hordeum vulgare* L.) PhD thesis. Colorado State University, Fort Collins, p. 92.

Hang, A., C. S. Burton, and K. Satterfield. 1998. Nearly compensating diploids involving chromosome 6 (6H) in barley (*Hordeum vulgare* L.). *J. Genet. & Breed.* 52: 161–165.

Hang, A., and T. Tsuchiya. 1992. Production and cytological studies of nine-paired barley (*Hordeum vulgare*). *Genome* 35: 78–83.

Hanna, W. W. 1991. Apomixis in crop plants—Cytogenetic basis and role in plant breeding. In *Chromosome Engineering in Plants: Genetics, Breeding, Evolution Part A*. P. K. Gupta, and T. Tsuchiya, Eds., Elsevier, Amsterdam, pp. 229–242.

Hanna, W. W. 1995. Use of apomixis in cultivar development. *Adv. Agron.* 54: 333–350.

Hanna, W. W., and E. C. Bashaw. 1987. Apomixis: Its identification and use in plant breeding. *Crop Sci.* 27: 1136–1139.

Hanna, W. W., M. Dujardin, P. Ozias-Akins, E. Lubbers, and L. Arthur. 1993. Reproduction, cytology, and fertility of pearl millet x *Pennisetum squamulatum* BC$_4$ plants. *J. Hered.* 84: 213–216.

Hanna, W. W., and J. B. Powell. 1973. Stubby head, an induced facultative apomict in pearl millet. *Crop Sci.* 13: 726–728.

Hanna, W. W., J. B. Powell, and G. W. Burton. 1976. Relationship to polyembryony, frequency, morphology, reproductive behavior, and cytology of autotetraploids in *Pennisetum americanum*. *Can. J. Genet. Cytol.* 18: 529–536.

Hanna, W. W., J. B. Powell, J. C. Millot, and G. W. Burton. 1973. Cytology of obligate sexual plants in *Panicum maximum* Jacq. and their use in controlled hybrids. *Crop Sci.* 13: 695–697.

Hanna, W. W., K. F. Schertz, and E. C. Bashaw. 1970. Apospory in *Sorghum bicolor* (L.) Moench. *Science* 170: 338–339.

Hansen, F. L., S. B. Anderson, I. K. Due, and A. Olesen. 1988. Nitrous oxide as a possible alternative agent for chromosome doubling of wheat haploids. *Plant Sci.* 54: 219–222.

Hanson, L., K. A. Mcmahon, M. A. Johnston, M. D. Bennett. 2001. First nuclear DNA C-values for 25 angiosperm families. *Ann. Bot.* 87: 251–258.

Harlan, J. R. 1976. Genetic resources in wild relatives of crops. *Crop Sci.* 16: 329–333.

Harlan, J. R. 1992. *Crops and Man*, 2nd edition. American Society of Agronomy and Crop Science Society of America, Madison, WI.

Harlan, J. R., and J. M. J. de Wet. 1963. The compilospecies concept. *Evolution* 17: 497–501.

Harlan, J. R., and J. M. J. de Wet. 1971. Toward a rational classification of cultivated plants. *Taxon* 20: 509–517.

Harlan, J. R., and J. M. J. deWet. 1975. On Ö Winge and a prayer: The origins of polyploidy. *Bot. Rev.* 41: 361–390.

Harlan, J. R., and J. M. J. de Wet. 1977. Pathways of genetic transfer from *Tripsacum* to *Zea mays*. *Proc. Natl. Acad. Sci. U.S.A.* 74: 3494–3497.

Harland, S. C. 1940. New polyploids in cotton by the use of colchicine. *Trop. Agri.* (Trinidad) 17: 53–54.

Hart, G. E., A. K. M. R. Islam, and K. W. Shepherd. 1980. Use of isozymes as chromosome markers in the isolation and characterization of wheat-barley chromosome addition lines. *Genet. Res.* (Camb.) 36: 311–325.

Hartman, J. B., and D. A. ST Clair. 1999a. Combining ability for beet armyworm, *Spodoptera exigua*, resistance and horticultural traits of selected *Lycopersicon pennellii*-derived inbred backcross lines of tomato. *Plant Breed.* 118: 523–530.

Hartman, J. B., and D. A. ST Clair. 1999b. Variation for aphid resistance and insecticidal acyl sugar expression among and within *Lycopersicon pennellii*-derived inbred backcross lines of tomato and their F$_1$ progeny. *Plant Breed.* 118: 531–536.

Hatasaki, M., T. Morikawa, and J. M. Leggett. 2001. Intraspecific variation of 18S-5.8S-26S rDNA sites revealed bt FISH and RFLP in wild oat, *Avena agadiriana*. *Gene Genet. Syst.* 76: 9–14.

Haunold, A. 1970. Fertility studies and cytological analysis of the progeny of a triploid x diploid cross in hop, *Humulus lupulus* L. *Can. J. Genet. Cytol.* 12: 582–588.

Haus, T. E. 1958. A linkage between two sections of a chromosome in barley. *J. Hered.* 49: 179–180.

Hawbaker, M. S., W. R. Fehr, L. M. Mansur, R. C. Shoemaker, and R. G. Palmer. 1993. Genetic variation for quantitative traits in soybean lines derived from tissue culture. *Theor. Appl. Genet.* 87: 49–53.

Hawkes, J. G. 1977. The importance of wild germplasm in plant breeding. *Euphytica* 26: 615–621.

Hazel, C. B., T. M. Klein, M. Anis, H. D. Wilde, and W. A. Parrott. 1998. Growth characteristics and transformability of soybean embryogenic cultures. *Plant Cell Rep.* 17: 765–772.

Heath, D. W., and E. D. Earle. 1995. Synthesis of high erucic acid rapeseed (*Brassica napus* L.) somatic hybrids with improved agronomic characters. *Theor. Appl. Genet.* 91: 1129–1136.

Heemert, C. Van, and J. Sybenga. 1972. Identification of the three chromosomes involved in the translocations which structurally differentiate the genomes of *Secale cereale* L. from those of *Secale montanum* Guss and *Secale vavilovii* Grossh. *Genetica* 43: 387–393.

Heijbroek, W., A. J. Roelands, and J. H. De Jong. 1983. Transfer of resistance to beet cyst nematode from *Beta patellaris* to sugar beet. *Euphytica* 32: 287–298.

Heinz, D. J., G. W. P. Mee, and L. G. Nickell. 1969. Chromosome numbers of some *Saccharum* species hybrids and their cell suspension cultures. *Am. J. Bot.* 56: 450–456.

Henderson, S. A. 1970. The time and place of meiotic crossing-over. *Annu. Rev. Genet.* 4: 295–324.

Hermann, F. J. 1962. A revision of the genus *Glycine* and its immediate allies. *U. S. Dep. Agric. Tech. Bull.* No. 1268. pp. 1–82.

Hermsen, J. G. Th. 1970. Basic information for the use of primary trisomics in genetic and breeding research. *Euphytica* 19: 125–140.

Hermsen, J. G. Th. 1994. Introgression of genes from wild species, including molecular and cellular approaches. In *Potato Genetics*. J. E. Bradshaw and G. R. MacKay, Eds., CAB International, U.K., pp. 515–538.

Hernandez-Soriano, J. M., and R. T. Ramage. 1974. Normal vs. desynapsis. *Barley Genet. Newsl.* 4: 137–142.

Hernandez-Soriano, J. M., and R. T. Ramage. 1975. Normal vs. desynapsis. *Barley Genet. Newsl.* 5: 113.

Hernandez-Soriano, J. M., R. T. Ramage, and R. F. Eslick. 1973. Normal vs. desynapsis. *Barley Genet. Newsl.* 3: 124–131.

Heslop-Harrison, J. S., and M. D. Bennett. 1983. The spatial order of chromosomes in root-tip metaphases of *Aegilops umbellulata*. *Proc. R. Soc. Lond. Ser. B.* 218: 225–239.

Heslop-Harrison, J. S., and T. Schwarzacher. 1996. Genomic Southern and *in situ* hybridization for plant genome analysis. In *Methods of Genome Analysis in Plants*. P. P. Jauhar, Ed., CRC Press, Boca Raton, FL, pp. 163–179.

Heun, M., A. E. Kennedy, J. A. Anderson, N. L. V. Lapitan, M. E. Sorrells, and S. D. Tanksley. 1991. Construction of a restriction fragment length polymorphism map for barley (*Hordeum vulgare*). *Genome* 34: 437–447.

Higashiyama, T., S. Yabe, N. Sasaki, Y. Nishimura, S. Migagishima, H. Kuroiwa, and T. Kuroiwa. 2001. Pollen tube attraction by the synergid cell. *Science* 293: 1480–1483.

Hinchee, M. A. W., D. V. Connor-Ward, C. A. Newell, R. E. McDonnell, S. J. Sato, C. S. Gasser, D. A. Fischhoff, D. B. Re, R. T. Fraley, and R. B. Horsch. 1988. Production of transgenic soybean plants using *Agrobacterium*-mediated DNA transfer. *Bio/Technol.* 6: 915–922.

Ho, K. M., and K. J. Kasha. 1975. Genetic control of chromosome elimination during haploid formation in barley. *Genetics* 81: 263–275.

Hoffmann, F., and T. Adachi. 1981. "Arabidobrassica": Chromosomal recombination and morphogenesis in asymmetric intergeneric hybrid cells. *Planta* 153: 586–593.

Hohmann, U., T. R. Endo, R. G. Herrmann, and B. S. Gill. 1995. Characterization of deletions in common wheat induced by an *Aegilops cylindrica* chromosome: Detection of multiple chromosome rearrangements. *Theor. Appl. Genet.* 91: 611–617.

Hoisington, D., M. Khairallah, T. Reeves, J. V. Ribaut, B. Skovmand, S. Taba, and M. Warburton. 1999. Plant genetic resources: What can they contribute toward increase crop productivity. *Proc. Natl. Acad. Sci. U.S.A.* 96: 5937–5943.

Hojsgaard, D., S. Klatt, R. Baier, J. G. Carman, and E. Hörandl. 2014. Taxonomy and viogeography of apomixis in angiosperms and associated biodiversity characteristics. *Crit. Rev. Plant Sci.* 33: 414–427.

Hollingshead, L. 1930a. Cytological investigations of hybrids and hybrid derivatives of *Crepis capillaris* and *Crepis tectorum*. *Univ. Calif. Publ. Agric. Sci.* 6: 55–94.

Hollingshead, L. 1930b. A cytological study of haploid *Crepis capillaris* plants. *Univ. Calif. Agric. Sci.* 6: 107–134.

Holm, G. 1960. An inversion in barley. *Hereditas* 46: 274–278.

Hougas, R. W., and S. J. Peloquin. 1957. A haploid plant of the potato variety Katahdin. *Nature* (London) 180: 1209–1210.

Hougas, R. W., S. J. Peloquin, and R. W. Ross. 1958. Haploids of the common potato. *J. Hered.* 49: 103–106.

Hu, C. H. 1968. Studies on the development of twelve types of trisomics in rice with reference to genetic study and breeding programme. *J. Agric. Assoc. China (N.S.)* 63: 53–71. (In Chinese with English summary.)

Huang, P.-L., K. Hahlbrock, and I. E. Somssich. 1988. Detection of a single-copy gene on plant chromosomes by *in situ* hybridization. *Mol. Gen. Genet.* 211: 143–147.

Huang, Q. C., and J. S. Sun. 1999. Autotriploid plants obtained from heteroploid rice crosses. *Acta Bot. Sinica.* 41: 741–746.

Hueros, G., J. M. Gonzalez, J. C. Sanz, and E. Ferrer. 1991. Gliadin gene location and C-banding identification of *Aegilops longissima* chromosomes added to wheat. *Genome* 34: 236–240.

Huskins, C. L., and L. M. Steinitz. 1948. The nucleus in differentiation and development. II. Induced mitoses in differentiated tissues of Rhoeo roots. *J. Hered.* 39: 67–77.

Hussain, S. W., W. M. Williams, C. F. Mercer, and D. W. R. White. 1997. Transfer of clover cyst nematode resistance from *Trifolium nigrescens* Viv. to *T. repens* L. by interspecific hybridization. *Theor. Appl. Genet.* 95: 1274–1281.

Hutchinson, J., and T. E. Miller. 1982. The nucleolar organisers of tetraploid and hexaploid wheats revealed by *in situ* hybridization. *Theor. Appl. Genet.* 61: 285–288.

Hyde, B. B. 1953. Addition of individual *Haynaldia villosa* chromosomes to hexaploid wheat. *Am. J. Bot.* 40: 174–182.

Hymowitz, T., and C. Newell. 1975. A wild relative of the soybean. *Illinois Res.* 17(4): 18–19.

Iglesias, V. A., E. A. Moscone, I. Papp, F. Neuhuber, S. Michalowski, T. Phelan, S. Spiker, M. Matzke, and A. J. M. Matzke. 1997. Molecular and cytogenetic analysis of stably and unstably expressed transgene loci in tobacco. *Plant Cell* 9: 1251–1264.

Iijma, K., and K. Fukui. 1991. Clarification of the conditions for the image analysis of plant chromosomes. *Bull. Natl. Inst. Agrobiol. Resour.* (Japan) 6: 1–58.

Imanywoha, J., K. B. Jensen, and D. Hole. 1994. Production and identification of primary trisomics in diploid *Agropyron cristatum* (crested wheatgrass). *Genome* 37: 469–476.

Inomata, N. 1982. Chromosome variation in callus cells of *Luzula elegans* Lowe with nonlocalized kinetochores. *Jpn. J. Genet.* 57: 59–64.

Ishida, Y., H. Saito, S. Ohta, Y. Hiei, T. Komari, and T. Kumashiro. 1996. High efficiency transformation of maize (*Zea mays* L.) mediated by *Agrobacterium tumefaciens*. *Nature Biotechnol.* 14: 745–750.

Ishiki, K. 1991. Cytological studies on African rice, *Oryza glaberrima* Steud. 3. Primary trisomics produced by pollinating autotriploid with diploid. *Euphytica* 55: 7–13.

Ising, G. 1969. Cytogenetic studies in *Cyrtanthus*. II. Aneuploidy and internal chromosome balance. *Hereditas* 61: 45–113.

Islam, A. K. M. R. 1980. Identification of wheat-barley addition lines with N-banding of chromosomes. *Chromosoma* (Berlin) 76: 365–373.

Islam, A. K. M. R., and K. W. Shepherd. 1990. Incorporation of barley chromosomes into wheat. In *Biotechnology in Agriculture and Forestry, Wheat*, vol. 13. Y. P. S. Bajaj, Ed., Springer-Verlag, Berlin, pp. 128–151.

Islam, A. K. M. R., K. W. Shepherd, and D. H. B. Sparrow. 1981. Isolation and characterization of euplasmic wheat-barley chromosome addition lines. *Heredity* 46: 161–174.

Iwata, N., and T. Omura. 1976. Studies on the trisomics in rice plants (*Oryza sativa* L.) IV. On the possibility of association of three linkage groups with one chromosome. *Jpn. J. Genet.* 51: 135–137.

Iwata, N., and T. Omura. 1984. Studies on the trisomics in rice plants (*Oryza sativa* L.) VI. An accomplishment of a trisomic series in *japonica* rice plants. *Jpn. J. Genet.* 59: 199–204.

Iwata, N., T. Omura, and M. Nakagahra. 1970. Studies on the trisomics in rice plants (*Oryza sativa* L.) I. Morphological classification of trisomics. *Jpn. J. Breed.* 20: 230–236.

Iyer, R. D. 1968. Towards evolving a trisomic series in Jute. *Curr. Sci.* 37: 181–183.

Jackson, J. A., and P. J. Dale. 1988. Callus induction, plant regeneration and an assessment of cytological variation in regenerated plants of *Lolium multiflorum* L. *J. Plant Physiol.* 132: 351–355.

Jackson, J. A., S. J. Dalton, and P. J. Dale. 1986. Plant regeneration from root callus in the forage grass *Lolium multiflorum*. In *Plant Tissue Culture and its Agricultural Applications*. L. A. Withers and P. J. Alderson, Eds., Butterworths, London, pp. 85–89.

Jacobsen, E. 1981. Polyploidization in leaf callus tissue and in regenerated plants of dihaploid potato. *Plant Cell Tiss. Org. Cult.* 1: 77–84.

Jacobsen, E., M. K. Daniel, J. E. M. Bergervoet-van Deelen, D. J. Huigen, and M. S. Ramanna. 1994. The first and second backcross progeny of the intergeneric fusion hybrids of potato and tomato after crossing with potato. *Theor. Appl. Genet.* 88: 181–186.

Jacobsen, E., J. H. de Jong, S. A. Kamstra, P. M. M. M. van den Berg, and M. S. Ramanna. 1995. Genomic *in situ* hybridization (GISH) and RFLP analysis for the identification of alien chromosomes in the backcross progeny of potato (+) tomato fusion hybrids. *Heredity* 74: 250–257.

Jahier, J., A. M. Chèvre, A. M. Tanguy, and F. Eber. 1989. Extraction of disomic addition lines of *Brassica napus–B. nigra*. *Genome* 32: 408–413.

Jain, S. K. 1959. Male sterility in flowering plants. *Bibliogr. Genet.* 18: 101–166.

Jain, S. K. 1960. Cytogenetics of rye (*Secale* spp.). *Bibliogr. Genet.* 19: 1–86.

Jan, C. C., J. M. Chandler, and S. A. Wagner. 1988. Induced tetraploidy and trisomic production of *Helianthus annuus* L. *Genome* 30: 647–651.

Jancey, R. C., and D. B. Walden. 1972. Analysis of pattern in distribution of breakage points in the chromosomes of *Zea mays* L. and *D. melanogaster Meigen*. *Can. J. Genet. Cytol.* 12: 429–442.

Janick, J., D. L. Mahoney, and P. L. Pfahler. 1959. The trisomics of *Spinacia oleracea*. *J. Hered.* 50: 47–50.

Janick, J., R. W. Schery, F. W. Woods, V. W. Ruttan. 1981. *Plant Science An Introduction to World Crops*, 3rd edition. W. H. Freeman, SF.

Janick, J., and E. C. Stevenson. 1955. The effects of polyploidy on sex expression in spinach. *J. Hered.* 46: 151–156.

Janse, J. 1985. Relative rate of development of aneuploid and euploid microspores in a tertiary trisomic of rye, *Secale cereale* L. *Can. J. Genet. Cytol.* 27: 393–398.

Janse, J. 1987. Male transmission of the translocated chromosome in a tertiary trisomic of rye: Genetic variation and relation to the rate of development of aneuploid pollen grains. *Theor. Appl. Genet.* 74: 317–327.

Jauhar, P. P. 1970. Chromosome behaviour and fertility of the raw and evolved synthetic tetraploids of pearl millet, *Pennisetum typhoides* Stapf et Hubb. *Genetica* 41: 407–424.

Jauhar, P. P. 1975. Genetic control of diploid-like meiosis in hexaploid tall fescue. *Nature* (London) 254: 595–597.

Jauhar, P. P. 1977. Genetic regulation of diploid-like chromosome pairing in *Avena. Theor. Appl. Genet.* 49: 287–295.

Jauhar, P. P. 1981. Cytogenetics of pearl millet. *Adv. Agron.* 34: 407–479.

Jauhar, P. P. 1990. Multidisciplinary approach to genome analysis in the diploid species, *Thinopyrum bessarabicum* and *Th. elongatum* (*Lophopyrum elongatum*), of the Triticeae. *Theor. Appl. Genet.* 80: 523–536.

Jauhar, P. P. 1996. *Methods of Genome Analysis in Plants.* CRC Press, Boca Raton, FL.

Jauhar, P. P., and C. F. Crane. 1989. An evaluation of Baum et al.'s assessment of the genomic system of classification in the Triticeae. *Am. J. Bot.* 76: 571–576.

Jauhar, P. P., O. Riera-Lizarazu, W. G. Dewey, B. S. Gill, C. F. Crane, and J. H. Bennett. 1991. Chromosome pairing relationships among the A, B, and D genomes of bread wheat. *Theor. Appl. Genet.* 82: 441–449.

Javornik, B., T. Sinkovic, L. Vapa, R. M. D. Koebner, and W. J. Rogers. 1991. A comparison of methods for identifying and surveying the presence of 1BL.1RS translocations in bread wheat. *Euphytica* 54: 45–53.

Jefferson, R. A., and R. Bicknell. 1996. The potential impacts of apomixis. A molecular genetic approach. In *The Impact of Molecular Genetics.* B. W. S. Sobral, Ed., Birkhäuser, Boston, pp. 87–101.

Jelodar, N. B., N. W. Blackhall, T. P. V. Hartman, D. S. Brar, G. Khush, M. R. Davey, E. C. Cocking, and J. B. Power. 1999. Intergeneric somatic hybrids of rice [*Oryza sativa* L. (+) *Porteresia coarctata* (Roxb.) Tateoka]. *Theor. Appl. Genet.* 99: 570–577.

Jena, K. K., and G. S. Khush. 1989. Monosomic alien addition lines of rice: Production, morphology, cytology, and breeding behavior. *Genome* 32: 449–455.

Jena, K. K., and G. S. Khush. 1990. Introgression of genes from *Oryza officinalis* Well ex Watt to cultivated rice, *O. sativa* L. *Theor. Appl. Genet.* 80: 737–745.

Jensen, W. A. 1998. Double fertilization: A personal view. *Sex Plant Reprod.* 11: 1–5.

Jewell, D. C. 1981. Recognition of two types of positive staining chromosomal material by manipulation of critical steps in the N-banding technique. *Stain Tech.* 56: 227–234.

Jewell, D. C., and C. J. Driscoll. 1983. The addition of *Aegilops variabilis* chromosomes to *Triticum aestivum* and their identification. *Can. J. Genet. Cytol.* 25: 76–84.

Jha, T. B., and S. C. Roy. 1982. Effect of different hormones on *Vigna* tissue culture and its chromosomal behaviour. *Plant Sci. Lett.* 24: 219–224.

Joersbo, M. 2001. Advances in the selection of transgenic plants using non-antibiotic marker genes. *Physiol. Plant.* 111: 269–272.

Johnson, B. L. 1975. Identification of the apparent B-genome donor of wheat. *Can. J. Genet. Cytol.* 17: 21–39.

Johnson, G. D., and, G. M. de C. Nogueira Araujo. 1981. A simple method of reducing the fading of immunofluorescence during microscopy. *J. Immunol. Meth.* 43: 349–350.

Johnson, S. S., R. L. Phillips, and H. W. Rines. 1987a. Meiotic behavior in progeny of tissue culture regenerated oat plants (*Avena sativa* L.) carrying near-telocentric chromosomes. *Genome* 29: 431–438.

Johnson, S. S., R. L. Phillips, and H. W. Rines. 1987b. Possible role of heterochromatin in chromosome breakage induced by tissue culture in oats (*Avena sativa* L.). *Genome* 29: 439–446.

Johnston, S. A., R. W. Ruhde, M. K. Ehlenfeldt, and R. E. Hanneman, Jr. 1986. Inheritance and microsporogenesis of a synaptic mutant (*sy*-2) from *Solanum commersonii* Dun. *Can. J. Genet. Cytol.* 28: 520–524.

Jones, B. L., G. L. Lookhart, A. Mak, and D. B. Cooper. 1982. Sequences of purothionins and their inheritance in diploid, tetraploid, and hexaploid wheats. *J. Hered.* 73: 143–144.

Jones, G. H. 1967. The control of chiasma distribution in rye. *Chromosoma* (Berlin) 22: 69–90.

Jones, G. H. 1974. Correlated components of chiasma variation and the control of chiasma distribution in rye. *Heredity* 32: 375–387.

Joppa, L. R., and S. S. Maan. 1982. A durum wheat disomic-substitution line having a pair of chromosomes from *Triticum boeticum*: Effect on germination and growth. *Can. J. Genet. Cytol.* 24: 549–557.

Joppa, L. R., and F. H. McNeal. 1972. Development of D-genome disomic addition lines of durum wheat. *Can. J. Genet. Cytol.* 14: 335–340.

Joppa, L. R., and N. D. Williams. 1977. D-genome substitution- monosomics of durum wheat. *Crop Sci.* 17: 772–776.

Jørgensen, R. B., and B. Anderson. 1989. Karyotype analysis of regenerated plants from callus cultures of interspecific hybrids of cultivated barley (*Hordeum vulgare* L.). *Theor. Appl. Genet.* 77: 343–351.

Kaeppler, H. E., S. M. Kaeppler, J.-H. Lee, and K. Arumuganathan. 1997. Synchronization of cell division in root tips of seven major cereal species for high yields of metaphase chromosomes for flow-cytometric analysis and sorting. *Plant Mol. Biol. Rep.* 15: 141–147.

Kakeda, K., K. Fukui, and H. Yamagata. 1991. Heterochromatic differentiation in barley chromosomes revealed by C- and N-banding techniques. *Theor. Appl. Genet.* 81: 144–150.

Kakeda, K., H. Yamagata, K. Fukui, M. Ohno, K. Fukui, Z. Z. Wei, and F. S. Zhu. 1990. High resolution bands in maize chromosomes by G-banding methods. *Theor. Appl. Genet.* 80: 265–272.

Kaltsikes, P. J., and L. E. Evans. 1967. Production and identification of trisomic types in *Beta vulgaris*. *Can. J. Genet. Cytol.* 9: 691–699.

Kamanoi, M., and B. C. Jenkins. 1962. Trisomics in common rye, *Secale cereale* L. *Seiken Zihô* 13: 118–123.

Kanda, M., S. Kikuchi, F. Takaiwa, and K. Oono. 1988. Regeneration of variant plants from rice (*Oryza sativa* L.) protoplasts derived from long term cultures. *Jpn. J. Genet.* 63: 127–136.

Kaneko, Y., Y. Matsuzawa, and M. Sarashima. 1987. Breeding of the chromosome addition lines of radish with single kale chromosome. *Jpn. J. Breed.* 37: 438–452. (Japanese with English summary.)

Kao, K. N., R. A. Miller, O. L. Gamborg, and B. L. Harvey. 1970. Variations in chromosome number and structure in plant cells grown in suspension cultures. *Can. J. Genet. Cytol.* 12: 297–301.

Kar, D. K., and S. Sen. 1985. Effect of hormone on chromosome behaviour in callus cultures of *Asparagus racemosus*. *Biol. Plant.* 27: 6–9.

Karp, A. 1986. Chromosome variation in regenerated plants. In *Genetic manipulation in Plant Breeding.* W. Horn, C. J. Jensen, W. Odenbach, and O. Schieder, Eds., Proc. Int. Symp. EUCARPIA. Walter de Gruyter, NY, pp. 547–554.

Karp, A., and S. E. Maddock. 1984. Chromosome variation in wheat plants regenerated from cultured immature embryos. *Theor. Appl. Genet.* 67: 249–255.

Karp, A., R. S. Nelson, E. Thomas, and S. W. J. Bright. 1982. Chromosome variation in protoplast-derived potato plants. *Theor. Appl. Genet.* 63: 265–272.

Karp, A., R. Risiott, M. G. K. Jones, and S. W. J. Bright. 1984. Chromosome doubling in monohaploid and dihaploid potatoes by regeneration from cultured leaf explants. *Plant Cell Tiss. Org. Cult.* 3: 363–373.

Karp, A., S. H. Steele, S. Parmar, M. G. K. Jones, P. R. Shewry, and A. Breiman. 1987a. Relative stability among barley plants regenerated from cultured immature embryos. *Genome* 29: 405–412.

Karp, A., Q. S. Wu, S. H. Steele, and M. G. K. Jones. 1987b. Chromosome variation in dividing protoplasts and cell suspensions of wheat. *Theor. Appl. Genet.* 74: 140–146.

Karpechenko, G. D. 1927. The production of polyploid gametes in hybrids. *Hereditas* 9: 349–368.

Karpen, G. H., M. -H. Lee, and H. Le. 1996. Centric heterochromatin and the efficiency of achiasmate disjunction in Drosophila female meiosis. *Science* 273: 118–122.

Karsenti, E., and I. Vernos. 2001. The mitotic spindle: A self-made machine. *Science* 294: 543–547.

Kasha, K. J. 1974. Haploids from somatic cells. In *Haploids in Higher Plants. Advances and Potential. Proc. Ist Int. Symp.* K. J. Kasha, Ed., Ainsworth Press, Canada, pp. 67–87.

Kasha, K. J., and K. N. Kao. 1970. High frequency haploid production in barley (*Hordeum vulgare* L.). *Nature* (London) 225: 874–876.

Kasha, K. J., and H. A. McLennan. 1967. Trisomics in diploid alfalfa I. Production, fertility and transmission. *Chromosoma* (Berlin) 21: 232–242.

Katayama, T. 1961. Cytogenetical studies on asynaptic rice plant (*Oryza sativa* L.) induced by X-ray. *La Kromosomo* 48: 1591–1601 (in Japanese with English summary.)

Katayama, T. 1963. Study on the progenies of autotriploid and asynaptic rice plants. *Jpn J. Breed.* 13: 83–87.

Katayama, T. 1964. Further review on the heritable asynapsis in plants. *La Kromosomo* 57–59: 1934–1942.

Katayama, T. 1982. Cytogenetical studies on the genus *Oryza*. XIII. Relationship between the genomes E and D. *Jpn. J. Genet.* 57: 613–621.

Katayama, T., and T. Ogawa. 1974. Cytogenetical studies on the genus *oryza*. VII. Cytogenetical studies on F$_1$ hybrids between diploid *O. punctata* and diploid species having C genome. *Jpn. J. Breed.* 24: 165–168.

Kato, S., T. Akiyama, C. M. O'Neill, and K. Fukui. 1997. Manual on the chromosome image analyzing system III, CHIAS III. *Res. Rep. Agr. Devel. Hokuriku Area, No. 36*: 1–76.

Katterman, F. R. H., and D. R. Ergle. 1970. A study of quantitative variations of nucleic acids in *Gossypium*. *Phytochemistry* 9: 2007–2010.

Kaul, M. L. H. 1988. Male sterility in higher plants. In *Monogr. Theor. Appl. Genet.* No. 10. Springer-Verlag, Berlin.

Kaul, M. L. H., and T. G. K. Murthy. 1985. Mutant genes affecting higher plant meiosis. *Theor. Appl. Genet.* 70: 449–466.

Kawata, M., A. Ohmiya, Y. Shimamoto, K. Oono, and F. Takaiwa. 1995. Structural changes in the plastid DNA of rice (*Oryza sativa* L.) during tissue culture. *Theor. Appl. Genet.* 90: 364–371.

Keathley, D. E., and R. L. Scholl. 1983. Chromosomal heterogeneity of *Arabidopsis thaliana* anther callus, regenerated shoots and plants. *Z. Pflanzenphysiol.* 112: 247–255.

Keen, N. T., R. L. Lyne, and T. Hymowitz. 1986. Phytoalexin production as a chemosystematic parameter within the genus *Glycine. Biochem. Syst. Ecol.* 14: 481–486.

Kejnoský, E., J. Vrána, S. Matsunaga, P. Souček, J., J. Široký, J. Doležel, and B. Vyakot. 2001. Localization of male-specifically expressed *MROS* genes of *Silene latifolia* by PCR on flow-sorted six chromosomes and autosomes. *Genetics* 158: 1269–1277.

Kerber, E. R. 1954. Trisomics in barley. *Science* 120: 808–809.

Kerber, E. R., and P. L. Dyck. 1990. Transfer to hexaploid wheat of linked genes for adult-plant leaf rust and seedling stem rust resistance from an amphiploid of *Aegilops speltoides* x *Triticum monococcum. Genome* 33: 530–537.

Kerby, K., and J. Kuspira. 1987. The phylogeny of the polyploid wheats *Triticum aestivum* (bread wheat) and *Triticum turgidum* (macaroni wheat). *Genome* 29: 722–737.

Kerby, K., and J. Kuspira. 1988. Cytological evidence bearing on the origin of the B genome in polyploid wheats. *Genome* 30: 36–43.

Kessel, R., and P. R. Rowe. 1975. Production of intraspecific aneuploids in the genus *Solanum*. 1. Doubling the chromosome number of seven diploid species. *Euphytica* 24: 65–75.

Khawaja, H. I. I., and J. R. Ellis. 1987. Colchicine-induced desynaptic mutations in *Lathyrus odoratus* L. and *L. pratensis* L. *Genome* 29: 859–866.

Kho, Y. O., and J. Baër. 1968. Observing pollen tubes by means of fluorescence. *Euphytica* 17: 298–302.

Khokhlov, S. S. 1976. Evolutionary-genetic problems of apomixis in angiosperms. In *Apomixis and Breeding*. S. S. Khokhlov, Ed., Amerind Publ. Co. Pvt. Ltd., New Delhi, pp. 1–102 (translated from the Russian).

Khush, G. S. 1962. Cytogenetic and evolutionary studies in *Secale*. II. Interrelationships of the wild species. *Evolution* 16: 484–496.

Khush, G. S. 1963. Cytogenetic and evolutionary studies in *Secale* IV. *Secale vavilovii* and its biosystematic status. *Z. Pflanzenüchtg.* 50: 34–43.

Khush, G. S. 1973. *Cytogenetics of Aneuploids*. Academic Press, NY.

Khush, G. S. 2010. Trisomics and addition lines in rice. *Breeding Sci.* 60: 469–474.

Khush, G. S., and C. M. Rick. 1966. The origin, identification and cytogenetic behavior of tomato monosomics. *Chromosoma* (Berlin) 18: 407–420.

Khush, G. S., and C. M. Rick. 1967a. Novel compensating trisomics of the tomato: Cytogenetics, monosomic analysis, and other applications. *Genetics* 56: 297–307.

Khush, G. S., and C. M. Rick. 1967b. Studies on the linkage map of chromosome 4 of the tomato and on the transmission of induced deficiencies. *Genetica* 38: 74–94.

Khush, G. S., and C. M. Rick. 1967c. Tomato tertiary trisomics: Origin, identification, morphology and use in determining position of centromeres and arm location of markers. *Can. J. Gent. Cytol.* 9: 610–631.

Khush, G. S., and C. M. Rick. 1968a. Cytogenetic analysis of the tomato genome by means of induced deficiencies. *Chromosoma* (Berlin) 23: 452–484.

Khush, G. S., and C. M. Rick. 1968b. Tomato telotrisomics: Origin, identification, and use in linkage mapping. *Cytologia* 33: 137–148.

Khush, G. S., and C. M. Rick. 1969. Tomato secondary trisomics: Origin, identification, morphology, and use in cytogenetic analysis of the genome. *Heredity* 24: 129–146.

Khush, G. S., and R. J. Singh. 1991. Chromosome architecture and aneuploidy in rice. In *Chromosome Engineering in Plants: Genetics, Breeding, Evolution*. Part A, P. K. Gupta and T. Tsuchiya, Eds., Elsevier, Amsterdam, pp. 577–598.

Khush, G. S., R. J. Singh, S. C. Sur, and A. L. Librojo. 1984. Primary trisomics of rice: Origin, morphology, cytology and use in linkage mapping. *Genetics* 107: 141–163.

Khush, G. S., and G. L. Stebbins. 1961. Cytogenetic and evolutionary studies in *Secale*. I. Some new data on the ancestry of *S. cereale. Am. J. Bot.* 48: 723–730.

Kihara, H. 1924. Cytologische und genetische studien bei wichtigen getreidearten mit besonderer rücksicht auf das verhalten der chromosomen und die sterilität in den bastarden. *Mem. Coll. Sci. Kyoto Imp. Uni. Ser. B* 1: 1–200. (in German.)

Kihara, H. 1930. Genomanalyse bei *Triticum* und *Aegilops. Cytologia* 1: 263–270 (in German).

Kihara, H. 1944. The discovery of the DD analyser, one of the ancestors of common wheat (preliminary report). *Agric. Hort.* 19: 889–890 (in Japanese).

Kihara, H. 1947. Entdeckung des DD-analysators beim weizen. *Seiken Zihô* 3: 1–15.

Kihara, H. 1951. Substitution of nucleus and its effects on genome manifestations. *Cytologia* 16: 177–193.

Kihara, H. 1958. Breeding for seedless fruits. *Seiken Zihô* 9: 1–7.

Kihara, H. 1963. Interspecific relationships in *Triticum* and *Aegilops*. *Seiken Zihô* 15: 1–12.

Kihara, H., and F. A. Lilienfeld. 1932. Genomanalyse bei *Triticum* und *Aegilops* IV. Untersuchungen an *Aegilops* x *Triticum* -und *Aegilops*-bastarden. *Cytologia* 3: 384–456 (in German).

Kihara, H., and I. Nishiyama. 1930. Genomanalyse bei *Triticum* und *Aegilops*. I. Genomeaffinitäten in tri-, tetra- und pentaploiden weizenbastarden. *Cytologia* 1: 270–284 (in German).

Kimber, G. 1961. Basis of the diploid-like meiotic behaviour of polyploid cotton. *Nature* (London) 191: 98–100.

Kimber, G. 1967. The addition of the chromosomes of *Aegilops umbellulata* to *Triticum aestivum* (var. Chinese Spring). *Genet. Res.* (Camb.) 9: 111–114.

Kimber, G., and L. C. Alonso. 1981. The analysis of meiosis in hybrids. III. Tetraploid hybrids. *Can. J. Genet. Cytol.* 23: 235–254.

Kimber, G., L. C. Alonso, and P. J. Sallee. 1981. The analysis of meiosis in hybrids. I. Aneuploid hybrids. *Can. J. Genet. Cytol.* 23: 209–219.

Kimber, G., and M. Feldman. 1987. Wild wheat: An introduction. Special report 353, College of Agriculture, University of Missouri, Columbia, MO, p. 142.

Kimber, G., B. S. Gill, J. M. Rubenstein, and G. L. Barnhill. 1975. The technique of Giemsa staining of cereal chromosomes. Res. Bull. 1012. University of Missouri, Columbia, College of Agriculture, Agric. Exp. Stn. pp. 3–6.

Kimber, G., and R. Riley. 1963. Haploid angiosperms. *Bot. Rev.* 29: 480–531.

Kimber, G., and E. R. Sears. 1969. Nomenclature from the description of aneuploids in the *Triticinae*. In *Proc. 3rd Int. Wheat Genet. Symp.* K. W. Finlay and K. W. Shepherd, Eds., Butterworth & Co Ltd., Australia, pp. 468–473.

Kimber, G., and K. Tsunewaki. 1988. Genome symbols and plasma types in the wheat group. In *Proc. 7th Int. Wheat Genet. Symp.* T. E. Miller and R. M. D. Koebner, Eds., Bath Press, Avon, U.K., pp. 1209–1211.

Kindiger, B., and J. B. Beckett. 1985. A hematoxylin staining procedure for maize pollen grain chromosomes. *Stain Tech.* 60: 265–269.

King, I. P., R. M. D. Koebner, S. M. Reader, and T. E. Miller. 1991. Induction of a mutation in the male fertility gene of the preferentially transmitted *Aegilops sharonensis* chromosome 4S[l] and its application for hybrid wheat production. *Euphytica* 54: 33–39.

Kitada, K., N. Kurata, H. Satoh, and T. Omura. 1983. Genetic control of meiosis in rice, *Oryza sativa* L. I. Classification of meiotic mutants induced by MNU and their cytogenetical characteristics. *Jpn. J. Genet.* 58: 231–240.

Kitada, K., and T. Omura. 1984. Genetic control of meiosis in rice, *Oryza sativa* L. IV. Cytogenetical analysis of asynaptic mutants. *Can. J. Genet. Cytol.* 26: 264–271.

Kitani, Y. 1963. Orientation, arrangement and association of somatic chromosomes. *Jpn. J. Genet.* 38: 244–256.

Kleckner, N. 1996. Meiosis: How could it work? *Proc. Natl. Acad. Sci. U.S.A.* 93: 8167–8174.

Klein, T. M., E. D. Wolf, R. Wu, and J. C. Sanford. 1987. High-velocity microprojectiles for delivering nucleic acids into living cells. *Nature* (London) 327: 70–73.

Knott, D. R. 1964. The effect on wheat of an *Agropyron* chromosome carrying rust resistance. *Can. J. Genet. Cytol.* 6: 500–507.

Knott, D. R., J. Dvořák, and J. S. Nanda. 1977. The transfer to wheat and homoeology of an *Agropyron elongatum* chromosome carrying resistance to stem rust. *Can. J. Genet. Cytol.* 19: 75–79.

Knox, R. B., and J. Heslop-Harrison. 1963. Experimental control of aposporous apomixis in a grass of the Andropogoneae. *Botaniska Notieser* 116: 127–141.

Koba, T., S. Takumi, and T. Shimada. 1997. Isolation, identification and characterization of disomic and translocated barley chromosome addition lines of common wheat. *Euphytica* 96: 289–296.

Kobayashi, S. 1987. Uniformity of plants regenerated from orange (*Citrus sinensis* Osb.) protoplasts. *Theor. Appl. Genet.* 74: 10–14.

Koch, J. E., S. Kølvraa, K. B. Petersen, N. Gregersen, and I. Bolunel. 1989. Oligonucleotide-priming methods for the chromosome-specific labelling if alpha satellite DNA *in situ*. *Chromosoma* 98: 259–265.

Kodani, M. 1948. Sodium ribose nucleate and mitosis. Induction of morphological changes in the chromosomes and of abnormalities in mitotic divisions in the root meristem. *J. Hered.* 39: 327–335.

Koduru, P. R. K., and M. K. Rao. 1981. Cytogenetics of synaptic mutants in higher plants. *Theor. Appl. Genet.* 59: 197–214.

Koller, O. L., and F. J. Zeller. 1976. The homoeologous relationships of rye chromosomes 4R and 7R with wheat chromosomes. *Genet. Res.* (Camb.) 28: 177–188.

Koller, P. C. 1938. Asynapsis in *Pisum sativum. J. Genet.* 36: 275–306.

Kollipara, K. P., R. J. Singh, and T. Hymowitz. 1997. Phylogenetic and genomic relationships in the genus *Glycine* Willd. based on sequences from the ITS region of nuclear rDNA. *Genome* 40: 57–68.

Koltunow, A. M., R. A. Bicknell, and A. M. Chaudhury. 1995. Apomixis: Molecular strategies for the generation of genetically identical seeds without fertilization. *Plant Physiol.* 108: 1345–1352.

Koltunow, A. M., T. Hidaka, and S. P. Robinson. 1996. Polyembryony in *Citrus. Plant Physiol.* 110: 599–609.

Kononowicz, A. K., K. Floryanowicz-Czekalska, J. Clithero, A. Meyers, P. M. Hasegawa, and R. A. Bressan. 1990a. Chromosome number and DNA content of tobacco cells adapted to NaCl. *Plant Cell Rep.* 8: 672–675.

Kononowicz, A. K., P. M. Hasegawa, and R. A. Bressan. 1990b. Chromosome number and nuclear DNA content of plants regenerated from salt adapted plant cells. *Plant Cell Rep.* 8: 676–679.

Konvička, O., and W. Gottschalk. 1974. Untersuchungen an der meiosis steriler mutanten von *Allium cepa. Angew. Botanik.* 48: 9–19 (in German with English summary).

Konzak, C. F., and R. E. Heiner. 1959. Progress in the transfer of resistance to bunt (*Tilletia caries* and *T. foetida*) from *Agropyron* to wheat. *Wheat Inf. Serv.* 9–10: 31.

Koo, H-D., S. K. Sehgal, B. Friebe, and B. S. Gill. 2015. Structure and stability of telocentric chromosomes in wheat. *PLOS One* 10(9): e0137747. DOI: 10.1371/JOURNAL.PONE.0137747 September 18, 2015.

Kooter, J. M., M. A. Matzke, and P. Meyer. 1999. Listening to the silent genes: Transgene silencing, gene regulation and pathogen control. *Trends Plant Sci.* 9: 340–347.

Kota, R. S., and J. Dvořák. 1988. Genomic instability in wheat induced by chromosome 6B^s of *Triticum speltoides. Genetics* 120: 1085–1094.

Kramer, H. H., R. Veyl, and W. D. Hanson. 1954. The association of two genetic linkage groups in barley with one chromosome. *Genetics* 39: 159–168.

Kramer, M. G. and S. M. Reed. 1988. An evaluation of natural nulli haploidy for *Nicotiana tabacum* L. Nullisomic production. II. A pollen irradiation and ovule culture approach. *J. Hered.* 79: 469–472.

Kramer, M. G., and K. Redenbaugh. 1994. Commercialization of a tomato with an antisense polygalacturonase gene: The FLAVR SAVR™ tomato story. *Euphytica* 79: 293–297.

Kranz, A. R. 1963. Beiträge zur cytologischen und genetischen evolutionsforschung an dem roggen. *Z. Pflazenzüchtg.* 50: 44–58.

Kranz, A. R. 1976. Karyotype analysis in meiosis: Giemsa banding in the genus *Secale* L. *Theor. Appl. Genet.* 47: 101–107.

Kreft, I. 1969. Cytological studies on an inversion in barley. *Hereditas* 62: 14–24.

Krishnaswamy, N., and K. Meenakshi. 1957. Abnormal meiosis in grain Sorghums-Desynapsis. *Cytologia* 22: 250–262.

Krolow, K. D. 1973. 4*x* Triticale, production and use in triticale breeding. In *Proc. 4th. Int. Wheat Genet. Symp.* E. R. Sears and L. M. S. Sears, Eds., Missouri Agric. Exp. Stn., Columbia, MO, pp. 237–243.

Krumbiegel, G., and O. Schieder. 1979. Selection of somatic hybrids after fusion of protoplasts from *Datura innoxia* Mill. and *Atropa belladonna* L. *Planta* 145: 371–375.

Kudirka, D. T., G. W. Schaeffer, and P. S. Baenziger. 1989. Stability of ploidy in meristems of plants regenerated from anther calli of wheat (*Triticum aestivum* L. em. Thell). *Genome* 32: 1068–1073.

Kumar, A., and E. C. Cocking. 1987. Protoplast fusion: A novel approach to organelle genetics in higher plants. *Amer. J. Bot.* 74: 1289–1303.

Kumar, P. S., and T. Hymowitz. 1989. Where are diploid (2*n* = 2*x* = 20) genome donors of *Glycine* Willd. (Leguminosae, Papilionoideae)? *Euphytica* 40: 221–226.

Künzel, G. 1976. Indications for a necessary revision of the barley karyogramme by use of translocations. In *Barley Genetics III. Proc. Int. Barley Genet. Symp.* H. Gaul, Ed., Thiemig Press, Munich, pp. 275–281.

Kurata, N., and T. Omura. 1978. Karyotype analysis in rice. I. A new method for identifying all chromosome pairs. *Jpn. J. Genet.* 53: 251–255.

Kushnir, U., and G. M. Halloran. 1981. Evidence for *Aegilops sheronensis* Eig as the donor of the B genome of wheat. *Genetics* 99: 495–512.

Kuspira, J., R. N. Bhambhani, R. S. Sadasivaiah, and D. Hayden. 1986. Genetic and cytogenetic analyses of the A genome of *Triticum monococcum*. III. Cytology, breeding behavior, fertility, and morphology of autotriploids. *Can. J. Genet. Cytol.* 28: 867–887.

Kuspira, J., R. N. Bhambhani, and T. Shimada. 1985. Genetic and cytogenetic analyses of the A genome of *Triticum monococcum* I. Cytology, breeding behaviour, fertility and morphology of induced autotetraploids. *Can. J. Cenet. Cytol.* 27: 51–63.

Lacadena, J. R., and M. Candela. 1977. Centromere co-orientation at metaphase I in interchange heterozygotes of rye, *Secale cereale* L. *Chromosoma* (Berlin) 64: 175–189.

La Cour, L. P., and B. Wells. 1970. Meiotic prophase in anthers of asynaptic wheat. A light and electron microscopical study. *Chromosoma* (Berlin) 29: 419–427.

Ladizinsky, G. 1974. Genome relationships in the diploid oats. *Chromosoma* (Berlin) 47: 109–117.

Ladizinsky, G., N. F. Weeden, and F. J. Muehlbauer. 1990. Tertiary trisomics in lentil. *Euphytica* 51: 179–184.

Ladizinsky, G., C. A. Newell, and T. Hymowitz. 1979. Wide crosses in soybeans: Prospects and limitations. *Euphytica* 28: 421–423.

Lakshmanan, K. K., and K. B. Ambegaokar. 1984. Polyembryony. In *Embryology of Angiosperms*. P. M. Johri, Ed., Springer, Amsterdam, pp. 445–474.

Lamm, R. 1944. Chromosome behaviour in a tetraploid rye plant. *Hereditas* 30: 137–144.

Lamm, R., and R. J. Miravalle. 1959. A translocation tester set in *Pisum*. *Hereditas* 45: 417–440.

Lander, E. C., and R. A. Weinberg. 2000. Genomic: Journey to the center of biology. *Science* 287: 1777–1782.

Landjeva, S., and G. Ganeva. 1999. Identification of *Aegilops ovata* chromosomes added to the wheat (*Triticum aestivum* L.) genome. *Cereal Res. Comm.* 27: 55–61.

Lange, W. 1971. Crosses between *Hordeum vulgare* L. and *H. bulbosum* L. II. Elimination of chromosomes in hybrid tissues. *Euphytica* 20: 181–194.

Lange, W., Th. S. M. De Bock, J. P. C. Van Geyt, and M. Oléo. 1988. Monosomic additions in beet (*Beta vulgaris*) carrying extra chromosomes of *B. procumbens*. *Theor. Appl. Genet.* 76: 656–664.

Lange, W., and G. Jochemsen. 1976. Karyotypes, nucleoli, and amphiplasty in hybrids between *Hordeum vulgare* L. and *H. bulbosum* L. *Genetica* 46: 217–233.

Lange, W., H. Toxopeus, T. H. Lubberts, O. Dolstra, and J. L. Harrewijn. 1989. The development of raparadish (x *Brasscoraphanus*, 2n = 38), a new crop in agriculture. *Euphitica* 40: 1–14.

Lapitan, N. L. V., M. W. Canal, and S. D. Tanksley. 1991. Organization of the 5S ribosomal RNA genes in the genomes of tomato. *Genome* 34: 509–514.

Lapitan, N. L. V., R. G. Sears, and B. S. Gill. 1984. Translocations and other karyotypic structural changes in wheat x rye hybrids regenerated from tissue culture. *Theor. Appl. Genet.* 68: 547–554.

Larkin, P. J., P. M. Banks, E. S. Lagudah, R. Appels, C. Xiao, X. Zhiyong, H. W. Ohm, and R. A. McIntosh. 1995. Disomic *Thinopyrum intermedium* addition lines in wheat with barley yellow dwarf virus resistance and with rust resistances. *Genome* 38: 385–394.

Larkin, P. J., and W. R. Scowcroft. 1981. Somaclonal variation—A novel source of variability from cell cultures for plant improvement. *Theor. Appl. Genet.* 60: 197–214.

Larson, R. I., and T. G. Atkinson. 1973. Wheat-*Agropyron* chromosome substitution lines as sources of resistance to wheat streak mosaic virus and its vector, *Aceria tulipae*. In *Proc. 4th. Int. Wheat Genet. Symp.* E. R. Sears and L. M. S. Sears, Eds., Missouri Agric. Exp. Stn., Columbia, MO, pp. 173–177.

Laurie, D. A. 1989. The frequency of fertilization in wheat x pearl millet crosses. *Genome* 32: 1063–1067.

Laurie, D. A., and M. D. Bennett. 1989. The timing of chromosome elimination in hexaploid wheat x maize crosses. *Genome* 32: 953–961.

Lavania, U. C., and S. Srivastava. 1988. Ploidy dependence of chromosomal variation in callus cultures of *Hyoscyamus muticus* L. *Protoplasma* 145: 55–58.

Law, C. N. 1963. An effect of potassium on chiasma frequency and recombination. *Genetica* 33: 313–329.

Lawrence, J. B., C. A. Villnave, and R. H. Singer. 1988. Sensitive, high-resolution chromatin and chromosome mapping *in situ*: Presence and orientation of two closely integrated copies of EBV in a lymphoma line. *Cell* 52: 51–61.

Lazar, G. B., D. Dudits, and Z. R. Sung. 1981. Expression of cycloheximide resistance in carrot somatic hybrids and their segregants. *Genetics* 98: 347–356.

Lazar, M. D., T. H. H. Chen, G. J. Scoles, and K. K. Kartha. 1987. Immature embryo and anther culture of chromosome addition lines of rye in Chinese Spring wheat. *Plant Sci.* 51: 77–81.

Lazzeri, P. A., D. F. Hildebrand, and G. B. Collins. 1987a. Soybean somatic embryogenesis: Effects of hormones and culture manipulations. *Plant Cell Tiss. Org. Cult.* 10: 197–208.

Lazzeri, P. A., D. F. Hildebrand, and G. B. Collins. 1987b. Soybean somatic embryogenesis: Effects of nutritional, physical and chemical factors. *Plant Cell Tiss. Org. Cult.* 10: 209–220.

Le, H. T., K. C. Armstrong, and B. Miki. 1989. Detection of rye DNA in wheat-rye hybrids and wheat translocation stocks using total genomic DNA as a probe. *Plant Mol. Biol. Rep.* 7: 150–158.

Leblanc, O., D. Grimanelli, D. González-de-León, and Y. Savidan. 1995a. Detection of the apomictic mode of reproduction in maize-*Tripsacum* hybrids using maize RFLP markers. *Theor. Appl. Genet.* 90: 1198–1203.

Leblanc, O., M. D. Peel, J. G. Carman, and Y. Savidan. 1995b. Megasporogenesis and megagametogenesis in several *Tripsacum* species (Poaceae). *Am. J. Bot.* 82: 57–63.

Lee, H. K., R. Kessel, and P. R. Rowe. 1972. Multiple aneuploids from interspecific crosses in *Solanum*: Fertility and cytology. *Can. J. Genet. Cytol.* 14: 533–543.

Lee, H. K., and P. R. Rowe. 1975. Trisomics in *Solanum chacoense*: Fertility and cytology. *Am. J. Bot.* 62: 593–601.

Lee, M., and R. L. Phillips. 1987. Genomic rearrangements in maize induced by tissue culture. *Genome* 29: 122–128.

Lee-Chen, S., and L. M. Steinitz-Sears. 1967. The location of linkage groups in *Arabidopsis thaliana. Can. J. Genet. Cytol.* 9: 381–384.

Leggett, J. M. 1984. Morphology and metaphase chromosome pairing in three *Avena* hybrids *Can. J. Genet. Cytol.* 26: 641–645.

Leitch, A. R., J. Maluszynska, I. J. Leitch, K. Anamthawat-Jónsson, T. Schwarzacher, and J. S. Heslop-Harrison. 1996. *In situ* hybridisation of chromosomes using DNA probes labelled with digoxigenin and detected by epifluorescence. In *Techniques of Plant Cytogenetics.* J. Jahier, Ed., Science Publishers, Lebanon, NH, pp. 82–86.

Leitch, I. J., A. R. Leitch, and J. S. Heslop-Harrison. 1991. Physical mapping of plant DNA sequences by simultanious *in situ* hybridization of two differently labelled fluorescent probes. *Genome* 34: 329–333.

Lesley, J. W. 1928. A cytological and genetical study of progenies of triploid tomatoes. *Genetics* 13: 1–43.

Lesley, J. W. 1932. Trisomic types of the tomato and their relation to the genes. *Genetics* 17: 545–559.

Levan, A. 1938. The effect of colchicine in root mitoses in *Allium. Hereditas* 24: 471–486.

Levan, A. 1942. The effect of chromosomal variation in sugar beets. *Hereditas* 28: 345–399.

Levan, A., K. Fredga, and A. A. Sandberg. 1964. Nomenclature for centromeric position on chromosomes. *Hereditas* 52: 201–220.

Li, H. W., W. K. Pao, and C. H. Li. 1945. Desynapsis in the common wheat. *Am. J. Bot.* 32: 92–101.

Liang, G. H. 1979. Trisomic transmission in six primary trisomics of sorghum. *Crop Sci.* 19: 339–344.

Libbenga, K. R., and J. G. Torrey. 1973. Hormone-induced endoreduplication prior to mitosis in cultured pea root cortex cells. *Am. J. Bot.* 60: 293–299.

Lichter, P., C.-J. C. Tang, K. Call, G. Hermanson, G. A. Evans, D. Housman, and D. C. Ward. 1990. High-resolution mapping of human chromosome 11 by *in situ* hybridization with cosmid clones. *Science* 247: 64–69.

Lilienfeld, F. A. 1951. H. Kihara: Genome-analysis in *Triticum* and *Aegilops.* X. Concluding review. *Cytologia* 16: 101–123.

Limin, A. E., and J. Dvořák. 1976. C-banding of rye chromosomes with cold SSC buffer. *Can. J. Genet. Cytol.* 18: 491–496.

Lin, B.-Y. 1984. Ploidy barrier to endosperm development in maize. *Genetics* 107: 103–115.

Lin, B.-Y. 1987. Cytological evidence of terminal deficiencies produced by the *r-X1* deletion in maize. *Genome* 29: 718–721.

Lin, B.-Y., and E. H. Coe, Jr. 1986. Monosomy and trisomy induced by the *r-X1* deletion in maize, and associated effects on endosperm development. *Can. J. Genet. Cytol.* 28: 831–834.

Lin, B.-Y., K. Marquette, and P. Sallee. 1990. Characterization of deficiencies generated by *r-X1* in maize. *J. Hered.* 81: 359–364.

Lin, W., C. S. Anuratha, K. Datta, I. Potrykus, S. Muthukrishnan, and S. K. Datta. 1995. Genetic engineering of rice for resistance to sheath blight. *Bio/Technol.* 13: 686–691.

Lin, Y. J., and J. F. Chen. 1981. Trisomics in diploid marigold, *Tagetes erecta. J. Hered.* 72: 441–442.

Lind, V. 1982. Analysis of the resistance of wheat-rye addition lines to powdery mildew of wheat (*Erysiphe graminis* F. sp. *tritici*). *Tag.-Ber., Akad. Landwirtsch. Wiss., DDR,* 198: 509–520.

Linde-Laursen, Ib. 1975. Giemsa C-banding of the chromosomes of 'Emir' barley. *Hereditas* 81: 285–289.

Linde-Laursen, Ib. 1978. Giemsa C-banding of barley chromosomes. II. Banding patterns of trisomics and telotrisomics. *Hereditas* 89: 37–41.

Liu, C. J., M. D. Atkinson, C. N. Chinoy, K. M. Devos, and M. D. Gale. 1992. Nonhomologous translocations between group 4, 5 and 7 chromosomes within wheat and rye. *Theor. Appl. Genet.* 83: 305–312.

Liu, D. J., P. D. Chen, G. Z. Pei, Y. N. Wang, B. X. Qiu, and S. L. Wang. 1988. Transfer of *Haynaldia villosa* chromosomes into *Triticum aestivum.* In *Proc. 7th Int. Wheat Genet. Symp.* T. E. Miller and R. M. D. Koebner, Eds., Bath Press, Avon, U.K., pp. 355–361.

Liu, W., R. S. Torisky, K. P. McAllister, S. Avdiushko, D. Hildebrand, and G. B. Collins. 1996. Somatic embryo cycling: Evaluation of a novel transformation and assay system for seed-specific gene expression in soybean. *Plant Cell Tiss. Organ Cult.* 47: 33–42.

Liu, Z.-W., R. R.-C. Wang, and J. C. Carman. 1994. Hybrids and backcross progenies between wheat (*Triticum aestivum* L.) and apomictic Australian wheatgrass [*Elymus rectisetus* (Nees in Lahm.) A. Löve & Connor]: Karyotype and genomic analysis. *Theor. Appl. Genet.* 89: 599–605.

Loidl, J. 1987. Synaptonemal complex spreading in *Allium ursinum*: Pericentric asynapsis and axial thickenings. *J. Cell Sci.* 87: 439–448.

Loidl, J. 1990. The initiation of meiotic chromosome pairing: The cytological view. *Genome* 33: 759–778.

Löve, V. 1992. *Origin and Geography of Cultivated Plants* (English translation of N. I. Vavilov's *Origin and Geography of Cultivated Plants*). Cambridge University Press, Cambridge.

Lu, C.-Y, S. F. Chandler, and I. K. Vasil. 1984. Somatic embryogenesis and plant regeneration from cultured immature embryos of rye (*Secale cereale* L.). *J. Plant Physiol.* 115: 237–244.

Lubbers, E. L., L. Arthur, W. W. Hanna, and P. Ozias-Akins. 1994. Molecular markers shared by diverse apomictic *Pennisetum* species. *Theor. Appl. Genet.* 89: 636–642.

Lucov, Z., S. Cohen, and R. Moav. 1970. Effects of low temperature on the somatic instability of an alien chromosome in *Nicotiana tabacum*. *Heredity* 25: 431–439.

Lucretti, S., J. Doležel, I. Schubert, and J. Fuchs. 1993. Flow karyotyping and sorting of *Vicia faba* chromosomes. *Theor. Appl. Genet.* 85: 665–672.

Lukaszewaski, A. J. 1995. Chromatid and chromosome type breakage-fusion-bridge cycles in wheat (*Triticum aestivum* L.). *Genetics* 140: 1069–1085.

Lukaszewski, A. J. 1990. Frequency of 1RS.1AL and 1RS. 1BL translocations in United States wheats. *Crop Sci.* 30: 1151–1153.

Lumaret, R. 1988. Cytology, genetics, and evolution in the genus *Dactylis*. *Crit. Rev. Plant Sci.* 7: 55–91.

Lundsteem, C., B. Bjerregaard, E. Granum, J. Philip, and K. Philip. 1980. Automated chromosome analysis. 1. A simple method for classification of B- and D-group chromosomes represented by band transition sequences. *Clin. Genet.* 17: 183–190.

Lutts, S., J. Ndikumana, and B. P. Louant. 1994. Male and female sporogenesis and gametogenesis in apomictic *Brachiaria brizantha, Brachiaria decumbens* and F_1 hybrids with sexual colchicine induced tetraploid *Brachiara ruziziensis*. *Euphytica* 78: 19–25.

Lutz. A. M. 1907. A preliminary note on the chromosomes of *Oenothera lamarkiana* and one of its mutants, *O. gigas. Science* 26: 151–152.

Lynch, P. T., J. Jones, N. W. Blackhall, M. R. Davey, J. B. Power, E. C. Cocking, M. R. Nelson et al. 1995. The phenotypic characterisation of R_2 generation transgenic rice plants under field and glasshouse conditions. *Euphytica* 85: 395–401.

Lyrene, P. M., and J. L. Perry. 1982. Production and selection of blueberry polyploids in vitro. *J. Hered.* 73: 377–378.

Ma, Y., M. N. Islam-Faridi, C. F. Crane, D. M. Stelly, H. J. Price, and D. H. Byrne. 1996. A new procedure to prepare slides of metaphase chromosomes of roses. *Hort. Sci.* 31: 855–857.

Maan, S. S. 1973a. Cytoplasmic and cytogenetic relationships among tetraploid *Triticum* species. *Euphytica* 22: 287–300.

Maan, S. S. 1973b. Cytoplasmic male-sterility and male-fertility restoration systems in wheat. In *Proc. Symp. Genet. Breed. Durum Wheat*, Bari (Italy), pp. 177–138.

Maessen, G. D. F. 1997. Genomic stability and stability of expression in genetically modified plants. *Acta Bot. Neerl.* 46: 3–24.

Magnard, J. L., Y. Yang, Y.-C. S. Chen, M. Leavy, and S. McCormick. 2001. The *Arabidopsis* gene *tardy asynchronous meiosis* is required for the normal pace and synchrony of cell division during male meiosis. *Plant Physiol.* 127: 1157–1166.

Maguire, M. P. 1978. Evidence for separate genetic control of crossing over and chiasma maintenance in maize. *Chromosoma* (Berlin) 65: 173–183.

Maguire, M. P., A. M. Paredes, and R. W. Riess. 1991. The desynaptic mutant of maize as a combined defect of synaptonemal complex and chiasma maintenance. *Genome* 34: 879–887.

Maheshwari, P. 1950. *An Introduction to the Embryology of Angiosperms*. McGraw-Hill, New York.

Maheshwari, S. C., A. K. Tyagi, and K. Malhotra. 1980. Induction of haploidy from pollen grains in angiosperms—The current status. *Theor. Appl. Genet.* 58: 193–206.

Mahfouz, M. N., M.-TH. De Boucaud, and J.-M. Gaultier. 1983. Caryological analysis of single cell clones of tobacco. Relation between the ploidy and the intensity of the caulogenesis. *Z. Pflanzenphysiol.* 109: 251–257.

Maizonnier, D., and A. Cornu. 1979. Preuve cytogenetique de la production de chromosomes lineaires remanies a partir d'un chromosome annulaire chez *Petunia hybrida* Hort. *Caryologia* 32: 393–412.

Makino, T. 1976. Genetic studies on alien chromosome addition to a durum wheat. I. some characteristics of seven monosomic addition lines of *Aegilops umbellulata* chromosomes. *Can. J. Genet. Cytol.* 18: 455–462.

Makino, T. 1981. Cytogenetical studies on the alien chromosome addition to *Durum* wheat. Bull. *Tohoku Natl. Agric. Exp. Stn.* 65: 1–58.

Maliga, P. 1981. Streptomycin resistance is inherited as a recessive Mendelian trait in a *Nicotiana sylvestris* line. *Theor. Appl. Genet.* 60: 1–3.

Maliga, P., Z. R. Kiss, A. H. Nagy, and G. Lázár. 1978. Genetic instability in somatic hybrids of *Nicotiana tabacum* and *Nicotiana knightiana*. *Mol. Gen. Genet.* 163: 145–151.

Malmberg, R. L., and R. J. Griesbach. 1980. The isolation of mitotic and meiotic chromosomes from plant protoplasts. *Plant Sci. Lett.* 17: 141–147.

Maluszynska, J., and D. Schweizer. 1989. Ribosomal RNA genes in B chromosomes of *Crepis capillaris* detected by non-radioactive *in situ* hybridization. *Heredity* 62: 59–65.

Maneephong, C., and K. Sadanaga. 1967. Induction of aneuploids in hexaploid oats, *Avena sativa* L., through X-irradiation. *Crop Sci.* 7: 522–523.

Manga, V., J. V. V. S. N. Murthy, P. Sukha dev, and M. V. Subba Rao. 1981. Meiosis in a double ditelosomic plant of pearl millet. *Caryologia* 34: 89–93.

Mannerlöf, M., S. Tuvesson, P. Steen, and P. Tenning. 1997. Transgenic sugar beet tolerant to glyphosate. *Euphytica* 94: 83–91.

Mano, Y., H. Takahashi, K. Sato, and K. Takeda. 1996. Mapping genes for callus growth and shoot regeneration in barley (*Hordeum vulgare* L.). *Breed. Sci.* 46: 137–142.

Manuelidis, L. 1990. A view of interphase chromosomes. *Science* 250: 1533–1540.

Marais, G. F., H. S. Roux, Z. A. Pretorius, and R. de V. Pienaar. 1988. Resistance to leaf rust of wheat derived from *Thinopyrum distichum* (Thunb.) Löve. In *Proc. 7th Int. Wheat Genet. Symp.* T. E. Miller and R. M. D. Koebner, Eds., Bath Press, Avon, U.K., pp. 369–373.

Marc, J. 1997. Microtubule-organizing centers in plants. *Trends Plant Sci.* 2: 223–230.

Marcker, K. A. and W. J. Stiekema. 1994. Long-range organization of a satellite DNA family flanking the beet cyst nematode resistance locus (*Hs1*) on chromosome-1 of *B. patellaris* and *B. procumbens. Theor. Appl. Genet.* 89: 459–466.

Marks, G. E. 1957. Telocentric chromosomes. *Am. Nat.* 91: 223–232.

Marsolais, A. A., G. Séguin-Swartz, and K. J. Kasha. 1984. The influence of anther cold pretreatments and donor plant genotypes on in vitro androgenesis in wheat (*Triticum aestivum* L.). *Plant Cell Tiss. Org. Cult.* 3: 69–79.

Martin, A. 1978. Aneuploidy in *Vicia faba* L. *J. Hered.* 69: 421–423.

Martins, L.A.-C.P. 1999. Did Sutton and Boveri propose the so-called Sutton-Boveri chromosome hypothesis? *Genet. Mol. Biol.* 22: 261–271.

Martínez, E. J., F. Espinoza, and C. L. Quarín. 1994. B_{III} progeny $(2n + n)$ from apomictic *Paspalum notatum* obtained through early pollination. *J. Hered.* 85: 295–297.

Martini, G., and A. Bozzini. 1966. Radiation induced asynaptic mutations in durum wheat (*Triticum durum* Desf.). *Chromosoma* (Berlin) 20: 251–266.

Mastenbroek, I., and J. M. J. de Wet. 1983. Chromosome C-banding of *Zea mays* and its closest relatives. *Can. J. Genet. Cytol.* 25: 203–209.

Mastenbroek, I., J. M. J. de Wet, and C.-Y. Lu. 1982. Chromosome behaviour in early and advanced generations of tetraploid maize. *Caryologia* 35: 463–470.

Matérn, B., and M. Simak. 1968. Statistical problems in karyotype analysis. *Hereditas* 59: 280–288.

Matsui, S., and M. Sasaki. 1973. Differential staining of nucleolus organizers in mammalian chromosomes. *Nature* (London) 246: 148–150.

Matsumura, S. 1958. Breeding of sugarbeets by means of triploidy. *Seiken Zihô* 91: 30–38.

Matthysse, A. G., and J. G. Torrey. 1967a. DNA synthesis in relation to polyploid mitoses in excised pea root segments cultured *in vitro. Expt. Cell. Res.* 48: 484–498.

Matthysse, A. G., and J. G. Torrey. 1967b. Nutritional requirements for polyploid mitoses in cultured pea root segments. *Physiol. Plant.* 20: 661–672.

Matzke, A. J., and M. A. Matzke. 1995. Trans-inactivation of homologous sequences in *Nicotiana tabacum. Current Top Microbiol. Immunol.* 197: 1–14.

Matzke, A. J., and M. A. Matzke. Position effects and epigenetic silencing of plant transgenes. *Curr. Opin. Plant Biol.* 1: 142–148.

Matzke, M. A., M. F. Mette, C. Kunz, J. Jakowitsch, and A. J. M. Matzke. 2000. Homology-dependent gene silencing in transgenic plants: Links to cellular defense responses and genome evolution. In *Genomes.* J. P. Gustafson, Ed., Kluwer Academic/Plenum Pub., New York, pp. 141–162.

Matzke, M. A., E. A. Moscone, Y.-D. Park, I. Papp, H. Oberkofler, F. Neuhuber, and A. J. M. Matzke. 1994. Inheritance and expression of a transgene insert in an aneuploid tobacco line. *Mol. Gen. Genet.* 245: 471–485.

Maughan, P. J., R. Philip, M.-J. Cho, J. M. Widholm, and L. O. Vodkin. 1999. Biolistic transformation, expression, and inheritance of bovine β-casein in soybean (*Glycine max*). *In Vitro Cell Dev. Biol.-Plant* 35: 344–349.

Mazzucato, A., G. Barcaccia, M. Pezzotti, and M. Falcinelli. 1995. Biochemical and molecular markers for investigating the mode of reproduction in the facultative apomict *Poa pratensis* L. *Sex Plant Reprod.* 8: 133–138.

Mazzucato, A., A. P. M. den Nijs, and M. Falcinelli. 1996. Estimation of parthenogenesis frequency in Kentucky bluegrass with auxin-induced parthenocarpic seeds. *Crop Sci.* 36: 9–16.

McClintock, B. 1929a. A 2N-1 chromosomal chimera in maize. *J. Hered.* 20: 218.

McClintock, B. 1929b. Chromosome morphology in *Zea mays. Science* 69: 629.

McClintock, B. 1929c. A cytological and a genetical study of triploid maize. *Genetics* 14: 180–222.

McClintock, B. 1930. A cytological demonstration of the location of an interchange between two non-homologous chromosomes of *Zea mays. Proc. Natl. Acad. Sci. U.S.A.* 16: 791–796.

McClintock, B. 1931. Cytological observations of deficiencies involving known genes, translocations and an inversion in *Zea mays. MO Agric. Exp. Stn. Res. Bull.* 163: 1–30.

McClintock, B. 1932. Cytological observations in *Zea* on the intimate association of non-homologous parts of chromosomes in the mid-prophase of meiosis and its relation to diakinesis configurations. *Proc. VI Int. Cong. Genet.* 2: 126–128.

McClintock, B. 1938a. The fusion of broken ends of sister half-chromatids following chromatid breakage at meiotic anaphases. *MO. Agric. Exp. Stn. Res. Bull.* 290: 1–48.

McClintock, B. 1938b. The production of homozygous deficient tissues with mutant characteristics by means of the aberrant mitotic behavior of ring-shaped chromosomes. *Genetics* 23: 315–376.

McClintock, B. 1941a. The association of mutants with homozygous deficiencies in *Zea mays. Genetics* 26: 542–571.

McClintock, B. 1941b. Spontaneous alterations in chromosome size and form in *Zea mays. Cold Spring Harbor Symp. on Quant. Biol.: Genes and chromosomes structure and organization.* 9: 72–81.

McClintock, B. 1941c. The stability of broken ends of chromosomes in *Zea mays. Genetics* 26: 234–282.

McClintock, B. 1950. The origin and behavior of mutable loci in maize. *Proc. Natl. Acad. Sci. U.S.A.* 36: 344–355.

McClintock, B., and H. E. Hill. 1931. The cytological identification of the chromosome associated with the *R-G* linkage group in *Zea mays. Genetics* 16: 175–190.

McComb, J. A. 1978. Isolation of hybrid lines after protoplast fusion. In *Proc. Symp. Plant Tissue Culture, Peking.* Science Press, Princeton, NJ, pp. 341–349.

McCoy, T. J., and E. T. Bingham. 1988. Cytology and cytogenetics of alfalfa. In *Alfalfa and Alfalfa Improvement.* A. A. Hanson, Ed., American Society of Agronomy/Crop Science Society of America/ Soil Science Society of America. Agron. Monogr. No. 29, pp. 737–776.

McCoy, T. J., and R. L. Phillips. 1982. Chromosome stability in maize (*Zea mays*) tissue cultures and sectoring in some regenerated plants. *Can. J. Genet. Cytol.* 24: 559–565.

McCoy, T. J., R. L. Phillips, and H. W. Rines. 1982. Cytogenetic analysis of plants regenerated from oat (*Avena sativa*) tissue cultures; high frequency of partial chromosome loss. *Can. J. Genet. Cytol.* 24: 37–50.

McFadden, E. S., and E. R. Sears. 1944 (1945). The artificial synthesis of *Triticum spelta.* Abstract of the papers prepared for the 1944 meetings of the Genetics Soc. of America. *Genetics* 30: 14.

McFadden, E. S., and E. R. Sears. 1946. The origin of *Triticum spelta* and its free-threshing hexaploid relatives. *J. Hered.* 37: 81–89, 107–116.

McGinnis, R. C. 1962. Aneuploids in common oats, *Avena sativa. Can. J. Genet. Cytol.* 4: 296–301.

McGinnis, R. C., and G. Y. Andrews. 1962. The identification of a second chromosome involved in chlorophyll production in *Avena sativa. Can. J. Genet. Cytol.* 4: 1–5.

McGinnis, R. C., G. Y. Andrews, and R. I. H. McKenzie. 1963. Determination of chromosome arm carrying a gene for chlorophyll production in *Avena sativa. Can. J. Genet. Cytol.* 5: 57–59.

McGinnis, R. C., and C. C. Lin. 1966. A monosomic study of panicle shape in *Avena sativa. Can. J. Genet. Cytol.* 8: 96–101.

McGinnis, R. C., and D. K. Taylor. 1961. The association of a gene for chlorophyll production with a specific chromosome in *Avena sativa. Can. J. Genet. Cytol.* 3: 436–443.

McGrath, J. M., and C. F. Quiros. 1990. Generation of alien chromosome addition lines from synthetic *Brassica napus*: Morphology, cytology, fertility and chromosome transmission. *Genome* 33: 374–383.

McIntosh, J. R., and M. P. Koonce. 1989. Mitosis. *Science* 246: 622–628.

McIntosh, R. A., P. L. Dyck, T. T. The, J. Cusick, and D. L. Milne. 1984. Cytogenetical studies in wheat XIII *Sr35*-a third gene from *Triticum monococcum* for resistance to *Puccinia graminis tritici. Z. Pflanzenzüchtg.* 92: 1–14.

McPheeters, K., and R. M. Skirvin. 2000. 'Everthornless' blackberry. *Hort. Sci.* 35: 778–779.

Mehra, R. C., M. G. Butler, and T. Beckman. 1986. N-banding and karyotype analysis of *Lens culinaris*. *J. Hered.* 77: 473–474.

Mehra, R. C., and K. S. Rai. 1970. Cytogenetic studies of meiotic abnormalities in *Collinsia tinctoria*. I: Chromosome stickiness. *Can. J. Genet. Cytol.* 12: 560–569.

Meijer, E. G. M., and B. S. Ahloowalia. 1981. Trisomics of ryegrass and their transmission. *Theor. Appl. Genet.* 60: 135–140.

Melchers, G., M. D. Sacristán, and A. A. Holder. 1978. Somatic hybrid plants of potato and tomato regenerated from fused protoplasts. *Carlsberg Res. Commun.* 43: 203–218.

Mello-Sampayo, T. 1972. Compensated monosomic 5B-trisomic 5A plants in tetraploid wheat. *Can. J. Genet. Cytol.* 14: 463–475.

Mello-Sampayo, T., and A. P. Canas. 1973. Suppressors of meiotic chromosome pairing in common wheat. In *Proc. 4th Int. Wheat Genet. Symp.* E. R. Sears and L. M. S. Sears, Eds., Missouri. Agric. Exp. Stn. Columbia, MO, pp. 709–713.

Melz, G., and R. Schlegel. 1985. Identification of seven telotrisomics of rye (*Secale cereale* L.). *Euphytica* 34: 361–366.

Melz, G., R. Schlegel, and J. Sybenga. 1988. Identification of the chromosomes in the 'Esto' set of rye trisomics. *Plant Breed.* 100: 169–172.

Melz, G., and A. Winkel. 1986. Evidence for monotelodisomics in rye (*Secale cereale* L.). *Plant Breed.* 97: 368–370.

Mendel, G. 1866. *Versuch über Pflanzen-Hybriden*. In Proc. Brünn National History Soc. Translated into English under the title *Experiments in Plant-Hybridization*. J. A. Peters, Ed., Prentice-Hall, Englewood Cliffs, NJ, pp. 1–20.

Menke, M., J. Fuchs, and I. Schubert. 1998. A comparison of sequence resolution on plant chromosomes: PRINS versus FISH. *Theor. Appl. Genet.* 97: 1314–1320.

Menzel, M. Y., and M. S. Brown. 1952. Viable deficiency-duplications from a translocation in *Gossypium hirsutum*. *Genetics* 37: 678–692.

Menzel, M. Y., and M. S. Brown. 1955. Isolating mechanisms in hybrids of *Gossypium gossypioides*. *Am. J. Bot.* 42: 49–57.

Menzel, M. Y., and M. S. Brown. 1978. Reciprocal chromosome translocations in *Gossypium hirsutum*. *J. Hered.* 69: 383–390.

Menzel, M. Y., and B. J. Dougherty. 1987. Transmission of duplication-deficiencies from cotton translocations is unrelated to map lengths of the unbalanced segments. *Genetics* 116: 321–330.

Menzel, M. Y., and D. W. Martin. 1970. Genome affinities of four African diploid species of *Hibiscus* sect. *Furcaria*. *J. Hered.* 61: 179–184.

Mercy-kutty, V. C., and H. Kumar. 1983. Studies on induced tetraploid in four diverse cultivars of pea (*Pisum sativum* L.). *Cytologia* 48: 51–58.

Merker, A. 1973. A Giemsa technique for rapid identification of chromosomes in triticale. *Hereditas* 75: 280–282.

Merker, A. 1979. The breeding behaviour of some rye wheat chromosome substitutions. *Hereditas* 91: 245–255.

Mesbah, M., T. S. M. deBock, and J. M. Sandbrink. 1997. Molecular and morphological characterization of monosomic additions in *Beta vulgaris*, carrying extra chromosomes of *B. procumbens* or *B. patellaris*. *Molec. Breed.* 3: 147–157.

Mettin, D., W. D. Blüthner, H. J. Schäfer, U. Buchholz, and A. Rudolph. 1977. Untersuchungen an samenproteinen in der Gattung *Aegilops*. *Tagundsber. Akad. Landwirtschaftswiss.* DDR, 158: 95–106.

Mettin, D., W. D. Blüthner, and G. Schlegel. 1973. Additional evidence on spontaneous 1B/1R wheat-rye substitutions and translocations. In *Proc. 4th Int. Wheat Genet. Symp.* E. R. Sears and L. M. S. Sears, Eds., Missouri. Agric. Exp. Stn. Columbia, MO, pp. 179–184.

Meyer, P. 1995. Variation of transgene expression in plants. *Euphytica* 85: 359–366.

Mian, H. R., J. Kuspira, G. W. R. Walker, and N. Muntjewerff. 1974. Histological and cytochemical studies on five genetic male-sterile lines of barley (*Hordeum vulgare*). *Can. J. Genet. Cytol.* 16: 355–379.

Michaelis, A. 1959. Über das verhalten eines ringchromosomes in der mitose und meiose von *Antirrhinum majus* L. *Chromosoma (Berlin)* 10: 144–162 (in German with English summary).

Michaelis, A., and R. Rieger. 1958. Cytologische und stoffwechselphysiologische untersuchungen am aktiven meristem der wurzelspitze von *Vicia faba* L. II. Präferentielle verteilung der chromosomalen bruch-und reunionspunkte nach anaerober quellung der samen. *Chromosoma (Berlin)* 9: 514–536 (in German with English summary).

Mies, D. W., and T. Hymowitz. 1973. Comparative electrophoretic studies of trypsin inhibitors in seed of the genus *Glycine*. *Bot. Gaz.* 134: 121–125.

Miller, O. L. Jr. 1963. Cytological studies in asynaptic maize. *Genetics* 48: 1445–1466.

Miller, T. E. 1984. The homoeologous relationship between the chromosomes of rye and wheat. Current status. *Can. J. Genet. Cytol.* 26: 578–589.

Miller, T. E., J. Hutchinson, and V. Chapman. 1982a. Investigation of a preferentially transmitted *Aegilops sharonensis* chromosome in wheat. *Theor. Appl. Genet.* 61: 27–33.

Miller, T. E., N. Iqbal, S. M. Reader, A. Mahmood, K. A., Cant, and I. P. King. 1997. A cytogenetic approach to the improvement of aluminium tolerance in wheat. *New Phytol.* 137: 93–98.

Miller, T. E., S. M. Reader, and V. Chapman. 1982b. The addition of *Hordeum chilense* chromosomes to wheat, *in Induced Variability in Plant Breedings.* EUCARPIA, Center for Agricultural Publication and Documentation, Wageningen, pp. 79–81.

Minocha, J. L., D. S. Brar, R. S. Saini, D. S. Multani, and J. S. Sidhu. 1982. A translocation tester set in pearl millet. *Theor. Appl. Genet.* 62: 31–33.

Mitra, J., M. O. Mapes, and F. C. Steward. 1960. Growth and organized development of cultured cells. IV. The behavior of the nucleus. *Am. J. Bot.* 47: 357–368.

Mix, G., H. M. Wilson, and B. Foroughi-Wehr. 1978. The cytological status of plants of *Hordeum vulgare* L. regenerated from microspore callus. *Z. Pflanzenzüchtg.* 80: 89–99.

Mizushima, U. 1950a. Karyogenetic studies of species and genus hybrids in the tribe *Brassiceae* of *Cruciferae.* *Tohoku J. Agr. Res.* 1: 1–14.

Mizushima, U. 1950b. On several artificial allopolyploids obtained in the tribe *Brassiceae* of *Cruciferae.* *Tohoku J. Agr. Res.* 1: 15–27.

Mochizuki, A. 1962. *Agropyron* addition lines of *Durum* wheat. *Seiken Zihô* 13: 133–138.

Moens, P. B. 1965. The transmission of a heterochromatic isochromosome in *Lycopersicon esculentum. Can. J. Genet. Cytol.* 7: 296–303.

Moens, P. B. 1969. Genetic and cytological effects of three desynaptic genes in the tomato. *Can. J. Genet. Cytol.* 11: 857–869.

Moens, P. B. 1973. Mechanisms of chromosome synapsis at meiotic prophase. *Int. Rev. Cytol.* 35: 117–134.

Moh, C. C., and R. A. Nilan. 1954. "Short" chromosome—A mutant in barley induced by atomic bomb irradiation. *Cytologia* 19: 48–53.

Mohandas, T., and W. F. Grant. 1972. Cytogenetic effects of 2, 4-D and amithole in relation to nuclear volume and DNA content in some higher plants. *Can. J. Genet. Cytol.* 14: 773–783.

Montelongo-Escobedo, H., and P. R. Rowe. 1969. Haploid induction in potato: Cytological basis for the pollinator effect. *Euphytica* 18: 116–123.

Morgan, Jr. D. T. 1950. A cytogenetic study of inversions in *Zea mays. Genetics* 35: 153–174.

Morgan, D. T. Jr. 1956. Asynapsis and plasmodial microsporocytes in maize following X-irradiation of the pollen. *J. Hered.* 47: 269–274.

Morgan, D. T. Jr. 1963. Asynapsis in pepper following X-irradiation of the pollen. *Cytologia* 28: 102–107.

Morgan, L. V. 1933. A closed X-chromosome in *Drosophila melanogaster. Genetics* 18: 250–283.

Morgan, R. N., P. Ozias-Akins, and W. W. Hanna. 1998. Seed set in an apomictic BC_3 pearl millet. *Int. J. Plant Sci.* 159: 89–97.

Morgan, W. G. 1991. The morphology and cytology of monosomic addition lines combining single *Festuca drymeja* chromosomes and *Lolium multiflorum. Euphytica* 55: 57–63.

Morikawa, H., T. Kumashiro, K. Kusakari, A. Iida, A. Hirai, and Y. Yamada. 1987. Interspecific hybrid plant formation by electrofusion in *Nicotiana. Theor. Appl. Genet.* 75: 1–4.

Morikawa, T. 1985. Identification of the 21 monosomic lines in *Avena byzantina* C. Koch cv. 'Kanota.' *Thore. Appl. Genet.* 70: 271–278.

Morinaga, T., and H. Kuriyama. 1959. A note on the cross results of diploid and tetraploid rice plants. *Jpn. J. Breed.* 9: 115–121.

Morrison, J. W. 1953. Chromosome behaviour in wheat monosomics. *Heredity* 7: 203–217.

Morrison, J. W., and T. Rajhathy. 1960a. Chromosome behaviour in autotetraploid cereals and grasses. *Chromosoma* (Berlin) 11: 297–309.

Morrison, J. W., and T. Rajhathy. 1960b. Frequency of quadrivalents in autotetraploid plants. *Nature* (London) 187: 528–530.

Morrison, R. A., and D. A. Evans. 1988. Haploid plants from tissue culture: New plant varieties in a shortened time frame. *Bio/Technol.* 6: 684–690.

Moseman, J. G., and L. Smith. 1954. Gene location by three-point test and telocentric half-chromosome fragment in *Triticum monococcum. Agron. J.* 46: 120–124.

Mouras, A., I. Negrutiu, M. Horth, and M. Jacobs. 1989. From repetitive DNA sequences to single-copy gene mapping in plant chromosomes by *in situ* hybridization. *Plant Physiol. Biochem.* 27: 161–168.

Mouras, A., G. Salessas, and A. Lutz. 1978. Sur L' utilisation des protoplastes en cytologie: Amelioration d'une methode recente en vue de L'identification des chromosomes mitotiques des genres *Nicotiana* et *Prunus*. *Caryologia* 31: 117–127.

Muehlbauer, G. J., O. Riera-Lizarazu, R. G. Kynast, D. Martin, R. L. Phillips, and H. W. Rines. 2000. A maize chromosome 3 addition line of oat exhibits expression of the maize homeobox gene liguleless 3 and alteration of cell fates. *Genome* 43: 1055–1064.

Mukai, Y. 1996. Multicolor fluorescence *in situ* hybridization: A new tool for genome analysis. In *Methods of Genome Analysis in Plants*. P. P. Jauhar, Ed., CRC Press, Boca Raton, FL, pp. 181–192.

Mukai, Y., T. R. Endo, and B. S. Gill. 1990. Physical mapping of the 5S rRNA multigene family in common wheat. *J. Hered.* 81: 290–295.

Mukai, Y., B. Friebe, and B. S. Gill. 1992. Comparison of C-banding patterns and *in situ* hybridization sites using highly repetitive and total genomic rye DNA probes of 'Imperial' rye chromosomes added to 'Chinese Spring' wheat. *Jpn. J. Genet.* 67: 71–83.

Mukai, Y., and B. S. Gill. 1991. Detection of barley chromatin added to wheat by genomic *in situ* hybridization. *Genome* 34: 448–452.

Muller, H. J. 1940. An analysis of the process of structural change in chromosomes of *Drosophila*. *J. Genet.* 40: 1–66.

Multani, D. S., K. K. Jena, D. S. Brar, B. G. de los Reyes, E. R. Angeles, and G. S. Khush. 1994. Development of monosomic alien addition lines and introgression of genes from *Oryza australiensis* Domin. to cultivated rice *O. sativa* L. *Theor. Appl. Genet.* 88: 102–109.

Müntzing, A. 1938. Note on heteroploid twin plants from eleven genera. *Hereditas*. 24: 487–491.

Müntzing, A. 1951. Cyto-genetic properties and practical value of tetraploid rye. *Hereditas* 37: 17–84.

Münzer, W. 1977. Zur identifizierung von roggenchromosomen inB/1R-weizen-roggen-substitutions- und translokationslinien mit hilfe der Giemsa-färbetechnik. *Z. Pflanzenzüchtg.* 79–74–78 (in German with English summary).

Murashige, T., and R. Nakano. 1965. Morphogenetic behavior of tobacco tissue cultures and implication of plant senescence. *Am. J. Bot.* 52: 819–827.

Murashige, T., and R. Nakano. 1967. Chromosome complement as a determinant of the morphogenic potential of tobacco cells. *Am. J. Bot.* 54: 963–970.

Murashige, T., and F. Skoog. 1962. A revised medium for rapid growth and bioassays with tobacco tissue cultures. *Physiol. Plant.* 15: 473–497.

Murata, M. 1983. Staining air dried protoplasts for study of plant chromosomes. *Stain Tech.* 58: 101–106.

Murata, M., and F. Motoyoshi. 1995. Floral chromosomes of *Arabidopsis thaliana* for detecting low-copy DNA sequences by fluorescence *in situ* hybridization. *Chromosoma* 104: 39–43.

Murata, M., and T. J. Orton. 1984. Chromosome fusions in cultured cells of celery. *Can. J. Genet. Cytol.* 26: 395–400.

Myers, W. M. 1944. Cytological studies of a triploid perennial ryegrass and its progeny. *J. Hered.* 35: 17–23.

Nagele, R., T. Freeman, L. McMarrow, and H.-Y. Lee. 1995. Precise spacial positioning of chromosomes during prometaphase: Evidence of chromosomal order. *Science* 270: 1831–1835.

Nagl, W., and M. Pfeifer. 1988. Karyological stability and microevolution of a cell suspension culture of *Brachycome dichromosomatica* (Asteraceae). *Pl. Syst. Evol.* 158: 133–139.

Nakamura, C., and W. A. Keller. 1982. Callus proliferation and plant regeneration from immature embryos of hexaploid triticale. *Z. Pflanzenzüchtg.* 88: 137–160.

Nakata, N., Y. Yasumuro, and M. Sasaki. 1977. An acetocarmine-Giemsa staining of rye chromosomes. *Jpn. J. Genet.* 52: 315–318.

Nandi, H. K. 1936. The chromosome morphology, secondary association and origin of cultivated rice. *J. Genet.* 33: 315–336.

Naranjo, C. A., L. Poggio, and P. E. Brandham. 1983. A practical method of chromosome classification on the basis of centromere position. *Genetica* 62: 51–53.

Naranjo, T. 1990. Chromosome structure of durum wheat. *Theor. Appl. Genet.* 79: 397–400.

Naranjo, T., and P. Fernández-Rueda. 1991. Homoeology of rye chromosome arms to wheat. *Theor. Appl. Genet.* 82: 577–586.

Naranjo, T., A. Roca, P. G. Goicoechea, and R. Giraldez. 1988. Chromosome structure of common wheat: Genome reassignment of chromosome 4A and 4B. In *Proc. 7th Int. Wheat Genet. Symp.* T. E. Miller and R. M. D. Koebner, Eds., Bath Press Avon, U.K., pp. 115–120.

Narasinga Rao, P. S. R. L., and J. V. Pantulu. 1982. Fertility and meiotic chromosome behaviour in autotetraploid pearl millet. *Theor. Appl. Genet.* 62: 345–351.

Nassar, N. M. A. 1994. Development and selection for apomixis in cassava, *Manihot esculenta* Crantz. *Can. J. Plant Sci.* 74: 857–858.

Nasuda, S., B. Friebe, and B. S. Gill. 1998. Gametocidal gene induce chromosome breakage in the interphase prior to the first mitotic cell division of the male gametophyte in wheat. *Genetics* 149: 1115–1124.

Natali, L., and A. Cavallini. 1987. Regeneration of pea (*Pisum sativum* L.) plantlets by *in vitro* culture of immature embryos. *Plant Breed.* 99: 172–176.

Natarajan, A. T., and G. Ahnström. 1969. Heterochromatin and chromosome aberrations. *Chromosoma* (Berlin) 28: 48–61.

Naumova, T. N. 1993. *Apomixis in Aangiosperms. Nucellar and Integumentary Embryony.* CRC Press, Inc., Boca Raton, FL.

Nayar, N. M. 1973. Origin and cytogenetics of rice. *Adv. Genet.* 17: 153–292.

Naylor, J., G. Sander, and F. Skoog. 1954. Mitosis and cell enlargement without cell division in excised tobacco pith tissue. *Physiol. Plant.* 7: 25–29.

Nel, P. M. 1973. Reduced recombination associated with the asynaptic mutant of maize. *Genetics* 74: s193–s194.

Nel, P. M. 1979. Effects of the asynaptic factor on recombination in maize. *J. Hered.* 70: 401–406.

Nelson, G. E., G. G. Robinson, and R. A. Boolootian. 1970. *Fundamental Concepts of Biology*, 2nd edition. John Wiley & Sons, Inc., New York.

Neuffer, M. G., and W. F. Sheridan. 1980. Defective kernel mutants in maize. I. Genetic and lethality studies. *Genetics* 95: 929–944.

Newell, C. A., M. L. Rhoads, and D. L. Bidney. 1984. Cytogenetic analysis of plants regenerated from tissue explants and mesophyll protoplasts of winter rape, *Brassica napus* L. *Can. J. Genet. Cytol.* 26: 752–761.

Newton, W. C. F., and C. D. Darlington. 1929. Meiosis in polyploids. I. *J. Genet.* 21: 1–16.

Newton, W. C. F., and C. Pellew. 1929. *Primula kewensis* and its derivatives. *J. Genet.* 20: 405–467.

Nezu, M., T. C. Katayama, and H. Kihara. 1960. Genetic study of the genus *Oryza*. I. Crossability and chromosomal affinity among 17 species. *Seiken Zihô* 11: 1–11.

Ninnemann, H., and F. Jüttner. 1981. Volatile substances from tissue cultures of potato, tomato and their somatic fusion products—Comparison of gas chromatographic patterns for identification of hybrids. *Z. Pflanzenphysiol.* 103: 95–107.

Nishiyama, I. 1970. Monosomic analysis of *Avena byzantina* C. Koch var. Kanota. *Z. Pflanzenzüchtg.* 63: 41–55.

Nishiyama, I. 1981. Trisomic analysis of genome B in *Avena barbata* Pott. (AABB). *Jpn J. Genet.* 56: 185–192.

Nishiyama, I., R. A. Forsberg, H. L. Shands, and M. Tabata. 1968. Monosomics of Kanota oats. *Can. J. Genet. Cytol.* 10: 601–612.

Nishiyama, I., and T. Taira. 1966. The effects of kinetin and indoleacetic acid on callus growth and organ formation in two species of *Nicotiana*. *Jpn. J. Genet.* 41: 357–365.

Nishiyama, I., T. Yabuno, and T. Taira. 1989. Genomic affinity relationships in the genus *Avena*. *Plant Breed.* 102: 22–30.

Noda, K., and K. J. Kasha. 1978. A modified Giemsa C-banding technique for *Hordeum* species. *Stain Technol.* 53: 155–162.

Noda, S. 1974. Chiasma studies in structural hybrids IX. Crossing-over in paracentric inversion of *Scilla scilloides*. *Bot. Mag.* (Tokyo) 87: 195–208.

Nogler, G. A. 1984. Gametophytic apomixis. In *Embryology of Angiosperms*. B. M. Johri, Ed., Springer, Amsterdam, pp. 475–518.

Noguchi, J., and K. Fukui. 1995. Chromatin arrangements in intact interphase nuclei examined by laser confocal microscopy. *J. Plant Res.* 108: 209–216.

Novák, F. J. 1974. The changes of karyotype in callus cultures of *Allium sativum* L. *Caryologia* 27: 45–54.

Novák, F. J., L. Ohnoutková, and M. Kubaláková. 1978. Cytogenetic studies of callus tissue of wheat (*Triticum aestivum* L.). *Cereal Res. Commun.* 6: 135–147.

Nuti Ronchi, V., S. Bonatti, and G. Turchi. 1986a. Preferential localization of chemically induced breaks in heterochromatic regions of *Vicia faba* and *Allium cepa* chromosomes-I. Exogenous thymidine enhances the cytological effects of 4-expoxyethyl-1, 2-epoxy-cyclohexane. *Environ. Exp. Bot.* 26: 115–126.

Nuti Ronchi, V., S. Bonatti, M. Durante, and G. Turchi. 1986b. Preferential localization of chemically induced breaks in heterochromatic regions of *Vicia faba* and *Allium cepa* chromosomes-II. 4-epoxyethyl-1, 2-epoxy cyclohexane interacts specifically with highly repetitive sequences of DNA in *Allium cepa*. *Environ. Exp. Bot.* 26: 127–135.

Nuti Ronchi, V., G. Martini, and M. Buiatti. 1976. Genotype-hormone interaction in the induction of chromosome aberrations: Effect of 2,4-Dichlorophenoxyacetic acid (2,4-D) and kinetin on tissue cultures from *Nicotiana* spp. *Mut. Res.* 36: 67–72.

Nuti Ronchi, V., M. Nozzolini, and L. Avanzi. 1981. Chromosomal variation on plants regenerated from two *Nicotiana* spp. *Protoplasma* 109: 433–444.

Nygren, A. 1954. Apomixis in the angiosperms. II. *Bot. Rev.* 20: 577–649.

Obermayer, R. 2000. Genome size variation in soybean (*Glycine max*). A flow cytometric analysis with reference to the postulated correlation between maturation rate and nuclear DNA content. Diploma thesis, Vienna, Austria.

Obermayer, R., and J. Greilhuber. 1999. Genome size in Chinese soybean accessions-stable or variable. *Ann. Bot.* 84: 259–262.

Ogawa, T., and T. Katayama. 1973. Cytogenetical studies on the genus *Oryza* VI. Chromosome pairing in the interspecific hybrids between *O. officinalis* and its related diploid species. *Jpn. J. Genet.* 48: 159–165.

Ogihara, Y. 1981. Tissue culture in *Haworthia*. Part 4: Genetic characterization of plants regenerated from callus. *Theor. Appl. Genet.* 60: 353–363.

Ogihara, Y., K. Hasegawa, and H. Tsujimoto. 1994. High-resolution cytological mapping of the long arm of chromosome 5A in common wheat using a series of deletion lines induced by gametocidal (Gc) genes of *Aegilops speltoides*. *Mol. Gen. Genet.* 244: 253–259.

Ogihara, Y., and K. Tsunewaki. 1979. Tissue culture in *Haworthia*. III. Occurrence of callus variants during subcultures and its mechanism. *Jpn. J. Genet.* 54: 271–293.

Ogura, H. 1990. Chromosomal variation in plant tissue culture. In *Biotechnology in Agriculture and Forestry. Somaclonal Variation in Crop Improvement I.* Y. P. S. Bajaj, Ed., Springer-Verlag, Berlin, 11: 49–84.

Ohad, N., L. Margossian, Y.-C. Hsu, C. williams, P. Repetti, and R. L. Fischer. 1996. A mutation that allows endosperm development without fertilization. *Proc. Natl. Acad. Sci. U.S.A.* 93: 5319–5324.

Ohkoshi, S., T. Komatsuda, S. Enomoto, M. Taniguchi, and K. Ohyama. 1991. Variations between varieties in callus formation and plant regeneration from immature embryos of barley. *Bull. Natl. Inst. Agrobiol. Resour.* 6: 189–207.

Okamoto, M. 1957. Asynaptic effect of chromosome V. *Wheat Inf. Sev.* 5: 6.

Okamoto, M. 1962. Identification of the chromosomes of common wheat belonging to the A and B genomes. *Can. J. Genet. Cytol.* 4: 31–37.

Okamoto, M., H. Asami, T. Shimada, and N. Inomata. 1973. Recent studies on variation of chromosomes in callus tissues from tetra-5A, -5B, and -5D of Chinese spring wheat. In *Proc. 4th Int. Wheat Genet. Symp.* E. R. Sears and L. M. S. Sears, Eds., Missouri. Agric. Exp. Stn. Columbia, MO, pp. 725–729.

Okamoto, M., and N. Inomata. 1974. Possibility of 5B like effect in diploid species. *Wheat Inf. Serv.* 38: 15–16.

Okamoto, M., and E. R. Sears. 1962. Chromosomes involved in translocations obtained from haploids of common wheat. *Can. J. Genet. Cyto.* 4: 24–30.

Olmo, H. P. 1936. Cytological studies of monosomic and derivative types of *Nicotiana tabacum*. *Cytologia* 7: 143–159.

O'Mara, J. G. 1940. Cytogenetic studies on triticale. I. A method for determining the effects of individual *Secale* chromosomes on *Triticum*. *Genetics* 25: 401–408.

O'Mara, J. G. 1947. The substitution of a specific *Secale cereale* chromosome for a specific *Triticum vulgare* chromosome. *Genetics* 32: 99–100.

Omara, M. K., and M. D. Hayward. 1978. Asynapsis in *Lolium perenne*. *Chromosoma* (Berlin) 67: 87–96.

Omielan, J. A., E. Epstein, and J. Dvořák. 1991. Salt tolerance and ionic relations of wheat as affected by individual chromosomes of salt-tolerant *Lophopyrum elongatum*. *Genome* 34: 961–974.

Oono, K. 1991. In vitro mutation in rice. In *Biotechnology in Agriculture and Forestry*, vol. 14. Y. P. S. Bajaj, Ed., Springer-Verlag, Berlin, Heidelberg, pp. 285–303.

Orel, V. 1996. *Gergor Mendel: The First Geneticist*. Oxford University Press, Oxford.

Orellana, J., M. C. Cermeño, and J. R. Lacadena. 1984. Meiotic pairing in wheat-rye addition and substitution lines. *Can. J. Genet. Cytol.* 26: 25–33.

Ortiz, J. P. A., S. C. Pessino, V. Bhat, M. D. Hayward, and C. L. Quarin. 2001. A genetic linkage map of diploid *Paspalum notatum*. *Crop Sci.* 41: 823–830.

Orton, T. J. 1980. Chromosomal variability in tissue cultures and regenerated plants of *Hordeum*. *Theor. Appl. Genet.* 56: 101–112.

Orton, T. J. 1985. Genetic instability during embryogenic cloning of celery. *Plant Cell Tiss. Org. Cult.* 4: 159–169.

Orton, T. J., and M. A. Browers. 1985. Segregation of genetic markers among plants regenerated from cultured anthers of broccoli (*Brassica oleracea* var. 'italica'). *Theor. Appl. Genet.* 69: 637–643.

Ozias-Akins, P., and I. K. Vasil. 1982. Plant regeneration from cultured immature embryos and inflorescences of *Triticum aestivum* L. (wheat): Evidence for somatic embryogenesis. *Protoplasma* 110: 95–105.

Padgette, S. R., K. H. Kolacz, X. Delannay, D. B. Re, B. J. La Vallee, C. N. Tinius, W. K. Rhodes et al. 1995. Development, identification, and characterization of glyphosate-tolerant soybean line. *Crop Sci.* 35: 1451–1461.

Pagliarini, M. S. 2000. Meiotic behavior of economically important plant species: The relationship between fertility and male sterility. *Genet. Mol. Biol.* 23: 997–1002.

Palmer, J. D. 1988. Intraspecific variation and multicircularity in *Brassica* mitochondrial DNAs. *Genetics* 118: 341–351.

Palmer, J. D., C. R. Shields, D. B. Cohen, and T. J. Orton. 1983. Chloroplast DNA evolution and the origin of amphidiploid *Brassica* species. *Theor. Appl. Genet.* 65: 181–189.

Palmer, R. G. 1971. Cytological studies of ameiotic and normal maize with reference to premeiotic pairing. *Chromosoma* (Berlin) 35: 233–246.

Palmer, R. G. 1974. Aneuploids in the soybean, *Glycine max. Can. J. Genet. Cytol.* 16: 441–447.

Palmer, R. G. 2000. Genetics of four male-sterile, female-fertile soybean mutants. *Crop Sci.* 40: 78–83.

Palmer, R. G., and H. H. Hadley. 1968. Interspecific hybridization in *Glycine*, subgenus *Leptocyamus*. *Crop Sci.* 8: 557–562.

Palmer, R. G., and H. Heer. 1973. A root tip squash technique for soybean chromosomes. *Crop Sci.* 13: 389–391.

Palmer, R. G., and H. Heer. 1976. Aneuploids from a desynaptic mutant in soybeans (*Glycine max* (L.) Merr.). *Cytologia* 41: 417–427.

Palmer, R. G., and M. L. H. Kaul. 1983. Genetics, cytology and linkage studies of a desynaptic soybean mutant. *J. Hered.* 74: 260–264.

Palmer, R. G., and T. C. Kilen. 1987. Qualitative genetics and cytogenetics. In *Soybeans: Improvement, Production, and Uses*, 2nd edition. J. R. Wilcox, Ed., Agronomy Monogr No. 16, ASA-CSSA-SSSA, Madison, WI, pp. 135–209.

Palmer, R. G., K. E. Newhouse, R. A. Graybosch, and X. Delannay. 1987. Chromosome structure of the wild soybean accessions from China and the Soviet Union of *Glycine soja* Sieb. & Zucc. *J. Hered.* 78: 243–247.

Panda, R. C., O. Aneil Kumar, and K. G. Raja Rao. 1987. Desynaptic mutant in Chili Pepper. *J. Hered.* 78: 101–104.

Pantulu, J. V. 1968. Meiosis in an autotriploid pearl millet. *Caryologia* 21: 11–15.

Pantulu, J. V., and G. J. Narasimha Rao. 1977. A pearl millet strain with $2n = 12 + 4$ telocentric chromosomes. *Curr. Sci.* 46: 390–392.

Pantulu, J. V., and M. K. Rao. 1982. Cytogenetics of pearl millet. *Theor. Appl. Genet.* 61: 1–17.

Parokonny, A. S., A. Kenton, Y. Y. Gleba, and M. D. Bennett. 1994. The fate of recombinant chromosomes and genome interaction in *Nicotiana asymmetria* somatic hybrids and their sexual progeny. *Theor. Appl. Genet.* 89: 488–497.

Parokonny, A. S., J. A. Marshall, M. D. Bennett, E. C. Cocking, M. R. Davey, and J. B. Power. 1997. Homoeologous pairing and recombination in backcross derivatives of tomato somatic hybrids [*Lycopersicon esculentum* (+) *L. peruvianum*]. *Theor. Appl. Genet.* 94: 713–723.

Partanen, C. R. 1963. The validity of auxin-induced divisions in plants as evidence of endopolyploidy. *Expt. Cell Res.* 31: 597–599.

Parthasarathy, N. 1938. Cytogenetical studies in *Oryzeae* and *Phalarideae*. I. Cytogenetics of some X-ray derivatives in rice (*Oryza sativa* L.). *J. Genet.* 37: 1–40.

Pathak, G. N. 1940. Studies in the cytology of cereals. *J. Genet.* 40: 437–467.

Patterson, E. B. 1973. Genic male sterility and hybrid maize production. Part 1. In *Proc. 7th Meeting Maize and Sorghum Sect.* Secretariat, Institute for Breeding and Production of Field Crops, Eds., Eur. Assoc. Res. Plant Breeding (EUCARPIA), Zagreb, Yugoslavia.

Patterson, E. B. 1978. Properties and uses of duplicate-deficient chromosome complements in maize. In *Maize Breeding and Genetics*. D. B. Walden, Ed., John Wiley & Sons, NY, pp. 693–710.

Pearson, O. H. 1972. Cytoplasmically inherited male sterility characters and flavor components from the species cross *Brassica nigra* (L.) Koch x *B. oleracea* C. *J. Am. Soc. Hort. Sci.* 97: 397–402.

Peffley, E. B., J. N. Corgan, K. E. Horak, and S. D. Tanksley. 1985. Electrophoretic analysis of *Allium* alien addition lines. *Theor. Appl. Genet.* 71: 176–184.

Peil, A., V. Korzun, V. Schubert, E. Schumann, W. E. Weber, and M. S. Röder. 1998. The application of wheat microsatellites to identify disomic *Triticum aestivum–Aegilops markgrafii* addition lines. *Theor. Appl. Genet.* 96: 138–146.

Pellestor, F., I. Quennesson, L. Coignet, A. Girardet, B. Andréo, and J. P. Charlieu. 1997. Direct detection of disomy in human sperm by the PRINS technique. *Human Genet.* 97: 21–25.

Pental, D., J. D. Hamill, A. Pirrie, and E. C. Cocking. 1986. Somatic hybridization of *Nicotiana tabacum* and *Petunia hybrida*. *Mol. Gen. Genet.* 202: 342–347.

Percival, A. E., and R. J. Kohel. 1991. Distribution, collection, and evaluation of *Gossypium*. *Adv. Agron.* 44: 225–256.

Percival, A. E., J. F. Wendel, and J. M. Stewart. 1999. Taxonomy and germplasm resources in cotton. In *Origin, History, Technology, and Production*. W. C. Smith, Ed., John Wiley & Sons, New York, pp. 33–66.

Person, C. 1956. Some aspects of monosomic wheat breeding. *Can. J. Bot.* 34: 60–70.

Peters, J. A. 1959. *Classic Papers in Genetics*. Prentice-Hall, Englewood Cliffs, NJ.

Pfeil, B. E., M. D. Tindale, and L. A. Craven. 2001. A review of *Glycine clandestina* species complex (Fabaceae: Phaseolae) reveals two new species. *Aust. J. Bot.* 14: 891–900.

Phillips, L. L. 1966. The cytology and phylogenetics of the diploid species of *Gossypium*. *Am. J. Bot.* 53: 328–335.

Phillips, L. L., and M. A. Strickland. 1966. The cytology of a hybrid between *Gossypium hirsutum* and *G. longicalyx*. *Can. J. Genet. Cytol.* 8: 91–95.

Phillips, R. and M. Rix. 1993. *Vegetables*. Toppan Printing, Singapore.

Phillips, R. L., C. R. Burnham, and E. B. Patterson. 1971. Advantages of chromosomal interchanges that generate haplo-viable deficiency-duplications. *Crop Sci.* 11: 525–528.

Pich, U., A. Meister, J. Macas, J. Doležel, S. Lucretti, and I. Schubert. 1995. Primed *in situ* labeling facilitates flow sorting of similar sized chromosomes. *Plant J.* 7: 1039–1044.

Pickering, R. A., A. M. Hill, M. Michel, and G. M. Timmerman-Vaughan. 1995. The transfer of a powdery mildew resistance gene from *Hordeum bulbosum* L. to barley (*H. vulgare* L.) chromosome 2 (2I). *Theor. Appl. Genet.* 91: 1288–1292.

Pickersgill, B. 1988. The genus *Capsicum*: A multidisciplinary approach to the taxonomy of cultivated and wild plants. *Biol. Zentralbl.* 107: 381–389.

Pietro, M. E., N. A. Tuleen, and G. E. Hart. 1988. Development of wheat-*Triticum searsii* disomic chromosome addition lines. In *Proc. 7th Int. Wheat Genet. Symp*. T. E. Miller, and R. M. D. Koebner, Eds., Bath Press, Avon, U.K. pp. 409–413.

Pijnacker, L. P., and M. A. Ferwerda. 1984. Giemsa C-banding of potato chromosomes. *Can. J. Genet. Cytol.* 26: 415–519.

Pijnacker, L. P., and M. A. Ferwerda. 1987. Karyotype variation in aminoethylcysteine resistant cell and callus cultures and regenerated plants of a dihaploid potato (*Solanum tuberosum*). *Plant Cell Rep.* 6: 385–388.

Pijnacker, L. P., and K. Sree Ramulu. 1990. Somaclonal variation in potato: A karyotypic evolution. *Acta Bot. Neerl.* 39: 163–169.

Pijnacker, L. P., K. Sree Ramulu, P. Dijkhuis, and M. A. Ferwerda. 1989. Flow cytometric and karyological analysis of polysomaty and polyploidization during callus formation from leaf segments of various potato genotypes. *Theor. Appl. Genet.* 77: 102–110.

Pijnacker, L. P., K. Walch, and M. A. Ferwerda. 1986. Behaviour of chromosomes in potato leaf tissue cultured in vitro as studied by BrdC-Giemsa labelling. *Theor. Appl. Genet.* 72: 833–839.

Pikaard, C. S. 1999. Nucleolar dominance and silencing of transcription. *Trends Plant Sci.* 4: 478–483.

Pinkel. D., T. Straume, and J. W. Gray. 1986. Cytogenetic analysis using quantitative, high-sensitivity, fluorescence hybridization. *Proc. Natl. Acad. Sci. U.S.A.* 83: 2934–2938.

Plewa, M. J., and D. F. Weber. 1975. Monosomic analysis of fatty acid composition in embryo lipids of *Zea mays* L. *Genetics* 81: 277–286.

Pohler, W., G. Schumann, M. Sulze, and I.-M. Kummer. 1988. Cytological investigation of callus tissue and regenerated plants from triticale anther culture. *Biol. Zentralbl.* 107: 643–652.

Poon, N. H., and H. K. Wu. 1967. Identification of involved chromosomes in trisomics of *Sorghum vulgare* Pers. *J. Agr. Ass. China* 58 (N.S.): 18–32 (in Chinese with English summary).

Potrykus, I. 2001. Golden rice and beyond. *Plant Physiol.* 125: 1157–1161.

Poulsen, G. B. 1996. Genetic transformation of *Brassica*. *Plant Breed.* 115: 209–225.

Powell, J. B., and R. A. Nilan. 1963. Influence of temperature on crossing over in an inversion heterozygote in barley. *Crop Sci.* 3: 11–13.

Powell, J. B., and R. A. Nilan. 1968. Evidence for spontaneous inversions in cultivated barley. *Crop Sci.* 8: 114–116.

Prakash, S., and K. Hinata. 1980. Taxonomy, cytogenetics and origin of crop *Brassicas*, a review. *Opera Bot.* 55: 1–57.

Prakken, R. 1943. Studies of asynapsis in rye. *Hereditas* 29: 475–495.

Prasad, G., and D. K. Tripathi. 1986. Asynaptic and desynaptic mutants in barley. *Cytologia* 51: 11–19.

Price, H. J., G. Hodnett, and J. S. Johnston. 2000. Sunflower (*Helianthus annuus*) leaves contain compounds that reduce nuclear propidium iodide fluorescence. *Ann. Bot.* 86: 929–934.

Pundir, R. P. S., N. K. Rao, and L. J. G. van der Maesen. 1983. Induced autotetraploidy in chickpea (*Cicer arietinum* L.) *Theor. Appl. Genet.* 65: 119–122.

Pupilli, F., S. Businelli, M. E. Caceres, F. Damiani, and S. Arcioni. 1995. Molecular, cytological and morpho-agronomical characterization of hexaploid somatic hybrids in *Medicago. Theor. Appl. Genet.* 90: 347–355.

Pupilli, F., P. Labomborda, M. E. Caceres, C. L. Quarin, and S. Arcioni. 2001. The chromosome segment related to apomixis in *Paspalum simplex* is homoeologous to the telomeric region of the long arm of rice chromosome 12. *Mol. Breed.* 8: 53–61.

Putievsky, E., and P. Broué. 1979. Cytogenetics of hybrids among perennial species of *Glycine* subgenus *Glycine. Austr. J. Bot.* 27: 713–723.

Qiren, C., Z. Zhenhua, and G. Yuanhua. 1985. Cytogenetical analysis on aneuploids obtained from pollen clones of rice (*Oryza sativa* L.). *Theor. Appl. Genet.* 71: 506–512.

Quarin, C. L. 1986. Seasonal changes in the incidence of apomixis of diploid, triploid, and tetraploid plants of *Paspalum cromyorrhizon. Euphytica* 35: 515–522.

Quiros, C. F., O. Ochoa, S. F. Kianian, and D. Douches. 1987. Analysis of the *Brassica oleracea* genome by the generation of *B. campestris-oleracea* chromosome addition lines: Characterization by isozymes and rDNA genes. *Theor. Appl. Genet.* 74: 758–766.

Rajhathy, T. 1975. Trisomics of *Avena strigosa. Can. J. Genet. Cytol.* 17: 151–166.

Rajhathy, T., and P. L. Dyck. 1964. Methods for aneuploid production in common oats, *Avena sativa. Can. J. Genet. Cytol.* 6: 215–220.

Rajhathy, T., and G. Fedak. 1970. A secondary trisomics in *Avena strigosa. Can. J. Genet Cytol.* 12: 358–360.

Rajhathy, T., and H. Thomas. 1972. Genetic control of chromosome pairing in hexaploid oats. *Nature New Biol.* 239: 217–219.

Rajhathy, T., and H. Thomas. 1974. Cytogenetics of oats (*Avena* L.) *Miss. Publ. Genet. Soc. Canada,* No. 2. p. 90.

Ramachandran, C., and V. Raghavan. 1992. Apomixis in distant hybridization. In *Distant Hybridization of Crop Plants,* vol. 16. G. Kalloo, and J. B. Chowdhury, Eds., Monograph Theor. Appl. Genet, pp. 106–121.

Ramage, R. T. 1955. The trisomics of barley. Ph.D. thesis, University of Minnesota, p. 90.

Ramage, R. T. 1960. Trisomics from interchange heterozygotes in barley. *Agron. J.* 52: 156–159.

Ramage, R. T. 1965. Balanced tertiary trisomics for use in hybrid seed production. *Crop Sci.* 5: 177–178.

Ramage, R. T. 1971. Mapping chromosomes from the phenotypes of trisomics produced by interchange heterozygotes. In *Barley Genet. II. Proc. 2nd Int. Barley Genet. Symp.* R. A. Nilan, Ed., Washington State University Press, Pullman, WA, pp. 89–92.

Ramage, R. T. 1985. Cytogenetics. In *Barley.* Agro. Monograph No. 26. D. C. Rasmusson, Ed., American Society of Agronomy/Crop Science Society of America/Soil Science Society of America, Madison, WI, pp. 127–154.

Ramage, R. T., and A. D. Day. 1960. Separation of trisomic and diploid barley seeds produced by interchange heterozygotes. *Agron. J.* 52: 590–591.

Ramanna, M. S., and M. Wagenvoort. 1976. Identification of the trisomic series in diploid *Solanum tuberosum* L., group Tuberosum. I. Chromosome identification. *Euphytica* 25: 233–240.

Ramanujam, S. 1937. Cytogenetical studies in the Oryzeae II. Cytological behaviour of an autotriploid in rice (*Oryza sativa*). *J. Genet.* 35: 183–221.

Ramanujam, S., and N. Parthasarathy. 1953. Autopolyploidy. *Indian J. Genet. Plant Breed.* 13: 53–82.

Ramulu, K. S., P. Dijkhuis, A. Pereira, G. C. Angenent, M. M. van Lookeren Campagne, and J. J. M. Dons. 1998. EMS and transposon mutagenesis for the isolation of apomictic mutants in plants. In *Somaclonal Variation and Induced Mutations in Crop Improvement.* S. M. Jain, D. S. Brar, and B. S. Ahloowalia. Eds., Kluwer, Dordrecht, The Netherlands, pp. 379–400.

Ramulu, K. S., V. K. Sharma, T. N. Naumova, P. Dijkhuis, and M. M. vanLookeren Campagene. 1999. Apomixis for crop improvement. *Protoplasma* 208: 196–205.

Randolph, L. F. 1928. Chromosome numbers in *Zea mays* L. *Cornell Univ. Agric. Exp. Stn. Mem.* 117: 1–44.

Rao, G. M., and M. V. Reddi. 1971. Chromosomal associations and meiotic behaviour of a triploid rice (*Oryza sativa* L.). *Cytologia* 36: 509–514.

Rao, P. N., A. Nirmala, and P. Ranganadham. 1988. A cytological study of triploidy in pearl millet. *Cytologia* 53: 421–425.

Ray, D. T., and J. E. Endrizzi. 1982. A tester-set of translocations in *Gossypium hirsutum* L. *J. Hered.* 73: 429–433.

Ray, D. T., A. C. Gathman, and A. E. Thompson. 1989. Cytogenetic analysis of interspecific hybrids in *Cuphea. J. Hered* 80: 329–332.

Rayburn, A. L., D. P. Birdar, D. G. bullock, R. L. Nelson, C. Gourmet, and J. B. Wetzel. 1997. Nuclear DNA content diversity in Chinese soybean introductions. *Ann. Bot.* 80: 321–325.

Rayburn, A. L., and B. S. Gill. 1985a. Molecular evidence for the origin and evolution of chromosome 4A in polyploid wheats. *Can. J. Genet. Cytol.* 27: 246–250.

Rayburn, A. L., and B. S. Gill. 1985b. Use of biotin-labeled probes to map specific DNA sequences on wheat chromosomes. *J. Hered.* 76: 78–81.

Rayburn. A. L., and J. R. Gold. 1982. A procedure for obtaining mitotic chromosomes from maize. *Maydica* 27: 113–121.

Reamon-Ramos, S. M., and G. Wricke. 1992. A full set of monosomic addition lines in *Beta vulgaris* from *Beta webbiana*: Morphology and isozyme markers. *Theor. Appl. Genet.* 84: 411–418.

Reddi, V. R., and V. Padmaja. 1982. Studies on aneuploids of *Petunia*. Part 1: Cytomorphological identification of primary trisomics. *Theor. Appl. Genet.* 61: 35–40.

Rédei, G. P. 1965. Non-Mendelian megagametogenesis in *Arabidopsis. Genetics* 51: 857–872.

Rees, H. 1958. Differential behaviour of chromosomes in *Scilla. Chromosoma* (Berlin) 9: 185–192.

Rees, H. 1961. Genotypic control of chromosome form and behaviour. *Bot. Rev.* 27: 288–318.

Rees, H., and G. M. Evans. 1966. A correlation between the localisation of chiasmata and the replication pattern of chromosomal DNA. *Expt. Cell Res.* 44: 161–164.

Reeves, A. F., G. S. Khush, and C. M. Rick. 1968. Segregation and recombination in trisomics: A reconsideration. *Can. J. Genet. Cytol.* 10: 937–940.

Rhoades, M. M. 1933a. An experimental and theoretical study of chromatid crossing over. *Genetics* 18: 535–555.

Rhoades, M. M. 1933b. A secondary trisome in maize. *Proc. Nat. Acad. Sci. U.S.A.* 19: 1031–1038.

Rhoades, M. M. 1936. A cytogenetic study of a chromosome fragment in maize. *Genetics* 21: 491–502.

Rhoades, M. M. 1940. Studies of a telocentric chromosome in maize with reference to the stability of its centromere. *Genetics* 25: 483–520.

Rhoades, M. M. 1942. Preferential segregation in maize. *Genetics* 27: 395–407.

Rhoades, M. M. 1955. The cytogenetics of maize. In *Corn and Corn Improvement.* G. F. Sprague, Ed., Academic Press, New York, pp. 123–219.

Rhoades, M. M. 1956. Genetic control of chromosomal behavior. *Maize Genet. Coop. Newsl.* 30: 38–42.

Rhoades, M. M., and E. Dempsey. 1953. Cytogenetic studies of deficient-duplicate chromosomes derived from inversion heterozygotes in maize. *Am. J. Bot.* 40: 405–424.

Rhoades, M. M., and E. Dempsey. 1966. Induction of chromosome doubling at meiosis by the *elongate* gene in maize. *Genetics* 54: 505–522.

Rhoades, M. M., and E. Dempsey. 1972. On the mechanism of chromatin loss induced by the B chromosome of maize. *Genetics* 71: 73–96.

Rhoades, M. M., and E. Dempsey. 1973. Cytogenetic studies on a transmissible deficiency in chromosome 3 of maize. *J. Hered.* 64: 125–128.

Rhoades, M. M., E. Dempsey, and A. Ghidoni. 1967. Chromosome elimination in maize induced by supernumerary B chromosomes. *Proc. Nat. Acad. Sci. U.S.A.* 57: 1626–1632.

Rhoades, M. M., and B. McClintock. 1935. The cytogenetics of maize. *Bot. Rev.* 1: 292–325.

Rhodes, C. A., R. L. Phillips, and C. E. Green. 1986. Cytogenetic stability of aneuploid maize tissue cultures. *Can. J. Genet. Cytol.* 28: 374–384.

Rick, C. M. 1943. Cyto-genetic consequences of X-ray treatment of pollen in *Petunia. Bot. Gaz.* 104: 528–539.

Rick, C. M. 1945. A survey of cytogenetic causes of unfruitfulness in the tomato. *Genetics* 30: 347–362.

Rick, C. M., and D. W. Barton. 1954. Cytological and genetical identification of the primary trisomics of the tomato. *Genetics* 39: 640–666.

Rick, C. M., and L. Butler. 1956. Cytogenetics of tomato. *Adv. Genet.* 8: 267–382.

Rick, C. M., W. H. Dempsey, and G. S. Khush. 1964. Further studies on the primary trisomics of the tomato. *Can. J. Genet. Cytol.* 6: 93–108.

Rick, C. M., and B. S. Gill. 1973. Reproductive errors in aneuploids: Generation of variant extra-chromosomal types by tomato primary trisomics. *Can. J. Genet. Cytol.* 15: 299–308.

Rick, C. M., and G. S. Khush. 1961. X-ray-induced deficiencies of chromosome 11 in the tomato. *Genetics* 46: 1389–1393.

Rick, C. M., and N. K. Notani. 1961. The tolerance of extra chromosomes by primitive tomatoes. *Genetics* 46: 1231–1235.

Rickards, G. K. 1983. Alternate-1 and alternate-2 orientations in interchange (Reciprocal translocation) quadrivalents. *Genetics* 104: 211–213.

Rieger, R. A. Michaelis, and M. M. Green. 1976. *Glossary of Genetics and Cytogenetics. Classical and Molecular.* Springer-Verlag, Berlin.

Rietveld, R. C., R. A. Bressan, and P. M. Hasegawa. 1993. Somaclonal variation in tuber disc-derived populations of potato. II. Differential effects of genotype. *Theor. Appl. Genet.* 87: 305–313.

Riggs, R. D., S. Wang, R. J. Singh, and T. Hymowitz. 1998. Possible transfer of resistance to *Heterodera glycines* from *Glycine tomentella* to *Glycine max*. *J. Nematol.* 30: 547–552.

Riley, R. 1955. The cytogenetics of the differences between some *Secale* species. *J. Agric. Sci.* 46: 377–383.

Riley, R. 1966. Genotype-environmental interaction affecting chiasma frequency in *Triticum aestivum*. In *Chromosomes Today*, vol. 1. C. D. Darlington and K. R. Lewis, Eds., Oliver and Boyd, Edinburgh, pp. 57–65.

Riley, R., and V. Chapman. 1958a. Genetic control of the cytologically diploid behaviour of hexaploid wheat. *Nature* (London) 182: 713–715.

Riley, R., and V. Chapman. 1958b. The production and phenotypes of wheat-rye chromosome addition lines. *Heredity* 12: 301–315.

Riley, R., and V. Chapman. 1967. Effect of 5B^S in suppressing the expression of altered dosage of 5B^L on meiotic chromosome pairing in *Triticum aestivum*. *Nature* (London) 216: 60–62.

Riley, R., V. Chapman, and R. Johnson. 1968. Introduction of yellow rust resistance of *Aegilops comosa* into wheat by genetically induced homoeologous recombination. *Nature* (London) 217: 383–384.

Riley, R., V. Chapman, and R. C. F. Macer. 1966a. The homoeology of an *Aegilops* chromosome causing stripe rust resistance. *Can. J. Genet. Cytol.* 8: 616–630.

Riley, R., V. Chapman, and T. E. Miller. 1973. The determination of meiotic chromosome pairing. In *Proc. 4th Int. Wheat Genet. Symp.* E. R. Sears and L. M. S. Sears, Eds., Missouri. Agric. Exp. Stn. Columbia, MO, pp. 731–738.

Riley, R., V. Chapman, R. M. Young, and A. M. Belfield. 1966b. Control of meiotic chromosome pairing by the chromosomes of homoeologous group 5 of *Triticum aestivum*. *Nature* (London) 212: 1475–1477.

Riley, R., and G. Kimber. 1961. Aneuploids and the cytogenetic structure of wheat varietal populations. *Heredity* 16: 275–290.

Riley, R., and C. N. Law. 1965. Genetic variation in chromosome pairing. *Adv. Genet.* 13: 57–114.

Riley, R., and R. C. F. Macer. 1966. The chromosomal distribution of the genetic resistance of rye to wheat pathogens. *Can. J. Genet. Cytol.* 8: 640–653.

Rines, H. W., and L. S. Dahleen. 1990. Haploid oat plants produced by application of maize pollen to emasculated oat florets. *Crop Sci.* 30: 1073–1078.

Rines, H. W., and S. S. Johnson. 1988. Synaptic mutants in hexaploid oats (*Avena sativa* L.). *Genome* 30: 1–7.

Röbbelen, G. 1960. Beiträge zur analyse des *Brassica*-genoms. *Chromosoma* (Berlin) 11: 205–228 (in German with English summary).

Röbbelen, G., and S. Smutkupt. 1968. Reciprocal intergeneric hybridizations between wheat and rye. *Wheat Inf. Serv.* 27: 10–13.

Robertson, D. W. 1971. Recent information of linkage and chromosome mapping. In *Barley Genetics II. Proc. 2nd Int. Barley Genet. Symp.* R. A. Nilan, Ed., Washington State University Press, Pullman, pp. 220–242.

Robertson, W. M. R. B. 1916. Chromosome studies. I. Taxonomic relationships shown in the chromosomes of Tettegidae and Acrididiae: V-shaped chromosomes and their significance in Acridae, Locustidae and Grillidae: Chromosomes and variations. *J. Morphol* 27: 179–331.

Roeder, G. S. 1997. Meiotic chromosomes: It takes two to tango. *Genes & Develop.* 11: 2600–2621.

Rogers, S. O., and A. J. Bendich. 1985. Extraction of DNA from milligram amount of fresh, herbarium and mummified plant tissues. *Plant Mol. Biol.* 5: 69–76.

Romagosa, I., L. Cistue, T. Tsuchiya, and J. M. Lasa. 1985. Cytological identification of acrotrisomy in sugar beets. *J. Hered.* 76: 227–228.

Romagosa, I., R. J. Hecker, T. Tsuchiya, and J. M. Lasa. 1986. Primary trisomics in sugarbeet. I. Isolation and morphological characterization. *Crop Sci.* 26: 243–249.

Roman, H., and A. J. Ullstrup. 1951. The use of A-B translocations to locate genes in maize. *Agron. J.* 43: 450–454.

Romero, C., and J. R. Lacadena. 1982. Effect of rye B-chromosomes on pairing in *Triticum aestivum* x *Secale cereale* hybrids. *Z. Pflanzenzüchtg.* 89: 39–46.

Rooney, W. L., D. M. Stelly, and D. W. Altman. 1991. Identification of four *Gossypium sturtianum* monosomic alien addition derivatives from a backcrossing program with *G. hirsutum*. *Crop Sci.* 31: 337–341.

Rosenberg, O. 1909. Cytologische und morphologische studien an *Drosera longifolia* x *rotundifolia*. *Kungl. Sv. Vet. Akad. Handlingar* 43: 1–64 (in German).

Roseweir, J., and H. Rees. 1962. Fertility and chromosome pairing in autotetraploid rye. *Nature (London)* 195: 203–204.

Rowe, P. R. 1974. Methods of producing haploids: Parthenogenesis following interspecific hybridization. In *Haploids in Higher Plants. Advances and Potential.* K. J. Kasha, Ed., Ainsworth Press, Canada, pp. 43–52.

Roy, S. C. 1980. Chromosomal variations in the callus tissues of *Allium tuberosum* and *A. cepa. Protoplasma* 102: 171–176.

Rubatzky, V. E. and M. Yamaguchi. 1997. *World Vegetables, Principles, Production, and Nutritive Values,* 2nd edition. Chapman & Hall, New York.

Rudorf-Lauritzen, M. 1958. The trisomics of *Antirrhinum majus* L. *Proc. 10th Int. Congr. Genet. Montreal* 2: 243–244.

Ruíz, M. L., J. Rueda, M. I. Peláez, F. J. Espino, M. Candela, A. M. Sendino, and A. M. Vázquez. 1992. Somatic embryogenesis, plant regeneration and somaclonal variation in barley. *Plant Cell Tiss. Org. Cult.* 28: 97–101.

Russell, W. A., and C. R. Burnham. 1950. Cytogenetic studies of an inversion in maize. *Sci. Agr.* 30: 93–111.

Rutger, J. N. 1992. Searching for apomixis in rice. In *Proceedings Apomixis Workshops (Feb. 11–12, 1992),* Athens, GA, pp. 36–39.

Ryan, S. A., and W. R. Scowcroft. 1987. A somaclonal variant of wheat with additional ß-amylase isozymes. *Theor. Appl. Genet.* 73: 459–464.

Sacristán, M. D. 1971. Karyotypic changes in callus cultures from haploid and diploid plants of *Crepis capillaris* (L.) Wallr. *Chromosoma* (Berlin) 33: 273–283.

Sacristán, M. D. 1982. Resistance responses to *Phoma lingam* of plants regenerated from selected cell and embryogenic cultures of haploid *Brassica napus. Theor. Appl. Genet.* 61: 193–200.

Sacristán, M. D., and G. Melchers. 1969. The caryological analysis of plants regenerated from tumorous and other callus cultures of tobacco. *Mol. Gen. Genet.* 105: 317–333.

Sacristán, M. D., and M. F. Wendt-Gallitelli. 1971. Transformation to auxin-autotrophy and its reversibility in a mutant line of *Crepis capillaris* callus culture. *Mol. Gen. Genet.* 110: 355–360.

Sadanaga, K. 1957. Cytological studies of hybrids involving *Triticum durum* and *Secale cereale.* I. Alien addition races in tetraploid wheat. *Cytologia* 22: 312–321.

Sadasivaiah, R. S., R. Watkins, and T. Rajhathy. 1969. Somatic association of chromosomes in diploid and hexaploid *Avena. Chromosoma* (Berlin) 28: 468–481.

Saini, R. S., and J. L. Minocha. 1981. A compensating trisomic in pearl millet. *J. Hered.* 72: 354–355.

Sakamoto, K., Y. Akiyama, K. Fukui, H. Kamada, and S. Satoh. 1998. Characterization, genome sizes and morphology of sex chromosomes in hemp (*Cannabis sativa* L.). *Cytologia* 63: 459–464.

Samoylova, T. I., A. Meister, and S. Miséra. 1996. The flow karyotype of *Arabidopsis thaliana* interphase chromosomes. *Plant J.* 10: 949–954.

Sampson, D. R., A. W. S. Hunter, and E. C. Bradley. 1961. Triploid x diploid progenies and the primary trisomics of *Antirrhinum majus. Can. J. Genet. Cytol.* 3: 184–194.

Sánchez-Morán, E., E. Benavente, and J. Orellana. 1999. Simultaneous identification of A, B, D and R genomes by genomic *in situ* hybridization in wheat-rye derivatives. *Heredity* 83: 249–252.

Sandfaer, J. 1975. The occurrence of spontaneous triploids in different barley varieties. *Hereditas* 80: 149–153.

Santarém, E. R., and J. J. Finer. 1999. Transformation of soybean [*Glycine max* (L.) Merrill] using proliferative embryogenic tissue maintained on semi-solid medium. *In vitro Cell Dev. Biol.-Plant* 35: 451–455.

Santos, J. L. 1999. The relationship between synapsis and recombination: Two different views. *Heredity* 82: 1–6.

Santos, R. F., W. Handro, and E. I. S. Floh. 1983. Ploidy in *Petunia hybrida* plants regenerated from tissue cultures. *Revta Brasil. Bot.* 6: 33–39.

Sarkar, P., and G. L. Stebbins. 1956. Morphological evidence concerning the origin of the B genome in wheat. *Am. J. Bot.* 43: 297–304.

Sasakuma, T., and S. S. Maan. 1978. Male sterility-fertility restoration systems in *Triticum durum. Can. J. Genet. Cytol.* 20: 389–398.

Satina, S., and A. F. Blakeslee. 1935. Cytological effects of a gene in *Datura* which causes dyad formation in sporogenesis. *Bot. Gaz.* 96: 521–532.

Satina, S., and A. F. Blakeslee. 1937. Chromosome behavior in triploids of *Datura stramonium.* The male gametophyte. *Am. J. Bot.* 24: 518–527.

Satina, S., A. F. Blakeslee, and A. G. Avery. 1938. Chromosomal behavior in triploid *Datura.* III. The Seed. *Am. J. Bot.* 25: 595–602.

Sauer, C. O. 1969. *Agricultural Origins and Dispersals.* MIT Press, Cambridge, MA.

Savitsky, H. 1975. Hybridization between *Beta vulgaris* and *B. procumbens* and transmission of nematode (*Heterodera schachtii*) resistance to sugarbeet. *Can. J. Genet, Cytol.* 17: 197–209.

Sax, K. 1959. The cytogenetics of facultative apomixis in *Malus* species. *J. Arnold Abor.* 40: 289–297.

Schank, S. C., and P. F. Knowles. 1961. Colchicine induced polyploids of *Carthamus tinctorius* L. *Crop Sci.* 1: 342–345.

Schenck, H. R., and G. Röbbelen. 1982. Somatic hybrids by fusion of protoplasts from *Brassica oleracea* and *B. campestris. Z. Pflanzenzüchtg.* 89: 278–288.

Schertz, K. F. 1962. Cytology, fertility, and gross morphology of induced polyploids of *Sorghum vulgare. Can. J. Genet. Cyto.* 4: 179–186.

Schertz, K. F. 1966. Morphological and cytological characteristics of five trisomics of *Sorghum vulgare* Pers. *Crop Sci.* 6: 519–523.

Schertz, K. F. 1974. Morphological and cytological characteristics of five additional trisomics of *Sorghum bicolor* (L.) Moench. *Crop Sci.* 14: 106–109.

Schertz, K. F., and J. C. Stephens. 1965. Origin and occurrence of triploids of *Sorghum vulgare* Pers. and their chromosomal and morphological characteristics. *Crop Sci.* 5: 514–516.

Scheunert, E.-U., Z. B. Shamina, and H. Koblitz. 1978. Karyological features of barley callus tissues cultured *in vitro. Biol. Plant.* 20: 305–308.

Schiemann, E., and U. Nürnberg-Krüger. 1952. Neue Untersuchungen an *Secale africanum* Stapf. II. *Secale africanum* und seine bastarde mit *Secale montanum* und *Secale cereale. Naturwissenschaften.* 39: 136–137 (in German with English summary).

Schlegel, R., G. Melz, and R. Nestrowicz. 1987. A universal reference karyotype in rye, *Secale cereale* L. *Theor. Appl. Genet.* 74: 820–826.

Schlegel, R., T. Werner, and E. Hülgenhof. 1991. Confirmation of a 4BL/5RL wheat-rye chromosome translocation line in the wheat cultivar 'Viking' showing high copper efficiency. *Plant Breed.* 107: 226–234.

Schlegel, R. H. J. 2010. *Dictionary of Plant Breeding,* 2nd edition. CRC Press, Boca Raton, FL.

Schmidt, H. 1977. Contributions on the breeding of apomictic apple stocks. IV. On the inheritance of apomixis. *Z. Pflanzenzüchtg.* 78: 3–12.

Schnabelrauch, L. S., F. Kloc-Bauchan, and K. C. Sink. 1985. Expression of nuclear-cytoplasmic genomic incompatibility in interspecific *Petunia* somatic hybrid plants. *Theor. Appl. Genet.* 70: 57–65.

Schneerman, M. C., W. S. Lee, G. Doyle, and D. F. Weber. 1998. RFLP mapping of the centromere of chromosome 4 in maize using isochromosomes for 4S. *Theor. Appl. Genet.* 96: 361–366.

Schubert, I., and R. Rieger. 1990. Deletions are not tolerated by the *Vicia faba* genome. *Biol. Zentralbl.* 109: 207–213.

Schubert, I., and U. Wobus. 1985. In situ hybridization confirms jumping nucleolus organizing regions in *Allium. Chromosoma* (Berlin) 92: 143–148.

Schulenburg, H. G. Von Der. 1965. Beiträge zur aufstellung von monosomen-sortimenten beim hafer. *Z. Pflanzenzüchtg.* 53: 247–265.

Schultz-Schaeffer, J. 1980. *Cytogenetics Plants, Animals, Humans.* Springer-Verlag, Berlin.

Schulze, J., C. Balko, B. Zellner, T. Koprek, R. Hänsch, A. Nerlich, and R. R. Mendel. 1995. Biolistic transformation of cucumber using embryogenic suspension cultures: Long-term expression of reporter genes. *Plant Sci.* 112: 197–206.

Schwartz, D. 1953a. The behavior of an X-ray induced ring chromosome in maize. *Am. Nat.* 87: 19–28.

Schwartz, D. 1953b. Evidence of sister-strand crossing over in maize. *Genetics* 38: 251–260.

Schwarzacher, T. and J. S. Heslop-Harrison. 1991. *In situ* hybridization to plant telomeres using synthetic oligomers. *Genome* 34: 317–323.

Schwarzacher, T., A. R. Leitch, M. D. Bennett, and J. S. Heslop-Harrison. 1989. *In situ* localization of parental genomes in a wide hybrid. *Ann. Bot.* 64: 315–324.

Schwendiman, J., E. Koto, and B. Hau. 1980. Reflections on the evolution of chromosome pairing in cotton allohexaploids (*G. hirsutum* x *G. stocksii* and *G. hirsutum* x *G. longicalyx*) and on the taxonomic position of *G. longicalyx. Cot. Fib. Trop.* 35: 269–275.

Seal, A. G., and M. D. Bennett. 1981. The rye genome in winter hexaploid triticales. *Can. J. Genet. Cytol.* 23: 647–653.

Sears, E. R. 1939. Cytogenetic studies with polyploid species of wheat. I. Chromosomal aberrations in the progeny of a haploid of *Triticum vulgare. Genetics* 24: 509–523.

Sears, E. R. 1941. Amphidiploids in the seven-chromosome *Triticinae. Mo. Agric. Exp. Stn. Res. Bull.* 336: 1–46.

Sears, E. R. 1944. Cytogenetic studies with polyploid species of wheat. II. Additional chromosomal aberrations in *Triticum vulgare. Genetics* 29: 232–246.

Sears, E. R. 1952a. Misdivision of univalents in common wheat. *Chromosoma* (Berlin) 4: 535–550.

Sears, E. R. 1952b. The behavior of isochromosomes and telocentrics in wheat. *Chromosoma* (Berlin) 4: 551–562.

Sears, E. R. 1953a. Addition of the genome of *Haynaldia villosa* to *Triticum aestivum. Amer. J. Bot.* 40: 168–174.

Sears, E. R. 1953b. Nullisomic analysis in common wheat. *Amer. Nat.* 87: 245–252.

Sears, E. R. 1954. The aneuploids of common wheat *MO. Agric. Exp. Stn. Res. Bull.* 572: 1–58.

Sears, E. R. 1956a. The B genome in wheat. *Wheat Inf. Serv.* 4: 8–10.

Sears, E. R. 1956b. The transfer of leaf-rust resistance from *Aegilops umbellulata* to wheat. *Brookhaven Symp. Biol.* 9: 1–22.

Sears, E. R. 1958. The aneuploids of common wheat. In *Proc. Ist Int. Wheat. Genet. Symp.* B. C. Jenkins, Ed., Public Press, Winnipeg, Canada, pp. 221–229.

Sears, E. R. 1962. The use of telocentric chromosomes in linkage mapping. *Genetics* 47: 983 (abstr.)

Sears, E. R. 1966a. Chromosome mapping with the aid of telocentrics. *Proc. 2nd Int. Wheat Genet. Symp. Hereditas.* Suppl. 2: 370–381.

Sears, E. R. 1966b. Nullisomic-tetrasomic combinations in hexaploid wheat. In *Chromosome Manipulations and Plant Genetics.* R. Riley and K. R. Lewis, Eds., Oliver and Boyd, Edinburgh and London. suppl. *Heredity* 20: 29–45.

Sears, E. R. 1968. Relationships of chromosomes 2A, 2B, and 2D with their rye homoeologue. In *Proc. 3rd. Int. Wheat Genet. Symp.* K. W. Finlay and K. W. Shepherd, Eds., Australian Acad. Sci., Canberra, pp. 53–61.

Sears, E. R. 1973. *Agropyron*—Wheat transfers induced by homoeologous pairing. In *Proc. 4th Int. Wheat Genet. Symp.* E. R. Sears and L. M. S. Sears, Eds., Missouri Agric. Exp. Stn., Columbia, MO, pp. 191–199.

Sears, E. R. 1975. The wheats and their relatives. In *Handbook of Genetics.* R. C. King, Ed., Plenum Press, New York, pp. 59–91.

Sears, E. R. 1976. Genetic control of chromosome pairing in wheat. *Ann. Rev. Genet.* 10: 31–51.

Sears, E. R. 1977. An induced mutant with homoeologous pairing in common wheat. *Can. J. Genet. Cytol.* 19: 585–593.

Sears, E. R., and M. Okamato. 1958. Intergenomic chromosome relationships in hexaploid wheat. *Proc. X Int. Cong. Genet.* 2: 258–259.

Sears, L. M. S., and S. Lee-Chen. 1970. Cytogenetic studies in *Arabidopsis thaliana. Can. J. Genet. Cytol.* 12: 217–223.

Selentijn, E. M. J., N. N. Sandal, R. Klein-Lankhorst, W. Lange, Th. S. M. De Bock, K. A. Marcker, and W. J. Stiekema. 1994. Long-range organization of a satellite NA family flanking the beet cyst nematode resistance locus (*HsI*) on chromosome-1 of *B. patellaris* and *B. procumbens. Theor. Appl. Genet.* 89: 459–466.

Sen, N. K. 1952. Isochromosomes in tomato. *Genetics* 37: 227–241.

Senior, I. J. 1998. Uses of plant gene silencing. *Bio/Technol. Genet. Engineer. Rev.* 15: 79–119.

Serizawa, N., S. Nasuda, F. Shi, T. R. Endo, S. Prodanovic, I. Schubert, and G. Künzel. 2001. Deletion-based physical mapping of barley chromosome 7H. *Theor. Appl. Genet.* 103: 827–834.

Seshu, D. V., and T. Venkataswamy. 1958. A monosome in rice. *Madras Agric. J.* 45: 311–314.

Sethi, G. S., K. S. Gill, and B. S. Ghai. 1970. Cytogenetics of induced asynapsis in barley. *Indian J. Genet. Plant Breed.* 30: 604–607.

Shahla, A., and T. Tsuchiya. 1983. Additional information on the Triplo 7S in barley. *Barley Genet. Newsl.* 13: 22–23.

Shahla, A., and T. Tsuchiya. 1986. Cytogenetics of the acrotrisomic 5S^{5L} in barley. *Can. J. Genet. Cytol.* 28: 1026–1033.

Shahla, A., and T. Tsuchiya. 1987. Cytogenetic studies in barley chromosome 1 by means of telotrisomic, acrotrisomic and conventional analysis. *Theor. Appl. Genet.* 75: 5–12.

Shahla, A., and T. Tsuchiya. 1990. Genetic analysis in six telotrisomic lines in barley (*Hordeum vulgare* L.). *J. Hered.* 81: 127–130.

Shang, X. M., R. C. Jackson, and H. T. Nguyen. 1988a. Heterochromatin diversity and chromosome morphology in wheats analyzed by the HKG banding technique. *Genome* 30: 956–965.

Shang, X. M., R. C. Jackson, and H. T. Nguyen. 1988b. A new banding technique for chromosomes of wheat (*Triticum*) and its relatives. *Cereal Res. Commun.* 16: 169–174.

Sharma, A. K., and A. Sharma. 1965. *Chromosome Techniques—Theory and Practice.* Butterworths, London.

Sharma, D., and D. R. Knott. 1966. The transfer of leaf-rust resistance from *Agropyron* to *Triticum* by irradiation. *Can. J. Genet Cytol.* 8: 137–143.

Sharma, D. C., and R. A. Forsberg. 1977. Spontaneous and induced interspecific gene transfer for crown rust resistance in *Avena. Crop Sci.* 17: 855–860.

Sharma, G. C., L. L. Bello, V. T. Sapra, and C. M. Peterson. 1981. Callus initiation and plant regeneration from triticale embryos. *Crop Sci.* 21: 113–118.

Sharma, H., H. Ohm, L. Goulart, T. Lister, R. Appels, and O. Benlhabib. 1995. Introgression and character-
    ization of barley yellow dwarf virus resistance from *Thinopyrum intermedium* into wheat. *Genome*
    38: 406–413.
Sharma, H. C., and B. S. Gill. 1983. Current status of wide hybridization in wheat. *Euphytica* 32: 17–31.
Sharma, R. K., and E. Reinbergs. 1972. Gene controlled abnormal cytokinesis causing male sterility in barley.
    Indian J. Genet. Plant Breed. 32: 408–410.
Sharp, L. W. 1934. *Introduction to Cytology.* McGraw-Hill, New York.
Shastry, S. V. S., and D. R. Ranga Rao. 1961. Timing imbalance in the meiosis of the F_1 hybrid *Oryza sativa* x
    O. australiensis. Genet. Res. (Camb.) 2: 373–383.
Shastry, S. V. S., S. D. Sharma, and D. R. Ranga Rao. 1961. Pachytene analysis in *Oryza*. III. Meiosis in an
    inter-sectional hybrid, *O. sativa* x *O. officinalis. The Nucleus* (Calcutta) 4: 67–80.
Shen, D., Z. Wang, and M. Wu. 1987. Gene mapping on maize pachytene chromosomes by *in situ* hybridiza-
    tion. *Chromosoma* (Berlin) 95: 311–314.
Shepard, J. F., D. Bidney, T. Barsby, and R. Kemble. 1983. Genetic transfer in plants through interspecific
    protoplast fusion. *Science* 219: 683–688.
Shepard, J. F., D. Bidney, and E. Shahin. 1980. Potato protoplasts in crop improvement. *Science* 208: 17–24.
Shepherd, K. W., and A. K. M. R. Islam. 1988. Fourth compendium of wheat-alien chromosome lines. In *Proc.
    7th Int. Wheat. Genet. Symp.* T. E. Miller and R. M. D. Koebner, Eds., Bath Press, U.K., pp. 1373–1395.
Sherwood, R. T., C. C. Berg, and B. Y. Young. 1994. Inheritance of apospory in buffelgrass. *Crop Sci.* 34:
    1490–1494.
Sherwood, R. T., B. A. Young, and E. C. Bashaw. 1980. Facultative apomixis in buffelgrass. *Crop Sci.* 20: 375–379.
Shewry, P. R., S. Parmar, and T. E. Miller. 1985. Chromosomal location of the structural genes for the M_r
    75,000 Γ-Secalins in *Secale montanum* Guss: Evidence for a translocation involving chromosomes 2R
    and 6R in cultivated rye (*Secale cereale* L.). *Heredity* 54: 381–383.
Shewry, P. R., A. S. Tatham, F. Barro, P. Barcelo, and P. Lazzeri. 1995. Biotechnology of breadmaking:
    Unraveling and manipulating the multi-protein gluten complex. *Bio/Technol.* 13: 1185–1190.
Shi, A. N., S. Leath, and J. P. Murphy. 1998. A major gene for powdery mildew resistance transferred to com-
    mon wheat from wild einkorn wheat. *Phytopathology* 88: 144–147.
Shi, F., and T. R. Endo. 1999. Genetic induction of structural changes in barley chromosomes added to common
    wheat by a gametocidal chromosome derived from *Aegilops cylindrica. Genes Genet. Syst.* 74: 49–54.
Shigyo, M., Y. Tashiro, S. Isshiki, and S. Miyazaki. 1996. Establishment of a series of alien monosomic
    addition lines of Japanese bunching onion (*Allium fistulosum* L.) with extra chromosomes from shallot
    (*A. cepa* L. Aggregatum group). *Genes Genet. Syst.* 71: 363–371.
Shimada, T. 1971. Chromosome constitution of tobacco and wheat callus cells. *Jpn. J. Genet.* 46: 235–241.
Shimada, T., and M. Tabata. 1967. Chromosome numbers in cultured pith tissue of tobacco. *Jpn. J. Genet.*
    42: 195–201.
Shimada, T., and Y. Yamada. 1979. Wheat plants regenerated from embryo cell cultures. *Jpn. J. Genet.*
    54: 379–385.
Shimamoto, K., R. Terada, T. Izawa, and H. Fujimoto. 1989. Fertile transgenic rice plants regenerated from
    transformed protoplasts. *Nature (London)* 337: 274–276.
Shin, Y.-B., and T. Katayama. 1979. Cytogenetical studies on the genus *Oryza* XI. Alien addition lines of
    O. sativa with single chromosomes of *O. officinalis. Jpn. J. Genet.* 54: 1–10.
Shishido, R., S. Apisitwanich, N. Ohmido, Y. Okinaka, K. Mori, and K. Fukui. 1998. Detection of specific
    chromosome reduction in rice somatic hybrids with the A, B, and C genomes by multi-color genomic
    in situ hybridization. *Theor. Appl. Genet.* 97: 1013–1018.
Shoemaker, R. C., L. A. Amberger, R. G. Palmer, L. Oglesby, and J. P. Ranch. 1991. Effect of 2,4-dichloro-
    phenoxyacetic acid concentration on somatic embryogenesis and heritable variation in soybean [*Glycine
    max* (L.) Merr.]. *In Vitro Cell Dev. Biol.* 27P: 84–88.
Shoemaker, R. C., K. Polzin, J. Labote, J. Specht, E. C. Brummer, T. Olson, N. Young et al. 1996. Genome
    duplication in soybean (*Glycine* subgenus *soja*). *Genetics* 144: 329–338.
Sibi, M. 1984. Heredity of epigenic-variant plants from culture in vitro. In *Efficiency in Plant Breeding.*
    W. Lange, A. C. Zeven, and N. G. Hogenboom, Eds., *Proc. 10th Congress European Assoc. Res. Plant
    Breed.* EUCAPIA, Wageningen, The Netherlands, pp. 196–198.
Siep, L. 1980. The telotrisomic for the short arm of chromosome 6 in barley. M.S. thesis, Colorado State
    University, Fort Collins, p. 49.
Simeone, R., A. Blanco, and B. Giorgi. 1985. Chromosome transmission in primary trisomics of *Durum* wheat
    (*Triticum durum* Desf.). *Cereal Res. Commun.* 13: 27–31.

Simmonds, D. H., and P. A. Donaldson. 2000. Genotype screening for proliferative embryogenesis and biolistic transformation of short-season soybean genotypes. *Plant cell. Rep.* 19: 485–490.

Singh, B. D. 1981. Origin of aneuploid variation in tissue cultures of *Haplopappus gracilis* and *Vicia hajastana. Caryologia* 34: 337–343.

Singh, B. D., and B. L. Harvey. 1975a. Cytogenetic studies on *Haplopappus gracilis* cells cultured on agar and in liquid media. *Cytologia* 40: 347–354.

Singh, B. D., and B. L. Harvey. 1975b. Selection for diploid cells in suspension cultures of *Haplopappus gracilis. Nature* (London) 253: 453.

Singh, B. D., B. L. Harvey, K. N. Kao, and R. A. Miller. 1975. Karyotypic changes and selection pressure in *Haplopappus gracilis* suspension cultures. *Can. J. Genet. Cytol.* 17: 109–116.

Singh, K., T. Ishii, A. Parco, N. Huang, D. S. Brar, and G. S. Khush. 1996. Centromere mapping and orientation of the molecular linkage map of rice (*Oryza sativa* L.). *Proc. Natl. Acad. Sci. U.S.A.* 93: 6163–6168.

Singh, K., D. S. Multani, and G. S. Khush. 1996. Secondary trisomics and telotrisomics of rice: Origin, characterization, and use in determining the orientation of chromosome map. *Genetics* 143: 517–529.

Singh, R. J. 1974. Cytogenetics of telotrisomics in barley. PhD thesis. Colorado State University, Fort Collins, p. 112.

Singh, R. J. 1977. Cross compatibility, meiotic pairing and fertility in 5 *Secale* species and their interspecific hybrids. *Cereal Res. Commun.* 5: 67–75.

Singh, R. J. 1986. Chromosomal variation in immature embryo derived calluses of barley (*Hordeum vulgare* L.). *Theor. Appl. Genet.* 72: 710–716.

Singh, R. J. 1993. *Plant Cytogenetics.* CRC Press, Boca Raton, FL.

Singh, R. J. 2003. *Plant Cytogenetics*, 2nd edition. CRC Press, Boca Raton, FL.

Singh, R. J., and T. Hymowitz. 1985a. Diploid-like meiotic behavior in synthesized amphidiploids of the genus *Glycine* Willd. subgenus *Glycine. Can. J. Genet. Cytol.* 27: 655–660.

Singh, R. J., and T. Hymowitz. 1985b. The genomic relationships among six wild perennial species of the genus *Glycine* subgenus *Glycine* Willd. *Theor. Appl. Genet.* 71: 221–230.

Singh, R. J., and T. Hymowitz. 1985c. Intra- and interspecific hybridization in the genus *Glycine*, subgenus *Glycine* Willd.: Chromosome pairing and genome relationships. *Z. Pflanzenzüchtg.* 95: 289–310.

Singh, R. J., and T. Hymowitz. 1987. Intersubgeneric crossability in the genus *Glycine* Willd. *Plant Breed.* 98: 171–173.

Singh, R. J., and T. Hymowitz. 1988. The genomic relationship between *Glycine max* (L.) Merr. and *G. soja* Sieb. and Zucc. as revealed by pachytene chromosome analysis. *Theor. Appl. Genet.* 76: 705–711.

Singh, R. J., and T. Hymowitz. 1999. Soybean genetic resources and crop improvement. *Genome* 42: 605–616.

Singh, R. J., and H. Ikehashi. 1981. Monogenic male-sterility in rice: Induction, identification and inheritance. *Crop Sci.* 21: 286–289.

Singh, R. J., H. H. Kim, and T. Hymowitz. 2001. Distribution of rDNA loci in the genus *Glycine* willd. *Theor. Appl. Genet.* 103: 212–218.

Singh, R. J., T. M. Klein, C. J. Mauvais, S. Knowlton, T. Hymowitz, and C. M. Kostow. 1998. Cytological characterization of transgenic soybean. *Theor. Appl. Genet.* 96: 319–324.

Singh, R. J., and F. L. Kolb. 1991. Chromosomal interchanges in six hexaploid oat genotypes. *Crop Sci.* 31: 726–729.

Singh, R. J., K. P. Kollipara, F. Ahmad, and T. Hymowitz. 1992a. Putative diploid ancestors of 80-chromosome *Glycine tabacina. Genome* 35: 140–146.

Singh, R. J., K. P. Kollipara, and T. Hymowitz. 1987a. Intersubgeneric hybridization of soybeans with a wild perennial species, *Glycine clandestina* Wendl. *Theor. Appl. Genet.* 74: 391–396.

Singh, R. J., K. P. Kollipara, and T. Hymowitz. 1987b. Polyploid complexes of *Glycine tabacina* (Labill.) Benth. and *G. tomentella* Hayata revealed by cytogenetic analysis. *Genome* 29: 490–497.

Singh, R. J., K. P. Kollipara, and T. Hymowitz. 1988. Further data on the genomic relationship among wild perennial species (2n = 40) of the genus *Glycine* Willd. *Genome* 30: 166–176.

Singh, R. J., K. P. Kollipara, and T. Hymowitz. 1989. Ancestors of 80- and 78-chromosome *Glycine tomentella* Hayata (Leguminosae). *Genome* 32: 796–801.

Singh, R. J., K. P. Kollipara, and T. Hymowitz. 1990. Backcross-derived progeny from soybean and *Glycine tomentella* Hayata intersubgeneric hybrids. *Crop Sci.* 30: 871–874.

Singh, R. J., K. P. Kollipara, and T. Hymowitz. 1992b. Genomic relationships among diploid wild perennial species of the genus *Glycine* Willd. subgenus *Glycine* revealed by cytogenetics and seed protein electrophoresis. *Theor. Appl. Genet.* 85: 276–282.

Singh, R. J., K. P. Kollipara, and T. Hymowitz. 1993. Backcross (BC_2–BC_4)—Derived fertile plants from *Glycine max* (L.) Merr. and *G. tomentella* Hayata intersubgeneric hybrids. *Crop Sci.* 33: 1002–1007.

Singh, R. J., K. P. Kollipara, and T. Hymowitz. 1998a. The genomes of *Glycine canescens* F. J. Herm., and *G. tomentella* Hayata of Western Australia and their phylogenetic relationships in the genus *Glycine* Willd. *Genome* 41: 669–679.

Singh, R. J., K. P. Kollipara, and T. Hymowitz. 1998b. Monosomic alien addition lines derived from *Glycine max* (L.) Merr. and *G. tomentella* Hayata: Production, characterization, and breeding behavior. *Crop Sci.* 38: 1483–1489.

Singh, R. J. and A. Lebeda. 2007. Landmark research in vegetable crops. In *Genetic Resources, Chromosome Engineering, and Crop Improvement, Vegetable Crops, Volume 3*. R. J. Singh, Ed., CRC Press, Boca Raton, FL, pp. 1–15.

Singh, R. J., and T. Lelley. 1975. Giemsa banding in meiotic chromosomes of rye, *Secale cereale* L. *Z. Pflanzenzüchtg.* 75: 85–89.

Singh, R. J., and R. L. Nelson. 2014. Methodology for creating alloplasmic soybean lines by using *Glycine tomentella* as a maternal parent. *Plant Breed.* 133: 624–631.

Singh, R. J., and R. L. Nelson. 2015. Intersubgeneric hybridization between *Glycine max* and *G. tomentella*: Production of F_1, amphidiploid, BC_1, BC_2, BC_3, and fertile soybean plants. *Theor. Appl. Genet.* 128: 1117–1136.

Singh, R. J., and G. Röbbelen. 1975. Comparison of somatic Giemsa banding pattern in several species of rye. *Z. Pflanzenzüchtg.* 75: 270–285.

Singh, R. J., and G. Röbbelen. 1976. Giemsa banding technique reveals deletions within rye chromosomes in addition lines. *Z. Pflanzenzüchtg.* 76: 11–18.

Singh, R. J., and G. Röbbelen. 1977. Identification by Giemsa technique of the translocations separating cultivated rye from three wild species of *Secale*. *Chromosoma* (Berlin) 59: 217–225.

Singh, R. J., G. Röbbelen, and M. Okamoto. 1976. Somatic association at interphase studied by Giemsa banding technique. *Chromosoma* (Berlin) 56: 265–273.

Singh, R. J., and T. Tsuchiya. 1975a. Hypertriploid plants in barley (*Hordeum vulgare* L.). *Caryologia* 28: 89–98.

Singh, R. J., and T. Tsuchiya. 1975b. Pachytene chromosomes of barley. *J. Hered.* 66: 165–167.

Singh, R. J., and T. Tsuchiya. 1977. Morphology, fertility, and transmission in seven monotelotrisomics of barley. *Z. Pflanzenzüchtg.* 78: 327–340.

Singh, R. J., and T. Tsuchiya. 1981a. Cytological study of the telocentric chromosome in seven monotelotrisomics of barley. *Bot. Gaz.* 142: 267–273.

Singh, R. J., and T. Tsuchiya. 1981b. Identification and designation of barley chromosomes by Giemsa banding technique: A reconsideration. *Z. Pflanzenzüchtg.* 86: 336–340.

Singh, R. J., and T. Tsuchiya. 1981c. A novel compensating partial tetrasomic diploid plant of barley. *Barley Genet. Newsl.* 11: 66–68.

Singh, R. J., and T. Tsuchiya. 1981d. Origin and characteristic of telotrisomic for 3S (Triplo 3S) in barley. *Barley Genet. Newsl.* 11: 69.

Singh, R. J., and T. Tsuchiya. 1981e. Ring chromosome in barley (*Hordeum vulgare* L.). *Chromosoma (Berlin)* 82: 133–141.

Singh, R. J., and T. Tsuchiya. 1982a. Identification and designation of telocentric chromosomes in barley by means of Giemsa N-banding technique. *Theor. Appl. Genet.* 64: 13–24.

Singh, R. J., and T. Tsuchiya. 1982b. An improved Giemsa N-banding technique for the identification of barley chromosomes. *J. Hered.* 73: 227–229.

Singh. R. J., and T. Tsuchiya. 1993. Cytogenetics of two novel compensating diploids in barley (*Hordeum vulgare*). *Genome* 36: 343–349.

Singh, R. M., and A. T. Wallace. 1967a. Monosomics of *Avena byzantina* C. Koch. I. Karyotype and chromosome pairing studies. *Can. J. Genet. Cytol.* 9: 87–96.

Singh, R. M., and A. T. Wallace. 1967b. Monosomics of *Avena byzantina* C. Koch. II. Breeding behavior and morphology. *Can. J. Genet. Cytol.* 9: 97–106.

Singh, U. P., R. Sai Kumar, R. M. Singh, and R. B. Singh. 1982. Tertiary trisomics of pearl millet (*Pennisetum americanum* (L.) K. Schum): Its cytomorphology, fertility and transmission. *Theor. Appl. Genet.* 63: 139–144.

Singleton, W. R., and P. C. Mangelsdorf. 1940. Gametic lethals on the fourth chromosome of maize. *Genetics* 25: 366–390.

Sinha, S. K., and B. K. Mohapatra. 1969. Compensatory chiasma formation in maize. *Cytologia* 34: 523–527.

Sinnott, E. W., L. C. Dunn, and Th. Dobzhansky. 1950. *Principles of Genetics*. McGraw-Hill Book Company, Inc., New York.

Sinnott, E. W., H. Houghtaling, and A. F. Blakeslee. 1934. The comparative anatomy of extra-chromosomal types in *Datura stramonium*. *Carnegie Inst. Wash. Publ.* 451: 1–50.

Sjödin, J. 1970. Induced asynaptic mutants in *Vicia faba* L. *Hereditas* 66: 215–232.

Sjödin, J. 1971. Induced paracentric and pericentric inversions in *Vicia faba* L. *Hereditas* 67: 39–54.

Skirvin, R. M. 1978. Natural and induced variation in tissue cultures. *Euphytica* 27: 241–266.

Skirvin, R. M., M. Coyner, M. A. Norton, S. Motoike, and D. Gorvin. 2000. Somaclonal variation: Do we know what causes it? *AgBiotechNet* 2: 1–4.

Skirvin, R. M., and J. Janick. 1976. Tissue culture-induced variation in scented *Pelargonium* spp. *J. Am. Soc. Hort. Sci.* 101: 281–290.

Skirvin, R. M., K. D. McPheeters, and M. Norton. 1994. Sources of frequency of somaclonal variation. *Hort. Sci.* 29: 1232–1237.

Skorupska. H., M. C. Albertsen, K. D. Longholz, and R. G. Palmer. 1989. Detection of ribosomal RNA genes in soybean, *Glycine max* (L.) Merr. by *in situ* hybridization. *Genome* 32: 1091–1095.

Skorupska, H., and R. G. Palmer. 1987. Monosomics from synaptic Ks mutant. *Soybean Genet. Newsl.* 14: 174–178.

Skovsted, A. 1934. Cytological studies in cotton. II. Two interspecific hybrids between Asiatic and new world cottons. *J. Genet.* 28: 407–424.

Skovsted, A. 1937. Cytological studies in cotton. IV. Chromosome conjugation in interspecific hybrids. *J. Genet.* 34: 97–134.

Smartt, J., W. C. Gregory, and M. Pfluge Gregory. 1978. The genomes of *Arachis hypogaea*. I. Cytogenetic studies of putative genome donors. *Euphytica* 27: 665–675.

Smith, F. J., J. H. de Jong, and J. L. Oud. 1975. The use of primary trisomics for the localization of genes on the seven different chromosomes of *Petunia hybrida*. 1. Triplo V. *Genetica* 45: 361–370.

Smith, H. H. 1943. Studies on induced heteroploids in *Nicotiana*. *Am. J. Bot.* 30: 121–130.

Smith, H. H. 1968. Recent cytogenetic studies in the genus *Nicotiana*. *Adv. Genet.* 14: 1–54.

Smith, H. H. 1971. Broadening the base of genetic variability in plants. *J. Hered.* 62: 265–276.

Smith, L. 1936. Cytogenetic studies in *Triticum monococcum* L. and *T. aegilopoides*. *Mo. Agric. Exp. Stn. Res. Bull.* 248: 1–38.

Smith, L. 1941. An inversion, a reciprocal translocation, trisomics and tetraploids in barley. *J. Agric. Res.* 63: 741–750.

Smith, L. 1942. Cytogenetics of a factor of multiploid sporocytes in barley. *Am. J. Bot.* 29: 451–456.

Smith, L. 1947. A fragmented chromosome in *Triticum monococcum* and its use in studies of inheritance. *Genetics* 32: 341–349.

Smith, M. U., and A. C. H. Kindfield. 1999. Teaching cell division Basics & recommendations. *Am. Biol. Teach.* 61: 366–371.

Smith, S. M., and H. E. Street. 1974. The decline of embryogenic potential as callus and suspension cultures of carrot (*Daucus carota* L.) are serially subcultured. *Ann. Bot.* 38: 223–241.

Smutkupt, S. 1968. Herstellung und eigenschaften von *Secalotricum* im vergleich mit *Triticale*. *Angew. Bot.* 42: 95–118.

Snow, R. 1963. Alcoholic hydrochloric acid—Carmine as a stain for chromosomes in squash preparations. *Stain Tech.* 38: 9–13.

Snow, R. 1964. Cytogenetic studies in *Clarkia*, section Primigenia. III. Cytogenetics of monosomics in *Clarkia amoena*. *Genetica* 35: 205–235.

Snowdon, R. J., H. Winter, A. Diestel, and M. D. Sacristàn. 2000. Development and characterisation of *Brassica napus–Sinapis arvensis* addition lines exhibiting resistance to *Leptosphaeria maculans*. *Theor. Appl. Genet.* 101: 1008–1014.

Solntseva, M. P. 1976. Basis of embryological classification of apomixis in angiosperms. In *Apomixis and Breeding*. S. S. Khokhlov, Ed., Amerind Publ., New Delhi. (Translated from Russian.) pp. 89–101.

Song, K. M., T. C. Osborn, and P. H. Williams. 1988. *Brassica* taxonomy based on nuclear restriction fragment length polymorphisms (RFLPs). 1. Genome evolution of diploid and amphidiploid species. *Theor. Appl. Genet.* 75: 784–794.

Songstad, D. D., D. A. Somers, and R. J. Griesbach. 1995. Advances in alternative DNA delivery techniques. *Plant Cell Tiss. Organ Cult.* 40: 1–15.

Soost, R. K. 1951. Comparative cytology and genetics of asynaptic mutants in *Lycopersicon esculentum* Mill. *Genetics* 36: 410–434.

Soriano, J. D. 1957. The genus *Collinsia*. IV. The cytogenetics of colchicine-induced reciprocal translocations in *C. heterophylla*. *Bot. Gaz.* 118: 139–145.

Sosnikhina, S. P., Y. S. Fedotova, V. G. Smirnov, E. I. Mikhailova, O. L. Kolomiets, and Y. F. Bogdanov. 1992. Meiotic mutants of rye *Secale cereale* L. 1. Synaptic mutant *sy-1*. *Theor. Appl. Genet.* 84: 979–985.

Southern, D. I. 1969. Stable telocentric chromosomes produced following centric misdivision in *Myrmeleotettix maculatus* (Thunb.). *Chromsoma* (Berlin) 26: 140–147.

Sparvoli, E., H. Gay, and B. P. Kaufmann. 1966. Duration of the mitotic cycle in *Haplopappus gracilis*. *Caryologia* 19: 65–71.

Speckmann, G. J., Th. S. M. De Bock, and J. H. De Jong. 1985. Monosomic additions with resistance to beet cyst nematode obtained from hybrids of *Beta vulgaris* and wild *Beta* species of the section Patellares. I. Morphology, transmission and level of resistance. *Z. Pflanzenzüchtg.* 95: 74–83.

Spetsov, P., D. Mingeot, J. M. Jacquemin, K. Samardjieva, and E. Marinova. 1997. Transfer of powdery mildew resistance from *Aegilops variabilis* into bred wheat. *Euphytica* 93: 49–54.

Spurr, A. R. 1969. A low-viscosity epoxy resin embedding medium for electron microscopy. *J. Ultrstruct. Res.* 26: 31–43.

Sree Ramulu, K., F. Carluccio, D. de Nettancourt, and M. Devreux. 1977. Trisomics from triploid-diploid crosses in self-incompatible *Lycopersicum peruvianum*. I. Essential features of aneuploids and of self-compatible trisomics. *Theor. Appl. Genet.* 50: 105–119.

Sree Ramulu, K., P. Dijkhuis, and S. Roest. 1983. Phenotypic variation and ploidy level of plants regenerated from protoplasts of tetraploid potato (*Solanum tuberosum* L. cv. 'Bintje'). *Theor. Appl. Genet.* 65: 329–338.

Sree Ramulu, K., P. Dijkhuis, Ch. H. Hanisch TenCate, and B. de Groot. 1985. Patterns of DNA and chromosome variation during *in vitro* growth in various genotypes of potato. *Plant Sci.* 41: 69–78.

Sree Ramulu, K., P. Dijkhuis, and S. Roest. 1984a. Genetic instability in protoclones of potato (*Solanum tuberosum* L. cv. 'Bintje'): New types of variation after vegetative propagation. *Theor. Appl. Genet.* 68: 515–519.

Sree Ramulu, K., P. Dijkhuis, S. Roest, G. S. Bokelmann, and B. de Groot. 1984b. Early occurrence of genetic instability in protoplast cultures of potato. *Plant Sci. Lett.* 36: 79–86.

Sree Ramulu, K., P. Dijkhuis, S. Roest, G. S. Bokelmann, and B. de Groot. 1986. Variation in phenotype and chromosome number of plants regenerated from protoplasts of dihaploid and tetraploid potato. *Plant Breed.* 97: 119–128.

Stack, S. M., and W. V. Brown. 1969. Somatic pairing: Reduction and recombination: An evolutionary hypothesis of meiosis. *Nature* (London) 222: 1275–1276.

Stack, S. M., and D. E. Comings. 1979. The chromosomes and DNA of *Allium cepa*. *Chromosoma* (Berlin) 70: 161–181.

Stalker, H. T. 1980. Utilization of wild species for crop improvement. *Adv. Agron.* 33: 111–147.

Stam, M., J. N. M. Mol, and J. M. Kooter. 1997. The silence of genes in transgenic plants. *Ann. Bot.* 79: 3–12.

Stamova, B. S., and R. T. Chetelat. 2000. Inheritance and genetic mapping of cucumber mosaic virus resistance introgressed from *Lycopersocon chilense* into tomato. *Theor. Appl. Genet.* 101: 527–537.

Stebbins, G. L. Jr. 1949. The evolutionary significance of natural and artificial polyploids in the family Gramineae. In *Proc. 8th Int. Congr. Genet.* July 7th–14th, 1948, Stockholm, Sweden, pp. 461–485.

Stebbins, G. L. Jr. 1950. *Variation and Evolution in Plants*. Columbia University Press, New York.

Stebbins, G. L. 1958. The inviability, weakness, and sterility of interspecific hybrids. *Adv. Genet.* 9: 147–215.

Stebbins, G. L. Jr. 1971. *Chromosomal Evolution in Higher Plants*. Addison-Wesley, Reading, MA.

Steinitz-Sears, L. M. 1963. Chromosome studies in *Arabidopsis thaliana*. *Genetics* 48: 483–490.

Steinitz-Sears, L. M. 1966. Somatic instability of telocentric chromosomes in wheat and the nature of the centromere. *Genetics* 54: 241–248.

Stelly, D. M., D. W. Altman, R. J. Kohel, T. S. Rangan, and E. Commiskey. 1989. Cytogenetic abnormalities of cotton somaclones from callus cultures. *Genome* 32: 762–770.

Stephens, J. C., and K. F. Schertz. 1965. Asynapsis and its inheritance in *Sorghum vulgare* Pers. *Crop Sci.* 5: 337–339.

Stephens, S. G. 1947. Cytogenetics of *Gossypium* and the problem of the origin of new world cottons. *Adv. Genet.* 1: 431–442.

Stephens, S. G. 1950. The internal mechanism of speciation in *Gossypium*. *Bot. Rev.* 16: 115–149.

Stewart, C. N. Jr., M. J. Adang, J. N. All, H. R. Boerma, G. Cardineau, D. Tucker, and W. A. Parrott. 1996. Genetic transformation, recovery, and characterization of fertile soybean transgenic for a synthetic *Bacillus thuringiensis cryIAc* gene. *Plant Physiol.* 112: 121–129.

Stewart, D. M., E. C. Gilmore Jr., and E. R. Ausemus. 1968. Resistance to *Puccinia graminis* derived from *Secale cereale* incorporated into *Triticum aestivum*. *Phytopathology* 58: 508–511.

Stino, K. R. 1940. Inheritance in *Nicotiana tabacum*, XV. Carmine-white variegation. *J. Hered.* 31: 19–24.

Storlazzi, A., L. Xu, L. Cao, and N. Kleckner. 1995. Crossover and noncrossover recombination during meiosis: Timing and pathway relationships. *Proc. Natl. Acad. Sci. U.S.A.* 92: 8512–8516.

Storlazzi, A., L. Xu, A. Schwacha, and N. Kleckner. 1996. Synaptonemal complex (SC) component Zip 1 plays a role in meiotic recombination independent of SC polymerization along the chromosomes. *Proc. Natl. Acad. Sci. U.S.A.* 93: 9043–9048.

Stout, J. T., and R. L. Phillips. 1973. Two independently inherited electrophoretic variants of the lysine-rich histones of maize (*Zea mays*). *Proc. Natl. Acad. Sci. U.S.A.* 70: 3043–3047.

Straus, J. 1954. Maize endosperm tissue grown in vitro. II. Morphology and cytology. *Am. J. Bot.* 41: 833–839.

Strid, A. 1968. Stable telocentric chromosomes formed by spontaneous misdivision in *Nigella doerfleri* (Ranunculaceae). *Bot. Notiser* 121: 153–164.

Strickberger, M. W. 1968. *Genetics*. Macmillan Co., New York.

Stringam, G. R. 1970. A cytogenetic analysis of three asynaptic mutants in *Brassica campestris* L. *Can. J. Genet. Cytol.* 12: 743–749.

Stupar, R. M., J. W. Lilly, C. D. Town, Z. Cheng, S. Kaul, C. R. Buell, and J. Jiang. 2001. Complex mtDNA constitutes an approximate 620-kb insertion on *Arabidopsis thaliana* chromosome 2: Implication of potential sequencing errors caused by large-unit repeats. *Proc. Natl. Acad. Sci. U.S.A.* 98: 5099–5103.

Sturm, W., and G. Melz. 1982. Telotrisome-*Secal cereale* L. *Tagundsber. Akad. Landwirtschaftswiss.* DDR 198: 217–224.

Sturtevant, A. H., and G. W. Beadle. 1936. The relations of the inversions in the X chromosome of *Drosophila melanogaster* to crossing over and non-disjunction. *Genetics* 21: 554–604.

Stutz, H. C. 1972. On the origin of cultivated rye. *Am. J. Bot.* 59: 59–70.

Subba Rao, M. V. S., P. Sukha dev, J. V. V. S. N. Murty, and V. Manga. 1982. Analysis of meiosis and non-random bivalent formation in desynaptic mutants of *Pennisetum americanum* (L.) Leeke. *Genetica* 59: 157–160.

Subrahmanyam, N. C. 1978. Meiosis in polyhaploid *Hordeum*; Hemizygous ineffective control of diploid-like behaviour in a hexaploid? *Chromosoma* (Berlin) 66: 185–192.

Subrahmanyam, N. C., and K. J. Kasha. 1973. Selective chromosomal elimination during haploid formation in barley following interspecific hybridization. *Chromosoma* (Berlin) 42: 111–125.

Subrahmanyam, N. C., and K. J. Kasha. 1975. Chromosome doubling of barley haploids by nitrous oxide and colchicine treatments. *Can. J. Genet. Cytol.* 17: 573–583.

Subrahmanyam, N. C., P. S. Kumar, and D. G. Faris. 1986. Genome relationships in pigeon pea and its implications in crop improvement. In *Genetics and Crop Improvement*. P. K. Gupta and J. R. Bahl, Eds., Rastogi, Meerut, India, pp. 309–320.

Suen, D. F., C. K. Wang, R. F. Lin, Y. Y., Kao, F. M. Lee, and C. C. Chen. 1997. Assignment of DNA markers to *Nicotiana sylvestris* chromosomes using monosomic alien addition lines. *Theor. Appl. Genet.* 94: 331–337.

Sun, Z.-X., C.-Z. Zhao, K.-L. Zheng, X.-F. Qi, and Y.-P. Fu. 1983. Somaclonal genetics of rice, *Oryza sativa* L. *Theor. Appl. Genet.* 67: 67–73.

Sunderland, N. 1974. Anther culture as a means of haploid induction. In *Haploids in Higher Plants. Advances and Potential*. K. J. Kasha, Ed., Ainsworth Press, Canada, pp. 91–122.

Sutka, J., G. Galiba, A. Vagujfalvi, B. S. Gill, and J. W. Snape. 1999. Physical mapping of the *Vrn-A1* and *Frl* genes on chromosome 5A of wheat using deletion lines. *Theor. Appl. Genet.* 99: 199–202.

Sutton, W. S. 1903. The chromosomes in heredity. *Biol. Bull.* 4: 231–251.

Sutton, E. 1939. Trisomics in *Pisum sativum* derived from an interchange heterozygote. *J. Genet.* 38: 459–476.

Swaminathan, M. S., M. L. Magoon, and K. L. Mehra. 1954. A simple propiono-carmine PMC smear method for plant with small chromosomes. *Indian J. Genet. Plant Breed.* 14: 87–88.

Swaminathan, M. S., and B. R. Murty. 1959. Aspects of asynapsis in plants. I. Random and non-random chromosome associations. *Genetics* 44: 1271–1280.

Swanson, C. P. 1940. The distribution of inversions in *Tradescantia*. *Genetics* 25: 438–465.

Swanson, C. P., T. Merz, and W. Young. 1981. *Cytogenetics the Chromosome in Division, Inheritance, and Evolution*, 2nd edition. Prentice Hall, Englewood Cliffs, NJ.

Swedlund, B., and I. K. Vasil. 1985. Cytogenetic characterization of embryogenic callus and regenerated plants of *Pennisetum americanum* (L.) K. Schum. *Theor. Appl. Genet.* 69: 575–581.

Sybenga, J. 1983. Rye chromosome nomenclature and homoeology relationships. Workshop report. *Z. Pflanzenzüchtg.* 90: 297–304.

Sybenga, J. 1996. Aneuploid and other cytological tester sets in rye. *Euphytica* 89: 143–151.

Sybenga, J., J. van Eden, Q. van der Meijs, and B. W. Roeterdink. 1985. Identification of the chromosomes of the ryr translocation tester set. *Theor. Appl. Genet.* 69: 313–316.

Sybenga, J., and A. H. G. Wolters. 1972. The classification of the chromosomes of rye (*Secale cereale* L.): A translocation tester set. *Genetica* 43: 453–464.

Symko, S. 1969. Haploid barley from crosses of *Hordeum bulbosum* (2x) x *Hordeum vulgare* (2x). *Can. J. Genet. Cytol.* 11: 602–608.

Tabushi, J. 1958. Trisomics of spinach. *Seiken Zihô* 9: 49–57 (in Japanese with English summary).

Takagi, F. 1935. Karyogenetical studies on rye I. A trisomic plant. *Cytologia* 6: 496–501.

Takahashi, R., T. Tsuchiya, and I. Moriya. 1964. Heritable mixoploidy in barley. III. On a dwarf mutant from a cultivar. *Taisho-mugi*. *Nogaku Kenkyu* 50: 123–132 (in Japanese).

Taketa, S., T. Nakazaki, S. Shigenaga, and H. Yamagata. 1991. Preferential occurrence of specific R-D chromosome constitutions in stable hexaploid progenies of the hybrid between hexaploid triticale and bread wheat. *Jpn. J. Genet.* 66: 587–596.

Tanaka, M. 1956. Chromosome pairing and fertility in the hybrid between the new amphidiploid-S^lS^lAA and emmer wheat. *Wheat Inf. Ser.* 3: 21–22.

Tanksley, S. D., 1983. Gene mapping. In *Isozymes in Plant Genetics and Breeding*. Part A. S. D. Tanksley, and T. J. Orton, Eds., Elsevier, Amsterdam, pp. 109–138.

Tanksley, S. D., and S. R. McCouch. 1997. Seed banks and molecular maps: Unlocking genetic potential from the wild. *Science* 277: 1063–1066.

Taylor, N. L., M. K. Anderson, K. H. Quesenberry, and L. Watson. 1976. Doubling the chromosome number of *Trifolium* species using nitrous oxide. *Crop Sci.* 16: 516–518.

Tempelaar, M. J., E. Jacobsen, M. A. Ferwerda, and M. Hartogh. 1985. Changes of ploidy level by *in vitro* culture of monohaploid and polyploid clones of potato. *Z. Pflanzenzüchtg.* 95: 193–200.

Temsch, E. M., and J. Greilhuber. 1999. Genome size variation in *Arachis hypogaea* and *A. monticola* re-evaluated. *Genome* 43: 449–451.

Teoh, S. B., J. Hutchinson, and T. E. Miller. 1983. A comparison of the chromosomal distribution of cloned repeated DNA sequences in different *Aegilops* species. *Heredity* 51: 635–641.

Thakare, R. G., D. C. Joshua, and N. S. Rao. 1974. Radiation induced trisomics in jute. *Indian J. Genet. Plant Breed.* 34: 337–345.

Therman, E., and G. Sarto. 1977. Premeiotic and early meiotic stages in the pollen mother cells of *Eremurus* and in human embryonic oocytes. *Hum. Genet.* 35: 137–151.

Thiebaut, J., K. J. Kasha, and A. Tsai. 1979. Influence of plant development stage, temperature, and plant hormones on chromosome doubling of barley haploids using colchicine. *Can. J. Bot.* 57: 480–483.

This, P., O. Ochoa, and C. F. Quiros. 1990. Dissection of the *Brassica nigra* genome by monosomic addition lines. *Plant Breed.* 105: 211–220.

Thomas, H. 1968. The addition of single chromosomes of *Avena hirtula* to the cultivated hexaploid oat *A. sativa*. *Can. J. Genet. Cytol.* 10: 551–563.

Thomas, H. 1973. Somatic association of chromosomes in asynaptic genotypes of *Avena sativa* L. *Chromosoma* (Berlin) 42: 87–94.

Thomas, H., J. M. Leggett, and I. T. Jones. 1975. The addition of a pair of chromosomes of the wild oat *Avena barbata* (2n = 28) to the cultivated oat *A. sativa* L. (2n = 42). *Euphytica* 24: 717–724.

Thomas, H., and T. Rajhathy. 1966. A gene for desynapsis and aneuploidy in tetraploid *Avena*. *Can. J. Genet. Cytol.* 8: 506–515.

Thomas, H. M. 1988. Chromosome elimination and chromosome pairing in tetraploid hybrids of *Hordeum vulgare* x *H. bulbosum*. *Theor. Appl. Genet.* 76: 118–124.

Thomas, H. M., and R. A. Pickering. 1988. The cytogenetics of a triploid *Hordeum bulbosum* and of some of its hybrid and trisomic derivatives. *Theor. Appl. Genet.* 76: 93–96.

Thomas, H. M., and B. J. Thomas. 1994. Meiosis in triploid *Lolium*. I. synaptonemal complex formation and chromosome configurations at metaphase I in aneuploid autotriploid *L. multiflorum*. *Genome* 37: 181–189.

Thomas, J. B., and P. J. Kaltsikes. 1974. A possible effect of heterochromatin on chromosome pairing. *Proc. Nat. Acad. Sci. U.S.A.* 71: 2787–2790.

Tindale, M. D. 1984. Two new eastern Australian species of *Glycine* Willd. (Fabaceae). *Brunonia* 7: 207–213.

Tindale, M. D. 1986. Taxonomic notes on three Australian and Norfolk Island species of *Glycine* Willd. (Fabaceae: Phaseolae) including the choice of a neotype for *G. clandestina* Wendl. *Brunonia* 9: 179–191.

Tindale, M. D., and L. A. Craven. 1988. Three new species of *Glycine* (Fabaceae: Phaseolae) from north-western Australia, with notes on amphicarpy in the genus. *Austr. Syst. Bot.* 1: 399–410.

Tjio, J. H., and A. Hagberg. 1951. Cytological studies on some X-ray mutants of barley. *An. Estacion Exp. Aula Dei* 2: 149–167.

Tjio, J. H., and A. Lavan. 1956. The chromosome number of man. *Hereditas* 42(1–2): 1–6.

Tjio, J. H., and A. Levan. 1950. The use of oxyquinoline in chromosome analysis. *An. Estacion Exp. Aula Dei.* 2: 21–46.

Toriyama, K., Y. Arimoto, H. Uchimiya, and K. Hinata. 1988. Transgenic rice plants after direct gene transfer into protoplasts. *Bio/Technol.* 6: 1072–1074.

Toriyama, K., K. Hinata, and T. Sasaki. 1986. Haploid and diploid plant regeneration from protoplasts of anther callus in rice. *Theor. Appl. Genet.* 73: 16–19.

Torp, A. M., A. L. Hansen, and S. B. Andersen. 2001. Chromosomal regions associated with green plant regeneration in wheat (*Triticum aestivum* L.) anther culture. *Euphytica* 119: 377–387.

Torres, A. M., Z. Satovic, J. Canovas, S. Cobos, and J. L. Cubero. 1995. Genetics and mapping of new isozyme loci in *Vicia faba* L. using trisomics. *Theor. Appl. Genet.* 91: 783–789.

Torrey, J. G. 1959. Experimental modification of development in the root. In *Cell Organism and Milieu.* D. Rudnick, Ed., Ronald Press, New York, pp. 189–222.

Torrey, J. G. 1967. Morphogenesis in relation to chromosomal constitution in long-term plant tissue cultures. *Physiol. Plant* 20: 265–275.

Torrezan, R., and M. S. Pagliarini. 1995. Influence of heterochromatin on chiasma localization and terminalization in maize. *Caryologia* 48: 247–253.

Tsuchiya, T. 1952. Cytogenetic studies of a triploid hybrid plant in barley. *Seiken Zihô* 5: 78–93.

Tsuchiya, T. 1960a. Cytogenetic studies of trisomics in barley. *Jpn. J. Bot.* 17: 177–213.

Tsuchiya, T. 1960b. Studies on cross compatibility of diploid, triploid and tetraploid barley, III. Results of 3x x 3x crosses. *Seiken Zihô* 11: 29–37.

Tsuchiya, T. 1961. Studies on the trisomics in barley, II. Cytological identification of the extra chromosomes in crosses with Burnham's translocation testers. *Jpn. J. Genet.* 36: 444–451.

Tsuchiya, T. 1967. Establishment of a trisomic series in a two-rowed cultivated variety of barley. *Can. J. Genet. Cytol.* 9: 667–682.

Tsuchiya, T. 1969. Cytogenetics of a new type of barley with 16 chromosomes. *Chromosoma* (Berlin) 26: 130–139.

Tsuchiya, T. 1971a. An improved aceto-carmine squash method, with special reference to the modified Rattenbury's method of making a preparation permanent. *Barley Genet. Newsl.* 1: 58–60.

Tsuchiya, T. 1971b. Telocentric chromosomes in barley. In *Barley Genetics II. Proc. 2nd Int. Barley Genet. Symp.* R. A. Nilan, Ed., Washington State University Press, Pullman, pp. 72–81.

Tsuchiya, T. 1972a. Cytogenetics of the telocentric chromosome of the long arm of chromosome 1 in barley. *Seiken Zihô* 23: 47–62.

Tsuchiya, T. 1972b. Revision of linkage map of chromosome 5 in barley by means of telotrisomic analysis. *J. Hered.* 63: 373–375.

Tsuchiya, T. 1973. A barley strain with $2n = 12 + 4$ telocentric chromosomes (2 pairs). *Cereal Res. Commun.* 1: 23–24.

Tsuchiya, T. 1991. Chromosome mapping by means of aneuploid analysis in barley. In *Chromosome Engineering in Plants: Genetics, Breeding, Evolution. Part A.* P. K. Gupta and T. Tsuchiya, Eds., Elsevier Science, Amsterdam, pp. 361–384.

Tsuchiya, T., and A. Hang. 1979. Telotrisomic analysis of *yst3* and *i* in barley. *Barley Genet. Newsl.* 9: 106–108.

Tsuchiya, T., A. Shahla, and A. Hang. 1986. Acrotrisomic analysis in barley. In *Barley Genetics V. Proc. 5th Int. Barley Genet. Symp.* S. Yasuda and T. Konishi, Eds., Okayama, Japan, pp. 389–395.

Tsuchiya, T., and R. J. Singh. 1982. Chromosome mapping in barley by means of telotrisomic analysis. *Theor. Appl. Genet.* 61: 201–208.

Tsuchiya, T., R. J. Singh, A. Shahla, and A. Hang. 1984. Acrotrisomic analysis in linkage mapping in barley (*Hordeum vulgare* L.). *Theor. Appl. Genet.* 68: 433–439.

Tsuchiya, T., and S. Wang. 1991. Cytogenetics of ditelotetrasomics for short arms of four chromosomes of barley (*Hordeum vulgare* L.). *Theor. Appl. Genet.* 83: 41–48.

Tsuji, S., Y. Mukai, and K. Tsunewaki. 1983. Syncyte formation in alloplasmic common wheat and triticale. *Seiken Zihô* 31: 20–26.

Tsujimoto, H., K. Shinotani, and H. Ono. 1984. Alien chromosome substitution of *durum* wheat: Substitution of the chromosome e_7 of *Elytrigia elongata* for chromosome 4A of Stewart *durum. Jpn. J. Genet.* 59: 141–153.

Tsujimoto, H., T. Yamada, K. Hasegawa, N. Usami, T. Kojima, T. R. Endo, Y. Ogihara, and T. Sasakuma. 2001. Large-scale selection of lines with deletions in chromosome 1B in wheat and applications for fine deletion mapping. *Genome* 44: 501–508.

Tsunewaki, K. 1959. A ring chromosome in an F_1 hybrid between wheat and *Agropyron. Can. J. Bot.* 37: 1271–1276.

Tsunewaki, K. 1963. An Emmer wheat with 15 chromosome pairs. *Can. J. Genet. Cytol.* 5: 462–466.

Tsunewaki, K., and E. G. Heyne. 1960. The transmission of the monosomic condition in wheat. *J. Hered.* 51: 63–68.

Tsunewaki, K., and Y. Ogihara. 1983. The molecular basis of genetic diversity among cytoplasms of *Triticum* and *Aegilops* species. II. On the origin of polyploid wheat cytoplasms as suggested by chloroplast DNA restriction fragment patterns. *Genetics* 104: 155–171.

Tsunewaki, K., T. Yoshida, and S. Tsuji. 1983. Genetic diversity of the cytoplasm in *Triticum* and *Aegilops*. IX. The effect of alien cytoplasms on seed germination of common wheat. *Jpn. J. Genet.* 58: 33–41.

Tuleen, N. A. 1971. Linkage data and chromosome mapping. In *Barley Genetics II. Proc. 2nd. Int. Barley Genet. Symp.* R. A. Nilan, Ed., Washington State University Press, Pullman, pp. 208–212.

Tuleen, N. A. 1973. Karyotype analysis of multiple translocation stocks of barley. *Can. J. Genet. Cytol.* 15: 267–273.

U N. 1935. Genome-analysis in *Brassica* with special reference to the experimental formation of *B. napus* and peculiar mode of fertilization. *Jpn. J. Bot.* 7: 389–452.

Uchimiya, H., and S. G. Wildman. 1978. Evolution of fraction I protein in relation to origin of amphidiploid *Brassica* species and other members of the Cruciferae. *J. Hered.* 69: 299–303.

Uhrig, H. 1985. Genetic selection and liquid medium conditions improve the yield of androgenetic plants from diploid potatoes. *Theor. Appl. Genet.* 71: 455–460.

Ullrich, S. E., J. M. Edmiston, A. Kleinhofs, D. A. Kudrna, and M. E. H. Maatougui. 1991. Evaluation of somaclonal variation in barley. *Cereal. Res. Commun.* 19: 245–260.

Unrau, J. 1950. The use of monosomes and nullisomes in cytogenetic studies of common wheat. *Sci. Agri.* 30: 66–89.

Valiček, P. 1978. Wild and cultivated cottons. *Cot. Fib. Trop.* 33: 363–387.

van den Bulk, R. W. 1991. Application of cell and tissue culture and *in vitro* selection for disease resistance breeding-a review. *Euphytica* 56: 269–285.

Vanderlyn L. 1948. Somatic mitosis in the root-tip of *Allium cepa*—A review and reorientation. *Bot. Rev.* 14: 270–318.

van der Maesen, L. J. G. 1986. *Cajanus DC and Atylosia W. & A. (Leguminosae). A revision of all taxa closely related to the pigeonpea, with notes on other related genera within the substribe Cajaninae.* Agricultural University, Wageningen, Agricultural University Wageningen paper 85-4 (1985), p. 225.

van Dijk, P. J., I. C. Q. Tas, M. Falque, and T. Baks-Schotman. 1999. Crosses between sexual and apomictic dendelions (*Taraxacum*). II. The breakdown of apomixis. *Heredity* 83: 715–721.

Van't Hof, J. 1965. Relationships between mitotic cycle duration, S period duration and the average rate of DNA synthesis in the root meristem cells of several plants. *Exp. Cell Res.* 39: 48–58.

Van't Hof, J., and B. McMillan. 1969. Cell population kinetics in callus tissues of cultured pea root segments. *Am. J. Bot.* 56: 42–51.

Van't Hof, J., and A. H. Sparrow. 1963. A relationship between DNA content, nuclear volume and minimum mitotic cycle time. *Proc. Natl. Acad. Sci. U.S.A.* 49: 897–902.

Van't Hof, J., G. B. Wilson, and A. Colon. 1960. Studies on the control of mitotic activity. The use of colchicine in the tagging of a synchronous population of cells in the meristem of *Pisum sativum*. *Chromosoma* (Berlin) 11: 313–321.

Vasek, F. C. 1956. Induced aneuploidy in *Clarkia unguiculata* (*Onagraceae*). *Am. J. Bot.* 43: 366–371.

Vasek, F. C. 1963. Phenotypic variation in trisomics of *Clarkia unguiculata*. *Am. J. Bot.* 50: 308–314.

Vasil, I. K. 1988. Progress in the regeneration and genetic manipulation of cereal crops. *Bio/Tech.* 6: 397–402.

Vasil, I. K. 1990. The realities and challenges of plant biotechnology. *Bio/Tech.* 8: 296–301.

Vavilov, N. I. 1950. *The Origin, Variation, Immunity and Breeding of Cultivated Crops.* Chronica Botanica, Waltham, MA, p. 13.

Vega, C., and J. R. Lacadena. 1982. Cytogenetic structure of common wheat cultivars from or introduced into Spain. *Theor. Appl. Genet.* 61: 129–133.

Venkateswarlu, J., and P. N. Rao. 1976. Effect of inbreeding and selection for vigour and fertility on meiotic behaviour in autotetraploid Job's Tears, *Coix lacryma*-jobi L. *Theor. Appl. Genet.* 47: 165–169.

Venkateswarlu, J., and V. R. Reddi. 1968. Cytological studies of Sorghum trisomics. *J. Hered.* 59: 179–182.

Venketeswaran, S. 1963. Tissue culture studies on *Vicia faba*. II. Cytology. *Caryologia* 16: 91–100.

Verma, S. C., and H. Rees. 1974. Giemsa staining and the distribution of heterochromatin in rye chromosomes. *Heredity* 32: 118–121.

Viegas-Pequignot, E., B. Dutrillaux, H. Magdelenat, and M. Coppey-Moisan. 1989. Mapping of single-copy DNA sequences on human chromosomes by *in situ* hybridization with biotinylated probes: Enhancement of detection sensitivity by intensified-fluorescence digital-imaging microscopy. *Proc. Natl. Acad. Sci. U.S.A.* 86: 582–586.

Vilhar, B., J. Greilhuber, J. D. Koce, E. M. Temsch, and M. Dermastia. 2001. Plant genome size measurement with DNA image cytometry. *Ann. Bot.* 87: 719–2001.

Villeneuve, A. M., and K. J. Hillers. 2001. Whence meiosis? *Cell* 106: 647–650.

Virmani, S. S., and I. B. Edwards. 1983. Current status and future prospects for breeding hybrid rice and wheat. *Adv. Agron.* 36: 145–214.

Visser, R. G. F., R. Hoekstra, F. R. van der Leij, L. P. Pijnacker, B. Witholt, and W. J. Feenstra. 1988. In situ hybridization to somatic metaphase chromosomes of potato. *Theor. Appl. Genet.* 76: 420–424.

Vogt, G. E., and P. R. Rowe. 1968. Aneuploids from triploid-diploid crosses in the series *Tuberosa* of the genus *Solanum. Can. J. Genet. Cytol.* 10: 479–486.

Voigt, P. W., and C. R. Tischler. 1994. Leaf characteristic variation in hybrid lovegrass populations. *Crop Sci.* 34: 679–684.

Vroh Bi, I., A. Maquet, J.-P. Baudoin, P. du Jardin, J. M. Jacquemin, and G. Mergeai. 1999. Breeding for "low-gossypol seed and high-gossypol plants" in upland cotton. Analysis of tri-species hybrids and backcross progenies using AFLPs and mapped RFLPs. *Theor. Appl. Genet.* 99: 1233–1244.

Waara, S., and K. Glimelius. 1995. The potential of somatic hybridization in crop breeding. *Euphytica* 85: 217–233.

Wagenaar, E. B. 1969. End-to-end chromosome attachments in mitotic interphase and their possible significance to meiotic chromosome pairing. *Chromosoma* (Berlin) 26: 410–426.

Wagenheim, K.-H. v., S. J. Peloquin, and R. W. Hougas. 1960. Embryological investigations on the formation of haploids in the potato (*Solanum tuberosum*). *Z. Vererbungslehre* 91: 391–399.

Wagenvoort, M. 1995. Meiotic behavior of 11 primary trisomics in diploid potato and its consequences for the transmission of the extra chromosome. *Genome* 38: 17–26.

Waines, J. G. 1976. A model for the origin of diploidizing mechanism in polyploid species. *Am. Nat.* 110: 415–430.

Wall, A. M., R. Riley, and M. D. Gale. 1971. The position of a locus on chromosome 5B of *Triticum aestivum* affecting homoeologous meiotic pairing. *Genet. Res.* (Camb.) 18: 329–339.

Wang, R. R.-C. 1989. An assessment of genome analysis based on chromosome pairing in hybrids of perennial Triticeae. *Genome* 32: 179–189.

Wang, S., N. L. V. Lapitan, and T. Tsuchiya. 1991. Characterization of telomeres in *Hordeum vulgare* chromosomes by *in situ* hybridization I. Normal barley. *Jpn. J. Genet.* 66: 313–316.

Wang, S., and T. Tsuchiya. 1990. Cytogenetics of four telotrisomics in barley (*Hordeum vulgare* L.). *Theor. Appl. Genet.* 80: 145–152.

Wang, X. E., P. D. Chen, D. J. Liu, P. Zhang, B. Zhou, B. Friebe, and B. S. Gill. 2001. Molecular cytogenetic characterization of *Roegneria ciliaris* chromosome additions in common wheat. *Theor. Appl. Genet.* 102: 651–657.

Wang, Z. W., and N. Iwata. 1996. The origin, identification, and plant morphology of five rice monosomics. *Genome* 39: 528–534.

Wang, Z. Y., G. Second, and S. D. Tanksley. 1992. Polymorphism and phylogenetic relationships among species in the genus *Oryza* as determined by analysis of nuclear RFLPs. *Theor. Appl. Genet.* 83: 565–581.

Ward, E. J. 1980. Banding patterns in maize mitotic chromosomes. *Can. J. Genet. Cytol.* 22: 61–67.

Warmke, H. E. 1954. Apomixis in *Panicum maximum. Am. J. Bot.* 41: 5–11.

Watanabe, H., and S. Noda. 1974. Chiasma studies in structural hybrids. XI. Pericentric inversion in *Allium thunbergii. The Nucleus* (Calcutta) 17: 114–117.

Watanabe, K. 1981. Studies on the control of diploid-like meiosis in polyploid taxa of *Chrysanthemum* I. Hexaploid *Ch. japonense* Nakai. *Cytologia* 46: 459–498.

Watanabe, Y., S. Ono, Y. Mukai, and Y. Koga. 1969. Genetic and cytogenetic studies on the trisomic plants of rice, *Oryza sativa* L. *Jpn. J. Breed.* 19: 12–18.

Watkins, A. E. 1932. Hybrid sterility and incompatibility. *J. Genet.* 25: 125–162.

Weaver, J. B. 1971. An asynaptic character in cotton inherited as a double recessive. *Crop Sci.* 11: 927–928.

Weber, D. F. 1978. Nullisomic analysis of neuclear formation in *Zea mays. Can. J. Genet. Cytol.* 20: 97–100.

Weber, D. F. 1983. Monosomic analysis in diploid crop plants. In *Cytogenetics of Crop Plants.* M. S. Swaminathan, P. K. Gupta, and U. Sinha, Eds., McMillan, India, pp. 352–378.

Weber, D. F. 1991. Monosomic analysis in maize and other diploid crop plants. In *Chromosome Engineering in Plants: Genetics, Breeding, Evolution. Part A.* P. K. Gupta and T. Tsuchiya, Eds., Elsevier, Amsterdam, pp. 181–209.

Weber, D. F., and L. Chao. 1994. The *R-X1* deficiency system induces nondisjunction in the early embryo in maize. *Maydica* 39: 29–33.

Weeden, N. F., J. D. Graham, and R. W. Robinson. 1986. Identification of two linkage groups in *Cucurbita palmata* using alien additions lines. *Hort. Sci.* 21: 1431–1433.

Wehling, P. 1991. Inheritance, linkage relationship and chromosomal localization of the glutamate oxaloac-etate transminase, acid phosphatase and diaphorase isozyme genes in *Secale cereale* L. *Theor. Appl. Genet.* 82: 569–576.

Wei, J.-Z., W. F. Campbell, G. J. Scoles, A. E. Slinkard, and R. R.-C. Wang. 1995. Cytological identification of some trisomics of Russian wild rye (*Psathyrostachys juncea*). *Genome* 38: 1271–1278.

Weimarck, A. 1973. Cytogenetic behaviour in octoploid *Triticale* I. Meiosis, aneuploidy and fertility. *Hereditas* 74: 103–118.

Weimarck, A. 1975. Heterochromatin polymorphism in the rye karyotype as detected by the Giemsa C-banding technique. *Hereditas* 79: 293–300.

Wendel, J. F. 1989. New World tetraploid cottons contain Old World cytoplasm. *Proc. Natl. Acad. Sci. U.S.A.* 86: 4132–4136.

Wenzel, G., F. Hoffmann, and E. Thomas. 1977. Increased induction and chromosome doubling of androge-netic haploid rye. *Theor. Appl. Genet.* 51: 81–86.

Wenzel, G., and H. Uhrig. 1981. Breeding for nematode and virus resistance in potato via anther culture. *Theor. Appl. Genet.* 59: 333–340.

Wersuhn, G. 1989. Obtaining mutants from cell cultures. *Plant Breed.* 102: 1–9.

Wersuhn, G., and U. Dathe. 1983. Zur karyotypischen stabilität pflanzlicher *in vitro*-kulturen. *Biol. Zentralbl.* 102: 551–557.

Wersuhn, G., and B. Sell. 1988. Numerical chromosomal aberrations in plants regenerated from pollen, proto-plasts and cell cultures of *Nicotiana tabacum* L. cv. 'Samsun'. *Biol. Zentralbl.* 107: 439–445.

Westergaard, M. 1958. The mechanism of sex determination in dioecious flowering plants. *Adv. Genet.* 9: 217–281.

Wetter, L. R., and K. N. Kao. 1980. Chromosome and isoenzyme studies on cells derived from protoplast fusion of *Nicotiana glauca* with *Glycine max–Nicotiana glauca* cell hybrids. *Theor. Appl. Genet.* 57: 273–276.

White, M. J. D., and F. H. W. Morley. 1955. Effects of pericentric rearrangements on recombination in grass-hopper chromosomes. *Genetics* 40: 604–619.

Wiebe, G. A., R. T. Ramage, and R. F. Eslick. 1974. Eight paired barley lines. *Barley Genet. Newsl.* 4: 93–95.

Wienhues, A. 1966. Transfer of rust resistance of *Agropyron* to wheat by addition, substitution and transloca-tion. *Proc. 2nd Int. Wheat Genet. Symp.* Lund. 1963. *Hereditas Suppl.*: 328–341.

Wienhues, A. 1971. Substitution von weizenchromosomen aus verschiedenen homoeologen gruppen durch ein fremdchromosom aus *Agropyron intermedium*. *Z. Pflazenzüchtg.* 65: 307–321.

Wienhues, A. 1973. Translocations between wheat chromosomes and an *Agropyron* chromosome condition-ing rust resistance. In *Proc. 4th Int. Wheat Genet. Symp.* E. R. Sears and L. M. S. Sears, Eds., Missouri Agric. Exp. Stn., Columbia, MO, pp 201–207.

Wilson, E. B. 1928. *The Cell in Development and Heredity.* Macmillan Co., New York.

William, M. D. H. M., and A. Mujeeb-Kazi. 1995. Biochemical and molecular diagnostics of *Thinopyrum bessarabicum* chromosomes in *Triticum aestivum* germ plasm. *Theor. Appl. Genet.* 90: 952–956.

Williams, E., and K. K. Pandey. 1975. Meiotic chromosome pairing in interspecific hybrids of *Nicotiana.* II. South American species hybrids: The influence of genotype on pairing. *New Zealand J. Bot.* 13: 611–622.

Williamson, C. J. 1981. The influence of light regimes during floral development on apomictic seed produc-tion and on variability in resulting seedling progenies of *Poa ampla* and *P. pratensis*. *New Phytol.* 87: 769–783.

Wimber, D. E. 1960. Duration of the nuclear cycle in *Tradescantia paludosa* root tips as measured with H^3 thymidine. *Am. J. Bot.* 47: 828–834.

Winkelmann, T., R. S. Sangwan, and H.-G. Schwenkel. 1998. Flow cytometric analyses in embryogenic and non-embryogenic callus line of *Cyclamen persicum* Mill.: Relation between ploidy level and compe-tence for somatic embryogenesis. *Plant Cell Rep.* 17: 400–404.

Winkler, H. 1908. Über parthenogensis und apogemie im pflanzenreiche. *Pogr. Rei. Bot.* 2: 292–454.

Wolters, A. M., E. Jacobsen, M. O'Connell, G. Bonnema, K. S. Ramulu, H. de Jong, H. Schoenmakers, J. W. Wijbrandi, M. Koornneef. 1994a. Somatic hybridization as a tool for tomato breeding. *Euphytica* 79: 265–277.

Wolters, A. M. A., H. C. H. Schoenmakers, S. Kamstra, J. van Eden, M. Koornneef, and J. H. de Jong. 1994b. Mitotic and meiotic irregularities in somatic hybrids of *Lycopersicon esculentum* and *Solanum tuberosum*. *Genome* 37: 726–735.

Wright, M. S., D. V. Ward, M. A. Hinchee, M. G. Carnes, and R. J. Kaufman. 1987. Regeneration of soybean (*Glycine max* L. Merr.) from cultured primary leaf tissue. *Plant Cell Rep.* 6: 83–89.

Wynne, J. C., and T. Halward. 1989. Cytogenetics and genetics of *Arachis*. *Critical Rev. Plant Sci.* 8: 189–220.

Xu, S. J., R. J. Singh, and T. Hymowitz. 2000a. Monosomics in soybean: Origin, identification, cytology, and breeding behavior. *Crop Sci.* 40: 985–989.

Xu, S. J., R. J. Singh, K. P. Kollipara, and T. Hymowitz. 2000b. Hypertriploid in soybean: Origin, identification, cytology, and breeding behavior. *Crop Sci.* 40: 72–77.

Xu, S. J., R. J. Singh, K. P. Kollipara, and T. Hymowitz. 2000c. Primary trisomics in soybean: Origin, identification, breeding behavior, and use in linkage mapping. *Crop Sci.* 40: 1543–1551.

Xu, X. and B. Li. 1994. Fertile transgenic Indica rice obtained by electroporation of the seed embryo cells. *Plant Cell Rep.* 13: 237–242.

Yanagino, T., Y. Takahata, and K. Hinata. 1987. Chloroplast DNA variation among diploid species in *Brassica* and allied genera. *Jpn. J. Genet.* 62: 119–125.

Yang, D. P., and E. O. Dodson. 1970. The amounts of nuclear DNA and the duration of DNA synthetic period(s) in related diploid and autotetraploid species of oats. *Chromosoma* (Berlin) 31: 309–320.

Yarnell, S. H. 1931. A study of certain polyploid and aneuploid forms in *Fragaria*. *Genetics* 16: 455–489.

Yazawa, S. 1967. Cytological and morphological observations on the callus tissue of *Crepis capillaris*. *Bot. Mag. Tokyo* 80: 413–420 (in Japanese with English summary).

Young, B. A., R. T. Sherwood, and E. C. Bashaw. 1979. Clear-pistil and thick-sectioning techniques for detecting aposporous apomixis in grasses. *Can J. Bot.* 57: 1668–1672.

Yu, C. W., and E. A. Hockett. 1979. Chromosome behavior, breeding, characteristics, and seed set of partially sterile barley, *Hordeum vulgare* L. *Z. Pflanzenzüchtg.* 82: 133–148.

Yu, H.-G., and R. K. Dawe. 2000. Functional redundancy in the maize mitotic kinetochore. *J. Cell Biol.* 151: 131–141.

Zamir, D. 2001. Improving plant breeding with exotic genetic libraries. *Nature Rev.* 2: 983–989.

Zeller, F. J. 1973. 1B/1R wheat-rye chromosome substitutions and translocations. *Proc. 4th Int. Wheat Genet. Symp.* E. R. Sears and L. M. S. Sears, Eds., Missouri Agric. Exp. Stn. Columbia, MO, pp. 209–221.

Zeller, F. J., and A. C. Baier. 1973. Substitution des weizenchromosomenpaares 4A durch das roggenchromosomenpaar 5R in dem weihenstephaner weizenstamm W 70 a 86 (Blaukorn). *Z. Pflanzenzüchtg.* 70: 1–10 (in German with English summary).

Zeller, F. J., M-C., Cermeño, and B. Friebe. 1987. Cytological identification of telotrisomic and double ditelosomic lines in *Secale cereale* cv. Heines Hellkorn by means of Giemsa C-banding patterns and crosses with wheat-rye addition lines. *Genome* 29: 58–62.

Zeller, F. J., M.-C. Cermeño, and T. E. Miller. 1991. Cytological analysis on the distribution and origin of the alien chromosome pair conferring blue aleurone color in several European commom wheat (*Triticum aestivum* L.) strains. *Theor. Appl. Genet.* 81: 551–558.

Zeller, F. J., and E. Fuchs. 1983. Cytologie und krankheitsresistenz einer 1A/1R-und Mehrerer 1B/1R-weizen-roggen-translokationssorten. *Z. Pflanzenzüchtg.* 90: 285–296.

Zeller, F. J., and S. L. K. Hsam. 1983. Broadening the genetic variability of cultivated wheat by utilizing rye chromatin. In *Proc. 6th Int. Wheat Genet. Symp.* S. Sakamoto, Ed., Kyoto University, Kyoto, Japan, pp. 161–173.

Zeller, F. J., G. Kimber, and B. S. Gill. 1977. The identification of rye trisomics by translocations and Giemsa staining. *Chromosoma* (Berlin) 62: 279–289.

Zeller, F. J., E. Lintz, and R. Kunzmann. 1982. Origin, occurrence, and transmission rates of trisomic and telocentric chromosomes in *Secale cereale* cultivar Heines Hellkorn. *Tagunder. Akad. Landwirtschaftswiss. DDR* 198: 165–180.

Zenkteler, M., and G. Melchers. 1978. In vitro hybridization by sexual methods and by fusion of somatic protoplasts. Experiments with *Nicotiana tabacum* x *Petunia hybrida*, *N. tabacum* x *Hyoscyamus niger*, *H. niger* x *P. hybrida*, *Melandrium album* x *P. hybrida*. *Theor. Appl. Genet.* 52: 81–90.

Zenkteler, M., and M. Nitzsche. 1984. Wide hybridization experiments in cereals. *Theor. Appl. Genet.* 68: 311–315.

Zeven, A. C. 1998. Landraces: A review of definitions and classifications. *Euphytica* 104: 127–139.

Zhang, D. and R. B. Nicklas. 1996. 'Anaphase' and cytokinesis in the absence of chromosomes. *Nature* (London) 382: 466–468.

Zhang, D. L., K. Q. Li, W. Gu, and L. F. Hao. 1987. Chromosome aberration and ploidy equilibrium of *Vicia faba* in tissue culture. *Theor. Appl. Genet.* 75: 132–137.

Zhang, P., B. Friebe, A. J. Lukaszewski, and B. S. Gill. 2001. The centromere structure in Robertsonian wheat-rye translocation chromosomes indicates that centric breakage-fusion can occur at different positions within the primary constriction. *Chromosoma* 110: 335–344.

Zheng, Y.-Z., R. R. Roseman, and W. R. Carlson. 1999. Time course study of the chromosome-type breakage-fusion-bridge cycle in maize. *Genetics* 153: 1435–1444.

Zhou, K. D., X. D. Wang, M. Luo, K. M. Gao, S. L. Zhou, Z. B. Yan, P. Li, F. Chen, and G. M. Zhou. 1993. Initial study on Sichuan apomixis rice (SAR -1). *Sci. China. Ser. B—Chemistry.* 36: 420–429.

Ziauddin, A., and K. J. Kasha. 1990. Long-term callus cultures of diploid barley (*Hordeum vulgare*). II. Effect of auxin on chromosomal status of cultures and regeneration of plants. *Euphytica* 48: 279–286.

Zickler, D. and N. Kleckner. 1999. Meiotic chromosomes: Integrating structure and function. *Ann. Rev. Genet.* 33: 603–754.

Zohary, D. 1955. Chiasmata in a pericentric inversion in *Zea mays. Genetics* 40: 874–877.

Index

Printed and bound by CPI Group (UK) Ltd, Croydon, CR0 4YY

17/10/2024

01775698-0011